Communications in Computer and Information Science 1142

More information about this series at http://www.springer.com/series/7899

Tom Gedeon · Kok Wai Wong · Minho Lee (Eds.)

Neural Information Processing

26th International Conference, ICONIP 2019 Sydney, NSW, Australia, December 12–15, 2019 Proceedings, Part IV

Editors
Tom Gedeon
Australian National University
Canberra, ACT, Australia

Minho Lee
Kyungpook National University
Daegu, Korea (Republic of)

Kok Wai Wong
Murdoch University
Murdoch, WA, Australia

ISSN 1865-0929 ISSN 1865-0937 (electronic)
Communications in Computer and Information Science
ISBN 978-3-030-36807-4 ISBN 978-3-030-36808-1 (eBook)
https://doi.org/10.1007/978-3-030-36808-1

This Springer imprint is published by the registered company Springer Nature Switzerland AG
The registered company address is: Gewerbestrasse 11, 6330 Cham, Switzerland

Preface

Welcome to the proceedings of the 26th International Conference on Neural Information Processing of the Asia-Pacific Neural Network Society (APNNS 2019), held in Sydney during December 12–15, 2019.

The mission of the Asia-Pacific Neural Network Society is to promote active interactions among researchers, scientists, and industry professionals who are working in Neural Networks and related fields in the Asia-Pacific region. APNNS 2019 had governing board members from 13 countries/regions – Australia, China, Hong Kong, India, Japan, Malaysia, New Zealand, Singapore, South Korea, Qatar, Taiwan, Thailand, and Turkey. The society's flagship annual conference is the International Conference of Neural Information Processing (ICONIP).

The conference had three main themes: "Theory and Algorithms," "Computational and Cognitive Neurosciences," and "Human Centred Computing and Applications." The two CCIS volumes 1142–1143 are organized in topical sections which were also the names of the 12-minute presentation sessions at the conference. The topics were "Adversarial Networks and Learning," "Convolutional Neural Networks," "Deep Neural Networks," "Embeddings and Feature Fusion," "Human Centred Computing," "Human Centred Computing and Medicine," "Human Centred Computing for Emotion," "Hybrid Models," "Artificial Intelligence and Cybersecurity," "Image Processing by Neural Techniques," "Learning from Incomplete Data," "Model Compression and Optimisation," "Neural Network Applications," "Neural Network Models," "Semantic and Graph Based Approaches," "Social Network Computing," "Spiking Neuron and Related Models," "Text Computing using Neural Techniques," "Time-series and Related Models," and "Unsupervised Neural Models."

A Special thanks in particular to the reviewers who devoted their time to our rigorous peer-review process. Their insightful reviews and timely feedback ensured the high quality of the papers accepted for publication. Finally, thank you to all the authors of papers, presenters, and participants at the conference. Your support and engagement made it all worthwhile.

October 2019

Tom Gedeon
Kok Wai Wong
Minho Lee

Bohan Li	Nanjing University of Aeronautics and Astronautics, China
Mengmeng Li	Zhengzhou University, China
Yaoyi Li	Shanghai Jiao Tong University, China
Yanjun Li	Beijing Institute of Technology, China
Ming Li	Latrobe University, Australia
Mingyong Li	Donghua University, China
Chengcheng Li	Tianjin University, China
Xia Liang	University of Science and Technology, China
Alan Wee-Chung Liew	Griffith University, Australia
Chin-Teng Lin	UTS, Australia
Zheng Lin	Chinese Academy of Sciences, China
Yang Lin	The University of Sydney, Australia
Wei Liu	University of Technology Sydney, Australia
Jiayang Liu	Tsinghua University, China
Yunlong Liu	Xiamen University, China
Yi Liu	Zhejiang University of Technology, China
Ye Liu	Nanjing University of Posts and Telecommunications, China
Zhilei Liu	Tianjin University, China
Zheng Liu	Nanjing University of Posts and Telecommunications, China
Cheng Liu	City University of Hong Kong, Hong Kong, China
Linfeng Liu	Nanjing University of Posts and Telecommunications, China
Baoping Liu	IIE, China
Guiping Liu	Hetao College, China
Huan Liu	Xi'an Jiaotong University, China
Gongshen Liu	Shanghai Jiao Tong University, China
Zhi-Yong Liu	Institute of Automation, Chinese Academy of Science
Fan Liu	Beijing Ant Financial Services Information Service Co., Ltd., China
Zhi-Wei Liu	Huazhong University of Science and Technology, China
Chu Kiong Loo	University of Malaya, Malaysia
Xuequan Lu	Deakin University, Australia
Huimin Lu	Kyushu Institute of Technology, Japan
Biao Lu	Nankai University, China
Qun Lu	Yancheng Institute of Technology, China
Bao-Liang Lu	Shanghai Jiao Tong University, China
Shen Lu	The University of Sydney, Australia
Junyu Lu	University of Electronic Science and Technology of China, China
Zhengding Luo	Peking University, China
Yun Luo	Shanghai Jiao Tong University, China
Xiaoqing Lyu	Peking University, China

Kavitha MS	Hiroshima University, Japan
Wanli Ma	University of Canberra, Australia
Jinwen Ma	Peking University, China
Supriyo Mandal	Indian Institute of Technology Patna, India
Sukanya Manna	Santa Clara University, USA
Basarab Matei	University Paris 13, France
Jimson Mathew	IIT Patna, India
Toshihiko Matsuka	Chiba University, Japan
Timothy McIntosh	La Trobe University, Australia
Philip Mehrgardt	The University of Sydney, Australia
Jingjie Mo	Chinese Academy of Sciences, China
Seyed Sahand Mohammadi Ziabari	Vrije Universiteit Amsterdam, The Netherlands
Rafiq Mohammed	Murdoch University, Australia
Bonaventure Molokwu	University of Windsor, Canada
Maram Monshi	The University of Sydney, Australia
Ajit Narayanan	Auckland University of Technology, New Zealand
Mehdi Neshat	Adelaide University, Australia
Aneta Neumann	The University of Adelaide, Australia
Frank Neumann	The University of Adelaide, Australia
Dang Nguyen	University of Canberra, Australia
Thanh Nguyen	Robert Gordon University, UK
Tien Dung Nguyen	University of Technology Sydney, Australia
Thi Thu Thuy Nguyen	Griffith University, Australia
Boda Ning	RMIT University, Australia
Roger Nkambou	Uqam, Canada
Akiyo Nomura	IBM Research - Tokyo, Japan
Anupiya Nugaliyadde	Murdoch University, Australia
Atsuya Okazaki	IBM Research, Japan
Jonathan Oliver	Trendmicro, Australia
Toshiaki Omori	Kobe University, Japan
Takashi Omori	Tamagawa University, Japan
Shih Yin Ooi	Multimedia University, Malaysia
Seiichi Ozawa	Kobe University, Japan
Huan Pan	Ningxia University, China
Paul Pang	Unitec Institute of Technology, New Zealand
Shuchao Pang	Macquarie University, Australia
Kitsuchart Pasupa	King Mongkut's Institute of Technology Ladkrabang, Thailand
Jagdish Patra	Swinburne University of Technology, Australia
Cuong Pham	Griffith University, Australia
Mukesh Prasad	University of Technology Sydney, Australia
Yu Qiao	Shanghai Jiao Tong University
Feno Heriniaina Rabevohitra	Chongqing University, China
Sutharshan Rajasegarar	Deakin University, Australia

Md Mashud Rana	CSIRO, Australia
Md Mamunur Rashid	CQUniversity, Australia
Pengju Ren	Xi'an Jiaotong University, China
Rim Romdhane	Devoteam, France
Yi Rong	Wuhan University of Technology, China
Leszek Rutkowski	Czestochowa University of Technology, Poland
Fariza Sabrina	CQU, Australia
Naveen Saini	Indian Institute of Technology Patna, India
Toshimichi Saito	Hosei University, Japan
Michel Salomon	University of Bourgogne Franche-Comté, France
Toshikazu Samura	Yamaguchi University, Japan
Naoyuki Sato	Future University Hakodate, Japan
Ravindra Savangouder	Swinburne University of Technology, Australia
Rafal Scherer	Czestochowa University of Technology, Poland
Erich Schikuta	University of Vienna, Austria
Fatima Seeme	Monash University, Australia
Feng Sha	The University of Sydney, Australia
Jie Shao	University of Electronic Science and Technology, China
Qi She	Intel Labs China, China
Michael Sheng	Macquarie University, Australia
Jinhua Sheng	Hangzhou Dianzi University, China
Iksoo Shin	Korea Institute of Science and Technology Information, South Korea
Mohd Fairuz Shiratuddin	Murdoch University, Australia
Hayaru Shouno	University of Electro-Communications, Japan
Sarah Ali Siddiqui	Macquarie University, Australia
Katherine Silversides	The University of Sydney, Australia
Jiri Sima	Czech Academy of Sciences, Czech Republic
Chiranjibi Sitaula	Deakin University, Australia
Marek Śmieja	Jagiellonian University, Poland
Ferdous Sohel	Murdoch University, Australia
Aneesh Srivallabh Chivukula	University of Technology Sydney, Australia
Xiangdong Su	Inner Mongolia University, China
Jérémie Sublime	ISEP, France
Liang Sun	University of Science and Technology Beijing, China
Laszlo Szilagyi	Obuda University, Hungary
Takeshi Takahashi	National Institute of Information and Communications Technology, Japan
Hakaru Tamukoh	Kyushu Institute of Technology, Japan
Leonard Tan	University of Southern Queensland, Australia
Gouhei Tanaka	The University of Tokyo, Japan
Maolin Tang	Queensland University of Technology, Australia
Selvarajah Thuseethan	Deakin University, Australia
Dat Tran	University of Canberra, Australia

Oanh Tran	Vietnam National University, Vietnam
Enmei Tu	Shanghai Jiao Tong University, China
Hiroaki Uchida	Hosei University, Japan
Shinsuke Uda	Kyushu University, Japan
Brijesh Verma	Central Queensland University, Australia
Chaokun Wang	Tsinghua University, China
Xiaolian Wang	Chinese Academy of Sciences, China
Zeyuan Wang	The University of Sydney, Australia
Dong Wang	Hunan University, China
Qiufeng Wang	Xi'an Jiaotong-Liverpool University, China
Chen Wang	Institute of Automation, Chinese Academy of Sciences, China
Jue Wang	BIT, China
Xiaokang Wang	Beihang University, China
Zhenhua Wang	Zhejiang University of Technology, China
Zexian Wang	Shanghai Jiao Tong University, China
Lijie Wang	University of Macau, Macau, China
Ding Wang	Chinese Academy of Sciences, China
Peijun Wang	Anhui Normal University, China
Yaqing Wang	HKUST, China
Zheng Wang	Southwest University, China
Shuo Wang	Monash University and CSIRO, Australia
Shi-Lin Wang	Shanghai Jiaotong University, China
Yu-Kai Wang	University of Technology Sydney, Australia
Weiqun Wang	Institute of Automation, Chinese Academy of Sciences, China
Yoshikazu Washizawa	University of Electro-Communications, Japan
Chihiro Watanabe	NTT Communication Science Laboratories, Japan
Michael Watts	Auckland Institute of Studies, New Zealand
Yanling Wei	University of Leuven, Belgium
Hongxi Wei	Inner Mongolia University, China
Kok Wai Wong	Murdoch University, Australia
Marcin Woźniak	Silesian University of Technology, Poland
Dongrui Wu	Huazhong University of Science and Technology, China
Huijun Wu	The University of New South Wales, Australia
Fei Wu	Nanjing University of Posts and Telecommunications, China
Wei Wu	Inner Mongolia University, China
Weibin Wu	Chinese University of Hong Kong, Hong Kong, China
Guoqiang Xiao	Shanghai Jiao Tong University, China
Shi Xiaohua	Shanghai Jiao Tong University, China
Zhenchang Xing	Australian National University, Australia
Jianhua Xu	Nanjing Normal University, China
Huali Xu	Inner Mongolia University, China
Peng Xu	Jiangnan University, China

Caiming Zhong	Ningbo University, China
Jinghui Zhong	South China University of Technology, China
Mingyang Zhong	CQU, Australia
Xinyu Zhou	Jiangxi Normal University, China
Jie Zhou	Shenzhen University, China
Yuanping Zhu	Tianjin Normal University, China
Lei Zhu	Lingnan Normal University, China
Chao Zhu	University of Science and Technology Beijing, China
Xiaobin Zhu	University of Science and Technology Beijing, China
Dengya Zhu	Curtin University, Australia
Yuan Zong	Southeast University, China
Futai Zou	Shanghai Jiao Tong University, China

Contents – Part IV

Adversarial Networks and Learning

Convolutional Neural Networks

Deep Neural Networks

Embeddings and Feature Fusion

Human Centred Computing

Human Centred Computing for Emotion

Hybrid Models

Adversarial Networks and Learning

Adversarial Deep Learning with Stackelberg Games

Aneesh Sreevallabh Chivukula(✉), Xinghao Yang, and Wei Liu

School of Computer Science, University of Technology Sydney, Ultimo, Australia
{Aneesh.Chivukula,Wei.Liu}@uts.edu.au,
Xinghao.Yang@student.uts.edu.au

Abstract. Deep networks are vulnerable to adversarial attacks from malicious adversaries. Currently, many adversarial learning algorithms are designed to exploit such vulnerabilities in deep networks. These methods focus on attacking and retraining deep networks with adversarial examples to do either feature manipulation or label manipulation or both. In this paper, we propose a new adversarial learning algorithm for finding adversarial manipulations to deep networks. We formulate adversaries who optimize game-theoretic payoff functions on deep networks doing multi-label classifications. We model the interactions between a classifier and an adversary from a game-theoretic perspective and formulate their strategies into a Stackelberg game associated with a two-player problem. Then we design algorithms to solve for the Nash equilibrium, which is a pair of strategies from which there is no incentive for either the classifier or the adversary to deviate. In designing attack scenarios, the adversary's objective is to deliberately make small changes to test data such that attacked samples are undetected. Our results illustrate that game-theoretic modelling is significantly effective in securing deep learning models against performance vulnerabilities attached by intelligent adversaries.

1 Introduction

Adversarial learning algorithms are designed to exploit vulnerabilities in a given machine learning algorithm. These vulnerabilities are studied under various attack scenarios [1] and attack policies [15] formulated by an intelligent adversary. Goodfellow et al. [7] observe that many imperceptible infinitesimal non-random changes to the deep networks input (a.k.a., adversarial examples) add up to an arbitrarily large change in their output. For example, adding targeted noise to the input of a deep network classifier misleads it into classifying the image of a "panda" as that of a "gibbon" [7].

In this paper, we design a two-player Stackelberg game from interactions between an adversary and target Convolutional Neural Network (CNN) classifier. In every game iteration, the attack objective is to generate adversarial examples that mislead target CNN's classification result. Target CNN is manipulated to rematch adversary's data distribution changes on original data.

We then design adversarial learning algorithms that arrive at a Nash equilibrium for the game. Here, Nash equilibrium is a balanced state of play where there is no incentive

T. Gedeon et al. (Eds.): ICONIP 2019, CCIS 1142, pp. 3–12, 2019.
https://doi.org/10.1007/978-3-030-36808-1_1

for either the adversary or the classifier to deviate. This state is determined by each player's payoff functions.

We do not assume the adversary knows the targeted network architecture as this assumption cannot be met in practical attack scenarios. Our adversary's payoff function is composed of an adversarial cost and a classification error. The adversarial cost is computed from the magnitude of additive pixel manipulation to labelled training data. The classification error is computed from a CNN model's performance.

Following are the major contributions of this paper:

- We formulate game-theoretic adversaries and develop new algorithms for adversarial learning that attack deep learning models for image classification.
- We propose to formulate the game strategy space for the adversary using autoencoder networks in a stochastic optimization problem. We propose a new simulated annealing algorithm in the game's randomized strategy space to optimize the adversary's payoff function.
- Using Nash equilibrium from our (variable-sum two-player sequential) Stackelberg game model, we build a secure CNN classifier immune to multi-label adversarial data manipulations. Upon game convergence the CNN classifier can find weights that are robust to targetted adversarial attacks.
- We demonstrate the effectiveness of the proposed adversarial manipulations in case of both convolutional neural networks and generative adversarial networks. We also successfully benchmark our adversarial manipulations against existing adversarial examples to mislead classifiers in deep learning, deep generative learning and game theoretical adversarial learning.

2 Related Work

In deep learning, adversarial examples have been generated using two types of attacks called whitebox attacks and blackbox attacks [4]. In a whitebox attack, adversary has full knowledge of targetted network architecture, including the input data, the output predictions, the medial parameters and the number of layers. By contrast, adversary has no knowledge of targetted network architecture in a blackbox attack. Our research belongs to the blackbox attack scenarios, which is more realistic in practice.

Lowd et al. [11] introduced adversarial classification to construct adversarial attacks learning vulnerabilities of classifiers. Adversary's goal is to learn adversarial cost functions manipulating decision boundaries between class labels without assuming a data distribution for training data. Bruckner et al. [3] generated test data in response to a predictive model where interaction between data generator and classifier is modelled as a Stackelberg prediction game. An optimization problem involving adversaries is then derived to determine game's solution. Bruckner et al. [2] defined adversarial learning Stackelberg games to combine adversarial cost functions with adversarial payoff functions solving for adversarial attacks. Liu and Chawla [10] further analyzed the effect of adversarial payoff functions on Nash equilibrium of a prediction game.

Similar to our payoff functions and blackbox attack scenarios, Chivukula and Liu [6] proposed adversarial payoff functions for attacking deep learning classifiers. They proposed game theoretical adversarial learning algorithm that uses genetic algorithms for attacking supervised deep learning. Two-player Stackelberg game in Chivukula and Liu [6] is extended into Multiplayer Stackelberg game in Chivukula and Liu [5].

3 Game Formulation

In this section, we discuss problem formulation for two-player sequential Stackelberg game which is the foundation of our proposed adversarial learning algorithm.

3.1 Stackelberg Game Formulation

The key ingredients in a Stackelberg game [13] are (a) modelling players as the decision makers, (b) modelling actions (or series of actions) taken by players, and (c) modelling payoffs motivating players. To model adversarial attack on CNNs, we assume an intelligent adversary as a leader player (L) interacting in a Stackelberg game with a follower player (F) – the CNN classifier. We then design data manipulations over a strategy space as the adversary's attack actions. Such data manipulations are performed on targetted classes. In response to each attack, the CNN is allowed to re-optimize weights on manipulated training data.

The leader L starts the game by making initial action (or move). The actions available to L and F are assumed to be over search spaces (or strategy spaces) A and W respectively. The outcome of an action is determined by L's and F's payoff functions $J_L \in \mathbb{R}$ and $J_F \in \mathbb{R}$, respectively. Once CNN has been trained to learn optimal parameters w^* for all $w \in W$ on training data, the adversary's best action is formulated as α^* for all $\alpha \in A$ in a sequential game. The payoff function J_L for adversary is formulated as solving for α^* in Eq. 1. The payoff function J_F for classifier is formulated as solving for w^* in Eq. 2. In this research, A is taken to be data space of pixels in an image database and W is taken to be feature space of weights in a CNN classifier.

$$\alpha^* = argmax_{\alpha \in A} J_L(\alpha, w^*) \tag{1}$$

$$w^* = argmax_{w \in W} J_F(\alpha^*, w) \tag{2}$$

Equations 1 and 2 can be combined into Eq. 3 to form a sequential game with (α^*, w^*) expressed over adversarial manipulations α.

$$(\alpha^*, w^*) = argmax_{\alpha \in A} J_L(\alpha, argmax_{w \in W} J_F(\alpha, w)) \tag{3}$$

We assume the leader L's payoff is proportional to the follower F's payoff. The relation between L's and F's payoff functions is determined by a constant profit Φ for the game, variable profit $cost_L(\alpha)$ for the adversary and variable profit $cost_F(w)$ for the classifier. Further, the relative importance of adversary's cost $cost_L(\alpha)$ compared to classifier's cost $cost_F(w)$ in determining game solutions is controlled by a weighting parameter λ.

$$J_L + J_F = \Phi + \lambda * cost_L(\alpha) + cost_F(w) \tag{4}$$

To formulate a Stackelberg game using Eq. 4, we express classifier's J_F in terms of adversary's J_L so that Eq. 3 can be rewritten as a two-player game in Eq. 5. By treating CNN as a blackbox model, we assume $cost_L(\alpha)$ is independent of $cost_F(w)$ in adversary's attack scenario. It is worth noting that the cost of re-optimizing a CNN, $cost_F(w)$, does not have a closed form expression for optimization. Then the overall objective of the game is:

$$(\alpha^*, w^*) = argmax_{\alpha \in A} J_L(\alpha, argmax_{w \in W}(\Phi + \lambda * cost_L(\alpha) + cost_F(w) - J_L(\alpha, w))) \tag{5}$$

In every game iteration, each player's move depends on the opponent's previous move. For a CNN learning optimal w^* for all $w \in W$ on training data X_{train}, the adversary generates adversarial manipulation α^* by searching over candidate solutions $\alpha \in A$ best suited for attacking w^* according to game model in Eq. 5. The classifier is then allowed to re-optimize w^* on manipulated data $X_{train} + \alpha^*$ to defend against adversarial manipulation α^*. Game ends if the adversary's payoff J_L stops increasing with the game iterations. At the end of the game, adversary converges onto optimal adversarial manipulation α^*.

Equation 5 is a bilevel stochastic optimization problem. We design adversary's payoff function J_L that can be solved in Eq. 5 for optimal attack policy (α^*, w^*) where $cost_L(\alpha) = \|\alpha\|_F$.

$$J_L(\alpha, w) = error_F(w) - \lambda * cost_L(\alpha) \tag{6}$$

We define an attack scenario where adversary's payoff function J_L is given in Eq. 6. In Eq. 6, $error_F(w)$ is CNN's classification error for targetted classes and $cost_L(\alpha)$ is the adversary's manipulation cost for finding optimal solutions in Eq. 5. Our intuition for the attack scenario with targetted classes is that an adversary aims to increase CNN's error $error_F(w)$ of misclassifying targetted classes while ensuring minimum change $cost_L(\alpha)$ to clean data. We measure $cost_L(\alpha)$ in terms of Frobenius norm $\|\alpha\|_F$ of manipulating tensor with same shape as encoded training data examples.

The $cost_L(\alpha)$ is enhanced by weighting term λ, which empirically evaluates relative importance of error $error_F(w)$ and cost $cost_L(\alpha)$ in adversarial attack scenarios. Setting low value to λ leads to low values for adversarial cost $\lambda * cost_L(\alpha)$ in comparison to (hopefully high) misclassification error $error_F(w)$. For training data X_{train}, λ allows us to control effect of $\|\alpha\|_F$ on adversarial data $X_{train} + \alpha$.

To optimize adversarial manipulation $\alpha \in A$, we use a simulated annealing algorithm to solve the game model in Eq. 5 with payoff function given in Eq. 6. In an autoencoder network, adversarial manipulations α are generated on encoded data $Enc(X_{train})$, where Enc is the encoder function of an autoencoder network. Equation 6 is evaluated by decoding the perturbed data $Dec(Enc(X_{train}) + \alpha)$ that has been subject to adversarial manipulation α, where Dec is the decoder function of a autoencoder network.

3.2 Stackelberg Game Illustration

Figure 1 is a flowchart for our adversarial autoencoder based Stackelberg game model. A multi-label classifier $CNN_{original}$ (henceforth shortened as CNN_o) with weights

$w^* \in W$. It is trained on labelled training data X_{train} and evaluated on labelled testing data X_{test} sourced from an image database. CNN_o participates in a two-player game with our game theoretical adversary. Adversary attacks CNN_o on a targetted positive label $target = pos$ by generating optimal attack $\alpha^* \in A$ at Nash equilibrium for every negative label $neg \in Neg$ that targetted positive label pos is manipulated into. In this research pos and Neg are class labels where $overall = pos \cup Neg$, and $A = Enc(X_{train})$ is determined by an autoencoder function Enc trained on X_{train}.

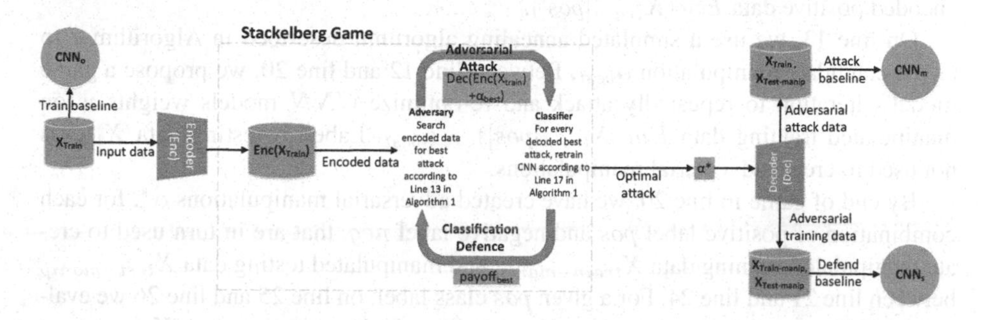

Fig. 1. A flowchart illustrating the adversarial autoencoder based Stackelberg game-theoretic modelling.

In each iteration of game, adversarial manipulation α_{best} is generated by the simulated annealing algorithm in Algorithm 2. For training data X_{train}, each α_{best} generates adversarial data $Enc(X_{train}) + \alpha_{best}$ in encoded space. It is then decoded as $Dec(Enc(X_{train}) + \alpha_{best})$ to be evaluated against CNN_o.

Upon convergence game outputs optimal α^* inferred for each pair of *pos* and *neg*. All α^*'s are then combined to effect a multi-label adversarial attack on CNN_o to output manipulated classifier $CNN_{manipulated}$ (henceforth shortened as CNN_m). CNN_m is finally retrained into secure classifier CNN_{secure} (henceforth shortened as CNN_s) that is robust to multi-label adversarial attacks.

4 Our Proposed Algorithms

In this section, we present the algorithms for a two-player Stackelberg game defined by Eq. 5.

Algorithm 1 is the training algorithm for Stackelberg game that solves for adversarial manipulations α^* in Eq. 5. As input, Algorithm 1 requires labelled training data X_{train}, labelled testing data X_{test}, target positive class *pos* for adversarial attack and negative classes Neg to mislead a CNN classifying *pos*. A key difference between adversarial examples in literature and those produced in every move of our game's iteration is that Algorithm 1 re-optimizes CNN's optimal weights w^* such that adversarial attacks α^* increase adversary's payoffs J_L across game moves. Moreover, various blackbox attack scenarios on CNN are controlled by λ weighting term on adversarial cost in Eq. 6. Algorithm 2 presents the simulated annealing method, which is used to determine best adversarial manipulation α_{best} in each game iteration.

4.1 Adversarial Learning Algorithm

Algorithm 1 initializes the game on line 3 by training CNN to get CNN_o(with optimal training weights w^*), and line 4 trains Autoencoder on X_{train} to get functions Enc and Dec. Adversary is assumed to target positive class pos to mislead classifier into misclassifying pos data into any negative class $neg \in Neg$. Two-player Stackelberg game loop from line 5 to line 20 creates adversarial data for attacking CNN. Between line 9 and line 11, we create adversarial data by adding random manipulation α to encoded positive data $Enc(X_{train}[pos])$.

On line 13, we use a simulated annealing algorithm described in Algorithm 2 to create candidate manipulation α_{best}. Between line 12 and line 20, we propose a game model's iteration to repeatedly attack and re-optimize CNN models weights w on manipulated training data $Enc(X_{train}[pos]) + \alpha_{best}$. Labelled testing data X_{test} is not used to create adversarial manipulations.

By end of game in line 20, we have created adversarial manipulations α^*, for each combination of positive label pos and negative label neg, that are in turn used to create manipulated training data $X_{train-manip}$ and manipulated testing data $X_{test-manip}$ between line 21 and line 24. For a given pos class label, on line 25 and line 26 we evaluate CNN_o on original testing data X_{test} and manipulated testing data $X_{test-manip}$

Algorithm 1. Adversarial Autoencoder based Stackelberg Game

Input:
1: Labelled training data X_{train}, Labelled testing data X_{test} (for final evaluations only), Positive class label pos, Negative class labels Neg, Cost weighting term λ, Maximum step size $maxstep$, Upper bound for simulated annealing iteration count $maxiter$
Output:
2: Original performance $error_o$, Attack performance $error_m$, Defence performance $error_s$

3: Train CNN on X_{train} to output model CNN_o, $CNN_m = CNN_o$
4: Train Autoencoder on X_{train} to get Encoder function Enc and Decoder function Dec
5: **for** $neg \in Neg$ **do**
6: $mask = mean(Enc(X_{train}[neg])) - mean(Enc(X_{train}[pos]))$
7: exitgame = False, $payoff_{prev} = payoff_{best} = 0, \alpha^*[neg] = 0$
8: **while** $\neg$ exitgame **do**
9: Randomly generate α, a tensor with values in [0,$maxstep$] and $mask$ size and shape
10: $\alpha = \alpha \odot mask$
11: Evaluate CNN on $Dec(Enc(X_{train}[pos]) + \alpha) \cup X_{train}[neg]$ to find error $error_{prev}$
12: $payoff_{prev} = error_{prev} - \lambda * \|\alpha\|_F$
13: Calculate α_{best} and $payoff_{best}$ from simulated annealing sa in Algorithm 2
14: **If** $payoff_{best} - payoff_{prev} > 0$ **then**
15: exitgame = False
16: $\alpha^*[neg] = \alpha_{best}$
17: Retrain CNN on $Dec(Enc(X_{train}[pos])) + \alpha_{best}) \cup X_{train}[neg])$
18: **else**
19: exitgame = True
20: $payoff_{prev} = payoff_{best}$
21: $X_{train-manip} = X_{train}, X_{test-manip} = X_{test}$
22: **for** $neg \in Neg$ **do**
23: $X_{train-manip} = X_{train-manip} \cup Dec(Enc(X_{train}[pos]) + \alpha^*[neg]) \cup X_{train}[neg]$
24: $X_{test-manip} = X_{test-manip} \cup Dec(Enc(X_{test}[pos]) + \alpha^*[neg]) \cup X_{test}[neg]$
25: Evaluate CNN_o on X_{test} to find error $error_o$
26: Evaluate CNN_m on $X_{test-manip}$ to find error $error_m$
27: Train CNN_m on $X_{train-manip}$ to output model CNN_s
28: Evaluate CNN_s on $X_{test-manip}$ to find error $error_s$
29: return $(error_o, error_m, error_s)$

Algorithm 2. Simulated Annealing with Random Restart

1: **function** SA($maxstep, mask, error_{prev}, payoff_{prev}, \alpha, maxiter, X_{train}, Enc, Dec$)
2: exitsearch = False, iter = 0, $payoff_{best} = error_{best} = 0$, jumpbound = 10
3: Initialize zeros tensor α_{best} with same shape as α
4: **while** $\neg$ exitsearch $\wedge$ iter<maxiter **do**
5: iter = iter + 1
6: Randomly generate manipulation increment δ, a tensor with with values in [0,$maxstep$]
7: $\delta = \delta \odot mask$
8: $\alpha = \alpha + \delta$
9: Evaluate CNN on $Dec(Enc(X_{train}[pos]) + \alpha) \cup X_{train}[neg]$ to find error $error_{curr}$
10: $payoff_{curr} = error_{curr} - \lambda * \|\alpha\|_F$
11: **If** $payoff_{curr} > payoff_{best}$ **then**
12: $payoff_{best} = payoff_{curr}$
13: $\alpha_{best} = \alpha$
14: $error_{best} = error_{curr}$
15: **If** $payoff_{curr} - payoff_{prev} > 0$ **then**
16: exitsearch = False
17: **else**
18: **If** $error_{curr} - error_{prev} \leq 0$ **then**
19: Randomly select $jump$ in [0,jumpbound]
20:
21: **If** $|error_{curr} - error_{prev}| > jump$ **then**
22: exitsearch = False
23: **else**
24: exitsearch = True
25: $payoff_{prev} = payoff_{curr}$
26: return ($payoff_{best}, \alpha_{best}$)

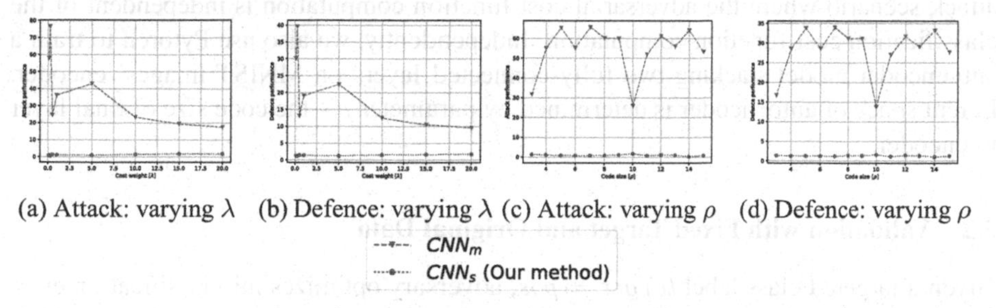

(a) Attack: varying λ (b) Defence: varying λ (c) Attack: varying ρ (d) Defence: varying ρ

Fig. 2. Testing performance (error) with variations in attack operators consisting of adversarial cost weight λ and autoencoder code size ρ.

to output errors $error_{original}, error_{manipulated}$ (shortened as $error_o$, $error_m$) respectively. A successful adversary misleads classifier into misclassifying *pos* as $neg \in Neg$ when $error_m$ is greater than $error_o$.

4.2 Simulated Annealing Algorithm

The simulated annealing algorithm *sa* in Algorithm 2 generates various candidate manipulations α on encoded positive data $Enc(X_{train}[pos])$. It targets to output the best changed α called α_{best}. From function arguments, *sa* initializes candidate manipulation α given by current iteration of game. Finally, *sa* returns adversary's payoff $payoff_{best}$ and classifier's error $error_{best}$ corresponding to α_{best}.

5 Experiments

In this section we discuss experimental validation and parameter settings for proposed adversarial learning algorithm. In a two-player Stackelberg game, various parameter settings correspond to various adversarial manipulations α^* that mislead CNN_o. Moreover we validate performance of various multi-label classifications produced by CNN models CNN_m trained on the MNIST database and tested on our data manipulations. We also compare defence performance of CNN models CNN_s retrained on data produced by our adversarial manipulations.

5.1 Classifier and Autoencoder Description

Pytorch[1] is used to create a CNN model that has deep representations of images using two convolution, one dropout and two fully connected layers. CNN has a rectified linear unit activation function. CNN has a log-softmax layer as loss function to make classification predictions. CNN is trained for 25 epochs on grayscale MNIST images to provide the baseline classification performance in our experiments. CNN's true positive rate for targeted class label is assumed to be adversary's target for Stackelberg game. We don't further fine tune the CNN learning processes since we assume a blackbox attack scenario where the adversarial cost function computation is independent of the classification cost function computation. Independently, we also use Pytorch to train a autoencoder model stacking two fully connected layers on MNIST images' encoder. Latent space of autoencoder is determined by parameter ρ – the code size of final layer in encoder.

5.2 Validation with Fixed Target and Original Data

Given a targeted class label $target = pos$, adversary optimizes misclassification error (a.k.a., attack performance) expressed in terms of true positive rate $tpr_{target}(w)$ while classifier optimizes classification error (a.k.a., defence performance) expressed in terms of f_1-score $fscore_{overall}(w)$ for all class labels $overall = pos \cup Neg$. Figure 2(a) and (c) compare CNN attack performance after adversarial manipulation with an example $pos = 7$. Figure 2(b) and (d) compare CNN defence performance after adversarial training. The x-axis in Fig. 2 gives parameter settings for adversarial attacks. The y-axis gives percentage error after adversarial attack. For attack parameters $\rho = 3$,

Table 1. Statistical t-tests before and after game by varying parameters for target class "7".

Attack parameter	p-values from t-statistics in Stackelberg games		
	CNN_o *vs* CNN_m	CNN_o *vs* CNN_s	CNN_m *vs* CNN_s
Cost weight (λ)	2.3×10^{-4}	3.3×10^{-2}	2.5×10^{-4}
Code size (ρ)	5.0×10^{-14}	6.9×10^{-6}	4.0×10^{-14}

[1] https://pytorch.org/docs/stable/index.html.

$maxstep = 1/10$ and $maxiter = 100$ in Algorithm 2, the weighting term λ in Eq. 6 is varied between 0.1 and 20. For same attack parameters, fixing $\lambda = 1$, Autoencoder's code size ρ is varied between 3 and 15.

Before the adversarial manipulation, the original performance of CNN_o, $error_o$, was found to be 0.86%. Table 1 compares p-values from pairwise t-tests comparing attack performances. Low p-values (< 0.05) allow us to reject null hypothesis that the attack performances are same before and after adversarial manipulations. Thus adversarial manipulations in our model are statistically significant across various parameter randomizations in comparing attack performances of original classifier CNN_o, manipulated classifier CNN_m and secure classifier CNN_s.

5.3 Validation with Varying Target and Generated Data

Table 2 gives classifier's defence performance as classification error expressed in terms of f_1-score $fscore_{overall}(w)$. In last column of Table 2, we change pos across various target labels by fixing attack parameters $\lambda = 0.25$, $\rho = 8$, $maxstep = 1/50$.

From Table 2, we observe that CNN_m after adversarial attack has higher defence error $error_m$ compared to original error $error_o$ of CNN_o before attack. Owing to our multi-step adversarial training, CNN_s achieves lower defence error $error_s$ than CNN_m defence error $error_m$. After game's convergence, classifier can be retrained on adversarial data to find secure model CNN_s that consistently has lower defence error $error_s$ than corresponding CNN_m defence error $error_m$ across varying target labels of both original data and generated data.

The low p-values (< 0.05) in Table 2 allow us to reject null hypothesis that the defence performances are the same before and after adversarial trainings. The low p-values also show that our game-theoretic attack settings are statistically significant across CNN models trained on datasets output by deep learning models, deep generative models and game theoretical adversarial learning models in literature.

Table 2. Comparisons on the defence to adversarial Nash equilibrium attacks

Classification error: Autoencoders attack in Stackelberg Game									
CNN_o	CNN_m								
	CNN [9]	$DCGAN$ [14]	$IWGAN$ [8]	$DeepFool$ [12]	$FGSM$ [7]	CNN_{GA} [5]	CNN_{SA} [5]	CNN_s (Our $method$)	Target class
1.16	27.71	28.63	27.38	24.71	27.41	3.91	2.12	0.77	0
0.52	39.24	33.74	40.11	31.21	37.16	6.38	2.98	0.42	1
1.59	22.89	22.53	21.68	26.61	21.11	5.62	5.31	1.01	2
1.43	17.56	26.61	20.06	26.99	15.01	5.08	3.73	0.53	3
0.91	46.47	43.54	35.46	36.44	33.02	3.81	16.0	0.92	4
1.45	38.81	40.51	43.03	32.33	33.24	51.35	32.78	0.91	5
1.04	28.12	31.55	35.02	27.79	21.81	9.03	7.09	0.86	6
1.98	32.24	39.18	36.78	33.75	29.39	22.49	9.73	1.05	7
1.34	24.82	33.77	26.18	28.52	15.54	4.08	4.17	1.16	8
2.00	26.16	43.31	26.78	25.64	21.18	30.71	10.92	1.51	9
t-statistics	3.5×10^{-9}	1.8×10^{-11}	3.9×10^{-10}	6.7×10^{-15}	7.6×10^{-9}	1.6×10^{-2}	8.9×10^{-3}	Base	

6 Conclusion and Future Work

We have formulated a Stackelberg game that models a stochastic optimization problem finding adversarial manipulations in convolutional neural networks. In each game iteration, adversary's strategy spaces and attack scenarios are determined by our payoff functions with evolutionary attack parameters. Optimal attack policy is found by a simulated annealing algorithm, which searches for attack parameters in encoded data. In Nash equilibrium, the game converges to adversarial manipulations affecting testing performance across targeted labels in multi-label classification models. Therefore, our proposed adversarial learning algorithm creates classifiers robust to targeted attacks. In future, we plan to experiment with deep generative networks and multiplayer games suitable for interclass discrimination in payoff functions.

References

1. Biggio, B., Roli, F.: Wild patterns: ten years after the rise of adversarial machine learning. Pattern Recogn. **84**, 317–331 (2018)
2. Brückner, M., Kanzow, C., Scheffer, T.: Static prediction games for adversarial learning problems. J. Mach. Learn. Res. **13**(1), 2617–2654 (2012)
3. Brückner, M., Scheffer, T.: Stackelberg games for adversarial prediction problems. In: Proceedings of ACM SIGKDD Conference on Knowledge Discovery and Data Mining (KDD) (2011)
4. Carlini, N., Wagner, D.: Adversarial examples are not easily detected: bypassing ten detection methods. In: Proceedings of the 10th ACM Workshop on Artificial Intelligence and Security (2017)
5. Chivukula, A., Liu, W.: Adversarial deep learning models with multiple adversaries. IEEE Trans. Knowl. Data Eng. **31**(6), 1066–1079 (2018)
6. Chivukula, A.S., Liu, W.: Adversarial learning games with deep learning models. In: Proceedings of 2017 International Joint Conference on Neural Networks (IJCNN) (2017)
7. Goodfellow, I., Shlens, J., Szegedy, C.: Explaining and harnessing adversarial examples. In: Proceedings of International Conference on Learning Representations (2015)
8. Gulrajani, I., Ahmed, F., Arjovsky, M., Dumoulin, V., Courville, A.C.: Improved training of wasserstein GANs. In: Advances in Neural Information Processing Systems 30 (2017)
9. Krizhevsky, A., Sutskever, I., Hinton, G.E.: Imagenet classification with deep convolutional neural networks. In: Advances in Neural Information Processing Systems (2012)
10. Liu, W., Chawla, S.: Mining adversarial patterns via regularized loss minimization. Mach. Learn. **81**(1), 69–83 (2010)
11. Lowd, D., Meek, C.: Adversarial learning. In: Proceedings of ACM SIGKDD Conference on Knowledge Discovery and Data Mining (KDD) (2005)
12. Moosavi-Dezfooli, S., Fawzi, A., Frossard, P.: Deepfool: a simple and accurate method to fool deep neural networks. In: Proceedings of Conference on Computer Vision and Pattern Recognition CVPR (2016)
13. Nisan, N., Roughgarden, T., Tardos, E., Vazirani, V.V.: Algorithmic Game Theory. Cambridge University Press, Cambridge (2007)
14. Radford, A., Metz, L., Chintala, S.: Unsupervised representation learning with deep convolutional generative adversarial networks. CoRR (2015)
15. Zhou, Y., Kantarcioglu, M., Xi, B.: A survey of game theoretic approach for adversarial machine learning. Wiley Interdisc. Rev. Data Min. Knowl. Discov. **9**(3), e1259 (2019)

Enhance Feature Representation of Dual Networks for Attribute Prediction

Yuchun Fang(✉), Yilu Cao, Wei Zhang, and Qiulong Yuan

School of Computer Engineering and Science, Shanghai University, Shanghai, China
ycfang@shu.edu.cn

Abstract. Traditional single branch CNN could not extract all the details of the input, which may lose some vital information, resulting in a decrease in recognition accuracy. In this paper, we propose a novel dual branch adversarial neural network named D-BANN. Inspired by adversarial learning, we drive parallel networks to extract complementary features and adopt a novel loss function to extend the application domain of the model. Moreover, we divide the network training procedure into multi-steps to alternatively optimize the loss functions. In order to evaluate the proposed method, we carry out comprehensive experiments on three attribute datasets. The results on facial attributes demonstrate that the proposed method can outperform other single task networks in face attribute recognition. Also, D-BANN achieves competitive results in two pedestrian datasets compared to the state-of-the-art multi-task methods. We visualize the D-BANN using Grad-CAM to verify the effectiveness of feature annotation.

Keywords: Adversarial training · Parallel network · Joint training · Attribute recognition

1 Introduction

Convolutional neural networks (CNN) have made significant breakthroughs in computer vision. Enhancing the representation ability is an essential problem in the design of network structure for handling complex missions such as attribute analysis.

Generally, the loss of CNN flows from the upper layer to the bottom layer in backpropagation, which may lose the critical information of objects in the middle layer. Therefore, it is difficult for a single network to extract all the details of the input. One direct solution is to use multiple sub-networks for task learning to handle the problem. The two-branch structure is effective in reaching a good balance in terms of computational cost and accuracy [4,5,8,10]. Bilinear CNN [5] set up local paired feature interactions in a translation-invariant manner. HD-CNN [10] embedded CNN into the hierarchical structure of the two-level classification. DDN [8] divided the data into disjoint classes and automatically built network structure. Dual Net [4] is the first network to focus on multi-CNN

T. Gedeon et al. (Eds.): ICONIP 2019, CCIS 1142, pp. 13–20, 2019.
https://doi.org/10.1007/978-3-030-36808-1_2

cooperation. Although these methods are instrumental in some tasks, they still have the problem of feature redundancy.

Recently, the adversarial learning based Generative Adversarial Network (GAN) [2] shows auspicious results in image generation. Yang and Peng [11,12] proposed the D-PCN model combing adversarial learning and Dual Net. It is expected that the feature representation of two branches of networks can be enriched by adversarial learning. D-PCN has excellent performance in several image classification datasets such as CIFAR-100 and ImageNet 32×32. However, due to the mode collapse, the training process is usually unstable, especially when dealing with fine-grained classification tasks, such as attribute recognition.

Inspired by the new adversarial loss of WGAN [1], we propose a novel model named Dual Branch Adversarial Neural Network (D-BANN). As shown in Fig. 1, the structure of D-BANN contains shared low-level convolutional layers, independent dual mid-level backbone networks, and fully connected layers for feature-level fusion. We also introduce the adversarial learning to enhance the feature representation of the two branches. Different from the D-PCN, we introduce the shared low-level layers to keep the model training more effective. In the proposed D-BANN, the adversarial loss of the discriminator denotes the competition between dual-branch. To handle the challenge of the instability of adversarial learning, we introduce the Wasserstein distance as the adversarial loss, which has been proved efficient in preventing gradient vanishes [1]. In realization, the losses of different branches are tuned alternatively to reach a stable network.

Through adversarial learning of the parallel networks, the proposed model can focus on different regions of the input image, and extract enhanced fusion features with less redundancy. The improved loss function can also serve to accelerate the training of the model. Experiments demonstrate that the proposed D-BANN can effectively enhance feature representation for visual attribute classification.

2 Method

The network structure of D-BANN is shown in Fig. 1. The shared low-level layers are the general convolutional layers that are used to form complex low-level feature maps of visual inputs, which can effectively decrease the number of parameters in the network and speed up the training. The subnet1 (S_1) and subnet2 (S_2) can be replaced by suitable backbone networks that might be decided by the tasks. Each subnet contains an independent classifier to supervise the training process. The features are respectively extracted from the two branches and fused in a cascade way as the inputs of the fully connected network. The discriminator (D) between the two subnets is the core of the D-BANN. Its parameters are tuned with the adversarial loss between the extracted features of the two subnets, guiding the two branches to extract complementary features. For simplicity, we adopt the three-layer fully connected network as classifiers for both subnets, the discriminator, and the fusion decision block.

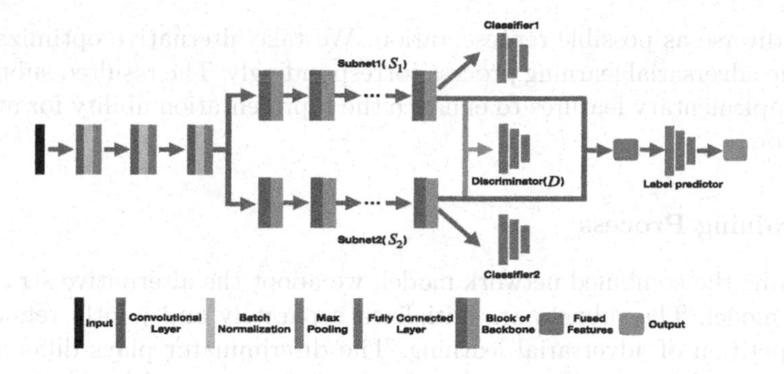

Fig. 1. Illustration of the proposed D-BANN. This architecture includes shared low-level layers, two independent subnets (S_1 and S_2), discriminator (D) and a fully connected feature fusion classifiers. Illustration of the proposed D-BANN. This architecture includes shared low-level layers, two independent subnets (S_1 and S_2), discriminator (D) and a fully connected feature fusion classifiers.

2.1 Loss Functions of the Discriminator

In the GAN model [2], the form of adversarial loss is a two-play minimax game between the discriminator D and the generator G as in Eq. (1). It aims at maximizing the distinction between the generated data and real data for D and minimizing the distribution distance between $G(z)$ and the real data for G.

$$\min_G \max_D V(D,G) = \mathbb{E}_{x \sim P_{data}(x)}[logD(x)] + \mathbb{E}_{z \sim P_z(z)}[log(1 - D(G(z)))] \quad (1)$$

In D-PCN [11,12], the adversarial loss is modified to a two-play max max set as denoted in Eq. (2). The discriminator is tuned to classify the features from Subnet1 as “real” and the features from Subnet2 as “false.”

$$\max_{S_1,S_2} \max_D V(S_1,S_2,D) = \mathbb{E}_x[logD(S_1(x))] + \mathbb{E}_x[log(1 - D(S_2(x)))] \quad (2)$$

where the S_1 and S_2 denote extractors of Subnet1 and Subnet2 respectively.

When training the dual-branch model with such adversarial loss on high-dimensional attribute images, the convergence of the model is hard to reach due to the vanishing gradient. To solve the problem, we introduce the Wasserstein distance in WGAN [1]. The overall objective function is denoted in Eq. (3). Different from the general GAN models, there is no generative component in the dual-branch architecture. The subnets of the network perform interactive learning with the discriminator through the adversarial loss.

$$\max_{S_1,S_2} \max_{D,\|D\|_L \le 1} V(D,S_1,S_2) = \mathbb{E}_{x \sim sub2}[D(S_2(x))] - \mathbb{E}_{x \sim sub1}[D(S_1(x))] \quad (3)$$

where D is a 1-Lipschitz function, by which, the calculation for the Wasserstein distance can be simplified.

To extract distinctive and discriminative features from dual branch network, the adversarial loss drives the dual-branches Subnet1 and Subnet2 to learn as diverse as possible representation. We take alternative optimization to adjust the adversarial learning process correspondingly. The resulted subnets can learn complementary features to enhance the representation ability for attribute recognition.

2.2 Training Process

For training the combined network model, we adopt the alternative strategy for a stable model. The subnets are initialized separately and jointly refined with the competition of adversarial learning. The discriminator plays different roles in different stages.

Low-Level Shared Layers. The shared layers serve to extract low-level features. They are trained through jointing separately with the two subnets, and the parameters are shared between the dual-branches. The design of low-level shared layers can be regarded as mimic the vision system of humans in sharing low-level neural paths.

Discriminator. The discriminator is trained to assign the correct labels to samples from Subnet1 and Subnet2, which makes the two extractors receive complementary features to avoid redundancy. The objective function of the discriminator is defined as in Eq. (4).

$$L_d = \min_{D, \|D\|_L \leq 1} -V(D, S_1, S_2) \tag{4}$$

here we employ Wasserstein distance as the loss function of the discriminator. In training, the parameters of discriminator are tuned alternatively jointing with Subnet1 or Subnet2 separately. The discriminator plays a key role in alternative training. The extractors from Subnet1 and Subnet2 use the discriminator as part of the loss function. Moreover, the Subnet1 extractor updates its parameters to extract features that look more 'real.' The discriminator, on the other hand, updates its parameters to make itself better at picking out 'false' features from 'real' features. We train the discriminator and the extractor form Subnet1 iteratively, and they are playing the game of 'cat and mouse' until the system reaches a balanced state. Besides, through the training of the discriminator, the weight from the Subnet2 has been further optimized.

Subnet1. The resulted feature maps of the shared layers are sent to Subnet1 for training. Taking the cross-entropy as loss function, the initial training of Subnet1 follows Eq. (5).

$$L_{sub1} = L_{cls1} \tag{5}$$

where the L_{cls1} denotes the error of classifier of Subnet1. After the initialization of Subnet2, the parameters of Subnet1 are fine-tuned with the supervision of the discriminator, and the objective function is denoted in Eq. (6).

$$L_{sub1} = L_{cls_1} + \alpha_1 L_{disc1}, \quad L_{disc1} = -\mathbb{E}_{x \sim sub1}[D(S_1(x))] \tag{6}$$

where α_1 is the hyperparameter to balance the contributions of the two parts of the loss, the L_{disc1} guides the features learned with the Subnet1 to be classified as "true" by the discriminator.

Subnet2. The training of Subnet2 is under the constraints of the discriminator to achieve both initial and fine-tuned parameters. The loss function of Subnet2 is defined as in Eq. (7).

$$L_{sub2} = L_{cls2} + \alpha_2 L_{disc2}, \quad L_{disc2} = -\mathbb{E}_{x \sim sub2}[D(S_2(x))] \tag{7}$$

where α_2 is a hyperparameter to balance the two parts of losses, and $D(S_2(x))$ denotes the output of discriminator for Subnet2. The discriminator acts as a regular term in the initialization of Subnet 2. In the alternative fine-tuned stage, the discriminator guides the dual-branches to learn different features.

Fusion Layers. With the trained the parallel networks, the features from Subnet1 and Subnet2 are integrated and fused in feature level as input to the final label predictor. The training of the fully connected layers is based on the loss function defined in Eq. (8).

$$L = L_{lp} + \lambda_1 L_{sub1} + \lambda_2 L_{sub2} \tag{8}$$

where the L_{lp} means the classification error of the final label predictor. L_{sub1} and L_{sub2} are regular terms balanced with two hyper-parameters λ_1 and λ_2 respectively.

3 Experiment

3.1 Datasets

To verify the validity of our model in attribute recognition, we have conducted sufficient experiments on three attribute datasets, including one face attribute dataset CelebA [7], two pedestrian attribute datasets Market-1501 [6] and Duke [6]. For pedestrian attributes datasets, we observe that several attributes are either more positive or more negative correlated with each other. As shown in

Table 1. Comparison of mean recognition accuracy on Market-1501, Duke and CelebA datasets. Numbers in brackets are +/=/−, which indicates D-BANN performs better/equally well/worse compared other approaches.

Datasets	Market-1501		Duke		Datasets	CelebA	
Methods	AVG	+/=/−	AVG	+/=/−	Methods	AVG	+/=/−
PedAtrNet [6]	88.19	**7/0/5**	82.39	**5/0/5**	Baseline [13]	80.03	**40/0/0**
APR [6]	88.16	**6/0/6**	86.42	**4/0/6**	Liu et al. [7]	87.33	**39/0/1**
D-PCN [11]	88.13	**9/1/2**	85.54	**8/0/2**	Independent [3]	91.06	**24/5/11**
Ours	**88.29**		**86.63**		Ours	**91.14**	

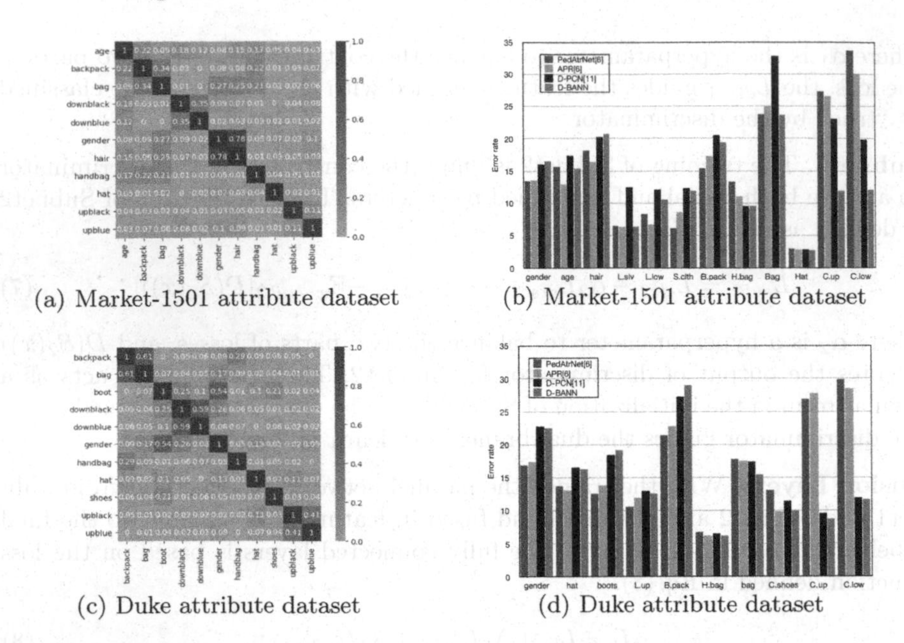

(a) Market-1501 attribute dataset (b) Market-1501 attribute dataset

(c) Duke attribute dataset (d) Duke attribute dataset

Fig. 2. (a) and (c) shows examples of attribute correlations on the Market-1501 and Duke datasets. (b) and (d) are performance comparison on Market-1501 and Duke attribute dataset.

Fig. 2(a) and (c), a high positive correlation exists between the attribute "gender" and "hair" in Market-1501 dataset, and a high negative attribute exists between the attribute "downblue" and "downblack" in Duke attribute datasets. Such correlation may deteriorate the identification ability of a single branch model. While with the dual-branch structure of D-BANN, more abundant representation can be learned to promote the overall performance.

3.2 Experimental Analysis and Comparison

In order to evaluate the effect of D-BANN, we conduct the attribute classification on Market-1501 and Duke attribute datasets. We compare our model with D-PCN [11] and two multi-task learning models, i.e. PedAtrNet [6] and APR [6]. We take the ResNet50 as the backbone of Subnet1 and Subset2 for a fair comparison.

In Table 1, D-BANN significantly outperforms D-PCN achieves accuracy by par with the multi-task methods on two pedestrian datasets. The comparison also demonstrates the stability of our method. For counting of ranks on a single attribute, D-BANN shows obvious superiority over the compared methods on the Market-1501 and Duke attribute datasets. The recognition accuracy on attributes "C.up" and "C.low" is greatly improved with our method.

We also compare the recognition results of 40 attributes on the CelebA dataset with the other attributes recognition methods, including Independent

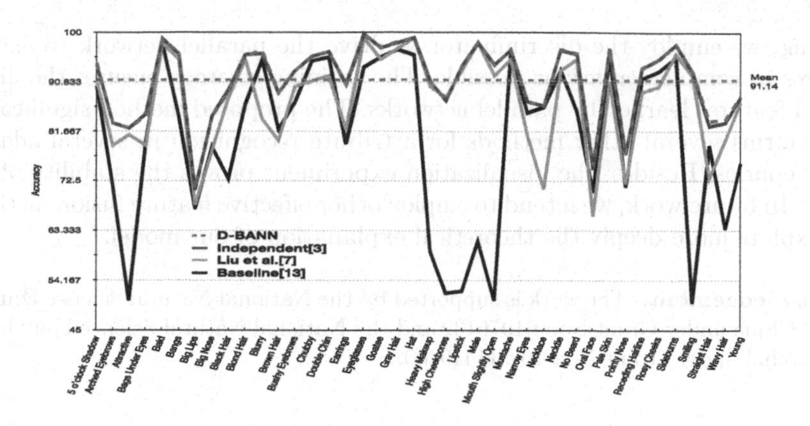

Fig. 3. Attribute recognition accuracy on CelebA dataset.

CNNs [3], Liu et al. [7], and the baseline [13]. We use VGG16 as the backbone of Subnet1 and Subset2. The results are summarized in Fig. 3, in which D-BANN outperforms other compared single-task models and achieves the best result in rank counting.

Further, for intuitive observation of the effectiveness of our model, we use Grad-CAM [9] to visualize D-BANN. The thermal maps are shown in the Fig. 4. In the response map of Grad-CAM, the gradual process from red to blue represents the gradual process of D-BANN attention from strong to weak. We can observe the obvious correlation between the D-BANN feature and human visual perception. People can quickly locate the part of the object that they want to observe while ignoring other things that are not related to the task. Similarly, the attention of D-BANN also responds strongly to the task-related features. The results of the visualization also verify the effectiveness of D-BANN.

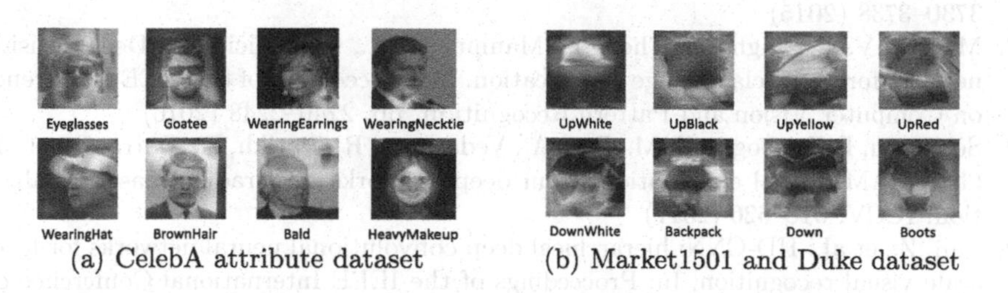

(a) CelebA attribute dataset (b) Market1501 and Duke dataset

Fig. 4. Grad-CAM visualization of D-BANN on CelebA dataset, Market-1501 attribute and Duke attribute dataset.

4 Conclusion

In this paper, we propose a dual adversarial neural network that can enhance the representation ability of CNN features. Inspired by the idea of adversarial learning, we employ the discriminator to drive the parallel network to extract features as complementary as possible. The training strategy ensures the difference of features learned by parallel networks. The proposed method significantly outperforms several other methods for attribute recognition in several adaptation scenarios. Besides, the visualization experiment proves the stability of our model. In future work, we intend to employ other effective feature fusion methods and explore more deeply the theoretical explanation of our model.

Acknowledgement. The work is supported by the National Natural Science Foundation of China under Grant No.: 61976132 and the National Natural Science Foundation of Shanghai under Grant No.: 19ZR1419200.

References

1. Arjovsky, M., Chintala, S., Bottou, L.: Wasserstein GAN. arXiv preprint arXiv:1701.07875 (2017)
2. Goodfellow, I., et al.: Generative adversarial nets. In: Advances in Neural Information Processing Systems, pp. 2672–2680 (2014)
3. Hand, E.M., Chellappa, R.: Attributes for improved attributes: a multi-task network utilizing implicit and explicit relationships for facial attribute classification. In: AAAI, pp. 4068–4074 (2017)
4. Hou, S., Liu, X., Wang, Z.: DualNet: learn complementary features for image recognition. In: 2017 IEEE International Conference on Computer Vision (ICCV), pp. 502–510. IEEE (2017)
5. Lin, T.Y., RoyChowdhury, A., Maji, S.: Bilinear CNN models for fine-grained visual recognition. In: Proceedings of the IEEE International Conference on Computer Vision, pp. 1449–1457 (2015)
6. Ou, X., Ma, Q., Wang, Y.: Improving person re-identification. Multimed. Tools Appl. **78**, 28257–28283 (2019)
7. Liu, Z., Luo, P., Wang, X., Tang, X.: Deep learning face attributes in the wild. In: Proceedings of the IEEE International Conference on Computer Vision, pp. 3730–3738 (2015)
8. Murthy, V.N., Singh, V., Chen, T., Manmatha, R., Comaniciu, D.: Deep decision network for multi-class image classification. In: Proceedings of the IEEE Conference on Computer Vision and Pattern Recognition, pp. 2240–2248 (2016)
9. Selvaraju, R.R., Cogswell, M., Das, A., Vedantam, R., Parikh, D., Batra, D., et al.: Grad-CAM: visual explanations from deep networks via gradient-based localization. ICCV, 618–626 (2017)
10. Yan, Z., et al.: HD-CNN: hierarchical deep convolutional neural networks for large scale visual recognition. In: Proceedings of the IEEE International Conference on Computer Vision, pp. 2740–2748 (2015)
11. Yang, S., Gang, P.: D-PCN: parallel convolutional neural networks for image recognition in reverse adversarial style (2017)
12. Yang, S., Peng, G.: D-PCN: parallel convolutional networks for image recognition via a discriminator (2017)
13. Zeiler, M.D., Fergus, R.: Visualizing and understanding convolutional networks. In: Fleet, D., Pajdla, T., Schiele, B., Tuytelaars, T. (eds.) ECCV 2014. LNCS, vol. 8689, pp. 818–833. Springer, Cham (2014). https://doi.org/10.1007/978-3-319-10590-1_53

Data Augment in Imbalanced Learning Based on Generative Adversarial Networks

Zhuocheng Zhou[1], Bofeng Zhang[1(✉)], Ying Lv[1], Tian Shi[1], and Furong Chang[1,2]

[1] School of Computer Engineering and Science, Shanghai University, Shanghai, China
{bearing512,bfzhang,lvying2016}@shu.edu.cn

[2] School of Computer Science and Technology, Kashi University, Kashi, Xinjiang, China

Abstract. Imbalanced learning is a traditional problem in machine learning and widely occurs in many applications. Most of the methods apply simple geometric transformation for data augment to imbalanced datasets. Due to those methods learn from local information, they might generate noisy samples in the dataset with high dimension and special complexity. To solve the problem, we propose an improved Generative Adversarial Networks with modification function (GAN-MF) to approximate the true distribution of the minority class of the dataset. The model could generate data from an overall perspective to overcome the limitation of the simple geometric transformation. The performance of GAN-MF is compared against multiple standard oversampling algorithms on several imbalanced learning tasks. Experiments demonstrate that the model has an improvement in data augment for imbalanced learning.

Keywords: Imbalanced learning · Generative Adversarial Networks (GAN) · Data augment · Modification function

1 Introduction

Learning from the imbalanced dataset is challenging and meaningful in many common areas including fraud detection, healthcare and medical diagnosis and many other applications. The reason why the performance of classifier drops sharply when dataset is imbalanced is that most standard algorithms assume or expect that the class distribution is balanced. So that, features of the minority class might be missed or neglected.

Imbalanced learning refers to the dataset in which one or several classes are outnumbered than the others. The gap in number of instances among the classes is defined as the imbalanced ratio (IR) [7].

Supported by National Key R&D Program of China grant (NO. 2017YFC0907505) and the Xinjiang Natural Science Foundation grant (NO. 2016D01B010).

T. Gedeon et al. (Eds.): ICONIP 2019, CCIS 1142, pp. 21–30, 2019.
https://doi.org/10.1007/978-3-030-36808-1_3

Devoted to improving the performance of imbalanced learning, different methods have been proposed, those could be summarized into several categories. The first is from the data perspective, focusing on reinforcing the learning on the minority by the means of sampling and feature selection [2]. Besides, synthetic sampling for data augment is also widely used. The second is to encourage the classifiers to minimize the cost errors by introduced cost-sensitive, ensemble learning and kernel-based methods [3,9]. The third is to restructure the classifier to suit the task according to the background of the applications. For instance, the algorithm of transfer learning and genetic algorithm are integrated in imbalanced learning [1].

However, with the rise in data complexity, methods mentioned above might be insufficient, especially in terms of data augment. Inspired by the fact that Deep Generative Models (DGMs) can synthesize new samples based on the distribution captured from the overall class rather than local information [8]. We try to synthesize sampling based on Generative Adversarial Networks (GAN) for data augment to improve the imbalanced binary classification. Whereas, the vanilla GAN model is restricted to continuous derivable variables for the gradient policy and the instability in training for model collapse and vanishing gradient.

To settle the matters, we proposed a novel GAN model (GAN-MF) based on a modification function $f(\mathrm{x})$ to approximate the true data distribution of the minority class. With the help of the modification function, the numeric discrete detests in imbalanced learning are converted into datasets with approximate Gaussian distribution that could be accepted by the GAN model and be trained in a stable way. The performance of the model is compared against multiple standard over-sampling algorithms and another generative model of Variational Auto-Encoder (VAE) based on 6 classifies. Experiments show GAN-MF has improved the results in imbalanced learning tasks.

The sections in the paper are organized as follows. In Sect. 2, an overview of related previous works regarding to GAN models and imbalanced learning are described. In Sect. 3, the model of the GAN-MF and application to the imbalanced learning is stated. In Sect. 4, the experiments and the results are addressed in detail. Finally, conclusions are provided in Sect. 5.

2 Related Works

Because of the simplicity and effectiveness of the algorithm, algorithms based on data augment are most widely used [6]. They offer additional minority-class instances derived by applying simple geometric transformations for the training. As the most classic one, Synthetic Minority Over-sampling Technique (SMOTE) provides a mechanism in creating artificial data based on the feature space similarities among the existing minority in the d-dimension dataspace X. The new instance x_{new} is created by $(x_i + \lambda(x_j - x_i))$, where $x_{i,j}$ is the minority instance in X, and x_j is selected considering to the k-nearest neighbor for x_i. Therefore, x_{new} is created in the vector between $x_{i,j}$, located in a random percent of way from $x_{i,j}$ as $\lambda \in [0, 1]$.

However, SMOTE has a vague understanding of the boundary and might generate noisy samples. To modify the algorithm, several rules including Edited Nearest Neighbor, balanced and weight level have been introduced into the algorithm which is summarized in [5]. Since they learn from local information, they might be ineffective in dealing with data in high dimensions.

DMGs have been gradually introduced to data augment for the excellent capability to represent multidimensional and complex data. Neural augment is firstly proposed in imbalanced picture classification in [11]. After that, Balancing GAN (BAGAN) [10] goes further more by taking attention mechanism to the training. The method based on Conditional generative adversarial networks (CGAN) [4] has also been addressed in learning numeric imbalanced data where additional space Y, as the label of the instance, is introduced to extra valuable information from latent space.

Although many efforts have been made, little research has been conducted in using GANs in learning the numerical variables dataset, and there is hardly no evidence suggests whether it is effective for GANs to generate discrete skewed data in dealing with imbalance learning. Meanwhile, it is unknown whether GANs have a shortage of capacity and training time when compared with standard over-sampling methods.

3 GAN-MF Model for Imbalanced Learning

3.1 The GAN-MF Model

The aim of generative model is to learn the data probability distribution $p_{data}(x)$ over the real space R^d. Although GANs have shown excellent ability to capture the distribution in many applications, the vanilla GAN model has been proved to be unsuitable to deal with discrete data for the model has hardly no gradient in generation process [12]. In addition, the model has a problem in training for model collapse and gradient vanish.

Thus, we introduced a modification function to figure out the limitation of the GAN model. The modification function $f(x)$ serves the role to convert problems of discrete data into an approximate continuous variable one that can be served by the GAN model. In other words, the d-dimensional real space is R^d is mapped to a special vector space $R^{d'}$ where numerical differences in features are relatively smooth and representative features of the dataset are preserved.

As the result, the two-player minimax game between the discriminator D and the generator G is improved. As G acts the role of producing fake data with striking resemblance from the latent variable z, D tells the data from sampled from the true data distribution $p_{data}(f(x))$ apart from those forged by G, where z is defined on the latent space Z.

The value function of the GAN-MF model is described in (1), where $E()$ represents the calculated expectation. From the view of D, it will maximize the outs if given data from real data and minimize the output if given data from G. Thus, D is optimized followed as $\log(1 - D(G(z)))$. At the same time, G tries the best to maximize the output of G when the fake is presented to D. G is optimized by $\log D(f(x))$. Finally, the generator's distribution $p_g(x)$ approaches to $p_{data}(f(x))$. The distribution of discrete dataset is related to $G(z, \theta)$, where θ is tuning parameters of the G.

$$\min_G \max_D V = E_{x \sim p_{data}(f(x))}(\log D(f(x))) + E_{z \sim p_z(Z)}[\log(1 - D(G(z)))] \quad (1)$$

3.2 Modification Function

With the help of modification function, the model has the ability to approximate the data distribution of the minority and generate augmented datasets that can present characteristics in a much smaller size than the simple geometric transformation.

Jensen-Shannon divergence and Wasserstein distance are widely used as the way to measure the difference in data distribution and optimizer for the network. We defined a vector $x = (x_1, x_2, x_3, ...x_n)$ as a discrete multivariate random variable where values of x_i are from fractions and integers. When we try to evaluate the Wasserstein distance between two probability distributions P_a and P_b, where $P_{(a,b)}$ is over the set of values for x, we find that it is a Linear Program (LP) problem. Therefore, the runtime reflects exponential growth with the increase in dimensions of data and variety of variables.

$$W(P_a, P_b) = min_{\gamma \in \prod(P_a, P_b)} \sum_i \sum_j \gamma(x_i, x_j) d(x_i, x_j) \quad (2)$$

where $d(x_i, x_j)$ is the distance between x_i, x_j and $\prod(P_a, P_b)$ is defined as the set of joint probability distribution $\gamma(x_i, x_j)$ whose marginals are P_a and P_b.

The same problem also occurs in the JSD which is used in most GAN models. As a consequence, it is clear that learning directly from difference in discrete mathematical distribution is not easy. Since the fact that it is difficult to measure the difference in discrete data distribution, the modification $f(x)$ become significant to GAN-MF. The modification we proposed is shown in (3), where μ_i is the mean of the feature x_i is and σ_i is the standard deviation of x_i.

$$max(0, \frac{x_i - \mu_i}{\sigma_i}) \quad (3)$$

Suppose that the networks is defined as $U = Wx + b$, $Z = F(U)$, where $F()$ is the activation function and W, b is the vector of weights and bias. When the modification is worked to the algorithm, the networks is transformed into (4):

$$U(f(x)) = W[max(0, \frac{x_i - \mu_i}{\sigma_i})] + b \quad (4)$$

Therefore, if $x_i > \mu_i$, $U(f(x)) = W(\frac{x_i - \mu_i}{\sigma_i}) + b$. All the features has been transformed to an approximate Gaussian distribution $\mathbf{N}(0,\ 1)$ which could be accepted by the GAN model and positive to the convergence of the networks. If $x_i < \mu_i$, x_i would be 0 in x. The vector would become sparse and the features would be more independent.

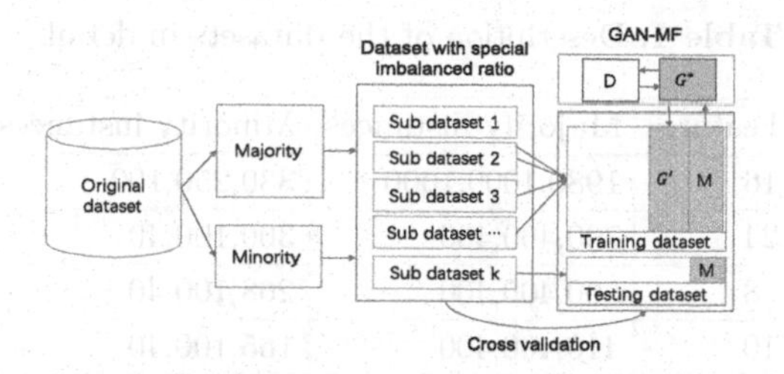

Fig. 1. GAN-MF for imbalanced learning. M refers to the minority of the dataset, $G^{'}$ is the augmented dataset generated by G^{*} for training. (Color figure online)

4 Experiments and Results

The framework of the GAN-MF Model in imbalanced learning we proposed is shown in Fig. 1.

(1) k-fold cross validation is applied with $k = 5$, the dataset is factitiously divided into 5 parts. Each part has approximately equal instances for both classes.
(2) All the hyperparameters of classifiers are performed with maximum accuracy under original dataset and used in subsequent experiments.
(3) The minority examples M colored in blue in Fig. 1 are isolated for training the GAN model G^{*} with tuning parameters.
(4) G^{*} as a generative model could generate artificial dataset $G^{'}$ by receiving random noise as input. Hence, the dataset used for training the classifies is composed of $G^{'}$ (colored green in Fig. 1) and sampled from real ones in M.

In this work, we rebalanced the dataset to the equal IR to the traditional methods. It made sure that classifies could learn unbrokenly. We doubled or tripled the number of minority classes in training for the methods based on deep generative models since they learn from an overall view.

Datasets. Several datasets from the Machine Learning Repository UCI and a credit card detection dataset were chosen for experiments. Aiming to objectively test the performance of the GAN-MF model, by the means of the under-sampling and random-sampling, the datasets from UCI were generated into additional dataset according to IR of 4, 10 on purpose. This procedure was applied only when the instances of the minority in the sub-dataset is no less than 5. Table 1 shows the datasets in detail. Values separated by comma in the table cells are related to the same dataset over original status and different IR in 4 and 10.

Architecture of the GAN-MF Model. In this work, both G and D used a module of multilayer perceptron with one single hidden layer. No convolution layers was need. Binary cross-entropy was served as the loss function. Relu was selected as the activation function in the output layer for G when Sigmoid was used in D. As Adam optimizer was used as the optimizer in G, Stochastic

Table 1. Description of the datasets in detail.

Dataset	Features	Majority instances	Minority instances	IR
Segment	16	1980,1000,1000	330,250,100	6,4,10
German	24	700,400,400	300,100,40	2.333,4,10
Pima	8	500,400,400	268,100,40	1.8656,4,10
Liver	10	416,400,400	165,100,40	2.491,4,10
Haberman	3	255,200,200	81,50,20	2.778,4,10
Ionosphere	34	255,200,200	126,50,20	1.786,4,10
Breastcancer	16	458,400,400	241,100,40	6,4,10
Credit card	29	284315	492	577.876

Gradient Descent(SGD) was chosen in D. Dropouts was used in G with a probability of 0.5. The input random noise followed a normal distribution. The other hyperparameters of the networks are described in Table 2. The optimal range for the numbers of epochs shoule be 5000–15000, much smaller than the one in the picture. The batch size should be set carefully to ensure that the final number of minority class instances is sufficient for the training. No dimensionality reduction methods were used. All samples with missing values were deleted.

Table 2. Parameters for GAN-MF model in detail. Including dimension d_z, number of hidden units for G and D, learning rate and batch size. The values in the same cell refers to the parameters under the IR of 4 and 10. N_G and N_D is defined as the number of units for hidden layer of G, D.

Dataset	d_z	N_G	N_D	Learning rate	Batch size
Segment	80,120	100,50	30,130	0.0005,0.0005	20,10
German	150,100	90,80	50,50	0.0005,0.0005	16,8
Pima	20,8	50,45	80,80	0.0005,0.0005	8,8
Liver	50,25	35,50	20,30	0.0001,0.0005	5,5
Haberman	10,10	20,20	10,15	0.0005,0.0005	10,8
Ionoshere	200,120	30,25	90,90	0.0005,0.0005	8,8
Breastcancer	70,70	90,90	30,30	0.0005,0.0005	10,10
Creditcard	200	36	100	0.0001,0.0001	10

Assessment Metric. F-measure, the geometric mean of specificity and sensitivity (G-mean) and Area Under the ROC Curve (AUC) were chosen as the assessment criteria. k-Nearest Neighbors (KNN), Logistic Regression (LR), Decision Trees (DT), AdaBoosting classifier, Nave Bayes (NB) and an ensemble learning method based on the simple voting (Vote) method were chosen as classifies.

Furthermore, a ranking score and the Friedman test were given for more holistic evaluation of the results. The ranking score was applied to cach data augment method for the experiments of 14 datasets under different assessment metrics and classifiers. In the rank, the best performing method ranks 1 and the worst one ranks 6. Besides, we defined the under-fitting as the situation that F-measure was under 50% and G-mean was under 40%. The under-fitting methods were set 6 in the rank. The Friedman test is a non-parametric statistical test, and widely used to detect the difference between treatments across multiple research attempts. The null hypothesis in the work is whether GAN-MF model is as effective as traditional over-sampling methods for data augment in imbalanced learning.

Results. The meaning ranking results are summarized in Fig. 2, where each plot is related to three assessment metrics and a classifier. Each mean rank is the result of 14 datasets based on the same classify. From a macro perspective, the model of GAN-MF has shown the improvement in most classifiers and datasets.

With fewer training data for data augment, we observe that the GAN-MF outperforms all other data augment methods when the voting algorithm

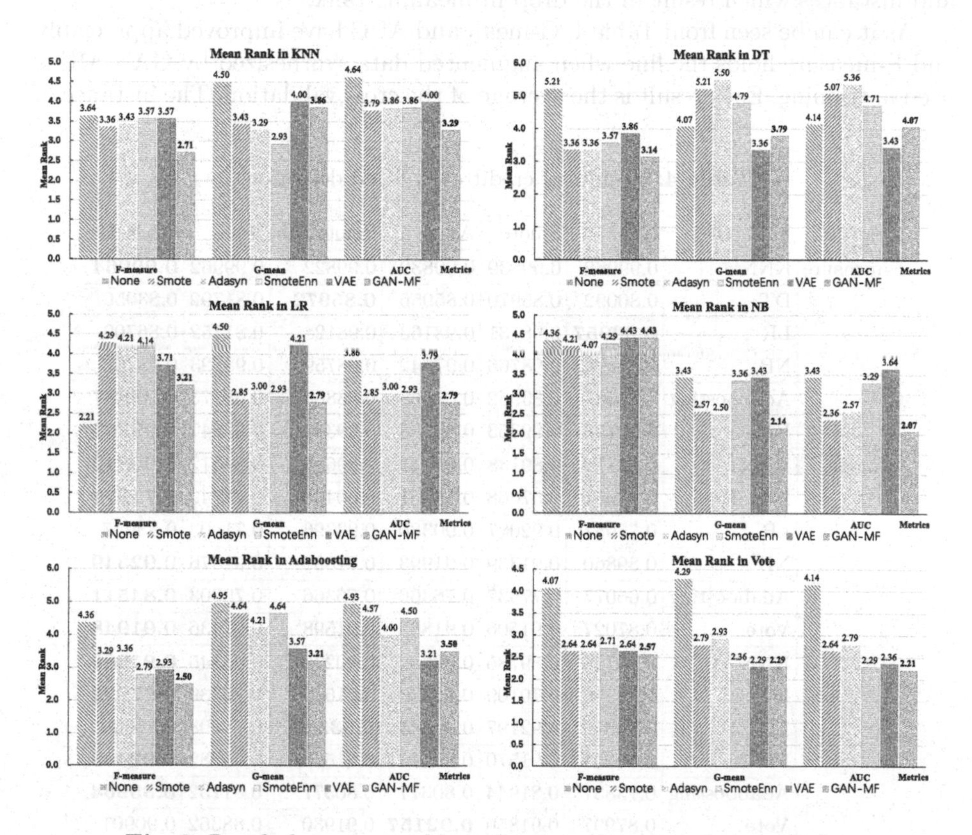

Fig. 2. Result for mean ranking of various data augment methods.

Table 3. Result of Friedman test. If $p < \alpha$, reject the hypothesis.

$\alpha = 0.05$	KNN	DT	LR	NB	Adaboost	Vote
X_γ^2	6.81	3.857	5.857	4.333	9.95	4.904
p	0.235	0.570	0.320	0.502	0.077	0.427

is selected as the classifier. It is also clear that the GAN-MF has an advantage to the metric of G-mean and AUC in more than four-fifths of cases.

The result of the Friedman test is shown in Table 3. All the p-values are more than the given standard value and the hypothesis are all not rejected where $\alpha = 0.05$. It means that the performance of the classifies show no bias in different methods and GAN-MF model is superior to traditional methods in data augment for imbalanced learning.

In the terms of the vibration in the mean rank for the GAN-MF, it should be noted that F-measure might be sick since the classifies would be favor to the majority and mark a high score for original imbalanced data. Both GAN and VAE have done a bad performance especially in the dataset with fewer features and instances which result in the drop in meaning rank.

As it can be seen from Table 4, G-mean and AUC have improved appreciably and F-measure holds the line when augmented data synthesized by GAN-MF is used in training. Each result is the average of the cross validation. The instance of

Table 4. Results of credit card fraud detection.

Metric	Methods	None	Smote	Adasyn	SmoteEnn	VAE	GAN-MF
F-measure	KNN	0.99962	0.99839	0.99839	0.99822	0.99962	**0.99964**
	DT	0.80092	0.85970	0.85956	**0.85972**	0.81792	0.83056
	LR	**0.99957**	0.98131	0.93163	0.98128	0.81752	0.86709
	NB	**0.98873**	0.98766	0.97642	0.98759	0.98723	0.98765
	Adaboosting	0.85081	0.83962	0.83865	0.83865	0.98773	**0.99887**
	Vote	0.99778	0.99353	0.98388	0.99331	0.99947	**0.99965**
G-mean	KNN	0.84812	0.89138	0.89013	0.89686	0.84812	**0.90812**
	DT	0.67981	0.57068	0.58051	0.60438	0.71749	**0.71979**
	LR	0.77057	0.92087	0.90380	0.93396	0.71803	0.83957
	NB	0.89860	0.91339	0.91993	0.91332	0.91576	**0.92349**
	Adaboosting	0.66077	0.77537	0.76366	0.76366	0.79003	**0.84511**
	Vote	0.87027	0.91506	0.91865	0.91598	0.87436	**0.91948**
AUC	KNN	0.86145	0.89785	0.89682	0.90276	0.86145	**0.91389**
	DT	0.76956	0.70109	0.70920	0.71636	0.76339	**0.77974**
	LR	0.80050	0.92197	0.90644	**0.93503**	0.78298	0.85602
	NB	0.90253	0.91570	0.92111	0.91563	0.91782	**0.92469**
	Adaboosting	0.73837	0.81944	0.80377	0.80377	0.81732	**0.85964**
	Vote	0.87997	0.91850	**0.92157**	0.91930	0.88362	0.90901

the augmented data generated by GAN-MF for training is about 1100, about one hundredth of the ones based on simple geometric transformation. Since GAN-MF learn the screwed dataset from an overall way, the classify can capture the representative feature in a more effective way. In the terms of G-mean and AUC, the GAN-MF model has outperformed in two-thirds of classifies. The classify of Adaboosting and KNN have been obviously improved by the GAN-MF.

5 Conclusion

In this work, we proposed a GAN-MF model for data augment to improve the imbalanced learning. Since the model learns the dataset from an overall view, it can generate data for augmentation based on the learned distribution. Modification function is employed to converts the numeric discrete detests into the one that could be train in a stable way. The model has been evaluated on several datasets, with much fewer augmented data, the model has done a good performance for most classifies, especially in dataset with high dimension.

More work should be taken to overcome the limitation of the model in stabilization, capacity and training time. The whole model still suffers from collapse problem. Our future work will try more different networks as well as take more other deep generative models into practice.

References

1. Al-Stouhi, S., Reddy, C.K.: Transfer learning for class imbalance problems with inadequate data. Knowl. Inf. Syst. **48**(1), 201–228 (2016)
2. Chen, H., Li, T., Fan, X., Luo, C.: Feature selection for imbalanced data based on neighborhood rough sets. Inf. Sci. **483**, 1–20 (2019)
3. Ding, S., et al.: Kernel based online learning for imbalance multiclass classification. Neurocomputing **277**, 139–148 (2018)
4. Douzas, G., Bacao, F.: Effective data generation for imbalanced learning using conditional generative adversarial networks. Expert Syst. Appl. **91**, 464–471 (2018)
5. Galar, M., Fernandez, A., Barrenechea, E., Bustince, H., Herrera, F.: A review on ensembles for the class imbalance problem: bagging-, boosting-, and hybrid-based approaches. IEEE Trans. Syst. Man Cybern. Part C (Appl. Rev.) **42**(4), 463–484 (2011)
6. Haixiang, G., Yijing, L., Shang, J., Mingyun, G., Yuanyue, H., Bing, G.: Learning from class-imbalanced data: review of methods and applications. Expert Syst. Appl. **73**, 220–239 (2017)
7. He, H., Garcia, E.A.: Learning from imbalanced data. IEEE Trans. Knowl. Data Eng. **21**(9), 1263–1284 (2008)
8. Hong, Y., Hwang, U., Yoo, J., Yoon, S.: How generative adversarial networks and their variants work: an overview. ACM Comput. Surv. (CSUR) **52**(1), 10 (2019)
9. Lu, H., Yang, L., Yan, K., Xue, Y., Gao, Z.: A cost-sensitive rotation forest algorithm for gene expression data classification. Neurocomputing **228**, 270–276 (2017)
10. Mariani, G., Scheidegger, F., Istrate, R., Bekas, C., Malossi, C.: BAGAN: data augmentation with balancing GAN. arXiv preprint arXiv:1803.09655 (2018)

11. Perez, L., Wang, J.: The effectiveness of data augmentation in image classification using deep learning. arXiv preprint arXiv:1712.04621 (2017)
12. Srivastava, A., Valkov, L., Russell, C., Gutmann, M.U., Sutton, C.: VEEGAN: reducing mode collapse in GANs using implicit variational learning. In: Advances in Neural Information Processing Systems, pp. 3308–3318 (2017)

A Deep Learning Scheme for Extracting Pedestrian-Parcel Tuples from Videos

Tongtong Wu[1], Xu Zhang[2], and Fuqing Duan[1(✉)]

[1] College of Artifical Intelligence, Beijing Normal University, 19 Xinjiekouwai Street, Haidian, Beijing 100875, People's Republic of China
fqduan@bnu.edu.cn

[2] National Engineering Laboratory for Intelligent Video Analysis and Application, 1 Shoutinanlu Street, Haidian, Beijing 100048, People's Republic of China

Abstract. Pedestrian parcel inspection is a common security measure in some public places like railway entrances. Automatic identification of the affiliation between pedestrians and parcels is an important task in an intelligent security inspection system. However, it is very challenging due to the high pedestrian volume in these places. In this paper, we propose a deep learning scheme for extracting pedestrian-parcel tuples from camera videos, which includes three modules, i.e. detection, interaction and re-identification of pedestrians and parcels. We first detect pedestrians and parcels in each frame, and then discriminate the affiliation between pedestrians and parcels by interaction behavior analysis, finally discard the redundant affiliations by re-identification of pedestrians and parcels. In the interaction module, we propose a lightweight interaction model for discriminating the affiliation between pedestrians and parcels in a single RGB image. Experiments on a video data at a subway entrance validate the proposed approach.

Keywords: Deep learning · Detection · Intelligent security inspection

1 Introduction

With the successful application of computer vision technology in visual surveillance, intelligent security inspection gradually attracts researcher's attention. In subway entrances, pedestrian parcel inspection is the most common scenario for security inspection. Every day, thousands of parcels pass security inspection. Once a dangerous parcel has been found out, it is necessary to look for the corresponding potential criminal which carries the dangerous parcel. Therefore, how to accurately identify and record the affiliation between pedestrians and parcels in real time from subway entrance video is essential to subway security.

Given a video, our goal is to capture the coordinate of pedestrians and parcels from it and calibrate the affiliation between the pedestrians and the parcels. But due to the high pedestrian volume in subway entrances, it comes a big challenge. Fortunately, the power of deep learning makes it possible. In recent years, deep

T. Gedeon et al. (Eds.): ICONIP 2019, CCIS 1142, pp. 31–39, 2019.
https://doi.org/10.1007/978-3-030-36808-1_4

learning has achieved great success in various fields such as object detection [12], classification [3], and re-identification [8]. By leveraging deep learning, we can tackle many challenging task like the one we just mentioned.

In this paper, we propose a deep learning scheme for extracting pedestrian-parcel tuples from camera videos, which includes three modules, i.e. detection, interaction and re-identification of pedestrians and parcels. We first detect pedestrians and parcels in each frame, and then discriminate the affiliation between pedestrians and parcels by interaction behavior analysis, finally discard the redundant affiliations by re-identification of pedestrians and parcels.

The major contributions of our work are two folds:

- We propose a lightweight interaction model in the interaction module to discriminate the affiliation between pedestrians and parcels in a single RGB image.
- We propose a deep learning frame for extracting pedestrian-parcel tuples from camera videos and obtain a powerful performance.

2 Method

In this section, we present our deep learning scheme for extracting pedestrian-parcel tuple from a video, where we use three deep learning modules, i.e. detection, interaction and re-identification. As show in Fig. 1, for each frame, we first detect pedestrians and parcels by detection module, and then crop the joint area of the candidate pedestrian-parcel pairs. After that, we use the interaction module to discriminate the affiliation of candidate pedestrian-parcel pairs and then use the Re-ID module to extract the feature of associated pedestrian-parcel pair. Finally, the algorithm merges the two information streams with previous results, and then updates the current results.

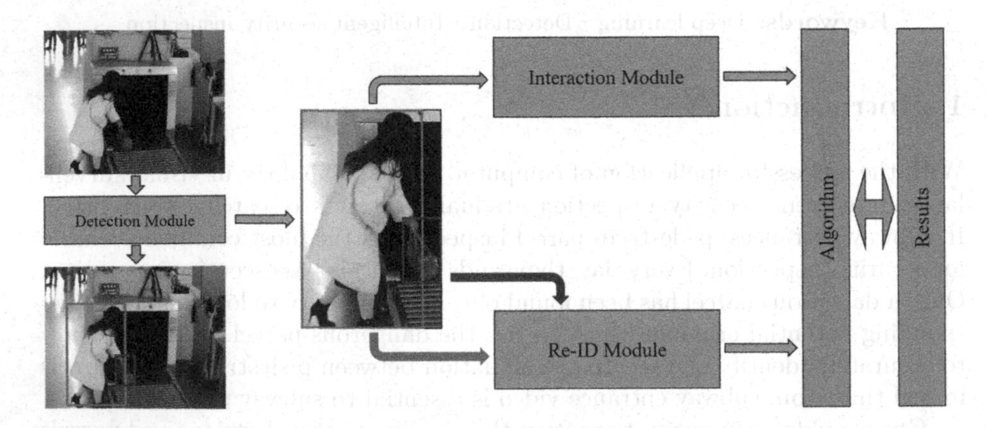

Fig. 1. Overview of the proposed scheme.

2.1 Detection Module

Object detection is a fundamental task in computer vision. The aim of object detection is to locate all pre-specified objects. Existing deep learning based methods are mainly divided into two categories, single-stage and two-stage detectors. Single-stage detectors, such as YOLO [9–11] and SSD [7], have a simple structure and is fast. R-CNN [1] and its derived structure Faster R-CNN [12] are the representative two-stage detectors. Compared to single-stage detectors, two-stage detectors is slower but more accurate.

Due to the high pedestrian volume in subway entrances, two-stage detectors with higher precision and recall is more suitable than one-stage detectors. Thus, in our work, we choose Faster R-CNN as our detection network to accurately detect pedestrians and parcels in each frame. In the Faster R-CNN network, a convolutional backbone is used to extract the features of image first, then a region proposal network (RPN) is used to score thousands of region proposals. After that, a small part of high-scoring region proposals is picked out and resized to a same size by a RoI pooling layer. Finally, all picked region proposals are sent into the R-CNN network, and the R-CNN network outputs a vector whose length is proportional to the number of categories for each region proposal. For each category, the R-CNN network outputs a 4-dimensional vector representing the size and offset and a score for each region proposal. After that, we use a hyperparameter H_1 as threshold to filter region proposals whose pedestrian score is lower than H_1 while another hyperparameter H_2 is used to filter region proposals whose parcel score is lower than H_2. In our work, H_1 is set to 0.8 and H_2 is set to 0.7. One more step is to use non-maximum suppression algorithm to filter most repeated regions. Finally, we obtain pedestrians detection results $A_i = \{a_{i0}, a_{i1}, ...\}$ and parcels detection results $B_i = \{b_{i0}, b_{i1}, ...\}$ for frame i.

2.2 Interaction Module

Understanding human-object interaction detection is an important task in visual analysis. The aim of the interaction module is to discriminate the affiliation between pedestrians and parcels in a single RGB image.

When interactions between pedestrians and parcels happen, it is impossible to have a large distance between them. So for the detected pedestrians and parcels in one frame, we first extract possible pedestrian-parcel pairs according to the boundary relationship between pedestrian bounding box and parcel bounding box. For each pedestrian bounding box $a_{im} = (a^x_{im}, a^y_{im}, a^w_{im}, a^h_{im})$ and parcel bounding box $b_{in} = (b^x_{in}, b^y_{in}, b^w_{in}, b^h_{in})$, compute an extension box Eb_{in}.

$$Eb_{in} = (b^x_{in}, b^y_{in}, 2 \times b^w_{in}, 2 \times b^h_{in}) . \tag{1}$$

After that, compute $IOU(a_{im}, Eb_{in})$. If it's positive, then (a_{im}, b_{in}) is a candidate pair. Note that $box_1 \cap box_2$ represents the overlapping area of the two boxes while $box_1 \cup box_2$ represents the joint area.

$$IOU(box_1, box_2) = \frac{box_1 \cap box_2}{box_1 \cup box_2} . \tag{2}$$

This step will remove many impossible combinations of pedestrians and parcels, so that the load of the interaction module can be decreased.

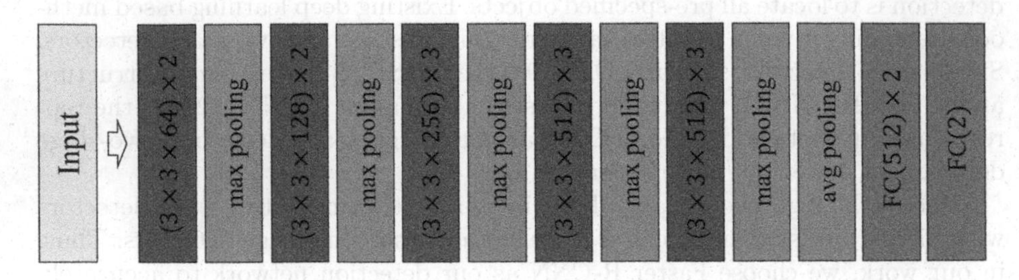

Fig. 2. Interaction model. The blue block represents a convolutional layer, while the orange block represents a pooling layer and the green block represents a fully connected layer. The $(3\times3\times64)$ means that the convolutional layer has 64 channels, and its kernal size is 3×3. The $FC(512)$ represents that the fully connected layer has 512-dimension. And the tail $\times 2$ means the current layer is repeated twice. (Color figure online)

We propose a lightweight interaction model to discriminate the affiliation of the pedestrian-parcel pairs. Figure 2 shows our interaction network structure. For each pedestrian-parcel pair, we crop joint area, zero the pixels outside the pedestrian and parcel bounding box, and resize the croped image to 224 * 224 as the input of the interaction network. At the last layer, we apply a softmax function to standardize output. The output of the network is a 2-dimensional vector that represents the associated score and unassociated score. A parcel b_{in} is considered to be associated with a pedestrian a_{im} if the associated score is higher than a hyperparameter H_3 which is set to 0.5 in our work. For one parcel b_{in}, if there are multiple pedestrians associated with it, then we consider the pedestrian with the highest associated score as its associated target.

In train process, we use a focal loss [5] to optimize our network. Focal loss is an advanced method to balance training weight between hard samples and easy samples. As shown in Eq. 3, the parameter t represents the class while p_t represents the confidence of class t. The α_t is calculated by Eq. 4. The parameters α and γ are hyperparameters selected by experience. In our work, α is set to 0.25 and γ is set to 2.

$$FL(p_t) = -\alpha_t(1 - p_t)^\gamma \log(p_t), \quad t \in \{0, 1\} . \tag{3}$$

$$\alpha_t = t \times \alpha + (1 - t) \times (1 - \alpha) . \tag{4}$$

2.3 Re-ID Module

Re-identification is to match the same subject in the images of different scenes or different times. Due to the large amount of duplicate information among frames, we need to discard redundant affiliations by re-identification of pedestrians and parcels. We use a standard baseline [8] as our re-identification network.

The structure of this network consists of a ResNet-50 [3] backbone, an average pooling layer, a 512-dimensional fully connected layer and a 751-dimensional fully connected layer. Note that there is a batch normlization operation after the 512-dimensional fully connected layer. The output of the last layer is 751 dimension because the baseline network was trained as a classifier on a dataset of 751 categories. Each pedestrian or parcel image is resized to 256×128, then sent into the re-identification network. The output of the 512-dimensional fully connected layer is used as the feature f of the input image.

For each pedestrian pair (a_1, a_2), its similarity is measured by $f_{a_1} \cdot f_{a_2}$, which is a number between 0 and 1. Because pedestrian re-identification is a similarity problem, the model can also be used to parcel re-identification.

2.4 Algorithm

In this section, we give the algorithm diagram of the whole system for an easy implementation. Since the frame changes very little in a short time, we only process 1 frame every 5 observed frames to reduce the burden of system. Our algorithm procedure is showed as Algorithm 1.

Algorithm 1

1: $Sequences \leftarrow \{\}, Results \leftarrow \{\}$
2: **for** frame $i = 0 \rightarrow N$ **do**
3: $A_i, B_i = Detection(), CandidatePairs = \{\}$
4: **for** a_{im} in A_i **do**
5: **for** b_{in} in B_i **do**
6: **if** $IOU(a_{im}, Eb_{in}) > 0$ and $Interaction(a_{im}, b_{in}) > H_3$ **then**
7: $CandidatePairs.add((a_{im}, b_{in}))$
8: **end if**
9: **end for**
10: **end for**
11: **for** $j = 0 \rightarrow len(CandidatePairs)$ **do**
12: **if** $max_k(Avesim(Seq_k, a_{im})) < H_4$ or $Sequences\, is\, empty$ **then**
13: $generate\, new\, Seq$
14: $Seq \leftarrow \{(a_{ij}, b_{ij})\}$
15: $Sequences.add(Seq)$
16: **else**
17: $Seq_k.add((a_{im}, b_{in}))$
18: **end if**
19: **end for**
20: **for** $j = 0 \rightarrow len(Sequences)$ **do**
21: **if** $Interval(Seq_j, i) > H_5$ or $len(Seq_j) > H_6$ **then**
22: $Results.add(Seq_j), Sequences.remove(Seq_j)$
23: **end if**
24: **end for**
25: **end for**

In this algorithm diagram, we first detect pedestrians and parcels in current frame, and then extract candidate pedestrian-parcel pairs from the detection results, finally match the same pedestrians or parcels between current frame and previous multiple frames by re-identification to discard the redundant affiliations. *Sequences* is a set for caching unfinished pedestrians sequence with corresponding parcels, while *Results* is used for saving finished ones. For current frame, we define a set named *CandidatePairs* for saving all candidate pedestrian-parcel pairs. Function *Avesim*() calculates an average similarity between a_{im} and the pedestrians stream in Seq_k. Another function *Interval*() returns an integer representing the frame interval between current frame i and the latest pair in Seq_j. H_4 is a hyperparameter which is set to 0.7 in our work while H_5 is set to 25 and H_6 is set to 100.

3 Experiments

3.1 Experimental Setup

Dataset. For the detection module, we labeled pedestrians and parcels from 3000 images extracted from a video. Any two images have an interval of more than one second. 80% of the images are used as training set while the rest as testing set.

For the interaction module, we labeled 5000 associated and 3756 unassociated pedestrian-parcels pairs from about 4000 different images. 90% of the pairs are used for training and the rest is used for testing.

For the Re-ID module, we use a public dataset called Market1501. It contains 751 individuals for training and 750 individuals for testing.

Evaluation Metrics. We evaluate the detection module performance using the commonly used mean average precision (mAP) [4]. For the interaction module, we use F1 score [2] to make an evaluation. And Rank-1, Rank-5, Rank-10 and mAP [6] results are used to evaluate the Re-ID module.

Implementation Details. For the detection module, we use a learning rate of 0.001, a weight decay of 0.0005 and a momentum of 0.9 to train network for 70 K iterations. For the Re-ID module, we use a learning rate of 0.005, a weight decay of 0.0005 and a momentum of 0.9 to train network for 60 epochs. And the interaction module is trained for 70 K iterations with a learning rate of 0.0001 and an Adam optimizer. All experiments are done on a single NVIDIA GTX1080Ti GPU.

3.2 Results

In the detection module, the AP of pedestrians is 0.890 while parcels' is 0.503, and the mAP is 0.697. In the interaction module, the results are 450 TP (True Positive), 90 FP (False Positive), 261 TN (True Negative) and 75 FN (False Negative). So the F1 score of interaction module is 0.845. In the Re-ID module, Rank-1 is 0.884, Rank-5 is 0.955, Rank-10 is 0.971 and mAP is 0.723.

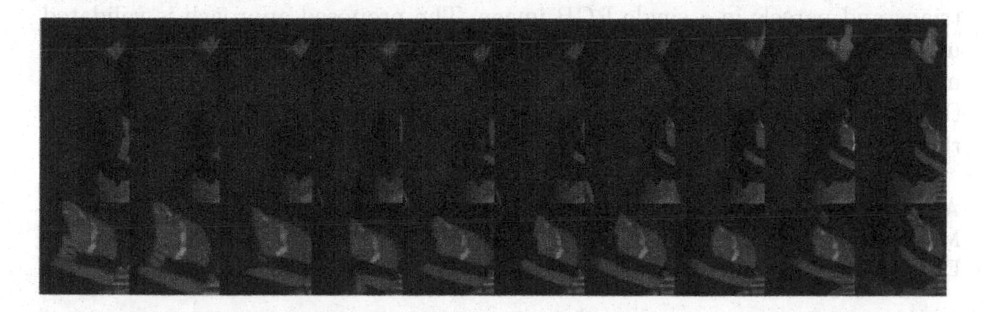

Fig. 3. Visualization. The image is a sequence of one pedestrian and its parcel extracted from a continuous frame stream.

From the results of detection module, we can see that the AP of parcel detection results is lower than the one of pedestrian detection results. The main reason is that the appearance, shape and scale of parcel images change dramatically with variations of camera viewpoints and we have not enough training data of parcels, while the pedestrian detection is a more common task solved using Faster RNN. Moreover, occlusions of parcels are more serious than pedestrians with a high pedestrian volume. From the results of interaction module, we can see that our proposed interaction model have an excellent performance in discriminating the affiliation between pedestrians and parcels. The accuracy is $450/(90+450) = 83.3\%$, and the recall is $450/(75+450) = 85.7\%$. Considering the small size of the training dataset, the error rate of the model is acceptable. The results of Re-ID module shows that the module has a powerful performance. It provides a strong support for our work.

Although the results of detection module and interaction module are not so satisfied, they will not bring about a significant impact on final result of the pedestrian-parcel tuple extraction. This is because the results are only for images, while the ultimate goal of the whole system is to extract pedestrian-parcel tuples from whole video. The correct pedestrian-parcel tuples can be obtained, provided that there exists one frame handled well during the interaction process between pedestrians and parcels. We visualize results of our scheme at Fig. 3. The visualization shows that our scheme has a powerful performance in extracting pedestrian-parcel tuples from videos.

4 Conclusions

Automatic identification of the affiliation between pedestrians and parcels is an important task in an intelligent security inspection system. In this paper, we propose a deep learning scheme for extracting pedestrian-parcel tuples from camera videos, which includes three modules, i.e. detection, interaction and re-identification of pedestrians and parcels. In the interaction module, we propose a lightweight interaction model for discriminating the affiliation between pedestrians and parcels in a single RGB image. The proposed approach is validated using the video data captured at subway entrances, and shows a good performance. However, some issues like parcel detection still need to be improved. In the future work, we will annotate more training data, and continue to improve the performance of the whole system.

Acknowledgment. This work was supported by National Science and Technology Major Project of China (grant 2018AAA0100800), and Opening Foundation of National Engineering Laboratory for Intelligent Video Analysis and Application.

References

1. Girshick, R., Donahue, J., Darrell, T., Malik, J.: Rich feature hierarchies for accurate object detection and semantic segmentation. In: Proceedings of the IEEE Conference on Computer Vision and Pattern Recognition, pp. 580–587 (2014)
2. Goutte, C., Gaussier, E.: A probabilistic interpretation of precision, recall and *F*-Score, with implication for evaluation. In: Losada, D.E., Fernández-Luna, J.M. (eds.) ECIR 2005. LNCS, vol. 3408, pp. 345–359. Springer, Heidelberg (2005). https://doi.org/10.1007/978-3-540-31865-1_25
3. He, K., Zhang, X., Ren, S., Sun, J.: Deep residual learning for image recognition. In: Proceedings of the IEEE Conference on Computer Vision and Pattern Recognition, pp. 770–778 (2016)
4. Henderson, P., Ferrari, V.: End-to-end training of object class detectors for mean average precision. In: Lai, S.-H., Lepetit, V., Nishino, K., Sato, Y. (eds.) ACCV 2016. LNCS, vol. 10115, pp. 198–213. Springer, Cham (2017). https://doi.org/10.1007/978-3-319-54193-8_13
5. Lin, T.Y., Goyal, P., Girshick, R., He, K., Dollár, P.: Focal loss for dense object detection. In: Proceedings of the IEEE International Conference on Computer Vision, pp. 2980–2988 (2017)
6. Lin, Y., et al.: Improving person re-identification by attribute and identity learning. Pattern Recogn. (2019)
7. Liu, W., et al.: SSD: single shot multibox detector. In: Leibe, B., Matas, J., Sebe, N., Welling, M. (eds.) ECCV 2016. LNCS, vol. 9905, pp. 21–37. Springer, Cham (2016). https://doi.org/10.1007/978-3-319-46448-0_2
8. Luo, H., Gu, Y., Liao, X., Lai, S., Jiang, W.: Bags of tricks and a strong baseline for deep person re-identification. arXiv preprint arXiv:1903.07071 (2019)
9. Redmon, J., Divvala, S., Girshick, R., Farhadi, A.: You only look once: Unified, real-time object detection. In: Proceedings of the IEEE Conference on Computer Vision and Pattern Recognition, pp. 779–788 (2016)

10. Redmon, J., Farhadi, A.: Yolo9000: better, faster, stronger. In: Proceedings of the IEEE Conference on Computer Vision and Pattern Recognition, pp. 7263–7271 (2017)
11. Redmon, J., Farhadi, A.: Yolov3: an incremental improvement. arXiv preprint arXiv:1804.02767 (2018)
12. Ren, S., He, K., Girshick, R., Sun, J.: Faster R-CNN: towards real-time object detection with region proposal networks. In: Advances in Neural Information Processing Systems (NIPS) (2015)

Support Matching: A Novel Regularization to Escape from Mode Collapse in GANs

Yinghua Yao[1,2], Yuangang Pan[2], Ivor W. Tsang[2], and Xin Yao[1](✉)

[1] Southern University of Science and Technology, Shenzhen, China
xiny@sustech.edu.cn
[2] CAI, University of Technology Sydney, Sydney, Australia
{yinghua.yao,yuangang.pan}@student.uts.edu.au,
ivor.tsang@uts.edu.au

Abstract. Generative adversarial network (GAN) is an implicit generative model known for its ability to generate sharp images. However, it is poor at generating diverse data, which refers to the mode collapse problem. It turns out that GAN is prone to emphasizing the quality of samples but ignoring their diversity. When mode collapse happens, the support of the generated data distribution is not aligned with that of the real data distribution. We thus propose Support Regularized-GAN (SR-GAN) to address such a mode collapse issue by matching their support. Our experiments on synthetic and real-world datasets show that our regularization can mitigate the mode collapse and also improve the data quality.

Keywords: GANs · Mode collapse · Support matching

1 Introduction

Generative adversarial networks (GANs) [3] implicitly model the statistical distributions for real data and have distinguished abilities in generating sharp images. Regarding the architecture of GANs, there are two networks – one is called the generator, which aims to generate samples like real data; another is called the discriminator, which examines input samples whether they are real or fake. GANs do not need an explicit form of the data distribution and instead train the generator by a binary classification of the discriminator.

GAN, however, suffers from several challenging problems. Mode collapse is one of the major challenges, which refers to poor mode diversity in generated samples [7]. There are two types of mode collapse being observed: entire modes of the input data are never generated, or the generator just generates some of the modes [7]. Like shown in Fig. 1a, the vanilla GAN only generates one mode of the data. Arjovsky et al. [1] derive that the unsaturated objective in the vanilla GAN is equivalent to minimize the reverse Kullback-Leibler (KL) divergence and

T. Gedeon et al. (Eds.): ICONIP 2019, CCIS 1142, pp. 40–48, 2019.
https://doi.org/10.1007/978-3-030-36808-1_5

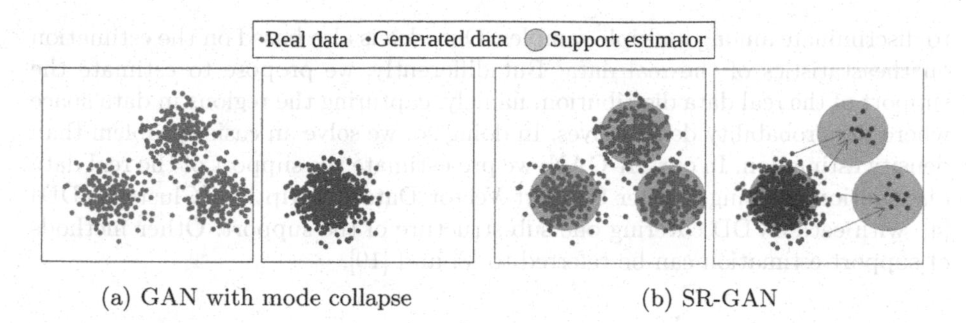

(a) GAN with mode collapse (b) SR-GAN

Fig. 1. (a) Problem: GAN suffers from the mode collapse, i.e., only generating one out of three modes. (b) Our solution (SR-GAN): red panels are the estimated support of real data. They are used to penalize the generated modes with no/scarce samples and guide the generator to disperse samples with all modes. (Color figure online)

maximize Jensen-Shannon (JS) divergence between the real data distribution and the generated data distribution simultaneously. The KL term assigns an extremely high cost to generating fake looking samples, and an extremely low cost to missing modes, which results in GAN's mode collapse problem.

Since the GAN's objective causes the mode collapse, we propose to add regularization to improve it. In Fig. 1a, we can observe that when mode collapse arises, the support of the real data distribution, namely, the domain of data space, and the support of the generated data distribution cannot be matched, which motivates us to use such matching as a regularization for GANs (Fig. 1b).

Our Contributions. We propose Support Regularized-GAN (SR-GAN) to improve mode diversity. Our main idea is to align the support of the generated data distribution with that of the real data distribution. To be specific, a support estimator is used to capture the structure of the real data support. Then an extra support matching regularization term is introduced to enforce the generator to cover all sub-structures of the data support. Experimental results on synthetic and real-world datasets show that the support matching indeed avoids the mode collapse and also improves the data quality for GANs.

2 Related Work

Many works are proposed to solve the mode collapse problem. In particular, there are two works most related to ours [2,6], which is to add an explicit regularization in the GAN's objective. DAN-S/2S [6] uses Maximum Mean Discrepancy to tell the difference between the real data distribution and the generated data distribution and use it as a regularization for GAN while LBT-GAN [2] utilizes the likelihood of the real data to guide the generator to cover all modes of the data through the density estimator. These two works are based on the statistics of the real data. However, LBT-GAN defines a bilevel optimization problem and has a high computational cost. DAN-S/2S needs to define one more discriminator to discriminate among multiple samples. Our idea is also based on the estimation on the statistics of the real data. But differently, we propose to estimate the support of the real data distribution, namely, capturing the regions in data space where the probability density lives. In doing so, we solve an easier problem than density estimation. In our SR-GAN, we pre-estimate the support of the real data distribution by using Cluster Support Vector Data Description (ClusterSVDD) [4], with each SVDD covering one sub-structure of the support. Other methods of support estimation can be referred to [8] and [10].

3 Support Matching as a Mode Regularizer

When mode collapse happens in the GANs, the support, which refers to the regions in data space where the probability density is larger than zero [8], of the generated data distribution cannot align with that of the real data distribution. So we propose to use support matching between the real data distribution and the generated data distribution as a mode regularizer. Through this regularizer, we push the generator to cover all modes of the real data; otherwise, the mismatching will occur and cause the penalty. Specifically, we capture the sub-structures of the support by using ClusterSVDD [4], which fits multiple hyperspheres on the support of the data distribution. Each sphere in ClusterSVDD will cover one sub-structure of the support, referred to one mode of the data. Therefore, the estimated spheres can be regarded as mode indicators for the generator to tell whether there is data generating in a certain mode.

3.1 Support Estimation on Real Data Distribution

We use ClusterSVDD to estimate the support of the real data distribution. This method unifies SVDD [10] and k-means clustering, which fits K hyperspheres that can be defined by its centers and radius $\{\mathbf{c}_k, R_k\}_{k=1}^K$, on the support of the real data distribution. With $\{\mathbf{c}_k, R_k\}_{k=1}^K$, we calculate the cluster label y_i for each sample and collect samples with regard to each cluster X_k as follows:

$$y_i = \text{argmin}_{k \in \{1,\dots,K\}} \|\mathbf{c}_k - \phi(\mathbf{x}_i)\|^2 - R_k^2, \; \forall i = 1, \dots, N, \tag{1}$$

$$X_k = \{\mathbf{x}_i | y_i = k, i = 1, \dots, N\}, \; \forall k = 1, \dots, K. \tag{2}$$

Each X_k is then used to solve one SVDD optimization problem [10].

3.2 Support Matching as a Regularizer

We use the estimated spheres $\{\mathbf{c}_k, R_k\}_{k=1}^K$ to evaluate the support of the generated data distribution and align it with that of the real data distribution. We apply Eq. (2) to divide generated data into K groups $\{X_k\}_{k=1}^K$ and then match the size of the groups between the real data and the generated data. K is set to the number of the modes in the data. If some mode is missed by the generator, the size of its corresponding group would be zero, i.e., $|X_k| = 0$, which causes

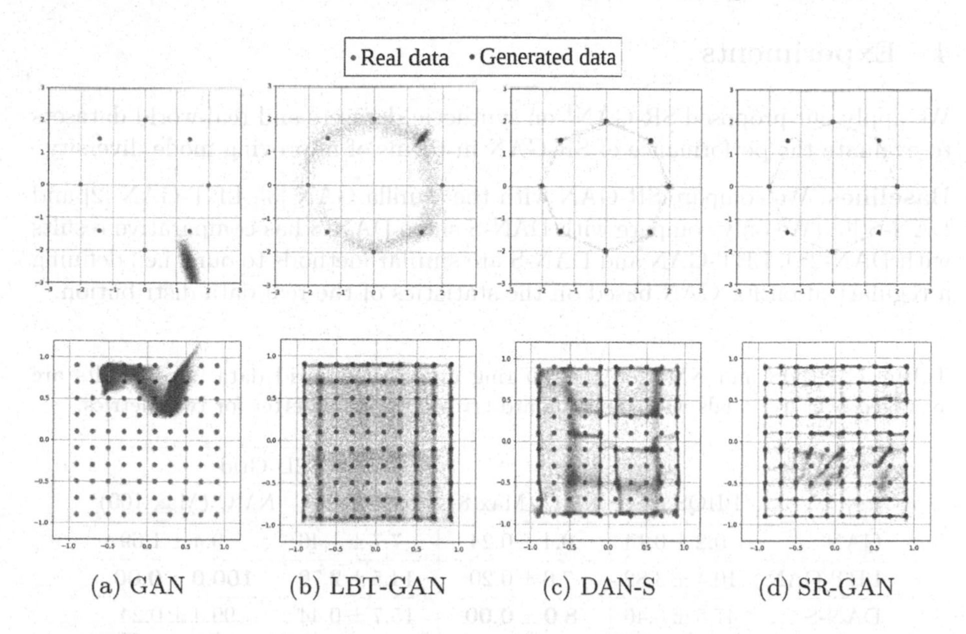

(a) GAN (b) LBT-GAN (c) DAN-S (d) SR-GAN

Fig. 2. Visual comparison of generated samples on the 2D ring data (Upper) and the 2D grid data (Lower). More overlapping between the generated samples and the real samples denotes a better generation.

the difference between the real data and the generated data. Such difference will guide the generator to generate data that is not covered currently. Since argmin function has no derivative, we instead replace it with softmax function for calculating the regularization term in GAN's objective as follows:

$$f_k(\mathbf{x}_i) = \frac{\exp(-\beta(\|\mathbf{c}_k - \phi(\mathbf{x}_i)\|^2 - R_k^2))}{\sum_j \exp(-\beta(\|\mathbf{c}_j - \psi(\mathbf{x}_i)\|^2 - R_j^2))}. \tag{3}$$

The matching of the support between the real data distribution and the generated data distribution is defined as a regularization for GAN. In short, the objective of SR-GAN consists of two terms, i.e., the discriminator and the generator:

$$\min_{\mathcal{D}} \mathbb{E}_{p_{data}(\mathbf{x})}[-\log \mathcal{D}(\mathbf{x})] + \mathbb{E}_{p(\mathbf{z})}[-\log(1 - \mathcal{D}(\mathcal{G}(\mathbf{z})))], \tag{4a}$$

$$\min_{\mathcal{G}} \mathbb{E}_{p(\mathbf{z})}[-\log \mathcal{D}(\mathcal{G}(\mathbf{z}))] + \lambda * \sum_k (\mathbb{E}_{p_{data}(\mathbf{x})}[f_k(\mathbf{x})] - \mathbb{E}_{p(\mathbf{z})}[f_k(\mathcal{G}(\mathbf{z}))])^2, \tag{4b}$$

where $\mathcal{G}$ is denoted as the generator network. $\mathcal{D}$ is denoted as the discriminator network. $p(\mathbf{z})$ is the distribution of the input noise. $p_{data}(\mathbf{x})$ is the distribution of the real data. λ balances the image quality and the mode diversity.

4 Experiments

We apply our proposed SR-GAN[1] on synthetic datasets and real-world datasets to evaluate the performance of SR-GAN in terms of improving mode diversity.

Baselines. We compare SR-GAN with the vanilla GAN [3], LBT-GAN [2] and DAN-S [6] (We only compare with DAN-S since DAN-S has comparative results with DAN-2S). LBT-GAN and DAN-S are similar methods to ours, i.e., defining a regularization for GAN based on the statistics of the real data distribution.

Table 1. PHQS and NMC on the 2D ring and the 2D grid data. The results are averaged over five trials with the standard error. Higher is better for two metrics.

	2D Ring		2D Grid	
	PHQS (%)	NMC (Max 8)	PHQS (%)	NMC (Max 100)
GAN	0.2 ± 0.14	0.4 ± 0.24	7.7 ± 1.46	9.4 ± 1.69
LBT-GAN	10.4 ± 3.82	7.8 ± 0.20	14.5 ± 2.70	$\mathbf{100.0 \pm 0.00}$
DAN-S	47.5 ± 5.46	$\mathbf{8.0 \pm 0.00}$	15.7 ± 0.44	99.4 ± 0.24
SR-GAN	$\mathbf{91.7 \pm 0.97}$	$\mathbf{8.0 \pm 0.00}$	$\mathbf{45.3 \pm 2.46}$	$\mathbf{100.0 \pm 0.00}$

Model Architectures and Hyperparameters. Following [7], fully connected networks (FCNs) are used for the generator network and the discriminator network on the synthetic datasets and the SatImage dataset. A recurrent neural network (RNN) is used for the generator and a convolutional neural network (CNN) for the discriminator (RNN-CNN) on the MNIST dataset. Furthermore, the FCNs are also applied for the generator and the discriminator (FCN-FCN) on MNIST following [6]. We keep the architectures similar for all GANs in order to make a fair comparison. In terms of the hyperparameters, the number of clusters K in CSVDD is set to the number of classes in the datasets. The trade-off factor λ is set to 1 for synthetic datasets and the SatImage dataset, 10 for MNIST with the RNN-CNN architecture and 50 for MNIST with the FCN-FCN architecture. β is set to 10 for all datasets.

4.1 Synthetic Datasets

We construct two synthetic datasets following [2]: (1) 2D ring, i.e., mixture of eight 2D Gaussian distributions with covariance matrix (CM) $0.02I$ arranged in a ring; (2) 2D grid, i.e., mixture of 100 2D Gaussian distributions with CM $0.01I$ arranged in a 10-by-10 grid. Same in [9], we use the percentage of high quality Samples (PHQS), and the number of mode covered (NMC) to measure the quality and diversity of the generated data, respectively.

[1] https://github.com/EvaFlower/SR-GAN.

Figure 2 shows the visualization of samples generated by GAN and its variants on the 2D ring and the 2D grid datasets, respectively. We can observe that: (1) the vanilla GAN suffers from severe mode collapse problems on both datasets. Regarding the 2D ring data (Fig. 2 Upper), GAN only generates samples nearly one mode. Regarding the 2D grid data (Fig. 2 Lower), GAN covers few modes. (2) In terms of LBT-GAN, DAN-S and SR-GAN, all real samples are surrounded by the generated samples, which means that they can cover all modes of the data. (3) The density of the data generated by GAN's variants is not equal to that of the real data. However, SR-GAN can learn a closer distribution comparing to other baselines. Table 1's quantitative results are consistent with the visualization results. LBT-GAN, DAN-S, and our SR-GAN achieve the maximum NMC while GAN gains small NMC on both two datasets. In addition, our SR-GAN achieves the highest PHQS, which means that SR-GAN can generate more high quality samples than LBT-GAN and DAN-S.

Table 2. NMC and KL on SatImage. The results are averaged over five trials with the standard error. Higher is better for NMC; lower is better for KL.

	NMC (Max 6)	KL
GAN	2.0 ± 0.77	1.38 ± 0.390
LBT-GAN	$\mathbf{6.0 \pm 0.00}$	0.21 ± 0.018
DAN-S	$\mathbf{6.0 \pm 0.00}$	0.06 ± 0.019
SR-GAN	$\mathbf{6.0 \pm 0.00}$	$\mathbf{0.02 \pm 0.005}$

4.2 SatImage Dataset

We then apply SR-GAN on a simple real-world dataset, i.e., SatImage dataset[2]. This dataset contains 4,435 instances with 36 attributes and 6 classes, which each class is regarded as one mode.

We also use NMC to evaluate the mode diversity on SatImage. The number of modes here is estimated using a trained classifier. We do not count high quality samples since it is hard to evaluate it on real-world datasets. Instead we count a mode as a covered mode if the number of its samples is greater than $\alpha\% \times \frac{\#of samples}{\#of modes}$ ($\alpha = 10$). The KL divergence between the generated data and the real data over class [7] (KL) is used to evaluate the quality of the generation.

It shows in Table 2 that (1) GAN only generates around two out of six modes, which denotes that GAN also suffers from a severe mode collapse problem on the simple real-world dataset. (2) LBT-GAN, DAN-S and our SR-GAN can cover all modes of the SatImage data. (3) In addition, our SR-GAN achieves the lowest KL divergence, which means that it can learn a more accurate data distribution.

[2] https://www.csie.ntu.edu.tw/~cjlin/libsvmtools/datasets/multiclass/satimage.scale.

4.3 MNIST Dataset

We further explore the superiority of our SR-GAN in terms of improving mode diversity on a more complex dataset: MNIST. It consists of zero to nine digits, denoted as 10 modes. We adopt two architectures: FCN-FCN and RNN-CNN.

Instead of doing the support estimation on raw image data directly, we apply support estimation on the embedding space. Particularly, we use deep neural networks [5] as feature extractors and input the discrete embedding features into ClusterSVDD. The dimension of the features is set to that of the input noise.

We use NMC and KL same in Sect. 4.2 to measure the diversity and the quality of the generated data, respectively.

MNIST with the FCN-FCN Architecture. The upper panel of Fig. 3 shows that: (1) the vanilla GAN and LBT-GAN both suffer from a severe mode collapse issue on MNIST. The visualization shows that they only generate few of ten digits. (2) DAN-S and SR-GAN can significantly mitigate the mode collapse. However, the performance of SR-GAN is inferior to that of DAN-S. That is because, in SR-GAN, we train the support regularization independently from the GAN's objective for simplicity. A better result could be achieved through training them in a unified framework, which we leave for a future work. The results in Table 3 (Left) is consistent with the visualization results.

MNIST with the RNN-CNN Architecture. The RNN-CNN architecture is asymmetric, resulting in a more complex power balance [7]. Therefore, its training is much harder than the previous FCN-FCN architecture. The lower panel of Fig. 3 shows that: (1) the samples generated by GAN, LBT-GAN and

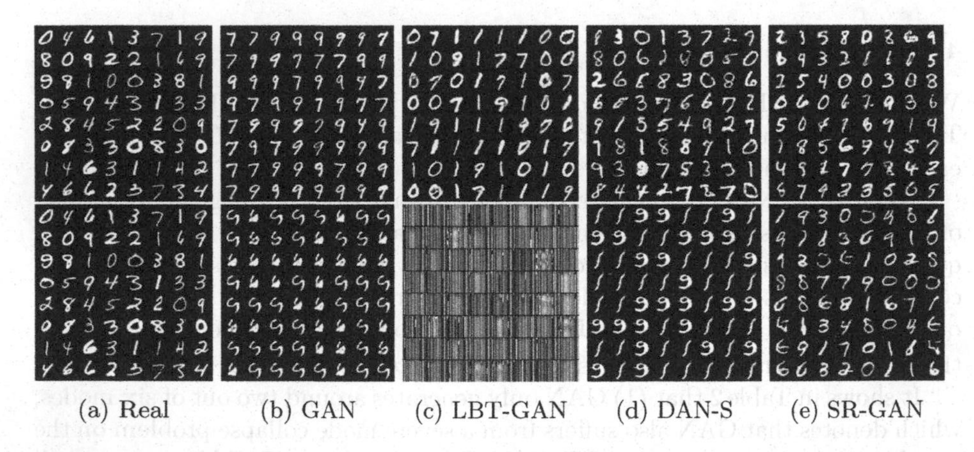

(a) Real (b) GAN (c) LBT-GAN (d) DAN-S (e) SR-GAN

Fig. 3. Visual comparison of generated samples on the MNIST data. **First column**: real samples from the MNIST dataset. **Upper column 2–5**: the generation with the FCN-FCN architecture. **Lower column 2–5**: the generation with the RNN-CNN architecture, which needs a more complex balance between the asymmetric architecture and thus is harder to train. The training of LBT-GAN with the RNN-CNN architecture is unstable and provides meaningless results.

Table 3. NMC and KL on MNIST. The results are averaged over five trials with the standard error. Higher is better for NMC; lower is better for KL. The results of LBT-GAN is unavailable since it generates meaningless samples.

	MNIST (FCN-FCN)		MNIST (CNN-RNN)	
	NMC (Max 10)	KL	NMC (Max 10)	KL
GAN	2.6 ± 0.25	1.57 ± 0.038	2.8 ± 0.80	1.70 ± 0.240
LBT-GAN	4.0 ± 0.00	1.28 ± 0.107	–	–
DAN-S	$\mathbf{10.0 \pm 0.00}$	$\mathbf{0.01 \pm 0.001}$	1.8 ± 0.20	1.86 ± 0.110
SR-GAN	8.2 ± 1.36	0.50 ± 0.259	$\mathbf{10.0 \pm 0.00}$	$\mathbf{0.12 \pm 0.025}$

DAN-S with the RNN-CNN architecture are less sharper compared to the results with the FCN-FCN architecture. This is because the asymmetric RNN-CNN architecture is harder to train. LBT-GAN even generates meaningless samples. (2) The images generated by GAN and DAN-S share only one single style within one mode. (3) SR-GAN achieves the best performance and covers all modes. It can generate diverse styles for the digits. The results in Table 3 (Right) are consistent with the visualization.

5 Conclusions

In this paper, we address the mode collapse problem by aligning the support of the generated data distribution with that of the real data distribution. The experiments show that our SR-GAN can avoid the mode collapse and also improve the data quality. SR-GAN introduces a simple extra regularization for GAN and does not modify any paradigm of GAN. Therefore the proposed support regularization term can be easy to be applied to other GANs, like conditional GAN, to solve the mode collapse problem.

Acknowledgement. This work was supported by the National Key R&D Program of China (Grant No. 2017YFC0804003), the Program for Guangdong Introducing Innovative and Enterpreneurial Teams (Grant No. 2017ZT07X386), Shenzhen Peacock Plan (Grant No. KQTD2016112514355531), the Program for University Key Laboratory of Guangdong Province (Grant No. 2017KSYS008), and Australian Research Council (Grant No. LP150100671 and No. DP180100106).

References

1. Arjovsky, M., Bottou, L.: Towards principled methods for training generative adversarial networks. In: ICLR (2017)
2. Du, C., Xu, K., Li, C., Zhu, J., Zhang, B.: Learning implicit generative models by teaching explicit ones. In: arXiv (2018)
3. Goodfellow, I.J., et al.: Generative adversarial nets. In: NeurIPS, pp. 2672–2680 (2014)

4. Görnitz, N., Lima, L.A., Müller, K., Kloft, M., Nakajima, S.: Support vector data descriptions and k-means clustering: one class? TNNLS **29**, 3994–4006 (2018)
5. Hu, W., Miyato, T., Tokui, S., Matsumoto, E., Sugiyama, M.: Learning discrete representations via information maximizing self-augmented training. In: ICML, pp. 1558–1567 (2017)
6. Li, C., Alvarez-Melis, D., Xu, K., Jegelka, S., Sra, S.: Distributional adversarial networks. In: ICLR (2018)
7. Metz, L., Poole, B., Pfau, D., Sohl-Dickstein, J.: Unrolled generative adversarial networks. In: ICLR (2017)
8. Schölkopf, B., Platt, J.C., Shawe-Taylor, J., Smola, A.J., Williamson, R.C.: Estimating the support of a high-dimensional distribution. Neural Comput. **13**(7), 1443–1471 (2001)
9. Srivastava, A., Valkov, L., Russell, C., Gutmann, M.U., Sutton, C.: VEEGAN: reducing mode collapse in gans using implicit variational learning. In: NeurIPS, pp. 3308–3318 (2017)
10. Tax, D.M., Duin, R.P.: Support vector data description. Mach. Learn. **54**(1), 45–66 (2004)

Patch-Based Generative Adversarial Network Towards Retinal Vessel Segmentation

Waseem Abbas[1], Muhammad Haroon Shakeel[2(✉)], Numan Khurshid[2], and Murtaza Taj[2]

[1] Cloud Application Solutions Division, Mentor, A Siemens Business, Lahore, Pakistan
muhammad_waseem@mentor.com

[2] Department of Computer Science, Syed Babar Ali School of Science and Engineering, Lahore University of Management Sciences (LUMS), Lahore, Pakistan
{15030040,15060051,murtaza.taj}@lums.edu.pk

Abstract. Retinal blood vessels are considered to be the reliable diagnostic biomarkers of ophthalmologic and diabetic retinopathy. Monitoring and diagnosis totally depends on expert analysis of both thin and thick retinal vessels which has recently been carried out by various artificial intelligent techniques. Existing deep learning methods attempt to segment retinal vessels using a unified loss function optimized for both thin and thick vessels with equal importance. Due to variable thickness, biased distribution, and difference in spatial features of thin and thick vessels, unified loss function are more influential towards identification of thick vessels resulting in weak segmentation. To address this problem, a conditional patch-based generative adversarial network is proposed which utilizes a generator network and a patch-based discriminator network conditioned on the sample data with an additional loss function to learn both thin and thick vessels. Experiments are conducted on publicly available STARE and DRIVE datasets which show that the proposed model outperforms the state-of-the-art methods.

Keywords: Deep Learning · Generative Adversarial Network · Segmentation · Retinal Vessels

1 Introduction

Deep learning has influenced image analysis in various important application areas including remote-sensing, autonomous vehicles, and specially medical [1,15]. Usually, diagnostic analysis and treatment which covers medical disorders, diabetic retinopathy, and glaucoma have been carried on retinal vessels of opthalmologic fundus images [18]. Morphological attributes and retinal vascular structures including vascular tree patterns, vessels thickness, color, density,

T. Gedeon et al. (Eds.): ICONIP 2019, CCIS 1142, pp. 49–56, 2019.
https://doi.org/10.1007/978-3-030-36808-1_6

crookedness, and relative angles are the key features used in the diagnostic process by ophthalmologists [7]. Thickness and visibility of retinal vessels are the core attributes for diabetic retinopathy analysis [2]. However, conventional and manual approaches for vascular analysis are time-consuming and prone to human error. Therefore, computer-assisted detection of retinal vessels is inevitable to segment out the retinal vessels from fundus images and mainly categorized into three approaches: *unsupervised learning*, *supervised learning*, and *deep learning*.

Unsupervised learning approaches extracts vessel pattern without class labels, mainly thorough filter-based approach. Image filters, such as Gaussian blur, are utilized to enhance vascular features. Zhang *et al.* combined the matched filter and first-order derivative of Gaussian filter to extract retinal vessel features [24]. Similarly, Fraz *et al.* applied the same approach in four directions along with multi-directional morphological top-hat operator to detect retinal vessels [6]. Yin *et al.* worked on orientation invariant approach and used Fourier transform to extract energy maps, consequently detecting retinal vessels using an orientation-aware detector [22]. A similar approach of using Wavelet transform to map images into a 3D lifted-domain was proposed in [25]. They utilized the Gaussian filter to detect retinal vessels.

Supervised approaches intend to assign a class label to each pixel of the image. These approaches can be based on either traditional *machine learning* or *deep learning*. The former utilizes classifiers that learn decision boundaries on handcrafted features e.g. Support Vector Machine (SVM) and K-nearest neighbor classifier (KNN).

However, more recently, Convolutional Neural Networks (CNNs) got popularity for segmentation problems since they learn the features directly from the input data. In the case of retinal vessel segmentation, CNNs surpass traditional handcrafted approaches [14]. However, the problem of blurred output images and false positives around indistinct tiny vessel branches is persistent in these methods. The main reason behind this limitation is use of a unified loss function in a pixel-wise manner to segment both thin and thick vessels. This led to blurred thin vessels resulting in non acceptable segmentation maps during binarization of generated probability maps. To address these issues a novel approach for retinal vessel segmentation using Generative Adversarial Networks (GAN) has been proposed in this research. Basically, a patch-based discriminator is utilized to learn inconsistencies and sharp edges of high-resolution blood vessels. Additionally, a loss term is integrated within the main objective function to learn low-frequency edges. We show that the proposed method is able to effectively segment out thick and thin vascular pixels form non-vascular pixels on two benchmark datasets namely DRIVE [16] and STARE [10]. Our results show a significant boost in the performance as compare to the state-of-the-art methods.

2 Methodology

Adversarial learning approach combines generator and discriminator networks in such a way that conditional input provides a head start to the overall learning

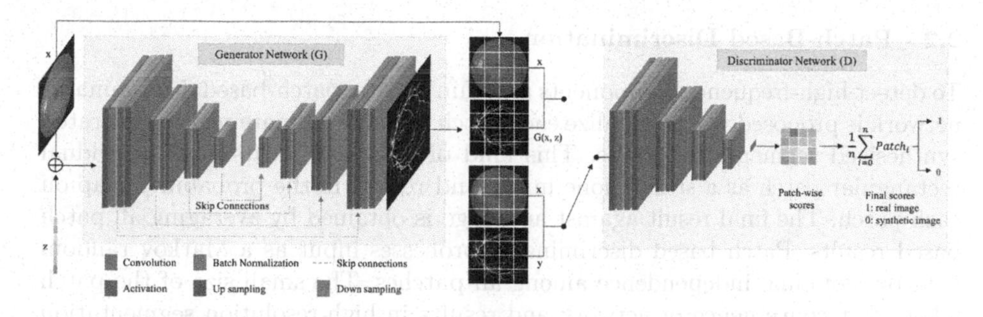

Fig. 1. The proposed Conditional Patch-based Generative Adversarial Network.

process. Generator network G expects noise and conditional input sample and learn to generate the synthetic retinal map. Discriminator network D takes two sets of input: conditional input and a generated synthetic map (fake sample) and conditional input and actual segmentation map (/real sample/ground truth). Segmentation maps generated from generator network are divided into rectangular patches and discriminator tries to discriminate each patch. This patch based discriminator is deployed to discriminate between actual or synthetically generated segmentation maps as shown in Fig. 1.

2.1 Objective Function

In adversarial learning, generator network G tries to map input conditional sample $x^{w \times h}$ and noise vector to its corresponding segmentation map $y^{w \times h}$ in encoder/decoder arrangement such that the difference between $y^{w \times h}$ and synthetic vessel map $G(x, z)$ is reduced (w:width and h:height of image). Discriminator network takes two pairs $\{x, y\}$ and $\{x, G(x, z)\}$ as input and predicts the score between 0 or 1 as $\{0, 1\}^n$ where n is a hyperparameter of the model and represents the total number of patches fed to the discriminator D. n could be selected between 0 and the total number of pixels of image. The objective function of the proposed model can be formulated as:

$$J = \arg\min_{G} \max_{D} \mathbb{E}_{x,y}\big(- log(D(x, y)) + \mathbb{E}_{x,z}\big(- log(1 - D(x, G(x, z)))\big) \\ + \lambda \mathcal{L}_{L_1}(G) \tag{1}$$

where $\mathbb{E}_{x,z}$ and $\mathbb{E}_{x,y}$ are loss functions for generator G and discriminator D respectively. To handle the problem of blurred outputs caused by L_2 norm reported in literature, we integrated L_1 norm in the main objective function to capture low and high-frequency components of the fundus retinal images. L_1 norm can be formulated as:

$$\mathcal{L}_{L_1}(G) = \mathbb{E}_{x,y,z}\Big[||y - G(x, z)||_1\Big] \tag{2}$$

where λ is a hyperparameter. The noise vector z is selected from a normal distribution to ensure that the generator learns a random sample at each training step. This also prevents the local minima problems.

2.2 Patch-Based Discrimination

To detect high-frequency components and thin vessels, patch-based discriminator network is proposed which penalize each patch and discriminate real or generated synthesized segmentation maps. This kind of approach treats each individual rectangular patch as a stand-alone image and results in the probability map on each patch. The final result against an image is obtained by averaging all patch based results. Patch based discriminator processes input as a Markov random field by assuming independence among all patches. The small size of the patch allows fast convergence of network and results in high-resolution segmentation maps.

2.3 Model Architecture

To extract low level features, proposed model followed inspiration form UNet [19] and SegNet [4]. The generative network is comprised of encoder and decoder networks. Encoder network extracts the hidden features and reduces the size of the input image whereas the decoder network reconstructs and up-sample at each stage till the last layer. Each input image passes through the entire generator network and skip connections between encoder and decoder network acts as bridge to semantic gap between the feature maps of the encoder and decoder prior to fusion.

2.4 Hyperparameter Tuning

The proposed model is trained in a manner such that a the generator network is trained and generates random output regardless of input sample and tries to learn the patterns of input as per given ground truths. Meanwhile, the discriminator network tries to discriminate the generated outputs of generator network and the ground truths. Further, discriminator learns to utilize the conditional samples (x) to learn the pattern of blood vessels in fundoscopic images. Stochastic gradient decent with Adam optimizer is used to train the generator network. All the hyperparameters used in training of the proposed model are selected empirically, which include trade-off coefficient ($\lambda = 10$), learning rate (lr = 0.002), learning rate decaying factor ($\eta = 0.75$) and momentum (m = 0.002).

2.5 Datasets and Evaluation Metrices

The proposed model is trained and evaluated on two publicly available datasets include DRIVE [16], STARE [10]. We followed the same evaluation metrics and protocols to conduct a fair evaluation with the reported state-of-the-art methods. Accuracy (*Acc*), sensitivity (*Se*) and specificity (*Sp*) are used as a benchmark for quantitative evaluation and area under the receiving operating curve (*AUC*) is used for the qualitative evaluation.

Table 1. Performance comparison of the proposed model with sate-of-the-art methods.

Scheme	Methods	Year	DRIVE				STARE			
			Acc	*Sp*	*Se*	*AUC*	*Acc*	*Sp*	*Se*	*AUC*
	Human Observer		0.9472	0.9724	0.7760	–	0.9349	0.9384	0.8952	–
Unsupervised	zhang [24]	2010	0.9382	0.9724	0.7120	–	0.9484	0.9753	0.7177	–
	Fraz [6]	2012	0.9430	0.9759	0.7152	–	0.9442	0.9686	0.7311	–
	Roychowdhuray [20]	2015	0.9494	0.9782	0.7395	0.9672	0.9560	0.9842	0.7317	0.9673
	Azzopardi [3]	2015	0.9442	0.9704	0.7655	0.9614	0.9497	0.9701	0.7716	0.9563
	Yin [22]	2015	0.9403	0.9790	0.7246	–	0.9325	0.9419	**0.8541**	–
	Zhang [25]	2016	0.9476	0.9725	0.7743	0.9636	0.9554	0.9758	0.7791	0.9748
Supervised	You [23]	2011	0.9434	0.9751	0.7410	–	0.9497	0.9756	0.7260	–
	Marin [13]	2011	0.9452	0.9801	0.7067	0.9588	0.9526	0.9819	0.6944	0.9769
	Fraz [8]	2012	0.9480	0.9807	0.7406	0.9747	0.9534	0.9763	0.7548	0.9768
	Orlando [17]	2017	–	0.9684	**0.7897**	–	–	0.9738	0.7680	–
	Dasgupta [5]	2017	0.9533	0.9801	0.7691	0.9744	–	–	–	–
Deep Learning	Melin [14]	2015	0.9466	0.9785	0.7276	0.9749	–	–	–	–
	Li [11]	2016	0.9527	0.9816	0.7569	0.9738	0.9628	0.9844	0.7726	0.9879
	Liskowski [12]	2016	0.9515	0.9806	0.7520	0.9710	**0.9696**	0.9866	0.8145	0.9880
	Fu [9]	2016	0.9523	–	0.7603	–	0.9585	–	0.7412	–
	Zengqiang [21]	2018	0.9538	0.9820	0.7631	0.9750	0.9636	0.9857	0.7735	0.9833
	Proposed	2019	**0.9562**	**0.9824**	0.7746	**0.9753**	0.9647	**0.9869**	0.7940	**0.9885**

3 Results and Discussion

Evaluation of the proposed model is conducted on DRIVE and STARE datasets and categorized into unsupervised, supervised and deep learning schemes as summarized in Table 1. *Acc*, *Se*, *Sp* and *AUC* values are mentioned against the proposed method and reported in the literature.

In unsupervised techniques, the most recent findings were reported in [25] where researchers used left invariant rotating derivative to get enhanced retinal vessels and obtained binary segmentation using thresholding. Their method outperformed the previous unsupervised techniques on DRIVE and STARE datasets [3,8,20,22,24]. In supervised learning techniques, [5] achieved best results on DRIVE dataset by deploying a combination of convolutional neural network and structured predictions. The second best method [6] in supervised learning used conditional random field model with a fully connected method and achieved comparable performance on DRIVE and STARE datasets as compared to other supervised learning schemes [8,13,23].

For the DRIVE dataset, the proposed model achieves 0.9562, 0.9824, 0.7746 and 0.9753 for *Acc*, *Sp*, *Se* and *AUC* respectively, where the model achieves better results for *Acc*, *Sp* and *AUC* as compared to the all current state-of-the-art unsupervised, supervised and deep learning techniques. However Orlando [17] outperforms all the methods in terms of *Se* as shown in Table 1, the only *Se* norm is not conclusive. In contrast, the performance of the proposed model is much better than all the compared methods.

On the STARE fundoscopic image dataset, the proposed model achieves 0.9647, 0.9862, 0.7940 and 0.9885 for *Acc*, *Sp*, *Se* and *AUC* respectively.

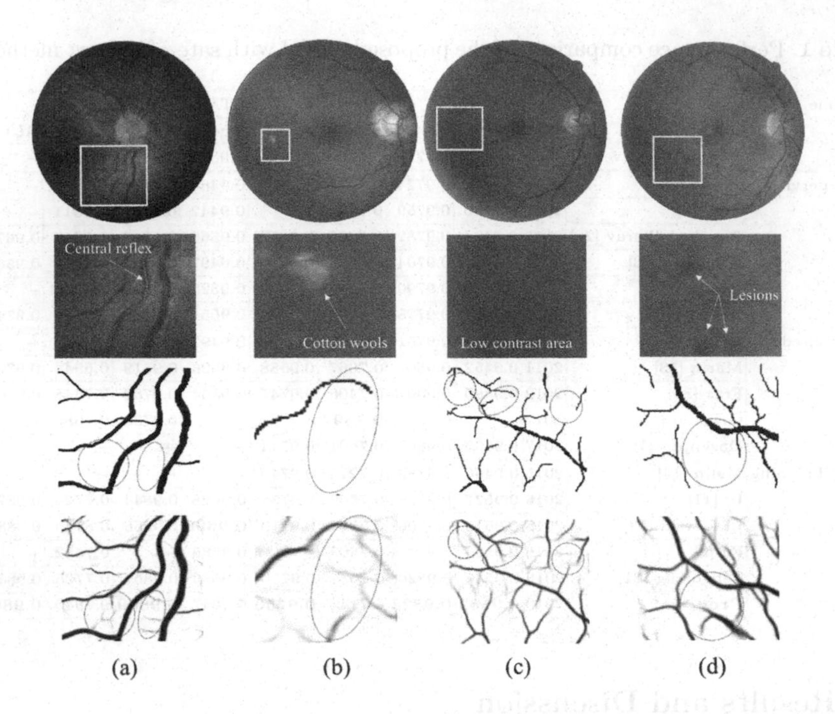

Fig. 2. Exemplar results of the proposed model on challenging cases: (a): central reflex vessels, (b): cotton wools, (c): low contrast, (d) lesions. From top to bottom: input fundus image, enlarged target patch of fundus image, corresponding manual annotation and the predicted probability maps.

In terms of Sp and AUC, the proposed model outperforms all the compared techniques. However, Yin [22] reported the best performance in terms of Se by achieving 0.0601 more sensitivity but obtains 0.030 lesser specificities. Similarly, Zengqiang [21] achieves better results in terms of Acc by achieving 0.9696 accuracy but lags in other evaluation benchmarks as compared to the proposed method.

Presence of lesions and cotton wools in fundoscopic images mainly affect the local features and the thick vessels. Other challenges are the presence of central reflex vessels and low contrasts. To address four types of challenges (central reflex vessels, cotton wools, low contrast, and lesions) in segmentation of fundoscopic images, the proposed model is able to segment out the retinal vessels in these challenging scenarios. By integrating L_1 norm in the main objective function, the generator network is able to detect low contrast and thin retinal vessels as shown in Fig. 2. The generator network learns the low contrast vessels whereas the discriminator network forces the model to learn the non-vessel pixels too by predicting a zero score. In this way, the entire model learns the structure and appearance of vessels simultaneously and the model is able to address the central reflex vessel problem. Patch based discriminator network allows the model to capture thin vessels in the presence of lesions and cotton wools. In summary, the proposed generative adversarial network can effectively address the main challenging cases by learning generator and discriminator network alternatively and integrating a custom loss term.

4 Conclusion

A novel generative adversarial network based deep learning model has been proposed, that can potentially address segmentation of retinal blood vessels in fundoscopic images. Training the generator network to learn small transitions in thin vessels and allowing the patch based discriminator to discriminate vascular and non-vascular pixels. Results on publicly available datasets showed that the proposed model is competitive with current state-of-the-art techniques. Averaging the patch based results over small patches of fundoscopic image and integration of additional loss term into the main objective function leverage and enhances the effectiveness of the proposed model. The model has the potential to probe the different patch sizes so that the influence of patch-based discriminator on segmentation performance can be better analyzed.

References

1. Abbas, W., Taj, M.: Adaptively weighted multi-task learning using inverse validation loss. In: IEEE International Conference on Acoustics, Speech and Signal Processing, pp. 1408–1412 (2019)
2. Abràmoff, M.D., Garvin, M.K., Sonka, M.: Retinal imaging and image analysis. IEEE Rev. Biomed. Eng. **3**, 169–208 (2010)
3. Azzopardi, G., Strisciuglio, N., Vento, M., Petkov, N.: Trainable cosfire filters for vessel delineation with application to retinal images. Med. Image Anal. **19**(1), 46–57 (2015)
4. Badrinarayanan, V., Kendall, A., Cipolla, R.: Segnet: a deep convolutional encoder-decoder architecture for image segmentation. IEEE Trans. Pattern Anal. Mach. Intell. **39**(12), 2481–2495 (2017)
5. Dasgupta, A., Singh, S.: A fully convolutional neural network based structured prediction approach towards the retinal vessel segmentation. In: IEEE International Symposium on Biomedical Imaging, pp. 248–251 (2017)
6. Fraz, M.M., et al.: An approach to localize the retinal blood vessels using bit planes and centerline detection. Comput. Methods Programs Biomed. **108**(2), 600–616 (2012)
7. Fraz, M.M., et al.: Blood vessel segmentation methodologies in retinal images-a survey. Comput. Methods Programs Biomed. **108**(1), 407–433 (2012)
8. Fraz, M.M., et al.: An ensemble classification-based approach applied to retinal blood vessel segmentation. IEEE Trans. Biomed. Eng. **59**(9), 2538–2548 (2012)
9. Fu, H., Xu, Y., Wong, D.W.K., Liu, J.: Retinal vessel segmentation via deep learning network and fully-connected conditional random fields. In: IEEE International Symposium on Biomedical Imaging, pp. 698–701 (2016)
10. Hoover, A., Kouznetsova, V., Goldbaum, M.: Locating blood vessels in retinal images by piecewise threshold probing of a matched filter response. IEEE Trans. Med. Imaging **19**(3), 203–210 (2000)

11. Li, Q., Feng, B., Xie, L., Liang, P., Zhang, H., Wang, T.: A cross-modality learning approach for vessel segmentation in retinal images. IEEE Trans. Med. Imaging **35**(1), 109–118 (2016)
12. Liskowski, P., Krawiec, K.: Segmenting retinal blood vessels with deep neural networks. IEEE Trans. Med. Imaging **35**(11), 2369–2380 (2016)
13. Marín, D., Aquino, A., Gegúndez-Arias, M.E., Bravo, J.M.: A new supervised method for blood vessel segmentation in retinal images by using gray-level and moment invariants-based features. IEEE Trans. Med. Imaging **30**(1), 146 (2011)
14. Melinščak, M., Prentašić, P., Lončarić, S.: Retinal vessel segmentation using deep neural networks. In: International Conference on Computer Vision Theory and Applications (2015)
15. Nazir, U., Khurshid, N., Ahmed Bhimra, M., Taj, M.: Tiny-inception-resnet-v2: using deep learning for eliminating bonded labors of brick kilns in South Asia. In: Proceedings of the IEEE Conference on Computer Vision and Pattern Recognition Workshops, pp. 39–43 (2019)
16. Niemeijer, M., Staal, J., Ginneken, B., Loog, M., Abramoff, M.: Drive: digital retinal images for vessel extraction. In: Methods for Evaluating Segmentation and Indexing Techniques Dedicated to Retinal Ophthalmology (2004)
17. Orlando, J.I., Prokofyeva, E., Blaschko, M.B.: A discriminatively trained fully connected conditional random field model for blood vessel segmentation in fundus images. IEEE Trans. Biomed. Eng. **64**(1), 16–27 (2017)
18. Patton, N., et al.: Retinal image analysis: concepts, applications and potential. Prog. Retinal Eye Res. **25**(1), 99–127 (2006)
19. Ronneberger, O., Fischer, P., Brox, T.: U-net: convolutional networks for biomedical image segmentation. In: Navab, N., Hornegger, J., Wells, W.M., Frangi, A.F. (eds.) MICCAI 2015. LNCS, vol. 9351, pp. 234–241. Springer, Cham (2015). https://doi.org/10.1007/978-3-319-24574-4_28
20. Roychowdhury, S., Koozekanani, D.D., Parhi, K.K.: Iterative vessel segmentation of fundus images. IEEE Trans. Biomed. Eng. **62**(7), 1738–1749 (2015)
21. Yan, Z., Yang, X., Cheng, K.T.: A three-stage deep learning model for accurate retinal vessel segmentation. IEEE J. Biomed. Health Inform. 1427–1436 (2018)
22. Yin, B., et al.: Vessel extraction from non-fluorescein fundus images using orientation-aware detector. Med. Image Anal. **26**(1), 232–242 (2015)
23. You, X., Peng, Q., Yuan, Y., Cheung, Y.M., Lei, J.: Segmentation of retinal blood vessels using the radial projection and semi-supervised approach. Pattern Recogn. **44**(10–11), 2314–2324 (2011)
24. Zhang, B., Zhang, L., Zhang, L., Karray, F.: Retinal vessel extraction by matched filter with first-order derivative of gaussian. Comput. Biol. Med. **40**(4), 438–445 (2010)
25. Zhang, J., Dashtbozorg, B., Bekkers, E., Pluim, J.P., Duits, R., ter Haar Romeny, B.M.: Robust retinal vessel segmentation via locally adaptive derivative frames in orientation scores. IEEE Trans. Med. Imaging **35**(12), 2631–2644 (2016)

A Gradient-Based Algorithm to Deceive Deep Neural Networks

Tianying Xie[1] and Yantao Li[2(✉)]

[1] College of Computer and Information Sciences, Southwest University, Chongqing 400715, China

[2] College of Computer Science, Chongqing University, Chongqing 400044, China
yantaoli@cqu.edu.cn

Abstract. Deep neural networks have achieved high performance in a variety of image recognition tasks. However, it is reported that the performance on image recognition of these networks is unstable to slight perturbations of images. To verify this weakness, we propose *DeceiveDeep*, a gradient-based algorithm for deceiving deep neural networks in this paper. There exists a lot of gradient-based attack methods, such as the L-BFGS, FGSM, and Deepfool. Specifically, based on an original method, L-BFGS, we exploit the Euclid norm of the gradient to update the space vector in an image to generate a deceivable image for fooling deep neural networks. We construct three types of deep neural network models and one convolutional neural network for testing the proposed algorithm. Based on the MNIST dataset and the Fashion-MNIST dataset, we evaluate the effectiveness of *DeceiveDeep* in terms of accuracy on training and testing data, and CNN model, respectively. The experimental results show that, comparing with L-BFGS, *DeceiveDeep* dramatically decreases the accuracy of the deep models on image recognition.

1 Introduction

Deep Neural Networks (DNNs) have achieved high performance, and supervised learning shows essential part in this area. However, it is similar to a black-box that we can not understand the processing procedure exactly. In [1,2], it suggests that the space, rather than the individual unit, contains the semantic information in the high layers of neural networks [3–6].

There are some works investigating how to fool deep neural networks. From [7], we know that deep neural network has two counter-intuitive properties. Due to the works [8–11], activating a given cell by looking for the maximum input set to analyze the semantics of each cell. In [12], the authors propose a method about computing continuous vector representations of words from very large datasets. Specifically, evolutionary algorithms or grandient ascent to generate images that revealed high accuracy by convolutional neural networks in [13,14]. The authors in [15] show a way that DNNs and human vision differ, and use evolutionary algorithms to do generated fooled images to do prediction [5]. Although the state-of-the-art deep neural networks can recognize natural images fast and accurately

T. Gedeon et al. (Eds.): ICONIP 2019, CCIS 1142, pp. 57–65, 2019.
https://doi.org/10.1007/978-3-030-36808-1_7

[16–19], they are also easily fooled into declaring familiar objects. There exists a lot of gradient-based attack methods, such as the L-BFGS [4], FGSM [20], and Deepfool [21].

The main purpose of this paper is to generate deceivable images that fool the neural network models without causing human visual errors, to protect users' privacy and security. Therefore, based on the original method L-BFGS [4], we propose *DeceiveDeep*, a gradient-based algorithm for deceiving deep neural networks.

The remainder of this paper is organized as follows: Sect. 2 details the proposed gradient-based algorithm. In Sect. 3, we construct three types of deep neural networks and on convolutional neural network for performance evaluation. In Sect. 4, we introduce the datasets for the performance evaluation and then evaluate the accuracy on the proposed deep neural networks with input images generated by *DeceiveDeep*. We conclude this work in Sect. 5.

2 Gradient-Based Algorithm

2.1 Original Method

We denote $f : \mathbb{R}^m \rightarrow \{1 \cdots k\}$ as a classifier mapping input image vectors to a discrete label set. The loss function is denoted by $loss_f : \mathbb{R}^m \times \{1 \cdots k\} \rightarrow \mathbb{R}^+$. Taking a given image $x \in \mathbb{R}^m$, target label $l \in \{1 \cdots k\}$ and perturbation r into consideration, the goal is to solve the optimization problem in Eq. (1) [5]:

$$\begin{aligned} &\min \|r\|_2 \\ &\text{s.t.} \quad f(x+r) = l \\ &\qquad x + r \in [0,1]^m \end{aligned} \tag{1}$$

where minimizer gradient r might not be unique, but we denote $x + r$ for an arbitrarily chosen minimizer by $D(x, l)$, where $D(x, f(x)) = f(x)$. We can find the minimum c of an approximation closing to a function $D(x, l)$ by performing line-search, as shown in Eq. (2):

$$\begin{aligned} &\min c|r| + loss_f(x + r, l) \\ &\text{s.t. } x + r \in [0,1]^m \end{aligned} \tag{2}$$

We denote the space information by d_x for fooling images, and we imply the following rule, as shown in Eq. (3):

$$x = x + d_x \tag{3}$$

2.2 Improved Method

A deceivable image is produced by the gradient ascent and an image is a space vector including information. However, deep neural networks learned by backpropagation have non-intuitive characteristics and intrinsic blind spots [5]. Nevertheless, gradient plus original image without processing to generate fooling images, which does not have a robust way to avoid semantic error. The original meaning of gradient is a vector, which indicates that the directional derivative of a function at this point is the maximum value along this direction. That is, the function changes rapidly along this direction at this point, and the change rate is the maximum (the modulus of the gradient).

To calculate an image space, we introduce Euclid norm which is often used to measure the length or size of each vector in a vector space (or matrix), as shown in Eq. (4):

$$||r||_2 = \sqrt{\sum_{N}^{i} r_i^2} \tag{4}$$

Thus, keeping the basic concepts to some extent, we use Euclid norm to upgrade Eq. (3) for space information:

$$d_x = \frac{\lambda \times r}{||r||_2} \tag{5}$$

Then, Eq. (3) can be updated as Eq. (6) via gradient ascend:

$$x = x + \frac{\lambda \times r}{||r||_2} \tag{6}$$

2.3 Gradient-Based Algorithm (*DeceiveDeep*)

There is a well-known algorithm proposed to produce novel images, evolutionary algorithm inspired by Darwinian evolution [7]. They contain a population of images that alternately face selection and then random perturbation. However, due to [8], this method has a series of limits and problems, such as only performing well on one subject in an image rather than multi-subjects.

There are heaps of useless space in an image where we can add some noise, and then the image will change in the unit's area. We try to add minimum noise to generate the least effect on an image. With an image as the input, neural network models can predict the category of the image with a prediction probability, which is used for determination. Then, we adjust the probability by backpropagation and update the parameters of the network to fool the neural network model. By varying the final probability value, we can calculate the increase of an image gradient based on the original one, and then add the image gradient to the original image to generate an image for fooling deep neural networks.

Based on the above discussion and Eq. (6), we propose *DeceiveDeep*, the gradient-based algorithm to generate deceivable images that can fool deep neural networks in Algorithm 1.

Algorithm 1. Gradient-based Algorithm for Generating Deceivable Images (*DeceiveDeep*)

Input: image x, neural model f, label l_x, target label l_t, learning rate λ
Output: fooling image x_f
1: for loop in 100 iterations:
2: do
3: true scores $= f(x)$; get target label $l_t = Random()$;
4: calculate target score $s_t = s[l_x, l_t]$;
5: calculate back propagation in term of hidden layer neurons $s_t.backward()$;
6: calculate gradient r about $x \rightarrow x_f$;
7: get $d_x = \frac{\lambda \times r}{||r||_2}$;
8: generate fooling image $x_f = x + d_x$;
9: end for
10: until STOP

3 Deep Neural Network Models

Based on deceivable images by *DeceiveDeep*, we test the image recognition accuracy of deep nerual networks. Therefore, we construct three types of deep neural networks and one convolutional neural network.

3.1 Proposed Deep Neural Networks

We build three types of architectures for deep neural networks, to compare with [5]: FC network model, FC100 network model, and FC200 network model. We refer to our network as "FC", which has a simple fully connected network with one or more hidden layers, and a classifier including Softmax or ReLU. The number of image pixels is 784; thus, we set all the first layer as 784 neurons to accept image, thereby preventing loss of images. The pixel intensities are scaled to be in the range [0, 1].

3.2 Convolutional Neural Network

Convolutional Neural Network (CNN) is composed of input layers, output layers, and multiple hidden layers, which can be divided into convolution layer, pooling layer, ReLU layer, and fully connected layer. Compared with traditional neural networks, CNN has three major features: local connectivity pattern, weight sharing, and multiple convolution kernel.

4 Experiments and Results

In this section, we first introduce the datasets for the image recognition, and then evaluate the accuracy on the proposed deep neural network models with input images generated by *DeceiveDeep*.

4.1 Dataset

In our experiments, we select two datasets: MNIST and Fashion-MNIST for the image recognition of *DeceiveDeep*.

MNIST[1] dataset is a handwriting dataset, and all images are 28×28.

Fashion-MNIST[2] is an image dataset that replaces the MNIST handwritten digital set, and has the same size to MNIST.

4.2 Accuracy on Testing Data

To evaluate the effectiveness of *DeceiveDeep*, we conduct experiments on the accuracy of the proposed deep neural network models with deceivable images generated by *DeceiveDeep* as testing data. For the proposed deep neural network models, we use the original image data to train them and exploit the deceivable images generated by *DeceiveDeep* for testing. In the testing step, we exploit the deceivable image data generated by the gradient-based algorithm to test them.

The results upon MNIST dataset with deceivable images as testing data are listed in Table 1. As shown in Table 1, we evaluate the accuracy of image recognition on three models with different λ values.

Table 1. Accuracy on MNIST dataset with deceivable images as testing data

Model name	Description	Original accuracy	Our accuracy
FC	Softmax with $\lambda = 10^{-5}$	98%	10%
FC 100-100-10	Sigmoid network with $\lambda = 10^{-5}$	90%	9%
FC 100-100-10	Sigmoid network with $\lambda = 10^{-6}$	85%	9%
FC 200-200-10	Sigmoid network with $\lambda = 10^{-5}$	90%	9%
FC 200-200-10	Sigmoid network with $\lambda = 10^{-6}$	85%	9%

4.3 Accuracy on Training and Testing Data

To further evaluate the efficiency of *DeceiveDeep*, we conduct experiments on the accuracy of image recognition of the proposed deep neural network models with deceivable images generated by *DeceiveDeep* as training and testing data. Half of the images are used to train the models and the rest are used to test them.

[1] http://yann.lecun.com/exdb/mnist/.

[2] https://github.com/zalandoresearch/fashion-mnist.

Based on the MNIST dataset, the results of the accuracy of deceivable images based on the MNIST dataset are listed in Table 2. As illustrated in Table 2, we list the original accuracy, the accuracy based on [5], and that on *DeceiveDeep* under different models with different λ values. From the Table, *DeceiveDeep* shows the best fooling effect with the average accuracy 6.9% compared with the original algorithm (average accuracy of 90.12%) and [5] (average accuracy of 10.46%). Specifically, the FC model with $\lambda = 10^{-3}$ reaches the highest accuracy 97.27% based on the original MNIST dataset. Based on the algorithm in [5], the accuracy drops to 9.85% while it further decreases to 5.4% based on *DeceiveDeep*.

Based on the Fashion-MNIST dataset, the results of the accuracy of deceivable image recognition based on the Fashion-MNIST dataset are listed in Table 3. In Table 3, we list the same accuracy under different models with different λ values as Table 2. As shown in Table 3, *DeceiveDeep* shows the best fooling effect with the average accuracy 8.44% compared to the original algorithm with an average accuracy of 82.3% and [5] with an average accuracy of 10.7%. More specifically, the FC model with $\lambda = 10^{-3}$ achieves the highest accuracy 93.27% based on the original Fashion-MNIST dataset. Based on the algorithm in [5], the accuracy decreases to 9.13% while it further drops to 6.72% based on *DeceiveDeep*. In addition, the FC100-100-10 model shows the lowest accuracy 5.85% based on *DeceiveDeep*.

Table 2. Accuracy on deceivable images based on MNIST dataset

Model Name	Description	Original Accuracy	Accuracy in [5]	Our Accuracy
FC	Softmax (with $\lambda = 10^{-4}$)	93.32%	9.42%	6.2%
FC	Softmax (with $\lambda = 10^{-3}$)	97.27%	9.85%	5.4%
FC	Softmax (with $\lambda = 10^{-2}$)	90.97%	11.71%	9.2%
FC	Softmax (with $\lambda = 1$)	80.08%	14.00%	10.3%
FC100-100-10	Sigmoid network (with $\lambda = 10^{-5}$)	90.11%	9.12%	5.87%
FC100-100-10	Sigmoid network (with $\lambda = 10^{-6}$)	86.85%	9.38%	5.84%
FC200-200-10	Sigmoid network (with $\lambda = 10^{-5}$)	91.56%	10.45%	6.28%
FC200-200-10	Sigmoid network (with $\lambda = 10^{-6}$)	90.78%	9.75%	6.12%

Table 3. Accuracy on deceivable images based on Fashion-MNIST dataset

Model Name	Description	Original Accuracy	Accuracy in [5]	Our Accuracy
FC	Softmax (with $\lambda = 10^{-4}$)	92.75%	9.21%	7.4%
FC	Softmax (with $\lambda = 10^{-3}$)	93.27%	9.13%	6.72%
FC	Softmax (with $\lambda = 10^{-2}$)	90.01%	10.58%	10.01%
FC	Softmax (with $\lambda = 1$)	78.81%	16.24%	14.57%
FC100-100-10	Sigmoid network (with $\lambda = 10^{-5}$)	90.25%	9.01%	5.85%
FC100-100-10	Sigmoid network (with $\lambda = 10^{-6}$)	71.82%	12.42%	9.98%
FC200-200-10	Sigmoid network (with $\lambda = 10^{-5}$)	70.58%	9.89%	6.51%
FC200-200-10	Sigmoid network (with $\lambda = 10^{-6}$)	70.87%	9.12%	6.48%

4.4 Accuracy on CNN

In order to further verify the effectiveness of the proposed algorithm, we conduct experiments on the image recognition accuracy of the Convolutional Neural Network (CNN) model, as shown in Fig. 1.

For the MNIST dataset, the CNN model achieves 99.25% accuracy based on the original algorithm, L-BFGS; however, it greatly drops to 9% based on the proposed algorithm. For the Fashion-MNIST dataset, the CNN model reaches 90.11% accuracy, but with data generated by our algorithm, it achieves 12% accuracy.

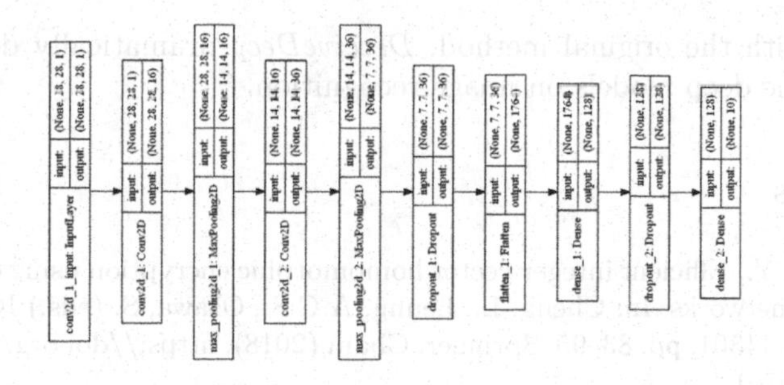

Fig. 1. Convolutional Neural Network Model

4.5 Discussion

Human eyes are unable to distinguish the difference between the processed image and the original image, mainly because we cannot see the difference directly from the appearance by adding minimal noise to the image. This approach does address privacy and security issues. For example, the content of an image can be easily distinguished by humans eyes, whereas a very robust model cannot because of the difference in the underlying composition of the image. In today's society, the rapid development of machine learning has spawned a large number of applications of artificial intelligence. For privacy and security, we need to guard against not only autonomous attack means from human beings, but also efficient attack means from machines.

The experimental results show that the image recognition accuracy of the neural network models is greatly reduced, which shows the validity and rationality of *DeceiveDeep*. It also proves that the proposed algorithm is effective and reasonable.

5 Conclusion

Through this paper, we can find that deep neural network is easy to be deceived, and the main reason is derived from its property and attribute. To protect user privacy and avoid privacy leakage problem, we propose *DeceiveDeep*, a gradient-based algorithm to deceive deep neural networks. Then, We construct three types of deep neural network models and one convolutional neural network for testing *DeceiveDeep* on image recognition. Based on the MNIST dataset and the Fashion-MNIST dataset, we evaluate the effectiveness of *DeceiveDeep* in terms of accuracy on training and testing data, and CNN model, respectively. The experimental results show that the deep neural networks achieve the lowest accuracy 9% on the MNIST dataset with deceivable images produced by *DeceiveDeep* as testing data for image recognition, 5.4% on that dataset with *DeceiveDeep*-generated deceivable images as training and testing data, 5.85% on the Fashion-MNIST dataset. CNN reaches 9% accuracy on the MNIST dataset processed by *DeceiveDeep* and 12% accuracy on the Fashion-MNIST dataset.

Comparing with the original method, *DeceiveDeep* dramatically decreases the accuracy of the deep models on image recognition.

References

1. Xie, T., Li, Y.: Efficient integer vector homomorphic encryption using deep learning for neural networks. In: Cheng, L., Leung, A.C.S., Ozawa, S. (eds.) ICONIP 2018. LNCS, vol. 11301, pp. 83–95. Springer, Cham (2018). https://doi.org/10.1007/978-3-030-04167-0_8
2. Szegedy, C., et al.: Intriguing properties of neural networks. arXiv preprint, arXiv:1312.6199 (2013)
3. Bengio, Y.: Learning deep architectures for AI. Found. Trends® Mach. Learn. **2**, 1–127 (2009). https://doi.org/10.1561/2200000006
4. Hinton, G.E.: Learning multiple layers of representation. Trends Cogn. Sci. **11**, 428–434 (2007). https://doi.org/10.1016/j.tics.2007.09.004
5. Yuan, X., He, P., Zhu, Q., Li, X.: Adversarial examples: attacks and defenses for deep learning. IEEE Trans. Neural Netw. Learn. Syst. **30**, 2805–2824 (2019). https://doi.org/10.1109/TNNLS.2018.2886017
6. Felzenszwalb, P., McAllester, D., Ramanan, D.: A discriminatively trained, multi-scale, deformable part model. In: 2008 IEEE Conference on Computer Vision and Pattern Recognition, pp. 1–8. IEEE Press, New York (2008). https://doi.org/10.1109/CVPR.2008.4587597
7. Floreano, D., Mattiussi, C.: Bio-inspired Artificial Intelligence: Theories, Methods, and Technologies. MIT Press, Cambridge (2008)
8. Cully, A., Clune, J., Tarapore, D., Mouret, J.B.: Robots that can adapt like animals. Nature **521**, 503–507 (2015). https://doi.org/10.1038/nature14422
9. Girshick, R., Donahue, J., Darrell, T., Malik, J.: Rich feature hierarchies for accurate object detection and semantic segmentation. In: 2014 IEEE Conference on CVPR, pp. 580–587. IEEE Press, New York (2014). https://doi.org/10.1109/CVPR.2014.81
10. Goodfellow, I., Lee, H., Le, Q.V., Andrew, Y.N.: Measuring invariances in deep networks. In: Proceedings of the 22nd International Conference on NIPS, pp. 646–654. ACM (2009). https://doi.org/10.5555/2984093.2984166
11. Zeiler, M.D., Fergus, R.: Visualizing and understanding convolutional networks. In: Fleet, D., Pajdla, T., Schiele, B., Tuytelaars, T. (eds.) ECCV 2014. LNCS, vol. 8689, pp. 818–833. Springer, Cham (2014). https://doi.org/10.1007/978-3-319-10590-1_53
12. Mikolov, T., Chen, K., Corrado, G., Dean, J.: Efficient estimation of word representations in vector space. arXiv preprint, arXiv:1301.3781 (2013)
13. Krizhevsky, A., Sutskever, I., Hinton, G.E.: ImageNet classification with deep convolutional neural networks. Commun. ACM **60**, 84–90 (2012). https://doi.org/10.1145/3065386
14. LeCun, Y., Bottou, L., Bengio, Y., Haffner, P.: Gradient-based learning applied to document recognition. Proc. IEEE **86**, 2278–2324 (1998). https://doi.org/10.1109/5.726791
15. Nguyen, A., Yosinski, J., Clune, J.: Deep neural networks are easily fooled: High confidence predictions for unrecognizable images. In: CVPR, pp. 427–436. IEEE Press (2015).https://doi.org/10.1109/CVPR.2015.7298640

16. Simonyan, K., Vedaldi, A., Zisserman, A.: Deep inside convolutional networks: visualising image classification models and saliency maps. arXiv preprint, arXiv:1312.6034 (2013)
17. Luo, C., Li, Z., Huang, K., Feng, J., Wang, M.: Zero-shot learning via attribute regression and class prototype rectification. IEEE Trans. Image Process. **27**, 637–648 (2018). https://doi.org/10.1109/TIP.2017.2745109
18. Hu, G., Peng, X., Yang, Y., Hospedales, T.M., Verbeek, J.: Frankenstein: learning deep face representations using small data. IEEE Trans. Image Process. **27**, 293–303 (2018). https://doi.org/10.1109/TIP.2017.2756450
19. Zhou, H., Wornell, G.: Efficient homomorphic encryption on integer vectors and its applications. In: 2014 Information Theory and Applications Workshop, pp. 1–9. IEEE Press, New York (2014). https://doi.org/10.1109/ITA.2014.6804228
20. Goodfellow, I.J., Shlens, J., Szegedy, C.: Explaining and harnessing adversarial examples. arXiv preprint, arXiv:1412.6572 (2014)
21. Moosavi-Dezfooli, S.M., Fawzi, A., Frossard, P.: Deepfool: a simple and accurate method to fool deep neural networks. In: Proceedings of the IEEE Conference on CVPR, pp. 2574–2582. IEEE (2016). https://doi.org/10.1109/CVPR.2016.282

Writing Style Adversarial Network for Handwritten Chinese Character Recognition

Huan Liu, Shujing Lyu(✉), Hongjian Zhan, and Yue Lu

Shanghai Key Laboratory of Multidimensional Information Processing, Department of Computer Science and Technology, East China Normal University, Shanghai 200062, China
{hliu,hjzhan}@stu.ecnu.edu.cn, {ylu,sjlv}@cs.ecnu.edu.cn

Abstract. The performance of handwritten Chinese character recognition (HCCR) has been greatly improved by using deep learning methods in recent years. But few people pay attention to the influence of writing style on it. In this paper, we aim to improve the performance of HCCR further by weakening the influence of different writing styles. We propose a writing style adversarial network (WSAN) which includes three parts: feature extractor, character classifier and writer classifier. In the training process, we first preprocess raw image with feature extractor. Afterwards, the learned features are fed into both the character classifier and the writer classifier. We apply joint optimization on the top of these two classifiers. Specifically, we minimize the loss value of the character classifier to achieve character recognition function. At the same time, we maximize the loss value of the writer classifier to reduce the influence of writing style in HCCR. The experimental results on CASIA-HWDB1.1 prove that the proposed WSAN has a promoting effect on HCCR. And the experiments on the offline HCCR competition dataset of ICDAR-2013 also give competitive results compared with other methods.

Keywords: Handwritten chinese character recognition · Style adversarial network · Gradient reversal layer

1 Introduction

The importance of HCCR is well recognized in both information retrieval and text recognition. The development of HCCR system has a long history and many methods have been proposed. However, some of the previously proposed methods, including both deep learning and machine learning based methods, were carefully designed for a specific distribution of training data, but ignore the effect of writing styles.

Figure 1 shows four examples of the same characters written by two writers. It is obvious that the writing styles of the two writers are different, and that

T. Gedeon et al. (Eds.): ICONIP 2019, CCIS 1142, pp. 66–74, 2019.
https://doi.org/10.1007/978-3-030-36808-1_8

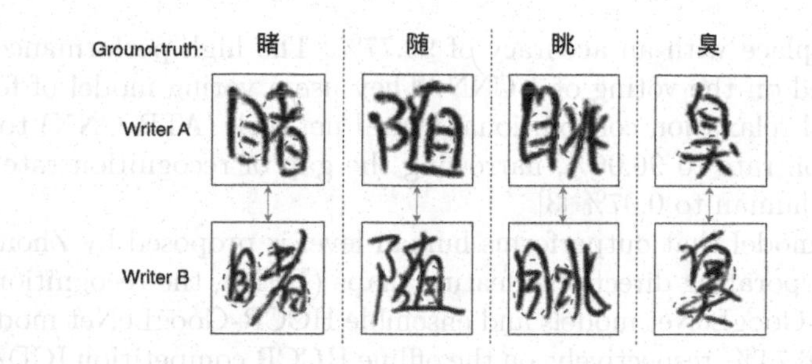

Fig. 1. Different handwriting styles from the two writers.

will produce different features for the same character. So we aim to weaken the impact of writing style on HCCR.

In this paper, we propose a novel domain adversarial based method, named writing style adversarial network(WSAN), to recognize new writing styles. On the one hand, we minimize the loss of character classifier to reach the purpose of character recognition. On the other hand, we maximize the loss of writer classifier to suppress writing styles.

In conclusion, the contributions of this work are:

1. We propose a domain adversarial neural network for handwritten Chinese character recognition, which is designed to reduce the variety between people's writing styles.
2. We propose a productive learning strategy by jointly optimizing the loss of recognition layer and write style discriminator. The result shows that it can improve both the recognition accuracy and training coverage speed.
3. We conduct two experiments on public datasets: CASIA-HWDB1.0, CASIA-HWDB1.1 and ICDAR-2013, the result demonstrates the effectiveness of our proposed methods. We have released the source code along with the paper[1].

2 Related Work

2.1 HCCR

The earlier classifiers used on HCCR are the traditional ones, including KNN, SVM, MQDF [1], etc., in which MQDF achieved comparable performance, but still far from application requirements.

Benefited from the improvement of computational performance and the enlargement of dataset, convolution neural networks (CNN) greatly promotes the performance of character recognition. The first CNN successfully applies to HCCR is multi-column deep neural network (MCDNN) [2], which is composed of multiple CNN, and its recognition rate approaches human performance. In the offline HCCR competition held by ICDAR in 2013, Fujitsu R&D Center

[1] https://github.com/qq2294011886/WSAN_HCCR.

win the first place with an accuracy of 94.77%. The high performance of their model is based on the voting of 4 CNN. They use a voting model of four alternately trained relaxation convolutional neural network (ATR-CNN) to increase the recognition rate to 96.06%, narrowing the gap of recognition rate between machine and human to 0.07% [3].

The first model that outperforms human-level is proposed by Zhong [4]. By properly incorporating directional feature maps (DFM), the recognition rates of single HCCR-GoogLeNet models and ensemble HCCR-GoogLeNet models reach 96.35% and 96.74%, respectively, on the offline HCCR competition ICDAR-2013. Zhang et al. [5] combine the traditional normalization-cooperated direction-decomposed feature map (directMap) with the deep convolutional neural network (convNet) to obtain a correct rate of 96.95%. They add an adaptation layer to the pre-trained convNet of this model, that increases the recognition rate to 97.37%. In the HCCR-CNN12Layer model proposed by Xiao et al. [6], the parametric rectifier linear unit (PReLU) is used instead of the rectifier linear unit (ReLU), and batch normalization layer is added after the convolution layer, which makes the recognition rate as high as 97.59%.

2.2 Domain Adversarial Network

The domain adversarial network (DAN) is successfully applied to the domain adaptation by Ganin et al. [7] for the first time. This network reduces the variation of feature distribution between the source domain and the target domain by learning the feature mapping between them. The model proposed by Ganin et al. achieves excellent performance in domain adaptation tasks such as sentiment analysis and image classification.

Domain adversarial network is widely used in many fields. Park et al. [8] applies the domain adversarial method to image-text multimodal learning for the first time. Compared to the previously proposed method, the domain adversarial method does not require the image-text pair to extract the semantic information of the image and the text, only needs the category label. The domain adversarial network is used by Kim et al. [9] to solve the problem of data transfer of the spoken language understanding (SLU). Data transfer involves two aspects, one is the transfer from synthetic data to live user data, and the other is the transfer from stale data to current data. The experimental results prove that the domain adversarial network has positive influence in both supervised and unsupervised scenarios. Liu et al. [10] propose an adversarial multi-task learning framework to alleviate the interference between shared and private feature spaces in text classification tasks.

Now using a single-domain discriminator to align the source and target domains has not met the requirements. Pei et al. [11] present a multi-adversarial domain adaptation (MADA) approach to capture multi-mode structures and achieve fine-grained alignment of multi-domain discriminators.

3 Proposed Method

Writing styles vary from person to person, which may have a certain impact on HCCR. In this paper, we propose a network, named WSAN, to reduce the influence of writing style on HCCR.

3.1 Model

The proposed network is shown in Fig. 2. It consists of three parts: feature extractor, character classifier, and writer classifier.

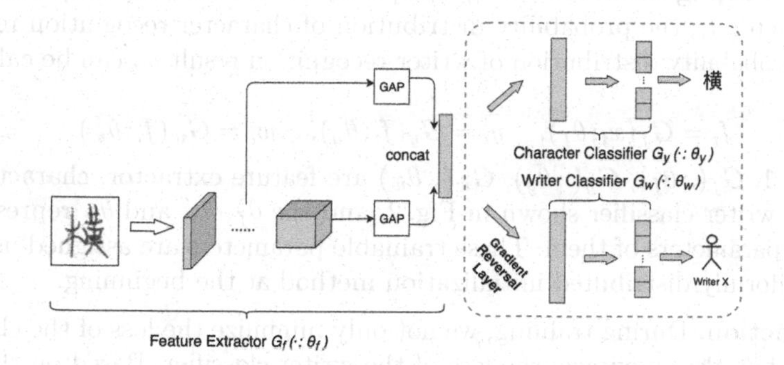

Fig. 2. The structure of WSAN model.

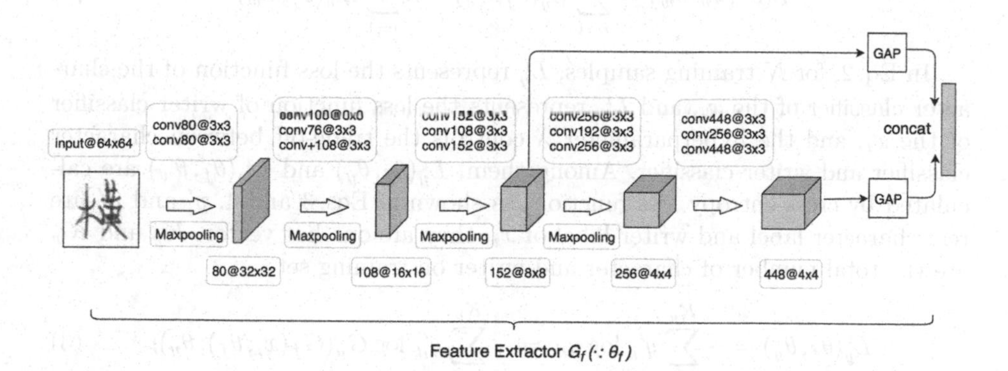

Fig. 3. The structure of feature extractor.

The feature extractor is a 14-layer CNN network, the specific structure is shown in Fig. 3. In order to speed up the training and improve the generalization ability of the network, we add the batch normalization layer behind each convolutional layer. To prevent overfitting, we add the dropout layer behind the max pooling layer and the global average pooling(GAP)[12] layer in the both middle and final layers of the feature extractor. The character classifier and the writer classifier have the same structure, including the fully connected layer and the softmax classifier. The softmax classifier of the character classifier outputs a 3755-dimensional vector. The softmax classifier of the writer classifier outputs a vector of m dimensions, which represents the number of writers in training dataset. In the experiment conducted on CASIA-HWDB1.0, m is 420, and m is 300 in another experiment.

3.2 Learning

Forward Propagation. For the i_{th} sample x_i, the feature map output from feature extractor f_i, the probability distribution of character recognition result y_i, and the probability distribution of writer recognition result w_i can be calculated by Eq. 1.

$$f_i = G_f(x_i;\theta_f), \quad y_i = G_y(f_i;\theta_y), \quad w_i = G_w(f_i;\theta_w) \tag{1}$$

In Eq. 1, $G_f(.;\theta_f)$, $G_y(.;\theta_y)$, $G_w(.;\theta_w)$ are feature extractor, character classifier and writer classifier shown in Fig. 2, and the θ_f, θ_y, and θ_w represent the trainable parameters of them. These trainable parameters are assigned using the glorot uniformly distributed initialization method at the beginning.

Loss Function. During training, we not only minimize the loss of the character classifier, but also maximize the loss of the writer classifier. Based on this idea, we propose a calculation method for the loss function. The specific formula is as shown in Eq. 2.

$$L(\theta_f,\theta_y,\theta_w) = \sum_{i=1}^{N} L_y^i(\theta_f,\theta_y) - \lambda \sum_{i=1}^{N} L_w^i(\theta_f,\theta_w) \tag{2}$$

In Eq. 2, for N training samples, L_y^i represents the loss function of the character classifier of the x_i, and L_w^i represents the loss function of writer classifier of the x_i, and the hyperparameter λ controls the trade-off between character classifier and writer classifier. Among them, $L_y^i(\theta_f,\theta_y)$ and $L_w^i(\theta_f,\theta_w)$ are calculated by cross-entropy loss function, as shown in Eqs. 3 and 4. y_i' and w_i' are real character label and writer label of x_i, they are one-hot vector. K_y and K_w are the total number of character and writer on training set.

$$L_y^i(\theta_f,\theta_y) = -\sum_{k=1}^{K_y} y_{ik}' \log y_{ik} = -\sum_{k=1}^{K_y} y_{ik}' \log G_y(G_f(x_i;\theta_f);\theta_y)_k \tag{3}$$

$$L_w^i(\theta_f,\theta_w) = -\sum_{k=1}^{K_w} w_{ik}' \log w_{ik} = -\sum_{k=1}^{K_w} w_{ik}' \log G_w(G_f(x_i;\theta_f);\theta_w)_k \tag{4}$$

We add a gradient reversal layer (GRL) [7] between the feature extractor and the writer classifier, as shown in Fig. 2. During back propagation, the GRL can update the feature extractor with the objective of maximizing the loss of writer classifier by reversing the gradient from writer classifier.

4 Experiments

4.1 Dataset

For evaluating the effectiveness of WSAN, we conduct the first experiment on the CASIA-HWDB1.1 test set compiled by the Chinese Academy of Sciences [13], where the training set is CASIA-HWDB1.1 training set. And the comparison experiment is conducted on the HCCR competition dataset ICDAR-2013, where the training set consists of CASIA-HWDB1.0 and CASIA-HWDB1.1.

4.2 The Evaluation of WSAN Effectiveness

To validate the superiority of our model, we conduct experiments on two datasets. The first experiment was done on CASIA-HWDB1.1 to verify the effectiveness of WSAN. The networks trained by Zhang et al. [14] is to verify the impact of network depth on the recognition accuracy. The networks can be combined with the style adversarial layer without changing the key parts of the original network. To save the cost and time of training, we select five of these networks and retrain them with WSAN. Then compare it with the network without WSAN.

From the experimental results in Table 1, we can see that after adding the WSAN, the top-1 and top-5 accuracy of the 3755-class classification is improved.

Table 1. The 3755-class classification results on the CASIA-HWDB1.1 test set.

Methods	Top 1 (%)		Top 5 (%)	
	w/o WSAN	w WSAN	w/o WSAN	w WSAN
DCNN-M6	94.60	95.02	98.90	99.04
DCNN-M6+	94.90	95.19	99.10	99.19
DCNN-M7-1	95.10	95.51	99.20	99.43
DCNN-M7-2	95.00	95.37	99.20	98.37
DCNN-Ensemble	95.50	95.86	99.30	99.53

In the experiment, we also find that the convergence rate of the model with WSAN is faster than that without WSAN. As shown in Fig. 4, the ensemble model without WSAN uses more than 7,000 mini-batch when the recognition rate reached 80%, while the ensemble model used less than 4,000 mini-batch after adding the WSAN. When the model is trained, if there is a tendency to overfit the writing style, the gradient reversal layer can correct the gradient descent from the wrong direction, which can accelerate the convergence of the model.

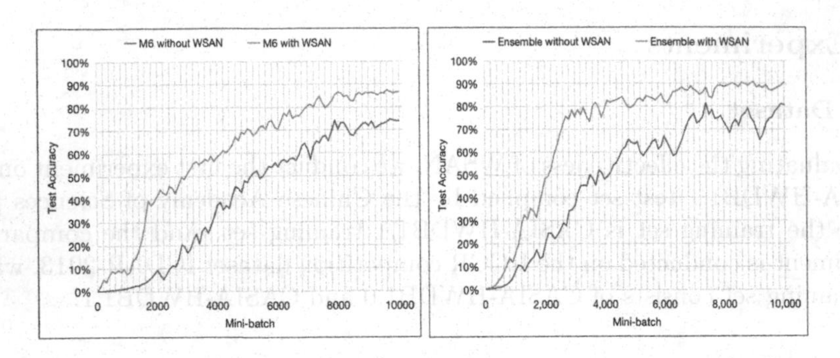

Fig. 4. Comparisons of convergence rates of models with and without WSAN.

4.3 Evaluation on ICDAR-2013

In the second experiment, we train the network on datasets CASIA-HWDB1.0 and CASIA-HWDB1.1 and test on the HCCR competition dataset ICDAR-2013. We compare our model with several representative models. The experiment results are shown in Table 2. It can be seen that the accuracy of our model reaches 97.27% in the ICDAR-2013 dataset.

Table 2. Classification results on the ICDAR-2013 database.

Methods	Top 1 accuracy (%)	Top 5 accuracy (%)
DCNN-M6 [14]	94.35	98.48
DCNN-M7-1 [14]	94.41	98.58
DCNN-Ensemble [14]	94.67	98.84
HCCR-AlexNet [15]	95.49	98.91
HCCR-GoogleNet [16]	96.26	99.58
Gabor+HCCR-GoogLeNet [4]	96.35	99.60
HCCR-Ensemble-GoogLeNet [4]	96.74	99.65
DirectMap+ConvNet+Adaptation [5]	97.37	n/a
HCCR-CNN12Layer [6]	97.59	n/a
Cascaded Model [17]	97.14	n/a
Ours Model(w/o WSAN)	96.89	99.59
Ours Model(w WSAN)	97.27	99.68

The reason why the accuracy of our model is lower than that of [5] and [6] is as follows: (1) The model in [5] combines domain knowledge with deep convolution neural network. However, our proposed model is an end-to-end model that does not require researchers to have prior knowledge. (2) The convolution kernel in [6] is deeper, and the model uses up to 48.7 MB of memory, while our model uses only 28.7 M of memory.

5 Conclusion

In this paper, we propose a novel adversarial network for handwritten Chinese character recognition called WSAN, which is designed to suppress the negative impact of writing style on character recognition. WSAN includes feature extractor, character classifier and writer classifier. We employ feature extractor to learn deep representations for raw image and then jointly optimize the network by minimize the loss of the character classifier and maximize the loss of the writer classifier. The experiments on two public datasets demonstrate that the proposed network achieves comparable performance on HCCR.

Acknowledgement. The work is supported by Shanghai Natural Science Foundation (No. 19ZR1415900).

References

1. Kimura, F., Takashina, K., Tsuruoka, S., Miyake, Y.: Modified quadratic discriminant functions and the application to chinese character recognition. IEEE Trans. Pattern Anal. Mach. Intell. **1**, 149–153 (1987)
2. Cireşan, D., Meier, U.: Multi-column deep neural networks for offline handwritten Chinese character classification. In: 2015 International Joint Conference on Neural Networks (IJCNN), pp. 1–6. IEEE (2015)
3. Wu, C., Fan, W., He, Y., Sun, J., Naoi, S.: Handwritten character recognition by alternately trained relaxation convolutional neural network. In: 2014 14th International Conference on Frontiers in Handwriting Recognition, pp. 291–296. IEEE (2014)
4. Zhong, Z., Jin, L., Xie, Z.: High performance offline handwritten Chinese character recognition using Googlenet and directional feature maps. In: 2015 13th International Conference on Document Analysis and Recognition (ICDAR), pp. 846–850. IEEE (2015)
5. Zhang, X.Y., Bengio, Y., Liu, C.L.: Online and offline handwritten chinese character recognition: a comprehensive study and new benchmark. Pattern Recogn. **61**, 348–360 (2017)
6. Xiao, X., Jin, L., Yang, Y., Yang, W., Sun, J., Chang, T.: Building fast and compact convolutional neural networks for offline handwritten chinese character recognition. Pattern Recogn. **72**, 72–81 (2017)
7. Ganin, Y., et al.: Domain-adversarial training of neural networks. J. Mach. Learn. Res. **17**(1), 2030–2096 (2016)
8. Park, G., Im, W.: Image-text multi-modal representation learning by adversarial backpropagation. arXiv preprint arXiv:1612.08354 (2016)
9. Kim, Y.B., Stratos, K., Kim, D.: Adversarial adaptation of synthetic or stale data. In: Proceedings of the 55th Annual Meeting of the Association for Computational Linguistics, vol. 1, Long Papers, pp. 1297–1307 (2017)
10. Liu, P., Qiu, X., Huang, X.: Adversarial multi-task learning for text classification. arXiv preprint arXiv:1704.05742 (2017)
11. Pei, Z., Cao, Z., Long, M., Wang, J.: Multi-adversarial domain adaptation (2018)
12. Lin, M., Chen, Q., Yan, S.: Network in network. arXiv preprint arXiv:1312.4400 (2013)

13. Liu, C.L., Yin, F., Wang, D.H., Wang, Q.F.: Online and offline handwritten chinese character recognition: benchmarking on new databases. Pattern Recogn. **46**(1), 155–162 (2013)
14. Zhang, Y.: Deep convolutional network for handwritten chinese character recognition. Computer Science Department, Stanford University (2015)
15. Krizhevsky, A., Sutskever, I., Hinton, G.E.: Imagenet classification with deep convolutional neural networks. In: Advances in Neural Information Processing Systems, pp. 1097–1105 (2012)
16. Szegedy, C., et al.: Going deeper with convolutions. In: Proceedings of the IEEE Conference on Computer Vision and Pattern Recognition, pp. 1–9 (2015)
17. Li, Z., Teng, N., Jin, M., Lu, H.: Building efficient cnn architecture for offline handwritten chinese character recognition. Int. J. Document Anal. Recogn. (IJDAR) **21**(4), 233–240 (2018)

Recovering Super-Resolution Generative Adversarial Network for Underwater Images

Yang Chen, Jinxuan Sun, Wencong Jiao, and Guoqiang Zhong(✉)

Department of Computer Science and Technology, Ocean University of China, 238 Songling Road, Qingdao 266100, China
gqzhong@ouc.edu.cn

Abstract. In this paper, we propose an end-to-end Recovering Super-Resolution Generative Adversarial Network (RSRGAN) to automatically learn super-resolution underwater images. RSRGAN mainly includes two parts. The first part is a Recovering GAN, aiming at color correction and removing noise in the images. The generator of Recovering GAN is based on an encoder-decoder network with self-attention on the global feature. The second part is a Super-Resolution GAN, which adopts the residual-in-residual dense block in its generator, to add details onto the results fed from the Recovering GAN. Both qualitative and quantitative experimental results show the advantage of RSRGAN over the state-of-the-art approaches for underwater image super-resolution.

Keywords: Underwater images · Super-resolution · Generative adversarial network

1 Introduction

Underwater images generally suffer from severe degradation, such as lack of contrast, color casting and noise. The poor visibility of underwater images limits the performance of subsequent vision tasks. Hence, high-resolution (HR) images are desirable for many underwater applications.

In this paper, we propose an end-to-end Recovering Super-Resolution Generative Adversarial Network (RSRGAN) to generate the super-resolution (SR) underwater images. We solve the problem in two stages. In the first stage, we use the first part of RSRGAN, called Recovering GAN, to correct the color and remove the noise of the underwater images. In the second stage, we use the second part of RSRGAN, called Super-Resolution GAN, to enrich the fine texture details of the images restored by the Recovering GAN. RSRGAN combines the generators of both GAN models and fine-tunes the entire model with the Super-Resolution GAN's discriminator. Experimental results show that RSRGAN outperforms the state-of-the-art methods for underwater image super-resolution.

T. Gedeon et al. (Eds.): ICONIP 2019, CCIS 1142, pp. 75–83, 2019.
https://doi.org/10.1007/978-3-030-36808-1_9

2 Related Work

As far as we know, there are very few super-resolution methods for underwater images. In this section, we mainly introduce the recent approaches for underwater image restoration and single image super-resolution (SISR).

2.1 Underwater Images Restoration

Typical underwater restoration algorithms, such as histogram equalization and automatic white balance [9], improve the visual quality to some extent, but they suffer from noise amplification and color deviations problems. The emergence of generative adversarial network (GAN) [4] provides a new chance for underwater image restoration problem. Fabbri et al. designed a GAN model with a fully convolutional encoder-decoder generator to restore underwater images [3]. However, this method cannot perform well on the heavy noise images.

2.2 Single Image Super-Resolution (SISR)

Interpolation and sparse representation learning are widely used SR methods [17]. However, the generated SR images generally lack detailed textures. With the development of deep learning, Dong et al. [2] used a three-layer fully convolutional network to get HR images. Kim et al. used a 20-layer VGG [13] to obtain SR images [6]. In [8], the authors proposed the enhanced deep residual networks for SISR. As GAN-based models boom, Ledig et al. [7] and Wang et al. [15] employed GAN with the perceptual loss for the applications of image SR. But unfortunately, these methods are only performed on the natural images. Particularly, Lu et al. [10] applied denoising and descattering methods to the SR underwater images. However, denoising and descattering brought additional blur to the SR images.

3 The Proposed Model

The goal of this work is to establish an underwater super-resolution system. The proposed RSRGAN includes two parts, Recovering GAN and Super-Resolution GAN. Figure 1 shows the architecture of RSRGAN.

3.1 Recovering GAN

To recover the clear image I^R from the noisy and degraded image I^U, we propose a GAN-based model, Recovering GAN, which can be formalized as:

$$\min_{G_R} \max_{D_R} V(G_R, D_R) = E_{I^T \sim p_{train}(I^T)}[\log D_R(I^T)] + E_{I^U \sim p_G(I^U)}[\log(1 - D_R(G_R(I^U)))], \quad (1)$$

where I^T is the clear image (ground-truth) corresponding to I^U.

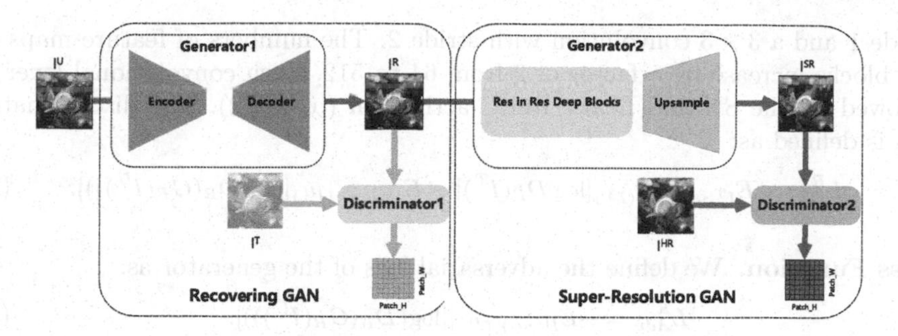

Fig. 1. The architecture of RSRGAN. The pretrained generators of both GAN models are combined as the generator of RSRGAN. The discriminator of the Super-Resolution GAN is used as the discriminator of RSRGAN. The light-colored parts are the discriminator of Recovering GAN only used for its pre-training.

Network Architecture. The generator of Recovering GAN is a fully convolutional encoder-decoder, as shown in Fig. 2. Every step in the encoder consists of a 3×3 convolutions with stride 1 and a 3×3 convolutions with stride 2 (in orange color). Every step in the decoder consists of an upsampling of the feature map followed by two 3×3 convolutions, a concatenation with the corresponding cropped feature map from the encoder, and a 3×3 convolution (in blue color). The other three 3×3 convolutions process the feature maps to an image (in purple color). Each convolutional layer is followed by the spectral normalization (SN) [11] and Leaky ReLU activation ($\alpha = 0.2$). Furthermore, we implement a self-attention block [14] on the global feature map before the decoder.

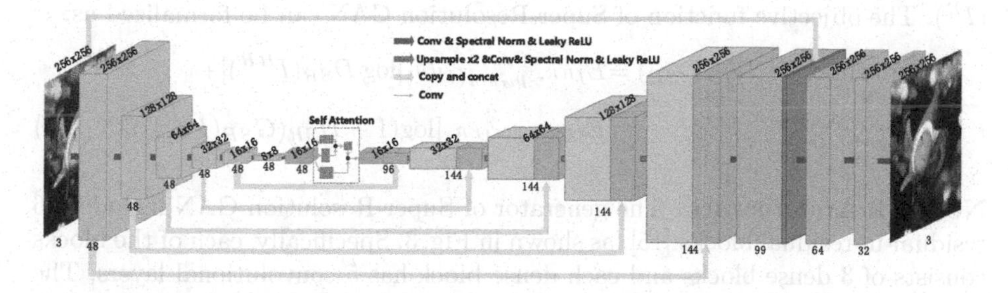

Fig. 2. The generator of Recovering GAN. Orange and blue boxes correspond to the feature maps in the encoder and decoder, respectively. The number of channels is denoted below the box. The size of the feature maps is denoted on the top of the box. Green boxes represent the global feature. Dark blue boxes represent copied feature maps, while purple boxes display the process from the feature maps to an image. (Color figure online)

The discriminator network D_R contains 4 convolutional blocks followed by a 3×3 convolution to obtain a 16×16 probability matrix for image patches classification. Here, each convolutional block consists of a 3×3 convolution with stride 1 and a 3×3 convolution with stride 2. The numbers of feature maps of the blocks increase by a factor of 2 from 64 to 512. Each convolutional layer is followed by the SN and Leaky ReLU activation ($\alpha = 0.2$). The discriminator loss is defined as:

$$L_D^R = -E_{I^T \sim p_{train}(I^T)}[\log D_R(I^T)] + E_{I^U \sim p_G(I^U)}[\log(D_R(G_R(I^U)))]. \quad (2)$$

Loss Function. We define the adversarial loss of the generator as:

$$L_{Adv}^R = -E_{I^U \sim p_G(I^U)}[\log(D_R(G_R(I^U)))]. \quad (3)$$

In Recovering GAN, we adopt the Mean Absolute Error (MAE) loss to measure the similarity between pixels:

$$L_{MAE}^R = \frac{1}{WH} \sum_{x=1}^{W} \sum_{y=1}^{H} |I_{x,y}^R - G_R(I^U)_{x,y}|. \quad (4)$$

Furthermore, we define the perceptual loss as:

$$L_{VGG}^R = \frac{1}{W_{i,j} H_{i,j}} \sum_{x=1}^{W_{i,j}} \sum_{y=1}^{H_{i,j}} |\phi_{i,j}(I^R)_{x,y} - \phi_{i,j}(G_R(I^U))_{x,y}|, \quad (5)$$

where $\phi_{i,j}$ is the feature map obtained by the j-th convolution (after activation) before the i-th max pooling layer within the pre-trained VGG-19 network [13], while $W_{i,j}$ and $H_{i,j}$ are the dimensions of the respective feature maps.

3.2 Super-Resolution GAN

The Super-Resolution GAN is trained to generate corresponding I^{SR} given I^{LR} (I^R). The objective function of Super-Resolution GAN can be formalized as:

$$\min_{G_{SR}} \max_{D_{SR}} V(G_{SR}, D_{SR}) = E_{I^{HR} \sim p_{train}(I^{HR})}[\log D_{SR}(I^{HR})] + E_{I^{LR} \sim p_G(I^{LR})}[\log(1 - D_{SR}(G_{SR}(I^{LR})))]. \quad (6)$$

Network Architecture. The generator of Super-Resolution GAN includes 16 residual-in-residual blocks [15], as shown in Fig. 3. Specifically, each of the blocks consists of 3 dense blocks and each dense block has 5 convolutional layers. The convolutions are 3×3 with stride 1 and the residual scaling parameter is 0.2. The pixel-shuffle layer increases the resolution of the input image. In addition, each convolutional layer in the generator is followed by the SN.

The discriminator's structure of Super-Resolution GAN is similar to that in Recovering GAN. However, we want to predict the probability that a real image I^{HR} is relatively more realistic than a fake one I^{SR}. The relativistic discriminator is formalized as:

$$D_{SR}(I^{HR}, I^{SR}) = \sigma(P_{SR}(I^{HR}) - E[P_{SR}(I^{SR})]), \quad (7)$$

$$D_{SR}(I^{SR}, I^{HR}) = \sigma(P_{SR}(I^{SR}) - E[P_{SR}(I^{HR})]), \quad (8)$$

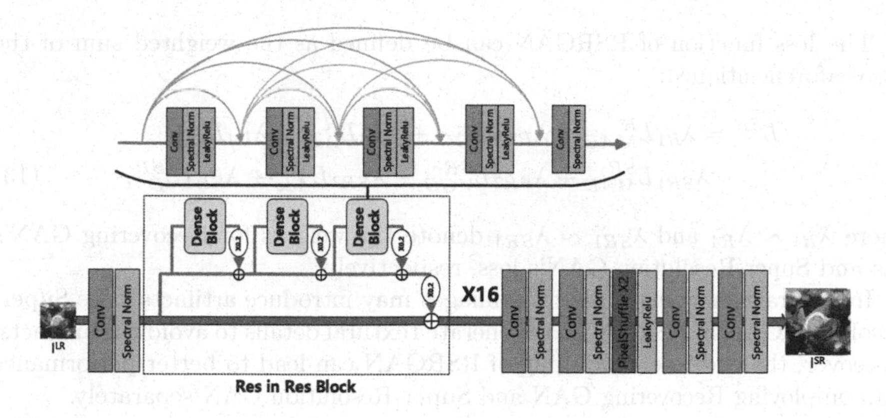

Fig. 3. The generator of Super-Resolution GAN.

where $P_{SR}(I)$ is the patch discriminator output, σ is the sigmoid function, $E[\cdot]$ takes an average for the images in the batch, and $I^{SR} = G_{SR}(I^{LR})$. The discriminator loss is then defined as:

$$L_D^{SR} = E_{I^{HR} \sim p_{train}(I^{HR})}[\log D_{SR}(I^{HR}, G_{SR}(I^{LR}))] + E_{I^{LR} \sim p_G(I^{LR})}[\log(1 - D_{SR}(G_{SR}(I^{LR}), I^{HR}))]. \quad (9)$$

Loss Function. Similar to the Recovering GAN, Super-Resolution GAN uses the MAE loss and perceptual loss to optimize the generator:

$$L_{MAE}^{SR} = \frac{1}{r^2 WH} \sum_{x=1}^{rW} \sum_{y=1}^{rH} |I_{x,y}^{HR} - G_{SR}(I^{LR})_{x,y}|, \quad (10)$$

$$L_{VGG}^{SR} = \frac{1}{W_{i,j} H_{i,j}} \sum_{x=1}^{W_{i,j}} \sum_{y=1}^{H_{i,j}} |\phi_{i,j}(I^{HR})_{x,y} - \phi_{i,j}(G_{SR}(I^{LR}))_{x,y}|. \quad (11)$$

Meanwhile, the adversarial loss for the generator is:

$$L_{Adv}^{SR} = E_{I^{HR} \sim p_{train}(I^{HR})}[\log(1 - D_{SR}(I^{HR}, G_{SR}(I^{LR})))] + E_{I^{LR} \sim p_G(I^{LR})}[\log D_{SR}(G_{SR}(I^{LR}), I^{HR})]. \quad (12)$$

We can see that, the adversarial loss for the generator benefits from the gradients from both I^{HR} and $I^{SR} = G_{SR}(I^{LR})$, while previous GAN generator is only benefited from the generated data.

3.3 Recovering Super-Resolution GAN (RSRGAN)

We combine the generators of pre-trained Recovering GAN and Super-Resoluiton GAN as the generator of RSRGAN. Furthermore, we use the discriminator of Super-Resolution GAN as the discriminator of RSRGAN. Finally, we fine-tune RSRGAN as an end-to-end network.

The loss function of RSRGAN can be defined as the weighted sum of the losses aforementioned:

$$L^{SR} = \lambda_{R1} L_{MAE}^{R} + \lambda_{R2} L_{VGG}^{R} + \lambda_{R3} L_{Adv}^{R} + \lambda_{R4} L_{D}^{R} + \\ \lambda_{SR1} L_{MAE}^{SR} + \lambda_{SR2} L_{VGG}^{SR} + \lambda_{SR3} L_{Adv}^{SR} + \lambda_{SR4} L_{D}^{SR}, \quad (13)$$

where $\lambda_{R1} \sim \lambda_{R4}$ and $\lambda_{SR1} \sim \lambda_{SR4}$ denote the weights for Recovering GAN's loss and Super-Resolution GAN's loss, respectively.

In general, removing noise from images may introduce artifacts. The Super-Resolution GAN in RSRGAN can generate textural details to avoid the artifacts. Moreover, the end-to-end training of RSRGAN can lead to better performance than employing Recovering GAN and Super-Resolution GAN separately.

4 Experiments

In this section, we compared RSRGAN with several state-of-the-art methods for underwater image super-resolution. For the parameters, we empirically set $\lambda_{R2} = \lambda_{SR2} = 2 \times 10^{-2}$, $\lambda_{R3} = \lambda_{SR3} = 1 \times 10^{-2}$. In addition, we set $\lambda_{R1} = \lambda_{R4} = \lambda_{SR1} = \lambda_{SR4} = 1$ as normal GAN-based model. The learning rate was set to 2×10^{-5}, and the Adam optimizer with $\beta_1 = 0.9$ and $\beta_2 = 0.999$ was employed for the network training.

4.1 Dataset

For this research, there are no available datasets yet, which contain pairs of clear ground-truth and corresponding low quality underwater images. Following [3], we used images from ImageNet to train a CycleGAN that learned the mapping from natural to underwater images. After that, we used the CycleGAN to generate underwater images with those containing marine creatures. Finally, we added marine snow noise [1] to the underwater images. Concretely, we used 5000 image pairs for training and 1100 image pairs for test.

4.2 Evaluation of the Underwater Image Restoration

Figure 4 shows some samples of the original images and the restored images obtained by Recovering GAN and some state-of-the-art image restoration methods, feeding them with the noisy underwater images. It is obvious that the restored images by CycleGAN [18] and UGAN [3] lack brightness. Pix2Pix [5] recovered the color of the images well, but the noise still remained. In contrast, Recovering GAN achieved the best performance. It not only recovered the color of the images, but also removed the noise in the images. Hence, Recovering GAN is beneficial for underwater image super-resolution.

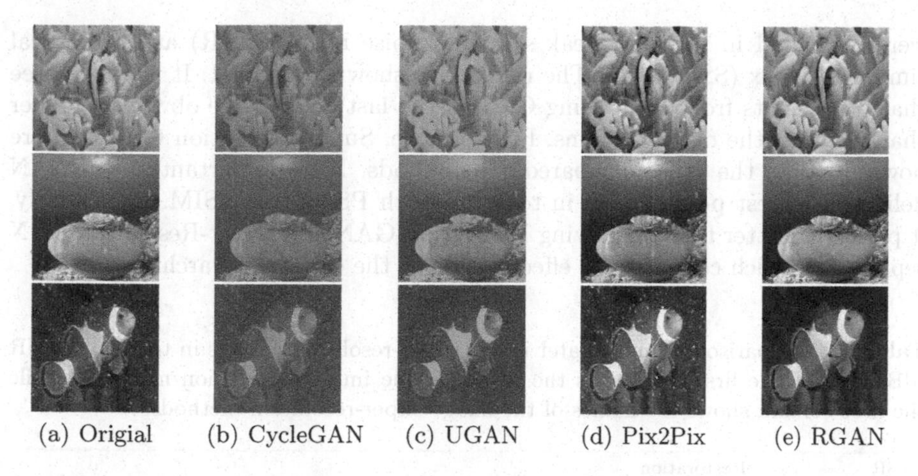

Fig. 4. Comparison between Recovering GAN (RGAN for short here) and the state-of-the-art methods for underwater image restoration.

4.3 Evaluation of the Underwater Image Super-Resolution

Figure 5 shows the super-resolution images generated by Super-Resolution GAN, VDSR [6], SRCNN [2] and ESRGAN [2], feeding them with the restored images by Recovering GAN. We can see that Super-Resolution GAN outperforms the compared SR methods in both sharpness and details. For instance, Super-Resolution GAN can produce sharper and more natural fins and contour line of the fish's face than the compared methods.

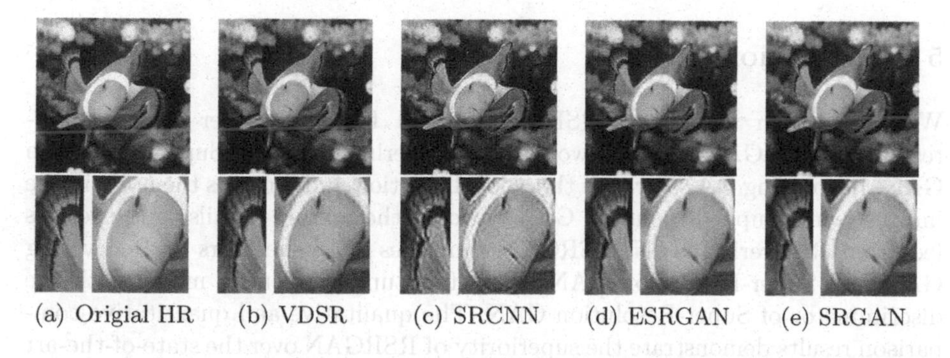

Fig. 5. Comparison between Super-Resolution GAN (SRGAN for short here) and the state-of-the-art methods for image SR.

4.4 Performance of RSRGAN

To quantitatively compare RSRGAN and the state-of-the-art methods, the performance of some image restoration methods followed by several SR methods were evaluated in terms of peak signal to noise ratio (PSNR) and structural similarity index (SSIM) [16]. The results are shown in Table 1. It is easy to see that, the results from Recovering GAN in the last column are obviously higher than those in the other columns. Furthermore, Super-Resolution GAN is more powerful than the other compared SR methods. More importantly, RSRGAN delivers the best performance in terms of both PSNR and SSIM. Particularly, it performs better than employing Recovering GAN and Super-Resolution GAN separately, which confirms the effectiveness of the end-to-end architecture.

Table 1. Comparison of underwater image super-resolution results in terms of PSNR (dB)/SSIM. The first row shows the names of the image restoration methods, while the first column shows the names of the image super-resolution methods.

SR	Restoration			
	CycleGAN [18]	UGAN [3]	Pix2Pix [5]	Recovering GAN
bicubic	16.70dB/0.5575	21.76dB/0.7195	18.87dB/0.6210	22.78dB/0.7322
VDSR [6]	16.85dB/0.5957	22.65dB/0.7811	19.21dB/0.6725	23.10dB/0.7689
EDSR [8]	16.84dB/0.5776	19.97dB/0.6404	18.67dB/0.5992	21.12dB/0.6378
ESPCN [12]	16.76dB/0.5735	21.47dB/0.6494	18.56dB/0.5978	21.12dB/0.6378
SRCNN [2]	16.72dB/0.5723	20.13dB/0.6462	18.55dB/0.5960	21.34dB/0.6440
SRGAN [7]	16.71dB/0.5747	19.68dB/0.6393	18.42dB/0.5945	20.71dB/0.6312
ESRGAN [15]	**16.87dB**/0.5906	22.54dB/0.7738	**19.23dB**/0.6633	**23.66dB/0.7806**
our SRGAN	-	-	-	**24.04dB/0.7832**
RSRGAN	**24.16dB/0.7886**			

5 Conclution

We propose an end-to-end RSRGAN model for underwater image super-resolution. RSRGAN includes two parts: Recovering GAN and Super-Resolution GAN. Recovering GAN corrects the color distortion and removes the noise in the images, while Super-Resolution GAN enriches the texture details to the results fed from Recovering GAN. RSRGAN combines the generators of Recovering GAN and Super-Resolution GAN, while fine-tunes the entire model with the discriminator of Super-Resolution GAN. The qualitative and quantitative comparison results demonstrate the superiority of RSRGAN over the state-of-the-art methods for underwater image super-resolution.

Acknowledgments. This work was supported by the National Key R&D Program of China under Grant No. 2016YFC1401004, the National Natural Science Foundation of China (NSFC) under Grant No. 41706010, the Science and Technology Program of Qingdao under Grant No. 17-3-3-20-nsh, the Joint Fund of the Equipments Pre-Research and Ministry of Education of China under Grand No. 6141A020337, and the Fundamental Research Funds for the Central Universities of China.

References

1. Cyganek, B., Gongola, K.: Real-time marine snow noise removal from underwater video sequences. J. Electron. Imaging **27**(04), 043002 (2018)
2. Dong, C., Loy, C.C., He, K., Tang, X.: Image super-resolution using deep convolutional networks. IEEE Trans. Pattern Anal. Mach. Intell. **38**(2), 295–307 (2016)
3. Fabbri, C., Islam, M.J., Sattar, J.: Enhancing underwater imagery using generative adversarial networks. In: ICRA, pp. 7159–7165 (2018)
4. Goodfellow, I.J., et al.: Generative adversarial networks. CoRR abs/1406.2661 (2014)
5. Isola, P., Zhu, J., Zhou, T., Efros, A.A.: Image-to-image translation with conditional adversarial networks. In: CVPR, pp. 5967–5976 (2017)
6. Kim, J., Lee, J.K., Lee, K.M.: Accurate image super-resolution using very deep convolutional networks. In: CVPR, pp. 1646–1654 (2016)
7. Ledig, C., et al.: Photo-realistic single image super-resolution using a generative adversarial network. In: CVPR, pp. 105–114 (2017)
8. Lim, B., Son, S., Kim, H., Nah, S., Lee, K.M.: Enhanced deep residual networks for single image super-resolution. In: CVPR, pp. 1132–1140 (2017)
9. Liu, Y.C., Chan, W.H., Chen, Y.Q.: Automatic white balance for digital still camera. IEEE Trans. Consum. Electron. **41**(3), 460–466 (1995)
10. Lu, H., Li, Y., Nakashima, S., Kim, H., Serikawa, S.: Underwater image super-resolution by descattering and fusion. IEEE Access **5**, 670–679 (2017)
11. Miyato, T., Kataoka, T., Koyama, M., Yoshida, Y.: Spectral normalization for generative adversarial networks. In: ICLR (2018)
12. Shi, W., et al.: Real-time single image and video super-resolution using an efficient sub-pixel convolutional neural network. In: CVPR, pp. 1874–1883 (2016)
13. Simonyan, K., Zisserman, A.: Very deep convolutional networks for large-scale image recognition. In: ICLR (2015)
14. Vaswani, A., et al.: Attention is all you need. In: NIPS, pp. 6000–6010 (2017)
15. Wang, X., et al.: ESRGAN: enhanced super-resolution generative adversarial networks. In: Leal-Taixé, L., Roth, S. (eds.) ECCV 2018. LNCS, vol. 11133, pp. 63–79. Springer, Cham (2019). https://doi.org/10.1007/978-3-030-11021-5_5
16. Wang, Z., Bovik, A.C., Sheikh, H.R., Simoncelli, E.P.: Image quality assessment: from error visibility to structural similarity. IEEE Trans. Image Processing **13**(4), 600–612 (2004)
17. Yang, W., Zhang, X., Tian, Y., Wang, W., Xue, J.: Deep learning for single image super-resolution: a brief review. CoRR abs/1808.03344 (2018)
18. Li, M., Huang, H., Ma, L., Liu, W., Zhang, T., Jiang, Y.: Unsupervised image-to-image translation with stacked cycle-consistent adversarial networks. In: Ferrari, V., Hebert, M., Sminchisescu, C., Weiss, Y. (eds.) ECCV 2018. LNCS, vol. 11213, pp. 186–201. Springer, Cham (2018). https://doi.org/10.1007/978-3-030-01240-3_12

Convolutional Neural Networks

Hierarchical Attention CNN and Entity-Aware for Relation Extraction

Xinyu Zhu, Gongshen Liu(✉), and Bo Su(✉)

Shanghai Jiao Tong University, Shanghai 200240, China
{jasonzxy,lgshen,subo}@sjtu.edu.cn

Abstract. Convolution neural network is a widely used model in the relation extraction (RE) task. Previous work simply uses max pooling to select features, which cannot preserve the position information and deal with the long sentences. In addition, the critical information for relation classification tends to present in a certain segment. A better method to extract feature in segment level is needed. In this paper, we propose a novel model with hierarchical attention, which can capture both local syntactic features and global structural features. A position-aware attention pooling is designed to calculate the importance of convolution features and capture the fine-grained information. A segment-level self-attention is used to capture the most important segment in the sentence. We also use the skills of entity-mask and entity-aware to make our model focus on different aspects of information at different stages. Experiments show that the proposed method can accurately capture the key information in sentences and greatly improve the performance of relation classification comparing to state-of-the-art methods.

Keywords: Relation extraction · Hierarchical attention · Entity-aware

1 Introduction

Relation extraction (RE) aims to obtain semantic relations between two given entities from plain text, such as the following examples: *contains*, *lives in*, *capital of*. It is an important task in natural language processing, particularly in knowledge graph construction, paragraph understanding and question answering.

Traditional RE suffered from the lack of training data. To solve this problem, distant supervision was proposed [8]. Distant supervision can easily generate a large amount of training data, but it also brings some challenges. Distant supervision is often used to address open corpora, such as Wikipedia and the New York Times. We counted the distribution of sentence length in a NYT dataset developed by Riedel et al. [11]. Over 70% of the sentences are longer than 30 words, and nearly half of the sentences are longer than 40 words. The performance of traditional methods decreases as the sentence length increases. We need to find more effective methods to capture features.

T. Gedeon et al. (Eds.): ICONIP 2019, CCIS 1142, pp. 87–94, 2019.
https://doi.org/10.1007/978-3-030-36808-1_10

In this study, we propose a novel model to address the limitations in feature extracting. Our model uses hierarchical attention to capture both local and global features. Also, entity-aware and position-aware is added to assist the classification prediction. The key contributions of this paper include:

- We apply hierarchical attention to better capture useful information. Position-aware attention is applied to capture the fine-grained features during the max pooling process. The sentences is divided into three segments according to the position of two entities. Self-attention is applied to these three segments to obtain the structural information in the high level.
- The entity mask is used in the input layer to help model focus on the global syntactic features. The entity-aware is applied in the output layer so that the semantic relation between two entities can be taken into account during the predict process.
- Experiments show that our methods can greatly improve the performance compared with the state-of-art models.

2 Related Work

Relation Extraction is an important work in NLP. Early methods proposed various features to identify different relations, particularly with supervised methods [1,4,10,13]. Recent years, neural networks were widely used in NLP. Various models were applied in RE task, including convolution neural network [6,9,12,17–19], recurrent neural networks [20] and long short-term memory network [21]. Attention mechanism is also applied to relation extraction task [3,6,21]. Some recent works try to capture more useful features with the help of side information [5,7,15].

Most methods based on CNN simply use max pooling to select convolution features, which ignores position information of the convolution features. Meanwhile, as the length of training sentences increases, the key information is often present in a certain segment. We need a better method to capture the structural information at segment level. To address these limitations, we propose a novel model which combines CNN with hierarchical attention.

3 Framework

We propose a novel framework for relation extraction, which uses hierarchical attention and entity-aware to help capture better semantic features. Figure 1 shows our neural network architecture.

Each sentence is transformed into a vector consisting of word embedding and position embedding. The specific entity words are masked and replaced with *Subject* and *Object*. Convolution Neural Network is used to extract features from the input sentence. Features are divided into three segments according to the entity position. The position-aware attention pooling is used to preserve the useful information, and then segment-level self-attention is used to capture the structural information. Finally, we combine the feature information with the entity information to get the predict output. These part is described in detail below.

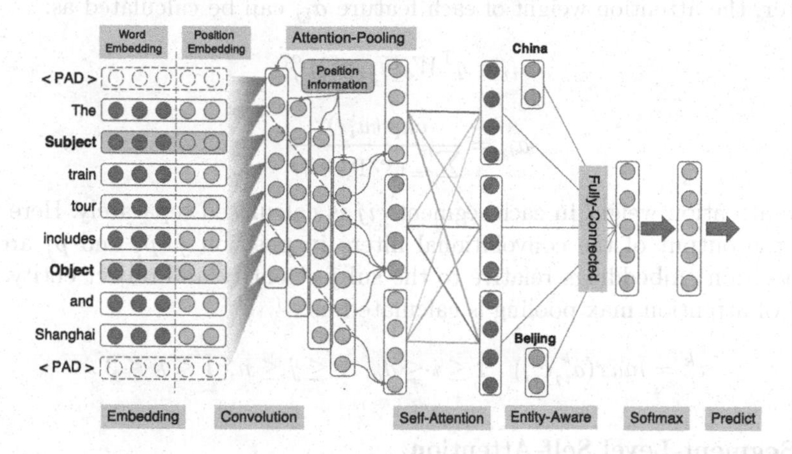

Fig. 1. The architecture of our method. Position-aware attention pooling and segment-level self-attention are used to capture both local and global features.

3.1 Vector Representation

The input of our model is the sequence of words in a sentence. Similar to previous papers [19], we use word embedding to capture the semantic information and position embedding to capture the structural information. Specifically, two specific tokens, *Sub* and *Obj*, are used to represent the entities in this phase, which we called entity masks. The dimension of each word in the input sentence is $d = d^w + 2 \times d^p$, where d^w is the word embedding dimension and d^p is the position embedding dimension.

3.2 Convolution

We use CNN to extract features from sentences. Given a input representation R, the convolution operation is applied to R with the sliding window of size k. We define the convolution matrix as $W_c \in \mathbb{R}^{d^c \times (k \times d)}$, where d^c is the number of filters. The output of the i-th convolutional filter can be expressed as:

$$c_i = [W_c q + b_c]_i \tag{1}$$

where $q_i = r_{i-k+1:i} (1 \leq i \leq N + k - 1)$ means the concatenation of k word embeddings. As for the boundary of sentences, $\frac{k-1}{2}$ padding tokens are placed at the beginning and the end of the sentence.

3.3 Position-Aware Attention Max Pooling

Similar to previous work [18], the output of each convolutional filter p_i is divided into three segments $\{c_i^1, c_i^2, c_i^3\}$ according to the position of two entities. For the i-th filter, the attention weight of each feature a_{ij} can be calculated as:

$$u_{ij} = q^\top W_a[h_j, p_j^s, p_j^o] \tag{2}$$

$$a_{ij} = \frac{\exp(u_{ij})}{\sum_{j=1}^{n} \exp(u_{ij})} \tag{3}$$

The attention weight in each segment cij is calculated separately. Here $h_j \in \mathbb{R}^{d_c}$ is the output of the convolutional filters in position j. p_j^s and p_j^o are the word position embeddings relative to the subject entity and object entity. The output of attention max pooling is calculated as:

$$z_i^k = max(a_{ij}^k c_{ij}^k) \quad 1 \leq i \leq d_c,\ 1 \leq j \leq n,\ 1 \leq k \leq 3 \tag{4}$$

3.4 Segment-Level Self-Attention

We calculate the relation between three segments based on the multi-head attention [16], which we called segment-level self-attention. We set the feature dimension $d_c = 600$ and employ $h = 6$ parallel attention heads. In each head we set the dimension of keys and values $d_k = d_v = d_c/h = 100$. This part is composed of two identical layers. Similar to [16], we add a fully connected feed-forward network after each self-attention layer. The dimension of the fully connected is $d_{ff} = 2048$.

3.5 Entity-Aware Output

Here we use the entity-aware softmax output, which concatenates the entity words and the feature vector, to help the relation prediction. The combined vector is fed into a softmax classifer:

$$o = \text{softmax}(W_c[h_s, e_1, e_2] + b_c) \tag{5}$$

where $W_c \in \mathbb{R}^{(3d_c+2d^e)\times r}$, $b_c \in \mathbb{R}^r$ and r is the number of possible relation types. e_1 and e_2 is the word embedding of the subject entity and object entity. Since each entity may have several words, we pad all the entity to the length of 5 words and $d^e = 5d^w$.

3.6 Loss Function

The softmax output can be interpreted as the probability score of different relations. We design a new loss function to help training:

$$J_c = \exp\left(\gamma\left(m - S_x^+ + S_x^-\right)\right) \tag{6}$$

where m is a margin and γ is a scaling factor. The margin gives extra penalization on the difference in scores and the scaling factor helps to magnifies the scores. S_x^+ refers to the score for the correct relation label and S_x^- refers to the highest score among all the wrong relations. By minimizing this loss function, we hope our model can give scores with a difference greater than m between positive label and negative label.

Table 1. Parameter settings

Word dimension d^w	300	Position dimension d^p	50	Windows size k	3
Learning rate λ	0.0003	margin m	1.0	scaling factor γ	2.0

4 Experiments

4.1 Experimental Setup

Dataset and Evaluation Metrics. We evaluate our method on a widely used dataset developed by Riedel [11]. This dataset is generated by aligning relation facts in Freebase with the New York Times (NYT) corpus. Training set contains 522611 sentences, and test set contains 172448 sentences. There are 53 relations including a special relation *NA* which indicates no relation between two entities.

Following previous work [8], we evaluate our model in two ways. We first compare the extracted relation facts with the Freebase data and report the precision/recall curves of the experiments. Then we manually check the precision of top N sentences in our experiments.

Parameter Settings. We use the word2vec model to pre-train the word representation. The detailed parameter settings are given in Table 1.

4.2 Precision/Recall Curve

The held-out evaluation provides an approximate measure of precision without consuming human evaluation. The relation facts extracted from the test data are automatically compared with those in Freebase.

To evaluate our model, we select several traditional models as baseline. Mintz [8], MultiR [2] and MIML [14] are feature-based models. PCNN+ONE [18] uses piecewise convolutional neural network and select one instance from each bag. PCNN+ATT [6] uses selective attention mechanism over instances to reduce the weights of noisy data.

As shown in Fig. 2, our method achieve the best performance in the entire interval. Our model simply select the sentence which get the highest scores in an entity bag. Even so, we still get better performance compared with PCNN+ATT, which demonstrates the effectiveness of our hierarchical attention and entity-aware for relation extraction.

4.3 Manual Evaluation

Distant supervision brings many noisy instances to the dataset. In order to better demonstrate the performance of our model, we select two types of specific data and conduct extra experiments on them.

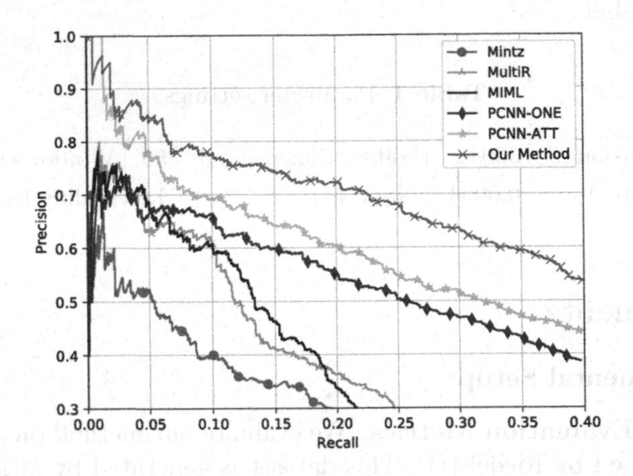

Fig. 2. Precision recall curves of our method and baselines.

Table 2. Top-N prediction accuracy for entity pairs present in freebase.

Model	100	500	1000	Avg
PCNN+ONE	99	98.6	97.1	98.23
PCNN+ATT	100	99	98.4	99.13
Our method	**100**	**99.8**	**99.6**	**99.8**

Freebase Evaluation. We select the test sentences whose entity pairs actually exist in the freebase. Although there may be some noisy instances, we can approximate that the labels are correct when we predict based on the entity bag. We make the relation prediction for these selected test sentences and the results are shown in Table 2.

From the results we can see that our method achieve the highest accuracy. Our method can keep the accuracy at a high level even in the top-1000 experiment, which indicates the effectiveness of our method.

NA Evaluation. In Riedel dataset, 166003 test sentences belong to *NA* relation, accounting for the majority of test set. However, many of them are false negative instances, which means there is actually a relation between two entities but it is missing in Freebase. We performed relation prediction for the sentences in which at least one of the participating entity is not present in freebase. We conducted manual evaluation for the top 100, top 200, and top 300 sentences which were predicted to have a certain relation.

Table 3 shows the Top-N prediction accuracy. We can observe that: (1) Many *NA* sentences are indeed false negative instances, which proves the necessity of this experiment. (2) The accuracy of relation prediction has been greatly improved with our method.

Table 3. Top-N prediction accuracy for entity pairs not present in freebase.

Model	100	200	300	Avg
PCNN+MAX	86	81	74.66	80.56
PCNN+ATT	89	84	79.33	84.11
Our method	**96**	**91**	**84**	**90.33**

5 Conclusion

In this paper, we propose a novel model for the relation extraction task. We use the hierarchical attention to capture both fine-grained and structural information. We introduce the position-aware attention pooling, which can obtain better features compared with traditional max pooling. We use the segment-level self-attention to capture structural features in high level. We also apply the entity mask and entity aware mechanism in the input layer and output layer respectively. As a result, our model can extract more useful information from the training sentences. Experiments demonstrate that our method can improve the performance of the state-of-the-art models.

Acknowledgements. This research work has been funded by the National Natural Science Foundation of China (Grant No. 61772337, U1736207), and the National Key Research and Development Program of China NO. 2016QY03D0604 and 2018YFC0830703.

References

1. Bunescu, R.C., Mooney, R.J.: A shortest path dependency kernel for relation extraction. In: Proceedings of HLT/EMNLP, pp. 724–731. Association for Computational Linguistics (2005)
2. Hoffmann, R., Zhang, C., Ling, X., Zettlemoyer, L., Weld, D.S.: Knowledge-based weak supervision for information extraction of overlapping relations. In: Proceedings of the 49th ACL: Human Language Technologies, vol. 1. pp. 541–550 (2011)
3. Ji, G., Liu, K., He, S., Zhao, J.: Distant supervision for relation extraction with sentence-level attention and entity descriptions. In: Thirty-First AAAI Conference on Artificial Intelligence (2017)
4. Kambhatla, N.: Combining lexical, syntactic, and semantic features with maximum entropy models for extracting relations. In: Proceedings of the ACL 2004 on Interactive Poster and Demonstration Sessions, p. 22 (2004)

5. Lei, K., et al.: Cooperative denoising for distantly supervised relation extraction. In: Proceedings of the 27th International Conference on Computational Linguistics, pp. 426–436 (2018)
6. Lin, Y., Shen, S., Liu, Z., Luan, H., Sun, M.: Neural relation extraction with selective attention over instances. In: Proceedings of the 54th ACL, vol. 1, Long Papers, pp. 2124–2133 (2016)
7. Liu, L., et al.: Heterogeneous supervision for relation extraction: a representation learning approach. arXiv preprint arXiv:1707.00166 (2017)
8. Mintz, M., Bills, S., Snow, R., Jurafsky, D.: Distant supervision for relation extraction without labeled data. In: Proceedings of the Joint Conference of the 47th Annual Meeting of the ACL and the 4th International Joint Conference on Natural Language Processing of the AFNLP: Volume 2-vol. 2. pp. 1003–1011. Association for Computational Linguistics (2009)
9. Nguyen, T.H., Grishman, R.: Relation extraction: perspective from convolutional neural networks. In: Proceedings of the 1st Workshop on Vector Space Modeling for Natural Language Processing, pp. 39–48 (2015)
10. Qian, L., Zhou, G., Kong, F., Zhu, Q., Qian, P.: Exploiting constituent dependencies for tree kernel-based semantic relation extraction. In: Proceedings of the 22nd International Conference on Computational Linguistics-Volume 1, pp. 697–704. Association for Computational Linguistics (2008)
11. Riedel, S., Yao, L., McCallum, A.: Modeling relations and their mentions without labeled text. In: Balcázar, J.L., Bonchi, F., Gionis, A., Sebag, M. (eds.) ECML PKDD 2010. LNCS (LNAI), vol. 6323, pp. 148–163. Springer, Heidelberg (2010). https://doi.org/10.1007/978-3-642-15939-8_10
12. Santos, C.N.D., Xiang, B., Zhou, B.: Classifying relations by ranking with convolutional neural networks. arXiv preprint arXiv:1504.06580 (2015)
13. Suchanek, F.M., Ifrim, G., Weikum, G.: Combining linguistic and statistical analysis to extract relations from web documents. In: The 12th ACM SIGKDD, pp. 712–717. ACM (2006)
14. Surdeanu, M., Tibshirani, J., Nallapati, R., Manning, C.D.: Multi-instance multi-label learning for relation extraction. In: Proceedings of the 2012 Joint Conference on Empirical Methods in Natural Language Processing and Computational Natural Language Learning, pp. 455–465. Association for Computational Linguistics (2012)
15. Vashishth, S., Joshi, R., Prayaga, S.S., Bhattacharyya, C., Talukdar, P.: Reside: improving distantly-supervised neural relation extraction using side information. In: Proceedings of EMNLP, pp. 1257–1266 (2018)
16. Vaswani, A., et al.: Attention is all you need. In: Advances in Neural Information Processing Systems, pp. 5998–6008 (2017)
17. Wang, L., Cao, Z., De Melo, G., Liu, Z.: Relation classification via multi-level attention CNNs. In: The 54th ACL (long papers), vol. 1, pp. 1298–1307 (2016)
18. Zeng, D., Liu, K., Chen, Y., Zhao, J.: Distant supervision for relation extraction via piecewise convolutional neural networks. In: EMNLP, pp. 1753–1762 (2015)
19. Zeng, D., Liu, K., Lai, S., Zhou, G., Zhao, J.: Relation classification via convolutional deep neural network. In: COLING: Technical Papers, pp. 2335–2344 (2014)
20. Zhang, D., Wang, D.: Relation classification via recurrent neural network. arXiv preprint arXiv:1508.01006 (2015)
21. Zhou, P., et al.: Attention-based bidirectional long short-term memory networks for relation classification. In: Proceedings of the 54th ACL (Volume 2: Short Papers), vol. 2, pp. 207–212 (2016)

Fault Tolerant Broad Learning System

Muideen Adegoke[1], Chi-Sing Leung[1(✉)], and John Sum[2]

[1] Department of Electronic Engineering, City University of Hong Kong, Kowloon Tong, Hong Kong
maadegoke2-c@my.cityu.edu.hk, eeleungc@cityu.edu.hk

[2] Institute of Technology Management, National Chung Hsing University, Taichung, Taiwan

Abstract. The broad learning system (BLS) approach provides low computational complexity solutions for training flat structure feedforward networks. However, many BLS algorithms deal with the faultless situation only. This paper addresses the fault tolerant ability of BLS networks. We call our approach fault tolerant BLS (FTBLS). First, we develop a fault tolerant objective function for BLS. Based on the developed objective function, we develop a training algorithm to construct a BLS network. The simulation results show that our proposed FTBLS is much better than the classical BLS.

Keywords: Broad learning system · Fault tolerance · Multiplicative noise

1 Introduction

Single hidden layered networks are well known for their universal approximation capability [1,2]. Inspired by feature extraction, the broad learning system (BLS) concept [3] was proposed to construct a flat structure network. It processes its input data using some feature mapped functions and uses the obtained features as its processed inputs. No iterative training procedures are required in the BLS concept. The desired output weights could be calculated easily by ridge regression. From this point of view, the BLS concept is computationally efficient.

Although the original BLS and its variants are proved to be effective in handling some benchmark datasets, existing BLS approaches focus on faultless situation only. In the implementation of a well-trained neural network, fault/noise occurrences are unavoidable [4,5]. To handle fault, such as weight and node failure in BLS networks, it is vital to understand how fault/noise affect the performance of a well trained BLS network. To best of our knowledge, there are few results on fault tolerant issues of BLS. This paper investigates the performance of BLS under noise situation, in which multiplicative noise occur at the feature nodes, the enhancement nodes and output weights concurrently. To mitigate the effect of such noise, we develop a fault tolerant objective function. Based on the developed objective function, we train a BLS network with the

T. Gedeon et al. (Eds.): ICONIP 2019, CCIS 1142, pp. 95–103, 2019.
https://doi.org/10.1007/978-3-030-36808-1_11

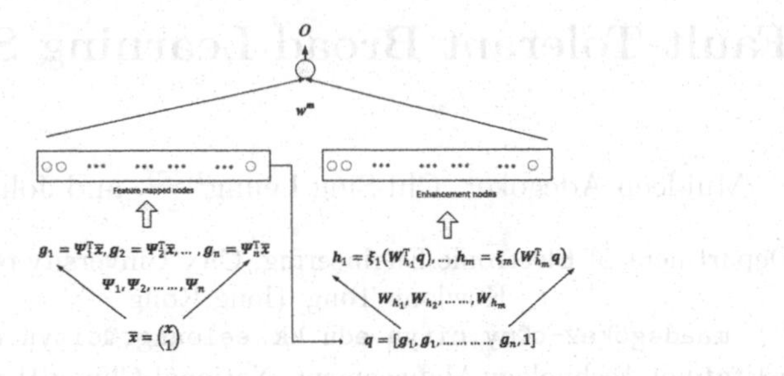

Fig. 1. The structure of the BLS network.

proposed objective function. We call our approach fault tolerant BLS (FTBLS) algorithm. Several simulations are carried out to show that our FTBLS algorithm can greatly suppress the effect of multiplicative noise.

The rest of this paper is organized as follows. The background of BLS is given in Sect. 2. The proposed FTBLS algorithm is developed in Sect. 3. Simulation results are provided in Sect. 4. Finally, conclusions are drawn in Sect. 5.

2 Background of Broad Learning System

Figure 1 shows a classical BLS network [3]. It has two types of nodes, calling feature mapped nodes and enhancement nodes. The details of the BLS model and its learning algorithm are given in the rest of this section.

2.1 Feature Mapped Nodes and Enhancement Nodes

Considering a regression problem, the input and output of the network are denoted as $\boldsymbol{x} \in \mathbb{R}^D$ and $o \in \mathbb{R}$, respectively. Let $\bar{\boldsymbol{x}} = [\boldsymbol{x}^{\mathrm{T}}, 1]^{\mathrm{T}}$ be the input augmented with 1.

Feature Mapped Nodes: There are n groups of feature nodes and there are r_i nodes in the i-th group. In order to explore hidden features of input data, for each group of feature nodes, a learned projection matrix, denoted as $\boldsymbol{\Psi}_i \in \mathbb{R}^{(D+1)\times r_i}$, is employed to project the input data to produce the i-th set of mapped features. The output $\mathfrak{g}_i$ of the i-th group of feature nodes is given by

$$\mathfrak{g}_i = [g_1, \cdots, g_{r_i}]^{\mathrm{T}} = \boldsymbol{\Psi}_i^{\mathrm{T}} \bar{\boldsymbol{x}} \quad \forall\, i = 1, \cdots, n. \tag{1}$$

The construction process of $\boldsymbol{\Psi}_i$'s is based on a sparse optimization process and will be presented in Sect. 2.3. Here, we follow the standard BLS system and do not apply the nonlinear activation function on $\mathfrak{g}_i$'s. We pack the outputs of all feature nodes together as $\boldsymbol{g} = [\mathfrak{g}_1^{\mathrm{T}}, \cdots, \mathfrak{g}_n^{\mathrm{T}}]^{\mathrm{T}} \in \mathbb{R}^{\sum_{i=1}^n r_i}$ and denote $\boldsymbol{q} = [\boldsymbol{g}^{\mathrm{T}}, 1]^{\mathrm{T}} \in \mathbb{R}^{\sum_{i=1}^n r_i + 1}$.

Enhancement Nodes: There are m groups of enhancement nodes. Each group has p_j nodes. The inputs of the enhancement nodes are taken from feature mapped nodes. The outputs of the j-th group of enhancement nodes are given by

$$\mathfrak{h}_j = [h_1, \cdots, h_{p_j}]^{\mathrm{T}} = \xi_j(\boldsymbol{W}_{\mathfrak{h}_j}^{\mathrm{T}} \boldsymbol{q}) \quad \forall\, j = 1 \cdots, m, \tag{2}$$

where $\xi_j(\cdot)$ is a nonlinear activation function for j-th group of enhancement nodes. In this paper, we use tanh as the activation function for all groups, given by

$$\xi_j(u) = \tanh(u) = \frac{2}{1 + \exp(-2u)} - 1. \tag{3}$$

We pack the outputs of all enhancement nodes together as $\boldsymbol{\eta} = [\mathfrak{h}_1^{\mathrm{T}}, \cdots, \mathfrak{h}_m^{\mathrm{T}}]^{\mathrm{T}} \in \mathbb{R}^{\sum_{j=1}^{m} p_j}$.

2.2 Network Output

We denote output of the network as o. We pack the outputs of features nodes and the outputs of enhancement nodes as $\boldsymbol{a} = [\boldsymbol{g}^{\mathrm{T}} | \boldsymbol{\eta}^{\mathrm{T}}]^{\mathrm{T}}$. The network output for a given input vector is given by $o = \boldsymbol{a}^{\mathrm{T}} \boldsymbol{w}^m$, where $\boldsymbol{w}^m$ is the output weight vector of the network.

2.3 Construction of the Projection Matrices: $\boldsymbol{\Psi}_i$ and the Weight Matrices of the Enhancement Nodes: $\boldsymbol{W}_{\mathfrak{h}_j}$

The way to construct the projection matrix is based on [3,6,7]. The training set is denoted as $\{(\boldsymbol{x}_1, y_1), \cdots, (\boldsymbol{x}_N, y_N)\}$, where $\boldsymbol{x}_k \in \mathbb{R}^D$ is the input vector of the k-th training sample and y_k is its the desire output. We first group all $\boldsymbol{x}_k$'s together to form training data matrices: $\boldsymbol{x} = [\boldsymbol{x}_1 | \cdots | \boldsymbol{x}_N]^{\mathrm{T}}$ and $\bar{\boldsymbol{x}} = [\boldsymbol{x} | \mathbf{1}]$, where $\mathbf{1}$ is a vector whose elements are equal to 1. The BLS first associates a random projection matrix $\bar{\boldsymbol{W}}_{f_i} \in \mathbb{R}^{(D+1) \times n}$ for each group of feature mapped nodes. The elements of $\bar{\boldsymbol{W}}_{f_i}$ are randomly generated. It should be noticed that the BLS do not use $\bar{\boldsymbol{W}}_{f_i}$'s as $\boldsymbol{\Psi}_i$'s.

With the random matrix $\bar{\boldsymbol{W}}_{f_i}$ [3,6,7], we obtain a random projection data matrix $\boldsymbol{Q}_i = \phi_i(\bar{\boldsymbol{X}} \bar{\boldsymbol{W}}_{f_i})$, where ϕ_i can be any activation function. In [3,6,7], the BLS considers linear function and then the random projection matrix is given by

$$\boldsymbol{Q}_i = \bar{\boldsymbol{x}} \bar{\boldsymbol{W}}_{f_i}. \tag{4}$$

The projection matrix $\boldsymbol{\Psi}_i$ is the solution of the following sparse approximation problem:

$$\min_{\boldsymbol{\Psi}_i} \|\boldsymbol{Q}_i \boldsymbol{\Psi}_i^{\mathrm{T}} - \bar{\boldsymbol{X}}\|_F^2 + \rho \|\boldsymbol{\Psi}_i\|_1 \tag{5}$$

where ρ in (5) is a regularization parameter for sparse regularization.

For each group of the enhancement nodes, weight matrix $\boldsymbol{W}_{\mathfrak{h}_j}$ is randomly generated [3,6,7].

2.4 Output Weight Vector

The next step is to construct the output weight vector $\boldsymbol{w}^m$. Given the n projection matrices $\boldsymbol{\Psi}_i$'s of the feature mapped nodes and the training data matrices $\bar{\boldsymbol{x}}$, for the i-th group of feature mapped nodes, we have the i-th set of features of the training data: $\boldsymbol{Z}_i = \bar{\boldsymbol{x}}\boldsymbol{\Psi}_i^{\mathrm{T}}$. We pack all the feature values of the training data together to get

$$\boldsymbol{Z}^n = [\boldsymbol{Z}_1, \cdots, \boldsymbol{Z}_n]. \tag{6}$$

Similarly, the outputs of enhancement nodes are $\boldsymbol{H}_j = \xi_j([\boldsymbol{Z}^n|1])\boldsymbol{W}_{\mathfrak{h}_j}$, for $j = 1, \cdots, m$. We pack all the outputs of enhancements nodes together as

$$\boldsymbol{H}^m = [\boldsymbol{H}_1, \cdots, \boldsymbol{H}_m]. \tag{7}$$

It should be noted that $\boldsymbol{w}^m$ can be computed with ease via ridge regression approximation of $[\boldsymbol{Z}^n|\boldsymbol{H}^m]^{\dagger}$, given by

$$\boldsymbol{w}^m = [\boldsymbol{Z}^n|\boldsymbol{H}^m]^{\dagger}\boldsymbol{y} \tag{8}$$

where $\boldsymbol{y} = [y_1, \cdots, y_N]^{\mathrm{T}}$ is the collection of the training outputs, and $[\boldsymbol{Z}^n|\boldsymbol{H}^m]^{\dagger}$ can be obtained by

$$[\boldsymbol{Z}^n|\boldsymbol{H}^m]^{\dagger} = \lim_{\lambda \to 0}(\lambda \boldsymbol{I} + [\boldsymbol{Z}^n|\boldsymbol{H}^m]^T[\boldsymbol{Z}^n|\boldsymbol{H}^m])^{-1}[\boldsymbol{Z}^n|\boldsymbol{H}^m]^T. \tag{9}$$

An alternative way is to formulate the training problem as an optimization problem, given by $\arg\min_{\boldsymbol{w}^m} \|\boldsymbol{A}\boldsymbol{w}^m - \boldsymbol{y}\|_2^2 + \lambda\|\boldsymbol{w}^m\|_2^2$, where $\boldsymbol{A} = [\boldsymbol{Z}^n|\boldsymbol{H}^m]$.

3 The Proposed Fault Tolerant BLS (FTBLS)

Let g_{l_i} be the output of the l_i-th node of the i-th group of the feature nodes. When it is affected by multiplicative noise, its value becomes

$$\tilde{g}_{l_i} = (1 + \delta_f)g_{l_i}, \tag{10}$$

where $l_i = 1, \cdots, r_i$. In (10) δ_f's are independent and identically distributed (i.i.d.) random variables with zero mean and variance equal to σ_ϕ^2. The variance σ_ϕ^2 describes the noise intensity. Hence, given the training data matrix, the n groups of feature nodes affected by noise become $\tilde{\boldsymbol{Z}}^n = \left[\tilde{\boldsymbol{Z}}_1, \cdots, \tilde{\boldsymbol{Z}}_n\right]$.

Let h_{l_j} be the output of the l_j-th node of the j-th group of enhancement nodes. Consider that the input weights of the enhancement nodes are affected by multiplicative noise. Hence, with the aids of Taylor series, the output of a faulty enhancement node can be modelled as

$$\tilde{h}_{l_j} = h_{l_j} + \delta_h \boldsymbol{w}_{\mathfrak{h}_{l_j}}^T \Delta\boldsymbol{h}_{l_j} \quad \forall\, l_j = 1, \cdots, p_j \tag{11}$$

where $\Delta \boldsymbol{h}_{l_j} = \frac{\partial \xi_j(\boldsymbol{w}^T_{\hbar_{l_j}} q)}{\partial \boldsymbol{w}_{\hbar_{l_j}}}$. Also, δ_h's are i.i.d. random variables with zero mean and variance equal to σ_ξ^2. Hence, the m groups of enhancement nodes affected by noise become $\tilde{\boldsymbol{H}}^m = \left[\tilde{\boldsymbol{H}}_1, \cdots, \tilde{\boldsymbol{H}}_m\right]$.

Furthermore, when the output weights of a BLS network are affected by noise, the weight value can be described as

$$\tilde{w}_l^m = (1 + \delta_w) w_l^m \tag{12}$$

for all l, where δ_w's are i.i.d. random variables with zero mean and variance equal to σ_w^2.

From the statistics properties of δ_f's, δ_h's and δ_w's, the first order statistics are given by

$$\langle \tilde{g}_{l_i} \tilde{w}_{l_i}^m \rangle = g_{l_i} w_{l_i}, \text{ and } \left\langle \tilde{h}_{l_j} \tilde{w}_j^m \right\rangle = h_{l_j} w_{l_j}. \tag{13}$$

The second order statistics are given by

$$\langle \tilde{g}_{l_i} \tilde{w}_{l_i}^m \rangle = g_{l_i} w_{l_i} \tag{14a}$$

$$\left\langle \tilde{h}_{l_j} \tilde{w}_j^m \right\rangle = h_{l_j} w_{l_j} \tag{14b}$$

$$\langle \tilde{g}_{l_i}^2 (\tilde{w}_{l_i}^m)^2) \rangle = \left(1 + \sigma_w^2\right) \left(1 + \sigma_\phi^2\right) g_{l_i}^2 (w_{l_i}^m)^2 \tag{14c}$$

$$\left\langle \tilde{h}_{l_j}^2 (\tilde{w}_{l_j}^m)^2 \right\rangle = \left(1 + \sigma_w^2\right) \left(h_{l_j}^2 w_{l_j}^2 + \sigma_\xi^2 \left(\boldsymbol{w}_{\hbar_{l_j}}^T \Delta \boldsymbol{h}_{l_j} \Delta \boldsymbol{h}_{l_j}^T \boldsymbol{w}_{\hbar_{l_j}}\right)\right) \tag{14d}$$

$$\left\langle \tilde{g}_{l_i} \tilde{w}_{l_i}^m \tilde{g}_{l_{i'}} \tilde{w}_{l_{i'}}^m \right\rangle = g_{l_i} w_{l_i} g_{l_{i'}} w_{l_{i'}}, \quad \forall l_i \neq l_{i'} \tag{14e}$$

$$\left\langle \tilde{h}_{l_j} \tilde{w}_{lj}^m \tilde{h}_{l_{j'}} \tilde{w}_{l_{j'}}^m \right\rangle = h_{l_j} w_{l_j} h_{l_{j'}} w_{lj'}, \quad \forall l_j \neq l_{j'} \tag{14f}$$

where $\langle \cdot \rangle$ is the expectation operator.

Based on the fault model given by (10)–(14), the training set error for a specific fault pattern (described as δ_f's, δ_h's, and δ_w's), is given by

$$\tilde{\zeta} = \| \tilde{\boldsymbol{A}} \tilde{\boldsymbol{w}}^m - \boldsymbol{y} \|_2^2 \tag{15}$$

It should be recalled that $\boldsymbol{A} = [\boldsymbol{Z}^n | \boldsymbol{H}^m]$. Hence, $\tilde{\boldsymbol{A}} = [\tilde{\boldsymbol{Z}}^n | \tilde{\boldsymbol{H}}^m]$.

In order to obtain the output weight vector that minimizes the training set error, we choose to minimize the expectation of the objective function stated in (15). Now, along with the statistics properties developed in (13) and simple manipulation, the following objective function is obtained:

$$J = \left\langle \| \tilde{\boldsymbol{A}} \tilde{\boldsymbol{w}}^m - \boldsymbol{y} \|_2^2 \right\rangle = \frac{1}{N} \| \boldsymbol{A} \boldsymbol{w}^m - \boldsymbol{y} \|_2^2 + \frac{1}{N} \left((\boldsymbol{w}^w)^{\mathrm{T}} \boldsymbol{R} \boldsymbol{w}^m \right), \tag{16}$$

where $\boldsymbol{R}$ is a $(\sum_{i=1}^n r_i + \sum_{j=1}^n p_j) \times (\sum_{i=1}^n r_i + \sum_{j=1}^n p_j)$ diagonal matrix, given by

$$\boldsymbol{R} = \sigma_w^2 \mathrm{diag}(\boldsymbol{A}^T \boldsymbol{A}) + (1 + \sigma_w^2) \mathrm{diag}(\boldsymbol{S}) \tag{17}$$

where diag(·) is the diagonal operator which extracts the diagonal elements of a matrix to form a diagonal matrix. The matrix $\boldsymbol{S}$ is a block diagonal matrix, given by

$$\boldsymbol{S} = \begin{bmatrix} \sigma_\phi^2 (\boldsymbol{Z}^n)^{\mathrm{T}} \boldsymbol{Z}^n & \boldsymbol{\emptyset} \\ \boldsymbol{\emptyset} & \sigma_\xi^2 \nabla \boldsymbol{F}^{\mathrm{T}} \nabla \boldsymbol{F}) \end{bmatrix}, \tag{18}$$

where $\nabla \boldsymbol{F} = [\nabla \boldsymbol{F}_1, \cdots, \nabla \boldsymbol{F}_m]$ and $\nabla \boldsymbol{F}_j = \boldsymbol{W}_{\mathfrak{h}_j}^{\mathrm{T}} \frac{\partial \xi([\boldsymbol{Z}^n|\mathbf{1}_{N\times 1}]\boldsymbol{W}_{\mathfrak{h}_j})}{\partial \boldsymbol{W}_{\mathfrak{h}_j}}$. By considering (3), $\nabla \boldsymbol{F}_j$ can be rewritten as

$$\nabla \boldsymbol{F}_j ! ([\boldsymbol{Z}^n|\mathbf{1}_{N\times 1}]\boldsymbol{W}_{\mathfrak{h}_j}) \odot \left(\mathbf{1}_{N\times p_j} - \xi([\boldsymbol{Z}^n|\mathbf{1}_{N\times 1}]\boldsymbol{W}_{\mathfrak{h}_j}) \odot \xi([\boldsymbol{Z}^n|\mathbf{1}_{N\times 1}]\boldsymbol{W}_{\mathfrak{h}_j})\right), \tag{19}$$

where $j = 1, \cdots, m$ and $\odot$ denotes Hadamard product.

Clearly, (16) is a convex function. Therefore, by setting its gradient to zero, we can obtain optimal output weight $(\boldsymbol{w}^m)^*$ which results in minimizing the training set error under the fault situation. The optimal output weight vector is given by

$$(\boldsymbol{w}^m)^* = (\boldsymbol{R} + [\boldsymbol{Z}^n|\boldsymbol{H}^m]^{\mathrm{T}}[\boldsymbol{Z}^n|\boldsymbol{H}^m])^{-1}\boldsymbol{Z}^n|\boldsymbol{H}^m]^{\mathrm{T}}\boldsymbol{y}. \tag{20}$$

One merit of our proposed learning algorithm (20) is that there is no tuning parameter. In the traditional BLS, we need to tune λ by trial and error (Table 1).

4 Numerical Experiments

In this section, the proposed FTBLS algorithm is compared with the original BLS algorithm. Four real life datasets from University of California Irvine (UCI) regression repository [8]. The datasets include Concrete Compressive Strength, Abalone, Wine Quality White (WQW), and Airfoil Self Noise (ASN). summarizes the properties of these datasets.

Table 1. Details of the data-sets

Data-set	Training set size	Test set size	Number of features
Concrete	500	530	9
Abalone	2000	2177	8
Whine Quality Qhite (WQW)	2000	2898	11
Airfoil Self Noise (ASN)	751	752	5

The data is pre-processed as follows. The input features are normalized to the range of $[-1, 1]$. The target outputs are normalized to the range of $[0, 1]$. In addition, the input weights and the biases of the enhancement nodes are generated randomly between the range $[-1, 1]$. For BLS, the main drawback of the original BLS algorithm is that there is no simple way to find an appropriate

λ. In our experiments, we try various λ values and select the value based on the training set. Table 2 shows those λ values.

We compare the traditional BLS algorithm with the proposed FTBLS under various fault levels. Three fault levels are considered. They are $\{\sigma_\phi^2 = \sigma_\xi^2 = \sigma_w^2 = 0.01\}$, $\{\sigma_\phi^2 = \sigma_\xi^2 = \sigma_w^2 = 0.09\}$, and $\{\sigma_\phi^2 = \sigma_\xi^2 = \sigma_w^2 = 0.25\}$. In the experiment, we set $r_i = 20, n = 20$. With this setting, we obtain 400 feature mapped nodes. In addition, we set $m = 1, p_j = 200$. Hence, we obtain 200 enhancement nodes. For fair comparison, we implement the same number of feature mapped nodes for the proposed FTBLS and the traditional BLS algorithm. Similarly, we use the same number of enhancement nodes for both algorithms.

In our numerical experiments, the simulation was ran for 20 times. In each trial, the samples of datasets were randomly split into training and testing set. Table 3 shows average test set mean square of error (MSE) over 20 trials.

Table 2. Tuning parameter settings of the traditional BLS algorithm

Data-set	Parameter
Concrete	$\lambda = \{0.01, 0.03, 0.06, 0.08, 0.1, 5, 8, 10\}$
Abalone	$\lambda = \{0.01, 0.05, 0.06, 0.08, 0.3, 5, 8, 10\}$
Wine Quality White (WQW)	$\lambda = \{0.001, 0.005, 0.01, 0.06, 0.3, 5, 8, 10\}$
Airfoil Self Noise (ASN)	$\lambda = \{0.005, 0.01, 0.06, 0.1, 0.3, 5, 8, 10\}$

From the result in Table 3, the average test set MSEs of the proposed FTBLS algorithm are smaller than those of the traditional BLS. For instance, in the Abalone dataset, when the fault level is small, i.e., $\sigma_\phi^2 = \sigma_\xi^2 = \sigma_w^2 = 0.01$, the average MSE of the traditional BLS is 0.0683, while the MSE of the proposed FTBLS is 0.0083 which is better than that of original BLS. Furthermore, when the fault level is increased to a larger value of $\sigma_\phi^2 = \sigma_\xi^2 = \sigma_w^2 = 0.25$, the average MSE of the traditional BLS is 1.0967 which is very large, while the average MSE of our proposed approach slightly increases to 0.0089. Obviously, the performance of FTBLS is better than that of original BLS. For other datasets, from Table 3, we have similar performance across. In addition, from the standard deviation (SD) in Table 3, it is observed that our proposed approach has smaller SD values. That means, the performance of our proposed FTBLS is more stable.

In order to further validate that our proposed FTBLS outperforms the traditional BLS, we carry out a paired t-test between the two algorithms. Table 4 summarizes the result of the paired t-test obtained.

For 20 trials and the one-tailed test with 95% level of confidence, the critical t-value is 1.729. From the Table, it is clear that all the p-values are smaller than 0.05 and all the test t-values are greater than 1.729. In other words, it is proven that on average the proposed FTBLS is better than the traditional BLS under faulty network. Furthermore, all confidence intervals in Table 4 do not include zero. Therefore, we have enough confidence to say that the improvement of our proposed algorithm is significant.

Table 3. Average MSE for test data-sets of the faulty network. The average values are taken over 20 trials. There are 400 feature mapped nodes and 200 enhancement nodes.

Data set	Fault level	BLS		FTBLS	
		Average MSE	Standard Deviation (SD)	Average MSE	Standard Deviation (SD)
Concrete	$\sigma_\phi^2 = \sigma_\xi^2 = \sigma_w^2 = 0.01$	0.1343	0.0152	0.0183	0.0008
	$\sigma_\phi^2 = \sigma_\xi^2 = \sigma_w^2 = 0.09$	0.7007	0.1226	0.0200	0.0009
	$\sigma_\phi^2 = \sigma_\xi^2 = \sigma_w^2 = 0.25$	2.0785	0.3849	0.0216	0.0009
Abalone	$\sigma_\phi^2 = \sigma_\xi^2 = \sigma_w^2 = 0.01$	0.0683	0.0125	0.0083	0.0003
	$\sigma_\phi^2 = \sigma_\xi^2 = \sigma_w^2 = 0.09$	0.3678	0.1017	0.0084	0.0003
	$\sigma_\phi^2 = \sigma_\xi^2 = \sigma_w^2 = 0.25$	1.0967	0.3200	0.0089	0.0003
WQW	$\sigma_\phi^2 = \sigma_\xi^2 = \sigma_w^2 = 0.01$	0.1238	0.0135	0.0175	0.0004
	$\sigma_\phi^2 = \sigma_\xi^2 = \sigma_w^2 = 0.09$	0.5937	0.1230	0.0181	0.0005
	$\sigma_\phi^2 = \sigma_\xi^2 = \sigma_w^2 = 0.25$	1.7365	0.3909	0.0188	0.0005
ASN	$\sigma_\phi^2 = \sigma_\xi^2 = \sigma_w^2 = 0.01$	0.1876	0.0223	0.0169	0.0005
	$\sigma_\phi^2 = \sigma_\xi^2 = \sigma_w^2 = 0.09$	1.1486	0.1676	0.0179	0.0006
	$\sigma_\phi^2 = \sigma_\xi^2 = \sigma_w^2 = 0.25$	3.4875	0.5221	0.0190	0.0006

Table 4. The paired t-test result between BLS and FTBLS

Data set	Fault level	BLS vs. FTBLS			
		AVG difference	t-value	p-value	Confidence interval
Concrete	$\sigma_\phi^2 = \sigma_\xi^2 = \sigma_w^2 = 0.01$	0.1160	34.01	1.74×10^{-18}	[0.1089 – 0.1231]
	$\sigma_\phi^2 = \sigma_\xi^2 = \sigma_w^2 = 0.09$	0.6807	24.84	5.99×10^{-16}	[0.6234 – 0.7381]
	$\sigma_\phi^2 = \sigma_\xi^2 = \sigma_w^2 = 0.25$	2.0585	23.92	1.20×10^{-15}	[1.8784 – 2.2386]
Abalone	$\sigma_\phi^2 = \sigma_\xi^2 = \sigma_w^2 = 0.01$	0.0600	21.3	1.04×10^{-14}	[0.0541 – 0.0659]
	$\sigma_\phi^2 = \sigma_\xi^2 = \sigma_w^2 = 0.09$	0.3593	15.78	2.25×10^{-12}	[0.3116 – 0.4070]
	$\sigma_\phi^2 = \sigma_\xi^2 = \sigma_w^2 = 0.25$	1.0879	15.20	4.36×10^{-12}	[0.9381 – 1.2377]
WQW	$\sigma_\phi^2 = \sigma_\xi^2 = \sigma_w^2 = 0.01$	0.1063	24.23	7.46×10^{-29}	[0.5929 – 0.6139]
	$\sigma_\phi^2 = \sigma_\xi^2 = \sigma_w^2 = 0.09$	0.5756	20.90	1.43×10^{-14}	[0.5180 – 0.6332]
	$\sigma_\phi^2 = \sigma_\xi^2 = \sigma_w^2 = 0.25$	1.7177	19.64	4.42×10^{-14}	[1.5347 – 1.9008]
ASN	$\sigma_\phi^2 = \sigma_\xi^2 = \sigma_w^2 = 0.01$	0.1708	33.75	2.01×10^{-18}	[0.1602 – 0.1814]
	$\sigma_\phi^2 = \sigma_\xi^2 = \sigma_w^2 = 0.09$	1.1307	30.10	1.70×10^{-17}	[1.0521 – 1.2093]
	$\sigma_\phi^2 = \sigma_\xi^2 = \sigma_w^2 = 0.25$	3.4685	29.69	2.20×10^{-17}	[3.2240 – 3.7131]

5 Conclusion

This paper aims at minimizing the influence of noise on a trained BLS network. In order to achieve this goal, we develop a new objective function to improve the robustness of a noisy BLS network, in which multiplicative noise exist in feature mapped nodes, the input weights of enhancement nodes, and the output weights. From the developed objective function, we propose a regularizer which does not need to be tuned to achieve the aforementioned goal. We train the BLS with the proposed fault tolerant objective function. The simulation experiments show that the proposed algorithm outperforms the original BLS.

Acknowledgments. The work was supported by a research grant from City University of Hong Kong (7005063).

References

1. Leshno, M., Lin, V.Y., Pinkus, A., Schocken, S.: Multilayer feedforward networks with a nonpolynomial activation function can approximate any function. Neural Netw. **6**(6), 861–867 (1993)
2. Pao, Y.H., Takefuji, Y.: Functional-link net computing: theory system architecture and functionalities. Computer **25**(5), 76–79 (1992)
3. Chen, C.P., Liu, Z.: Broad learning system: an effective and efficient incremental learning system without the need for deep architecture. IEEE Trans. Neural Netw. Learn. Syst. **29**(1), 10–24 (2017)
4. Feng, R.B., Han, Z.F., Wan, W.Y., Leung, C.S.: Properties and learning algorithms for faulty RBF networks with coexistence of weight and node failures. Neurocomputing **224**, 166–176 (2017)
5. Leung, C.S., Wan, W.Y., Feng, R.: A regularizer approach for RBF networks under the concurrent weight failure situation. IEEE Trans. Neural Netw. Learn. Syst. **28**(6), 1360–1372 (2017)
6. Jin, J., Liu, Z., Chen, C.P.: Learning system for image recognition. Sci. China Inf. Sci. **61**(11), 112209 (2018)
7. Jin, J.W., Chen, C.P.: Regularized robust broad learning system for uncertain data modeling. Neurocomputing **322**, 58–69 (2018)
8. Lichman, M.: UCI machine learning repository (2013)

Group Loss: An Efficient Strategy for Salient Object Detection

Yikai Hua and Xiaodong Gu(✉)

Department of Electronic Engineering, Fudan University,
Shanghai 200433, China
xdgu@fudan.edu.cn

Abstract. Deep convolutional neural networks (CNNs) have recently achieved great improvements in salient object detection. Most existing CNN-based models adopt cross entropy loss to optimize the networks for its capability in probability prediction. The function of cross entropy loss in salient object detection can be seemed as a pixel-wise label classification for images, which automatically predict whether the pixel is salient or non-salient. However, cross entropy loss pays attention to each single pixel of image when classifying the label, which doesn't consider the relationship with other pixels. In this paper, we propose an additional loss function, called group loss, to improve the above limitation of cross entropy loss. In our model, group loss as well as cross entropy loss work together to optimize the network for better saliency detection performance. The purpose of group loss is to make the difference between salient pixels smaller while the distance between salient and non-salient pixels as large as possible. Meanwhile, due to the large computation cost of pixel-wise comparisons, we design a superpixel pooling layer for computing group loss with no additional parameters, which converts the computation of group loss to superpixel level. The experimental results show that the introduction of group loss improves the performance of CNN network in salient object detection, which makes the boundaries of salient objects more distinct.

Keywords: Salient object detection · Convolutional neural network · Group loss · Superpixel pooling layer

1 Introduction

Salient object detection aims to automatically extract the most visually distinctive objects of an image from the rest part of background. It tends to focus on very few objects in an image, which actually attract human attention most. This attention-focused property means that salient object detection can discover the most critical information from the image, thus help to improve the performance of some follow-up computer vision tasks, including video compression, object tracking, image retrieval, object recognition, image semantic segmentation, etc.

Although many conventional saliency detection methods [1–6] combining with hand-crafted features have achieved great saliency performance with simple cases, they cannot handle images with complex scenes well. In order to obtain more robust features

T. Gedeon et al. (Eds.): ICONIP 2019, CCIS 1142, pp. 104–111, 2019.
https://doi.org/10.1007/978-3-030-36808-1_12

than the hand-crafted features, convolutional neural networks (CNNs) [7] are introduced to salient object detection. Many works [8, 9] have proved the priority of CNNs in mining high-level features and generating the saliency map representation.

Existing CNN based models usually use cross entropy loss to optimize the network, which does a pixel-wise label classification for images, and achieve great saliency prediction performance. However, cross entropy loss focuses most its attention on each pixel and doesn't take the relationship between pixels into account. To improve the limitation of cross entropy loss, we propose a new loss function, called group loss, for salient object detection.

The main highlights of our work are as follows.

(1) We propose a new loss function, called group loss, for salient object detection to take the relationship between pixels into account. The purpose is to make the difference between salient pixels smaller and the distance between salient and non-salient pixels as large as possible in specific feature space, thus making the boundaries of salient objects in saliency map more distinct. Group loss and cross entropy loss work together to optimize the CNN network for salient object detection.
(2) As the computation cost of group loss at pixel-wise level is large, we design a superpixel pooling layer for computing group loss. The superpixel pooling inputs a feature map and a superpixel map, and does max pooling operation to the input feature map to extract main features of each superpixel, taking the superpixel map as the mask of irregular pooling shape. This proposed layer is essentially a pooling layer, which means the model has no additional parameters to learn.

The rest part of this paper is organized as below. Section 2 gives a detailed description of our proposed group loss and the CNN based model for salient object detection. The experimental results and related analysis are discussed in Sect. 3. Finally, Sect. 4 makes an overall summary of this paper.

2 Proposed Model

2.1 Overview

The whole architecture of the propose model is shown in Fig. 1. The structure outside the red rectangle is a baseline network for salient object detection. The network consists of 2 parts, including feature extraction part and up-sampling part. The baseline network we use in our model is VGG-16 net. Cross entropy loss is used to train the network for salient object detection.

In order to improve the limitation of cross entropy loss mentioned above, we propose group loss for salient object detection and design a superpixel pooling layer to compute it, which is shown in the red rectangle in Fig. 1. The inputs of the superpixel pooling layer are the feature map exported from up-sampling part and the superpixel map of original image with k superpixels. The superpixel pooling layer do max pooling operation to the feature map and the superpixel map plays the role of a mask for irregular shape pooling. The output of this layer is a $k \times C$ pooling map for computing group loss. Group loss and cross entropy loss work together to optimize the network for better salient object detection performance.

A detailed description of each component in our proposed model is presented in the followed subsections.

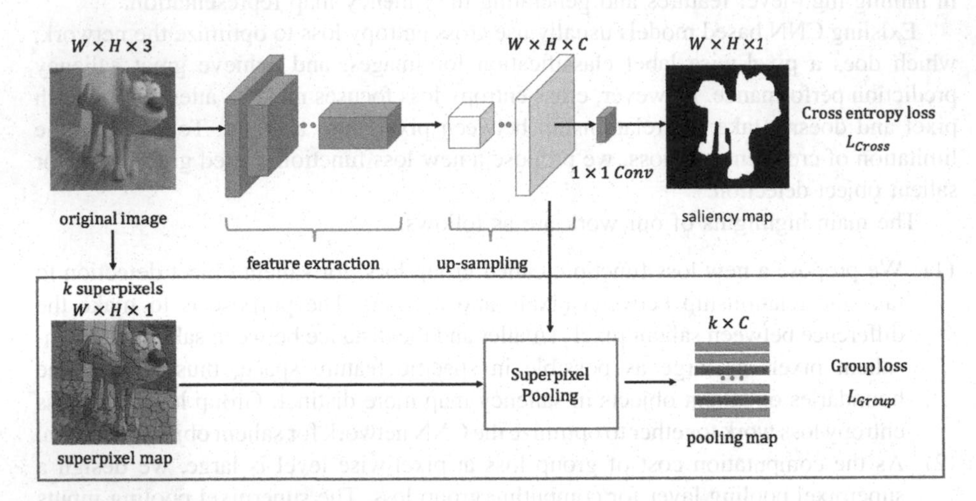

Fig. 1. The whole architecture of the proposed model. The model adopts cross entropy loss and proposed group loss together to optimize its parameters. The superpixel map is visualized for a more intuitive view. (Color figure online)

2.2 Superpixel Pooling Layer

It takes a large amount of computation cost for pixel-wise comparisons in group loss. In order to improve this problem, we design a superpixel pooling layer, which convert the computation to superpixel level and then extract the main features of superpixels. A superpixel is composed of a series of adjacent pixels with similar properties, which can represent these pixels on a larger dimension to some extent. Hence, using superpixels for group loss computation can greatly reduce the computation cost and achieve the purpose of separating the salient and non-salient pixels in the meantime.

The red rectangle in Fig. 1 shows the pipeline of how the superpixel pooling layer works in our model. The superpixel pooling layer has two inputs, including a feature map and a superpixel map. The feature map is the output of the up-sampling part in Fig. 1, whose size is $W \times H \times C$, while the superpixel pooling layer is a $W \times H \times 1$ map containing k superpixels in total, which is generated from Simple Linear Iterative Clustering (SLIC) method [10]. The superpixel map is divided into k areas, where each superpixel represents an irregular shape area, as the result of which the superpixel map is seen as a mask for the feature map in subsequent pooling operation. Different from the common way that we do pooling operation at each $d \times d$ regular rectangular area, we do max pooling operation for each channel of the feature map at k irregular areas, which are consistent with the corresponding superpixel areas in superpixel map, so that the output of the superpixel pooling layer is a $k \times C$ pooling map, which contains the main features of each superpixel. Finally, this pooling map is then used to compute our proposed group loss.

2.3 Group Loss

The purpose to design this group loss is to make up for cross entropy loss that it mainly focuses on each pixel, and take the relationship of different pixels into account. This group loss makes the difference between pixels in same class smaller and the difference between pixels in different classes larger, which makes the boundaries of salient objects in saliency map clearer. In practice, as the computation cost of group loss at pixel level is huge, we compute this group loss at superpixel level.

As described in Subsect. 2.2, the final output of superpixel pooling layer is a pooling map for computing group loss, of which each row vector C_i represents the features of corresponding superpixel. Group loss consists of two parts, $L_{salient}$ and L_{differ}. $L_{salient}$ measures the variance between superpixels which belongs to salient object, while L_{differ} measures the distance between non-salient superpixels and salient superpixels in specific feature space.

The $L_{salient}$ in group loss is defined as follows:

$$L_{salient} = \frac{1}{M}\sum_{i=1}^{M}\left(C_i - \overline{C}\right)^2 \tag{1}$$

where M denotes the number of salient superpixels in the superpixel map and C_i represents the corresponding feature vector of superpixel in the pooling map. $\overline{C}$ in Eq. (1) is the mean value of the feature vectors of these salient superpixels, which can be calculated by:

$$\overline{C} = \frac{1}{M}\sum_{i=1}^{M} C_i \tag{2}$$

The L_{differ} in group loss is given by:

$$L_{differ} = \frac{1}{N}\sum_{j=1}^{N} \max\left(0, \Delta - \left|\overline{C} - C_j\right|\right) \tag{3}$$

where N denotes the number of non-salient superpixels and Δ is a threshold judging whether the distance between this non-salient superpixels and other salient superpixels is far enough in this feature space.

Group loss is the combination of the above two loss functions:

$$L_{Group} = L_{salient} + L_{differ} \tag{4}$$

Therefore, the entire loss function of the proposed model is expressed as:

$$L = L_{cross} + L_{group} \tag{5}$$

The cross entropy loss L_{cross} in Eq. (5) can be represented by:

$$L_{cross} = -\frac{1}{S}\sum_{i=1}^{S}[g_i log y_i + (1 - g_i)\log(1 - y_i)] \quad (6)$$

where S is the number of pixels of the image. $g_i \in [0, 1]$ denotes the saliency ground truth of each pixel and y_i is the prediction score of each pixel.

3 Experimental Results

3.1 Experiment Settings

We evaluate the saliency detection performance of our model on 4 datasets, including MSRA-B [11], ECSSD [12], PASCAL-S [13] and SED2 [14]. All these four datasets own pixel-wise salient object detection ground truth and are widely adopted in visual saliency research.

In the experiments, the base learning rate is set to 10^{-4} and the threshold Δ in group loss is set to 1.0 during the training process. In superpixel pooling layer, the number of superpixels in each superpixel map is 196 and the salient pixel rate of each superpixel, which judges whether this superpixel is salient or not, is set to 0.5. In test phase, we just use the baseline network outside the red rectangle in Fig. 1 to generate saliency maps of images and it spends around 0.05 s processing one image.

3.2 Evaluation Metrics

In the experiments, we evaluate the saliency performance of all models by adopting two metrics, including F-measure and mean absolute error (MAE).

F-measure is the weighted combination of precision and recall, which evaluate the quality of saliency maps comprehensively. MAE measures the average pixel-wise error between the saliency map S and ground truth G.

3.3 Saliency Performance Comparison

We evaluate our model on four datasets with several state-of-the-art algorithms for comparison, including region-based contrast (RC) [5], graph-based manifold ranking (GMR) [2], discriminative regional feature integration (DRFI) [15], multiscale deep features (MDF) [9], multi-context deep learning (MCDL) [8], encoded low level distance (ELD) [16] and deep image saliency computing (DISC) [17].

Figure 2 gives an intuitive saliency performance comparison between our model and other compared models. As shown in the saliency maps in Fig. 2, our model is able to achieve favorable saliency detection results compared with other models, especially that the boundaries of salient objects in our model are relatively distinct.

More detailed quantitative comparisons are presented in Table 1. The comparison results show that our model, which is optimized by both cross entropy loss and proposed group loss, has a competitive saliency detection performance against other models. Especially, it is worth noting that we achieve high F_{avg} among all these four datasets, which shows the stability of our model on salient object detection.

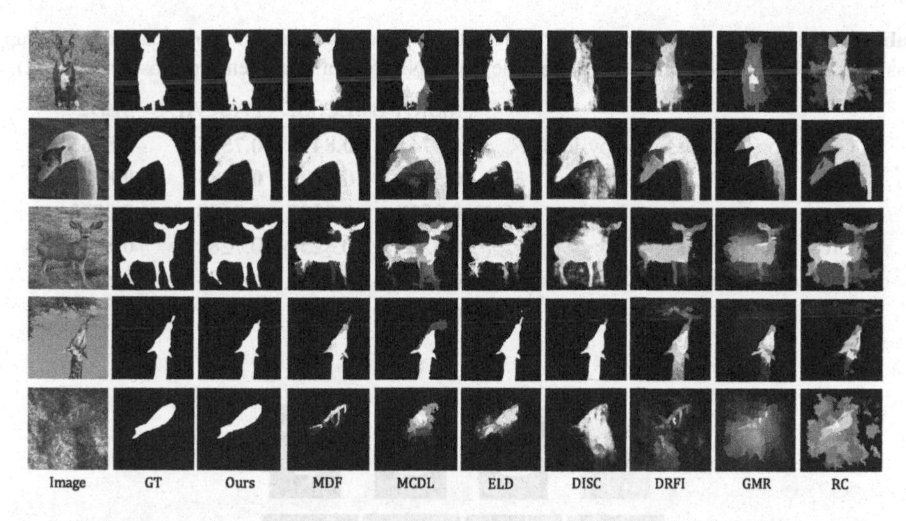

Fig. 2. Visual comparison between saliency maps of different models.

Table 1. Quantitatively comparison between our proposed approach and other methods on four datasets, involving F-measure and MAE.

	MSRA-B			ECSSD			PASCAL-S			SED2		
	F_{max}	F_{avg}	*MAE*	F_{max}	F_{avg}	*MAE*	F_{max}	F_{avg}	*MAE*	F_{max}	F_{avg}	*MAE*
RC	0.8323	0.8076	0.0544	0.7381	0.6771	0.1506	0.4670	0.3617	0.1607	0.7949	0.7467	0.1099
GMR	0.8396	0.8363	0.0371	0.7375	0.6423	0.1481	0.6448	0.5897	0.1011	0.7883	0.7324	0.1298
DRFI	0.8625	0.8360	0.0493	0.7860	0.6672	0.1185	0.6757	0.5942	0.0854	0.8597	0.7759	0.0945
MDF	0.8853	0.8534	0.0507	0.8316	0.8100	0.0225	0.7610	0.7134	0.0227	0.8828	0.8029	0.0245
MCDL	0.8720	0.8486	0.0247	0.8205	0.7809	0.0276	0.7256	0.6795	0.0425	0.7931	0.7432	0.0201
ELD	0.8805	0.8577	0.0303	0.8684	0.8177	0.0317	0.7775	0.7190	0.0482	0.8200	0.7331	0.0271
DISC	0.9054	0.8664	0.0283	0.8563	0.8127	0.0410	0.7583	0.6742	0.0555	0.8079	0.7249	0.0275
Ours	0.8741	0.8628	0.0258	0.8492	0.8398	0.0337	0.7525	0.7388	0.0638	0.8765	0.8583	0.0072

3.4 Contribution of Group Loss

In order to evaluate the effectiveness of group loss in our model, we compare the saliency performance of our model with a raw CNN model, which is just optimized by cross entropy loss. As the proposed superpixel pooling layer in our model does not introduce additional parameters, the learnable network structure in our model is consistent with that of the raw CNN model, which is same with the CNN structure outside the red rectangle in Fig. 1. Both two models start training under the same experiment settings and evaluate their saliency results.

Table 2 shows the detailed saliency detection performance of both two models on four datasets. The results of F-measure and MAE shows the overall improvement of our model in salient object detection brought by the introduction of group loss. A visual saliency comparison on several sample images of the two models is shown in Fig. 3. We can intuitively figure out that the introduction of group loss improves the capability of CNN model in detecting a more distinct boundary of salient object.

Table 2. The saliency detection performance comparison between our proposed model (group loss + cross entropy loss in Table 2) and the raw CNN model (cross entropy loss in Table 2).

		MSRA-B	ECSSD	PASCAL-S	SED2
Group loss + Cross entropy loss	F_{max}	**0.8741**	**0.8492**	**0.7525**	**0.8765**
	F_{avg}	**0.8628**	**0.8398**	**0.7388**	**0.8583**
	MAE	**0.0258**	**0.0337**	**0.0638**	**0.0072**
Cross entropy loss	F_{max}	0.8523	0.8258	0.7282	0.8480
	F_{avg}	0.8300	0.8080	0.7104	0.8280
	MAE	0.0330	0.0387	0.0645	0.0115

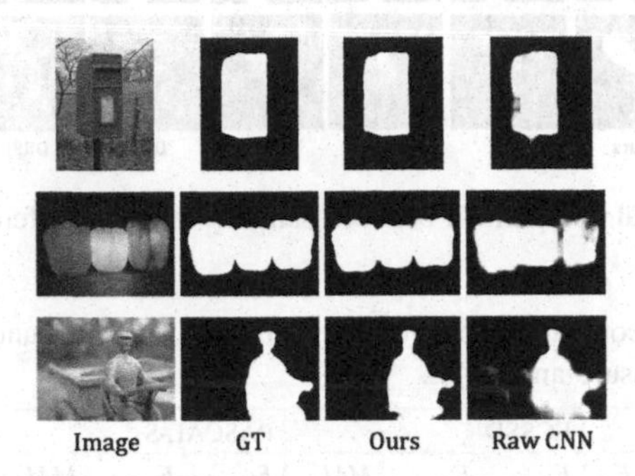

Fig. 3. The visual comparison between our proposed model and the raw CNN model.

4 Conclusion

Existing CNN based models for salient object detection usually adopt cross entropy loss to optimize the network and achieves great saliency performance. However, cross entropy loss pays attention to each pixel and does not take the relationship between pixels into account. In this paper, we propose group loss as supplement of cross entropy loss, which tends to make the difference between salient pixels smaller and difference between salient and non-salient pixels larger. Moreover, a superpixel pooling layer is designed to convert the computation of group loss to superpixel level for reducing computation cost with no additional learnable parameters. Experimental results show that the introduction of group loss is able to improve the saliency performance of CNN model and make the boundaries of salient objects more distinct.

Acknowledgments. This work was supported in part by National Natural Science Foundation of China under grant 61771145 and 61371148.

References

1. Zhu, W., Liang, S., Wei, Y., Sun, J.: Saliency optimization from robust background detection. In: 27th IEEE Conference on Computer Vision and Pattern Recognition, Columbus, pp. 2814–2821. IEEE Press (2014)
2. Yang, C., Zhang, L., Lu, H., Ruan, X., Yang, M.H.: Saliency detection via graph-based manifold ranking. In: 26th IEEE Conference on Computer Vision and Pattern Recognition, Portland, pp. 3166–3173. IEEE Press (2013)
3. Jiang, B., Zhang, L., Lu, H., Yang, C., Yang, M.H.: Saliency detection via absorbing Markov chain. In: 14th IEEE International Conference on Computer Vision, Sydney, pp. 1665–1672. IEEE Press (2013)
4. Cheng, M.M., Warrell, J., Lin, W.Y., Zheng, S., Vineet, V., Crook, N.: Efficient salient region detection with soft image abstraction. In: 14th IEEE International Conference on Computer Vision, Sydney, pp. 1529–1536. IEEE Press (2013)
5. Cheng, M.M., Zhang, G.X., Mitra, N.J., Huang, X., Hu, S.M.: Global contrast based salient region detection. IEEE Trans. Pattern Anal. Mach. Intell. **37**(3), 409–416 (2011)
6. Achanta, R., Hemami, S., Estrada, F., Susstrunk, S.: Frequency-tuned salient region detection. In: 22nd IEEE Conference on Computer Vision and Pattern Recognition, Miami, pp. 1597–1604. IEEE Press (2009)
7. Lecun, Y., Bottou, L., Bengio, Y., Haffner, P.: Gradient-based learning applied to document recognition. Proc. IEEE **86**(11), 2278–2324 (1998)
8. Zhao, R., Ouyang, W., Li, H., Wang, X.: Saliency detection by multi-context deep learning. In: The IEEE Conference on Computer Vision and Pattern Recognition on Proceedings, Boston, pp. 1265–1274. IEEE Press (2015)
9. Li, G., Yu, Y.: Visual saliency based on multiscale deep features. In: The IEEE Conference on Computer Vision and Pattern Recognition on Proceedings, Boston, pp. 5455–5463. IEEE Press (2015)
10. Achanta, R., Shaji, A., Smith, K., Lucchi, A., Fua, P., Süsstrunk, S.: SLIC superpixels compared to state-of-the-art superpixel methods. IEEE Trans. Pattern Anal. Mach. Intell. **34**(11), 2274–2282 (2012)
11. Liu, T., et al.: Learning to detect a salient object. IEEE Trans. Pattern Anal. Mach. Intell. **33** (2), 353–367 (2011)
12. Yan, Q., Xu, L., Shi, J., Jia, J.: Hierarchical saliency detection. In: The IEEE Conference on Computer Vision and Pattern Recognition on Proceedings, Portland, Oregon, pp. 1155–1162. IEEE Press (2013)
13. Li, Y., Hou, X., Koch, C., Rehg, J., Yuille, A.: The secrets of salient object segmentation. In: The IEEE Conference on Computer Vision and Pattern Recognition on Proceedings, Columbus, pp. 280–287. IEEE Press (2014)
14. Alpert, S., Galun, M., Basri, R., Brandt, A.: Image segmentation by probabilistic bottom-up aggregation and cue integration. In: The IEEE Conference on Computer Vision and Pattern Recognition on Proceedings, Minneapolis, pp. 1–8. IEEE Press (2007)
15. Jiang, H., Wang, J., Yuan, Z., Wu, Y., Zheng, N., Li, S.: Salient object detection: a discriminative regional feature integration approach. In: 26th IEEE Conference on Computer Vision and Pattern Recognition, Portland, pp. 2083–2090. IEEE Press (2013)
16. Lee, G., Tai, Y.W., Kim, J.: Deep saliency with encoded low level distance map and high level features. In: The IEEE Conference on Computer Vision and Pattern Recognition on Proceedings, Las Vegas, pp. 660–668. IEEE Press (2016)
17. Chen, T., Lin, L., Liu, L., Luo, X., Li, X.: DISC: deep image saliency computing via progressive representation learning. IEEE Trans. Neural Netw. Learn. Syst. **27**(6), 1135–1149 (2016)

PPGCN: A Message Selection Based Approach for Graph Classification

Xinyang Liu[1], Zheng Liu[2], and Yanwen Qu[1](✉)

[1] School of Computer Information and Engineering,
Jiangxi Normal University, Nanchang, China
{lxy,qu_yw}@jxnu.edu.cn

[2] Jiangsu Key Laboratory of BDSIP,
Nanjing University of Posts and Telecommunications, Nanjing, China
zliu@njupt.edu.cn

Abstract. Recently, the Graph Convolutional Networks (GCNs) have achieved state-of-the-art performance in many graph data related tasks. However, traditional GCNs may generate redundant information in the message passing phase. In order to solve this problem, we propose a novel graph convolution named Push-and-Pull Convolution (PPC), which follows the message passing framework. On the one hand, for each star-shaped subgraph, PPC uses a node pair based message generation function to calculate the message pushed by each local node to the central node. On the other hand, in the message aggregation substep, each central node pulls valuable information from the messages pushed by its local nodes based on a gate network with pre-perceiving function. Based on the PPC, a new network named Push-and-Pull Graph Convolutional Network (PPGCN) is proposed for graph classification. PPGCN stacks multiple PPC layers to extend the receptive field of each node, then applies a global pooling layer to get the graph embedding based on the concatenation of all PPC layers' outputs. The new network is permutation invariant and can be trained end-to-end. We evaluate the performance of PPGCN in 6 graph classification datasets. Compared with state-of-the-art baselines, PPGCN achieves the top-1 accuracy on 4 of 6 datasets.

Keywords: Graph classification · Graph Convolutional Networks · Node embedding · Message passing

1 Introduction

In recent years, there is a growing interest in developing Graph Neural Networks (GNNs) for graph learning. Among which, GCNs have achieved state-of-the-art performance in many graph data related tasks such as node classification, link prediction and graph classification [1,7,19]. GCNs methods fall into two classes, spectral-based and spatial-based. Spectral-based methods rely on the

T. Gedeon et al. (Eds.): ICONIP 2019, CCIS 1142, pp. 112–121, 2019.
https://doi.org/10.1007/978-3-030-36808-1_13

eigen-decomposition of the Laplacian matrix, thus they cannot be applied to graphs with different structures.

On the contrary, spatial-based methods, such as DCNN [1], MPNN [4] and DGCNN [19] directly operate convolution on the graph nodes and their neighbors, which imitates the convolution operation on image. For each node, spatial-based convolution operations usually construct a corresponding node-centered star-shaped subgraph through the adjacent matrix, where each edge associates the central node with a original neighbor node. Then the embeddings of all the nodes in the subgraph are aggregated to calculated the new embedding of the central node.

Most GCNs, as well as many GNNs, can be grouped into a generic framework called Message Passing Neural Networks [4]. The framework consists of two phases, the message passing phase and readout phase. In order to extend the receptive field of each node, multiple message passing steps (which corresponding to the convolution layers in GCNs), optionally alternate with pooling layers, are stacked in the message passing phase. Although MPNNs framework supports message selection in the message passing phase, most GCNs don't make full use of it. In those methods, each node either pass the same message to its neighbors, or accept messages from its neighbors fairly. In this case, information redundancy problem may appears in the message passing phase, which will be explained by an example in Sect. 3.1.

Inspired by the EdgeConv [15] and Graph Attention Network (GAT) [12], we propose a novel graph convolution named Push-and-Pull Convolution (PPC), and a corresponding graph neural network named Push-and-Pull Graph Convolutional Network (PPGCN). For the sake of discussion, we decompose each message passing step into three substeps: message generation, message aggregation and update. In the message generation substep, each local node in the subgraph pushes message to the central node base on the embeddings of the central node and of itself; in the message aggregation substep, based on the output of a gate network, the central node pulls valuable information from pushed messages sent by local nodes. In order to reduce information redundancy, the gate network pre-perceives the pushed messages in advance and assigns a corresponding input coefficient to each pushed message. Figure 1 illustrate the process of PPC. The reason why the new graph convolution named push-and-pull, is because we think that the message passing process is based on the push action performed by each local node and the pull action performed by central node, which is similar to the Label Propagation Algorithm in community detection related tasks [6]. We evaluate the effectiveness of our new network-PPGCN in 6 graph classification datasets, and the results show that it achieves the top-1 accuracy on 4 of 6 datasets.

2 Related Work

Here, we give a brief introduction to the MPNNs framework, EdgeConv and GAT.

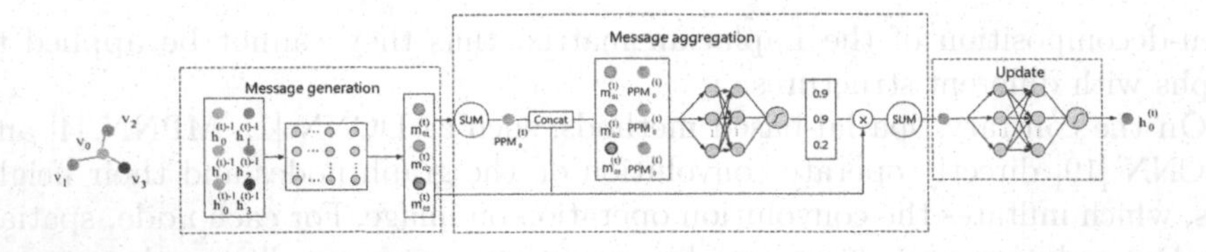

Fig. 1. The process of a PPC operation consists of three substeps: message generation, message aggregation and update. In message generation, we take the embedding of central node v_0 into account; in message aggregation, the pushed-messages $\{m_{01}^{(t)}, m_{02}^{(t)}, m_{03}^{(t)}\}$ are aggregated with different input coefficients at time step t. Besides, a neural network is used as the update function to make the embeddings more discriminative.

Definition. Given a graph $\mathbf{G}$ as (V, E, X), where $V = \{v_i\}, i \in \{1, ..., N_v\}$ is the set of nodes, N_v denotes the number of nodes, which equals to $|V|$; E is the set of edges, N_e denotes the number of edges, which equals to $|E|$, and $e_{ij} = (v_i, v_j) \in E$ denotes an edge; $X \in \mathbb{R}^{n \times c}$ is the node features matrix, where each row denotes the c-dimensional feature vector of a node.

The adjacent matrix of graph is denoted as $\mathbf{A} \in \mathbb{R}^{N_v \times N_v}$. In this paper, We only consider simple graph, where $\mathbf{A}$ is a symmetric binary matrix. The original graphs in datasets have no self-loops. For graph with node labels or node attributes, each row of X can be the one-hot encoding of the node label or be the attributes vector. For graph without node labels and node attributes, we use the one-hot encoding of the node degree as the node feature, where $degree(v_i) = \sum_{j=1}^{\mathbf{N}} A_{ij}$. For a node v, we use $N(v)$ to denote the neighbor nodes set.

2.1 Message Passing Neural Networks

Many graph neural networks for graph classification task abide by the principle of message passing framework [4]. The MPNNs consist of message passing phase and readout phase. Following the program architecture in [3], we decompose the message passing phase into message generation function $\mathcal{M}^{(t)}(\cdot)$, aggregation function $\square^{(t)}(\cdot)$ and update function $\mathcal{U}^{(t)}(\cdot)$ at each time step t.

$$a_v^{(t)} = \mathop{\square}_{w \in v \cup N(v)} (\{\mathcal{M}^{(t)}(h_v^{(t-1)}, h_w^{(t-1)}; e_{vw})\}) \tag{1}$$

$$h_v^{(t)} = \mathcal{U}^{(t)}(h_v^{(t-1)}, a_v^{(t)}) \tag{2}$$

where the $h_v^{(t-1)}$ is the embedding of node v at time step $(t-1)$, i.e the output of the $(t-1)$th graph convolution layer. The e_{vw} represents the optional edge feature associated with node v and w. The aggregation function can be MAX, MEAN, weighted SUM or a network.

The readout function READOUT$(\cdot)$ is used to get the graph embedding:

$$h_g = \text{READOUT}(\{h_v^{(T)} | v \in V\}) \tag{3}$$

where h_g is the graph embedding, T is the total time steps of message passing. READOUT(·) is a global pooling function, which can usually be a summation, mean, or max operation, even a neural network.

The Eqs. 1 and 2 show the powerful generalization capability of the MPNNs framework, which can support message selection, so our model also follows this framework.

2.2 EdgeConv and GAT

Recently, Wang et al. propose a novel graph convolution operation for point cloud data called EdgeConv [15], which can fit into the framework of MPNNs without readout phase. In the message generation substep, for each node v_c, EdgeConv performs a convolution-like operation on the node pairs $\{(v_c, v_i)\}, v_i \in v_c \cup N(v_c)$, to extract edge features. Then, the central node v_c aggregates the edge features to update its node embedding. The message passing process of EdgeConv is shown in Fig. 2(b).

There also has been a growing interest in using attention mechanism for graphs [12]. The key idea of graph attention models is assigning different weights to different nodes, walks and etc. It employs attention mechanism to aggregate the neighboring nodes features. The calculation process of GAT also can fit into the framework of MPNNs without readout phase, as shown in Fig. 2(c).

The above two models perform information selection either in the message generation substep or the message aggregation substep. In contrast, the PPC proposed in this paper will perform information selection in both substeps.

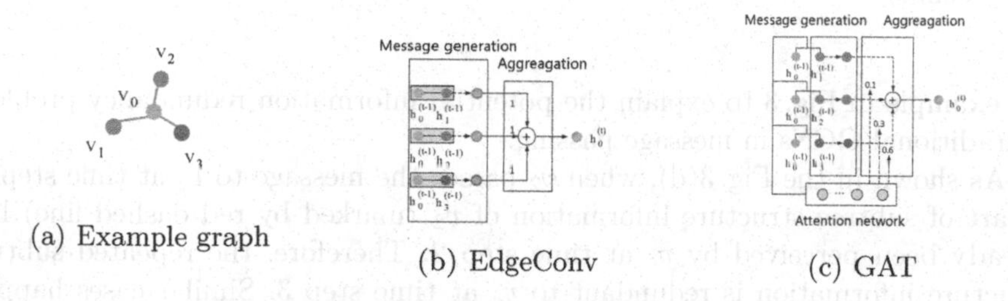

(a) Example graph (b) EdgeConv (c) GAT

Fig. 2. Figure 2(a) shows the sub-graph structure with v_0 as the central node. Figure 2(b) describes the corresponding message passing process of EdgeConv. Figure 2(c) illustrates the corresponding message passing process of single head GAT.

3 Proposed Method

3.1 Preliminaries

Since the neighborhood aggregation in GNNs is analogous to its 1-dimensional form of the Weisfeiler-Lehman subtree kernel [16], here we draw the subtree structure rooted at a fixed node to represent message passing process. We use

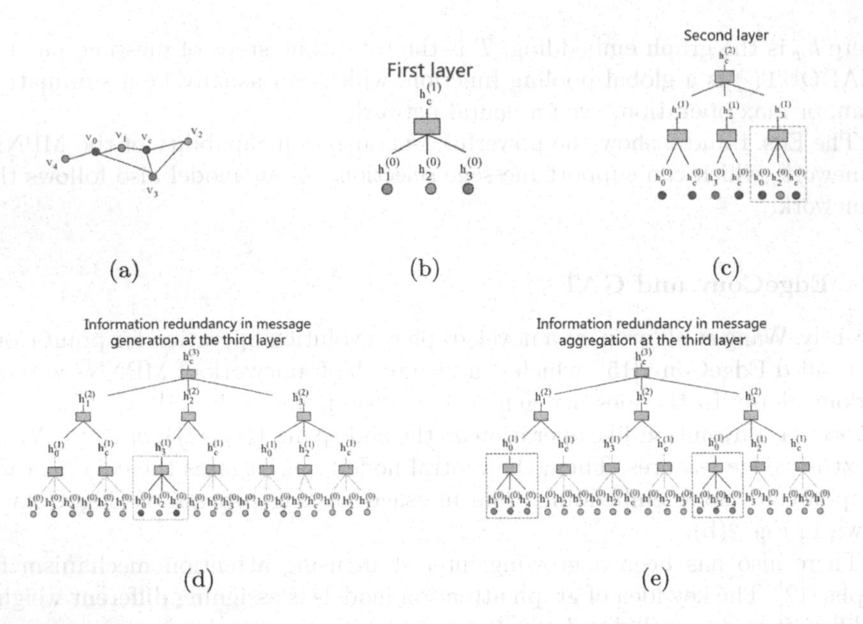

Fig. 3. We use the subtree structure rooted at the central node to illustrate the process of message passing. For better understanding, we remove the self loop. The structure information of the third layer consists of the structure information perceived from the first two steps. The $h_i^{(t)}$ presents the embedding of node v_i at time step t. The area marked by red dashed line in Fig. 3(d) represents redundant information produced in the message generation substep, and the areas marked by green dashed line in Fig. 3(e) represents redundant information produced in the message aggregation substep. (Color figure online)

the example in Fig. 3 to explain the potential information redundancy problem of traditional GCNs in message passing.

As shown in the Fig. 3(d), when v_2 passes the message to v_c at time step 3, a part of subtree structure information of v_2 (marked by red dashed line) has already been perceived by v_c at time step 2. Therefore, the repeated subtree structure information is redundant to v_c at time step 3. Similar cases happen to other neighbor nodes. Thus, in order to push informative message to v_c, the neighbor nodes should consider the embedding of central node.

On the other hand, in the message generation substep, even if each neighbor node considering the embedding of central node, redundant information may still be introduced when v_c aggregates all the messages passed to it. As shown in Fig. 3(e), v_1 and v_2 push the same message to v_c (marked by green dashed line). Therefore, it is necessary for the central node to pre-perceive all the messages in advance, so that it can filter the redundant information when doing aggregation operation.

In some sense, redundant information can be reduced by the information selection operation in the aggregation substep only. At this point, the messages passed to the central node will contain a lot of redundant information, which may increase the training difficulty of the model. An effective method is that each node performs message filtering beforehand, according to the embedding of the central node. Therefore, we suggest considering both the "push" and the "pull" operations in the message passing process.

3.2 Push-and-Pull Convolution

In order to reduce the information redundancy in the message passing phase, the Push-and-Pull Convolution is proposed based on the intuition derived from the above example. When calculating the pushed-message, both the embedding of central node and local node are considered by the message generation function; when aggregating the pushed messages from local nodes, the PPC assigns a corresponding input coefficient to each pushed message by a gate network which pre-perceives the messages pushed by all local nodes in advance. PPC aims to extract informative messages and reduce the information redundancy through the above two substeps.

In the message generation substep, the pushed-message is calculated as follows:

$$\begin{aligned} m_{ij}^{(t)} &= \mathcal{M}_{PPC}^{(t)}(h_i^{(t-1)}, h_j^{(t-1)}) \\ &= \mathrm{RELU}(\boldsymbol{W}_{\boldsymbol{1}}^{(t)}[h_i^{(t-1)} \| h_j^{(t-1)}] + b_1^{(t)}), v_j \in v_i \cup N(v_i) \end{aligned} \tag{4}$$

where $m_{ij}^{(t)}$ is the pushed-message from node j to node i, $\mathcal{M}_{PPC}^{(t)}(\cdot)$ is the message generation function at time step t, which is parameterized by $\{\boldsymbol{W}_{\boldsymbol{1}}^{(t)}, b_1^{(t)}\}$.

After obtaining the pushed-messages, PPC pulls informative messages by using aggregation function:

$$a_i^{(t)} = \underset{v_j \in v_i \cup N(v_i)}{\square_{PPC}}{}^{(t)}(\{m_{ij}^{(t)}\}) = \sum_{v_j \in v_i \cup N(v_i)} \alpha_{ij}^{(t)} \times m_{ij}^{(t)} \tag{5}$$

$$\begin{aligned} \alpha_{ij}^{(t)} &= gate^{(t)}(\{m_{ij}^{(t)}\}) \\ &= Sigmoid(\boldsymbol{A}_{\boldsymbol{2}}^{(t)} Relu(\boldsymbol{A}_{\boldsymbol{1}}^{(t)}[PPM_i^{(t)} \| m_{ij}^{(t)}]), v_j \in v_i \cup N(v_i) \end{aligned} \tag{6}$$

$$PPM_i^{(t)} = \sum_{v_j \in v_i \cup N(v_i)} m_{ij}^{(t)} \tag{7}$$

where $a_i^{(t)}$ is the pulled-message. $PPM_i^{(t)}$ is the pre-perceived message of central node, which is used by gate network $gate^{(t)}(\cdot)$ to filter the redundant information. The aggregation function $\square_{PPC}^{(t)}$ is a weighted summation over the messages pushed by local nodes. The input coefficients of pushed messages from local nodes which belongs to $v_i \cup N(v_i)$ are assigned by the gate mechanism $gate^{(t)}(\cdot)$, which is parameterized by $\{\boldsymbol{A}_{\boldsymbol{1}}^{(t)}, \boldsymbol{A}_{\boldsymbol{2}}^{(t)}\}$.

Finally, PPC use the update function to map node embedding into a more discriminative latent space:

$$h_i^{(t)} = \mathcal{U_{PPC}}^{(t)}(a_i^{(t)}) = RELU(\mathbf{W_3}^{(t)}(RELU(\mathbf{W_2}^{(t)} a_i^{(t)} + b_2^{(t)})) + b_3^{(t)}) \quad (8)$$

where the update function $\mathcal{U_{PPC}}^{(t)}(\cdot)$ is parameterized by $\{\mathbf{W_2}^{(t)}, \mathbf{W_3}^{(t)}, b_2^{(t)}, b_3^{(t)}\}$.

The whole process of a PPC layer is shown in Fig. 1. Obviously, the PPC is permutation invariant.

3.3 PPGCN Towards Graph Classification

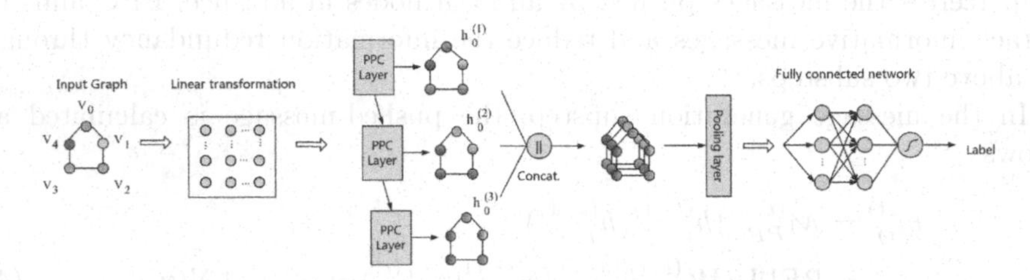

Fig. 4. The architecture of PPGCN. The v_i denotes the node i, The $h_i^{(t)}$ denotes the embedding of v_i at time step t.

Based on the MPNNs framework, a new network named Push-and-Pull Graph Convolution Network (PPGCN) is proposed in this paper. Firstly, PPGCN use a parameter matrix to transform original node embeddings into $\mathbb{R}^c$. Secondly, we stack multiple PPC layers to obtain node embeddings at different scales; as the number of layer increases, the receptive field of each node enlarges. Besides, to consider structural information under all scales, PPGCN combines node embeddings from all layers by using jumping network architecture [17]. Finally, PPGCN uses a global pooling layer to transform node embeddings into graph embedding and feed them into a fully connected network for graph classification. The whole architecture is shown in Fig. 4.

Since both the PPC layers and the global pooling layer are permutation invariant, the permutation invariance of PPGCN model can be easily derived.

4 Experiments

In order to evaluate the performance of our model, we compare it with eight baseline algorithms, including four graph kernels and four deep learning methods in 6 graph classification datasets.

4.1 Experiment Configuration

Datasets. The datasets including 3 bioinformatics datasets - MUTAG, PTC, PROTEINS, and 3 social network datasets IMDB-BINARY (IMDB-B), IMDB-MULTI (IMDB-M) and COLLAB. The bioinformatics datasets use categorical node labels as input features. The social network datasets do not have node labels, therefore we follow the convention, using one-hot encodings of node degrees as input features [3].

Baselines and Experimental Setup. We compare the graph classification accuracy of our model with four graph kernels: the graphlet kernel (GK) [11], the Weisfeiler-Lehman subtree kernel (WL) [10], the Shortest-Path kernel (SP) [2], the Random Walk kernel (RW) [14], and four other current the state-of-the-art deep learning approches for graph classification: Deep graph kernel (DGK) [18], Diffusion-CNN (DCNN) [1], DGCNN [19], PATCHYSAN [9]. To conduct the ablation study, we also evaluate the performance of EdgeConv and GAT on graph classification task. Considering the EdgeConv and GAT are not originally designed for graph classification, here we combine the EdgeConv and GAT with global sum pooling layer.

For our model, the Adam optimizer with L2 regularization is used for optimization, and the learning rate is decayed by half every 50 epochs. The drop out rate is set to 0.2, batch size is set to 50. We stack three PPC layers in the message passing phase, and batch normalization is applied before and after the update substep of every PPC layer. The other candidate hyper-parameters tuned during the experiment are as follows: the number of epoch $\in \{100, 200\}$; the dimension of pushed-message $\in \{32, 64, 128\}$; the dimension of pulled-message $\in \{16, 32\}$; the number of hidden units of update network $\in \{16, 32\}$; initial learning rate $\in \{0.001, 0.01\}$. All of the results are obtained under 10 fold cross validation.

4.2 Results for Graph Classification

We use the testing results reported in related literatures for baseline comparison. Table 1 lists the experiment results of PPGCN and baselines. We observe that our PPGCN model obtains the highest average performance among all social network datasets, which contains exceptionally dense graphs and rich node interaction information. Among the baseline methods, the kernel algorithm WL also performs quite well, achieving the second-best accuracy on the all social network datasets. But the other deep learning methods are all behind the PPGCN and WL.

Besides, in the biochemical dataset MUTAG, PPGCN also achieves the best performance. In PTC dataset, the accuracy of PPGCN is slightly behind PatchySan algorithm and ranks second. However, the PPGCN does not perform well on larger biochemical dataset - PROTEINS. PPGCN is not so effective in some biochemical datasets, probably because the edge features of biochemical datasets are more complex (e.g. molecular bonds), and may need to introduce

Table 1. Classification accuracies in percent, and the standard deviation (behind ±).

	Method	Data set					
		MUTAG	PTC	PROTEINS	IMDB-B	IMDB-M	COLLAB
Kernel	GK [5,8,13]	81.58 ± 2.11	57.26 ± 1.41	71.67 ± 0.55	65.87 ± 0.98	43.89 ± 0.38	72.84 ± 0.28
	WL [5,8,13]	80.72 ± 3.00	57.97 ± 0.49	74.68 ± 0.49	73.40 ± 4.63	49.33 ± 4.75	79.02 ± 1.77
	SP [8,13]	85.79 ± 2.51	58.24 ± 2.44	75.07 ± 0.54	–	–	–
	RW [8,13]	83.68 ± 1.66	57.85 ± 1.30	74.22 ± 0.42	–	–	–
GNN	PatchySan [8,13]	88.90 ± 4.37	**62.29 ± 5.68**	75.00 ± 2.51	71.00 ± 2.29	45.23 ± 2.84	72.60 ± 2.15
	DGK [8,13]	82.66 ± 1.45	60.08 ± 2.55	**75.68 ± 0.54**	66.96 ± 0.56	44.55 ± 0.52	73.09 ± 0.25
	DCNN [8,13]	66.98	56.60 ± 2.89	61.29 ± 1.60	49.06 ± 1.37	33.49 ± 1.42	52.11 ± 0.71
	DGCNN [8,13]	85.83 ± 1.66	58.59 ± 2.47	75.54 ± 0.94	70.03 ± 0.86	47.83 ± 0.85	73.76 ± 0.49
	EdgeConv-sum	69.02 ± 7.08	56.99 ± 1.99	59.57 ± 0.17	54.40 ± 8.98	34.00 ± 2.00	–
	GAT-sum	84.59 ± 6.30	58.14 ± 3.16	71.15 ± 7.12	72.50 ± 3.58	46.87 ± 3.17	71.84 ± 4.09
	PPGCN	**90.51 ± 4.81**	59.62 ± 5.42	72.30 ± 4.06	**73.70 ± 3.65**	**51.20 ± 3.54**	**80.82 ± 1.37**

additional labels and other attribute information of edges. But compared with other deep learning methods, PPGCN still shows comparable performance.

5 Conclusion

In this paper, we first analyze the feature redundancy problem that may occur in the message passing process of the traditional GCNs. In order to solve this problem, we propose a novel graph convolution named Push-and-Pull Convolution (PPC), which aims to generate and aggregate informative messages in the message passing phase. Further more, a new model named Push-and-Pull Graph Convolutional Network (PPGCN) is proposed for graph classification task.

In the future, we will consider the heterogeneous graph with additional edge labels and other attributes information. Furthermore, as a node embedding method, PPC also can be applied to other graph learning tasks, such as node classification task, link predicition task and etc.

Acknowledgement. This research is partially supported by the National Natural Science Foundation of China (Grant No. 61562041, Grant No. 61866018); Jiangsu Provincial Natural Science Foundation of China (Grant No. BK20171447); Jiangsu Provincial University Natural Science Research of China (Grant No. 17KJB520024).

References

1. Atwood, J., Towsley, D.: Diffusion-convolutional neural networks. In: Advances in Neural Information Processing Systems, pp. 1993–2001 (2016)
2. Borgwardt, K.M., Kriegel, H.P.: Shortest-path kernels on graphs. In: Fifth IEEE International Conference on Data Mining (ICDM 2005), pp. 8. IEEE (2005)
3. Fey, M., Lenssen, J.E.: Fast graph representation learning with pytorch geometric. arXiv preprint arXiv:1903.02428 (2019)
4. Gilmer, J., Schoenholz, S.S., Riley, P.F., Vinyals, O., Dahl, G.E.: Neural message passing for quantum chemistry. In: Proceedings of the 34th International Conference on Machine Learning, pp. 1263–1272 (2017)

5. Ivanov, S., Burnaev, E.: Anonymous walk embeddings. In: Proceedings of the 35th International Conference on Machine Learning, pp. 2191–2200 (2018)
6. Karp, R., Schindelhauer, C., Shenker, S., Vocking, B.: Randomized rumor spreading. In: Proceedings 41st Annual Symposium on Foundations of Computer Science, pp. 565–574 (2000)
7. Kipf, T.N., Welling, M.: Semi-supervised classification with graph convolutional networks. arXiv preprint arXiv:1609.02907 (2016)
8. Mallea, M.D.G., Meltzer, P., Bentley, P.J.: Capsule neural networks for graph classification using explicit tensorial graph representations. arXiv preprint arXiv:1902.08399 (2019)
9. Niepert, M., Ahmed, M., Kutzkov, K.: Learning convolutional neural networks for graphs. In: International Conference on Machine Learning, pp. 2014–2023 (2016)
10. Shervashidze, N., Schweitzer, P., Leeuwen, E.J.V., Mehlhorn, K., Borgwardt, K.M.: Weisfeiler-Lehman graph kernels. J. Mach. Learn. Res. **12**, 2539–2561 (2011)
11. Shervashidze, N., Vishwanathan, S., Petri, T., Mehlhorn, K., Borgwardt, K.: Efficient graphlet kernels for large graph comparison. In: Artificial Intelligence and Statistics, pp. 488–495 (2009)
12. Veličković, P., Cucurull, G., Casanova, A., Romero, A., Lio, P., Bengio, Y.: Graph attention networks. arXiv preprint arXiv:1710.10903 (2017)
13. Verma, S., Zhang, Z.L.: Graph capsule convolutional neural networks. arXiv preprint arXiv:1805.08090 (2018)
14. Vishwanathan, S.V.N., Schraudolph, N.N., Kondor, R., Borgwardt, K.M.: Graph kernels. J. Mach. Learn. Res. **11**, 1201–1242 (2010)
15. Wang, Y., Sun, Y., Liu, Z., Sarma, S.E., Bronstein, M.M., Solomon, J.M.: Dynamic graph CNN for learning on point clouds. arXiv preprint arXiv:1801.07829 (2018)
16. Weisfeiler, B., Lehman, A.A.: A reduction of a graph to a canonical form and an algebra arising during this reduction. Nauchno-Technicheskaya Informatsia **2**(9), 12–16 (1968)
17. Xu, K., Li, C., Tian, Y., Sonobe, T., Kawarabayashi, K.I., Jegelka, S.: Representation learning on graphs with jumping knowledge networks. arXiv preprint arXiv:1806.03536 (2018)
18. Yanardag, P., Vishwanathan, S.V.N.: Deep graph kernels. In: Proceedings of the 21th ACM SIGKDD International Conference on Knowledge Discovery and Data Mining, pp. 1365–1374 (2015)
19. Zhang, M., Cui, Z., Neumann, M., Chen, Y.: An end-to-end deep learning architecture for graph classification. In: Proceedings of the Thirty-Second AAAI Conference on Artificial Intelligence, pp. 4438–4445 (2018)

Multi-task Temporal Convolutional Network for Predicting Water Quality Sensor Data

Yi-Fan Zhang[1(✉)], Peter J. Thorburn[1], and Peter Fitch[2]

[1] Agriculture and Food, CSIRO, Brisbane, QLD 4067, Australia
{Yi-Fan.Zhang,Peter.Thorburn}@csiro.au
[2] Land and Water, CSIRO, Canberra, ACT 2601, Australia
Fitch.Peter@csiro.au

Abstract. Predicting the trend of water quality is essential in environmental management decision support systems. Despite various data-driven models in water quality prediction, most studies focus on predicting a single water quality variable. When multiple water quality variables need to be estimated, preparing several data-driven models may require unaffordable computing resources. Also, the changing patterns of several water quality variables can only be revealed by processing long term historical observations, which is not well supported by conventional data-driven models. In this paper, we propose a multi-task temporal convolution network (MTCN) for predicting multiple water quality variables. The temporal convolution offers one the capability to explore the temporal dependencies among a remarkably long historical period. Furthermore, instead of providing predictions for only one water quality variable, the MTCN is designed to predict multiple water quality variables simultaneously. Data collected from the Burnett River, Queensland is used to evaluate the MTCN. Compared to training a set of single-task TCNs for each variable separately, the proposed MTCN achieves the best RMSE scores in predicting both temperature and DO in the following 48 time steps but only requires 53% of the total training time of the TCN. Therefore, the MTCN is an encouraging approach for water quality management by processing a large amount of sensor data.

Keywords: Prediction model · Multi-task learning · Water quality

1 Introduction

Water quality is one of the major issues today because of its effects on human health and aquatic ecosystems. The water quality deterioration can be attributed to urbanisation, population growth, excessive water consumption, industrial wastewater discharge, and agricultural activities in the catchments [5]. An understanding of water quality dynamics is critical to the intelligent decision making in regards to ecological conservation [12, 13].

T. Gedeon et al. (Eds.): ICONIP 2019, CCIS 1142, pp. 122–130, 2019.
https://doi.org/10.1007/978-3-030-36808-1_14

Capturing long-term dependencies in time series data remains a fundamental challenge [17,18]. Despite advances in building models based on recurrent neural networks (RNNs), those models are still difficult to scale to very long data sequences. In the study proposed by Wang et al. [15], the maximum number of historical time steps used in their dissolved oxygen predictive model is five. Moon et al. [11] proposed an RNN-based model in forecasting electrical conductivity. Their model achieved the best performance when processing the inputs from 24 previous timesteps. Inputs with a small number of timesteps limits these RNN-based predictive models in identifying the long-range changing patterns, which is critical in numerous water quality variables.

Temporal Convolutional Networks (TCNs) overcome the previous shortcomings by capturing long-range patterns using a hierarchy of temporal convolutional filters [7]. Instead of using recurrent structure to maintain temporal dependencies, the TCN applies various sizes of convolutional filters to obtain the temporal dependencies at different time scale. Also, the dilated convolutions [14] increase the receptive field significantly so long historical data can be utilised.

Furthermore, most water quality researchers build predictive models for single water quality variable. For example, Alizadeh et al. [2] applied 30 different artificial neural network (ANN) models to predict daily values of salinity, temperature and DO separately. In the study conducted by Kim and Seo [6], an ANN ensemble model was developed to forecast the water quality variables such as pH, DO, turbidity, total nitrogen and phosphorus. The ANN ensemble model included 150 individual ANN models, with each of them needing to be trained and evaluated. In these studies, though all the models indeed deal with the same datasets, they cannot obtain benefits from each other's learning process.

Multi-task learning is an essential machine learning paradigm which aims at improving the generalisation performance of a task by using other related tasks [8]. It is prevalent in various applications ranging from computer vision [10] to speech recognition [4]. In the context of water quality prediction, each water quality variable interacts with and influences other variables in the same ecosystem. The temporal patterns of one water quality variable can, therefore, be precious in guiding us predicting other water constituents' values.

In this paper, we propose a multi-task temporal convolution network (MTCN) for predicting multiple water quality variables. The key contributions include:

- We develop a multi-variable predictive model to forecast various water quality constituents simultaneously. Applying a unified model in predicting multiple variables enables the knowledge sharing between multiple learning processes, and also reduces the necessities of computing resources significantly.
- We applied the temporal convolution network (TCN) to learn the long-term temporal dependencies for water quality data. Comparing to the RNN-based models, the TCN exhibits longer effective history data than the recurrent counterparts. The experiments demonstrate that the MTCN can obtain superior performance compared to equivalent separately trained models.

2 Proposed Multi-task Temporal Convolution Network

In this section, we propose a water quality multi-variable predictive model for forecasting various water quality constituents simultaneously. A TCN-based predictive model is built by following the multi-task learning paradigm. The model is designed to learn the temporal dependencies among various water quality monitoring properties within a long period of time. Each predictive task can benefit from the shared hidden representations. Moreover, task-specific layers are assigned to forecast the corresponded water quality variable concurrently.

We implement a temporal convolutional network similar to the one proposed by Bai et al. [3]. The TCN includes a stack of causal convolutional layers. Causal convolution is used to make sure the model will not capture information from the future time index to help the prediction task. In addition, the dilated convolutions and the residual connections are integrated into the TCN to enhance the utilization of long historical observations without the vastly deep structure.

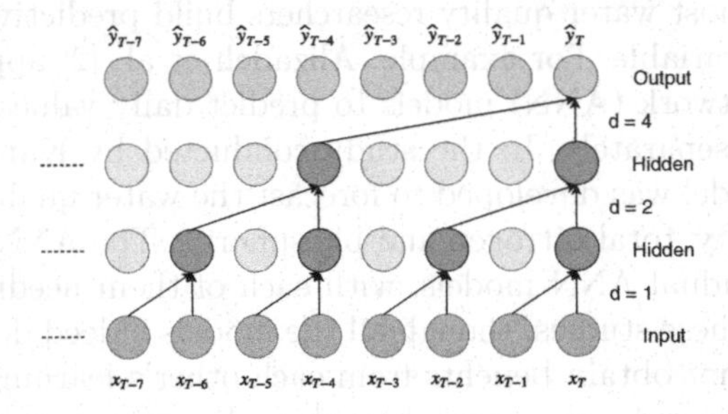

Fig. 1. The TCN with dilated convolutions. The dilated causal convolution is defined with dilation factors $d = [1, 2, 4]$, and filter size $k = 2$. In this case, the TCN is able to cover 8 numbers of historical observations.

2.1 Dilated Convolution

Figure 1 illustrates the way of applying dilated convolutions to increase the size of the receptive field. The dilated convolution operator can apply the same filter at different time scales using different dilation factors. 1D dilated convolution is defined as:

$$g[i] = \sum_{l=1}^{L} f[i + d \cdot l]h[l], \tag{1}$$

where $f[i]$ and $g[i]$ are the input and output time series, $h[l]$ denotes the filter of length L and d corresponds to the dilation rate.

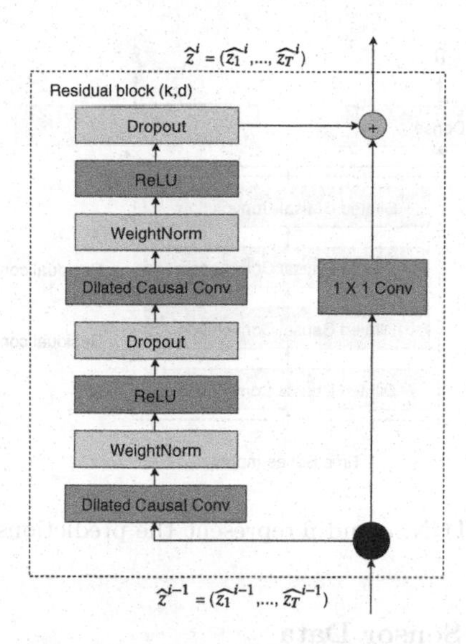

Fig. 2. Residual unit in the TCN. With the help of the skip connection within the residual unit, one can propagate larger gradients through the neural network.

2.2 Residual Unit

A residual unit defined in Bai et al.'s study [3] is implemented to improve the TCN's stability. The residual block (Fig. 2) includes two dilated causal convolutional layers. The weight normalization is applied to the convolutional filters and a spatial drop out is added after each dilated convolution for regularization. In addition, the input of the residual unit is added to the output through an additional 1×1 convolution.

2.3 Multi-task Temporal Convolution Network

In this subsection, we developed our multi-task temporal convolution network (MTCN) based on the TCN and the multi-task learning paradigm.

The proposed MTCN is illustrated in Fig. 3. By adjusting the dilation factors and filter size, the MTCN can cover a wide range of time series data by applying a hierarchy of filters with various sizes. In addition, the residual connections help to maintain the stability of the deep neural network by enhancing the information flow through the initial layer to the last layer in the deep neural network. The task-specific dense layers with the linear activation function are added on top of the shared convolutional layers.

3 Evaluation

In this section, we evaluate the effectiveness of the MTCN by using the water quality data collected by a water quality monitoring program in Australia.

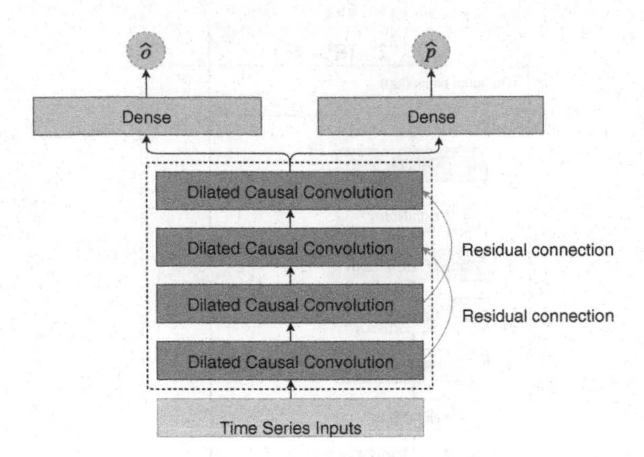

Fig. 3. The proposed MTCN. $\hat{o}$ and $\hat{p}$ represent the predictions of different variables.

3.1 Water Quality Sensor Data

The Burnett River is located on the southern Queensland coast and flows into the coral sea of the South Pacific Ocean. Cultivation of sugarcane and small crops are important lands uses in this region. A YSI model 6-Series Sonde is deployed in the river monitoring the water quality [1]. Temperature, electric conductivity (EC), pH, dissolved oxygen (DO), turbidity and chlorophyll-a (Chl-a) are recorded with half an hour time interval (Table 1).

Table 1. Water quality data during 1/3/2014 and 31/3/2018.

Variables	Unit	Min	Max	Mean	Std Dev
Temperature	°C	13.60	32.71	24.64	3.96
Electrical conductivity	μS/cm	2	50720	35931.09	14428.10
pH		6.62	8.63	7.85	0.63
Dissolved oxygen	mg L^{-1}	2.06	13.90	6.64	0.98
Turbidity	NTU	0.1	1850	19.85	87.18
Chlorophyll-a	μg L^{-1}	0.1	345.60	10.07	31.89

We choose the sensor data from 1/3/2014–31/3/2017 as training data and sensor data from 1/4/2017–31/3/2018 as testing data. During the training, 10% samples are selected as validation data. Considering the missing and abnormal measurements are inevitable in the monitoring network, we cleaned and normalized the chosen datasets first before feeding into the neural network models.

Beside this, studies [9,16] confirm that the concentration of DO in surface water is controlled by temperature and has both a seasonal and a daily cycle. We designed an MTCN to predict DO and temperature simultaneously. Two comparative TCNs were also designed to forecast the DO concentration and temperature separately.

3.2 Experimental Settings

To measure the performance of the predictive model, we used the mean absolute error (MAE) and the root mean square error (RMSE). Also, some of the optimised key hyperparameters are listed in Table 2.

Table 2. Key hyperparameters of the MTCN.

Hyperparameters	Value
No. of dense layers (per task)	2
No. of units in dense layers (per task)	[64, 48]
Dilated factors	[1, 2, 4, 8, 16, 32, 64]
Kernel size	3
Dropout rate	0.6

Based on the dilated factors and filter's kernel size (Table 2), the MTCN can cover 192 historical observations for predicting both the temperature and dissolved oxygen values in the future 48 time index. According to this experimental design, the MTCN is able to forecast the changing of the temperature and dissolved oxygen in the following 24 h.

3.3 Experimental Results and Discussion

We also compare the MTCN with the single-task TCN. The single-task TCN shares the same hyperparameter setting with MTCN, while it does not have the task-specific dense layer and multiple outputs. Hence, multiple TCNs have to be trained to meet the requirements of multi-variables prediction.

Table 3. Performance measurement.

Model	Metrics	Prediction accuracy		Training time
		Temperature	DO	
MTCN	RMSE	**0.59**	**0.49**	**9H:58M**
	MAE	**0.37**	0.27	
TCN	RMSE	0.60	0.49	18H:49M
	MAE	0.38	**0.26**	

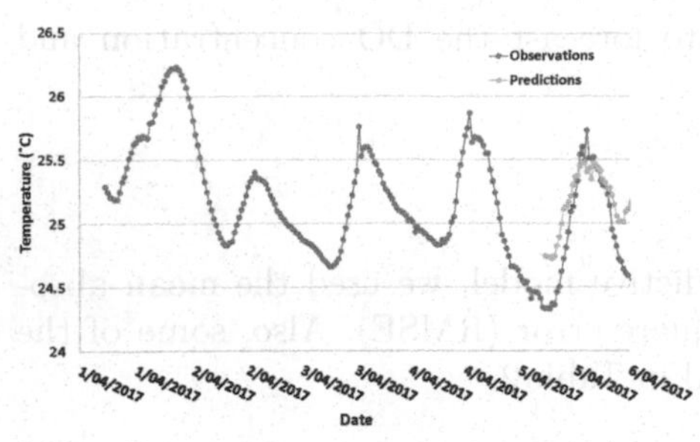

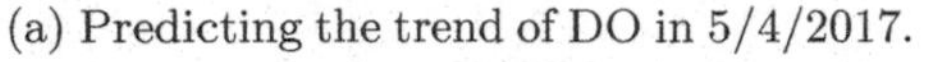
(a) Predicting the trend of DO in 5/4/2017.

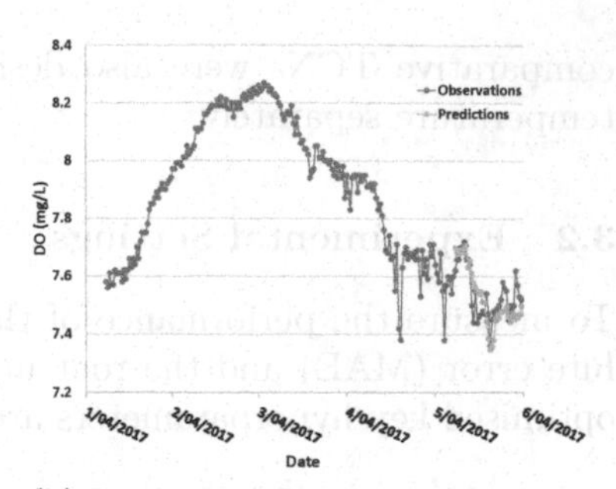

(b) Predicting the trend of temperature in 5/4/2017.

Fig. 4. Predicting the trend of DO and temperature by using the MTCN. 48 predictions are generated every half an hour based on historical inputs data.

Table 3 illustrates model performance for both MTCN and TCN. Benefiting from the temporal convolutional architecture, dilated convolution and the residual unit, both MTCN and TCN achieve remarkable predictive accuracy for both DO and temperature. As shown in Fig. 4a, the MTCN captures the trend of DO in the following 24 h, and also gives the proper estimation when the concentration of DO drops significantly. Similarly, the temperature predictions generated by the MTCN follow the expected daily temperature variation in Fig. 4b. Furthermore, the MTCN gains the best performance of both RMSE and MAE in predicting the change of temperature. Similarly, the MTCN also achieves the best RMSE scores in predicting the trend of DO.

In addition, the MTCN implemented in this experiment includes 1,304,096 trainable parameters, while the TCN with a single prediction task only has 710,064 parameters to train. To make a fair comparison, the training process was stopped after the 900th epoch for all the models listed in Table 3. Larger hidden parameters in the MTCN indicates that it requires more training epochs to converge, while the total training time is still much less than training separate TCNs for individual tasks. Therefore, the MTCN offers an efficient way in building predictive model for a number of water quality variables.

4 Conclusion

The development of reliable water quality predictions is critical to improve the management of aquatic ecosystems. This paper proposed a multi-task temporal convolutional network for predicting multiple water quality variables simultaneously. Experimental results were presented to demonstrate that the proposed model can achieve promising predictive accuracy for long term water quality prediction while requiring a significantly reduced training time.

Acknowledgement. This work was conducted within the CSIRO Digiscape Future Science Platform.

References

1. Ambient estuarine water quality monitoring data. https://data.qld.gov.au/dataset. Accessed 20 Nov 2017
2. Alizadeh, M.J., Kavianpour, M.R.: Development of wavelet-ANN models to predict water quality parameters in Hilo Bay, Pacific ocean. Mar. Pollut. Bull. **98**(1–2), 171–178 (2015)
3. Bai, S., Kolter, J.Z., Koltun, V.: An empirical evaluation of generic convolutional and recurrent networks for sequence modeling. arXiv preprint arXiv:1803.01271 (2018)
4. Huang, J.T., Li, J., Yu, D., Deng, L., Gong, Y.: Cross-language knowledge transfer using multilingual deep neural network with shared hidden layers. In: 2013 IEEE International Conference on Acoustics, Speech and Signal Processing (ICASSP), pp. 7304–7308. IEEE (2013)
5. Ji, X., Shang, X., Dahlgren, R.A., Zhang, M.: Prediction of dissolved oxygen concentration in hypoxic river systems using support vector machine: a case study of Wen-Rui Tang River, China. Environ. Sci. Pollut. Res. **24**(19), 16062–16076 (2017)
6. Kim, S.E., Seo, I.W.: Artificial neural network ensemble modeling with conjunctive data clustering for water quality prediction in rivers. J. Hydro-Environ. Res. **9**(3), 325–339 (2015)
7. Lea, C., Flynn, M.D., Vidal, R., Reiter, A., Hager, G.D.: Temporal convolutional networks for action segmentation and detection. In: 2017 IEEE Conference on Computer Vision and Pattern Recognition (CVPR), pp. 1003–1012. IEEE (2017)
8. Luong, M.T., Le, Q.V., Sutskever, I., Vinyals, O., Kaiser, L.: Multi-task sequence to sequence learning. arXiv preprint arXiv:1511.06114 (2015)
9. Manasrah, R., Raheed, M., Badran, M.I.: Relationships between watertemperature, nutrients and dissolved oxygen in the northern Gulf of Aqaba, Red Sea. Oceanologia **48**(2), 237–253 (2006)
10. Misra, I., Shrivastava, A., Gupta, A., Hebert, M.: Cross-stitch networks for multi-task learning. In: Proceedings of the IEEE Conference on Computer Vision and Pattern Recognition, pp. 3994–4003 (2016)
11. Moon, T., Ahn, T.I., Son, J.E.: Forecasting root-zone electrical conductivity of nutrient solutions in closed-loop soilless cultures via a recurrent neuralnetwork using environmental and cultivation information. Front. Plant Sci. **9**, 859 (2018)
12. Thorburn, P.J., et al.: Helping farmers mitigate nutrient losses to the great barrier reef through "digital agriculture". Occasional report, Fertiliser and Lime Research Centre, Massey University, 32, 6 (2019)
13. Thorburn, P., Wilkinson, S.: Conceptual frameworks for estimating the water quality benefits of improved agricultural management practices in large catchments. Agric. Ecosyst. Environ. **180**, 192–209 (2013)
14. Van Den Oord, A., et al.: Wavenet: a generative model for raw audio. CoRR abs/1609.03499 (2016)
15. Wang, Y., Zhou, J., Chen, K., Wang, Y., Liu, L.: Water quality prediction method based on LSTM neural network. In: 12th International Conference on Intelligent Systems and Knowledge Engineering (ISKE), pp. 1–5. IEEE (2017)
16. Zhang, Y., Fitch, P., Vilas, M.P., Thorburn, P.J.: Applying multi-layer artificial neural network and mutual information to the prediction of trends in dissolved oxygen. Front. Environ. Sci. **7**, 46 (2019)

17. Zhang, Y., Fitch, P., Vilas, M.P., Thorburn, P.J.: Predicting the trend of dissolved oxygen based on kPCA-RNN model in water quality monitoring. Water (2019, submitted)
18. Zhang, Y., Thorburn, P.J., Wei, X., Fitch, P.: SSIM - a deep learning approach for recovering missing time series sensor data. IEEE Internet Things J. **6**(4), 6618–6628 (2019)

CNN-LSTM Neural Networks for Anomalous Database Intrusion Detection in RBAC-Administered Model

Tae-Young Kim and Sung-Bae Cho(✉)

Department of Computer Science, Yonsei University, Seoul 03722, South Korea
{taeyoungkim, sbcho}@yonsei.ac.kr

Abstract. The relational database is designed to store and process large amount of information such as business records and personal data. There are many policies and access control techniques for database security, but they are not sufficient for detecting insider attacks. In order to detect threats for the database application, it is necessary to adopt role-based access control (RBAC) and classify the roles according to the authority of each user. In this paper, we propose a method of classifying user's role and authority using the CNN-LSTM neural networks by extracting features from SQL queries. In the anomaly detection method, CNN automatically extracts important features from database query and LSTM models the temporal information of the SQL sequence. The class activation map also identifies the SQL query features that affect the classification. Experiments with the TPC-E scenario-based benchmark query dataset show that the CNN-LSTM neural networks surpass other state-of-the-art machine learning methods, achieving an overall accuracy of 93.3% and recall of 88.7%. We also identify the characteristics of misclassification data through statistical analysis.

Keywords: Deep learning · Convolutional neural network · Long short-term memory · Database security · Access control

1 Introduction

The relational database management system (RDBMS) is the most popular for storing information of the company. It is based on a relational database model. Many companies require high security of the RDBMS because they store confidential information in the database for long periods of time [1]. Especially as the size of the company grows, many employees access the database. Therefore, system access should be controlled according to their authority. Role-based access control (RBAC) is the way to restrict database access based on the role of individual users within an enterprise [2]. RBAC allows employees to grant access only to the information they need to perform their tasks and prevent access to unrelated information [3].

Database intrusion detection systems identify and report access to unauthorized users by insider attacks through query patterns. Intrusion detection systems must accurately determine if a user's role matches to the database security [4]. In this paper, we use the RBAC in RDBMS, and classify roles according to user's SQL query based

T. Gedeon et al. (Eds.): ICONIP 2019, CCIS 1142, pp. 131–139, 2019.
https://doi.org/10.1007/978-3-030-36808-1_15

on TPC-E benchmark. This benchmark is gathered using the RBAC schema. The TPC-E benchmark is an online transaction processing (OLTP) workload from a brokerage firm [5]. There is a total of 11 roles in the TPC-E benchmark. User roles consist of brokers, customers, market transactions, and so on. Table 1 shows a virtual scenario of a TPC-E dataset generated by online transaction simulation. Figure 1(a) represents the connections among brokers, customers, and market transactions that make up the TPC-E benchmark schema. We preprocess SQL queries and extract features according to each role. Each feature represents the number of query elements that make up the select, from, where, order by, and group by clauses. Figure 1(b) shows the complex distribution of SQL query features according to the TPC-E benchmark role. Each role is difficult to classify because of the overlap of similar distributions.

Table 1. TPC-E benchmark scenario

Role	Transaction	Data manipulation	Authority
1	Broker-Volume	Select Only	Read-Only
2	Customer-Position		
3	Market-Watch		
4	Security-Detail		
5	Trade-Status		
6	Trade-Lookup		
7	Trade-Order	Select/Insert Only	Read/Write
8	Trade-Update	Select/Update Only	
9	Data-Maintenance		
10	Market-Feed	Select/Insert/Update/Delete	
11	Trade-Result		

In this paper, we propose CNN-LSTM networks that combine CNN and LSTM to extract features from relational database queries and perform intrusion detection. RBAC-based access control represents abnormal queries using 11 roles of the database query. The proposed CNN-LSTM networks transform preprocessed database queries

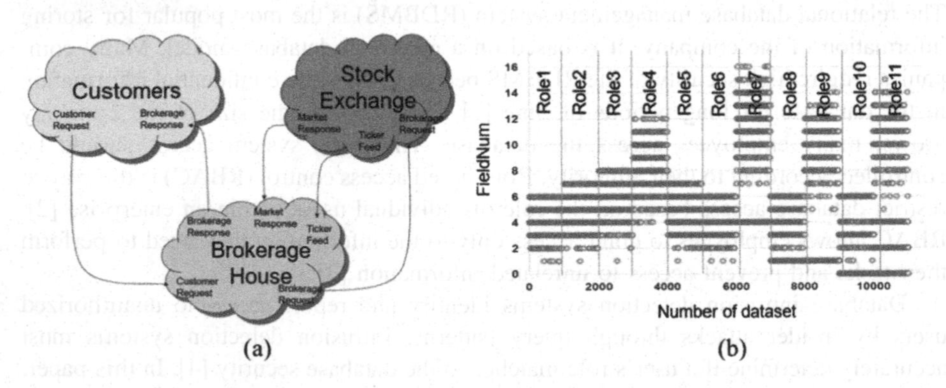

Fig. 1. TPC-E benchmark schema and distribution of SQL query data

using multiple CNN layers to reduce spectrum. The output of this CNN layer is used as input to the LSTM layer to model the sequence information among features. We can easily classify the role by mapping the function to a separate space. We also use the class activation map to identify the SQL query elements that affect the classification.

2 Related Works

In Table 2, there are many researchers to extract features from network packets or database queries and perform intrusion detection. Intrusion detection systems have been actively studied in the field of network systems, but these settings are inadequate for database security. Intrusion detection systems typically have three categories: statistical modeling, machine learning modeling, and neural network modeling.

Table 2. Related works on intrusion detection system

Category	Author	Year	Data	Method	Description
Statistical modeling	Ramachandran *et al.* [6]	2017	Database query	DBSCAN	Clustering using machine learning
	Kumar *et al.* [7]	2015	TCP/IP packet	k-means clustering	Gaussian similarity measure
	Horng *et al.* [8]	2011	TCP/IP packet	Hierarchical clustering	Feature selection using hierarchical clustering
Machine learning modeling	Ronao *et al.* [9]	2016	Database query	Random forest	Using weighted voting and PCA
	Rai *et al.* [10]	2016	TCP/IP packet	Decision tree	Improving the performance of DT
	Mulay *et al.* [11]	2010	TCP/IP packet	Decision tree	Decision tree and SVM integrated model
Neural network modeling	Kim *et al.* [12]	2017	Database query	Convolutional neural network	Deep learning based on learning classifier
	Qiu *et al.* [13]	2015	TCP/IP packet	BP neural network	BP neural network performance improvement
	Devaraju *et al.* [14]	2013	TCP/IP packet	Multi-layer perceptron	Intrusion detection using five classifiers

Horng *et al.* proposed intrusion detection system by extracting features from network traffic by combining hierarchical clustering algorithms with classification models [8]. Statistical modeling is the technique that identifies the characteristics of data with simple sampling. But as the amount of data should be sufficient, it is difficult to improve the performance. Mulay *et al.* performed intrusion detection in the TCP/IP packet using a model that incorporates a decision tree and SVM [11]. Machine learning methods can interpret variables that influence database intrusion detection. However, they make the simple decision boundary. It has the disadvantage of modeling discrete data rather than continuous data, and structurally converges to local optima, resulting in low performance. Devaraju and Ramakrishnan also performed intrusion detection in TCP/IP communication using multi-layer perceptron [14]. Neural network modeling generates more complex decision boundary than other machine learning techniques. However, overfitting problem occurs because of the slow learning and the difficulty of finding optimal parameter values. It is also difficult to model the spatial and temporal features of TCP/IP packet or database queries.

3 The Proposed CNN-LSTM Neural Networks

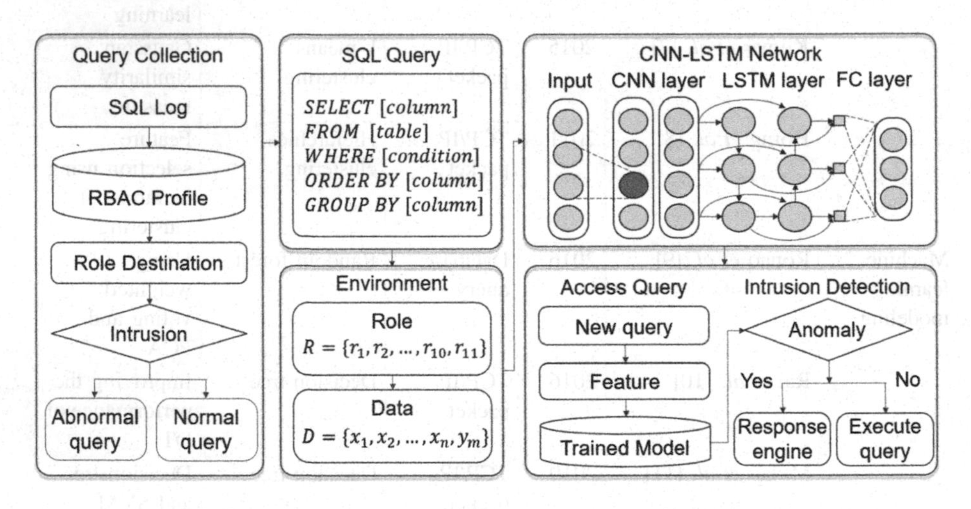

Fig. 2. The proposed CNN-LSTM intrusion detection structure

In order to learn the CNN-LSTM networks for classifying authority in DBMS, we need to understand the features of SQL query [15, 16]. Figure 2 shows the CNN-LSTM intrusion detection architecture for database access control. Input features extracted from the SQL query are composed of elements (SELECT, FROM, WHERE, GROUP BY, ORDER BY). We preprocess the clause elements of the SQL query to create a total of 277 feature vectors. We use the generated feature vectors as inputs to our proposed CNN-LSTM intrusion detection model. Figure 3 represents an example of the parsed features extracted from SQL queries.

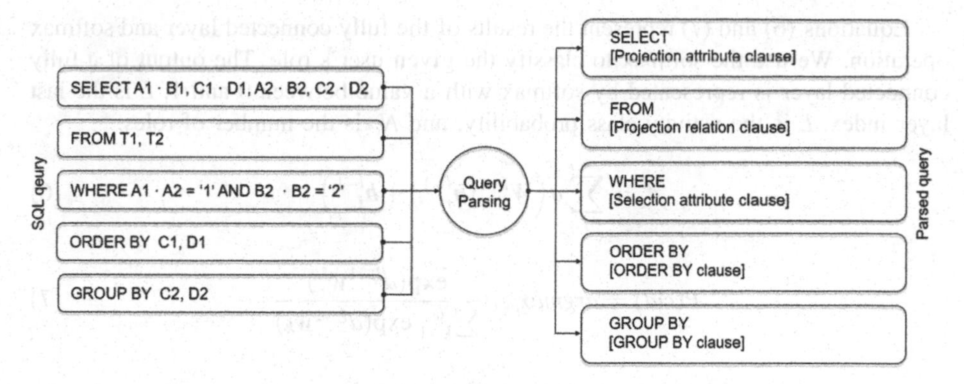

Fig. 3. An example of extracting parsed query from SQL queries

The CNN-LSTM network takes 277 parsed query log feature as input. First, we use CNN to extract features that have an important effect on role classification in parsed queries. Equation (1) represents the operation of l, the convolutional network. It consists of several m_1^{l-1} feature maps. The size of each feature map is $m_2^{l-1} \times m_3^{l-1}$. The ith feature map is represented by $\boldsymbol{Y}_i^l$. $\boldsymbol{B}_i^l$ represents a bias matrix. $\boldsymbol{K}_{i,j}^l$ represents a filter connecting the ith feature map of layer l and the jth feature map in layer $l-1$. Equation (2) represents the pooling layer that reduces the size of parsed query. It also increases computational efficiency. R represents the pooling size and T represents how much it strides the area. We adjust the settings of pooling according to performance.

$$\boldsymbol{y}_i^l = \boldsymbol{B}_i^l + \sum_{j=1}^{m_1^{l-1}} \boldsymbol{K}_{i,j}^l * \boldsymbol{Y}_j^{l-1} \tag{1}$$

$$\boldsymbol{p}_{ij}^l = \max_{r \in R} \boldsymbol{y}_{i \times T + r, j}^{l-1} \tag{2}$$

The LSTM layer uses memory cells to store the temporal sequence of the query feature vectors. It utilizes input $\boldsymbol{i}$, output $\boldsymbol{o}$ and forget $\boldsymbol{f}$ gate to efficiently store SQL query feature. It also controls the flow of data using hidden state $\boldsymbol{h}$ and cell states $\boldsymbol{c}$ for storage. Equations (3), (4) and (5) represent equations for calculating LSTM output.

$$\begin{pmatrix} \boldsymbol{i} \\ \boldsymbol{f} \\ \boldsymbol{o} \\ \boldsymbol{g} \end{pmatrix} = \begin{pmatrix} sigmoid \\ sigmoid \\ sigmoid \\ tanh \end{pmatrix} \boldsymbol{w}^l \begin{pmatrix} \boldsymbol{h}_t^{l-1} \\ \boldsymbol{h}_{t-1}^l \end{pmatrix} + \begin{pmatrix} \boldsymbol{b}_i \\ \boldsymbol{b}_f \\ \boldsymbol{b}_o \\ \boldsymbol{b}_c \end{pmatrix} \tag{3}$$

$$\boldsymbol{c}_t = \boldsymbol{f}_t \circ \boldsymbol{c}_{t-1} + \boldsymbol{i}_t \circ \boldsymbol{g} \tag{4}$$

$$\boldsymbol{h}_t = o_t \circ \sigma(c_t) \tag{5}$$

Equations (6) and (7) represent the results of the fully connected layer and softmax operation. We use the softmax to classify the given user's role. The output of a fully connected layer is represented by softmax with a value between 0 and 1. L is the last layer index, L is the activity class probability, and N_c is the number of roles.

$$d_i^l = \sum_j \sigma\left(W_{ji}^{l-1}\left(h_i^{l-1}\right) + b_i^{l-1}\right) \tag{6}$$

$$P(c|d) = argmax_{c \in C} \frac{\exp(d^{L-1} w^L)}{\sum_{k=1}^{N_c} \exp(d^{L-1} w_k)} \tag{7}$$

4 Experimental Results

4.1 TPC-E Benchmark Dataset

To evaluate the anomaly query classification in RBAC-based DBMS, we use the TPC-E benchmark, which simulates the online transaction processing (OLTP) workload. We have adopted standard transactions that correspond to the 11 roles. Each role consists of customer, broker, market, and so on. It also contains read-only and read/write transactions. To verify the proposed CNN-LSTM networks, we generated 11,000 SQL query data for 11 labels. It consists of 33 tables and 191 attributes. The SQL query vector is preprocessed for anomaly detection.

4.2 Performance Comparison

For the performance evaluation of the classifier, 10-fold cross validation is used. The proposed method achieves the best performance compared to other machine learning systems, followed by the random forest, decision tree, and k-nearest neighbor (KNN). Figure 4 is a box plot showing the accuracy achieved with 10-fold cross validation.

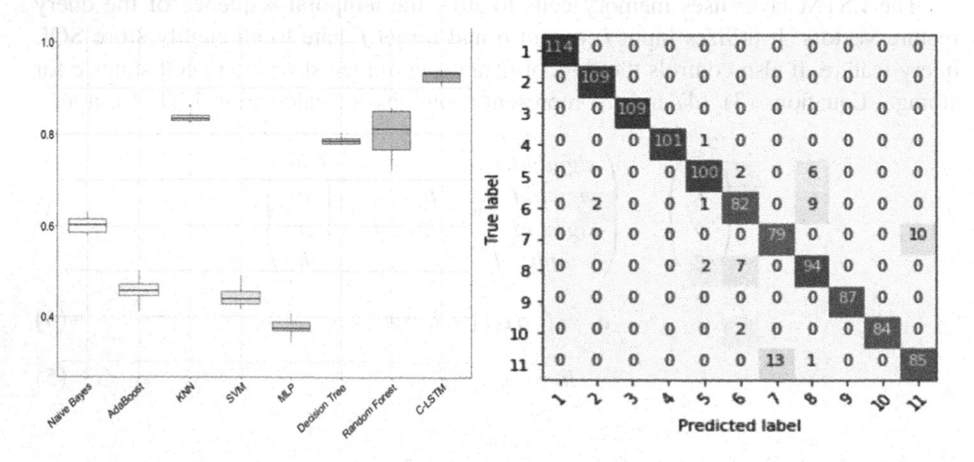

Fig. 4. Comparison of accuracy from 10-fold cross validation

4.3 Misclassification Data Analysis

The misclassification data are analyzed using the probability density function in Fig. 5. We select the features that have a large impact on the misclassification and then compare the probability density functions of the features of the classified data features with those of the misclassified data. We can see that the probability density is significantly different in several intervals of the probability density function. These features make it difficult to classify each role.

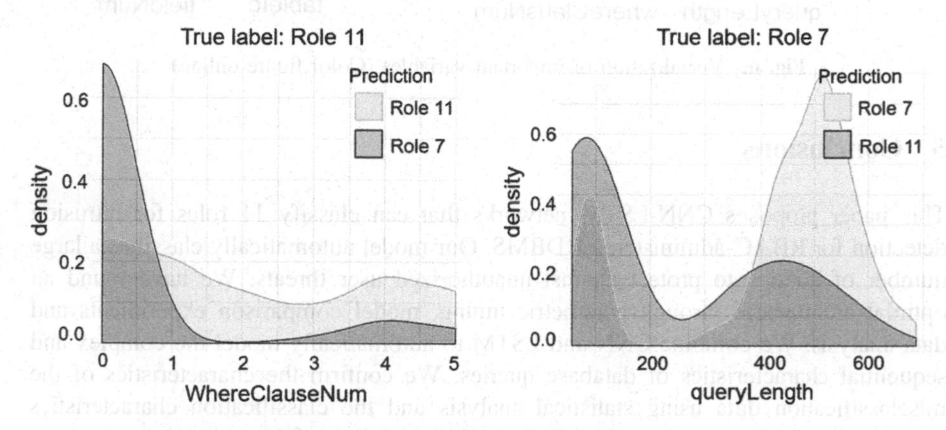

Fig. 5. Misclassification analysis using the probability density function

4.4 Analysis of Influential Variables

We can use the global average pooling layer, and shed light on influential SQL query variables. We analyze the variables that influenced the intrusion detection. A class activation map for a particular category indicates the discriminative feature regions used by the CNN to identify that role category [15]. We use a CNN-LSTM network and just before the final fully connected layer and softmax layer, we also use global average pooling on the convolutional feature maps and apply them as SQL query features. The model focuses on both queryLength and where ClausNum at the same time when misclassifying role 7 and role 11. The model focuses on tableId and fieldNum when misclassifying role 6 and role 8. The blue areas in the Fig. 6 represent variables that are important for role classification.

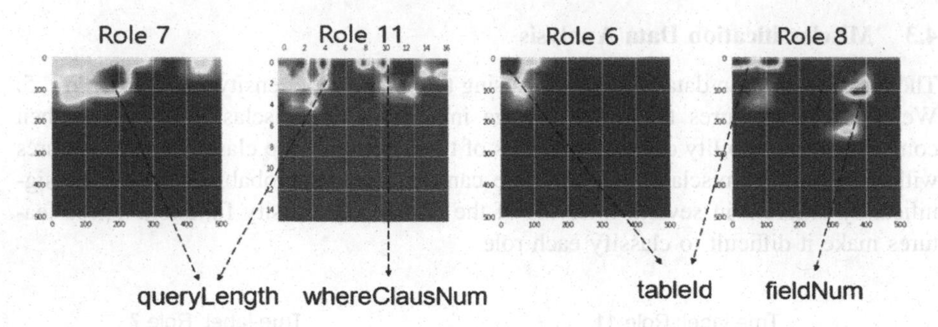

Fig. 6. Visualization of important variables (Color figure online)

5 Conclusions

This paper proposes CNN-LSTM networks that can classify 11 roles for intrusion detection for RBAC-administered RDBMS. Our model automatically classifies a large number of queries to protect against unauthorized user threats. We have found an optimal architecture through parametric tuning, model comparison experiments and data analysis. We combine CNN and LSTM to automatically model the complex and sequential characteristics of database queries. We confirm the characteristics of the misclassification data using statistical analysis and the classification characteristics using the class activation map. The CNN-LSTM model classifies and extracts the roles that could not be distinguished by using the conventional machine learning methods. However, we have manually optimized the CNN-LSTM neural network. Further research is needed to automatically find the optimal parameters of the CNN-LSTM model for intrusion detection.

Acknowledgements. This work was supported by the grant funded by 2019 IT promotion fund (Development of AI based Precision Medicine Emergency System) of the Korea government (Ministry of Science and ICT).

References

1. Bertino, E., Sandhu, R.: Database security-concepts, approaches, and challenges. IEEE Trans. Dependable Secure Comput. **1**, 2–19 (2005)
2. Ni, Q., et al.: Privacy-aware role-based access control. ACM Trans. Inf. Syst. Secur. (TISSEC) **13**(3), 24–34 (2010)
3. Li, D., Liu, C., Wei, Q., Liu, Z., Liu, B.: RBAC-based access control for SaaS systems. In: International Conference on Information Engineering and Computer Science (ICIECS), pp. 1–4 (2010)
4. Liao, H.J., Lin, C.H.R., Lin, Y.C., Tung, K.Y.: Intrusion detection system: a comprehensive review. J. Netw. Comput. Appl. **36**(1), 16–24 (2013)
5. Chen, S., et al.: TPC-E vs. TPC-C: Characterizing the new TPC-E benchmark via an I/O comparison study. ACM SIGMOD Rec. **39**(3), 5–10 (2011)

6. Ramachandran, R., Arya, P., Jayanthy, P.G.: A novel method for intrusion detection in relational databases. In: International Conference on Advances in Computing, Communications and Informatics (ICACCI), pp. 230–235 (2017)
7. Kumar, G.R., Mangathayaru, N., Narasimha, G.: An improved k-means clustering algorithm for intrusion detection using Gaussian function. In: International Conference on Engineering & MIS, pp. 69–79 (2015)
8. Horng, S.J., et al.: A novel intrusion detection system based on hierarchical clustering and support vector machines. Expert Syst. Appl. **38**(1), 306–313 (2011)
9. Ronao, C.A., Cho, S.-B.: Anomalous query access detection in RBAC-administered databases with random forest and PCA. Inf. Sci. **369**, 238–250 (2016)
10. Rai, K., Devi, M.S., Guleria, G.: Decision tree based algorithm for intrusion detection. Int. J. Adv. Netw. Appl. **7**(4), 2828–2838 (2016)
11. Mulay, S.A., Devale, P.R., Garje, G.B.: Intrusion detection system using support vector machine and decision tree. Int. J. Comput. Appl. **3**(3), 40–43 (2010)
12. Kim, J.-Y., Cho, S.-B.: Exploiting deep convolutional neural networks for a neural-based learning classifier system. Neurocomputing **354**, 61–70 (2019)
13. Qiu, C., Shan, J., Shandong, B.: Research on intrusion detection algorithm based on BP neural network. Int. J. Secur. Appl. **9**(4), 247–258 (2015)
14. Devaraju, S., Ramakrishnan, S.: Detection of accuracy for intrusion detection system using neural network classifier. Int. J. Emerg. Technol. Adv. Eng. **3**(1), 338–345 (2013)
15. Zhou, B., Khosla, A., Lapedriza, A., Oliva, A., Torralba, A.: Learning deep features for discriminative localization. In: Proceedings of the IEEE Conference on Computer Vision and Pattern Recognition (CVPR), pp. 2921–2929 (2016)
16. Kim, T.-Y., Cho, S.-B.: Particle swarm optimization-based CNN-LSTM networks for anomalous query access control in RBAC-administered model. In: Pérez García, H., Sánchez González, L., Castejón Limas, M., Quintián Pardo, H., Corchado Rodríguez, E. (eds.) HAIS 2019. LNCS (LNAI), vol. 11734, pp. 123–132. Springer, Cham (2019). https://doi.org/10.1007/978-3-030-29859-3_11

MC-HDCNN: Computing the Stereo Matching Cost with a Hybrid Dilated Convolutional Neural Network

Yunhong Liu(✉) and Yizhu Huang

School of Faculty of Electronic Information and Electrical Engineering, Dalian University of Technology, Dalian, China
lyunhong2003@aliyun.com, 136796495@qq.com

Abstract. Designing a model to quickly obtain an accurate matching cost is a vital problem in the stereo matching method. We present an algorithm called MC-HDCNN, which is based on hybrid dilated convolution neural network, for computing matching cost of two image patches. HDCNN uses the dilated convolution of the series to obtain a larger receptive field, while avoiding the "gridding" effect and ensuring the integrity of the receptive field. In addition, by adding batch normalization layer after each layer of the convolution, the gradient dispersion in the backward propagation and the generalization of the network can be improved effectively. We evaluate our method on the KITTI stereo data set. The results show that the proposed algorithm has certain advantages in accuracy and speed.

Keywords: Stereo vision · Matching cost · Similarity learning · HDC

1 Introduction

In recent years, stereo vision has been widely used in the areas of intelligent driving, robot navigation, and remote sensing measurement. Stereo matching can obtain disparity maps from stereo images, and how to efficiently obtain accurate and dense disparity maps is vital for stereo vision. Therefore, stereo matching is an important research direction for scholars.

The stereo matching algorithm mainly includes four steps: matching cost computation, cost (support) aggregation, disparity computation/optimization and disparity refinement [1]. The matching cost computation in the traditional method mainly includes (SAD), (NCC) and Census transform, etc. These algorithms have poor matching accuracy for areas where the texture is not obvious, and are susceptible to noise. With the development of deep learning, scholars began to use CNN to compute the cost of matching. In 2015, LeCun et al. [2] proposed an image patches matching method based on convolutional neural network, and proposed fast and accurate network structures. However, there are still shortcomings such as receptive field too small, low matching accuracy, and slow processing speed. Therefore, we propose the MC-HDCNN algorithm to increase the receptive field, and ensure the running speed, while the computation accuracy has a certain improvement.

T. Gedeon et al. (Eds.): ICONIP 2019, CCIS 1142, pp. 140–147, 2019.
https://doi.org/10.1007/978-3-030-36808-1_16

The contributions of this paper are:

(1) We enlarged the receptive field from 9×9 to 25×25 by using dilated convolution to replace the traditional convolution, improved the matching accuracy;
(2) In order to avoid the "gridding" effect, we used two series of hybrid dilated convolution and reasonably designed the dilation rate to ensure that all information of the receptive field is accepted;
(3) We added the corresponding batch normalization layer [3] after each convolutional layer, and improved the network training speed and generalization.
The convolutional network was trained in the KITTI2012 and KITTI2015 data sets and verified with the corresponding test sets. The results showed that the disparity map obtained by our algorithm is denser than the traditional matching method. Compared with MC-CNN, our algorithm has certain advantages in running time and matching accuracy.

2 Related Work

In this section, we introduce the traditional and CNN-based stereo matching algorithms.

The traditional algorithms mainly divided into local matching algorithm, global matching algorithm and semi-global matching algorithm between them. The local matching algorithm is computed based on the window, in order to avoid fixed-size windows blurring the edge details, Yoon et al. [4] proposed an adaptive weighting algorithm. The algorithm assigns different weights to the pixel points according to the gray and geometric distance between the pixel and the central pixel, so that the edge information is well preserved. The global matching algorithm solves the optimal disparity value by establishing and minimizing the global energy function. Dynamic programming method leads to horizontal band effect, so Lei et al. [5] proposed a dynamic programming algorithm based on tree structure to eliminate sideband effects. The graph cut method has a large computational complexity and a long running time. Kolmogorov et al. [6] added a unique constraint to the energy function to effectively reduce the amount of computation and achieved good results. The semi-global matching algorithm was proposed by Hirschmüller [7], and it uses the energy function to compute the matching cost, performs cost aggregation along different paths, and uses linear scan optimization to reduce the computational complexity.

LeCun [8] first used CNN to learn the similarity of stereo images and compute the cost of stereo matching. Lou et al. [9] proposed that regarding the computation of matching cost as a multi-classification problem, and improved the computational efficiency. Park et al. [10] proposed using the pyramid pool structure to increase the receptive field and improved the computation accuracy. However, the raw-disparity results of CNN have too much matching errors. Disparity regression using CNN integrates all steps of stereo matching into a network to form an end-to-end stereo matching convolutional neural network. Mayer et al. [11] first proposed the end-to-end network DispNet, which uses a codec structure to directly generate a disparity map. Kendall et al. [12] proposed using 3D convolution for semantic understanding to extract depth features, and performing disparity regression through a differentiable "Soft Argmin" operation to achieve disparity learning with sub-pixel precision. The EdgeStereo network proposed by Song [13] et al. consists of two sub-networks, CP-RPN and HED. The CP-RPN is responsible for generating the initial disparity map, and the HED is responsible for extracting the edge information. The outputs of the two sub-networks are fused by the residual network to obtain the final disparity map.

3 Architecture of MC-HDCNN

In this section, we propose a novel CNN architecture MC-HDCNN (Matching Cost by Hybrid Dilated Convolution Neural Network). It is an improvement based on the MC-CNN-fast algorithm and is a shared weight siamese network.

The proposed algorithm uses the dilated convolution to learn the similarity of stereo image patches, and the receptive field is nearly tripled. It can better acquire image information and improve the accuracy of disparity computation. At the same time, to ensure the continuity of the information, we divide three consecutive dilated convolution into a set of HDC, and the dilation rate of HDC is set to r = 1, 2, 3. We add two sets of HDC in our architecture. And we add a batch normalization layer after the convolutional layer to improve the training speed of the network. The network uses cosine similarity to measure the properties of the left and right input patches as the final output of the network. Figure 1 shows the architecture of MC-HDCNN.

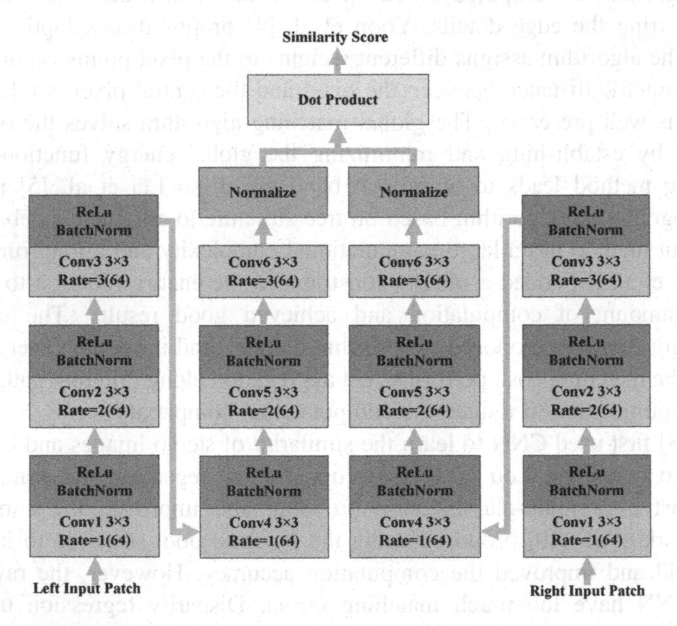

Fig. 1. Architecture of MC-HDCNN. Two sets of hybrid dilated convolution with dilation rates (1, 2 and 3) were added in the network.

3.1 Dilated Convolution

The original convolutional neural network uses four 3 × 3 convolution kernels with a stride of 1. The size of the receptive field is 9 × 9, so the received information is too small, and the mismatch rate is high. To enlarge the receptive field, there are three common methods: (1) using larger convolution kernels; (2) adding more convolution layers; (3) adding pooling layers.

However, the above methods will greatly increase the number of parameters of the CNN, reduce the efficiency of the algorithm, and the pooling layer will cause loss of feature information. To enlarge the receptive field under the premise of ensuring the running speed and computation accuracy, this paper replaces the ordinary convolution kernel with the same size dilated convolution kernel.

Dilated convolution is achieved by inserting zeros into a common convolution kernel. For a convolution kernel of size $k \times k$, when the dilation rate is r, the size of the dilated convolution kernel is $k_d \times k_d$, where $k_d = k + (k-1) \cdot (r-1)$. As shown in Fig. 2, the receptive field is enlarged from 3 × 3 to 5 × 5, when we replace the 3 × 3 convolution kernel with a dilated convolution kernel whose dilation rate is 2.

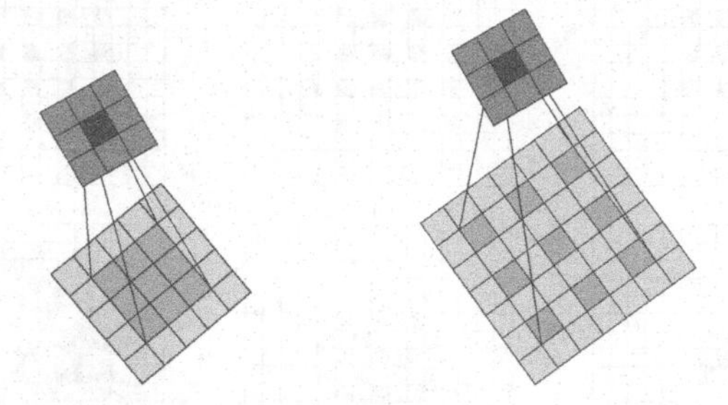

Fig. 2. The 3 × 3 convolutional layer with dilation rate 2. The receptive field is enlarged from 3 × 3 to 5 × 5.

3.2 Hybrid Dilated Convolution (HDC)

In multi-layer dilated convolution, for a pixel p in the layer n, the information conducing to it comes from a nearby $k_d \times k_d$ region in layer $n-1$ centered at p. Because the dilated convolution is used, the actual pixels taking part in the computation from the $k_d \times k_d$ region are just $k \times k$, and there will be a gap of $r-1$ between them. As shown in Fig. 3(a), if $k = 3$ and $r = 2$, then only 9 of the 25 pixels take part in the convolution operation. If the multi-layer dilated convolution uses the same dilation rate r, the top-level pixel p is affected by at most the bottom-layer $(w' \times h')/r^2$ pixels. When $r = 2$, at least 75% of the information will be lost. If a larger dilation rate is used, the actual number of pixels involved in the computation will be sparser, and the local information will be lost, resulting in the "gridding" effect (Fig. 3(a)).

In order to avoid the above problems, Wang et al. [14] proposed hybrid dilated convolution. HDC ensures that the final receptive field covers the entire area by designing reasonable dilation rate, avoiding voids or loss of edge information. Suppose there are N dilated convolution layers with dilation rates of $[r_1, \ldots, r_i, \ldots, r_n]$. And the size of the convolution kernels is $K \times K$. Define the "maximum distance between two non-zero values" as:

$$M_i = \max[M_{i+1} - 2r_i, M_{i+1} - 2(M_{i+1} - r_i), r_i] \tag{1}$$

with $M_n = r_n$. The design goal of hybrid dilated convolution is to let $M_2 \leq K$, and avoid using the same dilatation rate for each layer. Therefore, the HDC designed in this paper adopts the dilated convolution with the dilatation rates of 1, 2 and 3, as shown in Fig. 3(b). The size of the receptive field is the same as 3 convolution layers network with dilatation rate of 2. However, HDC guarantees the integrity of the receptive field and is more accurate. Another benefit of HDC is the ability to use any size of dilatation rate that meets the requirements, naturally expanding the receptive field without the need for additional modules.

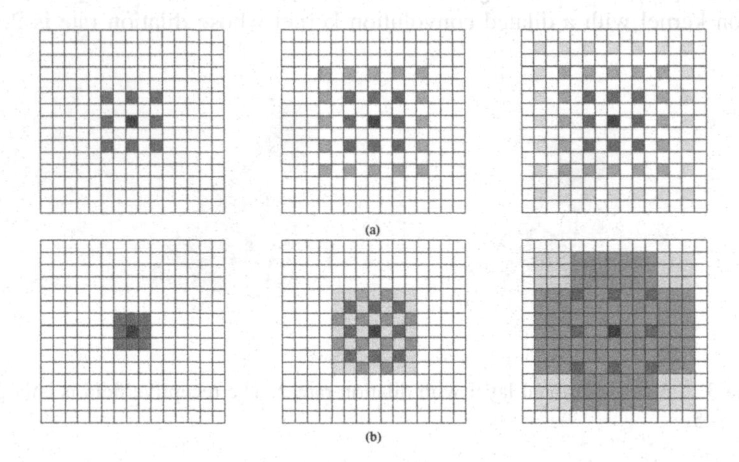

Fig. 3. Explanation of the gridding problem. (a) Three dilated convolutional layers with a dilation rate of 2. (b) Three dilated convolutional layers with the dilation rates of 1, 2, and 3.

4 Experiments and Results

We used the KITTI2012 and KITTI2015 stereo image datasets to train and verify the experimental results. 40 pairs of image data in two data sets are taken as verification sets and the remaining images are used as training sets. Considering that the color information has less influence on the algorithm of this paper, we converted all color images to grayscale.

4.1 Training Set Preparation

As described above, we extract a positive and a negative training example for each image position of the KITTI data set, whose true disparity is known. Each example is a pair of image patches from left and right images, respectively. We define p as the center of the left image patch and q as the center of the right image patch. Define d as the correct disparity for position p. When $p = (x, y)$, the positive example is obtained by setting q as:

$$q_{pos} = (x - d + o_{pos}, y) \quad (2)$$

where o_{pos} is a random number chose from (−1, 1). And the negative example is obtained by setting q as:

$$q_{neg} = (x - d + o_{neg}, y) \quad (3)$$

where o_{neg} is a random number chose from(−10, −4) and (4, 10). We generate examples for each pixel in pair of images according to the above rules, and finally get 40 million examples.

4.2 Comparison and Discussion

In order to evaluate the algorithm, the computation results of MC-HDCNN are compared with the original MC-CNN algorithm, and both algorithms use the same subsequent processing steps.

Table 1 shows the raw-disparity results of MC-HDCNN and MC-CNN without subsequent processing. We can see that the error rate computed by our algorithm on the KITTI2012 data set is 7.01%, which is 8.69% and 6.48% lower than the MC-CNN-fast algorithm and the MC-CNN-acrt algorithm, respectively; the error rate on the KITTI2015 dataset is 7.73%, which is 7.93% and 5.65% lower than the MC-CNN-fast algorithm and the MC-CNN-acrt algorithm, respectively. At the same time, our algorithm runtime is much less than the MC-CNN-acrt, and is close to the MC-CNN-fast.

Table 1. Comparison of MC-HDCNN、MC-CNN-acrt and MC-CNN-fast without subsequent processing. The "Matching error" is the percentage of miss-matching pixels with threshold 3.0. The "Runtime" is the time required for CNN to compute a pair of stereo images, in seconds.

Methods	Matching error		Runtime
	KITTI2012	KITTI2015	
MC-HDCNN	**7.01**	**7.73**	0.28
MC-CNN-acrt	13.49	13.38	35
MC-CNN-fast	15.70	15.66	0.20

In addition to comparison with the original algorithm, we also carefully compare the error results of the improved network with fast convolution network, weak supervised learning convolution network, MC-CNN-fast and MC-CNN-act algorithms. As shown in Tables 2 and 3, the values in the table are the rate of pixels where the true disparity differs from the predicted disparity by more than m (m = 2, 3, 4, 5) pixels.

The table shows that the improved method performs closely to the MC-CNN-acrt algorithm in the KITTI2012 dataset, and performs better in the KITTI2015 dataset.

Table 2. Error comparison of disparity with different algorithms (KITTI2012)

Methods	>2 pixel	>3 pixel	>4 pixel	>5 pixel
Fast CNN	4.98	3.07	2.39	2.03
MC-CNN-WS	4.76	3.02	2.33	1.96
MC-CNN-fast	4.81	2.97	2.26	1.91
MC-CNN-acrt	**4.28**	**2.63**	**2.02**	**1.72**
MC-HDCNN	4.67	2.70	2.08	1.79

Table 3. Error comparison of disparity with different algorithms (KITTI2015)

Methods	>2 pixel	>3 pixel	>4 pixel	>5 pixel
Fast CNN	6.78	4.38	2.56	2.03
MC-CNN-WS	6.75	3.78	2.91	2.35
MC-CNN-fast	7.46	3.95	2.80	2.30
MC-CNN-acrt	6.38	3.27	2.37	1.97
MC-HDCNN	**5.92**	**2.93**	**2.18**	**1.86**

The predicted disparity maps computed by our method in this paper are shown in Fig. 4, which are the test results of the KITTI2012 and KITTI2015 data sets respectively. Red pixels on error graphs represent miss-matching pixels with threshold 3.0. We can see that our method can obtain accurate and dense disparity maps, and the edge information is better preserved, such as the edges of vehicles and utility poles. The distinction between foreground and background is more obvious, the shaded area matches correctly, and the influence of illumination is less.

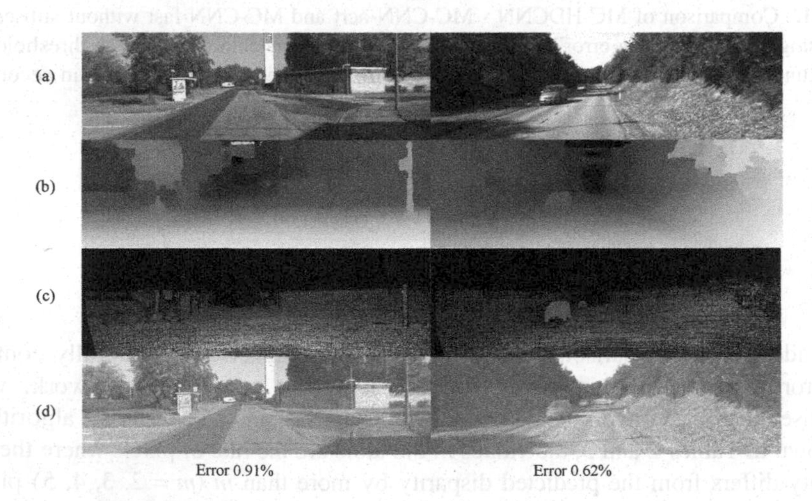

Fig. 4. The results of MC-HDCNN in KITTI2012 and KITTI2015 data sets. From top to bottom: (a) input image; (b) predicted disparity map; (c) true disparity map; (d) error graph (Color figure online)

5 Conclusions and Future Work

We propose a stereo matching method based on hybrid dilated convolution in this paper. Its receptive field has expanded to nearly 3 times without adding additional modules. At the same time, it avoids the "gridding" effect, and accepts all the information of the receptive field. The algorithm verification on the KITTI2012 and KITTI2015 datasets shows that the disparity computation results of our algorithm have certain advantage in accuracy compared with the similar deep learning methods, and the running speed can also meet the requirements of real-time computing. However the raw-disparity results of CNN have too much matching errors, we still need subsequent disparity optimization. In the future we will focus on improving the network to compute enough accurate disparity maps and adding more efficient disparity optimization algorithm.

References

1. Scharstein, D., Szeliski, R.: A taxonomy and evaluation of dense two-frame stereo correspondence algorithms. In: IEEE SMBV, pp. 131–140 (2001)
2. Žbontar, J., LeCun, Y.: Stereo matching by training a convolutional neural network to compare image patches. JMLR **17**(1), 2287–2318 (2016)
3. Ioffe, S., Szegedy, C.: Batch normalization: accelerating deep network training by reducing internal covariate shift. In: International Conference on Machine Learning, pp. 448–456 (2015)
4. Yoon, K.J., Kweon, I.S.: Adaptive support-weight approach for correspondence search. IEEE Trans. Pattern Anal. Mach. Intell. **28**, 650–656 (2006)
5. Lei, C., Selzer, J., Yang, Y.H.: Region-tree based stereo using dynamic programming optimization. In: IEEE CVPR, pp. 2378–2385 (2006)
6. Kolmogorov, V., Zabih, R.: Computing visual correspondence with occlusions using graph cuts. In: IEEE ICCV, pp. 508–515 (2001)
7. Hirschmüller, H.: Semi-global matching-motivation, developments and applications. Photogram. Week **11**, 173–184 (2011)
8. Žbontar, J., LeCun, Y.: Computing the stereo matching cost with a convolutional neural network. In: IEEE CVPR, pp. 1592–1599 (2015)
9. Luo, W., Schwing, A.G., Urtasun, R.: Efficient deep learning for stereo matching. In: IEEE CVPR, pp. 5695–5703 (2016)
10. Park, H., Lee, K.M.: Look wider to match image patches with convolutional neural networks. IEEE Signal Process. Lett. **24**, 1788–1792 (2017)
11. Mayer, N., Ilg, E., Häusser, P., et al.: A large dataset to train convolutional networks for disparity, optical flow, and scene flow estimation. In: IEEE CVPR, pp. 4040–4048 (2016)
12. Kendall, A., Martirosyan, H., Dasgupta, S., et al.: End-to-end learning of geometry and context for deep stereo regression. In: IEEE ICCV, pp. 66–75 (2017)
13. Song, X., Zhao, X., Hu, H., Fang, L.: EdgeStereo: a context integrated residual pyramid network for stereo matching. In: Jawahar, C.V., Li, H., Mori, G., Schindler, K. (eds.) ACCV 2018. LNCS, vol. 11365, pp. 20–35. Springer, Cham (2019). https://doi.org/10.1007/978-3-030-20873-8_2
14. Wang, P., Chen, P., Yuan, Y., et al.: Understanding convolution for semantic segmentation. In: IEEE WACV, PP. 1451–1460 (2018)

Convolutional Neural Network to Detect Thorax Diseases from Multi-view Chest X-Rays

Maram Mahmoud A. Monshi[1,2(✉)], Josiah Poon[1], and Vera Chung[1]

[1] School of Computer Science, University of Sydney, Sydney, Australia
mmon4544@uni.sydney.edu.au, {josiah.poon, vera.chung}@sydney.edu.au
[2] Department of Information Technology, Taif University, Taif, Saudi Arabia

Abstract. Chest radiography is the most common examination for a radiologist. This demands correct and immediate diagnosis of a patient's thorax to avoid life threatening diseases. Not only certified radiologists are hard to find, stress, fatigue and experience contribute to the quality of an examination. It is ideal that a chest X-ray can be interpreted by an automated deep learning algorithm. In this paper, we proposed a stage-wise model that is founded on a ResNet-50 based deep convolutional neural networks architecture to detect the presence and absence of twelve thorax diseases. This novel model has incorporated various recent techniques such as transfer learning, fine tuning, fit one cycle function and discriminative learning rates. The experiments were performed on 10% of the largest collection of chest X-rays to date, the MIMIC-CXR dataset. The model was trained for eight epochs using a subset of the available multi-view chest X-rays. The absolute labelling performance has achieved an encouraging average AUC of 0.779.

Keywords: Convolutional neural network · Thorax disease · Chest X-ray

1 Introduction

Currently, analyzing chest x-rays depends on the availability of professional radiologist. In some regions, access to such radiologists is limited [1]. Additionally, clinicians in emergency department and intensive care unit needs fast and accurate interpretations of medical images [2]. Globally, chest X-ray is the most common radiological examinations that required correct and fast analysis [1]. An automated and precise system that can flag potentially life-threatening diseases could allow care providers to handle emergency cases efficiently.

However, interpreting X-rays to detect thoracic diseases is still a challenging job. This is due to the highly diverse appearance of lesion areas on chest X-rays. Unlike the traditional computer-aided detection (CAD) systems that interpret medical images automatically to offer an objective diagnosis that assist radiologists [3], deep learning is able to learn useful features which are beyond the limit of radiology detection [4]. For example, deep learning has been applied on Mammography to discriminate breast cancer with microcalcification [5], on ultrasound to differentiate breast lesions and on

T. Gedeon et al. (Eds.): ICONIP 2019, CCIS 1142, pp. 148–158, 2019.
https://doi.org/10.1007/978-3-030-36808-1_17

CT lung scans to classify pulmonary [6]. Researchers [5, 6] showed a significant performance boost by their deep learning based models over the conventional CAD systems.

In this study, we present a supervised deep learning model using convolutional neural network to detect twelve thoracic diseases by reading a given chest X-ray. Residual network (ResNet-50) [7] is the backbone network for our model because it has clearly shown its outstanding performance on computer vision.

2 Related Work

Recently, several deep learning models that classify thorax diseases have been proposed as a result of the public release of a collection of large datasets namely Indiana Chest X-Ray [8], ChestX-ray14 [9], CheXpert [1], PadChest [10] and MIMIC-CXR [9]. For example, CheXNet [11], text-image embedding network (TieNet) [12], attention guided convolutional neural network (AG-CNN) [13], learning to diagnose from scratch network [14] classify thorax diseases from frontal chest x-rays using ChestX-ray14. However, [15] suggest that using lateral view enhances the performance for certain prediction tasks such as pleural effusion. Further, [2] proposed DualNet model to prove that simultaneous processing of both frontal and lateral chest X-ray inputs results in better classification performance. Unlike ChestX-ray14 [9] that only presents the frontal view of chest X-ray, MIMIC-CXR is a multi-view version of radiographs dataset. DualNet employed a limited released version of the MIMIC-CXR dataset to automate reading of frontal and lateral chest X-rays.

Convolutional neural network (CNN) which is a supervised deep learning model is the most common used deep learning technique for thoracic disease classification. It has also seen the widest variety in architectures, such as AlexNet [16], VGG-16 [17], DenseNet [18] and ResNet [7]. CNN-based classification model [19], for instance, adopt VGG-16 and ResNet-101 to classify X-rays based on nine chest diseases like emphysema and bronchitis. ResNet won the ImageNet large scale visual recognition challenge (ILSVRC) in 2015 with 3.6% top five error rate, which enables automated image classification to beat human brains with 5% error for the first time. ResNet is a feed forward network that contains several basic residual blocks, refer to Fig. 1, to handle the vanishing gradients [20] and the degradation issue.

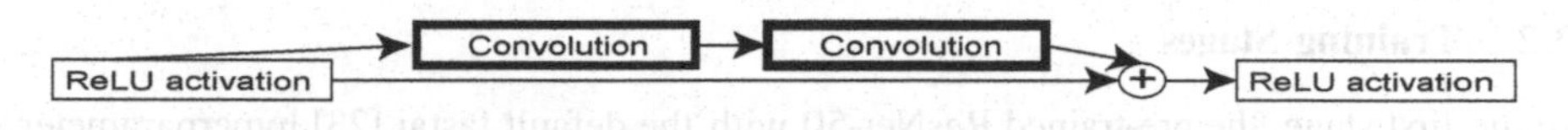

Fig. 1. A basic residual block

Consistent with recent proposed CNN models on automated chest x-rays classification [2, 11, 19], we focus on training CNN models to detect 12 common thoracic diseases namely enlarged cardiomediastinum, cardiomegaly, airspace opacity, lung lesion, edema, consolidation, pneumonia, atelectasis, pneumothorax, pleural effusion, pleural other and fracture (Fig. 2). Unique from past works, we propose a novel stage wise training approach to observe the model's performance and hence reduce training time and increase accuracy. We adopt a combination of recent techniques on multi-view chest X-rays including ResNet-50, transfer learning, fine tuning, fit one cycle function [21] and discriminative learning rates [22].

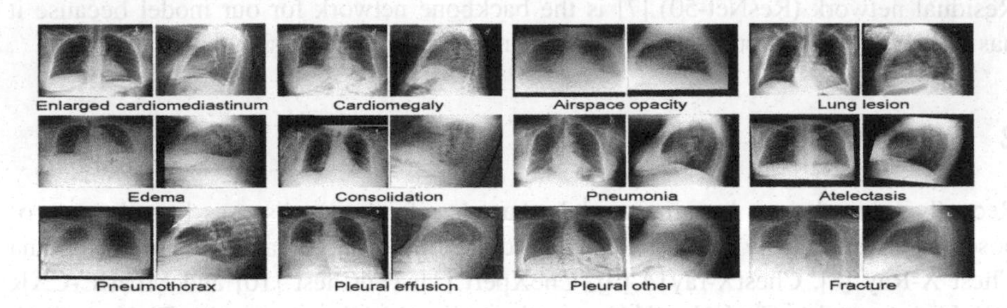

Fig. 2. Examples of Twelve Thoracic Diseases from MIMIC-CXR Dataset. Each disease is associated with frontal and lateral views of chest X-rays.

3 Proposed Model

3.1 Structure Overview

The task of detecting thorax diseases in chest x-rays is divided into 12 sub-tasks, where each task considers the presence and absence of a specific disease. Among the proposed variations of ResNet layers (i.e. 34, 50, 101, 152 and 1202), we adopt the popular ResNet-50 network which consists of 49 convolution layers and ends with 1 fully connected layer. Equation 1 defines the last output of residual unit x_l, where $F(x_{l-1})$ is the generated output after performing the convolution operations, batch normalization and activation function on x_{l-1}. Importantly, we use cyclical learning rates to enhance performance by decreasing the number of epochs required to accomplish the accuracy threshold. For each binary label problem, ResNet is used as the baseline CNN architecture in three main training stages (Fig. 3).

$$x_l = F(x_{l-1}) + x_{l-1} \tag{1}$$

3.2 Training Stages

In the first stage, the pre-trained ResNet-50 with the default fastai [23] hyperparameter values is trained for three epochs. That is setting all layers to frozen, excluding the final dense layer and examining each X-ray three times. In other words, the first stage embraces transfer learning approach to train faster with a model that is already trained to recognize 1000 categories of things in ImageNet. At the end of stage-1, model's weights were saved.

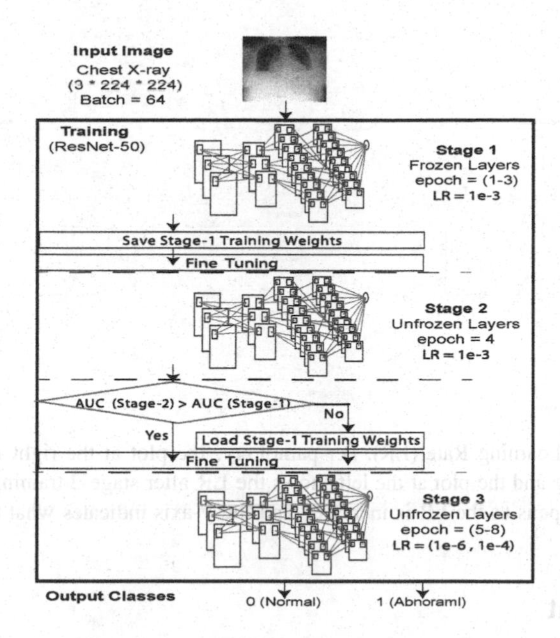

Fig. 3. Overall illustration of our model.

In the second stage, the whole model is trained again for one epoch by unfreezing the layers and calling the fit-one-cycle method. The objective of this stage is to observe the model's performance to reduce training time and increase accuracy. If the AUC is decreased at the end of this training stage, stage-1 weights are re-loaded.

In the third stage, the whole model is trained again for four epochs using the optimal learning rate finder. The learning rate is set by default to about 1e−3 at stage-1 and changed manually to a range of lower learning rates (1e−6 to 1e−4) at stage 3. Figure 4 illustrates the plotted learning rate after the first and second stages of the model, where the red dots on the graphs indicate the steepest gradient point. Using different learning rates for each layer at this stage is in line with the discriminate fine-tuning technique to tune each layer with various learning rates. In this case, the model's parameters θ and the learning rate η are split into $\{\theta^1, \ldots, \theta^L\}$ at time step t and $\{\eta^1, \ldots, \eta^L\}$ respectively, where L is the number of layers. This updated version of the regular stochastic gradient descent (SGD) with discriminative fine-tuning is defined in Eq. 2, where $\nabla_{\theta^l} J$ is the gradient of the model's objective function.

$$\theta_t^l = \theta_{t-1}^l - \eta^l \cdot \nabla_{\theta^l} j(\theta) \tag{2}$$

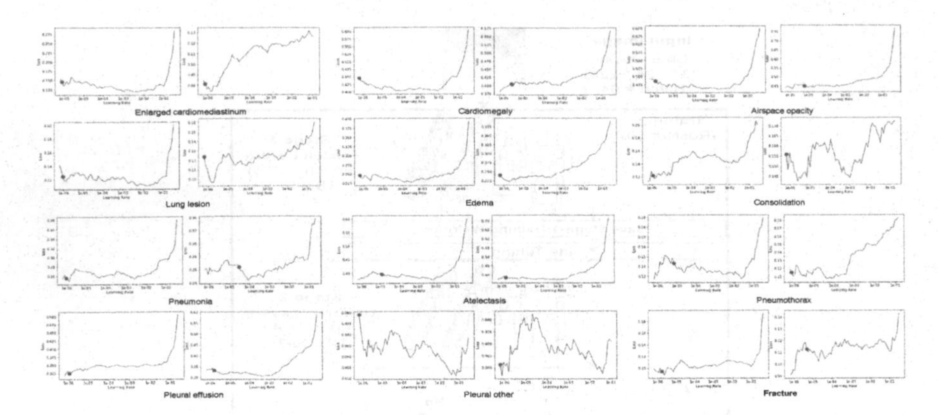

Fig. 4. Fluctuated Learning Rate (LR). Per pathology, the plot at the right represents the LR after stage-1 training and the plot at the left shows the LR after stage-2 training. Note the x-axis represents what happens as the LR is increased and the y-axis indicates what the loss is. (Color figure online)

4 Experiment

4.1 Dataset

MIMIC-CXR is the largest dataset of chest x-rays to date that consist of 371,920 images and relevant 227,943 studies derived from Beth Israel Deaconess Center [24]. Images are annotated with 14 labels, which overlap with those of the popular ChestX-ray14 dataset and match the co-released CheXpert dataset. Labels are extracted from the associated free-text radiology reports using the CheXpert labeler tool. The training labels for each observation are 0 for negative, 1 for positive, −1 for uncertain and blank for unknown. We organized a subset of 10% of the MIMIC-CXR v1.0.0 into training and validation sets that contains 33,195 and 3,688 images respectively.

Table 1. The MIMIC-CXR Dataset with 12 Labeled Pathologies. We account the number of positive and negative observations in %10 of the dataset.

Pathology	Positive (%)	Negative (%)
Enlarged cardiom.	1019 (2.8)	35367 (97.19)
Cardiomegaly	6932 (18.79)	29951 (81.2)
Airspace opacity	7582 (20.42)	29542 (79.57)
Lung lesion	1060 (2.82)	36472 (97.17)
Edema	3964 (11.06)	31859 (88.93)
Consolidation	1410 (3.8)	35634 (96.19)
Pneumonia	2738 (7.83)	32202 (92.16)
Atelectasis	6356 (17.54)	29876 (82.45)
Pneumothorax	1523 (4.05)	36059 (95.94)
Pleural effusion	7869 (21.34)	28994 (78.65)
Pleural other	425 (1.13)	37132 (98.86)
Fracture	805 (2.13)	36829 (97.86)

The validation set was selected at random. During training, the uncertain and unknown labels were ignored. Table 1 shows the positive and negative cases for each observation.

4.2 Pre-processing

Prior to models training, we employ several augmentation strategies (refer to Table 2) as data augmentation is a critical step of deep CNNs in medical imaging [25]. We crop each x-ray in both the training and validation sets to 224 by 224 pixels to reduces training time while maintaining robust model's performance. For example, training the model to diagnose cardiomegaly using 299 by 299 pixels would increase training time without improving the AUC per epoch (refer to Table 3). We perform a horizontal flip only for each image in the training set, since vertical flips often do not reflect chest x-rays (i.e. an upside-down chest x-ray may not improve training). The maximum lighting of the image is set to 0.3 with applying probability of 0.5. Note that no vertical flips, rotations, zooms or wraps were done on the images. In addition, uncertain and unknown labels were dropped.

Table 2. Data Augmentation for Chest X-rays. We applied a list of transforms parameters to the trained images.

Parameter	Value
Size	224
Flip (horizontally)	True
Lighting	0.3
Affine	0.5

Table 3. AUC per Epoch for Training ResNet-50 CNN. This model detects cardiomegaly using 299×299 or 224×224 pixels of chest X-rays.

Image size (pixels)	Epoch 1	2	3	4	5	6	7	8	Avg. AUC per Epoch
299	0.565	0.733	0.758	0.791	0.798	0.804	0.804	**0.807**	0.757
224	0.725	0.733	0.747	0.785	0.793	0.799	**0.802**	0.802	**0.773**

4.3 Training

The training algorithms were evaluated in twelve pathologies: enlarged cardiomediastinum, cardiomegaly, airspace opacity, lung lesion, edema, consolidation, pneumonia, atelectasis, pneumothorax, pleural effusion, pleural other and fracture. We used PyTorch software [26], fastai library, n1-highmem-8 (8 vCPUs, 52 GB memory) machine and 4 x NVIDIA Tesla P4 GPUs. This is in accordance with [27] work that demonstrate how time-per-epoch for the ResNet-50 architecture scale much better when training it on multiple GPUs. Table 4 records the time per epoch for training ResNet-50 based model to detect cardiomegaly using different number of GPUs, where parallel training on 4 GPUs reduce training time by around 20 min.

Table 4. Time per Epoch for Training ResNet-50 CNN. This model detects cardiomegaly using single NVIDIA Tesla P4 GPU or 4 x NVIDIA Tesla P4 GPUs in a parallel training. Note the batch size is set to 64 images and the image size is set to 224 pixels.

No. of GPUs	Epoch								Avg. time per Epoch (min)
	1	2	3	4	5	6	7	8	
1	32:42	32:26	32:36	34:34	33:40	33:52	33:58	34:00	33:28
4	13:32	12:54	13:01	13:05	13:07	13:08	13:07	13:06	**13:07**

4.4 Results

Table 5 shows the Area Under Curve (AUC) results of each pathology computed on the validation set for each of the eight training epochs. It can be seen that detection performance for each pathology fluctuate over epochs. For individual training epochs, the eighth unfrozen epoch accomplish a higher average AUC (0.777), compared to the first (0.670), second (0.704), third (0.718), forth (0.711), fifth (0.753), sixth (0.765) and seventh (0.776). Compared with stage-1 (epoch 1–3) and stage 2 (epoch 4), stage 3 (epoch 5–8) results in larger AUC values for all pathologies. This difference is likely due to the discriminative learning rates at the third stage of training.

Table 5. The Compression of AUC Scores in each Epoch. We trained each pathology for 8 epochs.

Pathology	Epoch							
	1	2	3	4	5	6	7	8
Enlarged cardiom.	0.670	0.694	0.700	0.544	0.702	0.705	0.708	**0.710**
Cardiomegaly	0.725	0.733	0.747	0.785	0.793	0.799	**0.802**	0.802
Airspace opacity	0.621	0.687	0.694	0.712	0.730	0.730	0.733	**0.737**
Lung lesion	0.520	0.638	0.612	0.638	0.651	0.688	**0.730**	0.729
Edema	0.816	0.848	0.857	0.887	0.892	0.894	0.896	**0.897**
Consolidation	0.748	0.758	0.769	0.778	0.788	0.797	0.797	**0.799**
Pneumonia	0.556	0.531	0.545	0.497	0.550	0.585	0.580	**0.587**
Atelectasis	0.706	0.706	0.743	0.827	0.830	0.835	0.837	**0.838**
Pneumothorax	0.710	0.786	0.817	0.839	0.853	0.862	**0.868**	0.860
Pleural effusion	0.837	0.869	0.881	0.891	0.903	**0.906**	0.905	0.899
Pleural other	0.585	0.637	0.676	0.533	0.707	0.736	**0.739**	0.727
Fracture	0.546	0.563	0.576	0.606	0.636	0.648	0.711	**0.741**
Average	0.670	0.704	0.718	0.711	0.753	0.765	0.776	**0.777**

Table 6 compares the per pathology AUC results between our proposed model and DualNet architecture using MIMIC-CXR dataset. We employed 10% of the dataset using all available frontal and lateral views of the chest X-rays. DualNet, on the other hand, considered a combination of posteroanterior (PA) and lateral as well as a composite of anteroposterior (AP) and lateral. In 5 out of 7 overlap pathologies, our model performs better than both DualNet models. Overall, it can be seen that average AUC is higher for our multi-view classifiers (0.779), compared to both PA-lateral (0.722) and AP-lateral (0.677).

Table 6. The Compression of AUC Scores. DualNet model used an older limited released version of the MIMIC-CXR dataset. Our model used 10% of the publicly released version of the dataset. Note that we ignored uncertain and unknown labels.

Pathology	DualNet [2]		Our model
	PA + Lateral	AP + Lateral	Multi-view
Enlarged cardiom.	–	–	0.710
Cardiomegaly	**0.840**	0.755	0.802
Airspace opacity	–	–	0.737
Lung lesion	–	–	0.730
Edema	0.734	0.749	**0.897**
Consolidation	0.632	0.623	**0.799**
Pneumonia	**0.625**	0.593	0.587
Atelectasis	0.766	0.671	**0.838**
Pneumothorax	0.706	0.621	**0.868**
Pleural effusion	0.757	0.733	**0.906**
Pleural other	–	–	0.739
Fracture	–	–	0.741
Average	0.722	0.677	**0.779**

4.5 Analysis

In DualNet model, chest X-rays labels were extracted from the associated radiology reports using an open source tool developed by the National Institute of Health (NIH), the NegBio labeler[1] [28]. This tool was used to annotate the popular ChestX-ray14 dataset. In contrast, our model follows the public released labels by [24] that utilized a different open source tool created by Stanford machine learning group, the CheXpert labeler[2]. Although the labeling algorithm of CheXpert is built upon the work of NegBio, it achieves a higher F1 score. Hence, our model is trained on a better annotated chest X-rays than DualNet. Interestingly, we reach improved results over those achieved by DualNet using small image sizes 224 by 224 pixels instead of 512 by 512 pixels.

[1] https://github.com/ncbi-nlp/NegBio.

[2] https://github.com/stanfordmlgroup/chexpert-labeler.

Nevertheless MIMIC-CXR is the largest open source X-ray images to date, the class labels in the training set are noisy because they were mined by natural language processing tool, rather than by experienced radiologist. Figure 5 visualizes the most incorrect predicted X-rays by our model with heatmaps, using the activations of the wrongly predicted class. In addition, the positive-negative subsets ratio was highly imbalanced in the enlarged cardiomediastinum, lung lesion, consolidation, pneumothorax, pleural other and fracture sets (Table 1). Yet, our model's AUC for each of these pathologies is above 0.7 (Table 6).

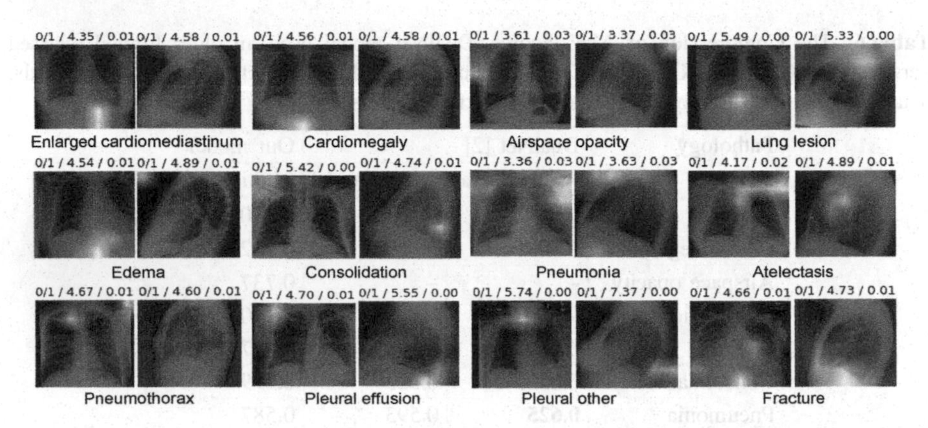

Fig. 5. Examples of the Most Confused Chest X-rays with Heatmaps. Each image is associated with the prediction, actual, loss and probability values after stage-1 training, where 0 and 1 represent negative and positive pathology respectively.

5 Conclusion

In this paper, ResNet-50 CNN based stage wise models have been proposed to detect twelve thorax diseases on 10% of the largest chest X-rays dataset to date, the MIMIC-CXR dataset. The absolute labelling performance with an average weighted AUC of 0.779 is encouraging, since we used only a subset of the available chest X-rays. In future work, we plan to improve our CNN model performance through utilizing common image-based classification techniques, in particular data augmentation. Importantly, we will incorporate useful information from the free-text radiology reports such as patient's history and clinical records to accurately recognize the presence and absence of thorax diseases.

References

1. Irvin, J., et al.: CheXpert: a large chest radiograph dataset with uncertainty labels and expert comparison. arXiv preprint arXiv:1901.07031 (2019)

2. Rubin, J., Sanghavi, D., Zhao, C., Lee, K., Qadir, A., Xu-Wilson, M.: Large scale automated reading of frontal and lateral chest x-rays using dual convolutional neural networks. arXiv preprint arXiv:1804.07839 (2018)
3. van Ginneken, B., Schaefer-Prokop, C.M., Prokop, M.: Computer-aided diagnosis: how to move from the laboratory to the clinic. Radiology **261**, 719–732 (2011)
4. Kohli, M., Prevedello, L.M., Filice, R.W., Geis, J.R.: Implementing machine learning in radiology practice and research. Am. J. Roentgenol. **208**, 754–760 (2017)
5. Wang, J., Yang, X., Cai, H., Tan, W., Jin, C., Li, L.: Discrimination of breast cancer with microcalcifications on mammography by deep learning. Sci. Rep. **6**, 27327 (2016)
6. Cheng, J.-Z., et al.: Computer-aided diagnosis with deep learning architecture: applications to breast lesions in US images and pulmonary nodules in CT scans. Sci. Rep. **6**, 24454 (2016)
7. He, K., Zhang, X., Ren, S., Sun, J.: Deep residual learning for image recognition. In: Proceedings of the IEEE Conference on Computer Vision and Pattern Recognition, pp. 770–778 (2016)
8. Demner-Fushman, D., et al.: Preparing a collection of radiology examinations for distribution and retrieval. J. Am. Med. Inform. Assoc. **23**, 304–310 (2015)
9. Wang, X., Peng, Y., Lu, L., Lu, Z., Bagheri, M., Summers, R.M.: Chestx-ray8: hospital-scale chest x-ray database and benchmarks on weakly-supervised classification and localization of common thorax diseases. In: 2017 IEEE Conference on Computer Vision and Pattern Recognition (CVPR), pp. 3462–3471. IEEE (2017)
10. Bustos, A., Pertusa, A., Salinas, J.-M., de la Iglesia-Vayá, M.: PadChest: a large chest x-ray image dataset with multi-label annotated reports. arXiv preprint arXiv:1901.07441 (2019)
11. Rajpurkar, P., et al.: Chexnet: radiologist-level pneumonia detection on chest x-rays with deep learning. arXiv preprint arXiv:1711.05225 (2017)
12. Wang, X., Peng, Y., Lu, L., Lu, Z., Summers, R.M.: TieNet: Text-image embedding network for common thorax disease classification and reporting in chest x-rays. In: Proceedings of the IEEE Conference on Computer Vision and Pattern Recognition, pp. 9049–9058 (2018)
13. Guan, Q., Huang, Y., Zhong, Z., Zheng, Z., Zheng, L., Yang, Y.: Diagnose like a radiologist: attention guided convolutional neural network for thorax disease classification. arXiv preprint arXiv:1801.09927 (2018)
14. Yao, L., Poblenz, E., Dagunts, D., Covington, B., Bernard, D., Lyman, K.: Learning to diagnose from scratch by exploiting dependencies among labels. arXiv preprint arXiv:1710.10501 (2017)
15. Bertrand, H., Hashir, M., Cohen, J.P.: Do lateral views help automated chest X-ray predictions? arXiv preprint arXiv:1904.08534 (2019)
16. Krizhevsky, A., Sutskever, I., Hinton, G.E.: ImageNet classification with deep convolutional neural networks. In: Advances in Neural Information Processing Systems, pp. 1097–1105 (2012)
17. Simonyan, K., Zisserman, A.: Very deep convolutional networks for large-scale image recognition. arXiv preprint arXiv:1409.1556 (2014)
18. Huang, G., Liu, Z., Van Der Maaten, L., Weinberger, K.Q.: Densely connected convolutional networks. In: 2017 IEEE Conference on Computer Vision and Pattern Recognition (CVPR), pp. 2261–2269. IEEE (2017)
19. Dong, Y., Pan, Y., Zhang, J., Xu, W.: Learning to read chest X-ray images from 16000+ examples using CNN. In: Proceedings of the Second IEEE/ACM International Conference on Connected Health: Applications, Systems and Engineering Technologies, pp. 51–57. IEEE Press (2017)

20. Glorot, X., Bengio, Y.: Understanding the difficulty of training deep feedforward neural networks. In: Proceedings of the Thirteenth International Conference on Artificial Intelligence and Statistics, pp. 249–256 (2010)
21. Smith, L.N.: A disciplined approach to neural network hyper-parameters: part 1–learning rate, batch size, momentum, and weight decay. arXiv preprint arXiv:1803.09820 (2018)
22. Howard, J., Ruder, S.: Universal language model fine-tuning for text classification. arXiv preprint arXiv:1801.06146 (2018)
23. https://docs.fast.ai
24. Johnson, A.E., et al.: MIMIC-CXR: a large publicly available database of labeled chest radiographs. arXiv preprint arXiv:1901.07042 (2019)
25. Hussain, Z., Gimenez, F., Yi, D., Rubin, D.: Differential data augmentation techniques for medical imaging classification tasks. In: AMIA Annual Symposium Proceedings, p. 979. American Medical Informatics Association (2017)
26. Ketkar, N.: Introduction to PyTorch. Deep Learning with Python: A Hands-on Introduction, pp. 195–208. Apress, Berkeley (2017)
27. Coleman, C., et al.: Analysis of dawnbench, a time-to-accuracy machine learning performance benchmark. arXiv preprint arXiv:1806.01427 (2018)
28. Peng, Y., Wang, X., Lu, L., Bagheri, M., Summers, R., Lu, Z.: NegBio: a high-performance tool for negation and uncertainty detection in radiology reports. In: AMIA Summits on Translational Science Proceedings 2017, p. 188 (2018)

Visual Speaker Authentication by a CNN-Based Scheme with Discriminative Segment Analysis

Jiahui Sun, Shilin Wang(✉), and Quanhai Zhang

School of Electronic Information and Electrical Engineering,
Shanghai Jiao Tong University, Shanghai, China
{sjh_717,wsl,qhzhang}@sjtu.edu.cn

Abstract. Recent research shows that the static and dynamic features of a lip utterance contain abundant identity-related information. In this paper, a new deep convolutional neural network scheme is proposed. The entire lip utterance is first divided into a series of overlapping segments; then an adaptive scheme is designed to automatically examine the discriminative power and assign a corresponding weight of each segment in the entire utterance. The final authentication result of the entire utterance is determined by weighted voting of the results for all the segments. In addition, considering the various lighting condition in the natural environment, an illumination normalization procedure is proposed. Experimental results show that different segments of the same utterance have different discriminative power for user authentication, and focusing on the discriminative details will be more effective. The proposed method has shown superior performance compared with two state-of-the-art lip authentication approaches investigated.

Keywords: Visual speaker authentication · 3DCNN · Lip feature · Discriminative weight

1 Introduction

In recent years, user authentication based on human biometric features has received much attention. In addition to the face [14], iris [15], and fingerprint [5], lip feature is also a popular biometric feature. Lip feature is a twin-biometric with a high discriminative power [2,11]. Speakers can be distinguished from different lip shapes and unique talking habits that are difficult to imitate [7,19].

Lip biometrics as a means of visual speaker authentication, was first introduced by Suzuki et al. [18]. Over the past decade, various ways have been proposed to verify human identity using lip features. Broun et al. [3] used the polynomial based approach [4] as the classifier. Based on the XM2VTS database, they achieved an FRR of 4.4% and an FAR of 8.2%. Chan et al. [6] proposed an ordinal contrast measure called Local Ordinal Contrast Pattern (LOCP). They obtained a very low $HTER$ of 0.36% on the XM2VTS dataset. In our

T. Gedeon et al. (Eds.): ICONIP 2019, CCIS 1142, pp. 159–167, 2019.
https://doi.org/10.1007/978-3-030-36808-1_18

previous work, Lai et al. [17] proposed a visual speaker authentication scheme which handles static lip appearance, lip movements during a specific word and lip movements during a word transition. Modeling dynamic and static segments using HMM-UBM and linear SVM, respectively.

The above methods have demonstrated that lip feature is effective and reliable in verifying the identity of the speaker. However, there are still two challenging tasks, one is how to highlight the speaker's unique speaking habits, the other is how to enhance the robustness in complex lighting environments. In order to handle the two challenges, a new visual speaker authentication scheme is proposed. The major contributions of this paper can be summarized as follows:

(1) A new deep convolutional neural network scheme is proposed, which can automatically examine the discriminative power of each segment in the utterance and provide more reliable authentication results.

(2) An illumination pre-processing method to overcome the influence of illumination is introduced to enhance the robustness of illumination variation.

(3) Experimental results show that the proposed approach achieves excellent authentication performance compared with two recent approaches investigated.

2 Motivation

In the previous work of our group [17], it has been demonstrated that the speaker's identity can be better recognized in some specific words or word transitions rather than the entire utterance. In this paper, we extend the above idea and propose a new deep neural network scheme for visual speaker authentication.

In our approach, the entire utterance is first divided into a series of overlapping segments. Then each segment is fed into a 3D convolutional neural network (3DCNN) to extract discriminative features. Meanwhile, a weighting measure which describes the discriminative power of the segment is automatically computed by examining the L-2 norm of the feature vector. The final authentication result can be obtained by weighted voting over the results for all the segments. With the above strategy, the discriminative segments will have more impact on the final authentication result and thus it will outperform the sentence level authentication approaches where all the segments have the same impact.

3 The Proposed Method

In general, visual speaker authentication is a two-class classification problem, i.e., a client or an imposter. In this section, considering the various lighting condition, an illumination normalization procedure is proposed. Then, the visual speaker authentication scheme based on discriminative segment analysis is proposed. Finally, the implementation procedure is introduced. The details are as follows.

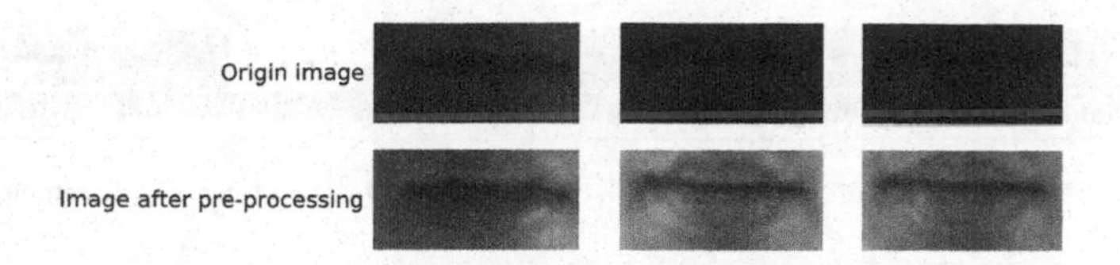

Fig. 1. The influence of illumination pre-processing

3.1 Illumination Pre-processing

In order to overcome the influence of illumination, referring to [1], an illumination normalization procedure is introduced as follows. The intensities of each channel of the input RGB image are stretched to [0, 255], the effects are shown in Fig. 1. The first row is the original image, and the second row is the image after the illumination pre-processing. It can be seen that this method can alleviate the situation where the light is too dark.

For R, G and B channels in the original image, the intensities of each channel are stretched to [0, 255]. For example, in the R channel, the maximum and minimum intensity are labeled as $I_{R,max}$, $I_{R,min}$, respectively. After processing, the intensity I_R in the original image is calculated as I'_R in Eq. 1. The pre-processing of G and B channels is the same as that of R channel.

$$I'_R = \frac{I_R - I_{R,min}}{I_{R,max} - I_{R,min}} \times 255 \tag{1}$$

3.2 The Visual Speaker Authentication Scheme Based on Discriminative Degment Analysis

A weighting measure is proposed to describe the discriminative power of the segment. The overall architecture of the proposed authentication system is shown in Fig. 2. The entire utterance is divided into a series of overlapping segments with predefined fixed time window size T and step size S. Then each segment is fed into a 3D convolutional neural network (3DCNN) [10] to extract discriminative feature. Meanwhile, a weighting measure which describes the discriminative power of the segment is automatically computed by examining the L-2 norm of the feature vector. The final authentication result, can be obtained by weighted voting over the authentication results for all the segments. The details are as follows:

(1) Given a predefined time window size T and step size S, the entire utterance contains V frames can be divided into K overlapping segments, where K can be obtained by $K = \lfloor (V - T)/S \rfloor + 1$. A series of segments in the same utterance are marked as $(E_1, E_2, ..., E_i, ..., E_k)$.

(2) For each segment E_i in the same entire utterance, 3DCNN is used for lip feature extraction. After the softmax layer, the predicted probability $\mathbf{p_i}$ is obtained. Meanwhile, the discriminative power w_i of the segment E_i is extracted

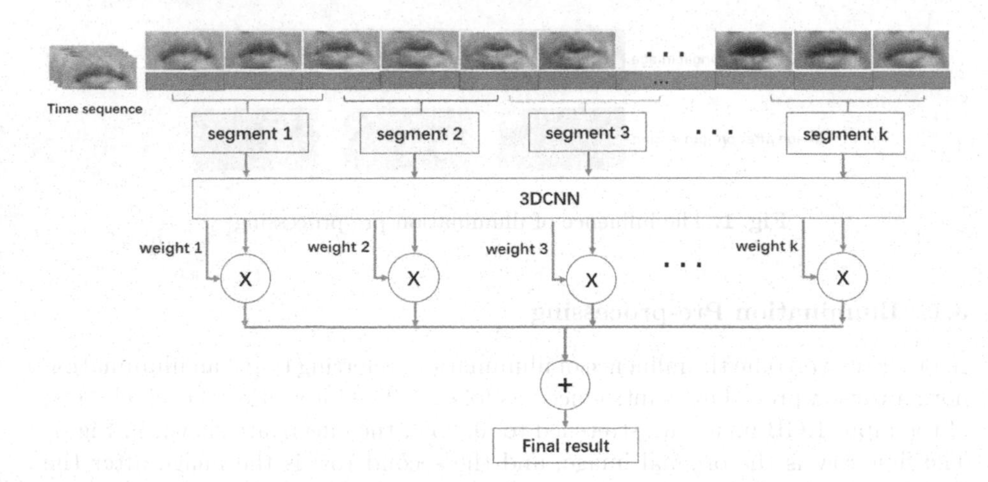

Fig. 2. Sketch of lip feature extraction procedure

from the last fully connected layer by calculating the L-2 norm of $\mathbf{f_i}$ as Eq. 2. Finally, the final result of the entire utterance $\mathbf{r}$ can be obtained by weighted voting over the authentication results for all the segments as Eq. 3.

$$w_i = \|\mathbf{f_i}\|_2 \tag{2}$$

$$\mathbf{r} = \sum_{i=1}^{k} w_i \times \mathbf{p_i} \tag{3}$$

(3) Note that when classifying with categorical crossentropy loss function, the network gives a very high probability to the label with higher probability, so that the majority of the maximum predicted values are distributed in [0.9, 1]. So a deformed sigmoid function in Eq. 4 is used to extend the difference, to highlight the effect of higher predicted values.

$$\mathbf{p_i'} = \frac{1}{1 + e^{(-\alpha \times (\mathbf{p_i} - \beta))}} \tag{4}$$

3.3 Implementation Procedure

The detailed network structure of the 3DCNN in Fig. 3 is given as follows: (1) A five-level pyramid is constructed, for conv1b, 2b, 3b, 4a and 4b, a stride of (1, 2, 2) is set (stride is 1 in the temporal domain and stride is 2 in the spatial domain) and a stride of (1, 1, 1) is set for the rest layers; (2) Batch normalization is applied after each pyramid to solve the gradient disappearance and explosion in training and speed up the convergence of the model; (3) All the 3D kernels are of the size $3 \times 3 \times 3$, which are set empirically to achieve the best performance compared with some other kernel sizes; (4) Except for the last layer using the

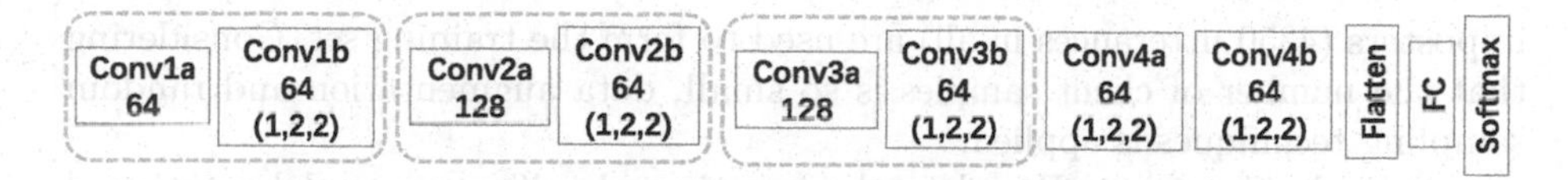

Fig. 3. The network structure of the 3DCNN

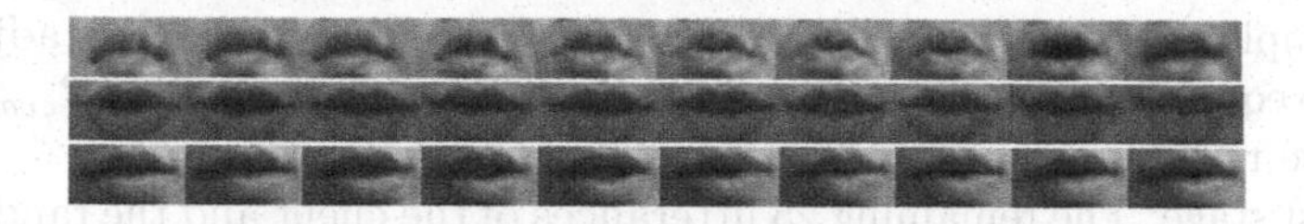

Fig. 4. Sample lip images in our dataset

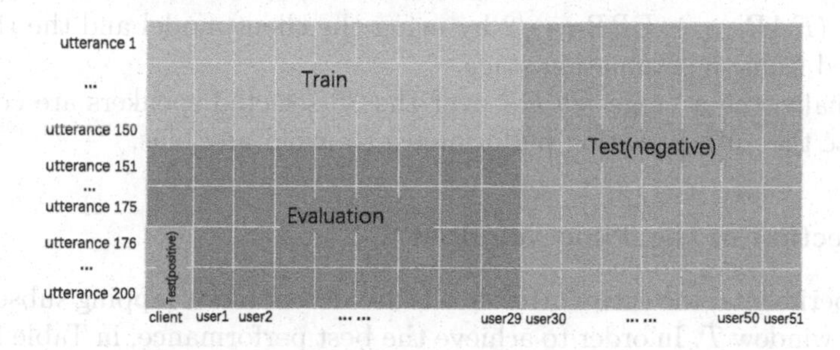

Fig. 5. The speaker authentication protocol in our experiment

softmax as activation functions, the activation functions of the other layers are ReLU, which alleviates over-fitting and reduces the cost of calculation; (5) In the training stage, the Adam optimizer is adopted to speed up training with an initial learning rate of 0.001, and categorical crossentropy loss is adopted as the loss function.

4 Experiments and Discussions

Since in most public speaker datasets, the lip region is of low resolution. To better evaluate the performance of our method, we have constructed a dataset containing 52 speakers under the natural environments. In the dataset, each speaker is required to read 200 four-digit utterances from "0000" to "9999", where the vocabulary is from 0 to 9. Among the 200 utterances, all speakers have the same 50 fixed utterances, the remaining 150 are random. The spatial resolution of the lip region is 50×100 pixels. Some sample are illustrated in Fig. 4.

In the visual speaker authentication scenario, the Lausanne protocol [12] is adopted in our experiment, which runs as follows and is illustrated in Fig. 5.

(1) Training stage: 30 users out of 52 users in the dataset are randomly selected to train 30 two-class classification model. 150 utterances (including 50 fixed and 100 random utterances) for the client and 29 other speakers as

imposters (4350 utterances in all) are used to form the training set. Considering that the number of client samples is so small, data augmentation and random sampling techniques is applied.

(2) Evaluation stage: Based on the two-class classification model mentioned in (1), for each authentication experiment, 25 utterances of the client and 1450 utterances of the other 29 speakers are used as the positive and negative evaluation samples, respectively. The threshold H for authentication is adjusted to obtain the equal error rate (EER), where the false accept rate (FAR_{eval}) equals to the false rejection rate (FRR_{eval}) in the evaluation set.

(3) Test stage: The remaining 25 utterances of the client and the random 1100 utterances of the other unused 22 speakers are used as the client and imposter's test samples. The Half Total Error Rate ($HTER$) in the test set is computed as $HTER = (FAR_{test} + FRR_{test})/2$ by using the client model and the threshold H obtained from the evaluation stage.

(4) Finally, the average $HTER$ over the 30 selected speakers are computed to evaluate the authentication performance to avoid any bias.

4.1 Selection of the Time Window

In our experiments, the entire utterance is divided into overlapping subsequences with time window T. In order to achieve the best performance, in Table 1, experiments are performed on T sizes of 8, 16, 24, 32, respectively.

Table 1 mainly includes the influence of window size on performance and calculation. When T is too large ($T = 32$), training samples are much reduced, and the authentication accuracy after voting is not significant, because each subsequence tends to be similar. When T is small ($T = 8$), there are more training samples, and get a very low EER but a high $HTER$, this is because each subsequence does not contain enough identity-related information. Therefore, considering the performance and calculation, the T is set to 16 in the following experiment to obtain the best authentication performance.

Table 1. Influence of the time window T ($EER/HTER$ in %)

T	8	16	24	32
EER	0.146	0.189	0.285	3.8298
HTER	0.454	0.402	0.411	4.2791
Params	24M	40M	56M	72M

4.2 Results Using Our Network

Illumination Pre-processing. In view of the diversity of illumination, the method mentioned in Sect. 3.1 is applied. The intensities of each channel of

Table 2. Influence of illumination pre-processing ($EER/HTER$ in %)

Methods	With pre-processing		Without pre-processing	
	Subsequence	Entire utterance	Subsequence	Entire utterance
EER	0.599	0.189	0.489	0.67
HTER	0.593	0.402	2.684	1.867

the input image are stretched and distributed at [0, 255]. Based on 3DCNN, experiments are performed to calculate the results of subsequence (prediction for each segment) and entire utterance with or without illumination pre-processing, respectively. As shown in Table 2, this pre-processing improves the performance of both the subsequence and the entire utterance.

Optimizing the Probability Distribution of a Segment. For each segment, as the majority of the maximum predicted values are distributed in [0.9, 1]. In order to enhances the higher probability distribution value, the $\mathbf{p_i}$ is optimized by a deformed sigmoid Eq. 4, setting $\alpha = 10$, $\beta = 0.97$, the weights originally distributed in [0.9, 1] are stretched to [0.33, 0.58]. The original method and the method above described is labeled as 'without nonlinear' and 'with nonlinear', respectively. Table 3 shows that this optimization can improve the performance.

Table 3. Optimizing the probability distribution of a segment ($EER/HTER$ in %)

Methods	Without nonlinear	With nonlinear
EER	0.211	0.189
HTER	0.498	0.402

4.3 Performance Comparison with Existing Approaches

In order to fully evaluate the proposed method, two recent methods of visual speaker authentication, namely Liao's [10] and Chan's [6], are adopted to compare. Table 4 shows the experimental results, it can be concluded that our method achieves the best performance compared to the other two methods.

Table 4. Use different authentication mechanisms ($EER/HTER$ in %)

Methods	Our method	Liao's	Chan's
EER	0.189	0.368	3.594
HTER	0.402	0.556	5.793

5 Conclusion

In this paper, a new deep neural network scheme for visual speaker authentication is proposed the segments. Experiments prove that comparing with analyzing the entire sequence indiscriminately, focusing on the discriminative details are more effective. It is effective to calculate the weight of each segment by examining the L-2 norm of the feature vector in the last fully connected layer. In addition, illumination normalization and optimized probability distribution are proposed to further improve performance. Experiments have shown that the proposed method has better performance than the lip authentication schemes investigated.

Acknowledgment. The work described in this paper is fully supported by NSFC Fund (No. 61771310).

References

1. van der Walt, S., et al., and the scikit-image contributors: scikit-image: image processing in Python. PeerJ, **2**, e453 (2014). https://doi.org/10.7717/peerj.453
2. Cheng, F., Wang, S.L., et al.: Visual speaker authentication with random prompt texts by a dual-task CNN framework. Pattern Recogn. **83**, 340–352 (2018)
3. Broun, C.C., Zhang, X., Mersereau, R.M., Clements, M.: Automatic speechreading with application to speaker verification. In: 2002 IEEE International Conference on Acoustics, Speech, and Signal Processing, vol. 1, p. 685. IEEE (2002)
4. Campbell, W.M.: Low-complexity speaker authentication techniques using polynomial classifiers. In: Applications and Science of Computational Intelligence II, vol. 3722, pp. 357–368. International Society for Optics and Photonics (1999)
5. Cao, K., Jain, A.K.: Automated latent fingerprint recognition. IEEE Trans. Pattern Anal. Mach. Intell. **41**(4), 788–800 (2019)
6. Chan, C.H., Goswami, B., Kittler, J., Christmas, W.: Local ordinal contrast pattern histograms for spatiotemporal, lip-based speaker authentication. IEEE Trans. Inf. Forensics Secur. **7**(2), 602–612 (2011)
7. Choraś, M.: The lip as a biometric. Pattern Anal. Appl. **13**(1), 105–112 (2010)
8. Kazemi, V., Sullivan, J.: One millisecond face alignment with an ensemble of regression trees. In: CVPR (2014)
9. Lai, J.Y., Wang, S.L., Liew, A.W.C., Shi, X.J.: Visual speaker identification and authentication by joint spatiotemporal sparse coding and hierarchical pooling. Inf. Sci. **373**, 219–232 (2016)
10. Liao, J., Wang, S., et al.: 3D convolutional neural networks based speaker identification and authentication. In: 2018 25th IEEE (ICIP), pp. 2042–2046. IEEE (2018)
11. Liu, X., Cheung, Y.M.: Learning multi-boosted HMMs for lip-password based speaker verification. IEEE Trans. Inf. Forensics Secur. **9**(2), 233–246 (2013)
12. Luettin, J., Maître, G.: Evaluation protocol for the extended M2VTS database (XM2VTSDB). Technical report, IDIAP (1998)
13. Marasco, E., Ross, A.: A survey on antispoofing schemes for fingerprint recognition systems. ACM Comput. Surv. (CSUR) **47**(2), 28 (2015)
14. Parkhi, O.M., Vedaldi, A., et al.: Deep face recognition. BMVC **1**, 6 (2015)

15. Raja, K.B., Raghavendra, R., Vemuri, V.K.: Smartphone based visible iris recognition using deep sparse filtering. Pattern Recogn. Lett. **57**, 33–42 (2015)
16. Schroff, F., Kalenichenko, D., Philbin, J.: Facenet: a unified embedding for face recognition and clustering. In: Proceedings of the IEEE Conference on Computer Vision and Pattern Recognition, pp. 815–823 (2015)
17. Shi, X.X., Wang, S.L., Lai, J.Y.: Visual speaker authentication by ensemble learning over static and dynamic lip details. In: 2016 IEEE (ICIP), pp. 3942–3946. IEEE (2016)
18. Suzuki, K., Tsuchihashi, Y.: A trial of personal identification by means of lip print II. Jap. J. Leg. Med. **23**, 324–325 (1970)
19. Wang, S.L., et al.: Physiological and behavioral lip biometrics: a comprehensive study of their discriminative power. Pattern Recogn. **45**(9), 3328–3335 (2012)
20. Zhang, K., et al.: Joint face detection and alignment using multitask cascaded convolutional networks. IEEE Signal Process. Lett. **23**(10), 1499–1503 (2016)

Intrusion Detection Using Temporal Convolutional Networks

Zhipeng Li, Zheng Qin(✉), Pengbo Shen, and Liu Jiang

School of Software, Tsinghua University, Beijing 100084, China
{lizp14,spb17,jiangl16}@mails.tsinghua.edu.cn,
qingzh@mail.tsinghua.edu.cn

Abstract. Intrusion detection system is an important network security facility. With the fast development of information technology, the information security is getting more serious. On the other side, making the IT equipment more intelligent via AI methods becomes a research hotpot. Recent studies show that temporal convolutional networks can outperform recurrent networks and convolutional architectures in sequence modeling problems. In this work, we propose a data processing method for intrusion detection. We conduct a systematic evaluation of temporal convolutional networks for intrusion detection with NSL_KDD data set. Compared with other standard baseline machine learning methods and some advanced deep learning architectures, the proposed model gives a promising performance in different level tests. With limited computational cost, TCN model converges fast and shows good performance. The proposed model can be easily adjusted to raw inputs and can be extended to large-scale online applications.

Keywords: Intrusion detection · Temporal convolutional networks · NSL_KDD

1 Introduction

With the fast growth of information technology, more and more electrical and electronic equipment are connected to the network, which brings convenience and efficiency. However, the endless incidents of network security result in huge losses and panic emotion to people and the popularity of security technology is making the situation worse. Intrusion detection system (IDS) is designed to recognize the attacks and treats from the network. IDS was first proposed by Denning in 1986 [3]. Intrusion detection is very important in nowadays' severe information security trend. In recent years IDS has evolved to take protection responding to attacks (not alarms only), which makes the facility more crucial and valuable.

In practice, most intrusion detection systems work with a set of rules which define the abnormal behaviors of users, systems and the network protocols. IDS detects attacks via rules that are called misuse mode. In contrast to misuse detection mode, IDS can also work in anomaly detection mode. Anomaly detection

T. Gedeon et al. (Eds.): ICONIP 2019, CCIS 1142, pp. 168–178, 2019.
https://doi.org/10.1007/978-3-030-36808-1_19

mode does not specify specific attack behavior, but gives a description of system running environment parameters by using statistic methods, machine learning or other AI methods. In this work, we using deep learning method to solve the anomaly detection problem.

Intrusion detection systems are applied in a variety of network structures. IDS can be divided into host-based or network-based according to its deploy mode. IDS which deploys on a host and do attacks recognition by monitoring the host events and network flow to host is so called host-based IDS. The other network-based IDS is deployed in a network, usually deploying on a mirror port of network switch and monitor the bypass network flow data. The two deploy methods have respective advantages and disadvantages. There are many famous important intrusion detection systems in history (e.g. Stanford's IDES, UCDavis' NSM). Snort and Bro are both open source software and widely used in both research and practice. In this work, we focus on the network-based intrusion detection system.

Temporal convolutional networks (TCN) is proposed for sequence modeling [1]. Traditional convolutional neural networks are not suitable for sequence modeling problem, because the convolutional kernel size limits the model's long time series gain ability. The temporal convolutional networks use causal convolution structure, dialed convolution structure and residual connections to learn the sequence model. By evaluation of multiple sequence modeling tasks, the temporal convolutional networks can achieve or even exceed various RNN structure.

Inspired by the temporal convolutional networks, we apply the model to intrusion detection scenario. To our best knowledge, there is still no work evaluating the TCN's performance on intrusion detection. We use NSL_KDD to test the model's performance, which is a successor data set of KDD99 and very popular in study work. The remainder of this paper is organized as follows: Sect. 2 introduces related literal intrusion detection works via various deep learning structures. Section 3 gives a detailed specification of temporal convolutional networks and the proposed method to use the model. Section 4 shows the evaluation experiment on the model with analysis and comparison to other methods. In the end, we give our conclusion and the future work in Sect. 5.

2 Related Work

Because the anomaly detection mode can find novel attack forms, there have been a lot of work that uses machine learning methods to achieve intrusion detection. Since deep learning has achieved great success in computer vision, audio processing, natural language processing, etc., literature work has introduced deep models into intrusion detection scenario [10]. Deep learning can do classification tasks without feature engineering. Some early work uses deep structures to extract feature automatically [5–7]. Some work uses different presentations of network data and introduces convolutional neural networks [12,13]. [11] uses semantic representation of the network data and does binary classification tests via LSTM. [19] gives a comprehensive test on recurrent neural network structure.

Some work tries to take the advantages of both CNN or RNN, using variant or hybrid architectures to recognize attacks [17,18,22].

3 Proposed Method

Temporal convolutional network (TCN) is a novel neural network architecture which was proposed in 2018 [1]. Aiming to solve time series prediction problem, TCN introduces 1D convolution, dilated convolution, causal convolution and residual block to make the model sequence sensitive, receptive field scalable, temporal casual sensitive and vanishing gradient immune. Temporal convolutional network is tested on several sequence modeling tasks (e.g. Seq.MNIST, Adding problem, Music JSB Chorales, Word-level PTB etc.) in that work. The performance is better than standard LSTM and GRU, which gives us inspiration to introduce the model into intrusion detection.

The overall architecture is shown in Fig. 1. Our intrusion detection model is based on the network. The raw data from network can be captured either from a network switch mirror port in bypass mode or simply from a network adapter in promiscuous mode. Some software can give a vivid illustration of the pcap format (e.g Wireshark). The raw data is binary bits sequence. Our model can directly process the data format without data preprocessing, and reasons for the advantage will be given in detail in following chapters. However, lots of work uses the abstract KDD99 format, which have statistical results of raw data features and some categorical features parsed by specific protocols. For fair comparison we test the model with this kind of data. We use ordinal encoding to convert the categorical features, with the numeric features with stand scale conversion and Boolean features remaining unchanged. Unlike some work which uses feature engineering, our method simply retains the data information without conversion or feature selection. We will give a detailed analysis of TCN's structure with each kind of feature types and show how the model can get the information of data presentation.

3.1 Temporal Convolutional Networks

There is a lot of work using recurrent neural network to solve intrusion detection issues as sequence modeling. And some work uses hybrid architecture and use RNN-like as independent part of the overall model. RNN-like network can capture temporal sequence information, but still suffers the attenuation problem of long signal. There is also some work trying to use convolutional neural network to solve sequence problem. 1-D convolutional operator is evolved from the original 2-D operator for image recognition and some special variant structures are also introduced from various CNN structure. Every specially designed structure makes the TCN more suitable for sequence modeling problem than RNN.

1-D Convolutions. Convolution operator can be seen as a signal multiplied with its effective function in a time period. It is a sum of weighted discrete

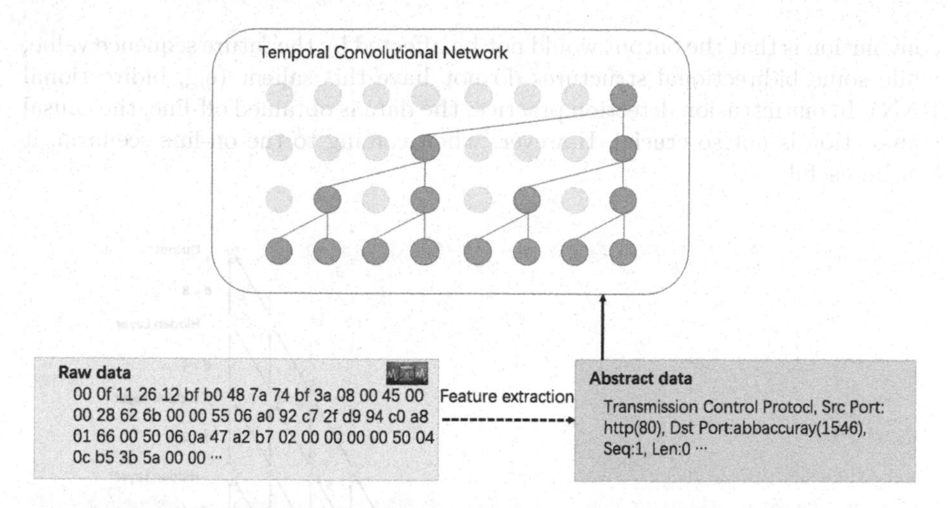

Fig. 1. Overview of Deep & Shallow model for intrusion detection

points. When the concept of convolutional neural networks (CNN) is introduced into neural network, CNN use a 2-D filter to weight the pix feature of the input image and move the filters by rows and columns. That is the 2-D convolution operator's original meaning. When the input is sequence data, the filter need not move by rows and columns, instead just in sequence direction. The comparison illustration is shown in Fig. 2.

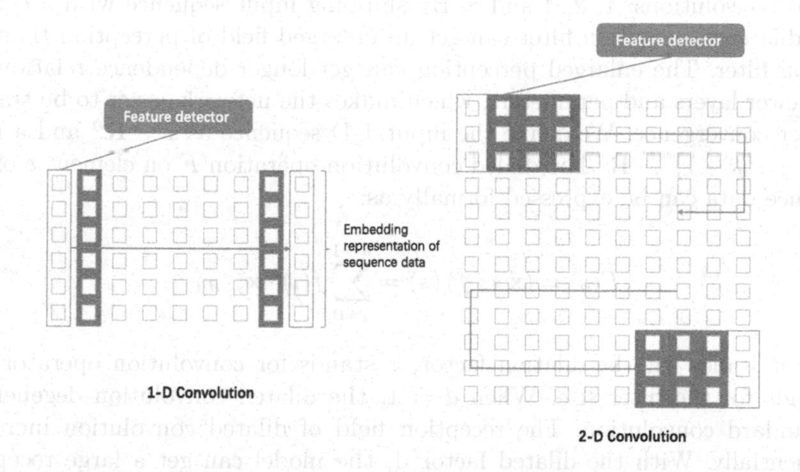

Fig. 2. Comparison of 1-D convolution & 2-D convolution

Causal Convolution. Causal convolution means that the output of time t only has relationship with the input data before it. The convolution operator has inputs from earlier time than time t. A major advantage of using causal convolution is that the output would not be affected by the future sequence value, while some bidirectional structures do not have this salient (e.g. bidirectional RNN). In our intrusion detection practice, the data is obtained off-line, the causal convolution is not so crucial. However, when coming to the on-line scenario, it can be useful.

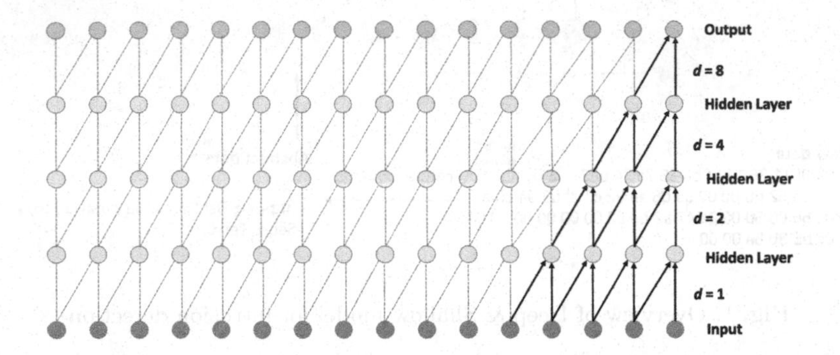

Fig. 3. Illustration of a stack of causal convolutional layers

Dilated Convolution. Dilated convolution is introduced to audio recognition [8,16], signal processing [4,9] and image segmentation [2,21]. Lots of work has shown that dilated convolution structure can get the reception field enlarged by stacking the dilated convolution filters. Figure 3 gives a detailed depict of dilated convolutions: 1, 2, 4 and 8. By skipping input sequence with a certain step, dilated convolution filter can get an enlarged field of perception than the original filter. The enlarged perception can get longer dependence relationship with fewer layers and parameters, which makes the network easier to be trained and get convergence. We define the input 1-D sequence as $x \in \mathbf{R}^n$ and a filter $f : \{0, \dots, k-1\} \rightarrow \mathbf{R}$. The dialed convolution operation F on element s of the sequence data can be expressed formally as:

$$F(s) = (\mathbf{x} *_d f)(s) = \sum_{i=0}^{k-1} f(i) \cdot \mathbf{x}_{s-d \cdot i} \tag{1}$$

where d stands for the dilation factor, $*$ stands for convolution operator and k stands for the filter size. When $d = 1$, the dilated convolution degenerates to standard convolution. The reception field of dilated convolution increases exponentially. With the dilated factor d, the model can get a large reception field through very few layers. E.g. in Fig. 3 the dilated filter was enlarged 8 times. The basic dilated filter structure can be stacked. If there are ns stacks of dilated filter structure, the reception is $ns \cdot kernel\_size \cdot last\_dilation$. By stacking dilated convolution layers, the model can get a very large receptive field.

Residual Connections. Residual Connections can provide a more flex feature expression. With different layer levels extracting different sizes of features, residual connections can recombine them. Formally, the residual block can be written as:

$$o = Activation(\mathbf{x} + \mathcal{F}(\mathbf{x})). \tag{2}$$

And the residual connections are friendly for training deep networks by speeding up convergence. The residual block for TCN is shown in Fig. 4(a), and the residual connection of TCN is shown in Fig. 4(b).

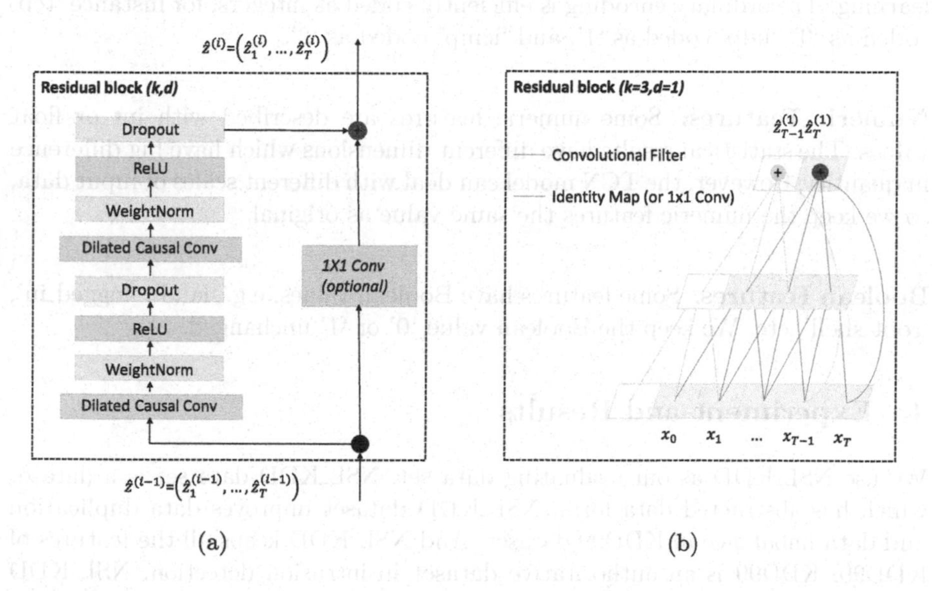

Fig. 4. TCN residual block and an example of residual connection in a TCN

3.2 Data Presentation

There are different kinds of data types in the abstracted data form, in which different features have meaningful physical or statistic information. The famous KDD99 data set and its successor NSL_KDD is abstracted data set. The data form is widely used in evaluation of intrusion detection. 2-D convolutional models need an image transformation of data [12,13], but the feature conversion process may cause precision loss. The RNN structure face the same problem: the quality of embedding operation will affect the learning effect [11,14].

Due to the salient of dilated convolution and the residual block, the different feature scale can be learned with recombination. And by stacking the dilated layers, the model can learn different scale of features. The TCN model can deal with the raw binary type of data or a simple presentation of abstracted data without feature engineering, which is also one of the most outstanding advantages of deep-learning.

Symbolic Features. There are several symbolic features in the abstracted form of data set. E.g. feature 'protocol_type' gives the protocol used in the connection, and its column has 'tcp, udp, icmp' in value. Feature 'service' gives the destination network service and feature 'flag' gives the status of the connection (normal or error). Lots of work uses one-hot encoding to this kind of categorical data and some work uses word2vec models or embedding functions to embedding the data. Due to the strong feature learning ability of TCN, we use an ordinary encoding to encode the category data. While the ordinary encoding retaining all the information, it also has a very short length, which is more conducive to model learning. The ordinary encoding is efficiently coded as integers, for instance 'tcp' coded as '0', 'udp' coded as '1', and 'icmp' coded as '2'.

Numeric Features. Some numeric features are described with int or float values. The statistical results have different dimensions which have big difference in quantity. However, the TCN model can deal with different scales of input data, so we keep the numeric features the same value as original.

Boolean Features. Some features have Boolean values, e.g. 'land', 'logged_in', 'root_shell', etc. We keep the Boolean value '0' or '1' unchanged.

4 Experiment and Results

We use NSL_KDD as our evaluating data set. NSL_KDD data set is a dataset which has abstracted data form. NSL_KDD dataset improves data duplication and data imbalance in KDD99 dataset. And NSL_KDD keeps all the features of KDD99. KDD99 is an authoritative dataset in intrusion detection. NSL_KDD has become a successor dataset of KDD99.

We evaluate the TCN model with different structures, using test dataset of different level to test the model's performance on 2 classifications and 5 classifications problems. And we give comparative analysis of other machine learning base-line methods and some advanced deep learning methods.

4.1 Implementation Detail

We evaluate the TCN model on a PC with a Intel i7-4790 CPU and 8 GB Memory. We use Keras with TensorFlow backend as our emulating software frame work. Three TCN models with different stacked layers are trained by 20 epochs with batch-size = 32 and evaluated by Test^{+} and Test^{-21}. We set the dropout factor = 0.05, dilated factor = 16 and filter numbers = 25 by comparisons of different tests. The training process uses Adam as optimizer and categorical cross entropy as cost function.

4.2 Performance Metric

We use accuracy as a main indicator in our comparative experiments. The precision and recall are also tested to give a description of the model's sensitivity and coverage capacity. And the harmonic mean of precision and recall is reflected by F1 score.

4.3 Results and Comparative Analysis

There are 5 category labels in the dataset: (normal, dos, U2R, R2L, probe) or simply 2 category species (normal, vicious). There are two test dataset Test^{+} and Test^{-21} in NSL_KDD. To evaluate the model's performance on recognizing novel attacks NSL_KDD includes different amount of novel attacks test dataset. Qualitatively, the Test^{-21} includes more difficult data than Test^{+}. We test TCN with different stacked layers for binary classification both on Test^{+} and Test^{-21}. Table 1 shows the details of these tests.

Table 1. Binary classification metrics of different layers of TCN

	Accuracy	Precision	Recall	F1 score
TCN (1 dilation convolutional layer on Test^{+})	**82.56%**	96.96%	71.60%	82.38%
TCN (1 dilation convolutional layer on Test^{-21})	**67.32%**	95.42%	63.11%	75.97%
TCN (2 dilation convolutional layers on Test^{+})	76.80%	95.81%	61.97%	75.26%
TCN (2 dilation convolutional layers on Test^{-21})	57.45%	94.54%	50.95%	66.22%
TCN (3 dilation convolutional layers on Test^{+})	75.50%	94.69%	94.68%	94.69%
TCN (3 dilation convolutional layers on Test^{-21})	48.62%	84.16%	84.13%	84.14%

We can see that TCN with 1 dilation convolutional layer have achieved the best performance on NSL_KDD dataset. We think the NSL_KDD training data sequence is not too long to need a more deeper structure. Shadow models already get to convergence very fast. Raw long sequence data flow may need deeper structure. On the other side, this can be a great advantage of TCN, it is effective and easy to train.

There is a lot of work use NSL_KDD dataset to evaluate deep model. However, some work uses feature engineering to improve performance and some work give their results by cross-validation treating the train set and test set as a whole data set. In fact, the test set (test^{+} and test^{-21}) include many novel attacks, which can effectively test the model's anomaly detection ability. The popular NSL_KDD is proposed by Tavallaee in [14], several machine learning are tested as base-line method. For fair comparisons, we list different models tested on NSL_KDD test dataset with binary classification.

As show in Table 2, our model achieved a accuracy 82.56% on test^{+} and 67.32% on test^{-21}. The performance on a relative easy test^{+} is the second best performance (weaker than RNN). We think the TCN can get more better performance than RNN, if the input sequence is longer. Considering our model is more fast convergence than RNN when the model achieves the same performance nearly, the result is very promising. When tested on a more harder test^{-21} data set, the model get a relatively good score like RNN. Our proposed method outperforms than the base-line machine learning methods

Besides the model's accuracy advantage, the TCN structure has more advantages. First, the TCN model is simple and has fast rate of convergence. Among all the deep learning models, the TCN has a shadow structure and is very easy to train. The effective structure can be applied to on-line scenario. Second, the TCN model can adapt to other input formats very easily. Unlike other method needing feature engineering or feature conversion, the input data of TCN can be raw and simple.

Table 2. Comparison of different models

Model	Test^{+}	Test^{-21}
J48 [15]	81.05%	63.97%
Naive Bayes [15]	76.56%	55.77%
NB Tree [15]	82.02%	66.16%
Random Forest [15]	80.67%	63.26%
Random Tree [15]	81.59%	58.51%
Muti-layer Perceptron [15]	77.41%	57.34%
SVM [15]	69.52%	42.29%
CNN (ResNet50) [12]	79.14%	81.57%
CNN (GoogLeNet) [12]	**77.04%**	**81.84%**
RNN [20]	**83.28%**	**68.55%**
Semantic LSTM [11]	82.21%	66.10%
TCN (proposed method)	82.56%	67.32%

5 Conclusion and Future Work

We evaluate the temporal convolutional network for intrusion detection in this paper. We give the details of how to use KDD99\NSL_KDD data format via TCN model. We compare the TCN model with different layers to find an optimal structure. By comparing with other base-line methods and other advanced deep learning methods, our model achieves a promising performance. With very limited computational cost, TCN model converges fast and gets relatively good accuracy. In addition, this model can easily adjust to raw inputs and can be extended to large-scale online applications.
As future work, we would have more intrusion detection dataset to test the model. And more variant model structures with the dilated convolution layer are also in consideration.

References

1. Bai, S., Kolter, J.Z., Koltun, V.: An empirical evaluation of generic convolutional and recurrent networks for sequence modeling. arXiv preprint arXiv:1803.01271 (2018)
2. Chen, L.C., Papandreou, G., Kokkinos, I., Murphy, K., Yuille, A.L.: Semantic image segmentation with deep convolutional nets and fully connected CRFs. arXiv preprint arXiv:1412.7062 (2014)
3. Denning, D.E.: An intrusion-detection model. IEEE Trans. Software Eng. **2**, 222–232 (1987)
4. Dutilleux, P.: An implementation of the "algorithme à trous" to compute the wavelet transform. In: Combes, J.M., Grossmann, A., Tchamitchian, P. (eds.) Wavelets. Inverse Problems and Theoretical Imaging, pp. 298–304. Springer, Heidelberg (1990). https://doi.org/10.1007/978-3-642-75988-8_29
5. Erfani, S.M., Rajasegarar, S., Karunasekera, S., Leckie, C.: High-dimensional and large-scale anomaly detection using a linear one-class SVM with deep learning. Pattern Recogn. **58**, 121–134 (2016)
6. Fiore, U., Palmieri, F., Castiglione, A., De Santis, A.: Network anomaly detection with the restricted Boltzmann machine. Neurocomputing **122**, 13–23 (2013)
7. Gao, N., Gao, L., Gao, Q., Wang, H.: An intrusion detection model based on deep belief networks. In: 2014 Second International Conference on Advanced Cloud and Big Data, pp. 247–252. IEEE (2014)
8. Haque, A., Guo, M., Verma, P., Fei-Fei, L.: Audio-linguistic embeddings for spoken sentences. arXiv preprint arXiv:1902.07817 (2019)
9. Holschneider, M., Kronland-Martinet, R., Morlet, J., Tchamitchian, P.: A real-time algorithm for signal analysis with the help of the wavelet transform. In: Combes, J.M., Grossmann, A., Tchamitchian, P. (eds.) Wavelets. Inverse Problems and Theoretical Imaging, pp. 286–297. Springer, Heidelberg (1990). https://doi.org/10.1007/978-3-642-97177-8_28
10. Lecun, Y., Bengio, Y., Hinton, G.: Deep learning. Nature **521**(7553), 436 (2015)
11. Li, Z., Qin, Z.: A semantic parsing based LSTM model for intrusion detection. In: Cheng, L., Leung, A.C.S., Ozawa, S. (eds.) ICONIP 2018. LNCS, vol. 11304, pp. 600–609. Springer, Cham (2018). https://doi.org/10.1007/978-3-030-04212-7_53
12. Li, Z., Qin, Z., Huang, K., Yang, X., Ye, S.: Intrusion detection using convolutional neural networks for representation learning. In: Liu, D., Xie, S., Li, Y., Zhao, D., El-Alfy, E.-S.M. (eds.) ICONIP 2017. LNCS, vol. 10638, pp. 858–866. Springer, Cham (2017). https://doi.org/10.1007/978-3-319-70139-4_87
13. Lin, S.Z., Shi, Y., Xue, Z.: Character-level intrusion detection based on convolutional neural networks. In: 2018 International Joint Conference on Neural Networks (IJCNN), pp. 1–8. IEEE (2018)
14. Sheikhan, M., Jadidi, Z., Farrokhi, A.: Intrusion detection using reduced-size RNN based on feature grouping. Neural Comput. Appl. **21**(6), 1185–1190 (2012)
15. Tavallaee, M., Bagheri, E., Lu, W., Ghorbani, A.A.: A detailed analysis of the KDD CUP 99 data set. In: 2009 IEEE Symposium on Computational Intelligence for Security and Defense Applications, pp. 1–6. IEEE (2009)
16. Van Den Oord, A., et al.: Wavenet: a generative modelfor raw audio. SSW **125** (2016)
17. Vinayakumar, R., Soman, K., Poornachandran, P.: Applying convolutional neural network for network intrusion detection. In: 2017 International Conference on Advances in Computing, Communications and Informatics (ICACCI), pp. 1222–1228. IEEE (2017)

18. Wei, W., Sheng, Y., Wang, J., Zeng, X., Ming, Z.: HAST-IDS: learning hierarchical spatial-temporal features using deep neural networks to improve intrusion detection. IEEE Access **6**(99), 1792–1806 (2018)
19. Yin, C.L., Zhu, Y.F., Fei, J.L., He, X.Z.: A deep learning approach for intrusion detection using recurrent neural networks. IEEE Access **5**(99), 21954–21961 (2017)
20. Yin, C., Zhu, Y., Fei, J., He, X.: A deep learning approach for intrusion detection using recurrent neural networks. IEEE Access **5**, 21954–21961 (2017)
21. Yu, F., Koltun, V.: Multi-scale context aggregation by dilated convolutions. arXiv preprint arXiv:1511.07122 (2015)
22. Yu, Y., Long, J., Cai, Z.: Network intrusion detection through stacking dilated convolutional autoencoders. Secur. Commun. Netw. **2017**, 1–10 (2017)

Empirical Study of Easy and Hard Examples in CNN Training

Ikki Kishida(✉) and Hideki Nakayama

Graduate School of Information Science and Technology, The University of Tokyo, Tokyo, Japan
{kishida,nakayama}@nlab.ci.i.u-tokyo.ac.jp

Abstract. Deep Neural Networks (DNNs) generalize well despite their massive size and capability of memorizing all examples. There is a hypothesis that DNNs start learning from simple patterns and the hypothesis is based on the existence of examples that are consistently well-classified at the early training stage (i.e., *easy examples*) and examples misclassified (i.e., *hard examples*). Easy examples are the evidence that DNNs start learning from specific patterns and there is a consistent learning process. It is important to know how DNNs learn patterns and obtain generalization ability, however, properties of easy and hard examples are not thoroughly investigated (e.g., contributions to generalization and visual appearances). In this work, we study the similarities of easy and hard examples respectively for different Convolutional Neural Network (CNN) architectures, assessing how those examples contribute to generalization. Our results show that easy examples are visually similar to each other and hard examples are visually diverse, and both examples are largely shared across different CNN architectures. Moreover, while hard examples tend to contribute more to generalization than easy examples, removing a large number of easy examples leads to poor generalization. By analyzing those results, we hypothesize that biases in a dataset and Stochastic Gradient Descent (SGD) are the reasons why CNNs have consistent easy and hard examples. Furthermore, we show that large scale classification datasets can be efficiently compressed by using *easiness* proposed in this work.

Keywords: Easy examples · Hard examples · Deep Neural Networks · Dataset compression

1 Introduction

From a traditional perspective of generalization, overly expressive models can memorize all examples and result in poor generalization. However, deep neural networks (DNNs) achieve an excellent generalization performance even if models are over-parameterized [16]. The reason for this phenomenon remains unclear. Arpit et al. [1] show that DNNs do not memorize examples, and propose a hypothesis that DNNs start learning from simple patterns. Their hypothesis is

T. Gedeon et al. (Eds.): ICONIP 2019, CCIS 1142, pp. 179–188, 2019.
https://doi.org/10.1007/978-3-030-36808-1_20

based on the existence of examples that are consistently well-classified at the early training stage (i.e., easy examples) and examples misclassified (i.e., hard examples). If DNNs memorize examples in brute force way, easy examples should not exist. Easy examples are the evidence that DNNs start learning from specific patterns and there is a consistent learning process. Therefore, we believe that analyzing easy and hard examples is one of the keys to understanding what kind of learning process DNNs have and how DNNs obtain generalization ability.

In this work, we study easy and hard examples, and their intriguing properties are shown. For our experiments, we introduce *easiness* as a metric to measure how early examples are classified correctly. In addition, we calculate the matching rates of easy and hard examples between different CNN architectures. As a result, we discover that both easy and hard examples are largely shared across CNNs, and easy examples are visually similar to each other and hard examples are visually diverse.

These results imply that CNNs start learning from a larger set of visually similar images and we hypothesize that easy and hard examples originate from biases in a dataset and Stochastic Gradient Descent (SGD). A dataset naturally contains various biases leading some images to appear as a majority or a minority. For instance, if there are many white dogs and rarely black dogs in dog images, the majority of visually similar images (i.e., white dogs) become easy examples and visually unique examples (i.e., black dogs) become hard examples. Since SGD randomly picks samples for training a model, discriminative patterns in easy examples tend to be focused more than those in hard examples. Thus, the gradient values of easy examples dominate the direction of the update at the beginning of training. Such intra-class biases are the reason why some examples are classified well at an early training stage.

According to this hypothesis, the gradient values of easy examples are thought to be redundant and we may be able to remove easy examples without significantly affecting generalization ability. To investigate how easy and hard examples contribute differently to generalization, we conduct ablation experiments. We find that hard examples contribute more to generalization than easy examples, however, removing a large number of easy examples leads to poor generalization. By using *easiness*, we show that datasets can be efficiently compressed than random selection even in the large-scale ImageNet-1k dataset [11]. Our contributions are as follows:

- We propose *easiness* to measure how early an example is classified correctly
- Empirical finding and analysis of easy and hard examples based on *easiness*. For instance, easy examples are visually similar to each other and hard examples are visually diverse, and both easy and hard examples are largely shared across different CNN architectures. We hypothesize such properties originate from the biases in the dataset and SGD.
- We demonstrate dataset compression by *easiness*. It is more efficient than random selection and works even for the large-scale ImageNet-1k dataset.

2 Method

2.1 Easiness

To measure how early an example is classified correctly, we introduce *easiness* $e_{x_i}^T \in \mathbb{R}$ as a criterion, where x_i represents one example and $T \in \mathbb{N}$ is the number of the model updates. For a criterion of how correctly a model classifies the example, the loss value is appropriate. However, since the model is stochastically updated, the loss value is uncertain in a single trial. To improve the certainty of the loss value, it is necessary to take an average of the loss value over several times. We propose *easiness* $e_{x_i}^T$ that is the averaged loss value as follows:

$$e_{x_i}^T = \frac{1}{M} \sum_{m=1}^{M} L(\mathbf{t}_i, f(\mathbf{x}_i, \mathbf{W}_m^T)), \tag{1}$$

where $f(\mathbf{x}_i, \mathbf{W}_m^T)$ is the prediction and t_i is the corresponding ground truth label. L is the loss function, for which we use the cross-entropy in this work since we focus on image classification. $M \in \mathbb{N}$ is the number of trials and we set M as 10 in this work.

In this work, we define **10% of the examples with the lowest *easiness* as easy examples** and **10% of the highest as hard examples**.

2.2 Matching Rate

It is important to know how large easy and hard examples are shared between various CNN architectures. If easy and hard examples are not shared, it means that the learning process depends on the architecture of CNN and model-dependent analysis would be required. To calculate the consistency of the set of examples, we use matching rate in this work. Let us consider two different sets of examples X_A and X_B. The matching rate $M_{AB} \in [0, 1]$ between X_A and X_B is calculated as

$$M_{AB} = \frac{|X_A \cap X_B|}{\max(|X_A|, |X_B|)}, \tag{2}$$

where $|\ |$ denotes the size of a set.

3 Experiments

3.1 Preparations

We use CIFAR-10 [5] and ImageNet 2012 dataset (ImageNet-1k) [11] for our experiments.

CIFAR-10. CIFAR-10 is the image classification dataset. There are 50000 training images and 10000 validation images with 10 classes. For data augmentation and preprocessing, translation by 4 pixels, stochastic horizontal flipping, and global contrast normalization are applied onto images with 32×32 pixels. We use three types of models of WRN 16-4 [15], DenseNet-BC 12-100 [4] and ResNeXt 4-64d [14].

ImageNet-1k. ImageNet-1k is the large scale dataset for the image classification. There are 1.28M training images and 50k validation images with 1000 classes. For data augmentation and preprocessing, resizing images with the scale and aspect ratio augmentation and stochastic horizontal flipping are applied onto images. Then, global contrast normalization is applied to randomly cropped images with 224×224 pixels. In this work, we use AlexNet [6], ResNet-18 [3], ResNet-50 and DenseNet-121 [4].

As the optimizer, we use Momentum SGD with 0.9 momentum and weight decay of 0.0001. The initial learning rate is 0.1 and it is divided by 10 at [150th, 250th] epochs and [100th, 150th, 190th] epochs on CIFAR-10 and ImageNet-1k, respectively.

3.2 Visual Property of Easy and Hard Examples

Figure 1 shows easy and hard examples in CIFAR-10 and ImageNet-1k dataset. Regardless of the size of the dataset, easy examples are visually similar to each other, and hard examples tend to be visually diverse.

In [2,10], the diversity of images is investigated by averaging the group of images. The more diverse the images are, the more uniform the average image is. The averaged images of easy and hard examples are shown in Fig. 2. The averaged image of hard examples is more uniform than the averaged easy or random examples, thus hard examples are the most diverse among three.

Those results imply that CNNs start learning from a large set of visually similar images.

3.3 Are Easy and Hard Examples Are Common Between Different CNN Architectures?

To investigate whether easy and hard examples are shared across different CNN architectures, we calculate matching rates according to *easiness*.

Results are shown in Fig. 3. The horizontal axis is the epoch and the vertical axis is the matching rate of easy and hard examples between different CNN architectures. Easy and hard examples are largely shared at an early epoch and the matching rate is high across any architectures compared to random case. These results indicate that the learning process is similar regardless of the difference in the architecture design of CNN.

(a) Easiest examples of dog (b) Hardest examples of dog

(c) Easiest examples of panda (d) Hardest examples of panda

Fig. 1. Easiest and hardest examples of CIFAR-10 and ImageNet-1k dataset. **(a–b)** are from CIFAR-10 with *easiness* of WRN 16-4. **(c–d)** are from ImageNet-1k with *easiness* of AlexNet.

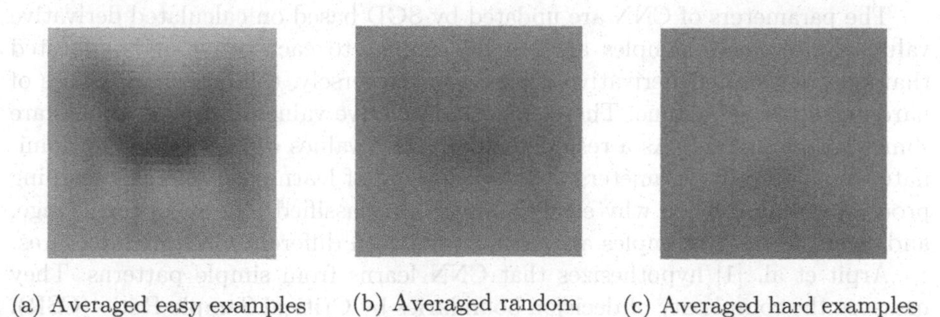

(a) Averaged easy examples of horse (b) Averaged random examples of horse (c) Averaged hard examples of horse

Fig. 2. Averaged easy, random and hard examples of the horse in CIFAR-10. The *easiness* is calculated by WRN 16-4. Each average image uses 500 images.

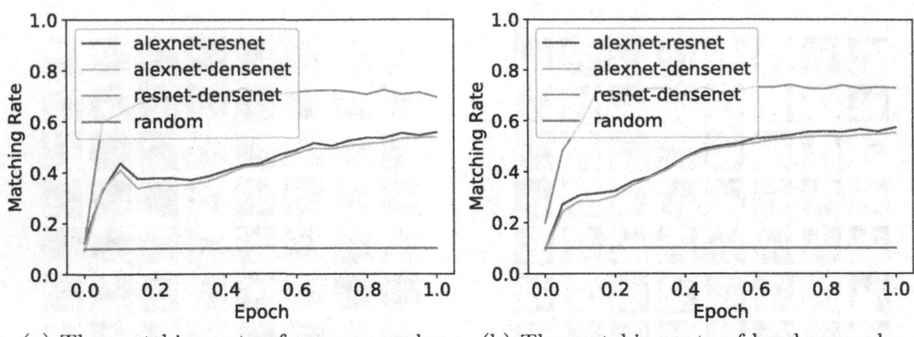

(a) The matching rate of easy examples (b) The matching rate of hard examples

Fig. 3. The matching rate of easy and hard examples between different CNNs in ImageNet-1k. "random" represents the chance rate of the case that 10% of images are randomly sampled.

3.4 Why Some Examples Are Consistently Easy or Hard?

Results in previous experiments show that

- CNNs start learning from a larger set of visually similar images,
- Easy and hard examples are largely shared across different CNN architectures.

We hypothesize that this phenomenon originates from dataset biases and Stochastic Gradient Descent (SGD).

There are many biases in the dataset and [13] mentions several biases in a dataset. *Selection bias* means that examples in a dataset tend to have particular kinds of images (example: there are many examples of a sports car in the car category). *Capture bias* represents the manner in which photos are usually taken (example: a picture of a dog is usually taken from the front with the dog looking at the photographer and occupying most of the picture). Easy examples are the result of such biases.

The parameters of CNN are updated by SGD based on calculated derivative values. Since easy examples are visually similar to each other, it is expected that they get similar derivative values, and conversely, the derivative values of hard examples are unique. Therefore, the derivative values of easy examples are somewhat redundant. As a result, the derivative values of easy examples dominate the update of parameters at the beginning of learning. From this learning process, we can explain why easy examples are classified well at an early stage, and easy and hard examples are common between different CNN architectures.

Arpit et al. [1] hypothesizes that CNN learns from simple patterns. They measure the complexity of decision boundaries by Critical Sample Ratio (CSR). CSR counts how many training examples are fooled by adversarial noises with radius r. The higher CSR is, the more complex decision boundaries are. Their results show that CSR becomes higher as CNN continues training, indicating that CNN firstly learns from simple patterns. However, their hypothesis does not explain our results well such as the question of "why the learned simple patterns are consistent between different CNNs".

Our hypothesis is the extension of Arpit et al. [1] in this respect. We argue that CNNs firstly learn from simple patterns and such patterns are affected by the intra-class biases in a dataset.

3.5 Generalization and Easiness

We perform ablation experiments on easy and hard examples to investigate if they equally contribute to the generalization ability. For this purpose, we decide which examples to ablate based on *easiness*. In detail, we first normalize *easiness* $e^T_{x_i}$ by dividing each $e^T_{x_i}$ by $\sum_{i=1}^N e^T_{x_i}$, where N is the size of examples. We randomly select which to ablate by using the normalized *easiness*.

The result is shown in Fig. 4(a, b). The horizontal axis is the ablation ratio and the vertical axis is accuracy. If ablation ratio is 0.3, then the size of the training dataset is 70%. "easy", "hard" and "random" on figures means easy, hard and randomly selected examples are mainly removed, respectively. "stepwise" is the gradual case of "easy".

As can be seen in Fig. 4a, removing hard examples consistently degrades the classification performance more drastically than other strategies. Therefore, we conclude that hard examples contribute more to generalization than easy examples do.

However, as can be seen in "easy" of Fig. 4a, if we remove too many easy examples, the accuracy starts degrading sharply. This phenomenon can be explained by our hypothesis. Since a dataset is randomly split into training and testing subsets, training and testing share the same biases. For example, if white dogs are easy examples and black dogs are hard examples in the training dataset, there are more white dogs than black dogs in the testing dataset too. Thus, if the trained model fails to learn white dogs (i.e., easy examples), the test accuracy will drop sharply since there are many white dogs in the testing dataset. Therefore, it is better to keep some easy examples even though redundant images can be ablated with less affecting generalization ability.

"stepwise" keeps some of the easy examples while ablating them. As can be seen in Fig. 4, "stepwise" gives the best performance. In addition, in Fig. 4b, "stepwise" outperforms "random" case even in the large-scale ImageNet-1k dataset. The difference in accuracy between "random" and "stepwise" is around 1.1% at 0.3 ablation ratio. It is approximately worth extra 100k images to achieve the comparable accuracy in "random".

4 Related Work

A dataset naturally contains various biases. For instance, Ponce et al. [10] shows some averaged images of Caltech-101 [9] are not homogeneous and recognizable. They claim that Caltech-101 may have *inter-class* variability but lacks *intra-class* variability. In this work, we find that easy examples lack *intra-class* variability, and hard examples are more diverse than easy examples.

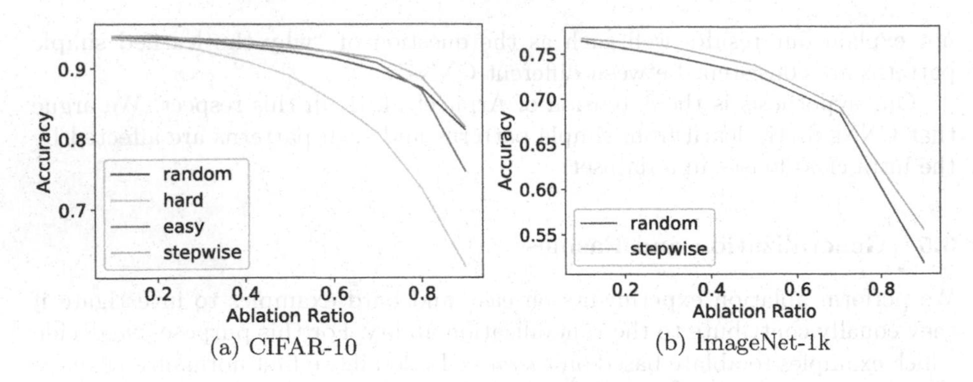

(a) CIFAR-10 (b) ImageNet-1k

Fig. 4. The result of the ablation experiments. The vertical axis is accuracy and the horizontal axis is ablation ratio. If ablation ratio is 0.3, it means that 30% of examples in the dataset are discarded. WRN 16-4 and ResNet-50 are used respectively for CIFAR-10 and ImageNet-1k dataset. "easy", "hard" and "random" on figures means easy, hard and randomly selected examples are mainly removed, respectively. "stepwise" is the gradual case of "easy". Unlike "easy" that ablating examples in one shot, every time "stepwise" ablates 10% of easy examples in the dataset and re-calculate *easiness* until reaching the target ablation ratio.

Arpit et al. [1] investigate the memorization of DNNs, and claims that DNNs tend to prioritize learning simple patterns first. They analyze the complexity of the decision boundary based on Critical Sample Ratio (CSR). CSR is the criterion of how many training examples change the predictions by adding adversarial noises with radius r. The high CSR means that CNN has complex decision boundaries. Arpit et al. [1] empirically show that CSR becomes higher as CNNs continue training and propose the hypothesis that CNN learns from simple patterns. However, their hypothesis does not explain why firstly learned simple patterns are consistent between different CNN architectures. Our hypothesis is the extension of [1] in this respect. We argue that CNN firstly learns from simple patterns and such patterns are affected by the intra-class biases in a dataset.

Lapedriza et al. [7] investigate well-classified examples and misclassified examples at the end of training based on SVMs with hand-crafted features in the small-scale datasets. They conclude that some examples are the reason to degrade the model's performances and examples with high loss values contribute to generalization well. In this work, we empirically investigate and analyze properties of examples at an early training stage in CNNs, and perform experiments on the large-scale ImageNet-1k dataset.

Toneva et al. [12] empirically investigate forgettable and unforgettable examples on small-scale MNIST [8] and CIFAR-10 dataset. The difference of the metric between [12] and our *easiness* is to use the loss values of the model with small updates unlike tracking the degradation of the accuracy across the whole training period in [12]. Table 1 shows the matching rates between easy and hard examples at the beginning and at the end of the training. As can be seen in Table 1, easy and hard examples are different, especially in ResNeXt. Therefore, we assume easy and hard examples are different from forgettable and unforgettable examples since [12] use records of the whole training period.

Table 1. The matching rates between easy and hard examples at the beginning and at the end of training in CIFAR-10. The chance rate is 0.1.

Model	Easy or hard	Matching rate
WRN 16-40	Easy	0.18
	Hard	0.255
DenseNet-BC 12-100	Easy	0.24
	Hard	0.279
ResNeXt 4-64d	Easy	0.129
	Hard	0.174

5 Conclusion

In this work, easy and hard examples are investigated to understand the learning process of DNNs.

Firstly, the metric of *easiness* is introduced to define easy and hard examples. Then, we discover that easy and hard examples are common among different CNN architectures, and easy examples are visually similar to each other and hard examples are visually diverse. To explain these phenomena, we propose the hypothesis that biases in the dataset and SGD make some examples easy or hard.

From this hypothesis, we consider easy examples are visually redundant and can be removed without significantly affecting the generalization ability of a model. In ablation experiments, we demonstrate that hard examples contribute more to generalization ability than easy examples in CIFAR-10 and the large-scale ImageNet-1k dataset. Therefore, the dataset can be efficiently compressed than random selections by using *easiness*.

For future work, further analysis of intra-class biases is fruitful directions. In addition, studying how to design biases in a dataset is promising directions to control the learning process of CNNs.

Acknowledgements. This work was supported by JSPS KAKENHI Grant Number JP19H04166. We would like to thank the two anonymous reviewers for their valuable feedback on this work.

References

1. Arpit, D., et al.: A closer look at memorization in deep networks. In: ICML (2017)
2. Deng, J., Dong, W., Socher, R., Li, L.J., Li, K., Fei-Fei, L.: ImageNet: a large-scale hierarchical image database. In: CVPR (2009)

3. He, K., Zhang, X., Ren, S., Sun, J.: Deep residual learning for image recognition. In: CVPR (2016)
4. Huang, G., Liu, Z., van der Maaten, L., Weinberger, K.Q.: Densely connected convolutional networks. In: CVPR (2017)
5. Krizhevsky, A.: Learning multiple layers of features from tiny images. Technical report (2009)
6. Krizhevsky, A., Sutskever, I., Hinton, G.E.: ImageNet classification with deep convolutional neural networks. In: NIPS (2012)
7. Lapedriza, À., Pirsiavash, H., Bylinskii, Z., Torralba, A.: Are all training examples equally valuable? arXiv (2013)
8. LeCun, Y., Cortes, C., Burges, C.J.: The MNIST database of handwritten digits (1999). http://yann.lecun.com/exdb/mnist/
9. Li, F., Fergus, R., Perona, P.: Learning generative visual models from few training examples: an incremental Bayesian approach tested on 101 object categories. In: CVPR Workshops (2004)
10. Ponce, J., et al.: Dataset issues in object recognition. In: Ponce, J., Hebert, M., Schmid, C., Zisserman, A. (eds.) Toward Category-Level Object Recognition. LNCS, vol. 4170, pp. 29–48. Springer, Heidelberg (2006). https://doi.org/10.1007/11957959_2
11. Russakovsky, O., et al.: ImageNet large scale visual recognition challenge. IJCV **115**(3), 211–252 (2015)
12. Toneva, M., Sordoni, A., des Combes, R.T., Trischler, A., Bengio, Y., Gordon, G.J.: An empirical study of example forgetting during deep neural network learning. In: ICLR (2019)
13. Torralba, A., Efros, A.A.: Unbiased look at dataset bias. In: CVPR (2011)
14. Xie, S., Girshick, R., Dollár, P., Tu, Z., He, K.: Aggregated residual transformations for deep neural networks. In: CVPR (2017)
15. Zagoruyko, S., Komodakis, N.: Wide residual networks. In: BMVC (2016)
16. Zhang, C., Bengio, S., Hardt, M., Recht, B., Vinyals, O.: Understanding deep learning requires rethinking generalization. In: ICLR (2017)

GCNDA: Graph Convolutional Networks with Dual Attention Mechanisms for Aspect Based Sentiment Analysis

Junjie Chen[1,2], Hongxu Hou[1(✉)], Jing Gao[2], Yatu Ji[1], Tiangang Bai[1], and Yi Jing[3]

[1] College of Computer Science, Inner Mongolia University, Hohhot, China
chenjj@imau.edu.cn, cshhx@imu.edu.cn, jiyatu0@126.com

[2] College of Computer Science and Information Engineering, Inner Mongolia Agricultural University, Hohhot, China
gaojing@imau.edu.cn

[3] Faculty of Science, The University of Sydney, Camperdown, Australia
yjin5439@uni.sydney.edu.au

Abstract. As the amount of user-generated content on the web continues to increase, a great interest has been shown in aspect-level sentiment analysis, which provides more detailed information than general sentiment analysis. In recent years, neural-based models have achieved success in this task because of their powerful representation learning capabilities. However, they ignore that the sentiment polarity of the target is related to the entire text structure. In this paper, we present a method based on graph convolutional neural networks named GCNDA, in which the given text is considered as a graph and the target is the specific region of the graph. Dual graph-based attention models are used to concentrate on the relation between words and certain regions of the graph. We conduct comprehensive experiments on publicly accessible datasets, and results demonstrate that our model outperforms the state-of-the-art baselines.

Keywords: Aspect Based Sentiment Analysis · Graph Convolutional Networks · Attention mechanisms

1 Introduction

Sentiment analysis [19], also known as opinion mining [13,20], is a vital task in text mining. For example, consumers want to know sentiment of existing users about products, meanwhile business want to obtain public opinions for their decision making. Due to its great value in practical applications, it has attracted widespread attention from both the industry and academic communities.

Supported by Inner Mongolia Natural Science Foundation of China (2018MS06005, 2015MS0628) and Inner Mongolia Autonomous Region Key Laboratory of Big Data Research and Application of Agriculture and Animal Husbandry.

T. Gedeon et al. (Eds.): ICONIP 2019, CCIS 1142, pp. 189–197, 2019.
https://doi.org/10.1007/978-3-030-36808-1_21

Typically, users write both positive and negative aspects in the same review, although the general sentiment may be positive or negative. Given the review *"the food is so good and so popular that waiting can really be a nightmare"*. It expresses negative sentiment towards "waiting" while holding positive sentiment towards "food". Aspect Based Sentiment Analysis (ABSA) is a fine-grained task in the field of sentiment classification [13,22]. The goal of this subtask is to predict the sentiment polarity of aspects that appear in a given text.

The existing deep learning models mainly rely on Recurrent Neural Networks (RNNs) [12,16,23,26,30], Memory Networks [4,6,14,24] and Convolutional Neural Networks (CNNs) [28]. However, these architectures run on a grid or sequential structure without using the entire graphical text structure. Therefore, they are difficult to obtain a structural relationship between words [29].

Text graph has been widely used in NLP tasks [1,7,18,27], which can grasp the entire text structure. Moreover, the Graph Convolutional Networks (GCNs) [9] have demonstrated the powerful ability to obtain hidden representation of nodes in a graphical structure. Inspired by this, we develop a deep learning framework based on graph convolutional networks to obtain structural information. To emphasis the relation between context and aspect words, we design dual graph-based attention models for the ABSA task. One is Graph Attention Mechanism (GAM), which learns attention weights for words to different neighbor words in the context. The other is an Aspect-based Structural Attention Mechanism (ABSAM), which focus on the part of aspect terms in the text graph.

2 Graph Convolutional Networks

Graph convolutional networks [3] have been proposed for learning over graphs. The majority of these methods do not scale to large graphs or are designed for whole-graph classification. Kipf and Welling [9] proposed a localized first-order approximation of spectral graph convolutions, which is very effective for the semi-supervised nodes classification. Since the GCNs succeeded in nodes classification, they have been introduced into NLP tasks such as semantic role labeling [17], machine translation [2].

3 Our Model

We assume that sentiment polarity of aspect terms is not only related to target words, but also to the whole text. Since it is necessary to take fully account of the relation between words in a given text, our model is based on the graph where the text is regarded as a graph and the aspect terms are considered part of the graph.

Our framework (GCNDA) can be divided into two parts, one is a text graph representation and the other is an aspect-based structural attention. We employ multiple GCN layers with GAM to get the text hidden state, and the aspect-based structural attention model to obtain the specific regions representation.

Then, two hidden states are fed forward to the fully connected (FC) layer. Finally, the representation vectors are put into the softmax layer to get the class label. In the following, we will explain each part of the framework in detail.

3.1 Graph Construction

In our approach, operations are performed on the text graph, so the structure of the text is important. To illustrate the effectiveness of our model, we construct undirected text graph in two ways, one based on co-occurrence information and the other on syntactic dependencies, which are widely used in the literature.

For the given text, each vertex corresponds to a word. For the co-occurrence graph, if two nodes v_i and v_j have a co-occurrence relation, the edge (v_i, v_j) is established. Where co-occurrence relation is defined as two nodes co-occur within the specific window size. The edge weight of (v_i, v_j) is the number of co-occurrences. For the syntactic dependency graph, establish connections for the dependencies where two nodes belong to a specific part of speech set. The adjacency matrix A is obtained by the undirected graph structure.

3.2 Text Graph Representation

After the graph construction, the given text is converted to a graph. The text graph representation is obtained by input nodes and adjacency matrix. Each node corresponds to the word w_{si} and is represented by the vector x_i of the dimension D after the embedding layer. Usually, the length N of words set in the corpus is larger than the length M of the set in the text. Since it is not necessary to build the adjacency matrix A$\in \mathbb{R}^{N\times N}$ for each given text, we covert nodes sequence to the current text words sequence, then the adjacency matrix A becomes $\mathbb{R}^{M\times M}$, and L represents the Laplacian matrix of A.

After embedding layer, the initial representation of nodes is defined as $H^{(0)} = X \in \mathbb{R}^{M\times D}$. The GAM produces the importance of node N_j to node N_i, and the attention coefficient is computed by

$$e_{ij} = score(H_i^{(0)}, H_j^{(0)}) \tag{1}$$

Where $score(\cdot)$ is the attention function. As mentioned in [15], the score function can be divided into "dot", "general" and "concat". In our model the "general" is used, the attention score is computed by following formula.

$$e_{ij} = H_i^{(0)}(W_{att}H_j^{(0)})^T \tag{2}$$

Where W_{att} is trainable parameters. The equation indicates the importance of the node to each node in the graph without any structural information. We perform masked attention similar to [25], injecting the graph structure into the mechanism as Eq. 6.

$$Att_{ij} = \begin{cases} \frac{\exp(e_{ij})}{\sum_{k\in N(i)}\exp(e_{ik})} & if\ j \in N(i) \\ 0 & others \end{cases} \tag{3}$$

Where $N(i)$ is the neighbor set of node i.

The output of the $l-th$ GCN hidden layer can be obtained by the following equation.

$$H^{(l+1)} = Relu(\beta \times LH^{(l)}W^{(l)} + (1-\beta) \times AttH^{(l)}W^{(l)}) \quad (4)$$

Where β is a hyperparameter between 0 and 1, $W^{(l)} \in \mathbb{R}^{F \times F'}$ is a weight matrix in $l-th$ GCN layer and, and $Att \in \mathbb{R}^{M \times M}$ is an attention matrix obtained from Eq. (6). F and F' are the input and output feature sizes, respectively. Compared with Eq. (3), the GCN layers in our model incorporate the graph attention mechanism. Unlike the attention method in the graph attention network [25], the attention weights in our model are shared among all GCN layers which is called GAM. Equation (6) describes the GCN hidden representation consists of two parts: one is calculated by the attention matrix and the other is the adjacency matrix. We assume the attention matrix represents the knowledge of the relationship between the nodes acquired by training, while the adjacent matrix is the relationship between nodes in the current context, both of which are proportionally combined output vectors. The output of the GCNs is forwarded to the pooling operations layers, which is the element operation on hidden vectors to get representation of the text graph.

3.3 Aspect-Based Structural Attention

For the general graph classification based on sequential methods, one challenge is to give the order of the recurrent neural networks [10], while for text mining, word sequences in the original text can be used naturally.

We extend the structural attention model [11] for ABSA task. See articles [11] for more details on the structural attention model. The sequences are produced through the aspect sequence generator. First the generator produces a set of nodes, referred to herein as aspect structural nodes, which are in the largest connected subgraph containing the aspect words. Then two agents are defined, one from the left and the other from right traversing to aspect terms. At each step, agents move along the text sequence to the word that are in aspect structural nodes. Obviously, in our approach the rank vector is the distance between the current node and aspect structural nodes. The ultimate goal of the agent is to collect enough information about the aspect terms. Finally, the generator generates a left sequence and a right sequence that are fed forward to the left and right LSTM layers, respectively. The hidden state of the aspect structural attention is obtained by concatenating the last hidden vectors of the left and right LSTM.

4 Experiments and Results

4.1 Experimental Setting

We test our model on three public datasets, two of them come from SemEval 2014 [22], and the third is a collection of twitters [5]. SemEval 2014 includes user-generated reviews of laptop and restaurant domains, following previous work [24],

Table 1. Statistics of aspects in different datasets

Datasets	Positive		Negative		Neutral	
	Train	Test	Train	Test	Train	Test
Restaurant	2164	728	805	196	633	196
Laptop	987	341	866	128	460	169
Twitter	1561	173	1560	173	3126	346

we removed a few examples having the "conflict" label. The statistics of the datasets are shown in Table 1.

Our models are performed on co-occurrence graph and on syntactic dependency graph, denoted as GCNDAc and GCNDAs, respectively. The window size is set to 2 for co-occurrence graph construction, and Stanford parser[1] is used for syntactic dependency graph construction.

In experiments, four-layer GCNs with RELU activation function is developed. We use Adam [8] optimizer with the learning rate 0.01, dropout 0.2, and the maximum number of epoch 50. 300-dimensional word embeddings pre-trained by GloVe [21] are utilized, which are not tuned during training time.

4.2 Compared Methods

We compare our model with following baseline methods:

TD-LSTM [23] is a model based on LSTM network, in which two LSTM models are used to model the preceding and following contexts surrounding the target string for sentiment classification.

MemNet [24] is a neural attention model over an external memory, which consists of multiple computational layers.

RAM [4] is a framework that adopts multiple-attention mechanism on recurrent neural network.

IAN [16] is an interactive attention networks model. It uses two attention networks to model the target and context interactively.

Cabasc [14] is based on the memory model, which can solve the semantic mismatch problem through two attention mechanisms, namely sentence-level content attention mechanism and context attention mechanism.

GCAE [28] is based on convolutional neural networks and gating mechanisms.

We note that in the different literature, different results are reported for the same model performed on the same dataset. We think that the results of the baseline methods are affected by text preprocessing and word embeddings, as mentioned in [16]. To reveal the capability of models, same word vectors used in our models are applied to all baselines.

[1] https://nlp.stanford.edu/software/lex-parser.shtml.

4.3 Main Results

For all methods accuracy evaluation is used as metric, and results are shown in Table 2. The best scores are highlighted in bold and the underlines indicate the second best performances. As the results show, our two models, GCNDAc and GCNDAs, consistently outperform all comparison methods on these three datasets. GCNDAs outperforms GCNDAc on Laptop and Restaurant datasets. This may be due to the fact that the syntactic dependency graph establishes a connection between two long distance words, shortening the distance between the aspect and the related words. However, the text in Twitter is irregular and short, the dependency parsing is not guaranteed to work well. Although the performance of GCNDA on the co-occurrence graph is not optimal, it is easier to construct a co-occurrence graph than to build a syntactic dependency graph, and its performance is superior to other baselines.

Table 2. Results of our model against baselines.

Methods	Restaurant	Laptop	Twitter
TD-LSTM	75.17	66.94	67.72
MemNet	76.88	68.18	69.63
IAN	76.96	67.86	68.63
RAM	76.87	67.24	<u>69.88</u>
Cabasc	77.05	68.65	67.33
GCAE	76.12	68.65	69.79
GCNDAs	**79.35**	**72.88**	**70.81**
GCNDAc	<u>78.93</u>	<u>70.21</u>	**70.81**

In LSTM-based models, TD-LSTM and IAN, IAN has better results than TD-LSTM because IAN uses context and target attention mechanisms, which make better use of important parts of a sentence for aspect words. MemNet is based on memory network, containing multiple attention layers, superior to LSTM-based models on Laptop and Twitter. RAM achieves the best performances on Twitter among baselines, which adopts not only the multi-hop attention mechanism but also deep bidirectional LSTM. Compared with RAM and MemNet, Cabasc enhances the ability to capture important information about a given aspect from a global perspective by sentence-level content attention mechanism and context attention mechanism, thus has a best performance in all baselines on Laptop and Restaurant. GCAE utilizes convolutional neural network with gating mechanisms, obtaining the best result as Cabasc on Laptop.

4.4 Effect of Hyperparameter β

The hyperparameter β is used in GCN layers, which represents the ratio obtained from the adjacency matrix in GCN output vectors. The effect of β on performance is shown in Fig. 1.

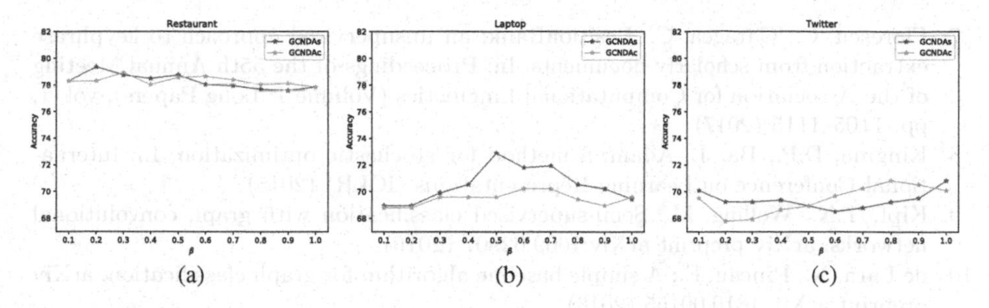

Fig. 1. Effect of β on GCNDA

5 Conclusion

In this paper, we present a novel method based on graph convolutional networks and two attention mechanisms for the aspect-based sentiment analysis task. Compared with baselines on public datasets, the experimental results show that our model outperforms the state-of-the-art baselines.

We performed our model on co-occurrence graph structure and syntactic graph structure, and the results demonstrate that although co-occurrence graph is simple in construction, it can achieve better performance datasets.

References

1. Angelova, R., Weikum, G.: Graph-based text classification: learn from your neighbors. In: Proceedings of the 29th Annual International ACM SIGIR Conference on Research and Development in Information Retrieval, pp. 485–492. ACM (2006)
2. Bastings, J., Titov, I., Aziz, W., Marcheggiani, D., Simaan, K.: Graph convolutional encoders for syntax-aware neural machine translation. In: Proceedings of the 2017 Conference on Empirical Methods in Natural Language Processing, pp. 1947–1957. Association for Computational Linguistics, Copenhagen, September 2017. https://www.aclweb.org/anthology/D17-1209
3. Bruna, J., Zaremba, W., Szlam, A., Lecun, Y.: Spectral networks and locally connected networks on graphs. In: International Conference on Learning Representations (ICLR 2014), CBLS, April 2014
4. Chen, P., Sun, Z., Bing, L., Yang, W.: Recurrent attention network on memory for aspect sentiment analysis. In: Proceedings of the 2017 Conference on Empirical Methods in Natural Language Processing, pp. 452–461 (2017)
5. Dong, L., Wei, F., Tan, C., Tang, D., Zhou, M., Xu, K.: Adaptive recursive neural network for target-dependent twitter sentiment classification. In: Proceedings of the 52nd Annual Meeting of the Association for Computational Linguistics (Volume 2: Short Papers), vol. 2, pp. 49–54 (2014)
6. Fan, C., Gao, Q., Du, J., Gui, L., Xu, R., Wong, K.F.: Convolution-based memory network for aspect-based sentiment analysis. In: The 41st International ACM SIGIR Conference on Research & Development in Information Retrieval, pp. 1161–1164. ACM (2018)

7. Florescu, C., Caragea, C.: PositionRank: an unsupervised approach to keyphrase extraction from scholarly documents. In: Proceedings of the 55th Annual Meeting of the Association for Computational Linguistics (Volume 1: Long Papers), vol. 1, pp. 1105–1115 (2017)
8. Kingma, D.P., Ba, J.: Adam: a method for stochastic optimization. In: International Conference on Learning Representations (ICLR) (2015)
9. Kipf, T.N., Welling, M.: Semi-supervised classification with graph convolutional networks. arXiv preprint arXiv:1609.02907 (2016)
10. de Lara, N., Pineau, E.: A simple baseline algorithm for graph classification. arXiv preprint arXiv:1810.09155 (2018)
11. Lee, J.B., Rossi, R., Kong, X.: Graph classification using structural attention. In: Proceedings of the 24th ACM SIGKDD International Conference on Knowledge Discovery & Data Mining, pp. 1666–1674. ACM (2018)
12. Li, L., Liu, Y., Zhou, A.: Hierarchical attention based position-aware network for aspect-level sentiment analysis. In: Proceedings of the 22nd Conference on Computational Natural Language Learning, pp. 181–189 (2018)
13. Liu, B.: Sentiment analysis and opinion mining. Synth. Lect. Hum. Lang. Technol. **5**(1), 1–167 (2012)
14. Liu, Q., Zhang, H., Zeng, Y., Huang, Z., Wu, Z.: Content attention model for aspect based sentiment analysis. In: Proceedings of the 2018 World Wide Web Conference on World Wide Web, pp. 1023–1032. International World Wide Web Conferences Steering Committee (2018)
15. Luong, M.T., Pham, H., Manning, C.D.: Effective approaches to attention-based neural machine translation. arXiv preprint arXiv:1508.04025 (2015)
16. Ma, D., Li, S., Zhang, X., Wang, H.: Interactive attention networks for aspect-level sentiment classification. In: Proceedings of the 26th International Joint Conference on Artificial Intelligence, pp. 4068–4074. AAAI Press (2017)
17. Marcheggiani, D., Titov, I.: Encoding sentences with graph convolutional networks for semantic role labeling. In: Proceedings of the 2017 Conference on Empirical Methods in Natural Language Processing, pp. 1507–1516. Association for Computational Linguistics, Copenhagen, September 2017. https://www.aclweb.org/anthology/D17-1159
18. Mihalcea, R., Tarau, P.: TextRank: Bringing Order into Texts. Association for Computational Linguistics, Stroudsburg (2004)
19. Nasukawa, T., Yi, J.: Sentiment analysis: capturing favorability using natural language processing. In: Proceedings of the 2nd International Conference on Knowledge Capture, pp. 70–77. ACM (2003)
20. Pang, B., Lee, L., et al.: Opinion mining and sentiment analysis. Founda. Trends® Inf. Retr. **2**(1–2), 1–135 (2008)
21. Pennington, J., Socher, R., Manning, C.: Glove: global vectors for word representation. In: Proceedings of the 2014 Conference on Empirical Methods in Natural Language Processing (EMNLP), pp. 1532–1543 (2014)
22. Pontiki, M., Galanis, D., Pavlopoulos, J., Papageorgiou, H., Androutsopoulos, I., Manandhar, S.: SemEval-2014 task 4: aspect based sentiment analysis. In: Proceedings of International Workshop on Semantic Evaluation, pp. 27–35 (2014)
23. Tang, D., Qin, B., Feng, X., Liu, T.: Target-dependent sentiment classification with long short term memory. CoRR, abs/1512.01100 (2015)
24. Tang, D., Qin, B., Liu, T.: Aspect level sentiment classification with deep memory network. In: Proceedings of the 2016 Conference on Empirical Methods in Natural Language Processing, pp. 214–224 (2016)

25. Veličković, P., Cucurull, G., Casanova, A., Romero, A., Liò, P., Bengio, Y.: Graph attention networks. In: International Conference on Learning Representations (2018). https://openreview.net/forum?id=rJXMpikCZ. Accepted as poster
26. Wang, Y., Huang, M., Zhao, L., et al.: Attention-based LSTM for aspect-level sentiment classification. In: Proceedings of the 2016 Conference on Empirical Methods in Natural Language Processing, pp. 606–615 (2016)
27. Wei, F., Li, W., Lu, Q., He, Y.: A document-sensitive graph model for multi-document summarization. Knowl. Inf. Syst. **22**(2), 245–259 (2010)
28. Xue, W., Li, T.: Aspect based sentiment analysis with gated convolutional networks. In: Proceedings of the 56th Annual Meeting of the Association for Computational Linguistics, ACL 2018, Melbourne, Australia, 15–20 July 2018, Volume 1: Long Papers, pp. 2514–2523 (2018). https://aclanthology.info/papers/P18-1234/p18-1234
29. Yang, Z., et al.: GLoMo: unsupervisedly learned relational graphs as transferable representations. arXiv preprint arXiv:1806.05662 (2018)
30. Zhang, M., Zhang, Y., Vo, D.T.: Gated neural networks for targeted sentiment analysis. In: AAAI, pp. 3087–3093 (2016)

A Wind Power Prediction Method Based on Deep Convolutional Network with Multiple Features

Shizhan Chen[1,2,3], Bo You[4], Xuewei Li[1,2,3], Mei Yu[1,2,3], Jian Yu[1,2,3](✉), Zhuo Zhang[4], Jie Gao[1,2,3], Zhiqiang Liu[1,2,3], and Ruiguo Yu[1,2,3]

[1] College of Intelligence and Computing, Tianjin University, Tianjin 300354, China
yujian@tju.edu.cn
[2] Tianjin Key Laboratory of Cognitive Computing and Application, Tianjin 300354, China
[3] Tianjin Key Laboratory of Advanced Networking, Tianjin 300354, China
[4] Tianjin International Engineering Institute, Tianjin University, Tianjin 300354, China

Abstract. As a clean and renewable energy, wind power plays an increasingly significant role in the power system. And wind power prediction is crucial for the operation planning and cost control of power plants. In wind power prediction, the wind-related information such as wind speed, wind direction, air pressure and temperature will affect the accuracy of prediction. However, most of the existing models either use only one kind of information or fail to effectively integrate a variety of information. Considering these problems, a new convolutional neural network model is proposed by integrating multiple information based on spatio-temporal features, called FB-CNN (Feature Block CNN). Obtain the output of each convolution layers in the whole neural network, then integrate these outputs, the prediction results are obtained through the full connection layers in the end. Compared with the current convolutional network with the highest accuracy, the proposed FB-CNN combining various features can excellently fit the actual change of wind power data. And the Mean Square Errors (MSE) on two data sets are reduced by 9.53% and 7.13% respectively.

Keywords: Wind power prediction · Multiple features · Convolutional neural network

1 Introduction

Wind power, as an emerging, renewable energy that can be exploited and utilized on a large scale, has developed rapidly in recent years [3]. However, due to the property of the wind, turbines are unavoidably burdened with randomness and fluctuation. Therefore, accurate wind power prediction is crucial to operate the power system safely and stably [11], for it can help control wind power, ensure

T. Gedeon et al. (Eds.): ICONIP 2019, CCIS 1142, pp. 198–206, 2019.
https://doi.org/10.1007/978-3-030-36808-1_22

the stable operation of the power grid, reduce the cost of power generation, and improve the ability of the grid to receive wind power. At present, the main methods used to predict wind power generation include: physical methods [5], statistical methods [13], and machine learning methods, like kNN [8], SVR [9] and LSTM [10]. Machine learning methods effectively simplify wind power prediction, but their prediction accuracy has failed to improve in recent years.

In this paper, a multi-feature driven model is proposed, which can effectively integrate various spatio-temporal features of wind-related information and improve the expressing ability of the features. Also, a complex convolutional network FB-CNN model is designed, which is suitable to predict various spatio-temporal features of the wind-related information in multi-feature driven models and enable to fit the trend of the change of wind power data in wind farms. The two data sets in this paper are all from National Renewable Energy Laboratory (NREL), 2009–2010. The results show that the accuracy of proposed method is higher than the existing prediction methods and significantly outperforms state-of-the-art methods. The contributions of this paper are as follows:

1. A multi-feature driven model is achieved to fully integrate various spatio-temporal features of wind-related information.
2. A complex convolutional network model is constructed, which can fit the changing trend of wind power data well, suitable for wind power prediction by integrating multiple features. Compared with the existing methods, the prediction accuracy is further improved.

2 Related Work

Machine learning is often used in short-term wind power prediction. Recently, many new methods have been proposed in recent years: Signal decomposition algorithm, which is mainly used to pre-process the original wind speed series is a popular idea to simplify complex problems. For example, wavelet transform [1], Ensemble Empirical Mode Decomposition (EEMD) [7]. Hybrid model combines multiple deep learning algorithms to improve the prediction ability of model [2]. Essentially, these methods get higher accuracy by using complex models. But complex models would greatly increase the computing costs. Also these feature extraction methods can not reflect the spatio-temporal changes of wind power.

How to extract features effectively is a important factor affecting the accuracy of prediction. The most basic method relies on features extracted from the target turbine's own data called Single Feature (SF) and feature extracted from the target turbine and several adjacent turbines, namely, Local-Feature (LF) [6]. The features extracted by these methods can obtain much information but no spatial information. In [12], two convolutional network models based on scene and spatio-temporal features called FC-CNN and E2E are proposed. According to the geographical coordinates of the turbine, the output power of the turbine at a certain moment is mapped to the grid to form a two-dimensional image, namely the scene. This method of feature extraction can effectively reflect the spatio-temporal features of wind. However, with simple structures and not take wind-related information into consideration, both of the two CNN models cannot predict the results accurately. So these methods cannot produce an accurate prediction of wind power.

3 Proposed Method

3.1 Multi-feature Driven Model

In the two data sets selected in this paper, in addition to the historical data of turbine power, wind-related information also includes wind speed, direction, pressure, density, temperature, etc. In the experiment, we orthogonalize the wind direction.

The features which are selected from a variety of wind-related information are the basis of wind power prediction based on deep learning. We use the method mentioned in reference [12], embedding the turbine into the grid as small as possible to construct scene. Since the scene represents the spatial distribution of wind power in a certain time, it can effectively reflect spatio-temporal features. So the series of multiple continuous scenes can convey the process of space changing with time. We choose different combinations of multiple features, we construct scene time series with each feature separately, and link them together as input of CNN model, called Feature Block (FB). The constructing process of multi-feature model is shown in Fig. 1. Finally, we choose the best multi-feature combination for wind power prediction through experiments.

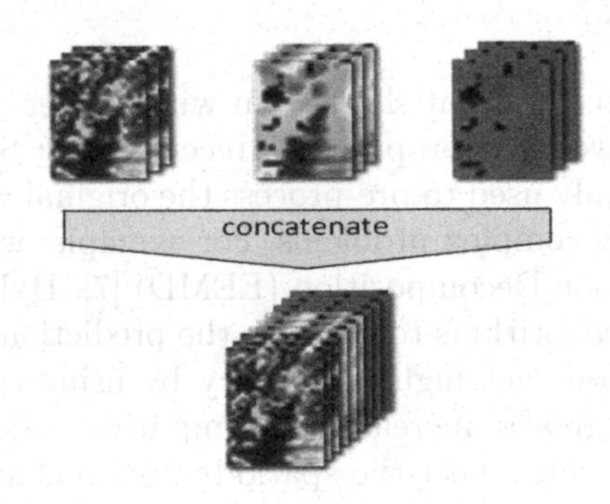

Fig. 1. The constructing process of the multi-feature driven model.

3.2 Feature-Block CNN

The advantage of DenseNet network is that it has a narrower network and fewer parameters [4]. It largely results from the design of dense block structure. This connecting mode makes the transmission of features and gradients more effective, further alleviating the problem of vanishing-gradient when deepening the network. It is especially suitable for wind power prediction. Therefore, we construct a CNN prediction model feature-block CNN (FB-CNN) based on partial dense block and the Feature Block we constructed.

Firstly, after receiving the input Feature Block, convolution is carried out to preserve the spatial information of the original input image. Because the main task at this stage is to extract features adequately, the number of channels in the feature image increases rapidly. Then, deeper features are extracted through a pooling layer and the size of the image is reduced. Similar multi-layer convolution is then performed. After the convolution of each layer being saved, the output of all convolution layers in the whole neural network is saved, and then these outputs are integrated. And next, these convolution layers are connected by a fully connected neural network. By fitting the complex function relationship of the fully connected layer, the deep features are mapped to the output of each turbine. Finally, the length of the output vector is equal to the number of pixels in the input image, and is reconstructed into a two-dimensional image. The pixels are mapped to the pixels of the input image one by one. The model structure is shown in Fig. 2.

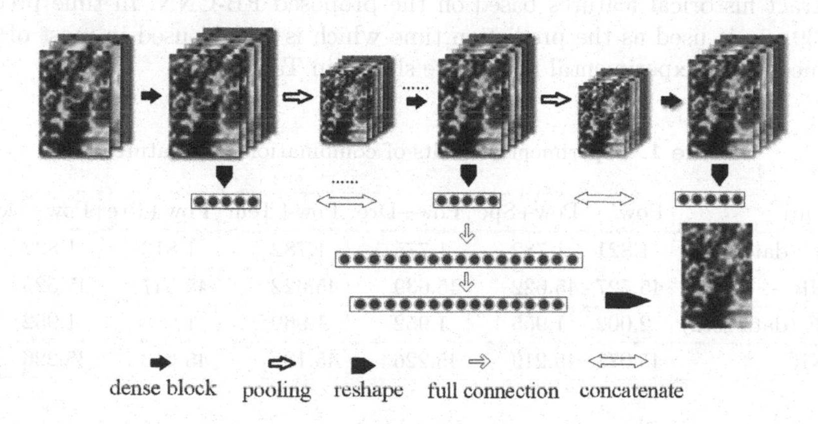

Fig. 2. The structure of Feature-Block CNN

4 Experiment and Analysis

4.1 Data Sets and Evaluating Criteria

The data sets used in this paper are the wind power data sets from NREL. We use the output values of the turbines working every 10 min from 2009 to 2010. 2009 as a training set, 2010 as a test set. Two regions respectively with dense and sparse turbine distribution are selected as data sets, which can verify the validity of our method more comprehensively. In the first data set, the number of turbines there reaches 559. In the second data set reaches 397 turbines.

We use Mean Square Error (MSE) to evaluate the accuracy of the prediction model. The calculation process of MSE is shown in Formula (1), where X denotes the sequence of true values, Y the sequence of predicted values, and n the length of the sequence. Peak signal-to-noise ratio (PSNR) is often used to measure the quality of signal reconstruction in the fields such as image compression. The formula is shown in (2), in which MAXI is the maximum value of the color of an image point. If each sampling point is represented by 8 bits, it would come to the number of 255.

$$MSE = \frac{1}{n}\sum_{i=0}^{n}(X_i - Y_i)^2 \tag{1}$$

$$PSNR = 10 \cdot log_{10}(\frac{MAX_I^2}{MSE}) \tag{2}$$

4.2 Experimental Results and Comparison

In order to verify the validity of multiple features, firstly in feature selection, we use power combined with other features, and select 10 min as the interval to extract historical features based on the proposed FB-CNN. In time prediction, 30 min is used as the prediction time which is widely used in most of the references. The experimental results are shown in Table 1.

Table 1. Experimental results of combination two features

Feature	Pow	Pow+Spe	Pow+Dre	Pow+Tem	Pow+Pre	Pow+Den
MSE (data set1)	1.821	1.782	1.775	1.782	1.813	1.822
PSNR	45.527	45.622	45.639	45.622	45.547	45.525
MSE (data set2)	2.002	1.955	1.952	1.969	1.958	1.952
PSNR	45.073	45.219	45.226	45.188	45.213	45.226

The calculation formulas for achieving the relevant wind power are formula (3), in which A denotes the sweep area, V the wind speed, and Cp the value of wind power conversion. Different technology of manufacturers would result in different values. The value of D represents the air density, which decreases with the increase of altitude. And η is a coefficient. By this formula, we select four related features for wind power prediction, gradually add other features, and finally use all the features for prediction. The prediction results are shown in Table 2.

$$P = \frac{1}{2}AV^3CpD\eta \tag{3}$$

To verify the effectiveness of the proposed method, the experimental results are compared with the existing time series prediction methods and two methods proposed in the reference [12] (E2E and FC-CNN). For LSTM, SVR, and kNN

Table 2. Prediction results of multiple feature combinations

Feature	Pow+Spe+ Pre+Den	Pow+Spe+Pre +Den+Dre	Pow+Spe+Pre +Den+Tem	All features
MSE (data set1)	1.787	1.736	1.781	1.756
PSNR	45.610	45.735	45.624	45.686
MSE (data set2)	1.953	1.916	1.951	1.958
PSNR	45.224	45.307	45.228	45.213

Table 3. The comparison between the experimental results of the proposed method and the existing methods.

Method			kNN	SVR	LSTM	E2E	FC-CNN	FB-CNN
MSE (data set1)	FW = 3	Max	5.745	5.512	5.335	4.578	4.250	3.941
		Min	0.053	0.052	0.053	0.163	0.078	0.073
		Ave	2.736	2.633	2.640	2.120	1.928	1.736
	FW = 4	Max	5.867	5.441	5.380	4.566	4.267	3.941
		Min	0.054	0.052	0.053	0.190	0.079	0.079
		Ave	2.793	2.599	2.658	2.123	1.919	1.748
	FW = 5	Max	5.999	*	5.442	4.677	4.262	3.929
		Min	0.055	*	0.053	0.178	0.073	0.074
		Ave	2.859	*	2.671	2.163	1.923	1.738
PSNR			*	*	*	44.867	45.300	45.735
MSE (data set2)	FW = 3	Max	4.637	4.831	4.741	4.410	3.703	3.562
		Min	1.236	1.203	1.269	0.979	0.917	0.839
		Ave	2.982	2.893	2.977	2.402	2.088	1.916
	FW = 4	Max	4.680	4.815	4.761	4.102	3.623	3.544
		Min	1.252	1.190	1.255	1.059	0.904	0.842
		Ave	3.040	2.867	2.991	2.374	2.063	1.924
	FW = 5	Max	4.713	*	4.782	4.052	3.623	3.530
		Min	1.275	*	1.278	1.060	0.911	0.851
		Ave	3.099	*	3.007	2.366	2.068	1.946
PSNR			*	*	*	44.391	44.985	45.307

methods, we selected local feature extraction features. The experimental results of feature windows of 3, 4 and 5 (the number of historical features extracted) in two data sets are shown in Table 3. Because the time cost of SVR when FW = 5 is too high, we only give the case of SVR when FW = 3 and FW = 4.

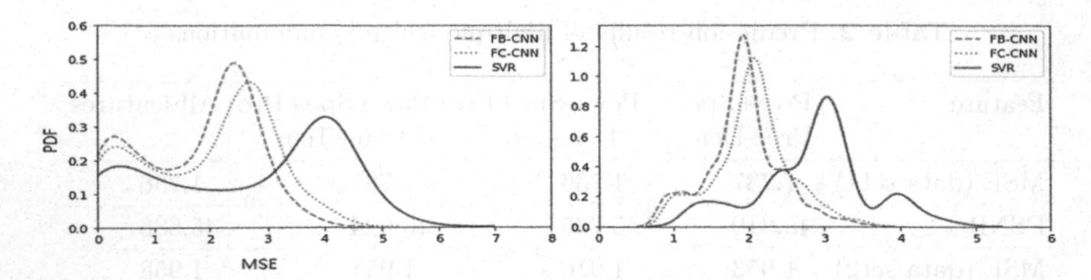

Fig. 3. The comparison of prediction error distribution of several methods: the left side is dataset 1, and the right side is dataset 2

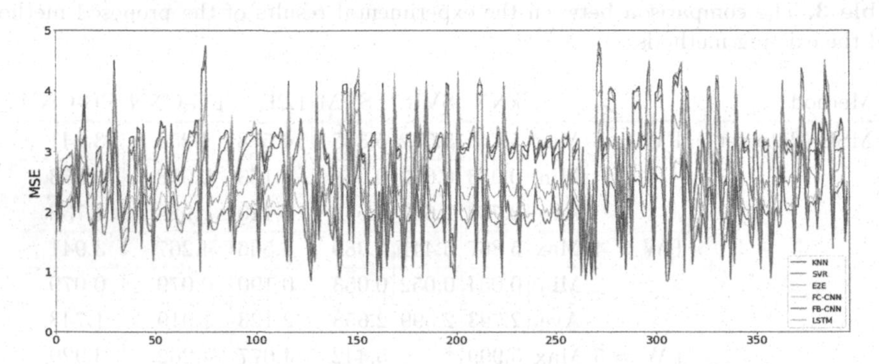

Fig. 4. Prediction error of each method at each turbines (data set 2)

4.3 Analysis of Experimental Results

From Table 1, it can be seen that the selection of different features has different improved on the prediction accuracy, among which the combination of power and wind direction has the best effect. Table 2 shows the prediction results of multiple feature combinations. And what should not be ignored is that more features may not come to a better prediction. The results show that the combination of power, wind speed, air pressure, density and wind direction achieves the best prediction accuracy. Compared with that of single feature, on the two data sets, the MSEs reduce by 4.67% and 4.30% respectively.

In Table 3, It can be clearly seen that compared with SVR which has the best single point prediction results at present, our prediction accuracy has increased by 33.21% and 33.17%, more than that of KNN, by 36.55% and 35.75%. Similarly, compared with the FC-CNN, the higher accuracy in the two methods proposed in the paper [12], increased by 9.53% and 7.13%. As an evaluating criteria of image quality, the PSNR value of the proposed method on two data sets is also the highest, 45.735 and 45.307. It is further proved that our method is effective and the best results are obtained in the regions with both dense and sparse turbines.

In the Fig. 3, we compare the SVR witch has the best single point prediction performance, the convolutional network FC-CNN and the method proposed method in this paper. The horizontal axis represents the value of MSE, and the vertical axis represents the Probability Density (PDF). Obviously, The probability density of FB-CNN is above FC-CNN and SVR in the region with small MSE value. In order to better demonstrate the effectiveness of the proposed method, we select all turbines from the data set and plot the MSE curves of all the methods on the turbines, as shown in Fig. 4. FB-CNN performs best among these methods.

5 Conclusion

In this paper, a method of wind power prediction is proposed using complex convolutional neural network based on multiple features. The Multi-feature driven model expressing spatio-temporal features and various wind related information effectively. The FB-CNN model could extract deep features and predict make more accurate prediction. The proposed method also performs better in terms of stability. It has crucial value and broad prospect in practical application.

Acknowledgements. This work is supported by National Natural Science Foundation of China (Grant No. 61877043, 61976155), Major Scientific and Technological Projects for A New Generation of Artificial Intelligence of Tianjin (Grant No. 18ZXZNSY00300), and Key Project for Science and Technology Support from Key R&D Program of Tianjin (Grant No. 18YFZCGX00960).

References

1. Chitsaz, H., Amjady, N., Zareipour, H.: Wind power forecast using wavelet neural network trained by improved clonal selection algorithm. Energy Convers. Manag. **89**, 588–598 (2015)
2. Feng, C., Cui, M., Hodge, B.M., Zhang, J.: A data-driven multi-model methodology with deep feature selection for short-term wind forecasting. Appl. Energy **190**, 1245–1257 (2017)
3. GWEC, Global Wind Energy Council: Global wind report. 2015. Brussels: GWEC (2017)
4. Huang, G., Liu, Z., Van Der Maaten, L., Weinberger, K.Q.: Densely connected convolutional networks. In: Proceedings of the IEEE Conference on Computer Vision and Pattern Recognition, pp. 4700–4708 (2017)
5. Jung, J., Broadwater, R.P.: Current status and future advances for wind speed and power forecasting. Renew. Sustain. Energy Rev. **31**, 762–777 (2014)
6. Treiber, N.A., Heinermann, J., Kramer, O.: Wind power prediction with machine learning. In: Lässig, J., Kersting, K., Morik, K. (eds.) Computational Sustainability. SCI, vol. 645, pp. 13–29. Springer, Cham (2016). https://doi.org/10.1007/978-3-319-31858-5_2
7. Wang, Y., Wang, S., Zhang, N.: A novel wind speed forecasting method based on ensemble empirical mode decomposition and GA-BP neural network. In: 2013 IEEE Power & Energy Society General Meeting, pp. 1–5. IEEE (2013)
8. Wen, Y., Song, M., Wang, J.: A combined AR-KNN model for short-term wind speed forecasting. In: 2016 IEEE 55th Conference on Decision and Control (CDC), pp. 6342–6346. IEEE (2016)

9. Woon, W.L., Kramer, O.: Enhanced SVR ensembles for wind power prediction. In: 2016 International Joint Conference on Neural Networks (IJCNN), pp. 2743–2748. IEEE (2016)
10. Wu, W., Chen, K., Qiao, Y., Lu, Z.: Probabilistic short-term wind power forecasting based on deep neural networks. In: 2016 International Conference on Probabilistic Methods Applied to Power Systems (PMAPS), pp. 1–8. IEEE (2016)
11. Xydas, E., Qadrdan, M., Marmaras, C., Cipcigan, L., Jenkins, N., Ameli, H.: Probabilistic wind power forecasting and its application in the scheduling of gas-fired generators. Appl. Energy **192**, 382–394 (2017)
12. Yu, R., et al.: Scene learning: Deep convolutional networks for wind power prediction by embedding turbines into grid space. Appl. Energy **238**, 249–257 (2019)
13. Ziel, F., Croonenbroeck, C., Ambach, D.: Forecasting wind power-modeling periodic and non-linear effects under conditional heteroscedasticity. Appl. Energy **177**, 285–297 (2016)

Simple ConvNet Based on Bag of MLP-Based Local Descriptors

Takumi Kobayashi(✉), Hidenori Ide, and Kenji Watanabe

National Institute of Advanced Industrial Science and Technology,
Umezono 1-1-1, Tsukuba, Ibaraki, Japan
{takumi.kobayashi,hidenori.ide,kenji.watanabe}@aist.go.jp

Abstract. Deep convolutional neural network (ConvNet) is applied to versatile image recognition tasks with great success, though demanding high computation cost. Toward efficient computation, we propose a simple ConvNet architecture based on local descriptors in the bag-of-features framework. The local descriptors are formulated in a simple form of MLP and thus are efficiently computed on various ROI in a flexible manner. The proposed method is effectively trained in an end-to-end manner by reformulating the MLP descriptor into the form of deep ConvNet stacking convolution layers *linearly*. Through projection-based visual word encoding, the local descriptors are aggregated and fed into a classifier for image recognition tasks, which enables us to compute the network forwarding pass by matrix-vector multiplication. In the experiments on image classification, the proposed method is analyzed thoroughly, exhibiting favorable generalization performance on various tasks.

1 Introduction

Hand-crafted local descriptors, such as SIFT [15], extracted from small image patches have played a key role on various computer vision tasks; image classification was enthusiastically addressed by utilizing the descriptors in the bag-of-features (BoF) framework [9,22]. In this decade, however, deep convolutional neural networks (ConvNets) [11,26] defeat them with promising performance, though the hand-crafted descriptor is practically useful due to the low computation cost [30]. While the ConvNet works on whole input image through deeply stacked convolution operations, it is internally dependent on local image feature extraction directed by the last convolution, *e.g.*, at so-called conv5 layer.

The local descriptors embedded in the deep ConvNets can be exposed and then combined with the traditional encoding schemes, such as Fisher kernel and bag of visual words, for image retrieval [16] and texture recognition [5]. There are also methods to leverage the ConvNet more directly to extract convolutional descriptors from image patches mainly on the task of patch matching [18,25]. Those ConvNet based descriptors are built on stacked convolution operations with computational burden [25], thus demanding sophisticated devices such as GPUs, unlike the hand-crafted SIFT. On the other hand, the hand-crafted descriptors are combined with neural network classifier of MLP through the Fisher kernel encoding

T. Gedeon et al. (Eds.): ICONIP 2019, CCIS 1142, pp. 207–215, 2019.
https://doi.org/10.1007/978-3-030-36808-1_23

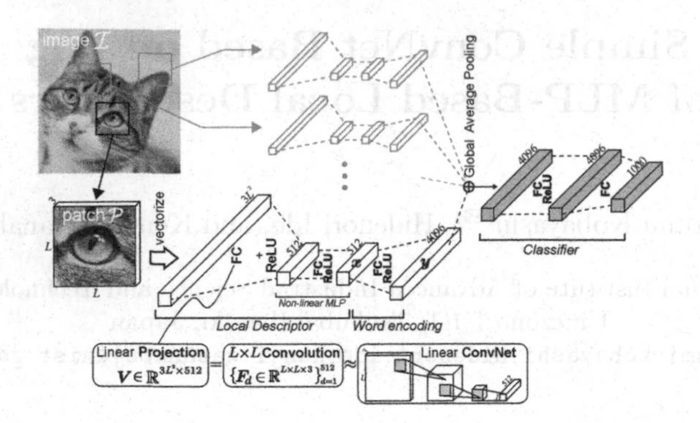

Fig. 1. Proposed network architecture based on MLP local descriptors.

Table 1. Baseline ConvNet architecture based on VGG16 [26].

	Block	Layers	Channel
Local descriptor ($L = 181$)	1	$\{3 \times 3$ Conv. + BatchNorm$\} \times 2$	64
		Down-sampling by 2-pixel stride	
	2	$\{3 \times 3$ Conv. + BatchNorm$\} \times 2$	128
		Down-sampling by 2-pixel stride	
	3	$\{3 \times 3$ Conv. + BatchNorm$\} \times 3$	256
		Down-sampling by 2-pixel stride	
	4	$\{3 \times 3$ Conv. + BatchNorm$\} \times 3$	512
		Down-sampling by 2-pixel stride	
	5	$\{3 \times 3$ Conv. + BatchNorm$\} \times 3$	512
		ReLU	512
	MLP	$\{1 \times 1$ Conv. + BatchNorm + ReLU$\} \times 0$	512
	BoW	1×1 Conv. + BatchNorm + ReLU	4096
		Global Average Pooling	4096
	FC1	FC + BatchNorm + ReLU	4096
	FC2	FC	1000
		SoftMax	1000

in [19]. The method improves performance of the SVM classification approach [22], though being slightly inferior to AlexNet [11], which reveals the less discriminativity of the hand-crafted descriptor than the learned one.

In this work, we formulate a simple ConvNet toward efficient computation by explicitly considering the bag-of-features approach in the end-to-end framework. In contrast to the hybrid method [19] incorporating the hand-crafted descriptors with neural network classifier, we propose a simple form of local descriptor followed by visual word encoding, all of which are trained in an end-to-end manner as in the standard deep ConvNets. The simple architecture in descriptor design is based on MLP which comprises a linear projection and a non-linear function, *i.e.*, ReLU [17]; as a result, our model only requires matrix-vector multiplication efficiently computed by well established technique on various devices [8].

In the case that local patches are sampled at regular grids over an input image, the computation of our local descriptors, especially at the first layer of the MLP, can be regarded as convolution operation, thereby exhibiting similarity to the deep ConvNets. From the architectural viewpoint of ConvNets, however, the proposed model contrasts with the standard ConvNets as follows. The model based on the local descriptors contains only *one* convolution layer followed by several matrix-vector multiplication in MLP; the spatial convolution operates only on an input RGB image. Thus, from this viewpoint, our model is *less* convolutional compared to the *deep* ConvNets [2]. Such a simple computational procedure enables us to efficiently compute the forwarding pass of the model. In addition, it is possible to efficiently compute the local descriptors at regular grids by leveraging the convolution theorem [4] to perform the only one convolution via FFT. The other research line toward lightweight ConvNet is found in recent years [23,29]. While those works focus on slimming ConvNet still heavily relying on convolutional operation, we simplify the form of local descriptor through breaking dependence on the convolution to achieve computational efficiency as well as generalization performance.

On the other hand, the proposed model based on local descriptors in the BoF framework flexibly deals with any shape of ROI beyond simple regular grids unlike the standard ConvNets usually working on the regular lattice. The MLP-based feature extraction for local descriptors is also found in PointNet [20] to cope with point cloud data for 3D recognition. In this work, we employ an MLP model for computational efficiency and show favorable performance on image recognition tasks in spite of the simple formulation.

2 MLP-Based BoF Network

We build the neural network based on bag of local descriptors which are computed by applying multilayer perceptron (MLP) to local image patches, as shown in Fig. 1. Thus, computation for this network is simply composed of ReLU [17] and matrix-vector multiplication which is well-established operation on various devices [8]. While the similar MLP architecture is found in small image classification such as for MNIST [12], in this work, we leverage it to extract features from local patches in the bag-of-feature framework. Following [10], the descriptor $\boldsymbol{x} \in \mathbb{R}^{512}$ is encoded into word representation $\boldsymbol{y} \in \mathbb{R}^{4096}$ via linear projection by the word vectors $\{\boldsymbol{w}_i\}_{i=1}^{4096}$ with ReLU;

$$y_i = \max[\boldsymbol{w}_i^\top \boldsymbol{x} - \rho_i, 0] = \texttt{ReLU}(\boldsymbol{w}_i^\top \boldsymbol{x} - \rho_i), \tag{1}$$

where ρ_i is a threshold for assigning the i-th word weight y_i to the descriptor $\boldsymbol{x}$ on the basis of inner-product similarity. The word representation aggregated across patches is then finally fed into the MLP classifier.

2.1 Training MLP Descriptor Through Linear ConvNet

The network (Fig. 1) can be trained end-to-end as in the deep ConvNets [11,26]. It, however, would be problematic to directly train the MLP descriptor which contains large projection matrix $\boldsymbol{V} \in \mathbb{R}^{3L^2 \times 512}$ in the first fully-connected layer; it depends on the patch size $L \times L$, say $L = 29$, which is larger than the standard convolution size, *e.g.*, 3×3. Thus, we reformulate the first fully-connected projection into a tractable form by means of *ConvNet*. It should be noted that our model is trained in a form of deep ConvNet but is deployed as the MLP-based form which is equivalent to the deep ConvNet.

In the descriptor MLP, the first fully-connected linear projection is viewed as *convolution* (without sliding) with the filters whose size corresponds to the patch size $L \times L \times 3$; this is just a transformation of the projection matrix $\boldsymbol{V}$ via unfolding. The moderately large $L \times L$ spatial filters are difficult to adequately learn due to the high degree of freedom (DoF). To mitigate it, we explicitly impose *decomposability into local convolutions* on the $L \times L$ convolution filters. This constraint is well validated by the Fractal structure, wavelet analysis and recent advances in deep ConvNet for image recognition. Thereby, the $L \times L$ convolution is approximated by stacking smaller convolutions, which results in the form of *linear* ConvNet (Fig. 1 & Table 1) without any non-linear functions, *e.g.*, ReLUs; a linear deep model is not *bad* even from the optimization viewpoint [7]. The linearly stacked convolution layers are compressed into a single convolution layer by enlarging the convolution filter as follows. Given two stacked convolutions whose filters are F of $l_F \times l_F$ and G of $l_G \times l_G$, we can describe the first convolution layer followed by down-sampling with factor s as

$$\tilde{I}(\boldsymbol{p}) = \sum_{\boldsymbol{\delta}\in\mathbb{Z}^2} F(\boldsymbol{\delta}) I(\boldsymbol{p} - \boldsymbol{\delta}), \ J(\boldsymbol{p}) = \tilde{I}(s\boldsymbol{p}), \tag{2}$$

where $I, \tilde{I}$ and J are input, intermediate and output feature maps, respectively, where the pixel position is denoted by $\boldsymbol{p}$. Then, the second convolution layer is

$$\begin{aligned}
\tilde{J}(\boldsymbol{p}) &= \sum_{\boldsymbol{\epsilon}\in\mathbb{Z}^2} G(\boldsymbol{\epsilon}) J(\boldsymbol{p} - \boldsymbol{\epsilon}) = \sum_{\boldsymbol{\epsilon}\in\mathbb{Z}^2} G(\boldsymbol{\epsilon}) \tilde{I}(s\boldsymbol{p} - s\boldsymbol{\epsilon}) \\
&= \sum_{\boldsymbol{\epsilon}\in\mathbb{Z}^2} \hat{G}(\boldsymbol{\epsilon}) \tilde{I}(s\boldsymbol{p} - \boldsymbol{\epsilon}) = \sum_{\boldsymbol{\delta},\boldsymbol{\epsilon}\in\mathbb{Z}^2} \hat{G}(\boldsymbol{\epsilon}) F(\boldsymbol{\delta}) I(s\boldsymbol{p} - \boldsymbol{\epsilon} - \boldsymbol{\delta}) \\
&= \sum_{\boldsymbol{\eta}\in\mathbb{Z}^2} \underbrace{\sum_{\boldsymbol{\delta}\in\mathbb{Z}^2} \hat{G}(\boldsymbol{\eta} - \boldsymbol{\delta}) F(\boldsymbol{\delta})}_{\text{Compressed filter } H(\boldsymbol{\eta})} I(s\boldsymbol{p} - \boldsymbol{\eta}),
\end{aligned} \tag{3}$$

where $\tilde{J}$ is the output feature map, and we use the dilated filter of $\hat{G}(\boldsymbol{\epsilon}) = G(\frac{\boldsymbol{\epsilon}}{s})$ if $\frac{\boldsymbol{\epsilon}}{s} \in \mathbb{Z}^2$ otherwise 0, and transform the variable as $\boldsymbol{\eta} = \boldsymbol{\delta} + \boldsymbol{\epsilon}$. The size l_H of the compressed filter H is $l_H = s(l_G - 1) + l_F$. Thus, the patch size L, hyper-parameter of the descriptor, is naturally determined according to the architecture of the linear ConvNet.

This linear ConvNet is followed by the *non-linear* MLP to extract discriminative descriptors. The MLP is implemented as NiN module [14] of 1×1 convolution + ReLU layers, and thus in the case of regularly sampling patches on an input image during training, we implement our network (Fig. 1) by deep ConvNet (*e.g.*, Table 1) to effectively train the local descriptors and BoW model in an end-to-end approach. Once the network is trained, the linear ConvNet part is compressed by (3) into the fully-connected layer to form MLP-based descriptors. And, for densely computing descriptors on an image as in training, the descriptor can be efficiently extracted by applying the convolution theorem [4] via FFT.

3 Experimental Results

We evaluate various configurations of the MLP-based local descriptor in our network by training the corresponding ConvNets on a ImageNet dataset of 1000 object classes. All the models are implemented by using MatConvNet [27] following the good practice provided; the stochastic gradient descent is applied with the learning rate decreasing in a log-scale from 0.1 to 0.0001 over 40 epochs, the momentum of 0.9, the weight decay of 0.0005 and the mini-batch size of 64 samples. We measure the performance of top-5 error rate by single center cropping [11] on the ImageNet validation set.

Table 2. Performance analysis on various configuration of the local descriptor. The performance is evaluated by top-5 error rate (%) on ImageNet validation set. The baseline architecture in Table 1 is sequentially updated by the one denoted in bold font from (a) to (f).

(a) Convolutions per block

Architecture	Error (%)
Table 1 [L=181]	31.18
{3 × 3 Conv. + BN}×2 [L=125]	29.31
{3 × 3 Conv. + BN}×1 [L=63]	**29.17**

(b) Down-sampling

Method	Error (%)
striding [L=63]	**29.17**
avg.-pool [L=78]	30.79

(c) Convolution Filter size

Filter size	Error (%)
3 × 3 [L=63]	29.17
5 × 5 [L=125]	**27.59**
7 × 7 [L=187]	28.12

(d) Depth of Linear ConvNet

# of block	Error (%)
5 [L=125]	27.59
4 [L=61]	**24.71**
3 [L=29]	**24.76**
2 [L=13]	30.61

(e) Degree of non-linearity

Depth in MLP	4 block [L=61]	3 block [L=29]
0	24.71	24.76
1	20.29	19.80
2	18.55	**18.00**

(f) Training form of descriptor

Form	Error (%)
linear ConvNet [L=29]	**18.00**
29 × 29 Conv. [L=29]	22.24

3.1 Quantitative Ablation Study

We modify the baseline ConvNet (Table 1), according to the following analyses with keeping the descriptor dimensionality as 512.

Number of Convolution. The baseline model (Table 1) contains 13 layers of 3 × 3 convolution, 2 or 3 layers per block, across five blocks. Table 2a shows that the performance is improved by decreasing the number of 3 × 3 convolutions per block in contrast to the non-linear ConvNet containing ReLUs [26]; only one 3 × 3 convolution per block works well.

Local Pooling. In the linear ConvNet (Table 1), the feature maps are simply down-sized by 2-pixel striding, since 2 × 2 local average pooling degrades performance as shown in Table 2b. The local pooling is composed of 2 × 2 average filtering and 2-pixel striding, which unfavorably increases the convolution layers harming performance as implied in Table 2a.

Convolution Filter Size. On the other hand, by moderately enlarging the convolution filter size, we can improve performance as shown in Table 2c; the 5 × 5 convolution produces the best performance. Note that at each block *one* 5 × 5 convolution is equivalent to *two* stacked 3 × 3 convolutions (Table 2a), which

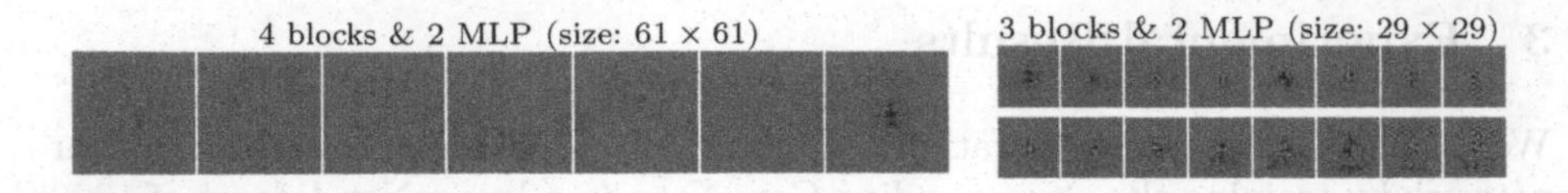

Fig. 2. The principal filters (columns of $\boldsymbol{U}_l$) by applying SVD to the learned filters.

conveys the insight that the larger-sized convolution in the shallower net is more effective than stacking smaller ones for a deep linear ConvNet.

Depth. Then, the number of blocks, *depth*, in the linear ConvNet stacking 5×5 convolutions is evaluated in Table 2d. The depth significantly affects the compressed filter size, *i.e.*, the patch size L. Compared to the larger patch descriptor, the moderate-sized ones produce the better performance; both the three ($L = 29$) and four ($L = 61$) blocks provide competitively good performance.

Non-linearity. The local descriptor is endowed with the non-linearity by the latter MLP part (Fig. 1 & Table 1) following the linear ConvNet part. Thus, the non-linearity is controlled by the depth of the MLP and Table 2e shows the performance improvement due to the higher non-linearity of the deeper MLP.

Training Form. As shown in Table 2f for training local descriptors, the form of linear ConvNet is superior to the naive MLP form, *i.e.*, one $L \times L$ convolution,

Based on the above analyses, we build the effective descriptor by stacking **three** 5×5 **convolution blocks** interlaced with the down-sampling of 2-pixel striding and **two-layer MLP**, which operates on a 29×29 patch with 4-pixel step for ImageNet classification. This configuration of the descriptor is closely related to the good practice [22] of the hand-crafted descriptor which extracts SIFT from 24×24 patches every 4 pixels on an image for image classification.

3.2 Qualitative Analysis

We qualitatively analyze the $L \times L$ spatial filter learned by the linear ConvNet.

For mining the principal characteristics in the spatial filters, we apply SVD to the (vectorized) filters $\boldsymbol{V} \in \mathbb{R}^{3L^2 \times 512}$ as $\boldsymbol{V} = \boldsymbol{U}_l \mathtt{diag}(\boldsymbol{s}) \boldsymbol{U}_r^\top$; the filters are decomposed into 512 components, the columns of $\boldsymbol{U}_l \in \mathbb{R}^{3L^2 \times 512}$. As shown in Fig. 2, the deeper linear ConvNet of 4 blocks activates the filter weights only on a small spatial region due to the larger patch size, while the filter weights by the shallower one are diversely distributed. Thus, we can conclude that for constructing the effective linear convolutional features, it is necessary to design moderately deep (linear) ConvNet to provide a proper receptive field, followed by the highly non-linear MLP.

3.3 Generality

The proposed simple network exhibits superior performance (18.00%) to AlexNet [11] which produces 19.29% on the ImageNet dataset. We further show

Table 3. The performance comparison on various image classification tasks. The performance is measured by classification accuracy (%).

Type	Object		Scene		Other	
Dataset	VOC2007 [1]	Caltech256 [6]	SUN397 [28]	MIT67 [21]	FMD [24]	Event8 [13]
Ours	78.22	66.71	50.78	66.48	79.23	95.14
AlexNet	77.87	73.79	48.36	63.96	72.75	95.07
Hand-craft [9]	63.83	57.4	46.1	63.4	57.3	92.6

the generality of the descriptor-based simple network across various image recognition tasks. For that purpose, the model trained on the ImageNet dataset (Sect. 3.1) is transferred to the other datasets. For simplicity, the pre-trained network is applied to extract a 4096-dimensional image feature vector at FC1 (Table 1) which is followed by the linear SVM classifier. It is noteworthy that in our model, the descriptors are computed on 29×29 local patches every 4 pixels and then encoded into the word representation in a quite similar manner to the hand-crafted methods in the BoF framework [9,22]. For comparison, we employ the same procedure for the pre-trained AlexNet and also show the performances reported by the hand-crafted method [9] on the datasets of various image recognition tasks.

Table 3 shows the performance results on various tasks of image classification. As mentioned in [3], the AlexNet exhibits favorable transferability on object recognition tasks which are closely related to ImageNet classification. On the other hand, the proposed model produces superior performance even to the AlexNet on the other types of tasks while working competitively with the AlexNet on the object classification tasks. The network simply relying on the MLP-based local descriptors is endowed with such a better generalization performance. And, our method consistently outperforms the hand-crafted one [9] based on the SIFT-based descriptors, demonstrating that our descriptor trained end-to-end on ImageNet dataset is well discriminative with favorable generality.

4 Conclusion

We have proposed a simple network architecture for image recognition toward efficient computation. The proposed method is explicitly built upon the bag-of-features procedure which leverages local descriptors and visual word based representation to extract image features. While the descriptor is formulated by means of simple MLP for efficiency, it is effectively trained in an end-to-end manner through transforming the MLP into a form of ConvNet, by utilizing standard techniques/procedures tailored for deep ConvNets on ImageNet dataset. The proposed network mainly composed of simple MLP computation exhibits favorable performance not only on the ImageNet classification task but also on the other various image recognition tasks.

References

1. The PASCAL Visual Object Classes Challenge 2007 (VOC2007). http://www.pascal-network.org/challenges/VOC/voc2007/index.html
2. Arandjelović, R., Gronat, P., Torii, A., Pajdla, T., Sivic, J.: NetVLAD: CNN architecture for weakly supervised place recognition. In: CVPR (2016)
3. Azizpour, H., Razavian, A.S., Sullivan, J., Maki, A., Carlsson, S.: Factors of transferability for a generic convnet representation. PAMI **38**(9), 1790–1802 (2016)
4. Bracewell, R.N.: The Fourier Transform and Its Applications. McGraw-Hill, New York (1999)
5. Cimpoi, M., Maji, S., Vedaldi, A.: Deep filter banks for texture recognition and segmentation. In: CVPR, pp. 3828–3836 (2015)
6. Griffin, G., Holub, A., Perona, P.: Caltech-256 object category dataset. Technical report 7694, Caltech (2007)
7. Kawaguchi, K.: Deep learning without poor local minima. In: NIPS, pp. 586–594 (2016)
8. Kestur, S., Davis, J.D., Chung, E.S.: Towards a universal FPGA matrix-vector multiplication architecture. In: FCCM, pp. 9–16 (2012)
9. Kobayashi, T.: Dirichlet-based histogram feature transform for image classification. In: CVPR, pp. 3278–3285 (2014)
10. Kobayashi, T.: Analyzing filters toward efficient convnets. In: CVPR, pp. 5619–5628 (2018)
11. Krizhevsky, A., Sutskever, I., Hinton, G.: ImageNet classification with deep convolutional neural networks. In: NIPS, pp. 1097–1105 (2012)
12. LeCun, Y., Bottou, L., Bengio, Y., Haffner, P.: Gradient-based learning applied to document recognition. Proc. IEEE **86**(11), 2278–2324 (1998)
13. Li, L.J., Fei-Fei, L.: What, where and who? Classifying events by scene and object recognition. In: ICCV (2007)
14. Lin, M., Chen, Q., Yan, S.: Network in network. In: ICLR (2014)
15. Lowe, D.G.: Distinctive image features from scale invariant features. IJCV **60**, 91–110 (2004)
16. Mohedano, E., McGuinness, K., O'Connor, N.E., Salvador, A., Marques, F., Giro-i-Nieto, X.: Bags of local convolutional features for scalable instance search. In: ICMR, pp. 327–331 (2016)
17. Nair, V., Hinton, G.: Rectified linear units improve restricted Boltzmann machines. In: ICML, pp. 807–814 (2010)
18. Paulin, M., Douze, M., Harchaoui, Z., Mairal, J., Perronnin, F., Schmid, C.: Local convolutional features with unsupervised training for image retrieval. In: ICCV, pp. 91–99 (2015)
19. Perronnin, F., Larlus, D.: Fisher vectors meet neural networks: a hybrid classification architecture. In: CVPR, pp. 3743–3752 (2015)
20. Qi, C.R., Su, H., Mo, K., Guibas, L.J.: PointNet: deep learning on point sets for 3D classification and segmentation. In: CVPR, pp. 77–85 (2017)
21. Quattoni, A., Torralba, A.: Recognizing indoor scenes. In: CVPR, pp. 413–420 (2009)
22. Sánchez, J., Perronnin, F., Mensink, T., Verbeek, J.: Image classification with the fisher vector: theory and practice. IJCV **105**(3), 222–245 (2013)
23. Sandler, M., Howard, A., Zhu, M., Zhmoginov, A., Chen, L.C.: MobileNETV2: inverted residuals and linear bottlenecks. In: CVPR, pp. 4510–4520 (2018)

24. Sharan, L., Rosenholtz, R., Adelson, E.: Material perception: what can you see in a brief glance? J. Vis. **9**(8), 784 (2009)
25. Simo-Serra, E., Trulls, E., Ferraz, L., Kokkinos, I., Fua, P., Moreno-Noguer, F.: Discriminative learning of deep convolutional feature point descriptors. In: ICCV, pp. 118–126 (2015)
26. Simonyan, K., Zisserman, A.: Very deep convolutional networks for large-scale image recognition. CoRR abs/1409.1556 (2014)
27. Vedaldi, A., Lenc, K.: MatConvNet - convolutional neural networks for MATLAB. In: ACM MM (2015)
28. Xiao, J., Hays, J., Ehinger, K.A., Oliva, A., Torralba, A.: Sun database: large-scale scene recognition from abbey to zoo. In: CVPR (2010)
29. Zhang, X., Lin, M., Sun, J.: ShuffleNet: an extremely efficient convolutional neural network for mobile devices. In: CVPR, pp. 6848–6856 (2018)
30. Zheng, L., Yang, Y., Tian, Q.: SIFT meets CNN: a decade survey of instance retrieval. PAMI **40**(5), 1224–1244 (2018)

Convolutional LSTM: A Deep Learning Method for Motion Intention Recognition Based on Spatiotemporal EEG Data

Zhijie Fang[1,2], Weiqun Wang[2](✉), and Zeng-Guang Hou[1,2,3]

[1] University of Chinese Academy of Sciences, Beijing 100049, China
{fangzhijie2018,zengguang.hou}@ia.ac.cn
[2] The State Key Laboratory of Management and Control for Complex Systems, Institute of Automation, Chinese Academy of Sciences, Beijing 100190, China
weiqun.wang@ia.ac.cn
[3] CAS Center for Excellence in Brain Science and Intelligence Technology, Beijing 100190, China

Abstract. Brain-Computer Interface (BCI) is a powerful technology that allows human beings to communicate with computers or to control devices. Owing to their convenient collection, non-invasive Electroencephalography (EEG) signals play an important role in BCI systems. Design of high-performance motion intention recognition algorithm based on EEG data under cross-subject and multi-category circumstances is a crucial challenge. Towards this purpose, a convolutional recurrent neural network is proposed. The raw EEG streaming is transformed into image sequence according to its location of the primary sensorimotor area to preserve its spatiotemporal features. A Convolutional Long Short-Term Memory (ConvLSTM) network is used to encode spatiotemporal information and generate a better representation from the obtained image sequence. The spatial features are then extracted from the output of ConvLSTM network by convolutional layer. The convolutional layer along with ConvLSTM network is capable of capturing the spatiotemporal features which enables the recognition of motion intention from the raw EEG signals. Experiments are carried out on the PhysioNet EEG motor imagery dataset to test the performance of the proposed method. It is shown that the proposed method can achieve high accuracy of 95.15%, which outperforms previous methods. Meanwhile, the proposed method can be used to design high-performance BCI systems, such as mind-controlled exoskeletons, prosthetic hands and rehabilitation robotics.

Keywords: Brain-Computer Interface · Convolutional LSTM EEG · Motion intention recognition

This work is supported in part by the National Key R&D Program of China (Grant 2018YFC2001700), the Strategic Priority Research Program of Chinese Academy of Science (Grant No. XDB32000000), and Beijing Natural Science Foundation (Grant L172050 and 3171001).

T. Gedeon et al. (Eds.): ICONIP 2019, CCIS 1142, pp. 216–224, 2019.
https://doi.org/10.1007/978-3-030-36808-1_24

1 Introduction

Brain science is one of the most challenging frontier research fields in the twenty-first century. The Brain-Computer Interface (BCI) is a kind of technology that helps human beings to communicate with computers or to control devices. Non-invasive Electroencephalography (EEG) is regarded as one of the most convenient signal sources for BCI systems in practice. When a person is doing mental preparations of motor activity without any muscular motion, appropriate motor related EEG rhythms fluctuate from their scalp [1]. Many promising EEG-based BCI systems have been developed in the literature, such as mind-controlled exoskeletons [2], prosthetic hands [3], and rehabilitation robotics [4]. Therefore, EEG-based intention recognition has become a significant topic because of its industrial and medical applications.

Although a large number of scientists are trying to recognize motion intentions by analyzing EEG signals, this technology is facing several challenges. The first challenge in EEG-based intention recognition is the collected EEG signals themselves because of the low signal-to-noise ratio, coupled with a large quantity of noise, including external noise and physiological noise. The noise definitely presents a severe difficulty for interpretation and analysis of the EEG signal. Also, a typical EEG-based BCI system suffers from the high price, tolerability of the end user, so there are limited public EEG datasets compared with audio, image and video data. More over, most EEG-based intention recognition mainly focuses on manual feature selection, which is time-consuming and highly relys on human experience. For examples, some methods use multiscale principal component analysis [5] to eliminate noise or discrete wavelet transform [6] to extract features followed by a classification model. Finally, many research projects have a terrible classification accuracy, though they just classify EEG signals under the intra-subject or binary circumstances.

Recently, deep learning [7] has shown strong capability when dealing with text, image, audio and video signals. Some researchers are trying to solve EEG-based intention recognition problem by using deep convolutional network or recurrent neural network. However, these methods only focus on spatial information [8] or temporal information. Thus, current approaches can't deal well with EEG signals. We formulate EEG-based intention recognition as a spatiotemporal sequence classification problem. In particular, we transform the spatially distributed EEG signals into 2-D images by projecting the corresponding location of electrodes from a 3-dimensional space onto a 2-D surface [9]. The ConvLSTM network is used to encode EEG signals from spatiotemporal EEG "movie". Several convolutional layers are applied to extract spatial features from the output of the ConvLSTM network. The major contributions of this paper can be outlined as follows:

- Firstly, we propose an end-to-end deep neural network model to recognize motion intentions based on raw spatiotemporal EEG data.
- It is shown that the proposed convolutional recurrent neural network is capable of encoding the spatiotemporal features from the raw EEG streaming and

recognizing motion intentions under cross-subject and multi-category classification circumstances.
- The experimental results demonstrate that the proposed method outperforms previous methods and achieves high accuracy of 95.15% for EEG-based intention recognition.

The remainder of this paper is organized as follows: The detail of the proposed framework is demonstrated in Sect. 2. The data processing, model training, and the result analysis are discussed in Sect. 3. Lastly, we conclude this paper in Sect. 4.

2 Methods

The goal of the proposed convolutional recurrent neural network is to recognize motion intentions based on spatiotemporal EEG data. Figure 1 shows an overview of the proposed method. The network is composed of a ConvLSTM layer for encoding spatiotemporal information and generating a better representation from raw EEG data and several convolutional layers for extracting spatial information.

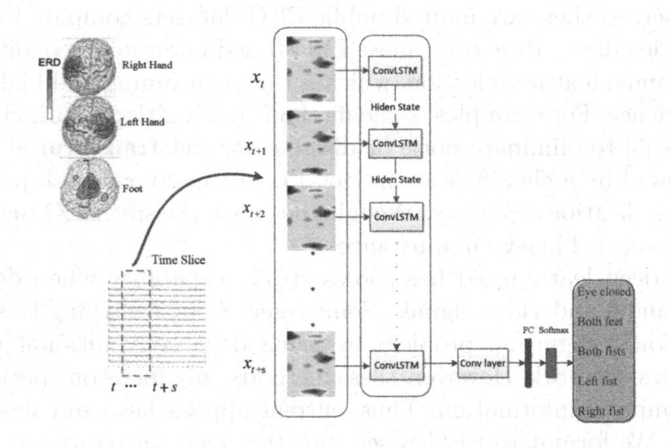

Fig. 1. The proposed convolutional recurrent neural network architecture.

2.1 Design of the Input Images from EEG Streaming

Neuroscience research found that the event-related desynchronization (ERD) starts before the motor imagery over the contralateral hemisphere then becomes bilaterally symmetrical with movement execution [10]. Specifically, when a person executes motor imagery, the specific area of the primary sensorimotor area is activated, in which the Rolandic mu and beta rhythms amplitude will decrease, resulting in event-related desynchronization [11]. The electrodes measure the EEG rhythms fluctuated from different areas of the brain. Hence, we transform the spatially distributed EEG signals into 2-D images by projecting the corresponding location of electrodes from a 3-dimensional space onto a 2-D surface [9]. Taking time into account, we can obtain a sequence of spatial information-preserving images. The detail will be discussed in Sect. 3.

2.2 Convolutional LSTM

By using the sliding window approach, the obtained image sequence can be divided into individual movie clips. The goal of the end-to-end deep neural network model is to classify motion intentions based on spatiotemporal features from EEG "movie" clips. For a model to recognize motion intentions based on EEG "movie" clips, it should be capable of identifying how the activated area of the primary sensorimotor is changing with time. Convolutional neural networks (CNN) is able to generate a spatial representation. Recurrent neural networks can encode temporal changes. Since the model should be able to deal with spatiotemporal information, ConvLSTM is a suitable option.

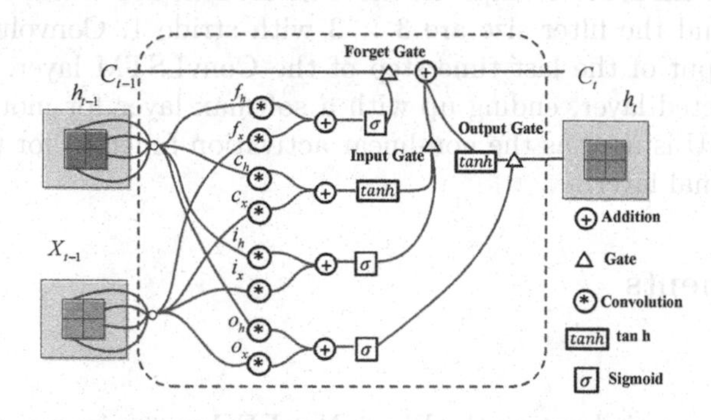

Fig. 2. The inner structure of a ConvLSTM cell.

ConvLSTM can encode spatiotemporal information and generate a better representation. The convLSTM model was first introduced to deal with precipitation nowcasting [12] due to its capacity of extracting spatiotemporal information. Figure 2 shows the inner structure of a ConvLSTM cell. Different from Long Short-Term Memory (LSTM) network, the input feature of a ConvLSTM cell is a 3-D spatiotemporal tensor, and the state-to-state and input-to-state transitions are related to convolutional operations. The key equations of the ConvLSTM are shown as follows:

$$f_t = \sigma\left(U_f * X_t + W_f * h_{t-1} + b_f\right) \tag{1}$$

$$i_t = \sigma\left(U_i * X_t + W_i * h_{t-1} + b_i\right) \tag{2}$$

$$o_t = \sigma\left(U_o * X_t + W_o * h_{t-1} + b_o\right) \tag{3}$$

$$C_t = f_t \circ C_{t-1} + i_t \circ \tanh\left(U_c * X_t + W_c * h_{t-1} + b_c\right) \tag{4}$$

$$h_t = o_t \circ \tanh\left(C_t\right) \tag{5}$$

In the equations, i_t, o_t, f_t are the outputs of input gate, output gate and forget gate at time step t. h_t stands for the hidden state of a cell at time step t. C_t stands for the cell output at time step t. The symbol "$*$" stands for the convolution operator, and "$\circ$" stands for the Hadamard product.

2.3 Network Architecture

After the spatial information-preserving image sequence is obtained, the end-to-end model is used to classify motion intentions based on the obtained image sequence. Figure 1 shows an overview of the proposed method. By using the sliding window approach, which can preserve valuable spatiotemporal information, we divide the obtained image sequence into individual movie clips. The length of each clip is fixed, and there are overlapping between nearby neighbors, avoiding losing significant information. Then the proposed model is used to recognize the motion intentions form the EEG "movie" clips. ConvLSTM network has the capability to encode spatiotemporal information in its memory cell based on the obtained EEG movie clips. In the ConvLSTM, 256 filters are applied in all the gates, and the filter size are 3×3 with stride 1. Convolutional layers receive the output of the last time step of the ConvLSTM layer, and feeds to the fully connected layer, ending up with a softmax layer for motion intention prediction. ReLU is used as the non-linear activation function for the output of each convolutional layer.

3 Experiments

3.1 Dataset

Experiments are carried out on the PhysioNet EEG motor imagery dataset [13], which contains 109 subjects. The dataset contains five motion intentions with eye closed, imagining moving both fists, both feet, right fist and left fist. And the dataset is collected by the BCI2000 instrumentation system, and this system has 64 channels and the sampling rate is 160 Hz. Each subject performed baseline runs and task runs.

3.2 Implementation Details

The collected EEG data has 64 channels, and we transform the EEG streaming into image sequence by projecting the corresponding location of electrodes from a 3-dimensional space onto a 2-D surface at each sampling moment. The obtained EEG image sequence is divided into clips with 10 sampling points and 5 sampling points overlap. Three-quarters data are chosen in random as the training set, and others are used as the validation set. The ConvLSTM layer is used to extract the spatiotemporal information, and several convolutional layers are used to extract spatial information. All experiments are established in Tensorflow framework with batch size 200. We adopt the Adam optimizer with 0.0005 learning rate.

3.3 Experiment Results

The performance of the proposed convolutional recurrent neural network is shown in this section. We compare the results with previous methods to evaluate the performance of the proposed model. Five convolutional recurrent neural network variants and the comparison models are shown in Table 1.

Table 1. Comparison between convolutional recurrent neural network and previous methods.

Method	Multi-class	Validation	Accuracy (%)
Wang [14]	Multi(3)	Intra-Sub	84.62
Pattnaik [6]	Binary	Cross-Sub	80.71
Bashivan [9]	Multi(4)	Cross-Sub	91.11
Kevric [5]	Binary	Intra-Sub	**92.80**
ConvLSTM + 2 Conv layers	Multi(5)	Cross-Sub	89.39
ConvLSTM + 3 Conv layers	Multi(5)	Cross-Sub	94.05
ConvLSTM + 4 Conv layers	Multi(5)	Cross-Sub	**95.15**
ConvLSTM + 2 Conv + 2 pooling layers	Multi(5)	Cross-Sub	83.18

As is shown in Table 1, the proposed convolutional recurrent neural network achieves high accuracy of 95.15% and outperforms the previous methods. ConvLSTM network along with four convolutional layers to extract spatiotemporal features can hit the best performance. Although Kevric [5] centers on the intra-subject and binary circumstance, the proposed model still achieves higher accuracy than their method. Their model requires decomposing raw EEG signals, which may lose significant features while extracting the higher order statistic features. What's more, we add a max-pooling layer after the convolutional layer, but the validation accuracy decreases. Max-pooling layer may make convolutional recurrent neural network achieve translation invariance. Thus, the proposed model can not distinguish which area of the primary sensorimotor is activated.

The accuracy of the proposed method lies in the range between 89% and 95.15%. A ConvLSTM layer with four convolutional layers to extract spatial information can reach the best performance, with an improvement of 2.35% over the previous methods [5]. The validation accuracy of ConvLSTM layer with different convolutional layers to extract spatial features are shown in Fig. 3.

It can be seen from Fig. 3 that the validation accuracy of three convolutional recurrent neural network variants increases rapidly from the first epoch; the validation accuracy increases slowly when the epoch is from 15 to 70; all model variants converge after several fluctuations. Although the ConvLSTM network

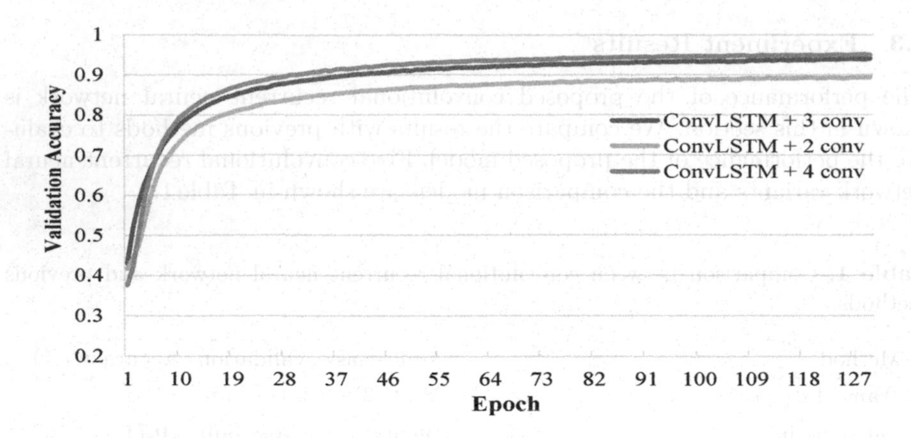

Fig. 3. The validation accuracy of three model variants based on the PhysioNet dataset. The horizontal axis stands for the number of epochs, and the left longitudinal axis stands for validation accuracy.

with four convolutional layers doesn't perform well after the first epoch, its convergence rate is faster than the other two model variants. With four convolutional layers to extract spatial features, the proposed model can achieve high accuracy of 95.15%.

The result of best model variant is used to calculate the confusion matrix, which is shown in Fig. 4. When distinguishing both feet and both fists classes or left fist and right fist classes, the proposed model may make mistakes. However, the proposed model outperforms previous methods. The results show that the proposed model is capable of recognizing motion intentions under cross-subject and multi-category classification circumstances.

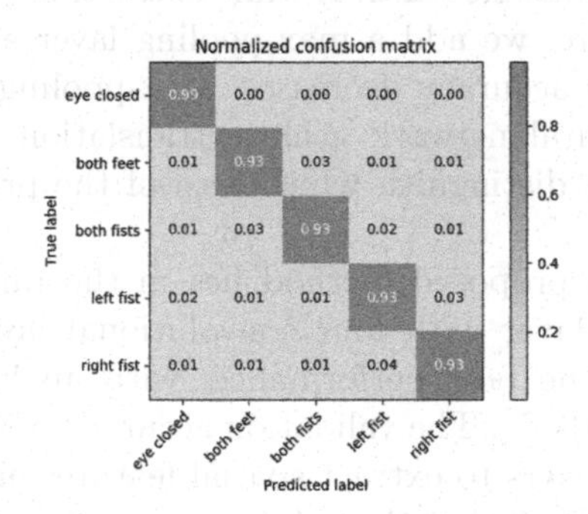

Fig. 4. Confusion matrix illustrating the per-class validation accuracy.

4 Conclusion

The work is motivated by the goal of achieving high-performance motion intention recognition algorithm under cross-subject and multi-category circumstances. The EEG streaming is transformed into image sequence according to its location of the primary sensorimotor area to preserve its spatiotemporal features. A convolutional recurrent neural network is proposed to learn features from raw EEG data. The proposed convolutional recurrent neural network is trained on PhysioNet EEG motor imagery dataset, and the results demonstrate that the proposed model outperforms the previous methods by achieving high accuracy of 95.15%. This results show that the proposed model can be used to design high-performance BCI systems, such as mind-controlled exoskeletons, prosthetic hands and rehabilitation robotics.

References

1. Kumar, S.U., Inbarani, H.H.: PSO-based feature selection and neighborhood rough set-based classification for BCI multiclass motor imagery task. Neural Comput. Appl. **28**(11), 3239–3258 (2017)
2. Chowdhury, A., Raza, H., Meena, Y.K., Dutta, A., Prasad, G.: Online covariate shift detection-based adaptive brain-computer interface to trigger hand exoskeleton feedback for neuro-rehabilitation. IEEE Trans. Cogn. Dev. Syst. **10**(4), 1070–1080 (2018)
3. Muller-Putz, G.R., Pfurtscheller, G.: Control of an electrical prosthesis with an SSVEP-based BCI. IEEE Trans. Biomed. Eng. **55**(1), 361–364 (2008)
4. Ang, K.K., et al.: A randomized controlled trial of EEG-based motor imagery brain-computer interface robotic rehabilitation for stroke. Clin. EEG Neurosci. **46**(4), 310–320 (2015)
5. Kevric, J., Subasi, A.: Comparison of signal decomposition methods in classification of EEG signals for motor-imagery BCI system. Biomed. Signal Process. Control **31**, 398–406 (2017)
6. Pattnaik, S., Dash, M., Sabut, S.: DWT-based feature extraction and classification for motor imaginary EEG signals. In: IEEE International Conference on Systems in Medicine and Biology (ICSMB), pp. 186–201. IEEE (2016)
7. LeCun, Y., Bengio, Y., Hinton, G.: Deep learning. Nature **521**(7553), 436 (2015)
8. Dai, M., Zheng, D., Na, R., Wang, S., Zhang, S.: EEG classification of motor imagery using a novel deep learning framework. Sensors **19**(3), 551 (2019)
9. Bashivan, P., Rish, I., Yeasin, M., Codella, N.: Learning representations from EEG with deep recurrent-convolutional neural networks. In: International Conference on Learning Representations (ICLR) (2016)
10. Stancák Jr., A., Pfurtscheller, G.: The effects of handedness and type of movement on the contralateral preponderance of μ-rhythm desynchronisation. Electroencephalogr. Clin. Neurophysiol. **99**(2), 174–182 (1996)
11. Pfurtscheller, G., Neuper, C.: Motor imagery activates primary sensorimotor area in humans. Neurosci. Lett. **239**(2–3), 65–68 (1997)

12. Xingjian, S., Chen, Z., Wang, H., Yeung, D.Y., Wong, W.K., Woo, W.c.: Convolutional LSTM network: a machine learning approach for precipitation nowcasting. In: Advances in Neural Information Processing Systems, pp. 802–810 (2015)
13. Goldberger, A.L., et al.: Physiobank, physiotoolkit, and physionet: components of a new research resource for complex physiologic signals. Circulation **101**(23), e215–e220 (2000)
14. Wang, Z., Du, X., Wu, Q., Dong, Y.: Research on the multi-classifier features of the motor imagery EEG signals in the brain computer interface. In: Tenth International Conference on Digital Image Processing (ICDIP 2018), vol. 10806, p. 108066Z. International Society for Optics and Photonics (2018)

Deep Neural Networks

A Deep Neural Network Model for Rating Prediction Based on Multi-layer Prediction and Multi-granularity Latent Feature Vectors

Bo Yang[1,2,3(✉)], Qilin Mu[2,3], Hairui Zou[1], Yancheng Zeng[1], Hau-San Wong[4], Zesong Li[3], and Peng Wang[3]

[1] School of Computer Science and Engineering, University of Electronic Science and Technology of China, Chengdu 611731, Sichuan, China
yangbo@uestc.edu.cn, 15002842735@163.com, zengyancheng@outlook.com

[2] Big Data Application on Improving Government Governance Capabilities National Engineering Laboratory, Guiyang 550022, China
muqilin@cetcbigdata.com

[3] CETC Big Data Research Institute Co., Ltd., Guiyang 550022, China
{muqilin,lizesongcd,wangpeng}@cetcbigdata.com

[4] Department of Computer Science, City University of Hong Kong, Tat Chee Avenue, Hong Kong, China
cshswong@cityu.edu.hk

Abstract. Recommender systems have attracted abundant research in the past decades. Side information is generally used besides the rating matrix to alleviate the data sparsity problem for recommendation models. To achieve better performance, in recent years deep learning (DL) technique has been introduced to recommendation models. It can be noted that most existing recommendation models incorporating DL technique only use one layer as the learned features; and the learned features for all users/items have the same dimension despite the fact that different users/items have different numbers of ratings. The aforementioned issues have negative impact on the performance of these recommendation models. To address the issues, in this paper we propose a Deep neural network model based on Multi-layer prediction and Multi-granularity latent feature vectors (DMM model). The DMM has two features: (1) A user or an item is represented by multiple latent vectors with different granularity, which can better describe the relationships between users and items. (2) Each layer in the DMM model produces a predicted rating for given user and item, then the overall rating is calculated by combining all these predicted values, which ensures fully use of the information in rating matrix and side information and thus may result in better performance. Experimental results on three widely used datasets demonstrate that the proposed DMM model outperforms the compared models.

Keywords: Rating prediction · Side information · Multi-granularity · Collaborative filtering

T. Gedeon et al. (Eds.): ICONIP 2019, CCIS 1142, pp. 227–236, 2019.
https://doi.org/10.1007/978-3-030-36808-1_25

1 Introduction

In recent years, recommender systems have attracted much attention and research [1]. Various recommendation models have been proposed in the literature, among which collaborative filtering (CF)-based models are popular ones [2, 3]. However, one drawback of CF-based models is that they suffer from the data sparsity problem. To mitigate the problem, side information such as user profile attributes or item descriptions is used in recommendation models besides rating matrix [4–6].

Recently, deep learning (DL) technique has been integrated into recommendation models due to its capability of capturing non-linear and non-trivial user-item relationships [7]. Some recommendation models incorporating DL technique have been proposed, e.g., DeepFM [8], mSDA-CF [9], AutoSVD++ [10], to improve the performance of CF-based models. However, current state-of-the-art recommendation models which incorporate DL technique still has two insufficiencies: (1) Some models make use of only rating matrix, and side information is not used in the model, e.g., [11, 12]. (2) For those models using both rating matrix and side information, only one layer, i.e., the last layer, e.g., [12, 13], or the middle layer, e.g., [4, 11], is used as the feature leaned, which may bring about information loss and consequently affect the performance of models.

In this paper, we propose a new CF-based recommendation model, which utilizes both rating matrix and side information as the inputs. The proposed model is based on DL technique and is referred to as Deep neural network model based on Multi-layer prediction and Multi-granularity latent feature vectors, abbreviated as DMM model hereafter. Note that recommendation models are either for rating prediction (RP) or for top-N recommendation [1], the proposed DMM model in this paper is for rating prediction.

The proposed DMM model has two features: (1) A user or an item is represented by multiple latent vectors with different granularity, which can better describe the relationships between users and items. Due to the fact that the number of observed ratings of a user/item can vary a lot in real world from that of other users/items, the ratings of a user/item thus may contain much less/more user-item interaction information than those of other users/items. However, most of existing rating prediction models have all users and items represented by same-dimensional latent vectors, which would hinder the expressing ability of the models, because the latent vectors of users/items with few ratings may suffer from the overfitting problem while the latent vectors of users/items with many ratings may suffer from the underfitting problem. In the proposed DMM model, the above-mentioned issue is resolved as the model has multi-granularity layers. (2) Each layer in the DMM model produces a predicted rating for given user and item, then the overall rating is calculated by combining all these predicted values. Since all layers in the DMM model contribute to the overall rating prediction, this ensures fully use of the information in rating matrix and side information and thus would result in better performance than existing rating prediction models that only use one layer as the feature learned.

The remainder of this paper is organized as follows. In Sect. 2, the details of the proposed models are presented. Experimental results are given in Sect. 3. Finally, concluding remarks are given in Sect. 4.

2 Methodology

2.1 Problem Definition

Similar to existing works [1, 10], the RP problem studied in this paper is defined as follows. Given M users and N items, $R \in \mathbb{R}^{M \times N}$ is the rating matrix, r_{ui} indicates the u-th user's rating (preference) on the i-th item. The partial observed vector $R_{u\cdot} = \{r_{u1}, r_{u2}, \ldots, r_{uN}\}$ is the u-th user's ratings on all the N items; and the vector $R_{\cdot i} = \{r_{1i}, r_{2i}, \ldots, r_{Mi}\}$ is all the M users' ratings on the i-th item. Denote by $\hat{r}_{ui}$ the predicted value of r_{ui}, the aim of RP is to predict the missing values in rating matrix R.

2.2 Feature Learning Network

Before we present the proposed DMM model, we first propose a feature learning network (FLN) that will be adopted in the DMM model. The FLN jointly makes use of the ratings and side information by learning latent features from both, as shown in Fig. 1.

It can be seen from Fig. 1 that the proposed FLN has two main features: (1) It uses both side information and ratings as inputs; (2) The hidden layers form a Multi-layer Perceptron (MLP) and thus are convenient to obtain the predicted ratings from each layer (details in Sect. 2.3), which is different from exiting works [4, 11–13].

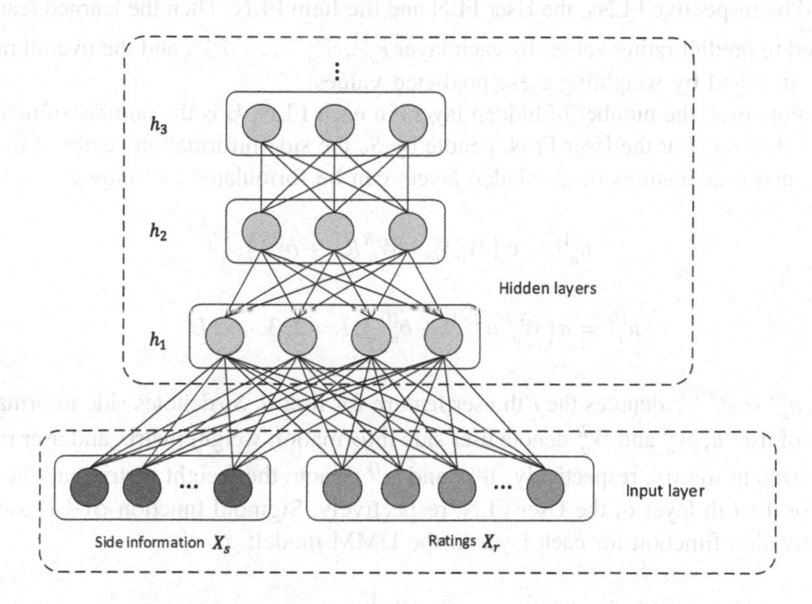

Fig. 1. The structure of the feature learning network

Formally, denote by X_s the side information, X_r is the rating vector (either $R_{u\cdot}$ or $R_{\cdot i}$), denote by h_i the hidden layers, the first hidden layer can be represented as follows:

$$h_1 = \varphi(W_s X_s + W_r X_r + b_1), \tag{1}$$

where W_s and W_r are side information weight matrix and ratings weight matrix, respectively; b_1 is the bias term; $\varphi(\cdot)$ is the activation function. Similarly, the rest hidden layers can be formulated as follows [14]:

$$h_i = \varphi(W_i h_{i-1} + b_i), i = 2, 3, \ldots, \tag{2}$$

where W_i is the weight matrix and b_i is the bias term for the *i-th* hidden layer.

Different from other deep neural networks, we can see from Fig. 1 that the proposed FLN does not contain an output layer. The reason is that each hidden layer can be considered to be a projecting function which projects rating vector and side information to a corresponding low-dimensional feature space, i.e., each layer learns corresponding feature from rating vector and side information. Through forward propagation, all layers of the FLN project rating vector and side information to respective low-dimensional spaces, thus no output layer is needed in the FLN.

2.3 The DMM Model

The proposed DMM model is illustrated in Fig. 2. User features and item features are learned by respective FLNs, the User FLN and the Item FLN. Then the learned features are used to predict rating values by each layer $\hat{r}_{ui}^{(1)}$, $\hat{r}_{ui}^{(2)}$, ..., $\hat{r}_{ui}^{(L)}$, and the overall rating $\hat{r}_{ui}$ is calculated by weighting these predicted values.

Denote by L the number of hidden layers in each FLN, k_l is the number of neurons in the l-th layer. For the User FLN, denote by S_u the side information vector of user u, the learned user features of all hidden layers can be formulated as follows:

$$u_u^{(1)} = \sigma\left(W_u^S S_u + W_u^R R_{u\cdot} + b_u^{(1)}\right), \tag{3}$$

$$u_u^{(l)} = \sigma\left(W_u^{(l)} u^{(l-1)} + b_u^{(l)}\right), l = 2, 3, \cdots, L, \tag{4}$$

where $u_u^{(l)} \in \mathbb{R}^{M \times k_l}$ denotes the l-th user feature for user u, S_u denotes side information vector of user u, W_u^S and W_u^R denote user side information weight matrix and user rating vector weight matrix, respectively. $W_u^{(l)}$ and $b_u^{(l)}$ denote the weight matrix and the bias term for the l-th layer of the User FLN, respectively. Sigmoid function $\sigma(\cdot)$ is used as the activation function for each layer in the DMM model:

$$\sigma(x) = \frac{1}{1 + e^{-x}} \tag{5}$$

Other kinds of activation functions such as ReLU and tanh are also used in neural networks. Here we chose sigmoid to be the activation function for the following reasons: (1) ReLU filters all negative values in forward propagation, which discards some latent information and therefore has negative impact on the effectiveness of learned features; (2) The output value of sigmoid falls between 0 and 1, unlike tanh's output value (between −1 and 1), which can be considered to be a probabilistic value, therefore it may better represent users' preferences and items' characteristics.

Similarly, we have item feature layers of the Item FLN formulated by:

$$v_i^{(1)} = \sigma\left(W_i^S S_i + W_i^R R_{\cdot i} + b_i^{(1)}\right), \tag{6}$$

$$v_i^{(l)} = \sigma\left(W_i^{(l)} v^{(l-1)} + b_i^{(l)}\right), l = 2, 3, \cdots, L \tag{7}$$

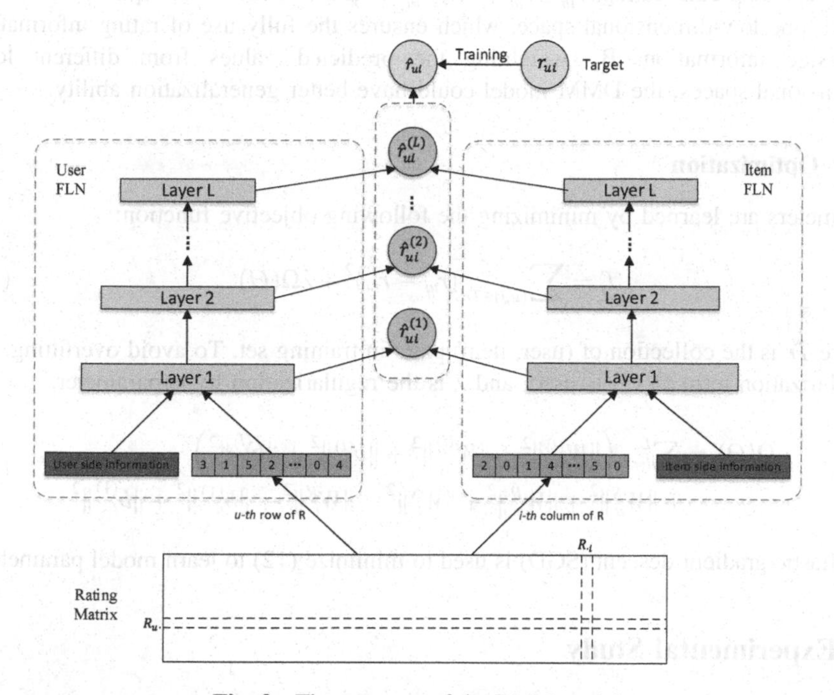

Fig. 2. The structure of the DMM model

Through the two FLNs, user preferences and item characteristics are mapped into L low-dimensional feature vectors $u_u^{(1)}, u_u^{(2)}, \ldots, u_u^{(L)}$ and $v_i^{(1)}, v_i^{(2)}, \ldots, v_i^{(L)}$, which are shown in Eqs. (8 and 9):

$$u_u^{(l)} = \sigma\left(W_u^{(l)}\left(\cdots\sigma\left(W_u^S S_u + W_u^R R_{u\cdot} + b_u^{(1)}\right)\cdots\right) + b_u^{(l)}\right), l = 1, 2, \cdots, L \tag{8}$$

$$v_i^{(l)} = \sigma\left(W_i^{(l)}\left(\cdots\sigma\left(W_i^S S_i + W_i^R R_{\cdot i} + b_i^{(1)}\right)\cdots\right) + b_i^{(l)}\right), l = 1, 2, \cdots, L \tag{9}$$

At the *l-th* layer of both FLNs, we use the learned user feature vector $u_u^{(l)}$ and item feature vector $v_i^{(l)}$ to generate the predicted value $\hat{r}_{ui}^{(l)}$ of user u on item i as follows:

$$\hat{r}_{ui}^{(l)} = u_u^{(l)\mathrm{T}} v_i^{(l)} \tag{10}$$

By combining the predicted values of L layers with respective weights, the predicted overall rating is calculated as follows:

$$\hat{r}_{ui} = \sum\nolimits_{l=1}^{L} \alpha_l \hat{r}_{ui}^{(l)}, \tag{11}$$

where α_l is a hyperparameter representing the weight of the *l-th* layer. As we have analyzed, we predict ratings $\hat{r}_{ui}^{(1)}$, $\hat{r}_{ui}^{(2)}$, ..., $\hat{r}_{ui}^{(L)}$ in L low-dimensional spaces rather than just in one low-dimensional space, which ensures the fully use of rating information and side information. By weighting the predicted values from different low-dimensional spaces, the DMM model could have better generalization ability.

2.4 Optimization

Parameters are learned by minimizing the following objective function:

$$\mathcal{L} = \sum\nolimits_{(u,i) \in Tr} (r_{ui} - \hat{r}_{ui})^2 + \lambda \Omega(\Theta), \tag{12}$$

where Tr is the collection of (user, item) pairs in training set. To avoid overfitting, the regularization term $\Omega(\Theta)$ is used, and λ is the regularization hyperparameter.

$$\begin{aligned} \Omega(\Theta) = \textstyle\sum_{l=2}^{L} \left(\|W_u^{(l)}\|^2 + \|W_i^{(l)}\|^2 + \|b_u^{(l)}\|^2 + \|b_i^{(l)}\|^2 \right) \\ + \|W_u^S\|^2 + \|W_u^R\|^2 + \|W_i^S\|^2 + \|W_i^R\|^2 + \|b_u^{(1)}\|^2 + \|b_i^{(1)}\|^2 \end{aligned} \tag{13}$$

Stochastic gradient descent (SGD) is used to minimize (12) to learn model parameters.

3 Experimental Study

We compare the DMM model with the following three state-of-the-art models which adopt DL technique to conduct rating prediction: Semi-AutoEncoder [4], AutoSVD++ [10], and ReDa [11]. In addition, the following two classical rating prediction models are compared with in the experimental study as well: Biased SVD [15], SVD++ [15].

Some other baseline models such as probabilistic rating autoencoder [16], marginalized stacked denoising autoencoder for collaborative filtering [9] have been outperformed by the compared models [10, 11] and therefore are not considered in our experiments. As the proposed DMM model is a rating prediction recommendation model, it is not suitable to conduct comparative studies of it and *top-N* recommendation models such as [5, 12, 13, 18].

MovieLens [17] are widely used for the evaluation of recommendation models, e.g., [10, 18]. Three real-world datasets from MovieLens, ML-100k, ML-1M and ML-HetRec, are used in our experiments. Moreover, user and item side information in ML-100k, ML-1M and ML-HetRec are used to construct S_u and S_i. User side information contains gender, age, occupation, etc., and item side information contains release date and a vector indicating genre of the movie.

The root mean squared error (RMSE) is adopted, similar to [2, 9, 10].

$$RMSE = \sqrt{\frac{1}{|Te|}\sum_{(i,j)\in Te}\left(r_{ij} - \hat{r}_{ij}\right)^2}, \tag{14}$$

where Te is the collection of (user, item) pairs in test set, and $|Te|$ denotes the number of pairs in Te collection.

3.1 Experiment #1: Performance Comparison

We construct five training sets with different percentages of ratings (50%, 60%, 70%, 80% and 90%, respectively) [11, 19] randomly selected from the original dataset, and the rest of the ratings are used as the test set. The following hyperparameters are adopted: $\lambda = 0.01$, $\eta = 0.001$. In each FLN, we have 6 hidden layers with different dimensions $k = (200, 100, 64, 50, 32, 16)$, and the weights to these hidden layers are $\alpha = (0.05, 0.05, 0.1, 0, 1, 0.2, 0.5)$. The experiment is repeated for 5 times on each constructed training set, and the average RMSE and standard deviation are summarized in Table 1, in which the smallest RMSE in each row is reported in bold.

As can be seen, all models have better performance when the sampling ratio of training data increases. For all sampling ratios, the proposed DMM model achieves the best performance on ML-100k, ML-1M and ML-HetRec datasets, which suggests that the learned features by the proposed DMM model can better describe the relationship between users and items than those of compared models.

Table 1. Performance comparison of the proposed DMM model and 5 other models.

Data set	Sampling ratio	Biased SVD	SVD++	Semi-autoencoder	ReDa	AutoSVD++	DMM	Improvement
ML-100k	50%	0.9329 ± 0.0000	0.9199 ± 0.0004	0.9247 ± 0.0003	0.9217 ± 0.0015	0.9142 ± 0.0004	**0.9049 ± 0.0007**	1.03%–3.09%
	60%	0.9328 ± 0.0001	0.9157 ± 0.0003	0.9181 ± 0.0004	0.9201 ± 0.0012	0.9118 ± 0.0003	**0.9039 ± 0.0003**	0.87%–3.20%
	70%	0.9268 ± 0.0001	0.9050 ± 0.0002	0.9064 ± 0.0011	0.9103 ± 0.0013	0.9016 ± 0.0002	**0.8911 ± 0.0006**	1.18%–4.01%
	80%	0.9243 ± 0.0001	0.8994 ± 0.0002	0.8959 ± 0.0010	0.9057 ± 0.0012	0.8957 ± 0.0004	**0.8846 ± 0.0004**	1.25%–4.49%
	90%	0.9112 ± 0.0001	0.8838 ± 0.0006	0.8810 ± 0.0004	0.8933 ± 0.0024	0.8804 ± 0.0004	**0.8664 ± 0.0001**	1.61%–5.17%

(continued)

Table 1. (*continued*)

Data set	Sampling ratio	Biased SVD	SVD++	Semi-autoencoder	ReDa	AutoSVD++	DMM	Improvement
ML-1M	50%	0.8709 ± 0.0001	0.8742 ± 0.0001	0.8828 ± 0.0002	0.8745 ± 0.0013	0.8626 ± 0.0001	**0.8596 ± 0.0003**	0.35%–2.70%
	60%	0.8633 ± 0.0000	0.8579 ± 0.0001	0.8746 ± 0.0002	0.8692 ± 0.0008	0.8553 ± 0.0001	**0.8522 ± 0.0002**	0.36%–2.63%
	70%	0.8582 ± 0.0000	0.8505 ± 0.0002	0.8650 ± 0.0004	0.8649 ± 0.0016	0.8479 ± 0.0001	**0.8440 ± 0.0003**	0.46%–2.49%
	80%	0.8566 ± 0.0001	0.8470 ± 0.0001	0.8614 ± 0.0002	0.8645 ± 0.0014	0.8447 ± 0.0002	**0.8414 ± 0.0002**	0.39%–2.75%
	90%	0.8510 ± 0.0001	0.8392 ± 0.0001	0.8505 ± 0.0002	0.8587 ± 0.0008	0.8368 ± 0.0002	**0.8301 ± 0.0002**	0.81%–3.45%
ML-HetRec	50%	0.8215 ± 0.0002	0.8169 ± 0.0002	0.8445 ± 0.0011	0.8237 ± 0.0013	0.8143 ± 0.0003	**0.8093 ± 0.0003**	0.62%–4.35%
	60%	0.8153 ± 0.0001	0.8119 ± 0.0003	0.8370 ± 0.0008	0.8203 ± 0.0007	0.8094 ± 0.0002	**0.8044 ± 0.0001**	0.62%–4.05%
	70%	0.8084 ± 0.0001	0.8044 ± 0.0002	0.8266 ± 0.0012	0.8148 ± 0.0013	0.8016 ± 0.0002	**0.7979 ± 0.0003**	0.46%–3.60%
	80%	0.8052 ± 0.0002	0.7991 ± 0.0004	0.8209 ± 0.0004	0.8141 ± 0.0008	0.7972 ± 0.0001	**0.7942 ± 0.0003**	0.38%–3.36%
	90%	0.8007 ± 0.0001	0.7906 ± 0.0002	0.8140 ± 0.0010	0.8094 ± 0.0015	0.7915 ± 0.0001	**0.7831 ± 0.0002**	0.95%–3.95%

3.2 Experiment #2: Is Multi-layer Structure Helpful?

As there is no existing work on using multiple learned features of users and items to predict ratings, it would be interesting to study whether the multi-layer structure can indeed help in making accurate rating predictions. Towards this end, we investigate the relationship between the performance of the DMM model and the layer weights α. We thus conduct a second experiment using four different settings of α: (1) Last-layer-only: $\alpha = (0, 0, 0, 0, 0, 1)$. (2) Mean: predicted ratings in different layers are combined with the same weight, $\alpha = \left(\frac{1}{6}, \frac{1}{6}, \frac{1}{6}, \frac{1}{6}, \frac{1}{6}, \frac{1}{6}\right)$. (3) Increasing: $\alpha = (0.05, 0.05, 0.1, 0, 1, 0.2, 0.5)$. (4) Decreasing, $\alpha = (0.5, 0.2, 0.1, 0, 1, 0.05, 0.05)$. Other hyperparameters such as λ and η stay the same as Experiment #1. The results are presented in Table 2.

Table 2. RMSE of the DMM model with different layer weight settings

Datasets	Last-layer-only	Mean	Decreasing	Increasing
ML-100k	0.9036	0.8754	0.8956	**0.8664**
ML-1M	0.8643	0.8467	0.8656	**0.8301**
ML-HetRec	0.8273	0.8011	0.8237	**0.7831**

It can be seen from Table 2 that the mean, decreasing and increasing settings outperform the last-layer-only setting, except for the value underlined. This suggests that the multi-layer structure be helpful yet a good weight setting is also important. From Table 2, we can observe that the increasing setting performs better than the mean setting, and the mean setting performs better than the decreasing setting. This indicates that the learned feature of deeper hidden layer is more effective in predicting ratings than that of shallower layer, therefore deeper layer deserves higher weight.

4 Conclusions

In this paper, we propose a deep neural network model based on multi-layer prediction and multi-granularity latent feature vectors. Different from existing models incorporating DL technique, the proposed model is capable to learn multiple features in different low-dimensional spaces for users and items from rating matrix and side information, which better describes the relationships between users and items; moreover, all layers in the proposed model contribute to the predicted overall rating, which ensures fully use of the information in rating matrix and side information. Experimental results on three datasets suggest that the proposed model have better performance than the compared models.

Acknowledgments. This work is supported by Big Data Application on Improving Government Governance Capabilities National Engineering Laboratory Open Fund Project (Contract No. w-2019006), National Natural Science Foundation of China (Project No. 61977013), and Sichuan Science and Technology Program (Project No. 2019YJ0164).

References

1. Ricci, F., Rokach, L., Shapira, B.: Recommender Systems Handbook, 2nd edn. Springer, Boston (2015). https://doi.org/10.1007/978-0-387-85820-3
2. Yuan, T., Cheng, J., Zhang, X., Qiu, S., Lu, H.: Recommendation by mining multiple user behaviors with group sparsity. In: 28th AAAI Conference on Artificial Intelligence, pp. 222–228. Association for the Advancement of Artificial Intelligence (2014)
3. Polatidis, N., Georgiadis, C.K.: A multi-level collaborative filtering method that improves recommendations. Expert Syst. **48**, 100–110 (2016)
4. Zhang, S., Yao, L., Xu, X., Wang, S., Zhu, L.: Hybrid collaborative recommendation via semi-autoencoder. In: Liu, D., Xie, S., Li, Y., Zhao, D., El-Alfy, E.S. (eds.) 24th International Conference on Neural Information Processing. Lecture Notes in Computer Science, vol. 10634, pp. 185–193. Springer, Cham (2017). https://doi.org/10.1007/978-3-319-70087-8_20
5. Chen, J., Zhang, H., He, X., Nie, L., Liu, W., Chua, T.: Attentive collaborative filtering. In: 40th International ACM SIGIR Conference on Research and Development in Information Retrieval, pp. 335–344 (2017)
6. Xiao, T., Shen, H.: Neural variational matrix factorization with side information for collaborative filtering. In: Yang, Q., Zhou, Z.H., Gong, Z., Zhang, M.L., Huang, S.J. (eds.) PAKDD 2019. LNCS (LNAI), vol. 11439, pp. 414–425. Springer, Cham (2019). https://doi.org/10.1007/978-3-030-16148-4_32
7. Zhang, S., Yao, L., Sun, A., Tay, Y.: Deep learning based recommender system: a survey and new perspectives. ACM Comput. Surv. **52**, 1–38 (2019)
8. Guo, H., Tang, R., Ye, Y., Li, Z., He, X.: DeepFM: a factorization-machine based neural network for CTR prediction. In: 26th International Joint Conference on Artificial Intelligence, pp. 1725–1731. International Joint Conferences on Artificial Intelligence Organization (2017)
9. Li, S., Kawale, J., Fu, Y.: Deep collaborative filtering via marginalized denoising auto-encoder. In: 24th ACM International Conference on Information and Knowledge Management, pp. 811–820. ACM Press (2015)

10. Zhang, S., Yao, L., Xu, X.: AutoSVD++: an efficient hybrid collaborative filtering model via contractive auto-encoders. In: 40th International ACM SIGIR Conference on Research and Development in Information Retrieval, pp. 957–960. ACM Press (2017)
11. Zhuang, F., Zhang, Z., Qian, M., Shi, C., Xie, X., He, Q.: Representation learning via dual-autoencoder for recommendation. Neural Netw. **90**, 83–89 (2017)
12. Xue, H., Dai, X., Zhang, J., Huang, S., Chen, J.: Deep matrix factorization models for recommender systems. In: 26th International Joint Conference on Artificial Intelligence, pp. 3203–3209. International Joint Conferences on Artificial Intelligence Organization (2017)
13. He, X., Liao, L., Zhang, H., Nie, L., Hu, X., Chua, T.: Neural collaborative filtering. In: 26th International Conference on World Wide Web, pp. 173–182. ACM Press (2017)
14. Goodfellow, I., Yoshua, B., Courville, A.: Deep Learning. MIT Press, Cambridge (2016)
15. Koren, Y., Bell, R., Volinsky, C.: Matrix factorization techniques for recommender systems. Computer **42**, 30–37 (2009)
16. Liang, H., Baldwin, T.: A probabilistic rating auto-encoder for personalized recommender systems. In: 24th ACM International Conference on Information and Knowledge Management, pp. 1863–1866. ACM Press (2015)
17. Harper, F.M., Konstan, J.A.: The MovieLens datasets: history and context. ACM Trans. Interact. Intell. Syst. **5**, 1–19 (2015)
18. Lee, J., Lee, D., Lee, Y., Hwang, W., Kim, S.W.: Improving the accuracy of top-N recommendation using a preference model. Inf. Sci. **348**, 290–304 (2016)
19. Zou, H., Chen, C., Zhao, C., Yang, B., Kang, Z.: Hybrid collaborative filtering with semi-stacked denoising autoencoders for recommendation. In: 4th IEEE Cyber Science and Technology Congress. IEEE Inc. (2019)

LSPM: Joint Deep Modeling of Long-Term Preference and Short-Term Preference for Recommendation

Jie Chen[1], Lifen Jiang[1](✉), Huazhi Sun[1], Chunmei Ma[1], Zekang Liu[1], and Dake Zhao[2]

[1] School of Computer and Information Engineering, Tianjin Normal University, Tianjin 300387, China
wxxjlf@sina.com

[2] Federation University Australia, Ballarat, Australia

Abstract. In the era of information, recommender systems are playing an indispensable role in our lives. A lot of deep learning based recommender systems have been created and proven to be good progress. However, users' decisions are determined by both long-term and short-term preferences, and most of the existing efforts study these two requirements separately. In this paper, we seek to build a bridge between the long-term and short-term preferences. We propose a **L**ong & **S**hort-term **P**reference **M**odel (LSPM), which incorporates LSTM and self-attention mechanism to learn the short-term preference and jointly model the long-term preference by a neural latent factor model. We conduct experiments to demonstrate the effectiveness of LSPM on three public datasets. Compared with the state-of-the-art methods, LSPM got a significant improvement in HR@10 and NDCG@10, which relatively increased by 3.875% and 6.363%. We publish our code at https://github.com/chenjie04/LSPM/.

Keywords: Deep learning · Collaborative filtering · Short-term preference · Long-term preference

1 Introduction

Recently, deep learning has made gigantic strides in many research areas [9]. It also brings a revolution to recommender systems [2]. A lot of deep learning based recommender systems have been created and proven to be good progress [3]. The latent factor techniques are the most effective methods for capturing the long-term preference which reflects users' inherent characteristics [8]. Session-based recommendations have shown a great advantage in extracting the short-term preference from the recent historical interactions [7]. However, the users' decisions are determined by both long-term and short-term preferences, a natural way to improve recommendation is to combine both of them. In this work, we seek to build a bridge between the long-term and short-term preferences.

T. Gedeon et al. (Eds.): ICONIP 2019, CCIS 1142, pp. 237–246, 2019.
https://doi.org/10.1007/978-3-030-36808-1_26

We focus on implicit feedback setting and propose a **L**ong & **S**hort-term **P**reference **M**odel (LSPM), which consists of a short-term preference module and a long-term preference module. More precisely, the short-term preference module is built on the LSTM network and self-attention mechanism to learn the short-term preference from recent historical interactions, and the long-term preference module is a neural latent factor model which is good at capturing long-term preference by considering the whole user-item matrix. By fusing the long-term preference and short-term preference greatly, LSPM can provide users with satisfactory recommendations. We conduct experiments on three public datasets. The results show that LSPM got a superior performance. Compared with the state-of-the-art methods, LSPM got a significant improvement in HR@10 and NDCG@10, which relatively increased by 3.875% and 6.363%.

The main contributions of this work are as follows.

1. We propose a **L**ong & **S**hort-term **P**reference **M**odel, which has an advantage in modeling user's long-term preference and short-term preference.
2. LSPM introduces the short-term preference into latent factor model, which improve the performance of the recommendations.
3. We conduct experiments on three public datasets to demonstrate the effectiveness of LSPM, and it outperforms the competitive baselines.

The structure of the paper is as follows. We start with related work in Sect. 2. In Sect. 3, we formalize the recommendation with implicit feedback and describe LSPM in detail. Experiments are presented in Sect. 4 to demonstrate its effectiveness. Finally, conclusions are presented in Sect. 5.

2 Related Work

In this section, we briefly review the related works from three perspectives: latent factor techniques, session-based recommendations and the combination of long & short-term preferences.

Latent factor techniques are the most effective methods for capturing long-term preference. Salakhutdinov *et al.* [14] use Restricted Boltzmann Machine to extract latent features of users' preference from user-item matrix. He *et al.* [6] replace the inner product with an MLP as the interaction function. ParVecMF fuses the textual user reviews along with the matrix factorization [1]. CoupledCF jointly learns explicit and implicit couplings within/between users and items for deep CF recommendation [19]. [18] proposes an expressive Deep Item-based Collaborative Filtering solution by accounting for the nonlinear and higher-order relationship among items.

Session-based recommendations have shown a great advantage in extracting short-term preference from the recent historical interactions. Hidasi *et al.* [7] take the lead in exploring GRUs for the prediction of the next user action in a session. Li *et al.* [10] argue that both the user's sequential behavior and the main purpose in the current session should be considered in recommendations.

Tuan *et al.* [15] describe a method that combines session clicks and content features to generate recommendations. Wu *et al.* [17] incorporate different kinds of user search behaviors to learn the session representation.

The combination of long & short-term preference is a natural way to improve the performances of recommender system. Liu *et al.* [13] explicitly take the effects of users' current actions on their next moves into account. BINN learns historical preference and present motivation of the target users by discriminatively exploiting users' behaviors [11]. Zhang *et al.* [20] propose an AttRec model that takes both short-term intention and long-term intention into consideration. In this work, we only compare AttRec with our work.

3 Long & Short-Term Preference Model

3.1 Preliminaries

Let $U = \{u_1, u_2, ..., u_M\}$ denote a set of users and $I = \{i_1, i_2, ..., i_N\}$ for items. We define the user-item interaction matrix $Y = R^{M*N}$ as Eq. 1.

$$y_{ui} = \begin{cases} 1 \text{ if interactions (click, view and so on) is observed;} \\ 0 \text{ otherwise.} \end{cases} \tag{1}$$

Let $H^{ui} = (x_0, x_1, x_2, ..., x_n)$ denote the historical interactions before user u interacts with the item i. The recommendation with implicit feedback can be thought as estimating how likely user u will interact with item i in the future, which can be expressed as $\hat{y}_{ui} = f(u, i, H^{ui}|\Theta)$.

3.2 Architecture

The Long & Short-term Preference Model proposed in this paper aims to jointly model the user's short-term preference and long-term preference for recommendations. The framework of LSPM is shown in Fig. 1.

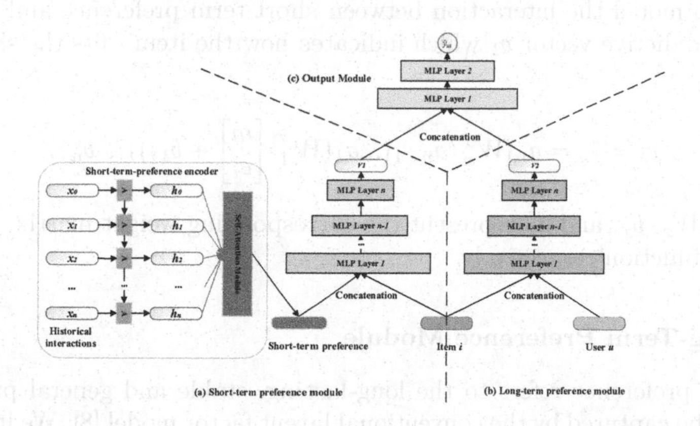

Fig. 1. The framework of Long & Short-term Preference Model.

LSPM comprises of three modules: short-term preference module, long-term preference module and an output module. The short-term preference module consists of a short-term preference encoder to extract the short-term preference from historical interactions H^{ui} and an MLP to learn how the item i fits the user's short-term preference. The long-term preference module is a neural latent factor model which learns how the item i fits the user's long-term preference. At last, the output module fuses the long-term and short-term preferences to obtain the final prediction.

3.3 Short-Term Preference Module

Recent historical interactions indicate users' special need and personalized taste in this short interval, which is what we call short-term preference. A short-term preference encoder which combines a LSTM network and a self-attention block is proposed to capture short-term preference, as shown in Fig. 1(a).

LSTM receives the user's recent interactions H^{ui} as input and outputs hidden states H, which can be expressed as,

$$H^{ui} = (x_0, x_1, ..., x_t, ..., x_n) \tag{2}$$

$$h_t = LSTM(x_t, h_{t-1}) \tag{3}$$

$$H = (h_0, h_1, ..., h_t, ..., h_n) \tag{4}$$

Where $x_t \in R^d$ is the embedding of item t, $h_t \in R^d$ and $H \in R^{n*d}$. We use a self-attention block [12] to learn the short-term preference,

$$a = softmax(W_2 tanh(W_1 H^T + b_1) + b_2) \tag{5}$$

$$m = aH \tag{6}$$

Where $W_1 \in R^{d_a * d}$, $W_2 \in R^{d_a}$, and d_a is a hyperparameter. The weight vector a indicates the importance of each hidden state. The hidden states of LSTM H are weighted sum to get the short-term preference m. Then, we use an MLP to model the interaction between short-term preference and item i, it output a predictive vector v_1 which indicates how the item i fits the short-term preference,

$$v_1 = l_n = a_n(W_n^T(a_{n-1}(...a_1(W_1^T \begin{bmatrix} m \\ e_i \end{bmatrix} + b_1))) + b_n) \tag{7}$$

Where W_x, b_x, and a_x represent the corresponding weight matrix, bias, and activation function respectively.

3.4 Long-Term Preference Module

Long-term preference refers to the long-lasting, stable and general preference, which can be captured by the conventional latent factor model [8]. We implement our long-term preference module base on MLP, as shown in Fig. 1(b).

Each item i or user u is associated with a vector, named as $q_i \in R^d$ or $p_u \in R^d$. Long-term preference module consumes q_i or p_u and outputs a predictive vector v_2, which indicate how the item i fits the long-term preference, defined as Eq. 8.

$$v_2 = l_n = a_n(W_n^T(a_{n-1}(...a_1(W_1^T \begin{bmatrix} p_u \\ q_i \end{bmatrix} + b_1))) + b_n) \tag{8}$$

3.5 Output Module

Due to the complexity of the real world, short-term preference is likely to be biased. It is necessary to introduce a long-term preference to ensure the recommendation is in the right direction. We use fully connected layers to merge this two preferences and make the final prediction, as shown in Eq. 9.

$$\hat{y}_{ui} = \sigma(h_T.relu(W_1^T \begin{bmatrix} v_1 \\ v_2 \end{bmatrix} + b_1) + b_2) \tag{9}$$

Where the final prediction score $\hat{y}_{ui}$ indicates how likely user u will interact with the item i in the future.

3.6 Model Learning

The recommendation problem with implicit feedback can be regarded as a two-class classification problem. And, we choose the binary cross-entropy as our loss function. It is defined as Eq. 10.

$$loss = - \sum_{(u,i) \in y \cup y^-} ylog\hat{y}_{ui} + (1-y)log(1-\hat{y}_{ui}) \tag{10}$$

Here y denotes the positive sample set and y^- means the negative sample set. We randomly sample 4 negative samples for each positive one and optimize our model by the adaptive gradient descent algorithm - Adam.

4 Experiments

4.1 Experimental Settings

BaseLines. In this section, to evaluate the effectiveness of LSPM, we compare it with three baseline methods that are

- **NeuMF** [6] is a composite matrix factorization jointly coupled with a multilayer perceptron model for item ranking.
- **AttRec** [20] takes both short-term and long-term intentions into consideration. We use an MLP network to replace the Euclidean distance metric.
- **LSPM-base**, we also implement a simple version of the LSPM model, in which the MLP network is replaced by an inner product function.

Table 1. Statistics of the datasets used in experiments

Dataset	Users	Items	Interactions
Movielens-1M	6040	3706	1000209
Movielens-10M	69878	10677	10000054
Movielens-20M	138493	26744	20000263
Amazon-Books	12252	5362	516486
Amazon-Electronics	3426	11777	107186
Taobao User-Behavior	197598	2215070	20039836

Datasets. The datasets employed in experiments are three public datasets, that are **(1) Movie Lens datasets** [4]. **(2) Amazon review dataset** [5]. **(3) Taobao User-Behavior dataset** [21]. After data preprocessing, detail statistics of the datasets are presented in Table 1.

Evaluation. We evaluate our model by the leave-one-out evaluation [6] and measure performance by HR [6] (Hit-Rate) and NDCG [16] (Normalized Discounted Cumulative Gain) metrics. Intuitively, HR measures the presence of the positive item and NDCG measures the item's position in the ranked list.

Parameters Setting. Hyperparameters are tuned according to a validation set. After a grid search was performed, the hyperparameters were set as follows, the $learning-rate$ was set to 0.001, $batch-size$ was set to 1024, and employed a 4-layer MLP in long-term and short-term module.

4.2 Performance Comparison

Table 2 lists the performance of LSPM and baselines for HR and NDCG with cut off at 10, in which the history size is set to 9, and embedding size is set to 64.

First, we can see that LSPM achieves the best performance on the all datasets, significantly outperforming the state-of-the-art method AttRec by a large margin. On average, the relative improvement of $HR@10$ and $NDCG@10$ over this baseline is 3.875% and 6.343%, respectively. The experimental results demonstrate that the combination of long-term preference and short-term preference gives an advantage to recommendations.

From the experimental results, we also have the following findings:

- NeuMF gives poor performance in most datasets. This indicates that latent factor models, which rely heavily on the long-term preference, may fail to provide accurate recommendations in some situation where the short-term preference can make a difference.

Table 2. Experimental results for different methods on public datasets.

Dataset	Metric	NeuMF	AttRec	LSPM-base	LSPM	Improv
Movielens-20m	HR@10	0.6330	0.7321	0.7485	**0.7634**	**4.28%**
	NDCG@10	0.3798	0.4823	0.4979	**0.5240**	**8.65%**
Movielens-10m	HR@10	0.8833	0.9223	0.9241	**0.9336**	**1.23%**
	NDCG@10	0.6298	0.7140	0.7221	**0.7426**	**4.30%**
Movielens-1m	HR@10	0.7018	0.8069	0.8255	**0.8303**	**2.90%**
	NDCG@10	0.4247	0.5788	0.6037	**0.6240**	**7.81%**
Amazon-books	HR@10	0.6947	0.7426	0.7355	**0.7489**	**0.85%**
	NDCG@10	0.4134	0.4895	0.4730	**0.5045**	**3.25%**
Amazon-electronics	HR@10	0.5394	0.5152	0.5406	**0.5552**	**7.76%**
	NDCG@10	0.3383	0.3254	0.3313	**0.3455**	**6.18%**
Taobao-User Behavior	HR@10	0.7461	0.7507	0.7775	**0.7975**	**6.23%**
	NDCG@10	0.5923	0.6034	0.6376	**0.6471**	**7.24%**

Note: *The best results are highlighted in* ***bold****. The* ***Improv*** *was computed compare with the AttRec model.*

- AttRec significantly outperform the NeuMF model, as it combines the benefits of both long-term and short-term preference. However, AttRec perform worse than our model, which might be caused by the lack of autoregression property in transformer.
- It is worth mentioning that the LSPM-base also achieves good performance than AttRec. This demonstrates the efficiency of the proposed model.

4.3 Impact of History Size

For a particular dataset, considering too many historical interactions will increase the complexity and introduce noise, but too few will fail to capture the useful dependencies in history. Here, we conduct experiments on the MovieLens-1m dataset to study the impact of history size on the performance. We perform a grid search over $\{5, 7, 9, 11, 13, 15\}$. Results are shown in Fig. 2.

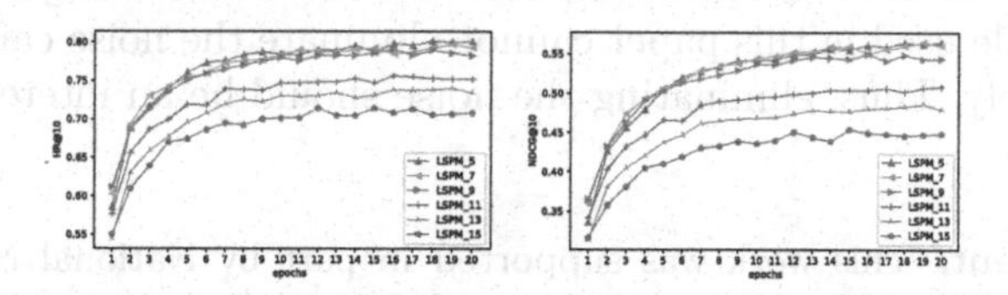

Fig. 2. Performances of LSPM *w.r.t* different history size.

We can see from Fig. 2 that performance begins to decrease when the history size is larger than 11. The decline of performance might be caused by the fact that too large a history size will bring the noise in, and our self-attention

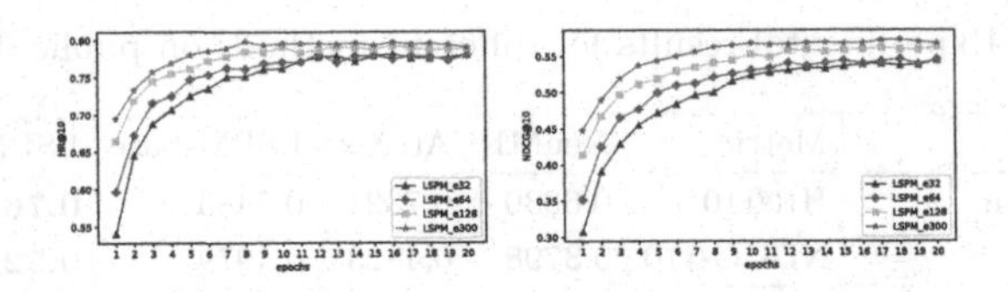

Fig. 3. Performances of LSPM *w.r.t* embedding size.

mechanism cannot effectively offset it. For different data sets, we should choose an appropriate history size.

4.4 Impact of Embedding Size

In general, the larger size of the embedding is, the more information can be carried, and the performance will improve. To study the impact of embedding size, we conduct experiments on MovieLens-1m dataset, in which the embedding size is searched on $\{32, 64, 128, 300\}$. The results are shown in Fig. 3.

In the first stage of training, the performance is better, as the embedding size growing. When the models converge, they tend to get the same performances. We think the embedding size of 32 is large enough to capture complex structures. When embedding size grows larger, the gain from increasing embedding size can't trade off the complexity it brings in.

5 Conclusion and Future Works

In this work, we propose a Long & Short-term Preference Model, which incorporates LSTM and self-attention mechanism to learn the short-term preference and jointly model the long-term preference by a neural latent factor model. By fusing the long-term and short-term preferences, LSPM can achieve more satisfactory recommendations. Comprehensive experiments on three public datasets demonstrate the effectiveness of the proposed model.

In the future, we will introduce more information into the model to provide users with more accurate recommendations. In addition, we argue that the self-attention module used in this paper cannot eliminate the noise caused by a long history effectively. Thus, eliminating the noise should be an interesting research direction.

Acknowledgment. This work was supported in part by National Natural Science Foundation of China (NO. 61702370); Natural Science Foundation of Tianjin (NO. 17JCYBJC16 400, 18JCQNJC70200, 18JCYBJC85900); Research project of Tianjin science and technology development strategy (NO. 17ZLZXZF00530); 131 three-level candidates of Tianjin Normal University (NO. 043/135305QS20); Doctoral Fund of Tianjin Normal University (NO. 043/135202XB1615, NO. 043/135202XB1705).

References

1. Alexandridis, G., Siolas, G., Stafylopatis, A.: ParVecMF: a paragraph vector-based matrix factorization recommender system. arXiv: Information Retrieval (2017)
2. Batmaz, Z., Yurekli, A., Bilge, A., Kaleli, C.: A review on deep learning for recommender systems: challenges and remedies. Artif. Intell. Rev. **52**, 1–37 (2018)
3. Ebesu, T., Shen, B., Fang, Y.: Collaborative memory network for recommendation systems. In: The 41st International ACM SIGIR Conference on Research and Development in Information Retrieval, pp. 515–524. ACM (2018)
4. Harper, F.M., Konstan, J.A.: The movielens datasets: history and context. ACM Trans. Interact. Intell. Syst. (TIIS) **5**(4), 19 (2016)
5. He, R., McAuley, J.: Ups and downs: modeling the visual evolution of fashion trends with one-class collaborative filtering. In: Proceedings of the 25th International Conference on World Wide Web, pp. 507–517. International World Wide Web Conferences Steering Committee (2016)
6. He, X., Liao, L., Zhang, H., Nie, L., Hu, X., Chua, T.S.: Neural collaborative filtering. In: Proceedings of the 26th International Conference on World Wide Web, pp. 173–182. International World Wide Web Conferences Steering Committee (2017)
7. Hidasi, B., Karatzoglou, A., Baltrunas, L., Tikk, D.: Session-based recommendations with recurrent neural networks. In: International Conference on Learning Representations (2016)
8. Koren, Y., Bell, R., Volinsky, C.: Matrix factorization techniques for recommender systems. Computer **8**, 30–37 (2009)
9. LeCun, Y., Bengio, Y., Hinton, G.: Deep learning. Nature **521**(7553), 436 (2015)
10. Li, J., Ren, P., Chen, Z., Ren, Z., Lian, T., Ma, J.: Neural attentive session-based recommendation. In: Conference on Information and Knowledge Management, pp. 1419–1428 (2017)
11. Li, Z., Zhao, H., Liu, Q., Huang, Z., Mei, T., Chen, E.: Learning from history and present: next-item recommendation via discriminatively exploiting user behaviors. In: Knowledge Discovery and Data Mining, pp. 1734–1743 (2018)
12. Lin, Z., Feng, M., Santos, C.N.D., Yu, M., Xiang, B., Zhou, B., Bengio, Y.: A structured self-attentive sentence embedding. In: International Conference on Learning Representations (2017)
13. Liu, Q., Zeng, Y., Mokhosi, R., Zhang, H.: STAMP: short-term attention/memory priority model for session-based recommendation, pp. 1831–1839 (2018)
14. Salakhutdinov, R., Mnih, A., Hinton, G.: Restricted Boltzmann machines for collaborative filtering. In: Proceedings of the 24th International Conference on Machine Learning, pp. 791–798. ACM (2007)
15. Tuan, T.X., Phuong, T.M.: 3D convolutional networks for session-based recommendation with content features, pp. 138–146 (2017)
16. Valizadegan, H., Jin, R., Zhang, R., Mao, J.: Learning to rank by optimizing NDCG measure, pp. 1883–1891 (2009)
17. Wu, C., Yan, M.: Session-aware information embedding for e-commerce product recommendation. In: Conference on Information and Knowledge Management, pp. 2379–2382 (2017)
18. Xue, F., He, X., Wang, X., Xu, J., Liu, K., Hong, R.: Deep item-based collaborative filtering for top-n recommendation. ACM Trans. Inf. Syst. **37**(3), 33 (2019)
19. Zhang, Q., Cao, L., Zhu, C., Li, Z., Sun, J.: CoupledCF: learning explicit and implicit user-item couplings in recommendation for deep collaborative filtering, pp. 3662–3668 (2018)

20. Zhang, S., Tay, Y., Yao, L., Sun, A.: Next item recommendation with self-attention. arXiv: Information Retrieval (2018)
21. Zhu, H., et al.: Learning tree-based deep model for recommender systems. In: Proceedings of the 24th ACM SIGKDD International Conference on Knowledge Discovery and Data Mining, pp. 1079–1088. ACM (2018)

How We Achieved a Production Ready Slot Filling Deep Neural Network Without Initial Natural Language Data

François Torregrossa, Nihel Kooli, Robin Allesiardo(✉), and Erwan Pigneul

Solocal/Pages Jaunes Digital/Search Tribe, Rennes, France
{ftorregrossa,nkooli,rallesiardo}@solocal.com

Abstract. Training deep networks requires large volumes of data. However, for many companies developing new products, those data may not be available and public data-sets may not be adapted to their particular use-case. In this paper, we explain how we achieved a production ready slot filling deep neural network for our new single-field search engine without initial natural language data. First, we implemented a baseline by using recurrent neural networks trained on expert defined templates with parameters extracted from our knowledge databases. Then, we collected actual natural language data by deploying this baseline in production on a small part of our traffic. Finally, we improved our algorithm by adding a knowledge vector as input of the deep learning model and training it on pseudo-labeled production data. We provide detailed experimental reports and show the impact of hyper-parameters and algorithm modifications in our use-case.

Keywords: Deep learning · Slot filling · Data generation · Pseudo-labeling · Knowledge database

1 Introduction: A Journey to a Single-Field Search Engine

PagesJaunes is a french search engine specialized in the search of local businesses. Historically, the search engine inputs are separated in two fields, the *Who What* and the *Where*. To be able to understand queries in natural language, we replaced the two fields with a single-field (see Fig. 1). This change is motivated by the democratization of dialogue systems available in smart-phones (Google Assistant, Siri), home-devices (Google Home, Amazon Echo) and the development of our own chat-bot. Being able to proceed queries in natural language is an

---

F. Torregrossa, N. Kooli and R. Allesiardo—The authors made equal contributions and are sorted at random.
François Torregrossa is also a member of The Research Institute on Informatics and Random Systems (IRISA) and of the Inria Rennes - Bretagne Atlantique research center.

T. Gedeon et al. (Eds.): ICONIP 2019, CCIS 1142, pp. 247–255, 2019.
https://doi.org/10.1007/978-3-030-36808-1_27

important challenge to tackle to achieve a presence on those supports. However, the single-field is only a proof of concept and our back-end still requires two separated fields. To be able to use the existing back-end, we needed to map the component of single-field queries to the *Who What* and *Where* fields. The task of extracting sub-concepts from a sentence is commonly known as slot filling.

Fig. 1. The PagesJaunes single-field search engine

In the following sections, we relate the challenges we faced and how we overcame them. In Sect. 2, we detail why we chose to implement our own slot filling solution instead of using a commercial one. In Sect. 3 we show how we used generated synthetic natural language data to train a recurrent neural network (RNN) and the results we obtained by deploying it on production on a small part of our traffic. Then, in Sect. 4, we present the improvements we made to our algorithm, by including features extracted from our knowledge data-bases and by auto-labeling the queries collected in production. Through the paper, we show how each modification impacted the performances of our solution. The complete results of our experiments are detailed in Sect. 5.

2 Slot Filling

2.1 Problem Setting

The Slot Filling problem consists of extracting sub-concepts, defined as entities, from a semantic frame, namely a user query. This is commonly performed as a supervised sequence classification problem where tokens are associated with corresponding slots. Usually, the slot filling task is proceeded after an intent detection task, whose purpose is to understand the general meaning of the user query. However, in our proof of concept, only one intention is supported by default: the search of local businesses. Table 1 shows an example of query where the slots are labeled following the In/Out/Begin (IOB) representation [13].

Table 1. Example of slot filling task with IOB representations of the slots.

Query	I	am	looking	for	a	vegan	restaurant	in	Paris
Slot	O	O	O	O	O	B-WhoWhat	I-WhoWhat	O	B-Where
Intent	FIND_BUSINESS								

2.2 Commercial Solutions and State of the Art

Several commercial solutions can be used on slot filling tasks, such as Google DialogFlow[1] or Microsoft Luis[2]. Those softwares are based on the definition of

[1] Google DialogFlow, https://dialogflow.com/.
[2] Microsoft Luis, https://www.luis.ai/.

intents and entities. Each intent is associated with several sentences, that provide examples of what a user could ask the system. Each of them can contains several slots that are filled with entities. For example, the sentence "I am looking for a restaurant in Paris" can be associated with the intent FIND_BUSINESS where the slots *WhoWhat* and *Where* are respectively filled with the entities *restaurant* and *Paris*.

DialogFlow and Luis can accept a limited number of entities by slot, in the order of ten thousands. Our solution must be able to manage several million entities by slot. As seen in Table 2, those commercial solutions failed to provide acceptable performances when challenged on our use-case[3]. The goal was to use the final solution inside our production environment, thus it had to achieve better performances. We chose to implement a customized solution based on state of the art research.

Table 2. Performances of commercial solutions on a slot filling task performed on hand labeled production queries. The accuracy is the proportion of queries where the slots have been successfully labeled.

Algorithm	Accuracy
Google DialogFlow	0.74
Microsoft Luis	0.64

The slot filling task is a challenging problem widely studied in the natural language understanding literature. One common method to address the slot filling task is to use expert knowledges, templates and dictionaries [14]. This approach has the advantage to be straightforward to implement but is domain dependent and can struggle to generalize on new domains. An alternative to those methods was to use conditional modeling algorithms, such as conditional random fields (CRF) [7,15]. Several works [8,9] challenged various network architectures against networks using recurrent layers. Models built with recurrent layers showed state of the art performances on the slot filling task and outperformed existing algorithms.

3 Deploying the First Model in Production

During this project, we implemented different algorithms from the literature: an algorithm based on our knowledge database; a CRF; several RNN based neural networks. Our experimentations confirmed the insight provided by the literature as the RNN based models outperformed the knowledge based algorithm and the CRF. In the following, we present the methodology used to obtain our first model without initial natural language data and the results obtained after deploying the RNN on a small traffic in our production environment.

3.1 Generation of Synthetic Training Data from Templates

The main challenge we faced when beginning this project was the absence of natural language data related to the search of local businesses, even unlabeled ones.

[3] Accuracies reported in Table 2 are obtained on our most recent testing set. At the beginning of the project, we challenged both solutions on a smaller set containing queries made by experts.

Nearly exhaustive databases of businesses, keywords and localities were available but were only used in the two fields search. Taking example on DialogFlow, we implemented a text generator based on sentence templates where the slots would be filled by entities from our databases (see Fig. 2).

(Give me|Tell me|What is) the (phone|number|num) of (BUSINESS_NAME).

Fig. 2. An example of template

Around fifty templates were defined by experts and added to the text generator. The main advantage of this approach is to know the label of each word by construction. Indeed, when generating a query, the indexes of each slot is known, as well as their type (*WhoWhat* or *Where*). By using those templates we generated a labeled dataset of several million queries.

3.2 The Initial Slot Filling Deep Neural Network

In this subsection we detail the steps used to process the queries before feeding it to the network.

Word Embedding is a technique used to map textual data to dense real valued vectors. When learned with appropriate algorithms they produce vector space where distance encode semantic similarity. Algorithms using word embeddings as inputs show good generalization properties on natural language processing tasks [12]. FastText library [2] is used here for its capacity to create vector for unseen word during training. This is particularly crucial in our use-case as users can make mistakes when writing queries.

Query Tokenization consits of, in our case, extracting words in the query and normalizing them by removing punctuation, accents and uppercase. In some cases, those special characters contain an important part of the information by highlighting the *Who What* and the *Where* from the rest of the sentence. However, most of the real-world queries do not contain this sort of characters, so we did not consider them. Moreover, if the RNN achieves a high accuracy with those generic conditions, it would be able to cope with every way to write words.

Recurrent Neural Networks [9] are used for their good performances on the slot filling task. Actually, their ability to capture individual words and their contextual information is beneficial for the slot filling task. More specifically, bi-GRUs [3,5] shown to be particularly promising among other tested RNN architectures on our use-case. The implementation details are listed in Sect. 5.

3.3 Labeling Production Data to Build a Validation/Test Dataset

We constituted a dataset of 6000 labeled single-field queries from production. Table 3 shows the accuracy of different baselines and of the retained RNN on this dataset. The RNN clearly outperformed the knowledge based algorithm and the CRF. The implementation of our knowledge based algorithm was quite naive and only looked for the largest matching expressions in our database. However, it showed unexpected performances and we decided to take advantage of those knowledge by combining them with the RNN. This improvement is detailed in the next section.

Table 3. Performances of several algorithms on a slot filling task performed on our use-case (details are provided in Sect. 5). The accuracy is the proportion of queries where the slots have been successfully labeled.

Algorithm	Accuracy
Knowledge Based Algorithm	0.764
Conditional Random Fields	0.778
RNN	**0.840**

4 Continuous Improvements

After the labelization of production data, we continued to improve our solution. We made two main improvements, adding a knowledge vector in input of the network (see Subsect. 4.1) and integrating pseudo-labeled production data to the training set (see Subsect. 4.2).

4.1 Taking Advantage of Our Existing Knowledge Databases

During the review of miss-classified queries we noticed that, for some of them, human experts had to use database queries to know if some sub-concepts of the query were business names and/or localities. As the current model did not achieve human-like performances, we had the intuition that to be able to perform better, we could include information from our databases as inputs of the network. Indeed, as those concepts are quite rare in Wikipedia documents, using the only information from the embedding may not be enough to solve the task.

N-Gram of Words. A list of every word N-gram is generated from the user query and each database is queried to see if it contains the N-gram. Each of them is thus associated with its frequency inside the *WhoWhat* and *Where* fields and its presence inside each database (businesses names and activities/keywords; the localities (cities/points of interest/addresses). See Fig. 3 for an example of N-gram decomposition.

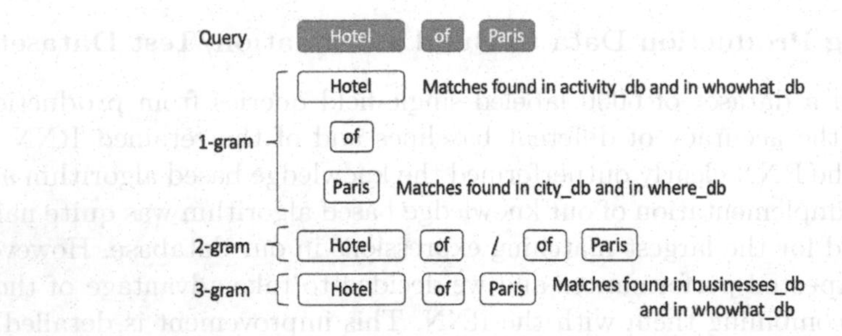

Fig. 3. Decomposition of a query in n-gram and search of matches in several database. In this example, activity_db contains the business activities, city_db the cities and whowhat/where_db the history of the inputs inside the *WhoWhat* and *Where* fields.

A Knowledge Matrix by token of the query is built by aggregating every information about it. The two dimensions of this matrix present: the size N of the N-grams ($N = 1, 2, 3, 4+$); the binning of frequencies inside the different fields (i.e. *WhoWhat* and *Where*) and the presence of the N-grams inside the different databases. Notice that features activated by an N-grams are activated on the knowledge matrix of all words compounding the N-grams. The knowledge matrix is finally flattened to form a **knowledge vector**. As shown in Fig. 4, the knowledge vector is forwarded through a dense layer before being fed to the RNN to expose non-linear relationships between the raw knowledge features. We call one output of this layer a **knowledge embedding**.

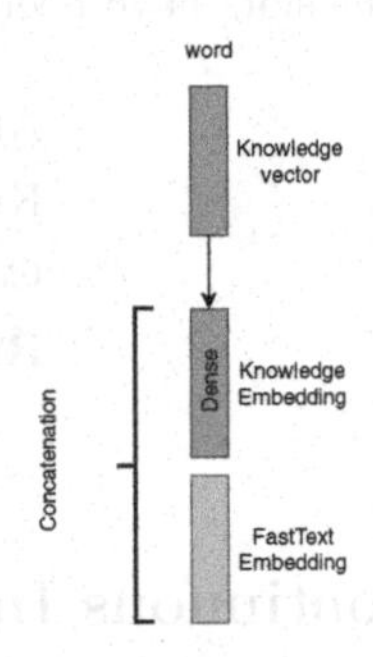

Fig. 4. The final architecture used by our RNN taking advantage of our knowledge database.

4.2 Achieving a Better Sentence Embedding with Pseudo-labeling

The main drawback of our model was the training set which is generated and does not include actual production data. Hand labeled data are not in sufficient quantities and are only used for validation and test. To include actual data, we create a pseudo-labeled data-set by labeling the production data with a RNN trained on the templates. This method is called pseudo-labeling [4] and achieves an effect similar to entropy regularization. It has the main advantage to allow the network to be trained on the actual distribution of queries. Indeed, while the sentences generated by the template are correctly labeled, the distribution of each type of queries may not reflect ground truth - actually, it does not. We train our network by incorporating pseudo-labeled data in the template dataset.

5 Experimentation

Optimizing Hyper-Parameters with Random Search. We optimize the hyper-parameters through a Random Search [1]: many networks are trained with sets of hyper-parameters sampled over distributions of Table 4. It has been shown that this approach is more efficient than grid-search or manual-search, especially when the hyper-parameter space is high dimensional.

Table 4. The sampling distribution of each hyper-parameter.

Parameter	Sampling distribution	Description
batch_size	choice([128, 256, 512])	Batch size
learning_rate	log_uniform(10^{-5}, 10^{-1})	Learning Rate
knowledge_emb_size	int.uniform(32, 512)	Size of the knowledge embedding layer
concat_dropout	float.uniform(0.01, 0.4)	Dropout value of the concatenation layer
rnn_layers	choice([1, 2])	Number of RNN layers (bi-GRU)
rnn_size	int.uniform(32, 1024)	Size of the RNN layers (bi-GRU)
rnn_dropout	float.uniform(0.1, 0.4)	Dropout value of the RNN outputs
dense_size	int.uniform(32, 1024)	Size of the dense layer after the RNN
dense_dropout	float.uniform(0.1, 0.4)	Dropout value of the dense layer after the RNN
pseudo_label_ratio	choice([0, 0.1, 0.2, 0.3, 0.4, 0.5])	Ratio of pseudo-labeled data in training
use_knowledge_vector	choice([True, False])	Availability of the Knowledge Vector

Implementation Details. The network is implemented in Pytorch [11]. Intermediate dense layers are activated through rectified linear units [10] and the prediction layer through a sigmoid function. The network is trained using the Adam optimizer [6]. The 6000 hand labeled queries are shuffled and separated into validation and test sets, each containing 3000 queries. Best models are selected using the accuracy on the validation set.

Results. More than 300 networks were trained using the random search. Table 5 presents the accuracies of the best networks obtained through the random search for each combination of improvements. Without surprise, **the baseline RNN** shows a lower accuracy than the improved networks. Indeed, the models suffer from the absence of actual natural language data and from the large number of entity values (several million). On another side, networks trained with a part of **pseudo-labeled data** or with **knowledge vectors** as input data achieve an higher accuracy. Combining both shows significant improvement over other networks. More than half of the best networks trained using both **Knowledge Vectors and Pseudo-Labeling** outperform the networks obtained by training the RNN alone or by using only one of the improvements.

Table 5. Recapitulation of the results obtained by the Recurrent Neural Network (RNN), the RNN with Knowledges Vectors (KV) as additional inputs, the RNN trained on templates and pseudo-labeled data and the combination of both.

Algorithm	Accuracy		
	Templates	Validation	Test
RNN	0.936	0.824	0.840
RNN+PL	0.928	0.856	0.867
RNN+KV	**0.939**	0.857	0.866
RNN+KV+PL	0.921	**0.875**	**0.886**

We also observe that pseudo-labeled data decrease the accuracy of the networks on the data generated from templates. This can be explained by the difference of distributions between generated and production data. However, as seen previously, including pseudo-labeled data increases the accuracy of the models significantly on validation and test.

6 Conclusion

During this work, many obstacles were faced, including the lack of real word labeled data. We showed that on the slot filling task, a baseline good enough (when manually tested by experts) to be used on a proof of concept can be achieved with data generated from templates. After manually labeling production data, we experimentally demonstrated that, on our use-case, combining pseudo-labeling with knowledge data allows to perform better. This work is an actual example of pseudo-labeling allowing to raise the performances of a network to a next level on a real world problem. Moreover, the use of random search was crucial in the search of a good combination of hyper-parameters. We think that the methodology implemented during this project will be useful to many practitioners trying to solve real world natural language understanding tasks.

References

1. Bergstra, J., Bengio, Y.: Random search for hyper-parameter optimization. J. Mach. Learn. Res. **13**, 2012 (2012)
2. Bojanowski, P., Grave, E., Joulin, A., Mikolov, T.: Enriching word vectors with subword information. arXiv preprint arXiv:1607.04606 (2016)
3. Cho, K., et al.: Learning phrase representations using RNN encoder-decoder for statistical machine translation. In: Conference on Empirical Methods in Natural Language Processing (2014)
4. Grandvalet, Y., Bengio, Y.: Semi-supervised learning by entropy minimization. In: NIPS (2004)
5. Graves, A., Schmidhuber, J.: Framewise phoneme classification with bidirectional LSTM and other neural network architectures. In: IJCNN (2005)
6. Kingma, D.P., Ba, J.: Adam: a method for stochastic optimization. In: 3rd International Conference on Learning Representations, ICLR 2015 (2015)
7. Lafferty, J.D., McCallum, A., Pereira, F.C.N.: Conditional random fields: probabilistic models for segmenting and labeling sequence data. In: ICML (2001)
8. Ma X., Hovy, E.: Investigation of recurrent-neural-network architectures and learning methods for spoken language understanding. In: End-to-End Sequence Labeling via Bi-directional LSTM-CNNS-CRF. Proceedings of the 54th Annual Meeting of the Association for Computational Linguistics (2016)
9. Mesnil G., He X., Deng, L., Bengio, Y.: Investigation of recurrent-neural-network architectures and learning methods for spoken language understanding. In: INTERSPEECH (2013)
10. Nair, V., Hinton, G.: Rectified linear units improve restricted Boltzmann machines. In: ICML 2010 (2010)
11. Paszke, A., et al.: Automatic differentiation in PyTorch. In: NIPS Workshop (2017)
12. Collobert, R., Weston, J., Bottou, L., Karlen, M., Kavukcuoglu, K., Kuksa, P.: Natural language processing (almost) from scratch. J. Mach. Learn. Res. **12**, 2493–2537 (2011)
13. Sang, E.F.T.K., Veenstra, J.: Representing text chunks. In: Proceedings of the Ninth Conference on European Chapter of the Association for Computational Linguistics, EACL 1999, pp. 173–179. Association for Computational Linguistics (1999)
14. Sun, A., Grishman, R., Xu, W., Min, B.: New York University 2011 system for KBP slot filling. In: TAC. NIST (2011)
15. Wang, Y., Deng, L., Acero, A.: Semantic frame based spoken language understanding. In: Tur, G., De Mori, R. (eds.) Spoken Language Understanding: Systems for Extracting Semantic Information from Speech, Chap. 3, pp. 35–80, January 2011

Swarm Intelligence Based Ensemble Learning of Deep Neural Networks

Tao Li and Jinwen Ma(✉)

Department of Information Science, School of Mathematical Sciences and LMAM, Peking University, Beijing 100871, China
jwma@math.pku.edu.cn

Abstract. Ensemble learning is a powerful tool in machine learning and it is very significant to utilize the mechanism of ensemble learning to improve the performance of deep learning for recognition and classification tasks. In this paper, we propose a general ensemble learning framework of deep neural networks based on swarm intelligence algorithms for solving the combination coefficients to the outputs of these component deep neural networks. We formulate the weights assigning problem for the deep neural networks as an optimization problem whose objective function is highly complicated and use swarm intelligence algorithms to solve it. We evaluate this ensemble learning framework on two real-world datasets, Market-1501 for person re-identification and CIFAR for image classification, and show that it outperforms a single deep neural network remarkably.

Keywords: Deep learning · Ensemble learning · Swarm intelligence

1 Introduction

Deep learning [13] has shown its great power in various research fields. In real-world applications, researchers and engineers often train multiple networks for a specific task and integrate the networks via simply averaging the outputs. However, due to the different characteristics of different networks, assigning equal weight to each network is unreasonable. Integrating multiple networks to obtain better performance falls into the category of ensemble learning [4], based on the philosophy that the whole is more than the sum of parts. We want to make the best use of the advantages and bypass the disadvantages of various networks, thus it is necessary and significant to consider how to integrate the networks more effectively.

In this paper, we propose a general ensemble learning framework of deep neural networks based on swarm intelligence. In the proposed framework, the objective function is a certain performance measure of the network, and the variables are combination coefficients of networks. Since the objective function is non-differentiable and highly complicated, we regard it as a black-box and solve the optimization problem via swarm intelligence algorithms. As a specific application, we apply our proposed ensemble learning framework to person re-identification task. To further claim the generality of the framework, we adapt it to image classification task and conduct experiments on

T. Gedeon et al. (Eds.): ICONIP 2019, CCIS 1142, pp. 256–264, 2019.
https://doi.org/10.1007/978-3-030-36808-1_28

CIFAR dataset [12]. The experimental results on CIFAR also confirm that the proposed method is very effective.

The rest of this paper is organized as follows. We review related work in Sect. 2. The ensemble learning framework is proposed in Sect. 3. We formulate the ensemble learning problem for re-identification task and describe the detailed algorithm in Sect. 4, including experimental results and analysis. Experiments on CIFAR are presented in Sect. 5 as an example of applying the framework to the image classification task. Finally, we give a brief conclusion in Sect. 6.

2 Related Work

Ensemble learning [4] is a widely used technique in statistical learning. Some previous works have been done in the area of ensemble learning methods for deep neural networks. Specifically, an efficient model averaging method for deep neural networks was proposed in [15]. The key point of this work is to group the hidden layer of a single network, while our work emphasizes on exploiting advantages of different networks. Some ensemble methods for deep neural networks have been proposed for specific tasks [3, 11], but our proposed framework is general and can be applied to various application scenes.

Person re-identification [24] is an active research topic in the field of computer vision. Most deep learning methods [14, 18, 25] train a convolutional neural network as a feature extractor, then calculate the similarity between a pair of probe and gallery images based on their features. Different network structures and loss functions influence the performance significantly. Metric learning methods [21, 22] are commonly used to improve performance on re-identification tasks, which share similarities with our method. However, metric learning usually aims at the output of a single network, while our method involves combining outputs of multiple networks.

3 Ensemble Learning Framework

3.1 Problem Formulation

For a given task, suppose we have trained n neural networks $\{\mathcal{N}_i\}_{i=1}^{n}$, which may have different network structures, loss functions or training datasets. The output of network $\mathcal{N}_i$ is denoted by $o(\mathcal{N}_i)$, which determines the performance of the network. Formally, we have a performance evaluation function f which depends on an evaluation dataset $\mathcal{E}$, then $f(o(\mathcal{N}_i); \mathcal{E})$ measures how well the network $\mathcal{N}_i$ performs on evaluation set $\mathcal{E}$.

It's reasonable to assume that a linear combination of the outputs $\{o(\mathcal{N}_i)\}_{i=1}^{n}$ leads to a better performance. More precisely, for certain weights $\omega = (\omega_1, \omega_2, \cdots, \omega_n)$, $f(\sum_{i=1}^{n} \omega_i o(\mathcal{N}_i); \mathcal{E})$ is large in value, indicating that combining the outputs of different networks yields better performance. Without loss of generality, we may assume $\sum_{i=1}^{n} \omega_i = 1$, since scaling the output has no influence on performance in most cases. Furthermore, we assume the performance of each single network is not too bad, thus we can constrain $\omega_i \geq 0, i = 1, 2, \cdots, n$. This means no network plays a negative role in the combination.

We need to find an ω such that $f(\sum_{i=1}^{n} \omega_i o(\mathcal{N}_i); \mathcal{E})$ is largest in order to obtain the best result. Ideally, $\mathcal{E}$ should be chosen as the test set, but the test set is not accessible in real applications, thus we need a proper strategy for choosing $\mathcal{E}$, which may be relevant to the task, and we will show how to choose $\mathcal{E}$ in Sects. 4 and 5. Now we can formulate the ensemble learning problem as a optimization problem as

$$\max_{\omega=(\omega_1,\omega_2,\cdots,\omega_n)} f(\sum_{i=1}^{n} \omega_i o(\mathcal{N}_i); \mathcal{E}) \text{ s.t. } \sum_{i=1}^{n} \omega_i = 1 \text{ and } \omega_i \geq 0, \forall i = 1, \cdots, n. \quad (1)$$

3.2 Optimization Procedure

In most cases, the performance evaluation function f is a black-box: we can calculate the value of f given input, but it is impossible to write f in an explicit form, let alone calculating the derivatives. Therefore, gradient-based optimization algorithms are not suitable for this problem. Instead, we propose to solve the optimization problem via swarm intelligence methods. Swarm intelligence algorithms are effective in optimization problems with highly complicated objective functions. We have a wide range of algorithms to choose in practice, such as particle swarm optimization algorithm (PSO), artificial bee colony algorithm (ABC), genetic algorithm (GA) and so on.

Algorithm 1. General ensemble learning framework of deep neural networks

Input: evaluation set $\mathcal{E}$, trained networks $\{\mathcal{N}_i\}_{i=1}^{n}$, evaluation function f
Output: ensemble weights $\omega = (\omega_1, \omega_2, \cdots, \omega_n)$

1: Set objective function as $f(\sum_{i=1}^{n} \omega_i o(\mathcal{N}_i) : \mathcal{E})$
2: Initialize swarm states
3: **for** each iteration **do**
4: Update swarm states via swarm intelligence algorithm
5: Calculate ω according to the final swarm state
6: Return $\omega = (\omega_1, \omega_2, \cdots, \omega_n)$

When we apply the framework in an application, we should clarify four issues: the output of the networks $\{o(\mathcal{N}_i)\}_{i=1}^{n}$, the evaluation set $\mathcal{E}$, the performance evaluation function f and the optimization algorithm. Once we have determined these issues, the general framework becomes a concrete algorithm. Since one can vary the settings according to the specific task, we claim the framework is suitable for a wide range of problems.

4 Experiment Study I: Person Re-identification

4.1 Formulation for Re-identification

In person re-identification task, a trained neural network can be viewed as a feature extractor. We use $\mathcal{N}$ to denote a neural network whose feature dimension is d, then for a given image q in query set $\mathcal{Q}$, we can extract the feature $\mathcal{N}(q) \in \mathbb{R}^d$. Similarly, suppose the gallery set is $\mathcal{G}$, for each $g \in \mathcal{G}$ the feature is $\mathcal{N}(g) \in \mathbb{R}^d$. Then neural network $\mathcal{N}$ induces a similarity metric function as $d_{\mathcal{N}}(q,g) = \langle \mathcal{N}(q), \mathcal{N}(g) \rangle$, where the last term is the standard inner product in $\mathbb{R}^d$. Suppose there are n trained neural networks $\{\mathcal{N}_i\}_{i=1}^n$ and correspondingly, n similarity metric functions $\{d_{\mathcal{N}_i}\}_{i=1}^n$. We then have $o(\mathcal{N}_i) = d_{\mathcal{N}_i}$, and use d_ω to denote $\sum_{i=1}^n \omega_i o(\mathcal{N}_i)$.

The training set can be randomly divided into training part and evaluation part. The training part is used to train the neural networks, and the evaluation part forms the evaluation set $\mathcal{E}$. For evaluation function, we set f to be the mAP **after re-ranking**. Furthermore, we choose particle swarm optimization (PSO) as the optimization method.

4.2 Experimental Settings

We use Market-1501 dataset [24] in this experiment. We train models on Market-1501 dataset with various network structures and loss functions. For network structures, we consider ResNet-50 [5] and DenseNet-121 [10]. For loss functions, we consider label smoothed cross entropy loss [19], center loss [20] and triplet loss [17]. Since the size of training set is relatively small, we use ResNet-50 and DenseNet-121 pre-trained on ImageNet dataset [2]. We use stochastic gradient descent (SGD) optimizer with batch equal to 32. Nesterov's acceleration, weight decay, and momentum are also used. In the training phase, the learning rates for network parameters are set to be small (0.01), except for the classifier block (0.1). The maximum epoch is set to be 60 and the learning rates are divided by 10 after 30 epochs. Random rotation, random crop, random erasing [27] and color jitter are used as data augmentation techniques.

Under the experimental settings above we train 6 neural networks on Market-1501. The performances of these models are reported in Table 1. Here, we consider Rank-1, Rank-5, Rank-10 accuracies and mean average precision (mAP) after re-ranking on the test set.

Table 1. Accuracies of various base models on Market-1501.

Model	Network	Loss	Rank-1	Rank-5	Rank-10	mAP
$\mathcal{N}_{RE}$	Resnet-50	Cross entropy	93.26%	94.71%	95.99%	84.35%
$\mathcal{N}_{RC}$	Resnet-50	Center loss	89.52%	94.12%	95.69%	84.32%
$\mathcal{N}_{RT}$	Resnet-50	Triplet loss	91.21%	95.22%	96.56%	86.58%
$\mathcal{N}_{DE}$	Densenet-121	Cross entropy	91.24%	95.10%	96.17%	86.39%
$\mathcal{N}_{DC}$	Densenet-121	Center loss	92.22%	95.99%	96.94%	87.18%
$\mathcal{N}_{DT}$	Densenet-121	Triplet loss	89.49%	94.09%	95.87%	84.06%

4.3 Results of Ensemble Learning

In this section, we investigate whether integrating different models leads to performance improvement. We consider three kinds of ensemble here.

- **Type 1:** Integrate models with a fixed network structure and different loss functions.
- **Type 2:** Integrate models with fixed loss function and different network structures.
- **Type 3:** All models are integrated.

For comparison, we also evaluate the performance of the simple averaging strategy, that is to say, we assign equal weight to each model. In the following, $\mathcal{N}_1 + \mathcal{N}_2$ means we integrate $\mathcal{N}_1$ and $\mathcal{N}_2$ together, and the weights are listed below. The experimental results are shown in Table 2.

According to Table 2, we have three observations. First, compared with Table 1, we can see that integrating multiple networks outperforms a single network. When we use a single network, $\mathcal{N}_{RT}$ has the best mAP 86.58%. However, when we consider more networks, almost all the models have higher mAP. This observation confirms the philosophy of ensemble learning: the whole is greater than the sum of parts. Second, weights assigned by the proposed ensemble learning framework lead to higher mAP and Rank-1 accuracy compared with simple averaging strategy. Therefore, we conclude that the ensemble learning framework is effective. Last but not least, the weights found by PSO are reasonable in the sense that better base learner has larger weight while worse base learner has a lower weight.

Table 2. Accuracies of different ensemble models on Market-1501.

Ensemble type	Models and weights	Rank-1	Rank-5	Rank-10	mAP
Type-1	$\mathcal{N}_{RE} + \mathcal{N}_{RC} + \mathcal{N}_{RT}$ $(0.0603, 0.0446, 0.8951)$	92.22%	95.90%	97.03%	88.86%
	$\mathcal{N}_{RE} + \mathcal{N}_{RC} + \mathcal{N}_{RT}(1/3, 1/3, 1/3)$	91.57%	95.01%	96.44%	86.99%
	$\mathcal{N}_{DE} + \mathcal{N}_{DC} + \mathcal{N}_{DT}$ $(0.2460, 0.6030, 0.1509)$	93.14%	96.26%	97.34%	89.44%
	$\mathcal{N}_{DE} + \mathcal{N}_{DC} + \mathcal{N}_{DT}$ $(1/3, 1/3, 1/3)$	93.08%	96.56%	97.51%	89.24%
Type-2	$\mathcal{N}_{RE} + \mathcal{N}_{DE}$ $(0.1726, 0.8274)$	92.37%	95.46%	96.67%	87.84%
	$\mathcal{N}_{RE} + \mathcal{N}_{DE}$ $(1/2, 1/2)$	91.95%	95.58%	96.62%	87.71%
	$\mathcal{N}_{RC} + \mathcal{N}_{DC}$ $(0.4208, 0.5792)$	92.43%	95.90%	97.27%	89.30%
	$\mathcal{N}_{RC} + \mathcal{N}_{DC}$ $(1/2, 1/2)$	92.46%	95.99%	97.27%	88.26%
	$\mathcal{N}_{RT} + \mathcal{N}_{DT}$ $(0.9307, 0.0693)$	92.01%	95.84%	97.09%	87.72%
	$\mathcal{N}_{RT} + \mathcal{N}_{DT}$ $(1/2, 1/2)$	90.62%	94.60%	96.14%	85.37%
Type-3	$\mathcal{N}_{RE} + \mathcal{N}_{RC} + \mathcal{N}_{RT} + \mathcal{N}_{DE} + \mathcal{N}_{DC} + \mathcal{N}_{DT}$ $(0.1817, 0.1027, 0.1692, 0.2392, 0.2135, 0.0937)$	**93.82%**	96.62%	**97.54%**	**89.70%**
	$\mathcal{N}_{RE} + \mathcal{N}_{RC} + \mathcal{N}_{RT} + \mathcal{N}_{DE} + \mathcal{N}_{DC} + \mathcal{N}_{DT}$ $(1/6, 1/6, 1/6, 1/6, 1/6, 1/6)$	93.65%	**96.67%**	**97.54%**	89.64%

4.4 Comparisons and Discussions

Market-1501 is a widely-used person re-identification dataset, and the performances of various methods are recorded on the home page of this dataset. Here, we compare the best model obtained in the last subsection with state-of-the-art methods[1] in Table 3.

Table 3. Comparisons with state-of-the-art methods on Market-1501 dataset.

Method	Rank-1	Rank-5	Rank-10	mAP
Zhang *et al.* [26]	88.79%	–	–	83.79%
Hermans *et al.* [7]	91.75%	95.78 %	–	87.18%
Sarfraz *et al.* [16]	90.30%	–	–	84.00%
Zhong *et al.* [27]	89.13%	–	–	83.93%
Li *et al.* [14]	93.80%	–	–	82.80%
Ours	**93.82%**	**96.62%**	**97.54%**	**89.70%**

Our best result outperforms state-of-the-art results. However, we confess that our best model requires higher computational cost and running time. In the training stage, we need to train multiple neural networks. In the test stage, we have to do the forward computation several times. Thus, the proposed framework may not be suitable for real-time tasks. Besides, one may ask, why we don't train the models jointly just like a mixture of expert model? Simultaneously training several models needs a large memory, which is unrealistic in certain scenes when the computational resources are scarce. On the other hand, training a single network is easy even on a personal computer. Thus, we choose to train the networks individually, then combine the outputs.

5 Experiment Study II: Image Classification

5.1 Formulation for Image Classification

To show the proposed framework is general, we consider image classification task in this part. Specifically, we use CIFAR dataset [12] in this experiment. Assume there are n trained neural networks $\{\mathcal{N}_i\}_{i=1}^n$. In a classification problem, given the input image x, the output $\mathcal{N}_i(x)$ is a probability distribution over classes. Sine $\sum_{i=1}^n \omega_i = 1, \omega_i \geq 0$, $\sum_{i=1}^n \omega_i \mathcal{N}_i(x)$ is still a probability distribution over classes. We use all the images in the training set to train the models. Then we randomly choose 2000 images from the standard test set as an evaluation set $\mathcal{E}$, and the remaining 8000 images are used for performance testing. In this way, we guarantee that the test set and evaluation set have no overlap. The performance evaluation function f is set to be the classification accuracy, and we consider PSO, ABC, and GA as optimization methods in this experiment.

[1] Data source: www.liangzheng.org/Project/state_of_the_art_market1501.html.

5.2 Experimental Settings

For the image classification task, we need the features to be separable rather than discriminative, thus we only use cross-entropy loss. Instead, we consider various network structures, including ResNet [5], PreAct ResNet [6], Wide ResNet [23], SeNet [9], MobileNet [8] and Dual Path Net [1]. For CIFAR-100 (CIFAR-10), we train the networks for 100 (50) epochs with batch size equals to 128 (128). The learning rate is set to be 0.1 (0.1) initially and divided by 10 (10) every 30 (20) epochs. When we integrate multiple networks, we use PSO, SA, and ABC as optimization algorithms.

5.3 Experimental Results

The experimental results are listed in Table 4. In this table, PSO, GA, and ABC mean that we combine all the models by the proposed ensemble learning framework, using PSO, GA, ABC as the optimization algorithm respectively. For comparison, we also assign equal weight to each network, and the corresponding results are shown in the AVE row. Due to the page limit, the weights assigned to networks are omitted here.

Table 4. Accuracies of different models on CIFAR-10 and CIFAR-100. The left table shows classification accuracies of baseline models, and the right table shows the performances obtained by ensemble methods.

	CIFAR-10	CIFAR-100
ResNet	93.13%	77.30%
PreAct ResNet	93.11%	77.26%
Wide ResNet	93.68%	78.99%
SeNet	92.59%	76.71%
MobileNet	90.34%	71.96%
Dual Path Net	91.84%	77.68%

	CIFAR-10	CIFAR-100
PSO	**94.46%**	81.54%
GA	94.18%	**81.70%**
ABC	94.18%	81.46%
AVE	94.06%	81.24%

With the proposed ensemble learning framework, we significantly improve the accuracies on CIFAR-10 and CIFAR-100 compared with a single network. All ensemble learning results outperform simple averaging strategy. This indicates that the proposed ensemble learning framework is more effective in solving the weight assigning problem. Different swarm intelligence algorithms lead to different results, among which PSO yields the best accuracy on CIFAR-10 while GA leads to best accuracy on CIFAR-100.

6 Conclusion

We have proposed a general ensemble learning framework of deep neural networks based on swarm intelligence. Compared to current ensemble methods in deep learning, our framework is universal and extensible, which can be applied in various applications.

Specifically, we adopt the proposed method to the person re-identification task and image classification task. The extensive experimental results on Market-1501 dataset and CIFAR dataset show that the proposed ensemble method is both effective and general.

References

1. Chen, Y., Li, J., Xiao, H., Jin, X., Yan, S., Feng, J.: Dual path networks. In: Advances in Neural Information Processing Systems (NeurIPs) (2017)
2. Deng, J., Dong, W., Socher, R., Li, L.J., Li, K., Fei-Fei, L.: ImageNet: a large-scale hierarchical image database. In: Proceedings of the IEEE Conference on Computer Vision and Pattern Recognition (CVPR). IEEE (2009)
3. Deng, L., Platt, J.C.: Ensemble deep learning for speech recognition. In: Fifteenth Annual Conference of the International Speech Communication Association (2014)
4. Friedman, J., Hastie, T., Tibshirani, R.: The Elements of Statistical Learning. Springer Series in Statistics, vol. 1. Springer, New York (2001). https://doi.org/10.1007/978-0-387-21606-5
5. He, K., Zhang, X., Ren, S., Sun, J.: Deep residual learning for image recognition. In: Proceedings of the IEEE Conference on Computer Vision and Pattern Recognition (2016)
6. He, K., Zhang, X., Ren, S., Sun, J.: Identity mappings in deep residual networks. In: Leibe, B., Matas, J., Sebe, N., Welling, M. (eds.) ECCV 2016. LNCS, vol. 9908, pp. 630–645. Springer, Cham (2016). https://doi.org/10.1007/978-3-319-46493-0_38
7. Hermans, A., Beyer, L., Leibe, B.: In defense of the triplet loss for person re-identification. arXiv preprint:1703.07737 (2017)
8. Howard, A.G., et al.: MobileNets: efficient convolutional neural networks for mobile vision applications. arXiv preprint:1704.04861 (2017)
9. Hu, J., Shen, L., Sun, G.: Squeeze-and-excitation networks. arXiv preprint:1709.01507 (2017)
10. Huang, G., Liu, Z., Weinberger, K.Q., van der Maaten, L.: Densely connected convolutional networks. In: IEEE Conference on Computer Vision and Pattern Recognition (CVPR) (2017)
11. Ensemble deep learning for biomedical time series classification: Jin, L.p., Dong. J. Comput. Intell. Neurosci. **2016** (2016). https://www.hindawi.com/journals/cin/2016/6212684/cta/. Article ID 6212684
12. Krizhevsky, A.: Learning multiple layers of features from tiny images. Technical report (2009)
13. LeCun, Y., Bengio, Y., Hinton, G.: Deep learning. Nature **521**(7553), 436 (2015)
14. Li, W., Zhu, X., Gong, S.: Harmonious attention network for person re-identification. In: Proceedings of the IEEE Conference on Computer Vision and Pattern Recognition (2018)
15. Opitz, M., Possegger, H., Bischof, H.: Efficient model averaging for deep neural networks. In: Lai, S.-H., Lepetit, V., Nishino, K., Sato, Y. (eds.) ACCV 2016. LNCS, vol. 10112, pp. 205–220. Springer, Cham (2017). https://doi.org/10.1007/978-3-319-54184-6_13
16. Sarfraz, M.S., Schumann, A., Eberle, A., Stiefelhagen, R.: A pose-sensitive embedding for person re-identification with expanded cross neighborhood re-ranking. arXiv preprint arXiv:1711.10378 (2017)
17. Schroff, F., Kalenichenko, D., Philbin, J.: FaceNet: a unified embedding for face recognition and clustering. In: The IEEE Conference on Computer Vision and Pattern Recognition (2015)
18. Sun, Y., Zheng, L., Deng, W., Wang, S.: SVDNet for pedestrian retrieval. In: 2017 IEEE International Conference on Computer Vision (ICCV). IEEE (2017)

19. Szegedy, C., Vanhoucke, V., Ioffe, S., Shlens, J., Wojna, Z.: Rethinking the inception architecture for computer vision. In: Proceedings of the IEEE Conference on Computer Vision and Pattern Recognition (CVPR) (2016)
20. Wen, Y., Zhang, K., Li, Z., Qiao, Y.: A discriminative feature learning approach for deep face recognition. In: Leibe, B., Matas, J., Sebe, N., Welling, M. (eds.) ECCV 2016. LNCS, vol. 9911, pp. 499–515. Springer, Cham (2016). https://doi.org/10.1007/978-3-319-46478-7_31
21. Xiong, F., Gou, M., Camps, O., Sznaier, M.: Person re-identification using kernel-based metric learning methods. In: Fleet, D., Pajdla, T., Schiele, B., Tuytelaars, T. (eds.) ECCV 2014. LNCS, vol. 8695, pp. 1–16. Springer, Cham (2014). https://doi.org/10.1007/978-3-319-10584-0_1
22. Yi, D., Lei, Z., Liao, S., Li, S.Z.: Deep metric learning for person re-identification. In: International Conference on Pattern Recognition (ICPR). IEEE (2014)
23. Zagoruyko, S., Komodakis, N.: Wide residual networks. arXiv preprint:1605.07146 (2016)
24. Zheng, L., Shen, L., Tian, L., Wang, S., Wang, J., Tian, Q.: Scalable person re-identification: a benchmark. In: IEEE International Conference on Computer Vision (ICCV) (2015)
25. Zheng, Z., Zheng, L., Yang, Y.: A discriminatively learned CNN embedding for person reidentification. ACM Trans. Multimedia Comput. Commun. Appl. (TOMM) **14**(1), 13 (2017). https://dl.acm.org/citation.cfm?id=3159171. Article No. 13
26. Zheng, Z., Zheng, L., Yang, Y.: Pedestrian alignment network for large-scale person reidentification. arXiv preprint arXiv:1707.00408 (2017)
27. Zhong, Z., Zheng, L., Kang, G., Li, S., Yang, Y.: Random erasing data augmentation. arXiv preprint:1708.04896 (2017)

DSMRSeg: Dual-Stage Feature Pyramid and Multi-Range Context Aggregation for Real-Time Semantic Segmentation

Mingdong Yang and Ying Shi(✉)

School of Automation, Wuhan University of Technology, Wuhan, China
231920@whut.edu.cn, a_laly@163.com

Abstract. Real-time semantic segmentation is a challenging task in computer vision. Many researches emphasize real-time inference speed while neglecting segmentation quality. To tackle this problem, we propose a framework called DSMRSeg to achieve high-speed with high-accuracy result after training on **only one GPU**. Overall, we accomplish this by three core components: (1) Dual-Stage Feature Pyramid Network structure is designed to obtain richer multi-scale information and enhance the entire features hierarchy by bidirectionally propagating features with strong semantics and accurate localization. (2) Multi-Range Context Module is developed to expand receptive fields by aggregating the local dense features and multi-range context information. (3) Light-weight Feature Fusion Module is proposed to merge dual-stage features effectively. We evaluate DSMRSeg on Cityscapes, CamVid and BDD100K datasets and produce competitive results compared with the state-of-the-art methods. Specifically, DSMRSeg achieves 75.5% mIoU on Cityscapes test set, with speed of 40 FPS on one NVIDIA GTX1080 card for 1024 × 512 high-resolution image.

Keywords: Deep neural networks · Real-time semantic segmentation

1 Introduction

Semantic segmentation performs pixel-level label prediction for images [1]. With the development of deep convolutional neural networks [2] has made notable progress in providing accurate segmentation results. However, then consist of extremely deep and wide layers, with huge parameters and computation complexity.

There are three methods accomplish real-time semantic segmentation. (1) *Based on light-weight networks*. [3, 4] uses light-weight networks of pre-training on ImageNet, such as ShuffleNet [5], as encoder part. Then skip connection or U-shape structure is used as decoder [1]. Though it is simple and effective, the receptive fields of light-weight networks is too small to cover large objects, resulting in low accuracy. (2) *Based on context module*. [6–8] adds context module at every stage of network, produces 1/8 feature map resolution and from scratch training. It not only expands receptive field, but also preserves spatial information. However, it discards decoder, lack multi-scale information and weaken feature discriminating ability. (3) *Based on multi-branch structure*. Some branches preserve the spatial information, while others

T. Gedeon et al. (Eds.): ICONIP 2019, CCIS 1142, pp. 265–273, 2019.
https://doi.org/10.1007/978-3-030-36808-1_29

obtain sufficient receptive field, finally, combine multi-branches results [9–12]. Nevertheless, it also can't exploit the multi-scale information effectively. Therefore, *these three methods achieve high-speed, but low-accuracy segmentation results.*

To achievement good trade-off on accuracy and speed, we first propose a novel Dual-Stage Feature Pyramid Network (DSFPN), which obtains richer multi-scale information and enhances the entire features hierarchy by bidirectionally propagating semantically strong and accurate localization features. Firstly, we choice light-weight network as encoder to achieve high-speed, like [9, 12]. Then, we design a *Inverted FPN* structure to propagate the accurate localization features to all levels. Finally, to strengthen the multi-scale features capability, we add extra top-down path on the Inverted FPN features. However, the decoder of DSFPN is much deep, resulting in vanishing gradient as introduced in [2]. To address it, we propose Long-Range Residual (LRR) unit, it can reuse previous features, enhances feature discriminating ability.

Furthermore, we propose Multi-Range Context Module (MRCM), which aggregates local dense features and multi-range context, to expand receptive fields of light-weight encoder tremendously. Small-range local feature expresses local texture and large-range feature can leverage more context information, learning long-range relationship between pixels. Meanwhile, DSFPN has different features representation, instead of simple summation them, we develop a light-weight Dual-stage Feature Fusion Module (DFFM) to merge these features.

Based on the above insight, we propose **D**ual-**S**tage feature pyramid and **M**ulti-**R**ange context aggregation network, named **DSMRSeg**. It achieves **75.5%** mIoU on *Cityscapes test set*, with speed of **40 FPS** on *GTX1080 card.*

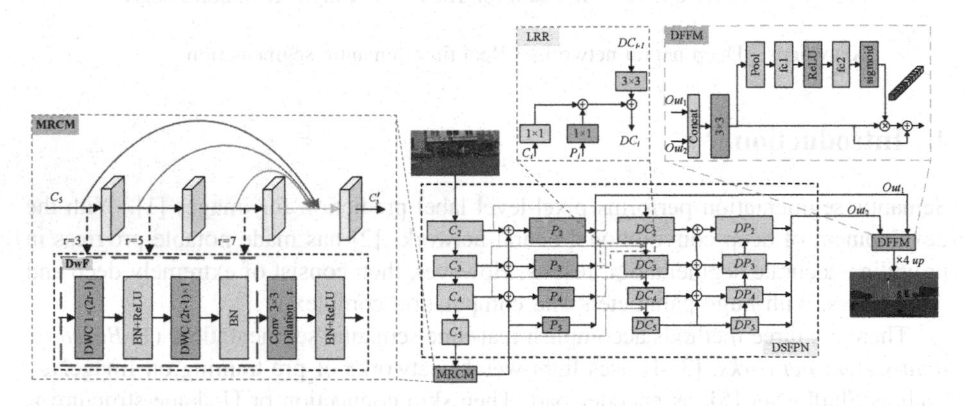

Fig. 1. DSMRSeg network architecture, 1 × 1 and 3 × 3 denote convolution kernels. MRCM: different color indicates multi-range context, red is local dense feature, the blue, and yellow is multi-range contxet feature. The "r" denote dilation rate, "DWC" denote depthwise convolution, "C_5" is output of last stage. (Color figure online)

2 Related Work

We introduce real-time semantic segmentation, with related work on context modules.

Real-Time Semantic Segmentation: Real-time semantic segmentation requires good trade-off on accuracy and speed. ENet [15] removes the last stage of the model to get extremely high speed, which the accuracy is very low. ERFNet designs factorized convnet to reduce computation cost. Based on PSPNet [16], ICNet [12] uses image cascade network to speed up the semantic segmentation. ESPNet [7] proposes efficient spatial pyramid to achieve real-time. [9–11] fused spatial path and context path to obtain better accuracy and speed. More recently, DFANet [17] utilizes deep feature aggregation network under resource constraints.

Context Module: Deeplabv2 [18] proposes ASPP module which uses dilated convolutions of different dilation rate to obtain rich context. Deeplabv3 [19] adds global average pooling with ASPP module to capture global context of the image. However, the dilated convolution based methods only obtain sparse context information, leads to low quality result. Furthermore, PSP module [16] uses different scales of average pooling layers to aggregate context, which lack adaptive context information. Recently, DANet [20] collects useful context information by using attention mechanism, which has high computation complexity, leads to non real-time speed.

3 DSMRSeg

3.1 Dual-Stage Feature Pyramid Network

We build a *effective* and *efficient* feature pyramid network, DSFPN, bidirectionally propagates features with strong semantics and accurate localization, as illustrated in Fig. 1. We first choice light-weight network as encoder to achieve high-speed, such as ResNet18 [2]. We define the four stages of encoder using $\{C_2, C_3, C_4, C_5\}$, and followed by FPN [14] to get valid features $\{P_2, P_3, P_4, P_5\}$. To propagate the accurate localization information of P_2 to all level P_3–P_5, we design a bottom-up path, named *Inverted FPN*, to obtain new features level $\{DC_2, DC_3, DC_4, DC_5\}$. Furthermore, we strengthen the features hierarchy to get richer multi-scale information by using new top-down path corresponding to $\{DC_2, DC_3, DC_4, DC_5\}$, and use $\{DP_2, DP_3, DP_4, DP_5\}$ denote these generated features. We bidirectionally propagating features with strong semantics and accurate localization by fusing top-down and bottom-up path.

Especially, we use 1×1 convolution and bilinear interpolation generate $\{DP_2, DP_3, DP_4, DP_5\}$, like FPN (we use 128 channels). However, DSFPN decoder is deeper, leads to vanishing gradient problem, the result is sub-optimize. Inspired by residual unit [2], we propose Long-Range Residual (LRR) unit, as shown in Fig. 1. It not only facilitates information flow, but also reuses $\{C_2, C_3, C_4, C_5\}$ and $\{P_2, P_3, P_4, P_5\}$ features, enhances feature discriminating ability across scales. Thus, we first downsampling DC_{i-1} by using a 3×3 convolution with stride 2, and utilize 1×1 convolutions in order to be aligned P_i, C_i, with DC_{i-1}, then sum up P_i, C_i, and DC_{i-1} to get feature DC_i. Repeat the process to obtain $\{DC_2, DC_3, DC_4, DC_5\}$.

3.2 Multi-Range Context Module

Motivation: Receptive fields is key for semantic segmentation. Large receptive field provides rich context information, and learn the long-range relationship between pixels. Philosophical ASPP [19] and PSP [16] context module expand receptive field by dilated convolutions and pooling operation. However, the former ignores dense context information. The latter collects context in a non-adaptive manner, missing the difference of local representation and context dependencies for different categories.

MRCM: Based on the above insight, we propose the Multi-Range Context Module (MRCM), aggregates local dense feature and multi-range context to expand receptive field tremendously in a adaptive manner. Especially, as shown in Fig. 1. We first cascade three 3×3 dilated convolution layers with different dilation rate (we empirically use 3, 5, 7 in experiments) to obtain rich multi-range features. Then we combine last features (e.g., C_5) of backbone (e.g., ResNet18) to generate the dense local features. Finally, we aggregate local and context features by concatenate.

However, dilated convolution can cause gridding artifacts [26], missing the local information and irrelevant of large range context. Inspired by [26], we design a novel *light-weight* DwF operation based on Depthwise [25] and Factorized [8] convolutions, it smoothes every dilated convolution to mitigate the gridding artifacts, as Fig. 1.

3.3 Dual-Stage Feature Fusion Module

DSFPN obtains dual-stage features Out_1 and Out_2 by upsample and summation multi-layers features $\{P_2, P_3, P_4, P_5\}$ and $\{DP_2, DP_3, DP_4, DP_5\}$, which belong to different feature representation. Instead of simple sum up Out_1 and Out_2, we introduce a light-weight Dual-stage Feature Fusion Module (DFFM) to combine them. Especially, Fig. 1 shows the details of this module, we first concatenate Out_1 and Out_2, then we use 3×3 convolution layer to fuse them. Next, following the SE block [27], we first learn a weight vector, then, re-weighting features by the rescaling operation which each of the features is enhanced or weakened by learned weight vector.

3.4 Network Architecture

The Fig. 1 shows our network architecture, **DSMRSeg**. We use pre-trained ResNet18 [2] with FPN [14] model as effective backbone. All network is end-to-end trained by cross-entropy loss. To stabilize the training process, main loss together with auxiliary loss are used to help optimization [16], where the main loss is defined on the final output of the network and the auxiliary loss is defined on Out_1 with weight of 0.4. Meanwhile, we build a simplified network architecture, **SSMRSeg**, which only employ our MRCM module.

4 Experiments

In this section, we evaluate the proposed DSMRSeg on Cityscapes [21], CamVid [22], BDD100K [23] road scene datasets. The performances are reported using the mIoU in all experiments. Especially, our implementation is based on PyTorch on only one GTX1080 desktop. We employ Adam with weight delay = 4e−4. The cosine learning rate policy is used, in which set base learning rate to 4e−4, the min learning rate is 1e−6. For data augmentation, we employ random mirror, mean subtraction and random scale on the input images. Due to the memory limitation, we employ 12 batch size and 448 × 448 fixed size for CamVid, 8 batch size and 640 × 640 size for other datasets. *Code will be made available upon publication.*

4.1 Ablation Study

Strong Baseline. We use the ResNet18-FPN [2, 14] as the backbone, and upsample the P3, P4, P5 feature map to P2 size. Then directly upsample output after summation of P2, P3, P4, P5 as original input image, like FCN [1]. The performance of the base model as our baseline, as shown in Table 1, achieves 71.712% mIoU performance.

Ablation for MRCM. (1) **MRCM.** It aggregate the local dense feature and multi-range context to expand receptive field tremendously. Therefore, it improve the performance from 71.712% to 75.229%, as shown in Table 1. (2) **ASPP** [19] and **PSP** [16]. The ASPP and PSP is widely used in the semantic segmentation, and achieve state-of-the-art results. But our MRCM improve the performance 0.505% and 1.017% relative to ASPP and PSP without extra parameters, as shown in Table 1. (3) **DwF.** The performance drops 0.462% if we remove the DwF operation.

Table 1. Validation of the different context module. All models were trained on **Cityscapes** *train* set, the evaluation is performed on Cityscapes *val* set.

Method	MRCM	w/o DwF	ASPP	PSP	Params (M)	mIoU (%)
Baseline					11.82	71.712
+ASPP [19]			√		12.10	74.724
+PSP [16]				√	12.21	74.212
+MRCM w/o DwF		√			12.30	74.767
+MRCM	√				12.30	**75.229**

Ablation for DSFPN. It obtains rich multi-scale information and enhance the entire feature hierarchy by fusing top-down and bottom-up path. As shown in Table 2, it improves the performance from 75.229% to 75.880%.

Ablation for LRR. To mitigate vanishing gradient problem caused by the deeper decoder, we propose novel LRR unit, it reuses previous stage features, enhance discriminating ability. The effect of the LRR is presented in Table 2.

Ablation for DSFFM. We introduce a light-weight DSFFM module to combine output features of DSFPN. First, we evaluate the performance of directly summation, then compare it with our proposed feature fusion module. The Table 2 shows our DSFFM outperform the summation method, from 75.930% to 76.835%.

Table 2. Validation of the proposed core components. All models were trained on **Cityscapes** *train* set, the evaluation is performed on Cityscapes *val* set.

Method	MRCM	DSFM	LRR	DFFM	Summation	mIoU (%)
Baseline						71.712
SSMRSeg	√					75.229
+DSFPN	√	√			√	75.880
+LRR	√	√	√		√	75.930
+DFFM	√	√	√	√		**76.835**

4.2 Speed Analysis

We measure inference speed on the GTX1080 and Jetson TX2 in different resolution input image of DSMRSeg and SSMRSeg, as shown in Fig. 2. Especially, for 1024 × 512 high-resolution input, DSMRSeg/SSMRSeg achieve 40/82 FPS on one GTX1080, and for 360 × 640 input, DSMRSeg/SSMRSeg get 8.6/14.5 FPS on Jetson TX2. Therefore, our model obtain good trade-off on speed and accuracy.

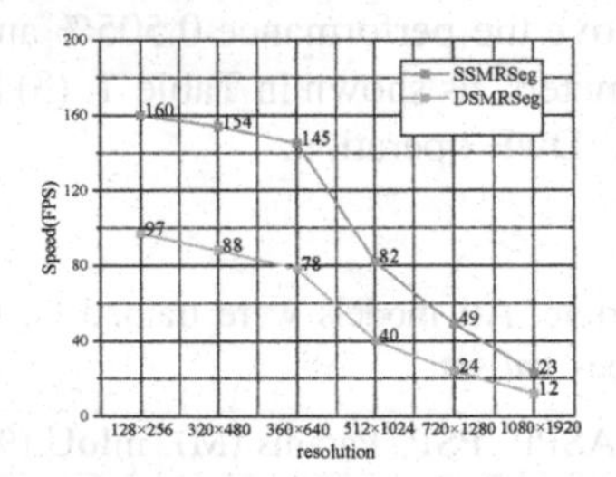

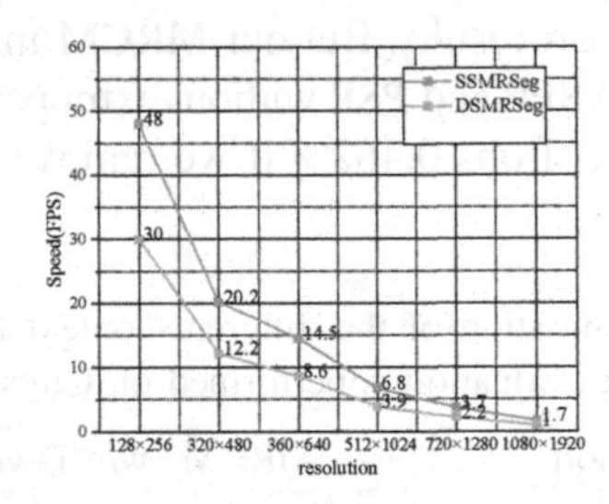

Fig. 2. Comparison of inference speed on the GTX 1080 GPU (left) and Jetson TX2 (right) in the different resolution input image.

4.3 Comparison with State-of-the-Arts

Cityscapes. We report the results of the proposed method on Cityscapes test set as shown in Table 3. Our SSMRSeg and DSMRSeg achieve large accuracy gain compare to the real-time methods which from scratch training, such as ENet [15], ERFNet [8], and CGNet [6] *et al.* For using pre-training model on the ImageNet, [12] and [17], DSMRSeg improve the performance 6.0% and 4.2%. Especially, compared with the previous state-of-the-art method BiseNet [9], DSMRSeg obtain better segmentation accuracy with less parameters. Figure 3 shows the Cityscapes test set visualization results where the DSMRSeg model performed better.

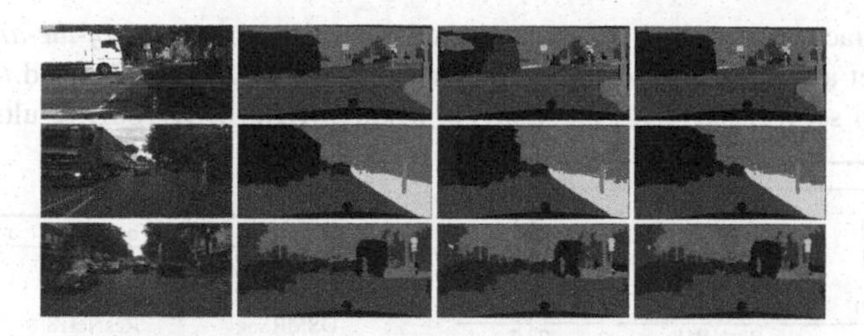

Fig. 3. The visualization results on **Cityscapes** *test* set. From left to right: original image, groundtruth, SSMRSeg, DSMRSeg.

Table 3. Accuracy comparison of our method against other *state-of-the-art* methods on **Cityscapes** *test* set. Our models were trained on **Cityscapes** *train* set. "–" indicates that the methods didn't give the corresponding result. "MS" indicates use Multi-Scale test of {0.5, 0.75, 1.0, 1.25}.

Method	Backbone	Params (M)	mIoU (%)	
			Test set	*Val set*
PSPNet [16]	ResNet	250.8	**78.4**	–
ENet [15]	From scratch	0.4	58.3	–
SegNet [13]	VGG16	29.5	56.1	–
ShuffleSeg [4]	ShuffleNet	–	58.3	–
ERFNet [8]	From scratch	2.03	68.0	–
BiseNet [9]	Xception39	5.8	68.4	69.0
BiseNet [9]	ResNet18	49.0	74.7	74.8
ICNet [12]	PSPNet50	26.5	69.5	67.7
ContextNet [10]	From scratch	0.9	66.1	65.9
CGNet [6]	From scratch	0.5	64.8	63.5
Fast SCNN [11]	From scratch	1.11	68.0	68.6
ESPNet [7]	From scratch	0.4	60.3	–
DFANet [17]	Xception	7.8	71.3	71.9
DSMRSeg	ResNet18	13.82	**75.5**	**76.8**
DSMRSeg-MS	ResNet18	–	–	**77.5**
SSMRSeg	ResNet18	12.30	74.5	75.2
SSMRSeg-MS	ResNet18	–	–	76.5
DSMRSeg	MobileNetv2	4.33	73.4	75.0

CamVid. The Table 4 shows the statistic result on CamVid test set. We adopt the 960 × 720 image resolution for training and testing, like as [9, 17]. Our methods get much higher segmentation accuracy than other state-of-the-art real-time methods.

BDD100K. As shown in Table 4, our model achieves 57.3% performance on val set. It's better than the strong baseline, DRN [24], on the BDD100K dataset.

Table 4. Accuracy comparison of our method against other *state-of-the-art* methods on **CamVid** *test* set and **BDD100K** *val* set. Our models were trained on CamVid *trainval* set and BDD100K *train* set respectively. Left: CamVid results, Right: BDD100K results.

Method	backbone	mIoU(%)
ENet[15]	From scratch	51.3
SegNet[13]	VGG16	55.6
BiseNet[9]	ResNet18	68.7
ICNet[12]	PSPNet50	67.1
DFANet[17]	Xception	64.7
DSMRSeg	ResNet18	**73.7**
SSMRSeg	ResNet18	72.7

Method	backbone	mIoU(%)
DRN-D-22[24]	DRN-D	53.2
DRN-D-38[24]	DRN-D	55.2
DSMRSeg	ResNet18	**57.3**

5 Conclusion and Future Work

In this paper, we propose SSMRSeg/DSMRSeg network for real-time Semantic Segmentation. Analysis and Quantitative experimental results on Cityscapes, CamVid, BDD100K dataset are presented to demonstrate the effectiveness of our method. Finally, we intend to achieve better trade-off on the speed and accuracy in future.

Acknowledgments. The work is supported by project grant of China (No. BE2016155).

References

1. Long, J., Shelhamer, E., Darrell, T.: Fully convolutional networks for semantic segmentation. In: Proceedings of CVPR, pp. 3431–3440 (2015)
2. He, K., Zhang, X., Ren, S., et al.: Deep residual learning for image recognition. In: Proceedings of CVPR, pp. 770–778 (2016)
3. Siam, M., Gamal, M., Abdel-Razek, M., et al.: RTSeg: real-time semantic segmentation comparative study. In: Proceedings of ICIP, pp. 1603–1607 (2018)
4. Gamal, M., Siam, M., Abdel-Razek, M.: ShuffleSeg: real-time semantic segmentation network (2018). arXiv preprint, arXiv:1803.03816
5. Zhang, X., Zhou, X., Lin, M., et al.: ShuffleNet: an extremely efficient convolutional neural network for mobile devices. In: Proceedings of ECCV, pp. 6848–6856 (2018)
6. Wu, T., Tang, S., Zhang, R., et al.: CGNet: a light-weight context guided network for semantic segmentation (2018). arXiv preprint, arXiv:1811.08201
7. Mehta, S., Rastegari, M., Caspi, A., Shapiro, L., Hajishirzi, H.: ESPNet: efficient spatial pyramid of dilated convolutions for semantic segmentation. In: Ferrari, V., Hebert, M., Sminchisescu, C., Weiss, Y. (eds.) ECCV 2018. LNCS, vol. 11214, pp. 561–580. Springer, Cham (2018). https://doi.org/10.1007/978-3-030-01249-6_34
8. Romera, E., Alvarez, J.M., Bergasa, L.M., et al.: Efficient ConvNet for real-time semantic segmentation. In: Proceedings of IV, pp. 1789–1794 (2017)
9. Yu, C., Wang, J., Peng, C., Gao, C., Yu, G., Sang, N.: BiSeNet: bilateral segmentation network for real-time semantic segmentation. In: Ferrari, V., Hebert, M., Sminchisescu, C., Weiss, Y. (eds.) ECCV 2018. LNCS, vol. 11217, pp. 334–349. Springer, Cham (2018). https://doi.org/10.1007/978-3-030-01261-8_20
10. Poudel, R.P.K., Bonde, U., Liwicki, S., et al.: ContextNet: exploring context and detail for semantic segmentation in real-time. In: Proceedings of BMVC (2018)

11. Poudel, R.P.K., Liwicki, S., Cipolla, R.: Fast-SCNN: fast semantic segmentation network (2019). arXiv preprint, arXiv:1902.04502
12. Zhao, H., Qi, X., Shen, X., Shi, J., Jia, J.: ICNet for real-time semantic segmentation on high-resolution images. In: Ferrari, V., Hebert, M., Sminchisescu, C., Weiss, Y. (eds.) ECCV 2018. LNCS, vol. 11207, pp. 418–434. Springer, Cham (2018). https://doi.org/10.1007/978-3-030-01219-9_25
13. Badrinarayanan, V., Kendall, A., Cipolla, R.: SegNet: a deep convolutional encoder-decoder architecture for image segmentation. IEEE Trans. Pattern Anal. Mach. Intell. **39**(12), 2481–2495 (2017)
14. Lin, T., Dollár, P., Girshick, R., et al.: Feature pyramid networks for object detection. In: Proceedings of CVPR, pp. 936–944 (2017)
15. Paszke, A., Chaurasia, A., Kim, S., et al.: ENet: a deep neural network architecture for real-time semantic segmentation (2016). arXiv preprint arXiv:1606.02147
16. Zhao, H., Shi, J., Qi, X., et al.: Pyramid scene parsing network. In: Proceedings of CVPR, pp. 6230–6239 (2017)
17. Li, H., Xiong, P., Fan, H., et al.: DFANet: deep feature aggregation for real-time semantic segmentation. In: Proceedings of CVPR, pp. 9522–9531 (2019)
18. Chen, L., Papandreou, G., Kokkinos, I., et al.: DeepLab: semantic image segmentation with deep convolutional nets, atrous convolution, and fully connected CRFs. arXiv (2016)
19. Chen, L.C., Papandreou, G., Schroff, F., Adam, H.: Rethinking atrous convolution for semantic image segmentation (2017). arXiv preprint arXiv:1706.05587
20. Fu, J., Liu, J., Tian, H., et al.: Dual attention network for scene segmentation. In: Proceedings of CVPR, pp. 3146–3154 (2019)
21. Cordts, M., Omran, M., Ramos, S., et al.: The cityscapes dataset for semantic urban scene understanding. In: Proceedings of CVPR, pp. 3213–3223 (2016)
22. Brostow, G.J., Shotton, J., Fauqueur, J., Cipolla, R.: Segmentation and recognition using structure from motion point clouds. In: Forsyth, D., Torr, P., Zisserman, A. (eds.) ECCV 2008. LNCS, vol. 5302, pp. 44–57. Springer, Heidelberg (2008). https://doi.org/10.1007/978-3-540-88682-2_5
23. Yu, F., Xian, W., Chen, Y., et al.: BDD100K: a diverse driving video database with scalable annotation tooling (2018). arXiv preprint arXiv:1805.04687
24. Yu, F., Koltun, V., Funkhouser, T.: Dilated residual networks. In: Proceedings of CVPR, pp. 636–644 (2017)
25. Sandler, M., Howard, A., Zhu, M., et al.: MobileNetV2: inverted residuals and linear bottlenecks. In: Proceedings of CVPR, pp. 4510–4520 (2018)
26. Wang, Z., Ji, S.: Smoothed dilated convolutions for improved dense prediction. In: Proceedings of KDD, pp. 2486–2495 (2018)
27. Hu, J., Shen, L., Sun, G.: Squeeze-and-excitation networks. In: Proceedings of CVPR, pp. 418–434 (2018)

Safety and Robustness of Deep Neural Networks Object Recognition Under Generic Attacks

Mallek Mziou Sallami[1(✉)], Mohamed Ibn Khedher[1(✉)], Asma Trabelsi[2], Samy Kerboua-Benlarbi[1], and Dimitri Bettebghor[2,3]

[1] IRT - SystemX, 8 Avenue de la Vauve, 91120 Palaiseau, France
{mallek.mziou-sallami,mohamed.ibn-khedher, samy.kerboua-benlarbi}@irt-systemx.fr

[2] Expleo, Yvelines, France
{asma.trabelsi,dimitri.bettebghor}@expleogroup.com

[3] ONERA, The French Aerospace Lab, Palaiseau, France
dbetteb@onera.fr

Abstract. Embedding machine or deep learning software into safety-critical systems such as autonomous vehicles requires software verification and validation. Such software adds non traceable hazards to traditional hardware and sensors failures, not to mention attacks that fool the prediction of a DNN and hampers its robustness. Formal methods from computer science are now applied to deep neural networks to assess the local and global robustness of a given DNN. Typically static analysis with Abstract Interpretation or SAT solvers approaches are applied to neural networks and leverages the important progress of formal methods over the last decades. Such approaches estimate bounds on the perturbation of the inputs and formally guarantee the same DNN prediction within these bounds. However formal methods over DNN for image perception system have only been applied to simple image attacks (2D rotation, brightness). In this work, we extend the definition of Lower and Upper Bounds to assess the robustness of a DNN perception system against more generic attacks. We propose a general method to verify object recognition systems using Abstract Interpretation theory. Another major contribution is the adaptation of Upper and Lower Bounds with the abstract intervals to support more complex attacks. We consider the three following classes: convolutional attacks, occlusion attacks and geometrical transformations. For the last one, we generalize the geometrical transformations with displacements in the three-dimensional space.

Keywords: AI Safety · Perception system · Object recognition · Classification · Deformation · Attacks · Neural networks · Abstract interpretation

T. Gedeon et al. (Eds.): ICONIP 2019, CCIS 1142, pp. 274–286, 2019.
https://doi.org/10.1007/978-3-030-36808-1_30

1 Introduction

1.1 AI Safety for Critical Systems

Despite significant success on image recognition tasks [17], automatic speech recognition [5], natural language processing [21] and many other AI-related tasks [1], deep learning models do not achieve sufficient confidence, explainability and transparency levels to be integrated into safety-critical systems, [11]. Amongst the very reasons that hamper DNN deployment in such systems (in transportation, energy production, military, medical...) the following aspects need to be rigorously addressed:

1. Train and deploy ML or DL models that are *consistent with specifications* (e.g. `Stop when a pedestrian crosses the road`)
2. *Test compliance* with respect to these specifications and *exhaustively detect worst cases* (e.g. find all possible inputs where the specification `Stop when a pedestrian crosses the road` fails)
3. *Demonstrate the consistency of the ML model with respect to specifications with formal methods* (e.g. for all possible inputs with a domain of definition verify that the software is consistent with respect to the specification `Stop when a pedestrian crosses the road`)

These aspects are related to the *correct-by-construction design principle*, identified in [16] as one of the 5 main challenges towards verification of Artificial Intelligence (*AI Safety* field), the other being environment modelling, formal specification, the modelling of the learning aspect of such systems and the computational engines. In this work, we hope to contribute to the *correct-by-construction design principle* aspect by enhancing robustness estimation of a general DNN object recognition system with respect to general and realistic attacks over inputs (images). Our work lie at the frontier between computer vision, deep learning and safety considerations and is aiming towards the validation of perception systems that are to be used in transportation systems (automotive, railroad, aircraft, ship).

1.2 Adversarial Attacks on DNN and Abstract Interpretation

In the mean time, there is both a growing general interest for AI-related products and an important societal pressure for risk mitigation of such products, especially for object detection in the automotive sector. This is where ensuring safety of perception systems becomes critical not only from safety consideration but also to guarantee general public acceptance. In this field, perception systems that recognize objects or persons poorly can significantly falsify enforcement actions dictated by the decision system. Therefore, it is important to ensure high operating safety as well as traffic safety. Recently, several researchers have demonstrated the lack of robustness of DNN against numerous attacks. Authors, in [9], have explained the sensibility causes of neural networks and have proposed an optimization (FGSM *Fast-Gradient Sign Method*) method to automatically generate adversarial examples. The adversarial examples correspond to images that are very close to the correctly classified images, while the DNN fails when classifying them. Authors have also demonstrated that training neural network along with adversarial examples may reduce the system sensitivity and improve the system robustness. To cope with this problem, researchers have proposed methods for automatically validate DNN and more generally Artificial Neural Networks. These methods can be broken down into three principal categories:

- **linear or mixed integer programming approaches**, [6,13] that focus on the *reachability* aspect of safety, i.e. determine whether some states (or bugs) can be attained by the neural network
- **algebraic approaches** [4] that focus more on the poorly understand theoretical properties of neural networks and aims at *representing the function* a given NN has actually learnt by means of more expressive and interpretable objects (typically kernels)
- **formal methods** [3] who aim to bring the rigorous and mathematical arsenal of formal proofs (static analysis with Abstract Interpretation, Boolean Satisfiability (SAT solvers), Boolean reasoning on Binary Decision Diagrams, Satisfiability Modulo Theories (SMT solvers)) already in use for Computer Aided Design of integrated circuits or in railway certification to neural networks seen as a software.

In this work, we focus on formal methods for NN-based object recognition systems and we propose a new formulation to assess the robustness of a given NN-based image classifier. We also highlight the importance of attack definition level. For example, authors, in [7,18,19], have introduced a certification method of neural network mainly based on the abstract interpretation. Experimental results on MNIST database have proven the capability of a such system to certify the robustness against attacks including simple contrast, FGSM noise and L_∞ attacks. Recent works (notably [19]) consider the robustness against geometric transformation attacks as a simple plane rotation. In this paper, we propose a generalization of Lower and Upper Bound concepts that allows us to verify the robustness of a NN against a larger class of attacks which is a mandatory requirement for critical safety systems.

The remaining of this paper is organized as follows. Section 2 is dedicated to state the fundamental concepts of the incomplete verification approach based on the Abstract Interpretation theory. We demonstrate in Sect. 3 the different possible attacks within a perception context and how to evaluate the robustness against each of these attacks. Our experimentation settings and results will be given in Sect. 4. Finally, in Sect. 5, we draw our conclusions and we discuss some future perspectives.

2 State of the Art and Related Works

2.1 Abstract Interpretations for Neural Networks Verification

Abstract Interpretation is an approximation approach to infer semantic properties from computer programs and demonstrate their soundness, see [2]. Static analysis by Abstract Interpretation allows to automatically extract information about all possible states of execution of a computer program and is used for automatic debugging, optimizing compilers and code execution and to certify programs against some classes of bugs. One of the first application of static analysis with Abstract Interpretation for NN can be found in [15] but focused at that time on shallow NN (MLP). Recently some authors in [7,18,19] have reused and adapted this method for verifying the robustness property of larger neural networks by proposing abstract transformers for each type of activation function. In what follows, we recall in a synthetic way some notions. Let $\bar{X}$ be a given input. $\bar{X}$ may undergo a deformation or even an attack. In such a case, $\bar{x} \in \bar{X}$ will be transformed into $\bar{x}_\epsilon$. The original inputs perturbed by ϵ are denoted by $R_{\bar{X},\epsilon}$. Verifying the robustness property for $R_{\bar{X},\epsilon}$ consists of checking the property over the whole possible perturbation of $\bar{X}$. Let C_L be the robustness condition that defines the output ensembles with the same label L. We denote $\bar{Y}$ the set of each prediction for each element in $R_{\bar{X},\epsilon}$.

$$C_L = \{\bar{y} \in \bar{Y} |\arg\max \bar{y}_i = L\} \tag{1}$$

The $(R_{\bar{X},\epsilon}, C_L)$ property is verified only if the outputs O_R of $R_{\bar{X},\epsilon}$ are included in C_L. However, in reality, we are not able to control the behavior of hidden layers. Accordingly, we have no knowledge about O_R. The Abstract Interpretation is a proposed alternative to face this shortcoming. In fact, it allows to determine an abstract domain thought transformers and verifies the inclusion condition in new abstract domains α_R, which is an abstraction of $\bar{X}$. Thanks to the neural network outputs, we denote the output abstract domain α_R^O.

The $(R_{\bar{X},\epsilon}, C_L)$ property is checked:

- If the outputs O_R of $R_{\bar{X},\epsilon}$ are included in C_L.
- If the outputs in α_R^O of the abstraction α_R of $R_{\bar{X},\epsilon}$ are included in C_L.

It seems necessary to define abstract transformers that are precise for the different existent activation functions. Authors in [7], have proposed a neural network analyzer known under the name of AI^2. This analyzer may automatically prove the robustness of different architectures neural networks, including convolutional neural networks. The test results demonstrate that AI^2 analyzer is fully accurate and may be used to certify the most recent defense efficiency for neural networks. Scalability is one of the major shortcoming of this approach.

Authors, in [18], have proposed an alternative solution, called DeepZ, for dealing with the scalability problem. DeepZ allows also to certify the robustness of neural network. It is characterized by its highly precision arithmetic in floating point and it manages several activation functions, including **ReLU** (*REctified Linear Unit*), **TanH**, and **Sigmoid**. It is worth noting that DeepZ is based on the abstract domains and more particularly the zonotopes [8]. Another analyzer, called DeepPoly, have been introduced in [7]. This approach relies on a novel abstract domain that merge polyhedron with floating point and intervals, see Fig. 1. Moreover, polygons live in a 2-dimensional space, while polyhedra live in a 3-dimensional one, and we have to take in account the fact that the generalized

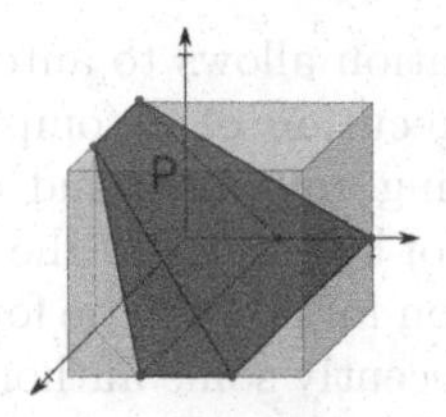

Fig. 1. Example of two abstract domains: intervals vs polyhedra on a set of 6 three-dimensional points

notion of the formers to a $n \geq 3$-dimensional space is the domain of polytopes. DeepPoly analyser supports refine transformation as well as modern activation functions such as **ReLU**, **sigmoid**, **TanH** and **maxpool**. According to authors, DeepPoly is the more precise comparatively with the AI^2 and DeepZ and manage also deep convolutional neural networks. This method has been used to check complex perturbation, including 2D rotation.

2.2 Lower and Upper Bound for Contrast and Geometrical Attacks

The definition of the abstract domain is a crucial step in the abstract interpretation verification process. The more precise the abstract domain is, the more complete is the verification. The three most used domains are: zonotopes, polytopes and abstract intervals. In the case of luminosity disturbance, the lower bound LB and the upper bound UB, which are the minimal brightness and its maximal value, are simple. We can approximate it to a brightness shift. Indeed, these two values allows us to define the abstract intervals that we need.

In the case of plane rotation, the contribution of the neighboring pixels for the intensity of the disturbed pixel is proportional to its distance from the initial pixel. This approximation lets us estimating the possible LB and UB, which give us the polytopes in which each rotated pixel is going to end. Combined with abstract intervals, they allow us to compute the needed abstract domain. It is desirable to add a tracing algorithm which subdivides the rotation interval into sub-intervals. Such procedure check whether the neural network is able to recognize the object when his orientation on the image change.

3 Image Attacks and Their Abstract Domains

Image attacks have been well-studied over the last few years for solving real world problems in several areas. Examples include the digital watermark. There are several kinds of attacks (see Fig. 2), in this work we focus on:

- **occlusion attacks** that mask some parts of the images
- **frequential attacks** that modify the spectral components of the images
- **geometric image transformation** that comes from sensor errors or limits (typically distortion)

Fig. 2. Some classical on images from [10] (a) JPEG (b) noise (c) occlusion

We identify two classes within occlusion attacks: additive and substractive. Malicious users try to detect the presence of an object as well as the location of this object to extract it from the target image. Analogically, one can define additive attacks that replaces a portion of the image by another portion.

Regarding the frequential attacks, they can be classified in two groups: filtering and compression. For example, when smoothing images in photo editing the high-frequency components are usually attenuated. Frequential attacks may also be due to compression as a format conversion or when setting different compression rates.

Concerning the geometric image transformations, the basic ones include rotation, uniform scale change, reflection and shearing.

3.1 Lower and Upper Bound for 3D Rotation

In this subsection, we consider a more complex geometrical transformation. This kind of transformations take into consideration the rigidity of the objects as well as the possible displacements in the real world (Fig. 3). To explain the relationship between the object displacement and its projection on the image, we refer to the pinhole camera model, which is very efficient and simple as it was described in [20]. Based on the pinhole camera model, we consider image plane that corresponds to a straight line $\{(x_1, x_2, x_3)^t \; \forall \; x_i \in \mathbb{R}\} \; \mathbb{R}^3$, with $x_2 = 1$. The image of a given point $(x_1, x_2, x_3)^t$ upon an image plane is given by: $j(x_1, x_2, x_3) = \frac{x_3 + ix_1}{x_2} = x_3 + ix_1$ with $(x_1, x_2, x_3)^t$, a point in the image plane and j a projection from $\mathbb{R}^3$ in $\mathbb{R}^2$.

The distortion of projected objects are generated through the applied rotation and translations followed by a projection upon an image plane. The image plane may be regarded as an extension of a complex line $\hat{C} = C \cup \{\infty\}$ with its affine part C defined by $x_2 = 1$. The special linear group $\mathcal{SL}(2, \mathbb{C})$, made up of all complex 2×2 matrix with determinant 1 defined as

$$\mathcal{SL}(2,\mathbb{C}) := \left\{ \begin{pmatrix} a \; b \\ c \; d \end{pmatrix} \mid a, b, c, d \in \mathbb{C}, ad - bc = 1 \right\} \tag{2}$$

$\mathcal{SL}(2, \mathbb{C})$ acts on the image plane $\hat{C}$ with the transformation g and the group of projective transformation is set to:

$$\mathcal{PSL}(2, \mathbb{C}) = \frac{\mathcal{SL}(2, \mathbb{C})}{\mp Id} \tag{3}$$

The special unitary group $\mathcal{SU}(2)$ defined as:

$$\mathcal{SU}(2) := \left\{ \begin{pmatrix} \alpha & -\overline{\beta} \\ \beta & \overline{\alpha} \end{pmatrix} : \alpha, \beta \in, |\alpha|^2 + |\beta|^2 = 1 \right\}$$

is a maximal compact subgroup of $\mathcal{SL}(2,\mathbb{C})$. Using these definitions, one can easily prove that there exist two elements for every rotation r of $\mathcal{SO}(3)$ such that $k = \pm \begin{pmatrix} \bar{a} & -\bar{b} \\ b & a \end{pmatrix}$ and consequently $k.z = \frac{az+b}{-\bar{b}+\bar{a}}$ with:

$$a = \pm\cos(\frac{\phi}{2})e^{\frac{i(\psi_1+\psi_2)}{2}}, \quad b = \pm i\sin(\frac{\phi}{2})e^{\frac{i(\psi_1-\psi_2)}{2}}$$

Consider Algorithm 1 coupled with Algorithm 2, which rotates an image by three angles. To compute the intensity of a given transformed pixel (x,y), we first computes the projection based on a and b values. By enumerating all possible integer values of c_{low}, c_{high}, v_{low} and v_{high}, we can identify a polygon where the pixel (x, y) is. The contribution of each pixel belonging to this polygon is proportional to its distance from the real position (x, y). Note that g and $-g$ have the same orbits, which shows that ψ_1 and ψ_2 have the same behavior. We can only consider one of the two parameters for our experiments.

To verify that for any image $I \in X$, for any angle $\phi \in [\phi_{min}, \phi_{max}]$ and any angle $\psi_1 \in [\psi^1_{min}, \psi^1_{max}]$ the neural network N classifies $I_{\epsilon=(\phi,\psi_1)}$ to the class of I, we cannot simply enumerate all possible rotations as done for simpler rotation algorithms and concrete images [14]. To refine UB et LB for large enough intervals, it is possible to subdivide intervals into several segments. This partitioning technique is coupled with batching to obtain precise enough output intervals. In the two algorithms, each temporary lower and upper bounds are initialized respectively to 255 and null matrices, and the T matrix is initialized to zeros too.

Fig. 3. Visualization of a projective effect resulting from a 3D rotation

3.2 Lower and Upper Bound for Convolutional Attacks

This kind of attacks includes filtering operations. Suppose that we have a filter with size $n \times n$. It consists of replacing a given pixel by the sum of the product of this pixel and its $n \times n - 1$ surrounding pixels by the filter's corresponding

Algorithm 1. Rotate Image I by 3D rotation

1: **procedure** PROCEDURE IMAGE_ROTATION
Input: $I \in [0, 255]^{m \times n}; \phi, \psi_1, \psi_2 \in [-\pi, \pi]; T, T_{LB}, T_{UB} \in [0, 255]^{m \times n}$
2: $(a, b) = (\cos(\frac{\phi}{2})e^{\frac{i(\psi_1+\psi_2)}{2}}, i\sin(\frac{\phi}{2})e^{\frac{i(\psi_1-\psi_2)}{2}})$
3: **for** $c \in \{1, \dots, m\}; v \in \{1, \dots, n\}$ **do**
4: $(x, y, z) = (c - \frac{m+1}{2}, \frac{n+1}{2} - v, x + iy)$
5: $z = \frac{(az+b)}{-\bar{b}z+\bar{a}}$
6: $(y', x') = (Im(z), Re(z))$
7: $(c'_{low}, c'_{high}) \leftarrow (\max(1, \frac{m+1}{2} - y'), \min(m, \frac{m+1}{2} - y')$
8: $(v'_{low}, v'_{high}) \leftarrow (\max(1, x' + \frac{n+1}{2}), \min(n, x' + \frac{n+1}{2})$
9: $R^{Low}_{c,v} \leftarrow \min(255, \min_{c' \in [c'_{low}, c'_{high}], v' \in [v'_{low}, v'_{high}]} I[c', v']$
10: $R^{Hight}_{c,v} \leftarrow \max(0, \max_{c' \in [c'_{low}, c'_{high}], v' \in [v'_{low}, v'_{high}]} I[c', v']$
11: $t \leftarrow \sum_{c'=c'_{low}, v'=v'_{low}}^{c'=c'_{high}, v'=v'_{high}} \max(0, 1 - \sqrt{(v' - x')^2 + (c' - y')^2})$
12: $t' \leftarrow \sum_{c'=c'_{low}, v'=v'_{low}}^{c'=c'_{high}, v'=v'_{high}} (\max(0, 1 - \sqrt{(v' - x')^2 + (c' - y')^2}) \times I[c', v'])$
13: **if** $t \neq 0$ **then**
14: $T[c, v] \leftarrow \frac{1}{t} \times t'$
15: $T_{LB}[c, v] \leftarrow min(T_{LB}[c, v], R^{Low}_{c,v})$
16: $T_{UB}[c, v] \leftarrow max(T_{UB}[c, v], R^{Hight}_{c,v})$
17: **else**
18: $T[c, v], T_{LB}[c, v], T_{UB}[c, v] \leftarrow 0$
19: **Return** T, T_{LB}, T_{UB}

Algorithm 2. Lower and Upper Bound for 3D Rotation (on a rotation interval)

1: **procedure** PROCEDURE ROTATION_LOWER_UPPER_BOUND
Input: $I, T^I_{LB}, T^I_{UB} \in [0, 255]^{m \times n}; bs_\phi, bs_{\psi_1}, bs_{\psi_2} \in \mathbb{N}$
$\phi_{min}, \phi_{max} \psi^1_{min}, \psi^1_{max}, \psi^2_{mmin}, \psi^2_{max}, \in [-\pi, \pi]$
2: $(step_\phi, step_{\psi 1}, step_{\psi 2}) = (\frac{|\phi_{max} - \phi_{min}|}{bs_\phi}, \frac{|\psi^1_{max} - \psi^1_{min}|}{bs_{\psi_1}}, \frac{|\psi^2_{max} - \psi^2_{min}|}{bs_{\psi_2}})$
3: Compute lists $\phi_{all}, \psi^1_{all}, \psi^2_{all}$ of all values using their respective steps
4: **for** $(\phi_0, \psi^1_0, \psi^2_0) \in (\phi_{all}, \psi^1_{all}, \psi^2_{all})$ **do**
5: $(T, T_{LB}, T_{UB}) \leftarrow IMAGE\_ROTATION(I, \phi_0, \psi^1_0, \psi^2_0)$
6: $T^I_{LB} = \min(T^I_{LB}, T_{LB})$
7: $T^I_{UB} = \max(T^I_{UB}, T_{UB})$
8: **Return** T^I_{LB}, T^I_{UB}

value. Our approach consists of defining a LB and an UB independently of the applied filter coefficients. The pixel on the filtered image is estimated according to the size of the filter. Each pixel, in his neighborhood, can be replaced by the nearest neighbor intensity value to find the LB. The final LB and UB correspond respectively to the min and the max values between the LB and the UB images relatives to the two dimensions (Fig. 4). Algorithm 3 describes in more details the different steps.

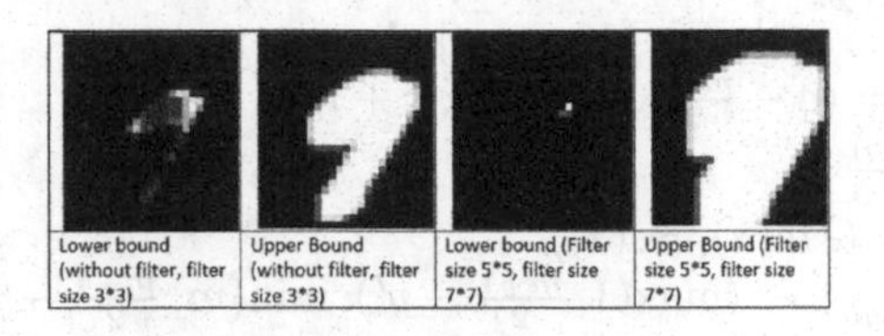

Fig. 4. Effects of filter size on UB and LB

Algorithm 3. Lower and Upper Bound for convolution

1: **procedure** PROCEDURE CONVOLUTION_LOWER_UPPER_BOUND
   Input: $I \in [0, 255]^{m\times n}; h, w \in [\![1, T]\!]$, T size of the filter
2: $I_{LB}, I_{UB} \leftarrow I$
3: **for** $c \in \{1, \dots, m\}; v \in \{1, \dots, n\}$ **do**
4: $L_1[c, v] \leftarrow \min(Neighbors(I, h, c, v))$
5: $U_1[c, v] \leftarrow \max(Neighbors(I, h, c, v))$
6: $L_2[c, v] \leftarrow \min(Neighbors(I, w, c, v))$
7: $U_2[c, v] \leftarrow \max(Neighbors(I, w, c, v))$
8: $I_{LB}[c, v] \leftarrow \min(L_1[c, v], L_2[c, v])$
9: $I_{UB}[c, v] \leftarrow \max(U_1[c, v], U_2[c, v])$
10: **Return** I_{LB}, I_{UB}

Algorithm 4. Lower and Upper Bound for occlusion

1: **procedure** PROCEDURE OCCLUSION_LOWER_UPPER_BOUND
   Input: $I \in [0, 255]^{m\times n}; h, w \in [\![1, T]\!], T$ filter size; $c_0 \in [1, m], v_0 \in [1, n]$
2: $I_{LB}, I_{UB} \leftarrow I$
3: $M1 \leftarrow Mask(h, c_0, v_0)$
4: $M2 \leftarrow Mask(w, c_0, v_0)$
5: $I_{M1}, I_{M2} \leftarrow I \times M1, I \times M2$
6: $I_{LB}, I_{UB} \leftarrow \min(I_{M1}, I_{M2}), \max(I_{M1}, I_{M2})$
7: **Return** I_{LB}, I_{UB}

3.3 Lower and Upper Bound for Occlusion Attack

With regards to the occlusion attacks, the image undergoes the disappearance of some pixels. It is a passage from gray level to black level through a given mask.

Let us denote by I the input image and let $M1$ and $M2$ be two given masks. We consider the lower bound ML and the upper bound MU of the mask M. Algorithm 4 describes the process to compute LB and UB with occlusion attack. The position of the mask center denoted by c_0 and v_0 are given as input. We have also the dimension h and w for the two masks. We apply the obtained masks for yielding the lower bound and the upper bound images.

4 Experimentation Settings and Results

This section is devoted to highlighting our experimentation settings and results for evaluating the effectiveness of our approach for verifying the robustness properties for complexes attacks including the convolution, the occultation and the 3D attacks.

4.1 Experimentation Settings

Herein, we point out the two main setting that allow us to carry out our experimentations. The first one is the evaluation dataset and the second one concerns the neural networks. Regarding the evaluation dataset, we relied on the MNIST dataset [12], one of the well commonly database within the fields of artificial intelligence and machine learning. It contains grayscale images of size 28 × 28 pixels. We have select, for the evaluation, the first 50 images as a test set. Our robustness criterion is then calculated as the number of verified image over the total number of well classified instance by the neural network. The robustness metric is set to: $Robustness = \frac{\text{Verified images}}{\text{Well classified images}}$

4.2 Experimentation Results

As described in earlier sections, we can apply our method to prove a neural network robustness against 3D rotations. Specifically, our analysis can prove if the MNIST network can classify a given image of a digit correctly even if each pixel is perturbed with three rotation using an arbitrary angle. Rotations according to ψ_1 and ψ_2 generate plane rotations in the plane of the image. So, to test the robustness of the neural network, just consider either ψ_1 or ψ_2 coupled with ϕ. Figure 5b shows example of robustness surface for $\phi \in [0, 10°], \psi_1 \in [0, 30°]$. We split the interval of ϕ [0, 10] into 50 batches, for ψ_1, we split the interval into 30 batches. To analyze a batch, we split the corresponding interval into 10 input intervals for interval analysis, resulting in 10 regions for each batch. We then run DeepPoly on the smallest common bounding boxes of all regions in each batch, 100 times in total. The results show that neural network models are more sensitive to rotations around the optical axis. This is expected since this rotation according to ϕ generates a greater deformation on the image than rotation according to ψ_1. Among the three studied attacks, convolutional attacks seem the strongest. Figure 5a shows the robustness results for different models of neural networks. On a basis of 50 images that have different classes, we represent the quotient between the number of image well classified and the number of image verified. For the different models, robustness does not exceed 8%. Occlusion attacks depend heavily on the surface of the hidden region (Fig. 6). In experiments, it was assumed that masks are positioned in the center of gravity of the image. Such simplification is possible for images with a single object on a black background. Otherwise, all possible positions and masks dimension must be considered. For the 3 models, the robustness decreases according to the number of deleted pixels.

5 Conclusion

We introduced a new approach to verify the object recognition system based on neural networks. The core idea is the adequate formulation of abstract intervals for each attack. These formula enable us to extend the deeppoly analyser and evaluate a wide range of attacks such as convolution and occlusion. We believe this work is a promising step towards a validation of a perception or object recognition system. As perspective, we plan to test more complex attacks such as attacks caused by weather conditions such as fog or snow.

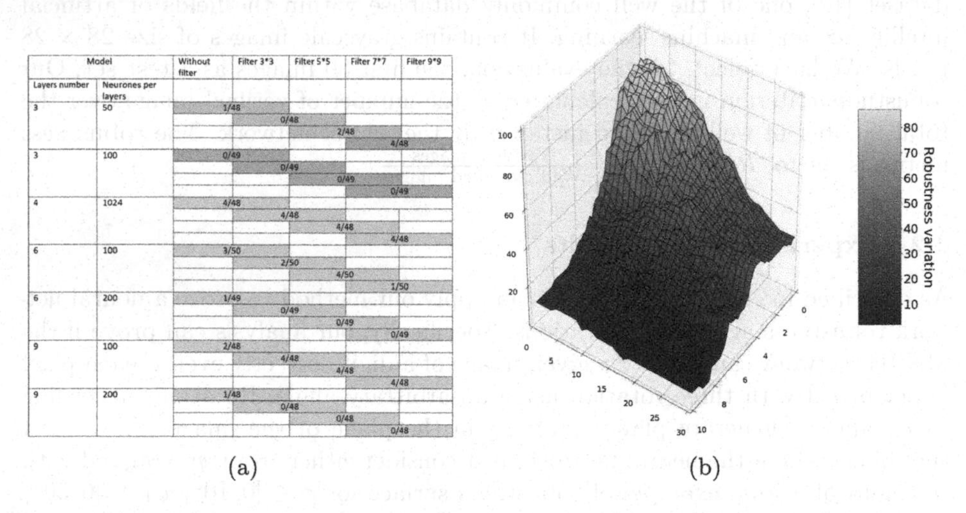

Model		Without filter	Filter 3*3	Filter 5*5	Filter 7*7	Filter 9*9
Layers number	Neurones per layers					
3	50	1/48				
			0/48			
				2/48		
					4/48	
3	100	0/49				
			0/49			
				0/49		
					0/49	
4	1024	4/48				
			4/48			
				4/48		
					4/48	
6	100	3/50				
			2/50			
				4/50		
					1/50	
6	200	1/49				
			0/49			
				0/49		
					0/49	
9	100	2/48				
			4/48			
				4/48		
					4/48	
9	200	1/48				
			0/48			
				0/48		
					0/48	

Fig. 5. (a) Robustness variation according to filters size for different fully connected neural networks, (b) Robustness variation according to 3D rotation

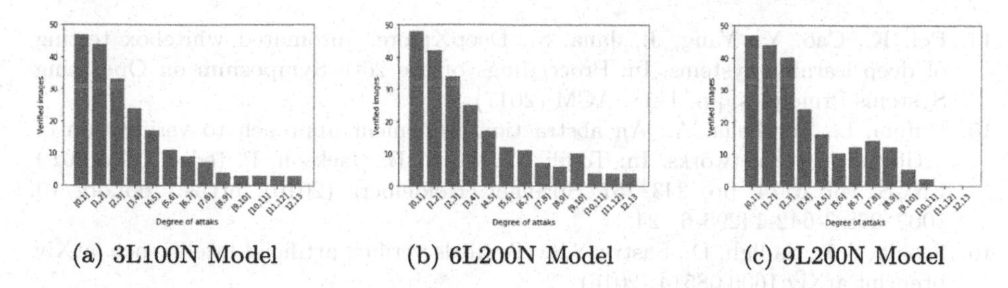

(a) 3L100N Model (b) 6L200N Model (c) 9L200N Model

Fig. 6. Robustness variation according to occlusion surface for different fully connected neural networks

References

1. Bengio, Y., et al.: Learning deep architectures for AI. Found. Trends® Mach. Learn. **2**(1), 1–127 (2009)
2. Cousot, P., Cousot, R.: Abstract interpretation and application to logic programs. J. Log. Program. **13**(2–3), 103–179 (1992)
3. Craven, M.W.: Extracting comprehensible models from trained neural networks. Technical report, University of Wisconsin-Madison Department of Computer Sciences (1996)
4. Daniely, A., Frostig, R., Singer, Y.: Toward deeper understanding of neural networks: the power of initialization and a dual view on expressivity. In: Advances In Neural Information Processing Systems, pp. 2253–2261 (2016)
5. Deng, L., et al.: Recent advances in deep learning for speech research at Microsoft. In: 2013 IEEE International Conference on Acoustics, Speech and Signal Processing, pp. 8604–8608. IEEE (2013)
6. Ehlers, R.: Formal verification of piece-wise linear feed-forward neural networks. In: D'Souza, D., Narayan Kumar, K. (eds.) ATVA 2017. LNCS, vol. 10482, pp. 269–286. Springer, Cham (2017). https://doi.org/10.1007/978-3-319-68167-2_19
7. Gehr, T., Mirman, M., Drachsler-Cohen, D., Tsankov, P., Chaudhuri, S., Vechev, M.: AI2: safety and robustness certification of neural networks with abstract interpretation. In: 2018 IEEE Symposium on Security and Privacy (SP), pp. 3–18. IEEE (2018)
8. Ghorbal, K., Goubault, E., Putot, S.: The zonotope abstract domain Taylor1+. In: Bouajjani, A., Maler, O. (eds.) CAV 2009. LNCS, vol. 5643, pp. 627–633. Springer, Heidelberg (2009). https://doi.org/10.1007/978-3-642-02658-4_47
9. Goodfellow, I.J., Shlens, J., Szegedy, C.: Explaining and harnessing adversarial examples. arXiv preprint arXiv:1412.6572 (2014)
10. He, M., Tan, Q., Cao, L., He, Q., Jin, G.: Security enhanced optical encryption system by random phase key and permutation key. Opt. Express **17**(25), 22462–22473 (2009)
11. Huang, X., Kwiatkowska, M., Wang, S., Wu, M.: Safety verification of deep neural networks. In: Majumdar, R., Kunčak, V. (eds.) CAV 2017. LNCS, vol. 10426, pp. 3–29. Springer, Cham (2017). https://doi.org/10.1007/978-3-319-63387-9_1
12. LeCun, Y., Bottou, L., Bengio, Y., Haffner, P., et al.: Gradient-based learning applied to document recognition. Proc. IEEE **86**(11), 2278–2324 (1998)
13. Lomuscio, A., Maganti, L.: An approach to reachability analysis for feed-forward ReLU neural networks. arXiv preprint arXiv:1706.07351 (2017)

14. Pei, K., Cao, Y., Yang, J., Jana, S.: DeepXplore: automated whitebox testing of deep learning systems. In: Proceedings of the 26th Symposium on Operating Systems Principles, pp. 1–18. ACM (2017)
15. Pulina, L., Tacchella, A.: An abstraction-refinement approach to verification of artificial neural networks. In: Touili, T., Cook, B., Jackson, P. (eds.) CAV 2010. LNCS, vol. 6174, pp. 243–257. Springer, Heidelberg (2010). https://doi.org/10.1007/978-3-642-14295-6_24
16. Seshia, S.A., Sadigh, D., Sastry, S.S.: Towards verified artificial intelligence. arXiv preprint arXiv:1606.08514 (2016)
17. Simonyan, K., Zisserman, A.: Very deep convolutional networks for large-scale image recognition. arXiv preprint arXiv:1409.1556 (2014)
18. Singh, G., Gehr, T., Mirman, M., Püschel, M., Vechev, M.: Fast and effective robustness certification. In: Advances in Neural Information Processing Systems, pp. 10825–10836 (2018)
19. Singh, G., Gehr, T., Püschel, M., Vechev, M.: An abstract domain for certifying neural networks. Proc. ACM Program. Lang. **3**, 41 (2019)
20. Turski, J.: Projective Fourier analysis for patterns. Pattern Recogn. **33**(12), 2033–2043 (2000)
21. Young, T., Hazarika, D., Poria, S., Cambria, E.: Recent trends in deep learning based natural language processing. IEEE Comput. Intell. Mag. **13**(3), 55–75 (2018)

Deep Neural Network Hyperparameter Optimization with Orthogonal Array Tuning

Xiang Zhang(✉), Xiaocong Chen, Lina Yao, Chang Ge, and Manqing Dong

University of New South Wales, Sydney, Australia
{xiang.zhang3,chang.ge,manqing.dong}@student.unsw.edu.au,
{xiaocong.chen,lina.yao}@unsw.edu.au

Abstract. Deep learning algorithms have achieved excellent performance lately in a wide range of fields (e.g., computer version). However, a severe challenge faced by deep learning is the high dependency on hyper-parameters. The algorithm results may fluctuate dramatically under the different configuration of hyper-parameters. Addressing the above issue, this paper presents an efficient Orthogonal Array Tuning Method (OATM) for deep learning hyper-parameter tuning. We describe the OATM approach in five detailed steps and elaborate on it using two widely used deep neural network structures (Recurrent Neural Networks and Convolutional Neural Networks). The proposed method is compared to the state-of-the-art hyper-parameter tuning methods including manually (e.g., grid search and random search) and automatically (e.g., Bayesian Optimization) ones. The experiment results state that OATM can significantly save the tuning time compared to the state-of-the-art methods while preserving the satisfying performance.

Keywords: Orthogonal array · Hyper-parameter · Deep learning

1 Introduction

Deep learning has been recently attracting much attention in both academia and industry, due to its excellent performance on various research areas such as computer vision, speech recognition, natural language processing, and brain-computer interface [11]. Nevertheless, deep learning faces an important challenge that the performance of the algorithm highly depends on the selection of hyper-parameters. Compared with traditional machine learning algorithms, deep learning requires hyper-parameter tuning more urgently because deep neural networks: (1) have more hyper-parameters to be tunned; (2) have higher dependency on the configuration of hyper-parameters. [10] reports the deep learning classification accuracy dramatically fluctuates from 32.2% to 92.6% due to the different selection of hyper-parameters. Therefore, an effective and efficient hyper-parameter tuning method is necessary.

T. Gedeon et al. (Eds.): ICONIP 2019, CCIS 1142, pp. 287–295, 2019.
https://doi.org/10.1007/978-3-030-36808-1_31

However, most of the existing hyper-parameter tuning methods have some drawbacks. In particular, grid search traverses all the possible combinations of different hyper-parameters, which is a time-consuming and ad-hoc process [2]. Random Search, which is developed based on grid research, set up a grid of hyper-parameter values and selects random combinations to train the algorithm [2]. Random search method oversteps some disadvantages of grid search such as time-consuming but meanwhile brings a major disadvantage which cannot converge to the global optimum [1]. The randomly selected hyper-parameter combinations cannot guarantee a steady and competitive result. Apart from the manually tuning methods, automated tuning methods being more popular in recent years [7]. Bayesian Optimization, a most widely-used automated hyper-parameter tunning approach, attempts to find the global optimum in a minimum number of steps. Nevertheless, the results of Bayesian optimization are sensitive to parameters of the surrogate model and the performance is highly depending on the quality of the learning model [3].

To address the aforementioned issue, we propose the Orthogonal Array Tuning Method (OATM) which can achieve a trade-off of the less tuning time and competitive performance. In detail, the OATM manner is proposed based on Taguchi Approach [8]. The OATM is a highly fractional orthogonal design method that is based on a design matrix and allows the user to consider a selected subset of combinations of multiple factors at multiple levels. Additionally, the OATM is balanced to ensure that all possible values of all hyper-parameters are considered equally. Moreover, OATM has been commonly used as an experimental design method in a wide variety of domains like mechanical engineering [6] and electrical engineering [5]. To our best knowledge, our work is *the first batch of work* adopting orthogonal array into parameter tuning in deep learning.

The proposed OATM adopts the orthogonal array to extract the most representative and balanced combinations from the whole set of possible combinations. The proposed OATM will be explained in detail in the context of two popular deep learning structures (Sect. 4). In addition, the OATM is evaluated over three datasets, which demonstrate the universality and adaptability. We notice that source codes performing grid search, random search, and especially Bayesian Optimization on deep learning are hard to online acquire. Thus, we provide the reusable source codes and datasets for reproduction[1].

2 Orthogonal Array Tuning

In this section, we first provide the background knowledge of orthogonal array, namely, the definition, the compose principles, and the terminology. Then, we report the working procedure of OATM.

2.1 Orthogonal Array Tuning Method

In this section, we propose the Orthogonal Array Tuning Method inspired by the basic principles of orthogonal array. Although deep learning algorithms can

[1] https://github.com/xiangzhang1015/OATM.

achieve good performance in many research areas, tuning the hyper-parameters (e.g., the number of layers, the number of nodes in each layer and the learning rate) is time-consuming and dependent on user's expertise.

In OATM, the hyper-parameters are regarded as factors and different values of each hyper-parameter are regarded as levels. The procedure is listed as follows.

- **Step 1:** Build the F-L (factor-level) table. Determine the number of to-be-tuned factors and the number of levels for each factor. The levels should be determined by experience and literature. We further suppose each factor has the same number of levels[2].
- **Step 2:** Construct Orthogonal Array Tuning table. The constructed table should obey the basic composition principles. Here[3] shows some commonly used tables. The Orthogonal Array Tuning table is marked as $L_M(h^k)$ which has k factors, h levels, and totally M rows.
- **Step 3:** Run the experiments with the hyper-parameters determined by the Orthogonal Array Tuning table.
- **Step 4:** Range analysis. This is the key step of OATM. Based on the experiment results in the previous step, range analysis method is employed to analyze the results and figure out the optimal levels and importance of each factor. The importance of a factor is defined by its influence on the results of the experiments. Note that range analysis optimizes each factor and combines the optimal levels together, which means that the optimized hyper-parameter combination is not restricted to the existing Orthogonal Array table.
- **Step 5:** Run the experiment with the optimal hyper-parameters setting.

3 Experimental Setting

To evaluate the proposed OATM, we design extensive experiments to tune the hyper-parameters of two most widely used deep learning structures, i.e., the Recurrent Neural Networks (RNNs) and the Convolutional Neural Networks (CNNs). Both of the two deep learning structures are employed on three real-world applications: (1) a human intention recognition task based on the Electroencephalography (EEG) signals [12]; (2) activity recognition based on wearable sensors like Inertial Measurement Unit (IMU); (3) activity recognition based on pervasive sensors like Radio Frequency IDentification (RFID).

3.1 Data Setting

The proposed OATM is evaluated over three different tasks on three benchmark datasets where each is divided into a training set (80%) and a testing set (20%).

[2] For the sake of simplicity, we consider all the factors with the same number of levels. More advanced knowledge can be found in [8] for more complex situations.

[3] https://www.york.ac.uk/depts/maths/tables/taguchi_table.htm.

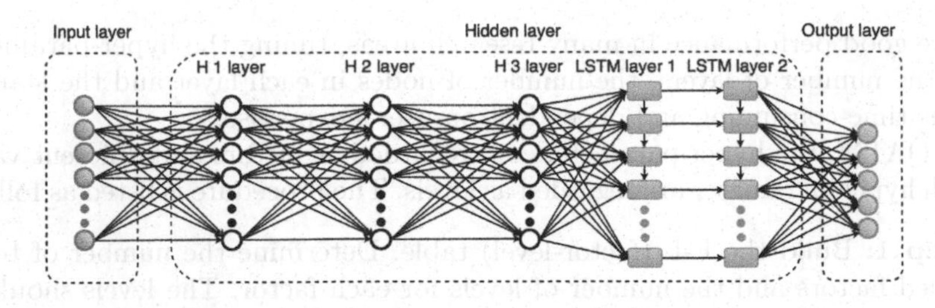

Fig. 1. The schematic diagram of RNN structure. 'H' denotes Hidden, where, for example, the *H 1 layer* denotes the first hidden layer.

EEG-Based Intention Recognition. We select the widely used EEG dataset from PhysioNet eegmmidb database[4] which contains 5 different categories. In this paper, we choose a subset of eegmmidb which contains 28,000 EEG samples. Every sample is a vector with 64 elements corresponding to 64 channels.

IMU-Based Activity Recognition. This dataset is collected by 9 participants [4], which contains 1200000 samples. 8 ADLs are selected as a subset of our paper. The activity is measured by 3 IMUs and each IMU collects sensor signal with 14 dimensions including two 3-axis accelerometers, one 3-axis gyroscopes, one 3-axis magnetometers, and one thermometer.

RFID-Based Activity Recognition. We collect the signals from passive RFID tags [9] and have 3100 samples in total. 21 activities, including 18 ADLs (Activity of Daily Living) and 3 abnormal falls, are performed by 6 subjects. Each sample has 12 dimensions corresponding to 12 RFID tags. RSSI measures the power present in a received radio signal, which is a convenient environmental measurement in ubiquitous computing.

3.2 Deep Learning Structures

In this section, we briefly describe RNN and CNN structures and then introduce the key hyper-parameters that will be tuned in the experiments.

RNN Structure. The RNN structure used in this paper is shown in Fig. 1. In the hidden layer, to implement the recurrent function, two LSTM (Long Short-Term Memory) layer is concentrated. LSTM is a simple cell structure which can be used to build a recurrent neural network. Different from other fully connected layers, LSTM layer is composed of cells (shown as rectangles) instead of neural nodes (shown as circles).

In this RNN structure, based on the deep learning hyper-parameters tuning experience, the learning rate, the regularization, and the number of nodes in each hidden layer are key factors affecting the algorithm performance. The loss is calculated by cross-entropy function, and the regularization method is ℓ_2

[4] https://www.physionet.org/pn4/EEGmmidb/.

norm with the coefficient λ, The loss is finally optimized by the AdamOptimizer algorithm. In summary, we choose four factors as to-be-tuned hyper-parameters: the learning rate lr, the regularization coefficient λ, the number of hidden layers n_l, and the number of nodes[5] in each hidden layer n_n.

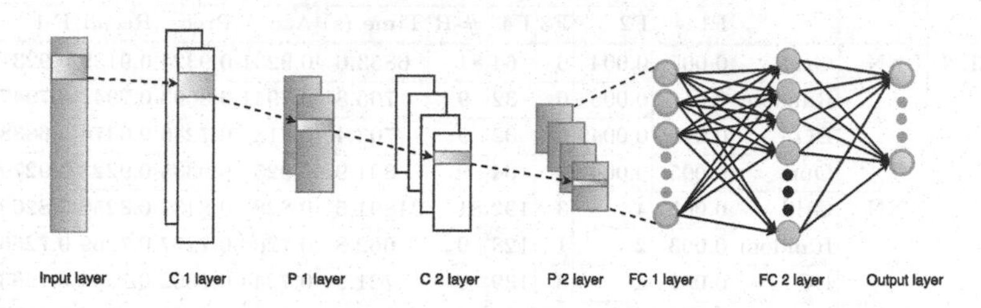

Fig. 2. The schematic diagram of CNN structure. *C*, *P*, and *FC* denote convolutional layer, pooling layer, and fully connected layer, respectively.

CNN Structure. The CNN diagram is presented in Fig. 2. The loss function, regularization method, and optimizer are the same as those in the RNN structure. Based on hyper-parameters tuning experience on CNN, we choose four most crucial factors to be tuned by OATM: the learning rate lr', the filter size f', the number of convolutional and pooling layers n_l'[6], and the number of nodes n_n' in the second fully connected layer.

4 Results and Analysis

4.1 Overall Comparison

In this section, we compare the proposed OATM with the most competitive state-of-the-art hyper-parameter tuning approaches including two manually methods (grid search and random search) and an automated one (Bayesian Optimization). It's easy to compute that there are $81 = 3^4$ exhausted combinations in grid search since we have four factors with three levels of the hyper-parameters. Thus, grid search requires 81 runnings to get the optimal hyper-parameters. On the other hand, our method requires only 9 runnings described in the corresponding orthogonal array table (detailed in Sect. 4.2). Due to the numbers of runnings in random search and Bayesian Optimization are manually set, they are set as 9 runnings which is same with our method in order to keep fair comparison. The baselines are introduced here: (1) Grid search simply goes through all the possible combinations according to the values provided which is exhaustive [2]; (2) Random search randomly picks combinations from all possible ones. It may not find a decent combination but is widely adopted in industry for the high-efficiency [1]; (3)Bayesian optimization uses a Gaussian process to minimize the loss function in order to maximize performance [7].

Table 1. Comparison with the state-of-the-art methods over three datasets and two deep learning architectures. The F1 $\sim$ F4 represent four tuning factors. Acc, Prec and F-1 denote accuracy, precision and F-1 score, respectively. #-R denotes the number of runnings.

Data	Models	Methods	Optimal factors				Metrics					
			F1	F2	F3	F4	#-R	Time (s)	Acc	Prec	Recall	F-1
EEG	RNN	Grid	0.005	0.004	6	64	81	6853.6	0.9251	0.9324	0.9139	0.9231
		Random	0.01	0.008	6	32	9	766.8	0.7941	0.8003	0.7941	0.7947
		BO	0.0135	0.0049	5	32	9	703.4	0.718	0.7246	0.6474	0.6838
		Ours	0.005	0.004	6	64	9	821.9	0.925	0.9335	0.9223	0.9279
	CNN	Grid	0.005	4	3	192	81	31891.5	0.828	0.8137	0.8256	0.8269
		Random	0.003	2	1	128	9	662.8	0.7268	0.7277	0.7269	0.7266
		BO	0.001	4	3	139	9	721.9	0.7244	0.7302	0.7244	0.7263
		Ours	0.003	4	1	128	9	680.4	0.797	0.7969	0.8112	0.8003
IMU	RNN	Grid	0.005	0.004	6	96	81	3027.2	0.9936	0.9909	0.9976	0.9971
		Random	0.015	0.004	4	32	9	1008.5	0.9139	0.9209	0.9412	0.9156
		BO	0.0132	0.0041	4	48	9	1078.8	0.9872	0.9877	0.9851	0.9863
		Ours	0.005	0.004	6	64	9	1138.2	0.9913	0.9924	0.9905	0.9919
	CNN	Grid	0.003	2	1	128	81	41804.9	0.9732	0.9708	0.9708	0.9707
		Random	0.003	2	2	128	9	7089.2	0.9692	0.9691	0.9692	0.9691
		BO	0.0012	2	2	192	9	6559.7	0.9696	0.9702	0.9701	0.9701
		Ours	0.003	2	2	128	9	6809.8	0.9702	0.9699	0.9703	0.9702
RFID	RNN	Grid	0.005	0.008	6	96	81	2846.1	0.9342	0.9388	0.9201	0.9252
		Random	0.005	0.012	4	32	9	642.3	0.8891	0.9138	0.8826	0.8895
		BO	0.0142	0.0093	6	79	9	452.2	0.9071	0.8556	0.8486	0.8436
		Ours	0.005	0.008	6	64	9	497.1	0.9134	0.9138	0.9029	0.9162
	CNN	Grid	0.005	4	2	192	81	7890.8	0.9316	0.9513	0.9316	0.9375
		Random	0.005	2	1	128	9	1210.3	0.8683	0.9113	0.8684	0.8779
		BO	0.005	5	3	64	9	872.9	0.9168	0.9058	0.9194	0.9086
		Ours	0.005	4	3	192	9	980.3	0.9235	0.9316	0.9188	0.9326

Table 2. Factor-Level table of RNN and CNN.

		Factor 1 (lr)	Factor 2 (λ)	Factor 3 (n_l)	Factor 4 (n_n)
RNN	Level 1	0.005	0.004	4	32
	Level 2	0.01	0.008	5	64
	Level 3	0.015	0.012	6	96
		Factor 1 (lr')	Factor 2 (f')	Factor 3 (n'_l)	Factor 4 (n'_n)
CNN	Level 1	0.001	[1, 2]	1	64
	Level 2	0.003	[1, 4]	2	128
	Level 3	0.005	[1, 6]	3	192

[5] Assume all the hidden layers have the same fixed number of nodes.

[6] We consider each convolutional layer and the following pooling layer as whole.

The hyper-parameter levels are selected based on empirical values. For grid search, random search, and our OATM, the empirical values are discrete as listed in Table 2 (take eegmmidb as an example). For Bayesian Optimization, the hyper-parameter ranges from the maximum and minimum of each factor.

The comparison results are shown in Table 1. It can be observed that: (1) under the same running numbers (9 runnings), our method outperforms the random search and Bayesian Optimization over all the datasets and deep learning architectures; (2) our method performs slightly lower than grid search but still competitive, however, take EEG dataset with RNN as an example, our approach saves 88% tuning time which is indicated from that the OATM only requires 9 runnings and costs 821.9 s while grid search requires 81 runnings and 6853.6 s; (3) the optimal factors selected by our method approximate to the global optimal factors selected by grid search.

4.2 Case Study in RNN and CNN

In this section, we take EEG classification as an example to present the detailed procedure of OATM in RNN and CNN architectures. The overall paradigm can be divided into five steps.

Step 1: Build the F-L Table. According to the description in Sect. 3.2, OATM will work on four different hyper-parameters (factors): the learning rate lr, the l-2 norm coefficient λ, the number of hidden layers n_l, and the number of nodes n_n. The number of levels h is set to be 3 which could be much larger in real-world applications. Based on the related work and tuning experience [10], the empirical values are shown in Table 2.

Table 3. Range analysis of RNN and CNN

(a) Range analysis of RNN

Row No.	Factor 1 (lr)	Factor 2 (λ)	Factor 3 (n_l)	Factor 4 (n_n)	Acc
1	0.005	0.004	4	32	0.875
2	0.005	0.008	5	64	0.8
3	0.005	0.012	6	96	0.521
4	0.01	0.004	5	96	0.888
5	0.01	0.008	6	32	0.797
6	0.01	0.012	4	64	0.451
7	0.015	0.004	6	64	**0.897**
8	0.015	0.008	4	96	0.335
9	0.015	0.012	5	32	0.471
R_{level1}	2.196	2.66	1.661	2.143	
R_{level2}	2.136	1.932	2.159	2.148	
R_{level3}	1.703	1.443	2.215	1.744	
A_{level1}	**0.732**	**0.887**	0.554	0.714	
A_{level2}	0.712	0.644	0.720	**0.716**	
A_{level3}	0.568	0.481	**0.738**	0.581	
Lowest Acc	0.568	0.481	0.554	0.581	
Highest Acc	0.732	0.887	0.738	0.716	
Range	0.164	0.406	0.184	0.135	
Importance	$lambda > n_l > lr > n_n$				
Best Level	**Level 1**	**Level 1**	**Level 3**	**Level 2**	
Optimal Value	**0.005**	**0.004**	**6**	**64**	**0.925**

(b) Range analysis of CNN

Row No.	Factor 1 (lr')	Factor 2 (f')	Factor 3 (n_l')	Factor 4 (n_n')	Acc
1	0.001	[1,2]	1	64	0.707
2	0.001	[1,4]	2	128	0.771
3	0.001	[1,6]	3	192	0.775
4	0.003	[1,2]	2	192	0.779
5	0.003	[1,4]	3	64	0.752
6	0.003	[1,6]	1	128	**0.797**
7	0.005	[1,2]	3	128	0.784
8	0.005	[1,4]	1	192	0.782
9	0.005	[1,6]	2	64	0.756
R_{level1}	2.253	2.27	2.993	2.215	
R_{level2}	2.328	2.305	2.306	2:352	
R_{level3}	2.322	2.328	2.311	2.336	
A_{level1}	0.751	0.757	**0.998**	0.738	
A_{level2}	**0.776**	0.768	0.769	**0.784**	
A_{level3}	0.774	**0.776**	0.770	0.779	
Lowest Acc	0.751	0.757	0.769	0.738	
Highest Acc	0.776	0.776	0.998	0.784	
Range	0.025	0.019	0.229	0.046	
Importance	$n_l' > n_n' > lr' > f'$				
Best Level	**Level 2**	**Level 3**	**Level 1**	**Level 2**	
Optimal Value	**0.003**	**[1,6]**	**1**	**128**	**0.797**

Step 2: OATM Table. Choose a suitable Orthogonal Array table with 4 factors and 3 levels for our experiments in this link[7] wich contains 9 combinations. The OATM table should satisfy two basic principles: (1) in each column, different levels have the same appear times; (2) in any two randomly-selected columns, nine differently-ordered element combinations are completed and balanced.

Step 3: Run the Experiments. Following the OATM table, run the 9 experiments and record the classification accuracy. In our case, each experiment runs 5 times with the corresponding average accuracy recorded. Each experiment is trained for 1,000 iterations to guarantee the convergence.

Step 4: Range Analysis. This is the key step of Orthogonal Array Tuning. The overall range analysis procedure and results are shown in Table 3a. The first 9 rows are measured and recorded in Step 3. R_{leveli} denotes the sum of accuracy under level i. For example, R_{level1} in factor 1 is the sum of the accuracy in the first 3 rows ($2.196 = 0.875 + 0.8 + 0.521$), where factor 1 is on level 1. A_{leveli} denotes the average accuracy of level i, calculated by $A_{leveli} = R_{leveli}/h$. In the above example, we calculate A_{level1} as $0.732 = 2.196/3$. Lowest and highest accuracy values, measuring the maximum and minimum of A_{leveli} respectively, are used to calculate the *range* of A_{leveli}. The importance denotes how important the factor is, which is ranked by the range value. *Best level* is the selected optimal level based on the *Highest Acc* while *Optimal Value* represents the corresponding value of the best level.

Step 5: Run the Optimal Setting. Run the experiment with the optimal hyper-parameters ($lr = 0.004$, $\lambda = 0.005$, $n_l = 6$, and $n_n = 64$) and finally achieve the optimal accuracy as 0.925. It can be observed that: (1) the optimal accuracy 0.925 is higher than the maximum of the accuracy (0.897) in the OATM experiments, which demonstrates that the OATM is enabled to approximate the global optimal instead of the local optimal; (2) the importance of each factor is ranked through the range analysis: $lambda > n_l > lr > n_n$, which can guide the researcher to grasp the dominating variable in the RNN structure and be helpful in the future algorithm development.

The OATM paradigm of CNN is similar to RNN. Here, we only report the F-L table (Table 2) and the range analysis table (Table 3b).

5 Discussion and Conclusion

One disadvantage of OATM is that it requires the empirical values as prerequisites. The values of the F-L table should be chosen appropriately. However, this is the common drawback of all the tuning methods. For instance, the hyper-parameter ranges in Bayesian Optimization are also pre-defined based

[7] https://www.york.ac.uk/depts/maths/tables/taguchi_table.htm.

on empirical values. In summary, we present an efficient OATM approach for hyper-parameter tuning in the context of deep learning. The proposed OATM is evaluated over two popular deep learning structures (RNN and CNN) over three real-world datasets. The experiment results show that our approach outperforms state-of-the-art hyper-parameter tuning methods.

References

1. Andradóttir, S.: A review of random search methods. In: Fu, M. (ed.) Handbook of Simulation Optimization, pp. 277–292. Springer, New York (2015). https://doi.org/10.1007/978-1-4939-1384-8_10
2. Bergstra, J.S., Bardenet, R., Bengio, Y., Kégl, B.: Algorithms for hyper-parameter optimization. In: NeurIPS 24, pp. 2546–2554 (2011)
3. Calandra, R., Gopalan, N., Seyfarth, A., Peters, J., Deisenroth, M.P.: Bayesian gait optimization for bipedal locomotion. In: Pardalos, P.M., Resende, M.G.C., Vogiatzis, C., Walteros, J.L. (eds.) LION 2014. LNCS, vol. 8426, pp. 274–290. Springer, Cham (2014). https://doi.org/10.1007/978-3-319-09584-4_25
4. Fida, B., Bibbo, D., Bernabucci, I., et al.: Real time event-based segmentation to classify locomotion activities through a single inertial sensor. In: MobiHealth, pp. 104–107 (2015)
5. Mahapatra, S., Patnaik, A.: Optimization of wire electrical discharge machining (wedm) process parameters using taguchi method. IJAMT **34**(9), 911–925 (2007)
6. Nalbant, M., Gökkaya, H., Sur, G.: Application of taguchi method in the optimization of cutting parameters for surface roughness in turning. Mater. Des. **28**(4), 1379–1385 (2007)
7. Snoek, J., Larochelle, H., Adams, R.P.: Practical Bayesian optimization of machine learning algorithms. In: NeurIPS 25, pp. 2951–2959. Curran Associates, Inc. (2012)
8. Taguchi, G., Taguchi, G.: System of experimental design; engineering methods to optimize quality and minimize costs. Technical report (1987)
9. Yao, L., et al.: Compressive representation for device-free activity recognition with passive RFID signal strength. IEEE Trans. Mob. Comput. **17**(2), 293–306 (2017)
10. Zhang, X., Yao, L., Huang, C., Sheng, Q.Z., Wang, X.: Intent recognition in smart living through deep recurrent neural networks. In: Liu, D., Xie, S., Li, Y., Zhao, D., El-Alfy, E.S. (eds.) ICONIP 2017. LNCS, vol. 10635, pp. 748–758. Springer, Cham (2017). https://doi.org/10.1007/978-3-319-70096-0_76
11. Zhang, X., Yao, L., Sheng, Q.Z., Kanhere, S.S., Gu, T., Zhang, D.: Converting your thoughts to texts: enabling brain typing via deep feature learning of EEG signals. In: PerCom 2018. IEEE (2018)
12. Zhang, X., Yao, L., Wang, X., Monaghan, J., Mcalpine, D., Zhang, Y.: A survey on deep learning based brain computer interface: Recent advances and new frontiers. arXiv preprint arXiv:1905.04149 (2019)

Improving the Identification of Code Smells by Combining Structural and Semantic Information

Mouna Hadj-Kacem(✉) and Nadia Bouassida

Mir@cl Laboratory, Sfax University, Sfax, Tunisia
mouna.hadjkacem@gmail.com, nadia.bouassida@isimsf.rnu.tn

Abstract. In software engineering, a code smell is an indication of a deeper problem in the source code, hindering the maintainability and evolvability of the system. In the literature, there is a significant emphasis on the detection of code smells because of its importance as a maintenance task. Most of previous studies focus in their analyses on one source of information, i.e. structural, historical or semantic information. However, some instances of bad smells could be identified by a type of information but missed by another one. In this paper, we propose an improved detection approach that combines structural and semantic information in order to fully exploit their complementarity in the identification of code smells. Both information are extracted separately using conventional and deep learning methods. For the evaluation, we have selected five open source projects which are JHotDraw, Apache Karaf, Freemind, Apache Nutch and JEdit. In order to optimize our performance results, we have set up four different experiments and compare between them. The obtained accuracy results confirm the effectiveness of combining structural and semantic information in improving the detection of code smells.

Keywords: Deep learning · Code smells · Variational auto-encoder · Semantic information · Structural information

1 Introduction

Over its evolution, the more the software system is affected by continuous changes, the more complex it becomes. Among the implications of this phenomenon, there exist potential problems that may hamper the software maintainability and evolvability. Code smells are examples of indicators that are associated with poor design and/or implementation problems [2]. They are generally induced because of an unwitting misunderstanding of developers in adapting with the continuous changes. Therefore, this implies the non-compliance with the software design principles, which in turn affects the software quality.

The impact of code smells on the software maintenance has been widely studied in many research studies [13,19,21]. They have demonstrated that there

T. Gedeon et al. (Eds.): ICONIP 2019, CCIS 1142, pp. 296–304, 2019.
https://doi.org/10.1007/978-3-030-36808-1_32

is a high diffusion of code smells over the software systems [13]. Also, they have indicated that the software quality is deteriorated which hinders its evolution. The refactoring is the appropriate technique devoted to deal with this type of problem. It is designed to reconstruct the internal software structure without affecting its external behaviour [2].

In the literature, there exist several researches to identify code smells [1,4, 6,10,11]. Most of the previous works exploit in their detection one source of information that could be structural, historical or semantic. Among these latter types, the structural information is more used than the others [5,18]. However, it has been proven that the historical and semantic information have the ability to identify additional instances of code smells [12]. This efficiency is due to the fact that each source of information has its own characteristics to identify specific aspects of code smells.

Theoretically, the complementarity has been studied between many types of information. In [11], the authors have shown that there are some complementarities between structural and textual information in order to obtain better results. Using the overlap metrics, the complementarity has been proven. In another work [12], there has been observed that the structural and historical information could be complementary. Based on these research observations, an opportunity of combination was apparent to improve the detection of code smells.

Given that there exist instances of code smells that are identified by a type of information but missed by another one and vice versa, we attempt in this paper to take advantage of two sources of information that are semantic and structural information by means of conventional and deep learning methods. In our previous study [6], a variational auto-encoder has been implemented to generate a deep representation that embeds the needed semantic information hidden into the Abstract Syntax Trees. In this paper, we propose an improved approach to identify code smells by combining different sources of information. Overall, we conduct 20 different experiments (four experiments over each of the five studied software projects) in order to determine the optimal way to combine information. Based on the reported findings, we have observed that the performance depends on some factors that we will discuss across the experiments.

The rest of the paper is organized as follows. Section 2 presents the background on code smells and reviews related work. Section 3 outlines our proposed research methodology. Section 4 shows the experimental set-up and evaluates the performance of our approach. Finally, Sect. 5 concludes the paper and outlines possible future research directions.

2 Code Smells and Existing Detection Approaches

In this section, we first introduce the target code smells. Then, we review the previous works related to the detection approaches based on more than one source of information.

2.1 Code Smells

Originally, the term code smell was coined by Fowler et al. [2] as an indication of poor design and/or implementation choices. The authors have provided a catalogue of 22 code smells with their appropriate refactoring operations. The purpose of refactoring is to reduce complexity and maintain the code simple and easier to evolve. In our study, we focus on three types of code smells: Blob, Feature Envy and Long Method.

- **Blob** is detected in the project where a class is monopolizing the most of the system's functionalities. In this class, there is a large number of attributes and methods that are depending on other classes [2].
- **Feature Envy** indicates a method that is more interested in exploiting the functionalities of a class other than its own. Thus, it accesses the data from other classes and causes a high coupling with them. Move Field and Move Method are the associated refactoring operations with this smell [2].
- **Long Method** is identified by its domination in implementing the main functionality of the class. This method includes a large number of data and causes complexity. Extract Method is the appropriate refactoring operation for this smell [2].

The selection of these three code smells is justified by their frequent detection in other studies [1,4,9–11]. Also, they belong to different levels of granularity that are class and method levels. Lastly, they are the target smells in the web-based platform Landfill [14] with which we intend to perform our experiments.

2.2 Existing Detection Approaches

Detecting code smells has attracted significant attention from both industry and academia as an important maintenance task. Most of previous approaches treated the detection problem based on one source of information [5], i.e. structural [1,4], historical [10] and semantic information [6,11]. Nonetheless, some instances of bad smells could be identified by a type of information but missed by another one. As stated by Palomba et al. [11,12,15], there exists a complementarity between different sources of information that could achieve better performance results. In the following, we discuss the existing detection approaches with respect to the nature of the source of information. We will emphasize our review mainly on the studies that are based on more than one source of information (multi-source) in their identification of code smells.

Very few researches have been conducted regarding the code smells identification based on different types of information. Fu and Shen [3] have used the evolutionary history of projects in order to extract the historical information by using the association rules. In their approach, they have combined the historical and structural information to find three types of bad smells which are Duplicated Code, Shotgun Surgery, and Divergent Change.

Liu et al. [9] have exploited both textual and structural information to detect Feature Envy. From the identifier names, the textual information has been extracted using Convolutional Neural Network (CNN). While for the structural information, the authors have used the distance metrics defined in [20] to identify the Feature Envy smell and its Move Method opportunities.

Our work differs from the aforementioned approaches in that we will extract the needed features from other sources of information. The semantic information is generated from the Abstract Syntax Tree, whereas the structural information is based on the object-oriented metrics. In addition, we will compare between different experiments in order to determine the best way of combination.

3 Methodology Description

As shown in Fig. 1, our approach consists of two separate phases. The first one extracts the semantic information using a variational auto-encoder. The second one extracts the structural information. Once the needed information are extracted, they will be combined by means of a classification algorithm to determine the nature of the testing instance.

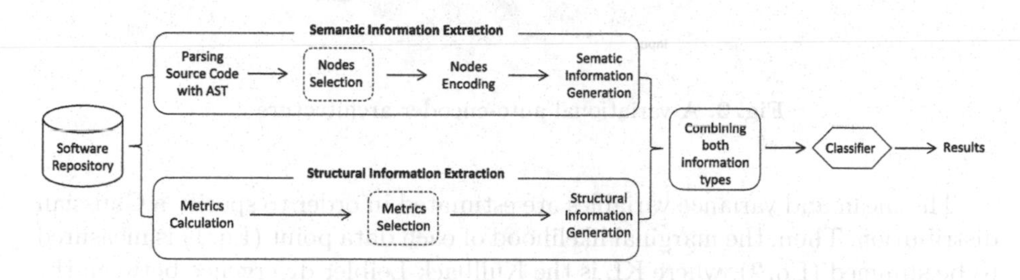

Fig. 1. The process of our improved approach to identify code smells

3.1 Semantic Information Extraction

In this phase, the source code is first parsed into the needed granularity of representation. According to Peng et al. [16], there are different granularities that are character-level, token-level, nodes of Abstract Syntax Trees, statement-level and higher. In our case, we need the nodes of Abstract Syntax Trees as the appropriate representation that preserves the hidden semantic in the source code [16,22].

The step of nodes selection is optional (framed by a dashed line in Fig. 1). This step will be applied or ignored according to the studied experiment (see Sect. 4), which means all nodes or a subset of them will be encoded. It is important to mention that due to space limitations, we do not list the details about the selected nodes.

As the variational auto-encoder accepts real-valued vectors as input, we will apply a conversion of the nodes vector to integer vectors. Each node will be identified by a unique integer and the length of vectors will be unified by padding with zero [22]. Afterwards, the vectors will be fed into the variational auto-encoder in order to generate the embedded semantic information.

The variational auto-encoder [7] has been selected because it is a generative algorithm that approximates the latent representation using a Bayesian inference approach. As shown in Fig. 2, the architecture of the variational auto-encoder is composed of two neural networks that are encoder and decoder. For a given input (x), the encoder network $q_{\emptyset}(z|x)$ produces a deep representation (z). Then, the decoder network $p_{\theta}(x|z)$ reconstructs the input by means of a reverse mapping to get the output $(\hat{x})$. The joint distribution is defined as $p_{\theta}(x, z) = p_{\theta}(x \mid z)p_{\theta}(z)$.

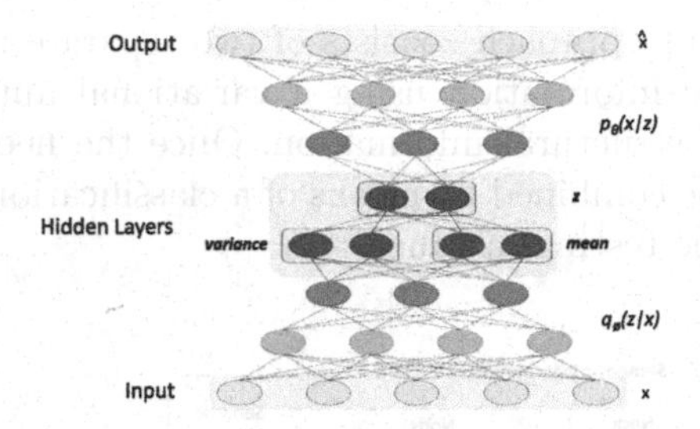

Fig. 2. A variational auto-encoder architecture

The mean and variance variables are estimated in order to specify a Gaussian distribution. Then, the marginal likelihood of each data point (Eq. 1) is measured to be summed (Eq. 2), where KL is the Kullback-Leibler divergence between the posterior and prior distributions.

$$log\, p_{\theta}(x^{(i)}) = D_{KL}(q_{\emptyset}(z|x)||p_{\theta}(z)) + \mathcal{L}(\theta, \emptyset; x^{(i)}) \tag{1}$$

$$log\, p_{\theta}(x^{(1)}, ..., x^{(N)}) = \sum_{i=1}^{N} log\, p_{\theta}(x^{(i)}) \tag{2}$$

3.2 Structural Information Extraction

A metric is a quantifiable characteristic that is measured to describe a particular aspect of a software. In this phase, the step of metrics calculation is accomplished by Metrics tool [17]. Metrics is a plug-in that calculates different metrics for Java projects. The calculated metrics belong to different quality dimensions, i.e. size (e.g. LOC), complexity (e.g. WMC) and inheritance (e.g. DIT).

As shown in Fig. 1, the step of metrics selection (or feature selection) is optional (framed by a dashed line). The purpose of the feature selection is to identify the relevant features from a large set of features [8]. The selected features are more correlated to the class distribution and the rest may not have an impact on the performance. To deal with, we use a feature selection technique that belongs to the feature-weighting algorithms. The Gain Ratio technique has been selected to measure the importance of metrics.

3.3 Code Smells Identification

At the end of the two phases, the semantic and structural information are combined. Afterwards, they will be fed into a classifier that is the Logistic Regression in order to determine the nature of the testing instance.

4 Experimental Evaluation

In this section, we evaluate the performance of our approach. To deal with, our evaluation aims at answering these two research questions (***RQs***):

- ***RQ1***: What is the optimal way to combine between semantic and structural information in order to enhance the performance results?
- ***RQ2***: To what extent does the nodes and metrics selection improve the code smells identification?

4.1 Dataset

To evaluate our combination method, we use a public dataset based on Java software projects. The dataset is a web-based platform named Landfill [14]. In this dataset, test smells and code smells are mined over 20 open source systems. In our experiments, the code smells are our target.

From the Landfill dataset, we have selected five projects that are JHotDraw, Apache Karaf, Freemind, Apache Nutch and JEdit. These projects belong to different domains and sizes. They are available on Github[1] and SourceForge[2]. For each project, we will apply four different experiments in order to answer the addressed research questions.

4.2 Experimental Results

In order to respond to the first research question ***RQ1***, we set up four different experiments over each software project. The experiments are defined as follows:

- **Exp1**: All Nodes with All Metrics
- **Exp2**: All Nodes with Selected Metrics
- **Exp3**: Selected Nodes with All Metrics
- **Exp4**: Selected Nodes with Selected Metrics

[1] https://github.com.
[2] https://sourceforge.net.

Table 1. Performance comparison of different experiments

Projects	Smells	Exp1			Exp2			Exp3			Exp4		
		P	R	F1	P	R	F1	P	R	F1	P	R	F1
JHotDraw	Blob	74,11%	76,11%	75,10%	75,45%	77.85%	76.63%	75.01%	80.04%	77.44%	78.99%	82.74%	80.82%
	FE	75.78%	77.82%	76.79%	75.91%	76.81%	76.36%	73.85%	78.93%	76.31%	76.23%	79.98%	78.06%
	LM	74.33%	77.01%	75.65%	75.79%	77.52%	76.65%	75.19%	78.97%	77.03%	79.23%	81.47%	80.33%
Apache Karaf	Blob	71.20%	76.53%	73.77%	70.78%	74.98%	72.82%	74.59%	78.23%	76.37%	76.75%	79.89%	78.29%
	FE	73.12%	78.18%	75.57%	73.88%	78.10%	75.93%	77.14%	80.07%	78.58%	79.12%	84.01%	81.49%
	LM	74.51%	78.97%	76.68%	75.97%	79.12%	77.51%	75.79%	81.47%	78.53%	80.09%	86.12%	83.00%
Freemind	Blob	70.56%	75.14%	72.78%	68.41%	70.38%	69.38%	75.41%	76.99%	76.19%	75.79%	78.93%	77.33%
	FE	76.66%	77.99%	77.32%	71.03%	73.95%	72.46%	77.52%	79.39%	78.44%	79.21%	82.18%	80.67%
	LM	76.15%	81.02%	78.51%	71.21%	73.92%	72.54%	79.23%	81.95%	80.57%	81.23%	87.47%	84.23%
Apache Nutch	Blob	74.37%	77.23%	75.77%	75.18%	77.58%	76.36%	78.21%	80.10%	79.14%	77.89%	79.21%	78.54%
	FE	75.79%	78.95%	77.34%	70.23%	72.13%	71.17%	76.10%	78.91%	77.48%	77.21%	79.92%	78.54%
	LM	76.71%	81.47%	79.02%	71.97%	73.85%	72.90%	78.26%	79.58%	78.91%	80.17%	85.48%	82.74%
JEdit	Blob	69.51%	75.62%	72.44%	70.01%	72.90%	71.43%	74.98%	76.92%	75.94%	76.92%	78.96%	77.93%
	FE	71.52%	75.69%	73.55%	74.00%	76.82%	75.38%	77.93%	79.22%	78.57%	78.26%	81.74%	79.96%
	LM	70.23%	74.91%	72.49%	71.76%	75.19%	73.43%	79.23%	80.02%	79.62%	81.35%	87.65%	84.38%

To evaluate the performance of the approach, we use True Positive (TP), True Negative (TN), False Positive (FP) and False Negative (FN). These four measurements are used to compute $Precision = TP/(TP + FP)$, $Recall = TP/(TP + FN)$, and $F-measure(F1) = 2*(Precision * Recall)/(Precision+Recall)$.

Table 1 reports the results of the four experiments. In Exp1 and Exp2, the results of F-measure were slightly convergent. The common factor between both first experiments was the use of all nodes. Although we have applied the metrics selection in the Exp2, there is not a remarkable change in the computed results. While in Exp3 and Exp4, we reached more better results than the first ones. In Exp3, the F-measure ranges between 76.19% and 80.57%. However, we found that Exp4 achieves better results. The recall ranges between 78.96% and 87.65% while the precision ranges between 75.79% and 81.35%. We have reached an average of F-measure equal to 80.42%.

Overall, to answer the ***RQ1***, it can be concluded that the selection of both nodes and metrics is a key factor for enhancing the results. Consequently, the optimal way to combine the two sources of information is insured by selecting the nodes prior to the encoding step and also by selecting the most relevant metrics.

To answer the second research question ***RQ2***, Fig. 3 shows, for each code smell, a comparison between the four experiments based on F-measure. The F-measure has been chosen as the comparative measurement because it is a harmonic mean between precision and recall values.

We found that the worst results were obtained from Exp1, where there is not a selection step neither for the nodes nor for the metrics. It is clearly observed from the results of Exp1 and Exp2 vs. Exp3 and Exp4 that the selection of nodes has more influence than the selection of metrics. The results make a remarkable enhancement starting from Exp3 and Exp4, where the common point between

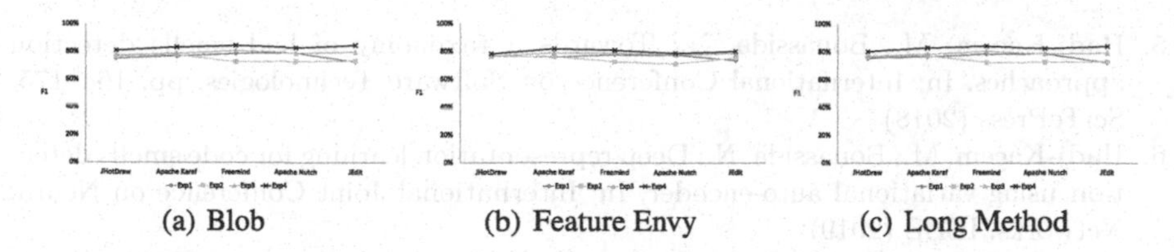

(a) Blob (b) Feature Envy (c) Long Method

Fig. 3. The comparison between the four experiments using F-measure for (a) Blob, (b) Feature Envy and (c) Long Method

them is the selected nodes. This is may be due to the large number of nodes compared to the number of metrics. As a consequence, this implies that the selection of nodes is more important than the metrics selection. Nonetheless, we cannot claim that the metrics selection serves no purpose, because it has shown a slight improvement. To this end, both selecting nodes and metrics are important in optimizing the code smells identification.

5 Conclusion

In this paper, we have proposed an improved approach to identify three target code smells. We have combined semantic and structural information to take full advantage of both these sources.

To evaluate our approach's performance, we have experimented different scenarios of combination. We have selected five open source projects from a public dataset. On each software, we have conducted four different experiments in order to define the best way of combination. Over the different experiments, we have reached significant performance results. We have observed that the application of nodes selection and metrics selection have improved the results. These latter steps have proved their efficiency by outperforming the results of the compared experiments.

In future work, we would like to expand our approach to identify other types of code smells that are scattered at different levels of granularity. Furthermore, we intend to explore other opportunities to combine other types of information.

References

1. Fontana, A.F., Mäntylä, M.V., Zanoni, M., Marino, A.: Comparing and experimenting machine learning techniques for code smell detection. Empir. Softw. Eng. **21**, 1143–1191 (2016)
2. Fowler, M., Beck, K., Brant, J., Opdyke, W., Roberts, D.: Refactoring: Improving the Design of Existing Code. Pearson Education India (1999)
3. Fu, S., Shen, B.: Code bad smell detection through evolutionary data mining. In: Symposium on Empirical Engineering and Measurement, pp. 1–9 (2015)
4. Hadj-Kacem, M., Bouassida, N.: A hybrid approach to detect code smells using deep learning. In: International Conference on Evaluation of Novel Approaches to Software Engineering, pp. 137–146 (2018)

5. Hadj-Kacem, M., Bouassida, N.: Towards a taxonomy of bad smells detection approaches. In: International Conference on Software Technologies, pp. 164–175. SciTePress (2018)
6. Hadj-Kacem, M., Bouassida, N.: Deep representation learning for code smells detection using variational auto-encoder. In: International Joint Conference on Neural Networks. IEEE (2019)
7. Kingma, D.P., Welling, M.: Auto-encoding variational bayes. arXiv (2013)
8. Kwak, N., Choi, C.-H.: Input feature selection for classification problems. IEEE Trans. Neural Netw. **13**, 143–159 (2002)
9. Liu, H., Xu, Z., Zou, Y.: Deep learning based feature envy detection. In: International Conference on Automated Software Engineering, pp. 385–396. ACM (2018)
10. Palomba, F., Bavota, G., Penta, M.D., Oliveto, R., Poshyvanyk, D., Lucia, A.D.: Mining version histories for detecting code smells. IEEE Trans. Softw. Eng. **41**, 462–489 (2015)
11. Palomba, F., Panichella, A., Lucia, A.D., Oliveto, R., Zaidman, A.: A textual-based technique for smell detection. In: International Conference on Program Comprehension, pp. 1–10. IEEE (2016)
12. Palomba, F.: Alternative sources of information for code smell detection: postcards from far away. In: International Conference on Software Maintenance and Evolution, pp. 636–640 (2016)
13. Palomba, F., Bavota, G., et al.: On the diffuseness and the impact on maintainability of code smells: a large scale empirical investigation. Empir. Softw. Eng. **23**, 1188–1221 (2018)
14. Palomba, F., Nucci, D.D., et al.: Landfill: an open dataset of code smells with public evaluation. In: Working Conference on Mining Software Repositories, pp. 482–485. IEEE (2015)
15. Palomba, F., Panichella, A., et al.: The scent of a smell: an extensive comparison between textual and structural smells. IEEE Trans. Softw. Eng. **44**, 977–1000 (2018)
16. Peng, H., Mou, L., Li, G., Liu, Y., Zhang, L., Jin, Z.: Building program vector representations for deep learning. In: Zhang, S., Wirsing, M., Zhang, Z. (eds.) KSEM 2015. LNCS (LNAI), vol. 9403, pp. 547–553. Springer, Cham (2015). https://doi.org/10.1007/978-3-319-25159-2_49
17. Sauer, F., Boissier, G.: Eclipse metrics plugin 1.3.8. http://metrics2.sourceforge.net/
18. Sharma, T., Spinellis, D.: A survey on software smells. J. Syst. Softw. **138**, 158–173 (2018)
19. Soh, Z., et al.: Do code smells impact the effort of different maintenance programming activities? In: International Conference on Software Analysis, Evolution, and Reengineering, pp. 393–402 (2016)
20. Tsantalis, N., Chatzigeorgiou, A.: Identification of move method refactoring opportunities. IEEE Trans. Softw. Eng. **35**, 347–367 (2009)
21. Tufano, M., Palomba, F., et al.: When and why your code starts to smell bad. In: International Conference on Software Engineering, pp. 403–414. IEEE (2015)
22. Wang, S., Liu, T., Tan, L.: Automatically learning semantic features for defect prediction. In: International Conference on Software Engineering, pp. 297–308 (2016)

Learnable Gabor Convolutional Networks

Guoqiang Zhong[1(✉)], Wei Gao[1], Wencong Jiao[1], Biao Shen[2], and Dongdong Xia[3]

[1] Department of Computer Science and Technology, Ocean University of China, 238 Songling Road, Qingdao 266100, China
gqzhong@ouc.edu.cn

[2] MOE Key Laboratory of Physical Oceanography, Ocean University of China, 238 Songling Road, Qingdao 266100, China

[3] State Oceanic Administration's State Key Laboratory of Marine Disaster Prediction Technology, National Marine Environmental Prediction Center, 8 Dahuisi Road, Beijing 100081, China

Abstract. Commonly used convolutional operation does not have the ability to learn invariant information of images. However, some hand-crafted image feature extractors, like Gabor wavelets, are robust to object's scale and orientation transformations. Hence, how to combine Gabor filters with convolutional kernels for image feature extraction is an interesting and urgent issue in recent research of image representation learning using deep convolutional neural networks (DCNNs). In this paper, we propose a new method, named learnable Gabor convolutional networks (LGCNs), to combine the Gabor filters and convolutional kernels together to form the Gabor convolutional filters (GCFs) for invariant information learning of images. The scale and orientation parameters in the Gabor function can be learned simultaneously with other parameters during the networks' training. Experimental results show that LGCNs perform better than the corresponding DCNNs and other related methods in image classification tasks.

Keywords: Gabor wavelets · Deep convolutional neural networks · Invariant information · Gabor convolutional filters · Image classification

1 Introduction

In recent years, deep convolutional neural networks (DCNNs) have attracted much attention in the areas related to deep learning. DCNNs learn the new representations of images with the convolutional kernels. However, the convolutional kernels cannot extract invariant information from images. Alternatively, Gabor wavelets are widely used for image processing tasks, as they can extract both scale and orientation invariant information from images.

In [7] and [16], the convolutional kernels in the shallow layers of DCNNs were demonstrated to perform similarly as the Gabor filters. Subsequently, Sarwar et al. applied the Gabor filters in several layers of DCNNs for fast learning [13]. Wang et al. used the Gabor transformations to replace the convolutional

T. Gedeon et al. (Eds.): ICONIP 2019, CCIS 1142, pp. 305–313, 2019.
https://doi.org/10.1007/978-3-030-36808-1_33

operation for images feature extraction [15]. Specifically, Luan et al. proposed the Gabor convolutional networks (GCNs), which integrated the Gabor filters into DCNNs, to enhance the resistance of the learned features to the orientation and scale changes [11]. However, GCNs simply use the same number of Gabor fitlers in each layer and can not update the parameters in the Gabor function.

In this paper, we propose a new model called learnable Gabor convolutional networks (LGCNs), which use Gabor filters to adaptively adjust convolutional kernels to form the Gabor convolutional filters (GCFs). Concretely, the adjusting method is element-wise multiplication of the convolutional kernels by the learnable Gabor filters. In general, the filters in the shallow layers of DCNNs can extract the low level information like edge and corner in an image. Furthermore, the kernels in the middle and deep layers of DCNNs can extract relatively higher level information, such as parts and semantic features. Correspondingly, we divide the DCNNs model into three stages corresponding to the levels of feature learning. In different stages, the convolutional kernels are manipulated by different numbers of Gabor filters: four Gabor filters in the low level stage, two and one in the middle and high level stages, respectively. Additionally, we derive the formulas to update the scale and orientation parameters in the Gabor functions.

The contributions of this work can be summarized as follows:

1. We propose the learnable Gabor convolutional networks (LGCNs), which use Gabor filters to adaptively adjust the convolutional filters in DCNNs.
2. The scale and orientation parameters of the Gabor functions can be learned together with others in LCGNs by the back-propagation (BP) algorithm during the model training.

2 Related Work

2.1 Gabor Filters

Gabor functions were proposed by Dennis Gabor in 1946 [3]. The 2D Gabor function is a product of an elliptical Gaussian and a complex plane wave. The Gabor function can be defined as follows [4]:

$$g(x, y; \lambda, \theta, \psi, \sigma, \gamma) = \exp(-\frac{x'^2 + \gamma^2 y'^2}{2\sigma^2}) \exp(i(2\pi \frac{x'}{\lambda} + \psi)), \tag{1}$$

where

$$x' = x \cos\theta + y \sin\theta, y' = -x \sin\theta + y \cos\theta. \tag{2}$$

Here, γ is spatial aspect ratio, which determines the ellipticity of the receptive field, λ is the wavelength and $1/\lambda$ is the spatial frequency of the cosine factor. In addition, θ is the orientation parameter, ϕ is the phase offset, and σ is the standard deviation of the Gaussian factor, while x and y indicate the position of the pixel on the x-axis and y-axis.

2.2 DCNNs

DCNNs have made a major breakthrough in the area of image classification [7,10]. Evidences show that the deeper the DCNNs, the better the classification results can be obtained [14]. However, traditional DCNNs cannot handle the problem of large spatial transformations in images.

To solve the above problem in DCNNs, several methods have been proposed. Among others, max-pooling is a down-sampling method [1]. It enables DCNNs to handle small spatial transformations in images. Transformation-invariant pooling (TI-pooling) utilizes parallel architectures to output the transformation invariant features before the fully-connected layers [8]. In addition, spatial transformer networks (STN) use an additional module to handle the spatial transformation problem [6]. Oriented response networks (ORNs) rotate the convolutional kernels to encode the rotation-invariant information into DCNNs [17]. However, ORNs are more suitable for small size convolutional kernels, i.e. 3×3.

2.3 Combination of Gabor Wavelets and DCNNs

Since the convolutional kernels perform similarly with Gabor filters in the shallow layers of DCNNs, [15] replaced the convolutional kernels with the Gabor filters for image feature learning. Sarwar et al. used the Gabor filters in several layers of DCNNs, which significantly reduced the storage requirements and training time with minimal degradation of the classification accuracy [13]. However, these methods only simply apply the Gabor filters as substitutes of the convolutional kernels in DCNNs.

Luan et al. proposed a convolutional kernel manipulation method named Gabor convolutional networks (GCNs) [11]. GCNs combine the Gabor filters and convolutional kernels to form the Gabor orientation filters (GoFs) in DCNNs. However, GCNs cannot learn the parameters of the Gabor functions in the training processes, which limits the performance of GCNs in image classification tasks.

3 Learnable Gabor Convolutional Networks (LGCNs)

In this section, we introduce a new deep architecture named learnable Gabor convolutional networks (LGCNs), which use the Gabor filters to adjust the convolutional kernels in DCNNs and form the Gabor convolutional filters (GCFs).

3.1 GCFs

To obtain the GCFs, we manipulate each convolutional kernel with the Gabor filters. Concretely, a GCF can be computed as:

$$M_k = C * G_k(\lambda, \theta), \tag{3}$$

where M_k denotes a GCF, C denotes a convolutional kernel, $G_k(\lambda, \theta)$ denotes the k_{th} Gabor filters, while λ and θ are the scale and orientation parameters in the Gabor function. In addition, $*$ stands for the element-wise multiplication, where C and G have the same size.

In LGCNs, we divide the convolutional layers into three stages and manipulate the convolutional kernels in each stage respectively. Suppose the convolutional layers can be equally divided into three stages. We use four Gabor filters in the first stage, two Gabor filters in the second stage and one Gabor filter in the third stage. Figure 1 illustrates the computation of the GCFs with the convolutional kernels and Gabor filters in the three stages. Additionally, when the convolutional layers cannot be divided into three stages equally, we have two solutions with respect to the two cases that the remainder of the layers is 2 or the remainder of the layers is 1. If the remainder is 2, we use the first number, which is a multiple of three and larger than the number of convolutional layers, to determine the split points. If the remainder is 1, we choose to neglect the first convolutional layer and only divide the remainders.

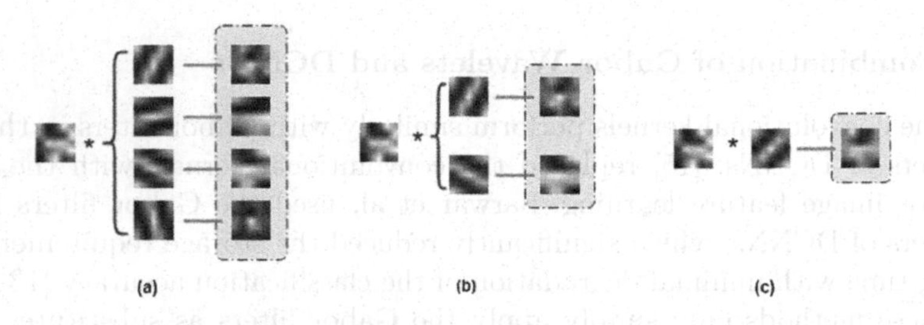

Fig. 1. Illustration of the computation of GCFs with the convolutional kernels and Gabor filters in the three stages of convolutional operations. (a) shows the computation in the first stage. A convolutional kernel is shown in the left column, which is adjusted by four Gabor filters as shown in the middle column. (b) and (c) shows the computation in the second and third stages, respectively.

3.2 Update of the Parameters in the Gabor Functions

In LGCNs, we only use the real part of the Gabor filters, which can be expressed as:

$$g_r(x, y; \lambda, \theta, \psi, \sigma, \gamma) = \exp(-\frac{x'^2 + \gamma^2 y'^2}{2\sigma^2}) \cos(2\pi\frac{x'}{\lambda} + \psi). \tag{4}$$

The gradient of this function with respect to the scale and orientation parameters can be written as:

$$\frac{\partial g_r}{\partial \lambda} = 2\pi\frac{x'}{\lambda^2} \exp(-\frac{x'^2 + \gamma^2 y'^2}{2\sigma^2}) \sin(2\pi\frac{x'}{\lambda} + \psi), \tag{5}$$

$$\begin{aligned} \frac{\partial g_r}{\partial \theta} = &- \exp(-\frac{x'^2 + \gamma^2 y'^2}{2\sigma^2})[\frac{x'\frac{\mathrm{d}x'}{\mathrm{d}\theta} + \gamma^2 y' \frac{\mathrm{d}y'}{\mathrm{d}\theta}}{\sigma^2} \cos(2\pi\frac{x'}{\lambda} + \psi) \\ &+ \frac{2\pi}{\lambda}\frac{\mathrm{d}x'}{\mathrm{d}\theta} \sin(2\pi\frac{x'}{\lambda} + \psi)], \end{aligned} \tag{6}$$

where

$$\frac{\mathrm{d}x'}{\mathrm{d}\theta} = y\cos\theta - x\sin\theta, \frac{\mathrm{d}y'}{\mathrm{d}\theta} = -x\cos\theta - y\sin\theta. \tag{7}$$

The gradient functions with respect to the scale and orientation parameters in the k_{th} Gabor filter are:

$$\delta_{\lambda}^{k} = \frac{1}{J}\frac{1}{N}\sum_{j=1}^{J}\sum_{n=1}^{N} C_{j,n}^{l} \frac{\partial L}{\partial M_{i,n}^{l}} \frac{\partial M_{i,n}^{l}}{\partial G_{k,n}(\lambda,\theta)} \frac{\partial G_{k,n}(\lambda,\theta)}{\partial \lambda}, \tag{8}$$

$$\delta_{\theta}^{k} = \frac{1}{J}\frac{1}{N}\sum_{j=1}^{J}\sum_{n=1}^{N} C_{j,n}^{l} \frac{\partial L}{\partial M_{i,n}^{l}} \frac{\partial M_{i,n}^{l}}{\partial G_{k,n}(\lambda,\theta)} \frac{\partial G_{k,n}(\lambda,\theta)}{\partial \theta}, \tag{9}$$

where N denotes the kernel size of the filters and J is the number of the convolutional kernels which are adjusted by G_k in layer l. The update functions of the scale and orientation parameters in the k_{th} Gabor filter can be expressed as:

$$\lambda^{k} = \lambda^{k} - \eta\delta_{\lambda}^{k}, \theta^{k} = \theta^{k} - \eta\delta_{\theta}^{k}, \tag{10}$$

where η denotes to the learning rate in the training processes. Particularly, the update functions of the convolutional kernels are the same as the conventional DCNNs, except the multiplication with the Gabor functions.

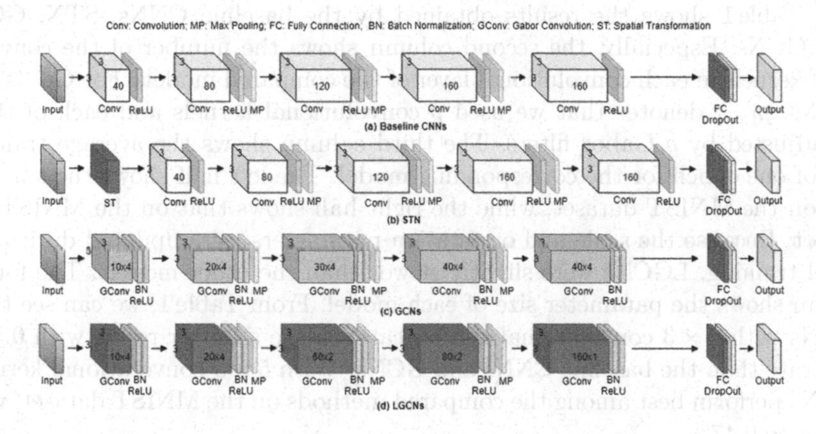

Fig. 2. The architectures of the baseline CNNs, STN [6], GCNs [11] and LGCNs.

4 Experiments

In our experiments, we evaluated LGCNs on the MNIST [9], MNIST-rot, SVHN [12] and CINIC-10 [2] datasets. All experiments were implemented in Pytorch with the NVIDIA GeForce GTX 1080Ti and Titan X GPU platforms.

4.1 Results Obtained on the MNIST and MNIST-rot Datasets

MNIST. We compared LGCNs with baseline CNNs, STN [6] and GCNs [11]. The architecture of baseline CNNs, STN, GCNs and LGCNs are shown in Fig. 2. We trained all these networks using the Stochastic Gradient Descent (SGD) algorithm with momentum 0.9, initial learning rate 0.01, a learning rate decay factor of 3e−05 per 10 epochs, and batch-size 128.

Table 1. Results obtained on the MNIST and the MNIST-rot datasets.

Model	Channels	Time (s) MNIST/ MNIST-rot	Parameters (M)	Error rate (%) MNIST/ MNIST-rot
Baseline CNNs (3 × 3)	40-80-120-160-160	4.96/12.05	0.69	0.73/1.39
STN (3 × 3)	40-80-120-160-160	10.80/26.11	0.70	0.56/1.25
GCNs (3 × 3)	$10_{(\times 4)}$-$20_{(\times 4)}$-$30_{(\times 4)}$-$40_{(\times 4)}$-$40_{(\times 4)}$	5.61/11.60	0.31	0.62/1.38
GCNs (5 × 5)	$10_{(\times 4)}$-$20_{(\times 4)}$-$30_{(\times 4)}$-$40_{(\times 4)}$-$40_{(\times 4)}$	8.12/12.17	0.54	0.58/1.40
LGCNs (3 × 3)	$10_{(\times 4)}$-$20_{(\times 4)}$-$60_{(\times 2)}$-$80_{(\times 2)}$-$160_{(\times 1)}$	12.00/12.12	0.54	0.52/1.24
LGCNs (5 × 5)	$10_{(\times 4)}$-$20_{(\times 4)}$-$60_{(\times 2)}$-$80_{(\times 2)}$-$160_{(\times 1)}$	18.60/18.39	1.20	**0.47/1.04**

In STN, a spatial transformation (ST) layer was implemented before the first convolutional layer, and the rest was the same as the baseline CNNs. In GCNs, each convolutional kernel was combined with four Gabor filters in each layer. Table 1 shows the results obtained by the baseline CNNs, STN, GCNs and LGCNs. Especially, the second column shows the number of the convolutional kernels in each convolutional layer of the compared models. For GCNs and LGCNs, $p_{(\times q)}$ denotes that we used p convolutional kernels and each of them was adjusted by q Gabor filters. The third column shows the average training time of one epoch for the corresponding model. The left half shows the training time on the MNIST dataset, while the right half shows that on the MNIST-rot dataset. Because the scale and orientation parameters were updated during the model training, LGCNs were slightly slower than the other models. The fourth column shows the parameter size of each model. From Table 1, we can see that, LGCNs with 3×3 convolutional kernels can achieve a better result with 0.52% error rate than the baseline CNNs and GCNs. With 5×5 convolutional kernels, LGCNs perform best among the compared methods on the MNIST dataset, with error rate 0.47%.

MNIST-rot. MNIST-rot was obtained by randomly rotating each digit image between $[0, 2\pi]$ in the MNIST dataset. From Table 1, we can see that, LGCNs with 3×3 convolutional filters and 5×5 convolutional filters can achieve 1.24% and 1.04% error rates respectively, which are better than that obtained by other compared methods. This demonstrates that LGCNs have the ability to extract the orientation invariant information of the digit images.

4.2 Results Obtained on the SVHN Dataset

The Street View House Numbers (SVHN) dataset [12] is a real-world image dataset, which is obtained from the house numbers in Google Street View images. The SVHN dataset contains more than 600,000 labeled digit images: 73,257 digits for training, 26,032 digits for test, and 531,131 additional digits. In our experiment, we only used the training and test sets.

Table 2. Classification results obtained on the SVHN dataset.

Model	ResNet-110	GCN-Res110	LGCN-Res110
Channels	16-32-64	$4_{(\times4)}$-$8_{(\times4)}$-$16_{(\times4)}$	$4_{(\times4)}$-$16_{(\times2)}$-$64_{(\times1)}$
Parameters (M)	1.73	0.44	1.52
Error rate (%)	6.03	6.43	**5.49**

We selected ResNet-110 [5] with 3 × 3 convolutional kernels as a baseline model. ResNet-110 includes a convolutional layer, 3 stages with 36 convolutional layers per stage and 16, 32, 64 channels per layer in the corresponding stage, and a fully connected layer. Hence, there are 109 convolutional layers in ResNet-110, which cannot be divided into three stages equally. We chose the second method to divide ResNet-110 to form the LGCNs-Res110 and GCN-Res110, neglecting the first convolutional layer. LGCNs-Res110 used 4, 2, 1 Gabor filters to adjust the convolutional kernels in the three stages, respectively. GCN-Res110 used 4 Gabor filters in each stage to keep the same number of channels with ResNet-110 and LGCNs-Res110. In the second row of Table 2, we show the number of channels per stage in ResNet-110, GCN-Res110 and LGCNs-Res110. From Table 2, we can see that LGCN-Res110 improves both ResNet-110 and GCN-Res110. Additionally, LGCN-Res110 has less parameters than ResNet-110.

4.3 Results Obtained on the CINIC-10 Dataset

The CINIC-10 dataset [5] contains the images from the CIFAR-10 data set and a selection of the ImageNet images (which are downsampled to 32 × 32). It is split into three equal subsets: train, validation, and test sets, each of which contains 90,000 images. We tested ResNet-110, GCN-Res110 and LGCN-Res110, and the error rate curves against the training epochs are shown in Fig. 3. It can be seen that LGCN-Res110 learns faster than GCN-Res110 and ResNet-110. Moreover, LGCN-Res110 achieved a better result compared with both ResNet-110 and GCN-Res110 (with error rate 29.48%, 32.77% and 32.19%, respectively).

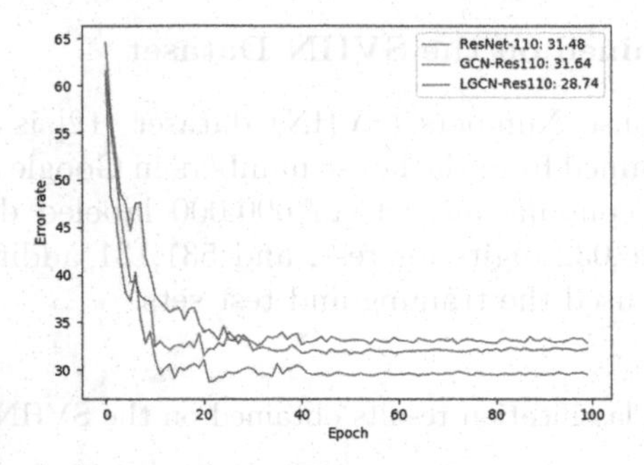

Fig. 3. The error rate curves of ResNet-110, GCN-Res110 and LGCN-Res110 on the CINIC-10 dataset.

5 Conclusion

Traditional convolutional kernels cannot learn the invariant information of images. In this paper, we propose a new method named learnable Gabor convolutional networks (LGCNs), to manipulate the convolutional kernels using Gabor filters for invariant information learning of images. The scale and orientation parameters in the Gabor functions can be learned simultaneously with other parameters during the training of LGCNs. Extensive experiments demonstrate the advantages of LGCNs over related state-of-the-art approaches.

Acknowledgments. This work was supported by the National Key R&D Program of China under Grant No. 2016YFC1401004, the National Natural Science Foundation of China (NSFC) under Grant No. 41706010, the Science and Technology Program of Qingdao under Grant No. 17-3-3-20-nsh, the CERNET Innovation Project under Grant No. NGII20170416, the Open Project Program of Key Laboratory of Research on Marine Hazards Forecasting, National Marine Environmental Forecasting Center, State Oceanic Administration (SOA), under Grand No. LOMF1802, and the Fundamental Research Funds for the Central Universities of China.

References

1. Boureau, Y., Ponce, J., LeCun, Y.: A theoretical analysis of feature pooling in visual recognition. In: ICML, pp. 111–118 (2010)
2. Darlow, L.N., Crowley, E.J., Antoniou, A., Storkey, A.J.: CINIC-10 is not ImageNet or CIFAR-10. CoRR abs/1810.03505 (2018)
3. Gabor, D.: Theory of communication. J. Inst. Electr. Eng. **93**, 429–457 (1946)
4. Grigorescu, C., Petkov, N., Westenberg, M.A.: Contour detection based on non-classical receptive field inhibition. IEEE Trans. Image Process. **12**(7), 729–739 (2003)
5. He, K., Zhang, X., Ren, S., Sun, J.: Deep residual learning for image recognition. In: CVPR, pp. 770–778 (2016)
6. Jaderberg, M., Simonyan, K., Zisserman, A., Kavukcuoglu, K.: Spatial transformer networks. In: NIPS, pp. 2017–2025 (2015)

7. Krizhevsky, A., Sutskever, I., Hinton, G.E.: ImageNet classification with deep convolutional neural networks. In: NIPS, pp. 1106–1114 (2012)
8. Laptev, D., Savinov, N., Buhmann, J.M., Pollefeys, M.: TI-POOLING: transformation-invariant pooling for feature learning in convolutional neural networks. In: CVPR, pp. 289–297 (2016)
9. LeCun, Y.: The MNIST Database of Handwritten Digits (1998). http://yann.lecun.com/exdb/mnist/
10. LeCun, Y., et al.: Backpropagation applied to handwritten zip code recognition. Neural Comput. **1**(4), 541–551 (1989)
11. Luan, S., Chen, C., Zhang, B., Han, J., Liu, J.: Gabor convolutional networks. IEEE Trans. Image Process. **27**(9), 4357–4366 (2018)
12. Netzer, Y., Wang, T., Coates, A., Bissacco, A., Wu, B., Ng, A.Y.: Reading digits in natural images with unsupervised feature learning. In: NIPS Workshop on Deep Learning and Unsupervised Feature Learning, p. 5 (2011)
13. Sarwar, S.S., Panda, P., Roy, K.: Gabor filter assisted energy efficient fast learning convolutional neural networks. In: ISLPED, pp. 1–6 (2017)
14. Szegedy, C., et al.: Going deeper with convolutions. In: CVPR, pp. 1–9 (2015)
15. Wang, L.-N., Liu, B., Wang, H., Zhong, G., Dong, J.: Deep Gabor scattering network for image classification. In: Lai, J.-H., et al. (eds.) PRCV 2018. LNCS, vol. 11257, pp. 332–343. Springer, Cham (2018). https://doi.org/10.1007/978-3-030-03335-4_29
16. Yosinski, J., Clune, J., Bengio, Y., Lipson, H.: How transferable are features in deep neural networks? In: NIPS, pp. 3320–3328 (2014)
17. Zhou, Y., Ye, Q., Qiu, Q., Jiao, J.: Oriented response networks. In: CVPR, pp. 4961–4970 (2017)

Deep Autoencoder on Personalized Facet Selection

Siripinyo Chantamunee(✉), Kok Wai Wong, and Chun Che Fung

Discipline of Information Technology, Mathematics and Statistics, College of Science, Health, Engineering and Education, Murdoch University, Perth, Australia
csiripin@wu.ac.th, {k.wong,l.fung}@murdoch.edu.au

Abstract. Information overloading leads to the need for an efficient search tool to eliminate a considerable amount of irrelevant or unimportant data and present the contents in an easy-browsing form. Personalized faceted search has been one of the potential tools to provide a hierarchical list of facets or categories that helps searchers to organize the information of the search results. Facet selection is one of the important steps to pursue a good faceted search. Collaborative-based personalization was introduced to facet selection. Previous studies have been performed on the use of Collaborative Filtering techniques for personalized facet selection. However, none of the study has investigated Artificial neural network techniques on personalized facet selection. Therefore, this study aims to investigate the possible use of deep Autoencoder on the prediction of facet interests. Autoencoder model was applied to address the association of collaborative interest in facets. The experiments were conducted on 100K and 1M rating records of Movielen dataset. Rating score was used to represent the explicit feedback on facet interests. The performance was reported by comparing the proposed technique and the state-of-the-art model-based Collaborative Filtering techniques in terms of prediction accuracy and computational time. The results showed that the proposed Autoencoder-based model achieved better performance and it was able to significantly improve the prediction of personal facet interests.

Keywords: Deep Autoencoder · Faceted search · Personalized facet selection

1 Introduction

With the exponential increasing of digital data, information retrieval system has been improved to provide greater accessing services for searchers to meet their information need. The progressive movement from query formation to information browsing can facilitate searchers to make direct access to their desirable information instead of surfing a sequential list of search results [14]. Faceted search is one of these potential features which provides a navigating category or *facet* of the knowledge underlying the search results [2]. With faceted search,

T. Gedeon et al. (Eds.): ICONIP 2019, CCIS 1142, pp. 314–322, 2019.
https://doi.org/10.1007/978-3-030-36808-1_34

facets are filtered from search results and represented in the form of a hierarchical list [2,20] which is particularly useful when searchers have to face with a large volume of search results [14]. Dynamic faceted search has replaced the traditional form of faceted search from the situation that search environment has been changed from searching on a fixed set of data to scalable volume. It has been applied to various applications such as E-commerce [18] and social media [12]. Facet selection is one of the main steps to create an appropriate faceted list [2,20]. Without selection process, dynamic faceted search becomes useless. A long list of facets can be produced and those items may not be interested by searchers [18]. Browsing through an irrelevant facet list does not help searching activity, but adding more effort to searchers. Facet selection needs to be personalized in order to produce a facet list which is relevant to the personal interests of each searcher. Personalization was first introduced to facet selection by [6].

Deriving searchers' interests from their own profiles may not be enough to develop personalization due to human information seeking behavior [2]. Interests from the past may not be adequate to predict current interests. Incorporating current interests from other people's opinions to predict the interests of individual person is an alternative way to develop personalization which called *collaborative approach*. Collaborative approach is based on the assumption that it is likely that people who have similar interests will also prefer the same items [17]. Collaborative approach has been one of the most successful techniques used in building recommender systems [10]. Collaborative Filtering (CF), one technique of collaborative approach, was first applied to personalize facet selection by Koren et al. [6]. The recent experimental study carried out by Chantamunee, et al. [3] showed that CF had the capability to significantly predict facet interests. The study reported that model-based CF achieved higher prediction accuracy than memory-based methods where Singular Value Decomposition (SVD) performed the best [10]. However, none of the studies has investigated Artificial Neural Network (ANN) techniques for personalized facet selection. ANN, deep learning in particular, is recently considered as one of the widely used model-based techniques for modeling collaborative information in recommender systems [19].

Several deep learning models have been applied in recommender systems [19]. In recent years, Autoencoder has attracted researchers in the area due to the capability of learning latent representation and predicting user preferences using content reconstruction technique [17]. The idea of Autoencoder is to operate dimensional reduction which is similar to the concept of SVD [7] where SVD is the best performer in the recent study by Chantamunee et al. [3]. Hence, it motivates this paper to investigate Autoencoder on the prediction of personal facet interests.

The main two contributions of this paper are presented as follows: (1) This paper aims to investigate deep Autoencoder in personalized facet selection. The performance will be measured by comparing to the state-of-the-art statistical CF methods where the performance is previously reported in [3]. (2) This paper further investigates the applicability of the deep Autoencoder to larger dataset by investigating whether the volume of training data has an effect on the prediction of personalized facets. The remainder of this paper is organized as follows: Sect. 2 provides the related works and backgrounds. Section 3 presents the proposed model. Section 4 describes the detail of the experiments, the dataset, and evaluation metrics. Section 4 gives the experimental results and discussions. Finally, Sect. 5 gives the conclusion.

2 Background and Related Works

2.1 Personalization on Facet Selection

Facet selection is one of the major processes to create faceted search where the task is to select a set of representative facets to construct a hierarchical search filter list [20]. The success of faceted search heavily relies on the performance of facet selection and ranking processes [1]. In recent years, faceted search has been changed to dynamic scheme. Facet list is automatically created and dynamically changed by certain criterion. In Kim et al. [5], information of the last searched facets was adopted to predict a current facet list for movie web engine. Count-based greedy algorithm was proposed to facet selection by [9]. The counting number was used to select a set of preferable facets. Vandic et al. [18] further used counting algorithm to count the frequency of products that were related to pre-defined facet values. The number of clicks was considered as a key factor to select a set of expected facets in [13].

However, these proposed works did not guarantee whether the selected facets satisfied the interests of an individual searcher. In general, searchers' satisfaction is the major purpose of a search engine. Searchers navigate the list to find their desirable facets where they are, in the worst situation, possibly shown at the bottom of the list. Personalization was then introduced to facet selection. Initially, user profile and search history were used to derive searchers' interests. In [8], the information of user activities and user-predefined interests was analysed to relate a set of preferable facets. However, past information may not be enough to generate current interests. The study by Tuong et al. [8] suggested that the frequency of current interaction was able to significantly select more representative facets than using search history.

The lack of user preferences's information is a major consideration for personalization task. Associating the interests from similar group people to generate user's own interests appears to be an alternative approach, known as *collaborative approach.* Collaborative approach provides the ability to relate collaborative interests to new users or users who have few interests in particular. The state-of-the-art statistical CF methods were applied to produce personalized facet selection in [3]. The study reported that model-based CF methods, SVD in particular, were able to predict more preferable facets than memory-based methods for most cases. However, none of the work has applied ANN, deep learning in particular, to personalized facet selection. Artificial neural network has been an active technique which is mostly used to model user preferences in recommender systems [19].

2.2 Autoencoder for Collaborative-Based Personalization

Personalization has been developed based on collaborative information by replacing the traditional use of personal profiles [3]. Collaborative-based personalization has achieved great success in the area of recommender systems [10]. Deep learning technique has recently received much attention and become an active model to develop personalization in recommender system [19] with the ability to learn the pattern of user preferences in collaborative environment. With deep learning, the accuracy of personalization has improved from past works, which used the traditional CF techniques [19].

In recent years, Autoencoder has attracted the attention of many researchers in the area of personalization due to its capability of predicting expected personal interests by reconstructing new user preferences from collaborative information [17]. Basically, low-level features are learned at the *bottleneck layer* (encoding process) and unknown user preferences are then generated during the last layer, called *reconstruction layer* (decoding process) [19]. A number of works have proposed variants of Autoencoder to personalize items' recommendation. Three-layer U-AutoRec and I-AutoRec models were proposed by Sedhain et al. [15] for predicting user-based and item-based recommendation respectively. Liu et al. [11] integrated Stacked Denoising Autoencoder (SDE) to Neural-based Collaborative Filtering to create a hybrid recommender system. SDE was used to learn the information of user and item prior to feeding to the ANN-based Collaborative Filtering.

3 Deep Autoencoder for Personalized Facet Selection

This paper investigates the application of deep Autoencoder to model user-facet interaction in order to develop personalization in facet selection. The proposed model is separated into two parts: (1) encoding part (feature learning) which is operated in the form of multiple hidden layers, and (2) decoding part upsamples the output which is situated at the output layer. The purpose of decoding part is to reconstruct the embedded user-facet ratings and predict unknown ratings. Increasing more layers for encoder part is examined in the experiments in order to investigate whether deeper layer has an influence on learning performance.

The proposed model aims to minimize the error from predicting personal facet interests. Mean Squared Error (MSE) was utilized as the loss function in the experiments. To prevent model overfitting, regularization method was adopted where L1 and L2 regularization were chosen. In addition, five activation functions including Sigmoid, Tanh, Relu, Elu, and linear function were tested while the choices of model optimizers were Adam (Adaptive Moment Estimation), SGD (Stochastic Gradient Descent), and Adagrad. Figure 1 presents the architecture.

4 Experiment Detail

4.1 Experimental Setup

Explicit user feedback was used to represent user preferences on facets. User rating was chosen in this experiment to represent explicit feedback. The range of rating score is within *1* to *5* where *1* denoted the least preference.

This paper compared the performance of the proposed model against the-state-of-the-art CF techniques. CF is classified into two main approaches: memory-based and model-based CF [10]. However, regarding to the experiment reported by Chantamunee et al. [3], model-based CF obtained better prediction than memory-based approach. Therefore, the techniques by model-based approach was only taken into the investigation. The comparison techniques includes SVD, SVD++, NMF (Normalised Matrix Factorisation), and PMF (Probabilistic Matrix Factorisation). The public toolkit, *Surprise*[1], was employed to implement CF techniques.

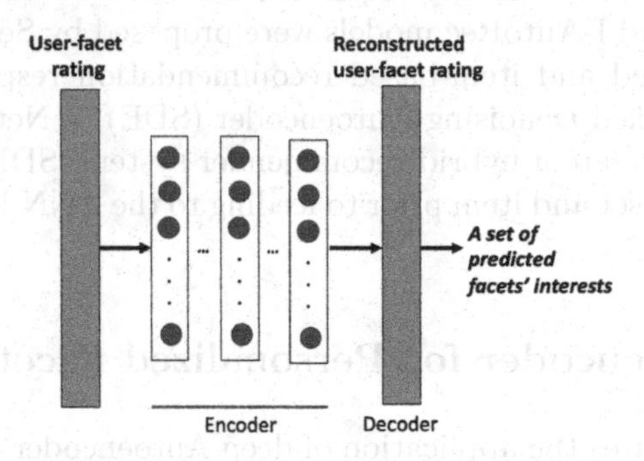

Fig. 1. The architecture of deep autoencoder model for personalized facet selection.

The comparison CF methods are not able to apply straight to personalized facet selection. The techniques are originally designed for recommending items in recommender systems. The prediction is made at item level. Facet, however, is the metadata of items. The example of facets for item *book* could be *author*, *book title*, and *year*. In order to perform the comparison, this paper therefore transformed the information of user-item ratings to user-facet ratings prior to feeding into the methods. The procedure of facet extraction is described in [3].

4.2 Dataset

The experiments were conducted based on the MovieLens-100K and MovieLens-1M datasets [4]. The datasets contained the information of movie metadata and user ratings. *User id* was used to represent individual user. The dataset only contains rating for users who rate more than 20 movies. From the metadata, facets

[1] http://surpriselib.com/.

were extracted and selected to four attributes including genre, actor/actress, keyword and production company. For the experiments, the dataset was randomly divided into 80% for training and 20% for testing.

4.3 Evaluation Metrics

The evaluation was measured by the accuracy of rating prediction and calculation time. Root Mean Squared Error (RMSE) was chosen for measuring the prediction accuracy as being done in [3]. RMSE is a common evaluational tool in recommendation task [16]. The lower value indicates that prediction error is small which gives a better prediction accuracy. Computation time includes training and prediction times.

5 Results and Discussions

This section presents the experimental results from the investigation of employing deep Autoencoder on predicting facet interests. The experiments were run on 2.9 GHz Intel Core i5 CPU and 8 GB RAM. In the experiments, four facets were tested including facet *genre*, *keyword*, *actor/actress*, and *production company*.

For parameters' setting, regularization was adopted to the proposed model in order to prevent model overfitting. L1 and L2 regularization methods were tested and the results were shown that L2 regularization outperformed another method. Hence, L2 regularization method was selected to run the model where the value of *beta* was set to *0.01*. In addition, Linear function was chosen as the activation function due to its performance above Sigmoid, Relu, Elu and Tanh in the experiment. The proposed model was trained for *500* epochs and learning rate was set to *0.01*. The number of nodes for each hidden layer was set to *100*. Adam optimizer was chosen as being outperformed another two optimizers.

With Movielen-100K dataset, the results were shown that the Autoencoder-based facet prediction models obtained higher prediction accuracy values for all facets. The best RMSE scores were reported at *0.0011* by the 1-hidden-layer Autoencoder model with L2 regularization (AE (1L) + L2), *0.2788* by the 2-hidden-layer model with L2 regularization (Deep-AE (2L) + L2), *0.6474* by the 1-hidden-layer Autoencoder model (AE (1L)), and *0.5401* by the 1-hidden-layer Autoencoder model (AE (1L)) for facet *genre*, *production company*, *keyword*, and *actor/actress* respectively. The prediction for personal facet interests was improved from the previous outperformer, SVD, did in [3]. It appeared that the Autoencoder model achieved the remarkable results on predicting the interests on facet *genre*. In the experiment, movie genre included only 20 values. The number of facet values was limited so that it may be easier for Autoencoder to learn the pattern and reconstruct the unknown ratings. The results on 100K-dataset are presented in Table 1.

The results were further confirmed by the experiment on larger dataset. The same four facets were then generated from 1M rating records of Movielen dataset. The experiment produced the results in the same direction that was done on

100K rating records. Table 1 shows the results of prediction accuracy carried out on Movielen-1M dataset. In conclusion, the proposed model outperformed the CF techniques. However, the Autoencoder-based model did not fit too well on the facets that were formed by name entity and free texts. In addition, facet *keyword* contained some special characters from foreign languages. It is likely to be difficult for pure Autoencoder model to learn the patterns. It may need text filtering process prior to feeding the content to the model. Alternatively, CNN can be another choice for learning facet preferences due to the feature of information filtering that can be done in the model. Dropout, another potential regularization technique [7], could be applied to the proposed model in order to drop noise nodes. These will be taken into consideration for future works.

In term of computation time, the Autoencoder-based models required much longer training time than CF methods, but it achieved less time on facet prediction. For instance, for facet *keyword* (on 100K records), training time was approximately ranged from 30 to 50 s while the model-based Collaborative Filtering techniques spent less time during 6 to 11 s. Prediction time for Autoencoder-based models was about 0.02 to 0.03 s while prediction time of the CF techniques was about 0.25 s in average. With spending shorter time for prediction, the Autoencoder-based model is rather appropriate for online real time prediction.

The experiments showed that modeling by deep layers achieved higher prediction accuracy in some cases. In addition, the experiments suggested that employing Adam optimizer in Autoencoder-based model was able to solve the problem of data sparsity in facet prediction where the percentage of data sparsity was high in facet *keyword*, *actor/actress*, and *production company.* When using Adagrad and SGD optimization for the Autoencoder models, their results are poorer when compared to the CF techniques.

Table 1. Facet prediction accuracy on Movielen-100K and Movielen-1M dataset measured by RMSE.

Techniques/facets	Genre		Product. com.		Keyword		Actor/actress	
	100K	1M	100K	1M	100K	1M	100K	1M
Proposed methods								
AE (1L)	0.0022	0.0048	0.2888	0.1195	**0.6474**	0.7194	**0.5401**	0.6752
AE (1L) + L2	**0.0011**	0.0026	0.2861	0.1194	0.6496	**0.7169**	0.5439	**0.6646**
Deep-AE (2L)	0.0016	0.0038	0.2813	0.1175	0.6660	0.7441	0.5517	0.6897
Deep-AE (2L) + L2	0.0033	**0.0024**	**0.2788**	0.1271	0.6566	0.7526	0.5563	0.6893
Deep-AE (3L)	0.0042	0.0026	0.2834	0.1588	0.6711	0.7870	0.5636	0.7140
Deep-AE (3L) + L2	0.0292	0.0025	0.2803	0.1266	0.6653	0.7770	0.5593	0.7197
Model-based CF								
SVD	0.6572	0.6841	0.6089	0.5940	0.8256	0.8009	0.8193	0.8602
SVD++	0.6502	0.6798	0.6586	0.5873	0.8262	0.8146	0.8208	0.8912
NMF	0.6837	0.7144	0.7811	0.7101	0.8713	0.9094	0.8812	0.9283
PMF	0.6824	0.6918	0.9899	1.8368	0.8501	0.8023	0.8514	0.8635

6 Conclusion

This paper presents a study of employing Autoencoder model to predict facet interests in personalized faceted search. The performance accelerated by the choices of multiple layers along with regularization methods, optimizers, and activation functions were reported. The proposed Autoencoder-based models were compared to the model-based CF techniques. The evaluation was measured in terms of prediction accuracy using RMSE and computation time. The experimental results showed that the proposed models obtained higher prediction accuracy than the models fitted by CF techniques for all facets. As expected, the proposed methods spent much time on training. However, the proposed models gave faster prediction. The study was extended on larger datasets and the results presented similar behaviour. Future work may consider the use of text filtering technique to improve the prediction on the facet formed by name entity and free texts.

References

1. Basu Roy, S., Wang, H., Das, G., Nambiar, U., Mohania, M.: Minimum-effort driven dynamic faceted search in structured databases. In: Proceedings of the 17th ACM Conference on Information and Knowledge Management, pp. 13–22. ACM (2008)
2. Chantamunee, S., Fung, C.C., Wong, K.W., Dumkeaw, C.: Knowledge discovery from thai research articles by solr-based faceted search. In: Unger, H., Sodsee, S., Meesad, P. (eds.) IC2IT 2018. AISC, vol. 769, pp. 337–346. Springer, Cham (2019). https://doi.org/10.1007/978-3-319-93692-5_33
3. Chantamunee, S., Wong, K.W., Fung, C.C.: Collaborative filtering for personalised facet selection. In: Proceedings of the 10th International Conference on Advances in Information Technology, p. 15. ACM (2018)
4. Harper, F.M., Konstan, J.A.: The movielens datasets: History and context. ACM Trans. Interact. Intell. Syst. (TiiS) **5**(4), 19 (2016)
5. Kim, H.J., Zhu, Y., Kim, W., Sun, T.: Dynamic faceted navigation in decision making using semantic web technology. Decis. Support. Syst. **61**, 59–68 (2014)
6. Koren, J., Zhang, Y., Liu, X.: Personalized interactive faceted search. In: Proceedings of the 17th International Conference on WWW, pp. 477–486. ACM (2008)
7. Kuchaiev, O., Ginsburg, B.: Training deep autoencoders for collaborative filtering. arXiv preprint arXiv:1708.01715 (2017)
8. Le, T., Vo, B., Duong, T.H.: Personalized facets for semantic search using linked open data with social networks. In: The 3rd International Conference on Innovations in Bio-Inspired Computing and Applications, pp. 312–317. IEEE (2012)
9. Liberman, S., Lempel, R.: Approximately optimal facet value selection. Sci. Comput. Program. **94**, 18–31 (2014)
10. Liu, H., Wu, Z., Zhang, X.: CPLR: collaborative pairwise learning to rank for personalized recommendation. Knowl. Based Syst. **148**, 31–40 (2018)
11. Liu, Y., Wang, S., Khan, M.S., He, J.: A novel deep hybrid recommender system based on auto-encoder with neural collaborative filtering. Big Data Min. Anal. **1**(3), 211–221 (2018)

12. Momeni, E., Braendle, S., Adar, E.: Adaptive faceted ranking for social media comments. In: Hanbury, A., Kazai, G., Rauber, A., Fuhr, N. (eds.) ECIR 2015. LNCS, vol. 9022, pp. 789–792. Springer, Cham (2015). https://doi.org/10.1007/978-3-319-16354-3_86
13. Nazi, A., Asudeh, A., Das, G., Zhang, N., Jaoua, A.: MobiFace: a mobile application for faceted search over hidden web databases. In: International Conference on Computer and Applications, pp. 13–17. IEEE (2017)
14. Niu, X., Fan, X., Zhang, T.: Understanding faceted search from data science and human factor perspectives. ACM Trans. Inf. Syst. **37**(2), 14:1–14:27 (2019)
15. Sedhain, S., Menon, A.K., Sanner, S., Xie, L.: Autoencoders meet collaborative filtering. In: Proceedings of 24th International Conference on WWW, pp. 111–112. ACM (2015)
16. Silveira, T., Zhang, M., Lin, X., Liu, Y., Ma, S.: How good your recommender system is? A survey on evaluations in recommendation. Inter. J. Mach. Learn. Cybern., 1–19 (2017)
17. Tran, D., et al.: Deep autoencoder for recommender systems: parameter influence analysis. CoRR (2019)
18. Vandic, D., Aanen, S., Frasincar, F., Kaymak, U.: Dynamic facet ordering for faceted product search engines. IEEE Trans. Knowl. Data Eng. **29**(5), 1004–1016 (2017)
19. Zhang, S., Yao, L., Sun, A., Tay, Y.: Deep learning based recommender system: a survey and new perspectives. ACM Comput. Surv. (CSUR) **52**(1), 5 (2019)
20. Zheng, B., Zhang, W., Feng, X.F.B.: A survey of faceted search. J. Web Eng. **12**(1&2), 041–064 (2013)

Attention-Based Deep Q-Network in Complex Systems

Kun Ni, Danning Yu, and Yunlong Liu(✉)

Department of Automation, Xiamen University, Xiamen 361005, China
ylliu@xmu.edu.cn

Abstract. In recent years, Deep Reinforcement Learning (DRL) has achieved great successes in many large scale applications, e.g., the Deep Q-Network (DQN) surpasses the level of professional human players in most of the challenging Atari 2600 games. As DQN transforms the whole input frames into some feature vectors by using convolutional neural networks (CNNs) at each decision step, all objects in the system are treated equally in the process of the feature extraction. However, in reality, for complex systems where many objects exist, the optimal action taken by the agent may only be affected by some important objects, which may lead to inefficiency or poor performance of DQN. In order to alleviate this problem, in this paper, we introduce two approaches that integrate global and local attention mechanisms respectively into the DQN model. For the approach with global attention, the agent is able to focus on all objects to varying degrees; for the approach with local attention, the agent is allowed to focus only on a few objects of great importance with the result that a better strategy can be learned by the agent. The performance of our proposed approaches are demonstrated on some benchmark domains. Source code is available at https://github.com/DMU-XMU/Attention-based-DQN.

Keywords: Deep Reinforcement Learning · Deep Q-Network · Atari 2600 games · Complex systems · Attention mechanisms

1 Introduction

Recent research has shown that Deep Reinforcement Learning (DRL) provides a power framework to solve the real-world problems where the input data is high dimensional. For example, the Deep Q-Network (DQN) model [11] is capable of learning control strategies directly from raw input frames to actions by combining deep neural networks with reinforcement learning. However, a possible drawback of DQN is that the agent is forced to consider all the information of the input frames in each decision-making, which makes the network hard to be trained

This work was supported by the National Natural Science Foundation of China (No. 61772438 and No. 61375077).

T. Gedeon et al. (Eds.): ICONIP 2019, CCIS 1142, pp. 323–332, 2019.
https://doi.org/10.1007/978-3-030-36808-1_35

well when many objects exist in the environment but some of them may have no effects on the agent's decision.

In the DRL literature, inspired by the human perception [2], attention mechanisms has been integrated with the related algorithms, e.g., the Deep Attention Recurrent Q-Network (DARQN) models [12], where Deep Recurrent Q-Network (DRQN) [4] is combined with soft and hard attention mechanisms respectively to make agents selectively focus only on some important objects of visual input. Both of them add attention to current frame at each time step. However, given only a single frame, most of games become a Partially-Observable Markov Decision Process (POMDP) rather than a Markov Decision Process (MDP). As a consequence, for the DARQN algorithm, to obtain the state of the moving object, attention must be paid to this object continuously, e.g., several frames, which will cause agents to ignore other objects, resulting in inefficiency of DQN.

In order to overcome this shortcoming of the DARQN model, in this paper, global and local attention-based DQN models are proposed. Given a history of the same length, in our approach, instead of only focusing on the current frame, the agent selectly focuses on the related regions of the input frames to directly obtain the states of all objects in regions of interest. Compared to soft DARQN [12], the global attention model can greatly improve the training efficiency without using a recurrent neural network. Due to the limitation of DARQN model, when hard attention is used, agents can only focus on one region at each decision time while our model can make agents focus on multiple regions (see Sect. 4.2 for details).

2 Related Work

In real-world applications, many problems are partially observable. The methods to solve them are divided to model-based and model-free methods. For model-based methods, the standard approaches are to establish environmental models firstly and then use environmental models to plan. An alternative interesting method of modeling is Predictive State Representations (PSRs) [6,8], which uses predictive vectors of action-observation series occurring in the future to represent the system states. Furthermore, agents can update the PSR model while planning through online learning [7].

For model-free methods, e.g., DQN uses the last four stacked frames as state to convert problems to a MDP. And for the problem of limited memory of DQN, DRQN [4] replaces the first fully-connected layer behind the convolutional networks by a recurrent LSTM network. Based on DRQN, DARQN [12] enables the agent to selectively focus on information of importance by adding visual attention mechanisms.

In other fields, visual attention mechanisms are widely used and have gained remarkable performance over many challenging problems. The recurrent neural network of visual attention (RAM) proposed in [10] outperforms a convolution neural network baseline significantly on image classification tasks and dynamic visual control problems. Subsequently, the deep recurrent visual attention model (DRAM), a extended version of RAM, is able to recognize multiple objects in images and be more accurate than the state-of-the-art convolutional networks by using fewer parameters and less computation [1]. But both of them are non-differentiable, which should be trained by using reinforcement learning methods. More recently, the Deep Recurrent Attentive Writer (DRAW) neural network [3], considered as "Differentiable RAM", demonstrates a significant improvement in test error on cluttered MNIST classification over the original RAM network.

3 Background

3.1 Deep Q-Network

Deep Q-Network uses deep neural networks parametered with θ as an approximator of state-action value functions defined as $Q(s, a; \theta)$. Given a current state s, DQN can output values for each possible action a to help the agent to make decisions. The loss function of DQN is defined as follows [11]:

$$L(\theta) = \mathbb{E}[(r + \gamma \max_{a'} Q(s', a'|\theta^-) - Q(s, a|\theta))^2] \tag{1}$$

where θ^- is the parameters of target network, which is a replication of θ every fixed steps to stabilise the learning and γ is a discount factor that makes a trade-off between immediate rewards and future rewards. Experiences stored as $\langle s, a, r, s' \rangle$ can be sampled repeatedly for training, which denotes the process that at some time step, the agent executes action a in the current state s and receives next state s' and an immediate reward r. And parameters are updated with a learning rate α as [11]:

$$\theta_{t+1} = \theta_t - \alpha \nabla_{\theta_t} L(\theta_t) \tag{2}$$

3.2 Attention Mechanisms

Attention mechanisms have been widely used in different fields, usually divided into soft attention and hard attention as [13].

The soft attention mechanism [9,12,13] is to assign attention weight between 0 and 1 to each pixel of the feature map. The hard attention mechanism [1, 10,12,13] is to select an attention area of the image, often called a glimpse, where the attention weight of this part is 1, and the attention weight of other areas is 0, e.g. image cropping, which is usually non-differentiable. However, a differentiable hard attention mechanism called the read mechanism is proposed in [3]. By applying a $N \times N$ grid of two-dimensional Gaussian filters consisting of the horizontal and vertical filterbank matrices $F_X \in \mathbb{R}^{N \times A}$ and $F_Y \in \mathbb{R}^{N \times B}$ to a input image with A columns and B rows, the read mechanism can extract a $N \times N$ glimpse g as [3]:

$$g = \gamma F_Y x F_X^T \tag{3}$$

where γ is a scalar intensity. Both γ and the parameters of filterbank matrices are determined by a linear transformation of the hidden state h of LSTM [3]:

$$\left(\tilde{g}_X, \tilde{g}_Y, \log\sigma^2, \log\tilde{\delta}, \log\gamma\right) = Linear(h) \tag{4}$$

where σ is standard deviation, $\left(\tilde{g}_X, \tilde{g}_Y\right)$ is the grid center and stride $\tilde{\delta}$ controls the scope of glimpses. To make the initial glimpse cover the entire input image, $\tilde{g}_X, \tilde{g}_Y, \tilde{\delta}$ are scaled as follows [3]:

$$g_X = \frac{A+1}{2}\left(\tilde{g}_X + 1\right) \tag{5}$$

$$g_Y = \frac{B+1}{2}\left(\tilde{g}_Y + 1\right) \tag{6}$$

$$\delta = \frac{\max(A, B) - 1}{N-1}\tilde{\delta} \tag{7}$$

and then the mean locations μ_X^i, μ_Y^i of filters at row i, column j are computed as [3]:

$$\mu_X^i = g_X + \left(i - \frac{N}{2} - 0.5\right) \cdot \delta \tag{8}$$

$$\mu_Y^j = g_Y + \left(j - \frac{N}{2} - 0.5\right) \cdot \delta \tag{9}$$

given the parameters μ_X^i, μ_Y^i, σ, filterbank matrices $F_X \in \mathbb{R}^{N \times A}$ and $F_Y \in \mathbb{R}^{N \times B}$ can be obtained via [3]:

$$F_X[i, a] = \text{Softmax}\left(\exp\left(-\frac{\left(a - \mu_X^i\right)^2}{2\sigma^2}\right)\right) \tag{10}$$

$$F_Y[j, b] = \text{Softmax}\left(\exp\left(-\frac{\left(b - \mu_Y^j\right)^2}{2\sigma^2}\right)\right) \tag{11}$$

4 Attention-Based Deep Q-Network

We integrate two attention mechanisms respectively with DQN so that the extended attention-based DQN models including global and local can perform well in many complex environments. Both of these two models consist of three components: the feature extraction network, the attention network and the action network.

Common to them is that they share a same feature network and a similar action network. The feature extraction network calculates a feature representation V with given stacked frames of the history as input. And its structure is the same as convolutional neural networks of DQN. The action network is simply composed of full-connected layers. Its output is state-action values evaluated, referred to as Q-values.

The difference between these two types of models is whether the attention of the agent is placed on the whole feature map or on only a part of the feature map. It is worth noting that both our models exert attention on four stacked frames instead of one frame at each step as DARQN. In the following subsections, we describe two different attention networks in detail.

4.1 Global Attention Model

The global attention model is to place a weight distribution on the feature map and does not need to use recurrent neural network, which is more efficient than the soft DARQN model that requires attention on each frame. The attention network of this model includes a regression network and a glimpse network. The regression network produces weights for each position of the feature map by using a fully-connected layer followed by a Softmax function, and then the glimpse network calculates the element-wise product of weight map and feature map.

In detail, firstly, the feature map is reshaped to $V = [v_1, v_2, \ldots, v_L], v_m \in \mathbb{R}^D, L = H \times W$, where H, W, D are respectively height, width and channel of V. Then we use the feature map V to product a positive attention weight α_m for each region v_m via a linear transformation. Below are the specific definitions of generating a weighted feature map $Z = [z_1, z_2, \ldots, z_L], z_m \in \mathbb{R}^D$:

$$\alpha = Softmax(Tanh(Linear(V))) \tag{12}$$

$$z_m = \alpha_m v_m \tag{13}$$

where *Linear* is a fully-connected layer in the regress network. At each decision step, the agent pays more attention to some important regions by giving a greater weight and suppresses the information of the unimportant regions by giving a weight of approximately 0.

4.2 Local Attention Model

The local attention model is an aggregation of read mechanism introduced in [3] and DQN. Different from [1,3,10], it extracts glimpses from the feature representation V instead of the input images, which reduces the amount of calculation. And then a LSTM network is used to integrate the information of these glimpses sequentially and produce attention parameters for next glimpse. Figure 1 shows the generation of k glimpses, where k is set according to the specific environment. We first transpose the feature map V to $V^T = [v_1, v_2, \ldots, v_D], v_n \in \mathbb{R}^{H \times W}$. By applying a two-dimensional Gaussian filter consisting of a group column Gaussian filters $F_X \in \mathbb{R}^{N \times W}$ and a group of row Gaussian filters $F_Y \in \mathbb{R}^{N \times H}$ to each v_n separately as formula (3), the read mechanism can extract a $N \times N \times D$ glimpse G. With glimpse G as input, the LSTM updates its hidden state dynamically as:

$$c_t, h_t = LSTM(G, c_{t-1}, h_{t-1}|W) \tag{14}$$

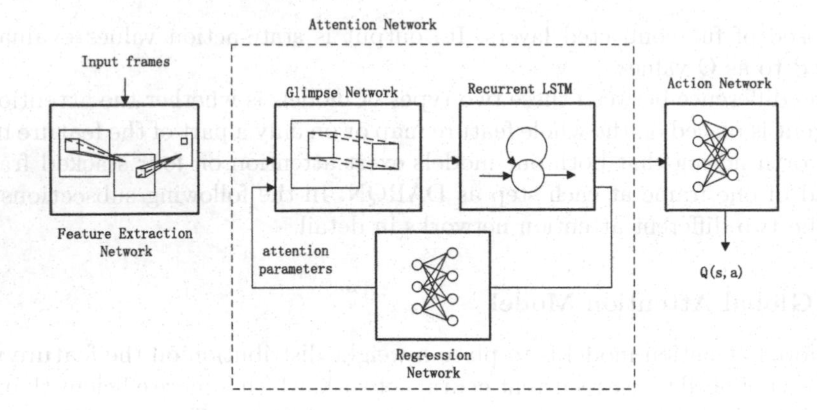

Fig. 1. The network architecture of the local attention model

where the memory state c and the hidden state h of the LSTM are both initialized to 0. And W is the parameters of the LSTM. Then h_t is used as input of the regression network to generate the next attention parameters through formula (4). Therefore, we can enable agents to focus on k regions in each decision, by setting the unroll step of LSTM to k.

An interesting part of this attention mechanism is that the larger a scope of the feature map V covered by glimpse, the lower the resolution of the glimpse will be, which requires the agent to make a trade-off between view and detail. As shown in Fig. 2, the left glimpse has a wider view but low resolution and the right is the opposite. To ensure that the agent has a global view at the beginning so that it can master roughly the motion states of each object, we use the normalization of Eqs. (5)–(7).

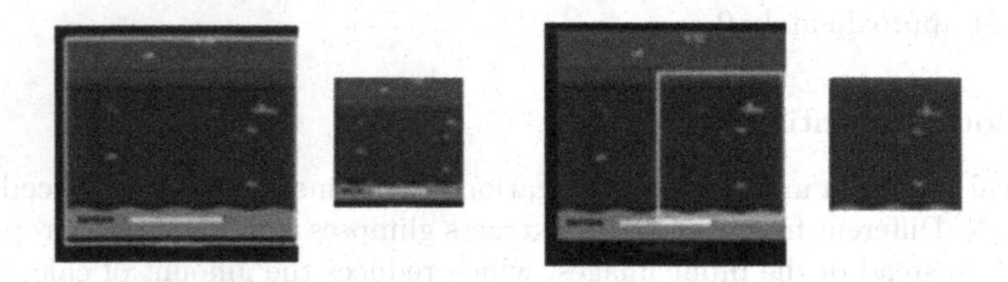

Fig. 2. Visualization of the results of extracting a 36×36 glimpse with different δ from a 84×84 gray frame of the game of Seaquest. The boxes in two big pictures are the areas covered by glimpses, and the two small pictures is the extracted glimpses. The left glimpse has a big δ while the right glimpse has a small δ.

Since DARQN needs to keep track of the same object in order to obtain the state of this object, the hard DARQN model will limit the agent to only focus on an area of input. For complex environments such as the game of Seaquest, if the agent only tracks the trajectory of an enemy, then it may suddenly lose a life because the agent does not know that the oxygen has been exhausted. Different from the hard DARQN model, at each unroll step of LSTM, as our agent focuses on an area of four stacked frames instead of current frame so that it can immediately get the state of objects in the area. Therefore, the agent can get the full information of k regions at each decision step. In the game of Seaquest, our agent with the local attention can sequentially select and focus on different objects such as the enemy and the oxygen with the local attention, which makes the agent survive more easily because of not ignoring the lack of oxygen. Another advantage of the local model over hard DARQN is that our model is differentiable.

5 Experiments

We selected several complex Atari 2600 games to test our models and compared the results to those of DQN [11] and soft DARQN [12] with 4 frames history. Simultaneously, we tested our models in two simple environments and compared their performance with that in complex environments to show which environments they are more suitable for.

Table 1. The best average reward per episode for the four models on 11 Atari 2600 games. The first nine rows are complex environments and the rest are simple environments.

	DQN	soft DARQN	gloal DQN	local DQN
Assault	1151.4	1155.2	1159.1	**1455.7**
Asterix	2640	2585	3025	**3360**
Breakout	151.2	143.3	187.2	**208.7**
DemonAttack	1397	1876	3418.5	**5432**
Jamesbond	255	**585**	555	525
Pooyan	2687.5	2443.5	2980	**3265.0**
Seaquest	1868	4739	6412	**7160**
SpaceInvaders	518.5	502	538.5	**589**
StarGunner	9000	5350	8940	**12200**
Pong	15.2	17.3	16.7	**17.9**
Boxing	94	94.3	94	**96.1**

In all our experiments, the hidden units and unroll steps of LSTM in the local attention model was set to 256 and 4 respectively. The size of memory was 200,000 and the discount factor γ was set to 0.99. At the first 50,000 steps, the agent used a random strategy in order to collect various experiences. After this,

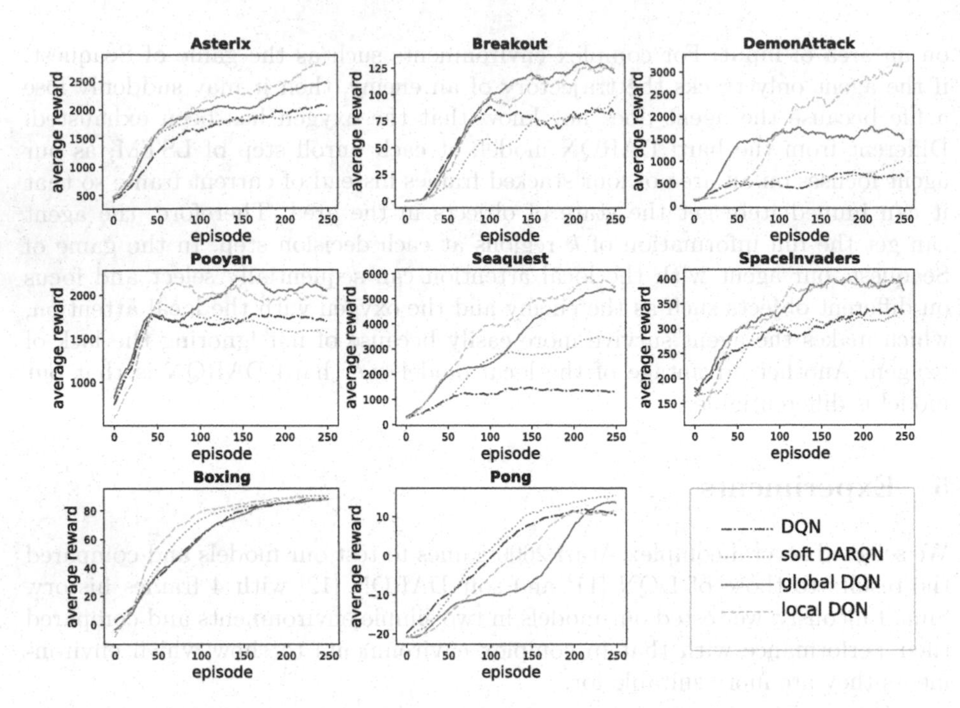

Fig. 3. Average reward curves for the four models on 8 Atari 2600 games, of which the first two rows are complex environments and the remaining are simple environments.

the agent was trained every 4 steps with experiences of batch size 32 sampled from the memory by running an ε-greedy strategy, where ε decayed linearly from 1 to 0.05 over 1 million steps. And models were assigned to the target network every 10,000 training steps. We used a ADAM optimization [5] to update the weights of the network with a learning rate of 0.0001. All models were trained on a GPU 2080TI for 2.5 million steps and evaluated every 10000 steps of training. We used the average rewards of 10 episodes as the evaluation results. And the main comparison results of the four models on 11 Atari games are presented in Table 1, which demonstrated the superiority of our models. We selected several environments to show their reward curves. In order to smooth the curves, we took the average value of 20 points before and after each point as the value of that point. The comparison results of the four models in two types of environment are presented in Fig. 3.

As shown in Fig. 3, both the global and local models are almost superior to DQN and the soft DARQN model in these 6 complex games, especially in the games of Seaquest and DemonAttack. And the local model tends to perform better. The performance of the soft DARQN model was not greatly improved compared with DQN in other games except for the games of Seaquest. Moreover, our global model requires only approximately 12 h to train for 2.5 million steps, which is more efficient than soft DARQN that needs to take about 16 h.

And in Pong and Boxing games, the final performance of the four models was almost the same. This can be well explained: in the game of Pong, there are only two objects in fact, a racket and a ball, so that whether or not attention is added will not make much difference to the final scores. In addition, our models can also be compared with other state-of-the-art DQN by setting their training mode and loss functions to the same.

6 Conclusion

In this paper, we integrate DQN with two efficient differentiable attention-based DQN models: global and local. The global attention model adds a weight to each location of the feature map, corresponding to a fixed size area in the original image. The local attention model only focuses on a part of the feature map at a time, which can make a trade-off between the size and resolution of regions of interest as needed. In experiments, we show that these two models are generally superior to DQN and the soft DARQN model in complex environments. And we will try to test our models in more challenging tasks in future work.

References

1. Ba, J., Mnih, V., Kavukcuoglu, K.: Multiple object recognition with visual attention. arXiv preprint arXiv:1412.7755 (2014)
2. Egeth, C.E., Connor, H.E.: Visual attention: bottom-up versus top-down. Curr. Biol. **14**, R850–R852 (2004)
3. Gregor, K., Danihelka, I., Graves, A., Wierstra, D.: DRAW: A recurrent neural network for image generation. CoRR abs/1502.04623 (2015). http://arxiv.org/abs/1502.04623
4. Hausknecht, M.J., Stone, P.: Deep recurrent q-learning for partially observable mdps. CoRR abs/1507.06527 (2015). http://arxiv.org/abs/1507.06527
5. Kingma, D., Ba, J.: Adam: a method for stochastic optimization. arXiv preprint arXiv:1412.6980 (2014)
6. Liu, Y., Tang, Y., Zeng, Y.: Predictive state representations with state space partitioning. In: Proceedings of the 2015 International Conference on Autonomous Agents and Multiagent Systems (AAMAS), pp. 1259–1266 (2015)
7. Liu, Y., Zheng, J.: Online learning and planning in partially observable domains without prior knowledge. In: Generative Modeling and Model-Based Reasoning for Robotics and AI at ICML 2019 (2019)
8. Liu, Y., Zhu, H., Zeng, Y., Dai, Z.: Learning predictive state representations via Monte-carlo tree search. In: Proceedings of the 25th International Joint Conference on Artificial Intelligence (IJCAI) (2016)
9. Luong, M., Pham, H., Manning, C.D.: Effective approaches to attention-based neural machine translation. CoRR abs/1508.04025 (2015). http://arxiv.org/abs/1508.04025
10. Mnih, V., Heess, N., Graves, A., Kavukcuoglu, k.: Recurrent models of visual attention. In: Ghahramani, Z., Welling, M., Cortes, C., Lawrence, N.D., Weinberger, K.Q. (eds.) Advances in Neural Information Processing Systems 27, pp. 2204–2212. Curran Associates, Inc. (2014). http://papers.nips.cc/paper/5542-recurrent-models-of-visual-attention.pdf

11. Mnih, V., et al.: Human-level control through deep reinforcement learning. Nature **518**(7540), 529 (2015)
12. Sorokin, I., Seleznev, A., Pavlov, M., Fedorov, A., Ignateva, A.: Deep attention recurrent q-network. CoRR abs/1512.01693 (2015). http://arxiv.org/abs/1512.01693
13. Xu, K., et al.: Show, attend and tell: neural image caption generation with visual attention. In: International Conference on Machine Learning, pp. 2048–2057 (2015)

Effect of Data Augmentation and Lung Mask Segmentation for Automated Chest Radiograph Interpretation of Some Lung Diseases

Peng Gang[1], Wei Zeng[1], Yuri Gordienko[2(✉)], Yuriy Kochura[2], Oleg Alienin[2], Oleksandr Rokovyi[2], and Sergii Stirenko[2]

[1] Huizhou University, Huizhou, China
peng@hzu.edu.cn

[2] National Technical University of Ukraine "Igor Sikorsky Kyiv Polytechnic Institute", Kyiv, Ukraine
yuri.gordienko@gmail.com

Abstract. The results of chest X-ray (CXR) analysis of 2D images to get the statistically reliable predictions of some lung diseases by computer-aided diagnosis (CADx) based on the convolutional neural network (CNN) are presented for the largest open CXR dataset with radiologist-labeled reference standard evaluation sets (CheXpert). The results demonstrate the lower validation loss and higher area under curve (AUC) values for the receiver operating characteristic curve (ROC) for the models with lung mask segmentation (for 4 from 14 lung diseases) and data augmentation (for 10 from 14 lung diseases) for small image sizes (320 × 320 pixels) and standard CNN (like DenseNet121) even. Moreover, the additional training leads to the lower validation loss and higher AUC values for the model with data augmentation. The further progress of CADx is assumed to be obtained for the big datasets with the bigger original image sizes by longer training with the tuned data augmentation.

Keywords: Deep learning · Convolutional neural network · Segmentation · Data augmentation · Chest X-ray · Computer-aided diagnosis

1 Introduction

Chest X-ray (CXR) imaging is the most common imaging technology used for screening, diagnosis, and management of many life threatening diseases, especially lung diseases like pneumonia, tuberculosis, cancer, etc. Manual CXR image

The work was partially supported by Huizhou Science and Technology Bureau and Huizhou University (Huizhou, P.R.China) in the framework of Platform Construction for China-Ukraine Hi-Tech Park Project.

T. Gedeon et al. (Eds.): ICONIP 2019, CCIS 1142, pp. 333–340, 2019.
https://doi.org/10.1007/978-3-030-36808-1_36

interpretation by expert radiologists is a long and complicated process. Moreover, the number of certified radiologists is not enough to organize nationwide screening for reasonable time. That is why the progress in automated CXR image interpretation close to the level of expert radiologists could provide substantial benefit for large-scale (nation-wide and worldwide) screening. The main prerequisites for such a progress are (a) large labeled datasets that follow strong reference standards and provide expert metrics for comparison, (b) reliable prediction models that are close to or outperform expert human performance, and (c) powerful computing infrastructures for training and updating the models with a purpose to provide the high-performance prediction. Availability of open datasets with labeled CXR images [1], new deep learning models [2], and new generation of general-purpose graphic processing cards (GPU) and tensor processing (TP) hardware [3] allowed data scientists to apply their deep learning algorithms for anatomical structure detection, segmentation, computer-aided detection (CADe) of suspicious regions, and computer-aided diagnosis (CADx).

2 Problem and Related Work

Recently, numerous important results were obtained in the field of CADx for an assessment of lung diseases by deep learning from CXR image analysis [3–8]. They become possible due to availability of various datasets with CXR images released for public domain recently: from small ones like JSRT dataset with 247 images of cancer [9]; LIDC dataset with $\sim 10^3$ images [10]; Montgomery County (MC) dataset with 138 images, Shenzhen Hospital (SH) dataset with 662 images [11] and up to the huge ones like ChestX-ray14 [12] with $>10^5$ images and CheXpert (Chest eXpert) [5] with $>2.2 * 10^5$ images. The new algorithms for medical image analysis [1], especially deep learning methods [2], have allowed scientists to detect automatically many lung diseases from CXR images at a level exceeding certified radiologists [4,5]. In addition to the newly available datasets and the better deep learning models, the progress of data processing techniques and development of have allowed researchers to get the better performance of prediction (lower loss and higher accuracy). For example, various segmentation methods were applied like the active shape models, active appearance models, and a multi-resolution pixel classification method [13,14], "internal segmentation" by exclusion of the effect of some body parts that shadow the lung, for example, ribs and clavicles [15]. The current and previous attempts to perform training for the tiny ($<10^3$ images) CXR datasets without any pre-processing were performed and failed, but "external lung segmentation" (exclusion of outside regions which are not pertinent to lungs) was demonstrated to be effective to provide successful training and, moreover, to increase the accuracy of predictions [6–8]. But the open question is how these techniques can be useful for the really big datasets ($>10^5$ images) for the multi-class and multi-label classification tasks and what is their impact on the accuracy of prediction of some classes. The main aim of this paper is to present the new results on application of some data processing technique (lung segmentation in combination with/without data augmentation) for CADx of 14 lung pathologies for the newly available biggest dataset CheXpert [4,5].

3 Dataset and Models

Dataset. In this work, CheXpert was used, which consists of 224,316 chest radiographs of 65,240 patients labeled for the presence of 14 common chest radiographic observations (no findings, diseases and other abnormalities) [5]. The images are available in low-quality and high-quality versions, where images can have a little bit various sizes and aspect ratios. In this work we used the low-quality version (CheXpert-v1.0-small), where all images where resized to the uniform size of 320 × 320 pixels, and the similar work on the high-quality version will be published elsewhere [16]. The training subset contains >200,000 images. The validation subset and test set were provided by creators of CheXpert dataset. The validation subset contains 234 images (from 200 patients) randomly sampled from the full dataset with no patient overlap with the report evaluation set. The test subset consists of 500 studies (from 500 patients) randomly sampled from the 1000 studies in the report test set. The details as to their labeling, radiologist annotations, and benchmarks of radiologist performance can be found in the relevant paper [5].

Data Processing Techniques. Recently, efficiency of some data processing techniques (like lung segmentation, bone shadow exclusion, and t-distributed stochastic neighbor embedding (t-SNE) for exclusion of outliers, etc.) was demonstrated by us for the small datasets (<10^3 images) for analysis of CXR 2D images to identify marks of lung cancer and tuberculosis [6–8]. Here some of these methods, actually lung segmentation and data augmentation were used. Lung segmentation was performed automatically by the previously trained model [6–8] based on U-Net [17] for the whole CheXpert dataset (Fig. 1). Data augmentation included the following image transformations with random parameters: horizontal flip, vertical flip, rotation (up to 5°), color jitter, resized crop (with a scale in the range from 0.9 to 1.0 and ratio from 0.9 to 1.1), and perspective distortion (with a scale = 0.1).

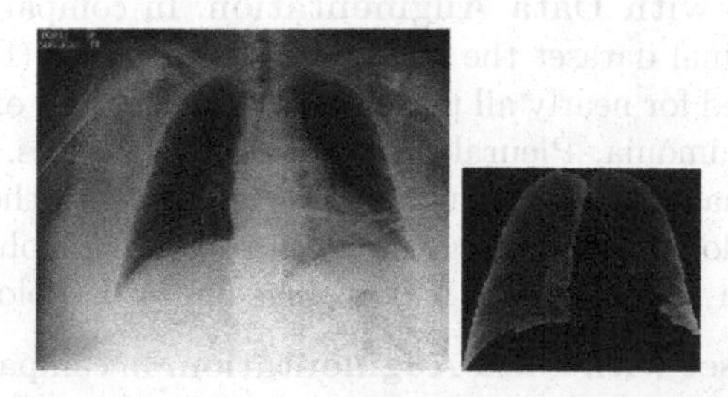

Fig. 1. The original CheXpert image (left) and its segmented image (right).

Model. In this work the deep learning model was used with the DenseNet121 convolutional neural network architecture [18], as far as the DenseNet121 architecture and its variations give the best results at the moment [5]. The dataset images were input into the network with size 320 × 320 pixels. The Adam optimizer was used with $\beta_1 = 0.9$, $\beta_2 = 0.999$, and learning rate 10^{-4} which was fixed during the training. The batch size was equal to 32 images. As far as some labels have uncertain labels, in this work the uncertain labels for any of the observations were replaced by the 1 label (so-called U-Ones model) [5].

4 Results

The model was trained on GPU card (NVIDIA Titan 1080) by means of TensorFlow machine learning framework [19] for: the original dataset (O); the segmented dataset (S), the original dataset with data augmentation (OA), and the segmented dataset with data augmentation (SA). The training was performed during 3 epochs (for O and S) and 18 epochs (for OA and SA) with checkpoints every 2240 iterations and calculation of training and validation losses. At each checkpoint the trained model was saved when the current validation loss was lower than the minimal loss at the previous checkpoints. For the model with the minimal validation loss the receiver operating characteristic curves (ROC) were created for all 14 chest radiographic observations (listed in Table 1, column "Pathologies"). Then the areas under curve (AUC) for ROCs were calculated.

Original Dataset. AUC values (Table 1, column O) demonstrate that some pathologies (No Finding, Cardiomegaly, Edema, Pleural Effusion, Support Devices) can be predicted with the better performance (AUC > 0.80) in comparison to others (Enlarged Cardiomegaly, Lung Opacity, Consolidation, Atelectasis) with the much lower performance (AUC < 0.70).

Segmented Dataset. AUC values (Table 1, column S) demonstrate the same division of pathologies to three groups that can be predicted with the better, medium, and lower prediction performance. In comparison to the model trained on the original dataset the increase of the AUC values for 4 from 14 pathologies only (like No Finding, Lung Opacity, Consolidation, Pleural Other) was observed, while the AUC values decreased for 10 others.

Original Dataset with Data Augmentation. In comparison to the model trained on the original dataset the increase of AUC values (Fig. 2; Table 1, column OA) is observed for nearly all pathologies (10 from 14), except for Enlarged Cardiomegaly, Pneumonia, Pleural Other, Support Devices. But it should be noted that additional training during 15 epochs has led to the increase of AUC values for all pathologies (they are given in parentheses in column OA, Table 1). Finally, OA allowed us to improve AUC values for 14 pathologies from 14.

Segmented Dataset with Data Augmentation. In comparison to the model trained on the original dataset the increase of AUC values (Table 1, column SA) was observed for some pathologies (6 from 14), namely for No Finding, Lung

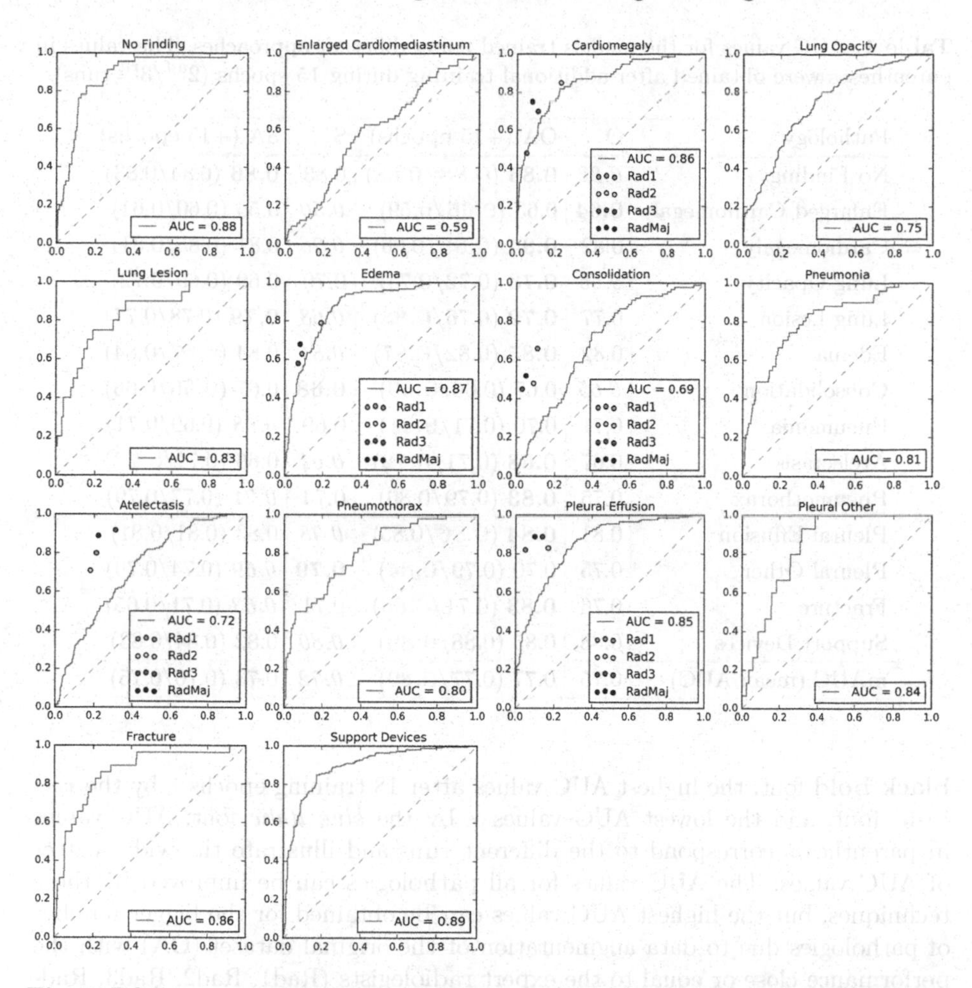

Fig. 2. ROC-curves and AUC values for the model trained using the original dataset with data augmentation (OA). The vertical axes are for true positive rate values, and horizontal axes are for false positive rate ones. The color (in the electronic version) dots are data by the 3 expert radiologists (Rad1, Rad2, Rad3) and their majority vote (RadMaj) given for comparison for the relevant diseases [5].

Opacity, Lung Lesion, Edema, Consolidation, and Pleural Effusion. The additional training during 15 epochs has led to the increase of AUC values (they are given in parentheses in column SA, Table 1) for 4 more pathologies like Cardiomegaly, Atelectasis, Pneumothorax, and Pleural Other. Finally, SA allowed us to improve AUC values for 10 pathologies from 14.

5 Discussion

The results obtained are summarized in Table 1 and emphasized by colors (in electronic version): the highest AUC values after 3 training epochs – by the

Table 1. AUC values for the models trained using different approaches. The values in parentheses were obtained after additional training during 15 epochs ($2^{nd}/3^{rd}$ runs).

Pathology	O	OA (+15 epochs)	S	SA (+15 epochs)
No Findings	*0.82*	**0.86** (0.88/0.88)	0.83	**0.86** (0.84/0.83)
Enlarged Cardiomegaly	**0.64**	0.63 (0.65/0.59)	*0.50*	0.56 (0.60/0.61)
Cardiomegaly	0.82	**0.84** (0.88/0.86)	*0.76*	0.81 (0.85/0.81)
Lung Opacity	*0.66*	**0.73** (0.72/0.75)	0.70	0.69 (0.69/0.72)
Lung Lesion	0.77	**0.79** (0.76/0.83)	*0.68*	**0.79** (0.78/0.77)
Edema	0.82	**0.85** (0.82/0.87)	*0.81*	0.84 (0.87/0.84)
Consolidation	*0.65*	0.67 (0.66/0.69)	**0.68**	0.67 (0.69/0.66)
Pneumonia	0.71	0.70 (0.71/0.81)	0.69	*0.68* (0.69/0.71)
Atelectasis	0.67	**0.68** (0.71/0.72)	*0.64*	0.66 (0.71/0.72)
Pneumothorax	0.75	**0.83** (0.79/0.80)	0.74	*0.71* (0.77/0.79)
Pleural Effusion	0.81	**0.84** (0.86/0.85)	*0.78*	0.83 (0.81/0.81)
Pleural Other	0.75	0.70 (0.79/0.84)	**0.79**	*0.69* (0.74/0.79)
Fracture	0.76	**0.83** (0.71/0.86)	0.71	*0.63* (0.71/0.65)
Support Devices	0.88	0.85 (0.88/0.89)	*0.80*	0.82 (0.84/0.83)
mAUC (mean AUC)	0.75	0.77 (0.77/0.80)	*0.72*	0.73 (0.76/0.75)

black bold font, the highest AUC values after 18 training epochs – by the red bold font, and the lowest AUC values – by the *blue italic* font. AUC values in parentheses correspond to the different runs and illustrate the wide scatter of AUC values. The AUC values for all pathologies can be improved by these techniques, but the highest AUC values can be obtained for the bigger number of pathologies due to data augmentation of the original dataset (OA) with the performance close or equal to the expert radiologists (Rad1, Rad2, Rad3, RadMaj points in Fig. 2) for small image sizes (320 × 320 pixels) even. This impact is visualized by the majority of the black bold (after 3 epochs) and the red bold AUC values (after 18 epochs) in column OA (Table 1). In reverse, segmentation mainly leads to the decrease of the AUC values nearly for all pathologies, and this impact is visualized by the majority of the blue italic AUC values in columns S and SA (Table 1). These results for the large CheXpert dataset contradict the results previously obtained by us for the very small datasets (662 images) [9,10], where the segmentation and augmentation improved the prediction of lung diseases [8]. It means that as far as segmentation allows to decrease the number of non-relevant features (by cropping out the regions outside of lungs) it is useful for the relatively small datasets. But for the large datasets the segmentation is not critically useful anymore, because such non-relevant features are effectively excluded after processing the sufficiently high number of images. Nevertheless, the data augmentation techniques remain important as far as they mimic variability of scanning conditions like different angles of scanning, slight distortions, small scaling up and down, etc. But the more careful investigation of data augmentation intensity, especially for the larger CXR images (like CheXpert), is of great importance to find the most appropriate set of data augmentation hyperparameters (angles, scales, etc.). It should be noted that high variability of AUC values for some pathologies on the same validation dataset after different training runs (Table 1 and [5]) remains a challenge for the real medical image applications of these methods and models. It is especially important aspect in the view of high variability of lungs for people of various ages, genders, geographical origin, substance abuse, professional activity, general health state, and other parameters of patients. In this context the better progress can be reached by further sharing the similar datasets around the world in the spirit of open science, volunteer data collection, processing and computing [20,21].

6 Conclusions

The results obtained here demonstrate the high efficiency of data augmentation technique and the limited usefulness of lung segmentation for CADx of some lung diseases for the models trained on the large (>10^5 images) lung image dataset, namely, CheXpert dataset, for small image sizes (320×320 pixels) and standard CNN (like DenseNet121) even. Lung mask segmentation has a subtle effect on the validation loss (for 4 from 14 lung diseases) for small image sizes (320×320 pixels) even in comparison to the original and other pre-processed datasets after data augmentation. For the large datasets the segmentation is not useful, because non-relevant features are effectively excluded after processing the sufficiently high number of images. The additional training for model with data augmentation results in the lower validation loss and the higher AUC values for 10 from 14 lung diseases. The matter is the data augmentation techniques allow to mimic variability of scanning conditions like different angles of scanning, slight distortions, small scaling up and down, etc. That is why, the more careful investigation of data augmentation intensity is of great importance to find the most appropriate set of data augmentation hyperparameters (angles, scales, etc). In conclusion, besides the more complex deep CNNs, the better progress of CADx for the big datasets (like CheXpert) could be obtained for some lung diseases by longer training after the tuned data augmentation, especially for larger CXR images.

References

1. Shen, D., Wu, G., Suk, H.: Deep learning in medical image analysis. Annu. Rev. Biomed. Eng. **19**, 221–248 (2017)
2. LeCun, Y., Bengio, Y., Hinton, G.: Deep learning. Nature **521**(7553), 436–444 (2015)
3. Smistad, E., et al.: Medical image segmentation on GPUs - a comprehensive review. Med. Image Anal. **20**(1), 1–18 (2015)
4. Rajpurkar, P., et al.: CheXNet: radiologist-level pneumonia detection on chest X-rays with deep learning. arXiv preprint arXiv:1711.05225 (2017)

5. Irvin, J., et al.: CheXpert: A Large Chest Radiograph Dataset with Uncertainty Labels and Expert Comparison. arXiv preprint arXiv:1901.07031 (2019)
6. Gordienko, Y., et al.: Deep Learning with lung segmentation and bone shadow exclusion techniques for chest X-ray analysis of lung cancer. In: Hu, Z., Petoukhov, S., Dychka, I., He, M. (eds.) ICCSEEA 2018. AISC, vol. 754, pp. 638–647. Springer, Cham (2019). https://doi.org/10.1007/978-3-319-91008-6_63
7. Peng, G., et al.: Dimensionality reduction in deep learning for chest X-ray analysis of lung cancer. In: Proceedings of 10th International Conference on Advanced Computational Intelligence, ICACI 2018, pp. 878–883. IEEE (2018)
8. Stirenko, S., et al.: Chest X-ray analysis of tuberculosis by deep learning with segmentation and augmentation. In: Proceedings of IEEE 38th International Conference on Electronics and Nanotechnology, pp. 422–428. IEEE (2018)
9. Shiraishi, J., et al.: Development of a digital image database for chest radiographs with and without a lung nodule: receiver operating characteristic analysis of radiologists' detection of pulmonary nodules. Am. J. Roentgenol. **174**, 71–74 (2000)
10. Armato, S.G., et al.: The lung image database consortium (LIDC) and image database resource initiative (IDRI): a completed reference database of lung nodules on CT scans. Med. Phys. **38**(2), 915–931 (2011)
11. Jaeger, S., et al.: Two public chest X-ray datasets for computer-aided screening of pulmonary diseases. Quant. Imaging Med. Surg. **4**(6), 475–477 (2014)
12. Wang, X., et al.: ChestX-ray8: hospital-scale chest x-ray database and benchmarks on weakly-supervised classification and localization of common thorax diseases. In: IEEE Conference on Computer Vision and Pattern Recognition (CVPR), pp. 2097–2106. IEEE (2017). arXiv preprint arXiv:1705.02315
13. van Ginneken, B., Stegmann, M.B., Loog, M.: Segmentation of anatomical structures in chest radiographs using supervised methods: a comparative study on a public database. Med. Image Anal. **10**(1), 19–40 (2006)
14. Hashemi, A., Pilevar, A.H.: Mass detection in lung CT images using region growing segmentation and decision making based on fuzzy systems. Int. J. Image Graph. Signal Process. **6**(1), 1–8 (2013)
15. Juhász, S., Horváth, Á., Nikházy, L., Horváth, G., Horváth, Á.: Segmentation of anatomical structures on chest radiographs. In: Bamidis, P.D., Pallikarakis, N. (eds.) XII Mediterranean Conference on Medical and Biological Engineering and Computing 2010. IFMBE Proceedings, vol. 29, pp. 359–362. Springer, Heidelberg (2010). https://doi.org/10.1007/978-3-642-13039-7_90
16. Kochura, Yu., Gordienko, Yu., Stirenko, S., et al.: Aggressive data augmentation and segmentation for lung disease diagnostics by deep learning (2019, submitted)
17. Ronneberger, O., Fischer, P., Brox, T.: U-Net: Convolutional networks for biomedical image segmentation. In: Navab, N., Hornegger, J., Wells, W.M., Frangi, A.F. (eds.) MICCAI 2015. LNCS, vol. 9351, pp. 234–241. Springer, Cham (2015). https://doi.org/10.1007/978-3-319-24574-4_28
18. Chollet, F.: Deep Learning with Python. Manning Publications, New York (2018)
19. Abadi, M., et al.: TensorFlow: large-scale machine learning on heterogeneous distributed systems. arXiv preprint:1603.04467 (2016)
20. Gordienko, N., Lodygensky, O., Fedak, G., Gordienko, Yu.: Synergy of volunteer measurements and volunteer computing for effective data collecting, processing, simulating and analyzing on a worldwide scale. In: Proceedings of the IEEE 38th International Convention on Information and Communication Technology, Electronics and Microelectronics (MIPRO), pp. 193–198. IEEE (2015)
21. Rather, N.N., Patel, C.O., Khan, S.A.: Using deep learning towards biomedical knowledge discovery. Int. J. Math. Sci. Comput. **3**(2), 1–10 (2017)

A Comparison Study of Deep Learning Techniques to Increase the Spatial Resolution of Photo-Realistic Images

Andrew M. Shackleton and Abdulrahman M. Altahhan(✉)

Leeds Beckett University, Leeds, UK
Andrew.shackleton1@ntlworld.com, A.Altahhan@leedsbeckett.ac.uk

Abstract. In this paper we present a perceptual and error-based comparison study of the efficacy of four different deep-learned super-resolution architectures, ESPCN, SRResNet, ProGanSR and LapSRN, all performed on photo-realistic images by a factor of 4x; adapting some of the current state-of-the-art architectures using Convolutional Neural Networks (CNNs). The resultant application and the implemented CNNs are tested with objective (Peak-Signal-to-Noise ratio and Structural Similarity Index) and perceptual metrics (Mean Opinion Score testing), to judge their relative quality and implementation within the program. The results of these tests demonstrate the effectiveness of super-resolution, showing that most network implementations give an average gain of +1 to +2 dB (in PSNR), and an average gain of +0.05 to +0.1 (in SSIM) over traditional Bicubic scaling. The results of the perception test also show that participants almost always prefer the images scaled using each CNN model compared to traditional Bicubic scaling. These findings also present a look into new diverging paths in super-resolution research; where the focus is now shifting from solely error-reduction, objective-based models to perceptually focused models that satisfy human perception of a high-resolution image.

1 Introduction

Traditional image scaling techniques such as nearest-neighbour, bilinear, and bicubic interpolation offer computationally quick methods of increasing the size of an image, but they do not provide any benefit to quality as they cannot construct or infer new data; able to only increase the scale of what is already present in the original image.

Nearest neighbour interpolation works by first enlarging the image by the desired factor and spreading the already available pixels within the newly defined space. The original pixels are surrounded by a 'grid' of blank space in which there are no original pixels from the image; the blank spaces are then filled by copying the 'nearest-neighbour' pixels to the blank space, turning one pixel to four identical pixels (for 4x scale). To perform bilinear interpolation, pixels are sampled in two directions. This type of scaling takes the closest 4 pixels located

T. Gedeon et al. (Eds.): ICONIP 2019, CCIS 1142, pp. 341–348, 2019.
https://doi.org/10.1007/978-3-030-36808-1_37

diagonally into account (2×2) and takes a weighted average, as opposed to nearest-neighbours singular sample. Bicubic interpolation further considers the weighted average of the nearest 16 pixels (in a grid of 4×4), which produces an overall smoother image and reduces artefacts. Because the region of sampling is greater for this algorithm compared to others, pixels closer to the chosen interpolated pixel are given a greater weighting in the calculation.

Whilst such image resampling techniques increase the actual 'resolution' of the image when upscaling, they do not present any added detail that contributes to the increase in spatial resolution of the final image. This results in an equal or less-than equally detailed output image, such that one might refer to the output as 'blurry' when compared to a similar image of native resolution. This issue has led to the research and development of machine learned models to improve upon traditional methods of image upscaling; a method known as super-resolution.

1.1 Motivation and Rationale

Super-Resolution can have applications in surveillance, medical imaging, astronomical observation, and so on (Yue et al.) [5]. Super-Resolution also has novel uses; a popular application of such techniques is upscaling textures from older video games to bring them into the modern era, as well as enhancing old low-resolution photographs, or enhancing complex drawings and diagrams. Image super-resolution by nature is an ill-posed problem as there is no true output to an image that does not have a corresponding high-resolution parent. There are a number of different approaches that have been taken using machine learning and convolutional neural networks (CNNs) for image super-resolution; such as SRCNN, SSResNet, Deep Image Prior and ESPCN. These all attempt, using different architectures, to up-scale an image while retaining/reconstructing fine image detail that is not found within the original low-resolution image (such as sharp edges on geometric shapes, or texture detail on small scale objects). Many of these methods for super resolution exist in a primitive form however; the majority being simply proposals that offer independent python command line implementations based on Linux, or working models built using and running within MATLAB.

1.2 Related Literature

ESPCN. The following method by Shi et al. [4] ESPCN, uses a shallow 3-layer convolutional neural network and avoids upscaling the low-resolution input like in (Dong et al.) [3]. A convolutional layer is applied directly on the low-resolution input to extract the feature maps, followed by a sub-pixel convolutional layer to upscale these feature maps to produce the super resolution output. This method differs from Dong et al. [3] in that it uses an efficient sub-pixel convolution layer instead of deconvolution layer (which recovers resolution from the max pooling layer, also known as backwards convolution). This pixel shuffle layer is faster than methods that use a deconvolution layer specifically in training, as well as being faster than methods performing upscaling or pre-processing before convolution is applied. In Shi et al. [4], ESPCN with ReLU activation trained with ImageNet data achieved significantly better performance compared to SRCNN models. Training the ESPCN model with more images saw a greater gain in PSNR than the values found with SRCNN. Interestingly, performance on this architecture is found to be high enough that it is capable of running on video without severe performance degradation.

SRResNet. Another architecture by Ledig et al. [6] presents a method of Super-resolution combining error reduction focused architectures with a GAN architecture. The authors pose that while performance and accuracy of current super-resolution models are a benefit, recovering fine-detail in the image has not yet been tackled successfully. Most methods (A+, SRCNN, ESPCN, and LapSRN for example) are based on Mean Squared Error (MSE) reduction during reconstruction. While the resultant PSNR values for these techniques are high, high-frequency details are missing and the images do not give the visual perception of being high-resolution to the human eye. By combining a CNN optimised for Mean Squared Error (SRResNet) with a Generative Adversarial Network-based model (SRGAN), this problem can be overcome. This architecture sees greater gains in PSNR and SSIM over both ESPCN (Shi et al.) [4] and SRCNN (Dong et al.) [3], however as the authors rightfully state, that these values are not representative of the fine detail reconstruction that SRGAN provides. The authors therefore take an extra step and use Mean Opinion Score testing to quantify the super resolution capabilities of each of these models (Fig. 1).

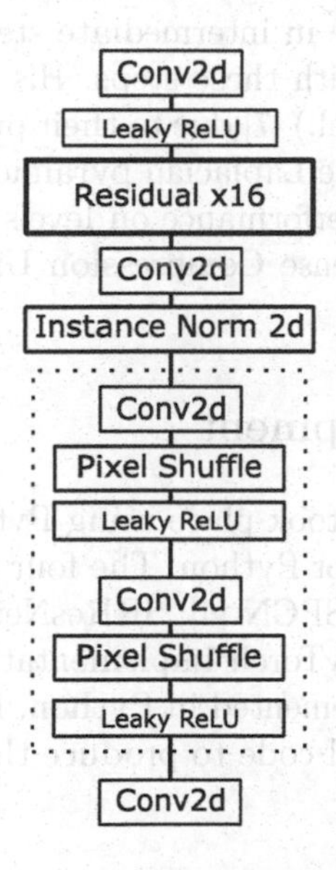

Fig. 1. The architecture of SRResNet

LapSRN. The architecture by Lai et al. [7] referred to as LapSRN provides an alternative process of super-resolution using Laplacian pyramids. The authors highlight drawbacks of using pre-processing methods found in other techniques, in that they increase the computational cost unnecessarily and do not provide any additional high frequency information for a HR output. Many techniques focus around MSE loss, resulting in overly smooth images (the same low-resolution patch may have multiple high-resolution output patches in correspondence). The authors propose a progressive approach which eliminates the single up-sampling step that most other models use (SRCNN, ESPCN use direct reconstruction in a single step), to progressively reconstruct images along the network. The Laplacian Pyramid structure of this network is a key concept; where weights are shared across pyramid levels to reduce network parameters. This subsequently allows for multi-scale training for different levels of super resolution at once (2x, 4x, 8x pyramids). The authors also state the LapSRN can be easily extended to incorporate adversarial training as a part of GAN, as found in Ledig et al. [6] and Wang et al. [8], however this is not provided in the paper.

ProSR. Taking the concept of progressive reconstruction a step further, Wang et al. [8] propose an architecture that combines two methods, ProSR; a progressive method to upscale images in intermediate steps, and ProGanSR which follows the same design principle but allows for more photo-realistic results to be generated using a GAN. This diverges from other traditional methods in that it takes a progressive approach with "curriculum learning" as opposed to direct methods which upsample in a single final step. The basis of this is that the network up-samples the image in intermediate steps while the learning process increases in difficulty along with these steps. His approach shares similarity in concept with LapSRN (Lai et al.) [7] due to their progressive approaches, but the authors of ProSR note that the Laplacian pyramid structure increases difficulty of optimisation and reduces performance on levels higher up the pyramid structure. The authors propose Dense Compression Units consisting of both Dense Blocks and Compression.

2 Design and Development

Neural Network development took place using Python 3 with PyTorch 1.0. The GUI was developed using Qt for Python. The four models mentioned above were chosen for implementation; ESPCN [4], SRResNet(w/o GAN) [6], LapSRN [7], ProGanSR [8]. Each has a PyTorch implementation officially provided by the author or independently implemented in Python. Each models code was further adapted to work with the GUI code to produce the resultant application.

2.1 Training

All models are trained for a desired resolution multiplier of 4x. Training was performed locally using an NVIDIA GeForce GTX 1080 Ti. Training was performed

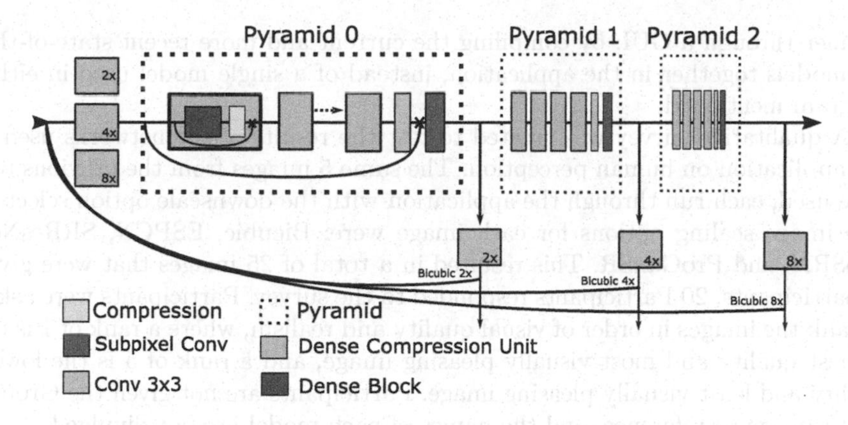

Fig. 2. The architecture of ProSR (without GAN) as found in [8]

using CUDA v9.0 to provide faster execution and training speeds. Datasets used for training include BSDS500 (Arbelaez et al.) [2], DIV2K (Agustsson et al.) [1]. In order to train, the data set images are first downscaled by 4x. An independent implementation of MATLAB's imresize function was used, as this provides the best results for bicubic downscaling compared to other methods found within Python. Training datasets were also augmented with random cropping, flipping, and transposing of each image. Each model was trained individually via said local machine, or via the provided model from the author for 100 Epochs (Fig. 2).

2.2 Testing

Two tests performed, a test validating output images from the application using PSNR and SSIM with a python script; and another evaluating human perception on the same set of test images to judge percieved quality via a survey. The Python implementations are not a perfect recreation of the models described in the relevant literature, as such the results for PSNR found within the literature are typically greater than those of the python versions when tested with similar images. The PSNR and SSIM testing for images within relevant literature is performed on the Y channel, and so for this test the image channels are separated, and testing is performed on the Y channel.

2.3 Similar Work

Applications such as Waifu2x and Topaz A.I. Gigapixel perform similar functions to the proposed application; Waifu2x works best on non-photoreal images such as drawings and cartoons at up to 2x factor scaling based around the (no longer state-of-the-art) SRCNN architecture, and Gigapixel is a proprietary piece of software in which the algorithms used are unknown. This prototype application differs from both of these in that it is a free application that makes practical use of more up-to-date, publicly available image scaling networks in a user-friendly manner through a GUI; by compiling the current and more recent state-of-the-art models together in the application, instead of a single model used in either program mentioned.

A qualitative survey was created to test the results of the networks used in the application on human perception. The same 5 images from the previous test were used, each run through the application with the downscale option selected. The image scaling options for each image were; Bicubic, ESPCN, SRResNet, LapSRN, and ProGanSR. This resulted in a total of 25 images that were given to participants. 20 Participants responded to the survey. Participants were asked to rank the images in order of visual quality and realism, where a rank of 1 is the highest quality and most visually pleasing image, and a rank of 5 is the lowest quality and least visually pleasing image. Participants are not given the Ground Truth image as reference, and the names of each model are not divulged.

3 Results and Evaluation

3.1 PSNR and SSIM

On the 'statuette' image set, bicubic scaling appears to give the highest value results for both SSIM and PSNR. It is unclear why this happens, but it is only the case on this image. This example is some justification as to why PSNR and SSIM alone are not a concrete metrics for judging image quality. SRResNet has the most occurrences of the highest values of PSNR and SSIM on the 5 sets of test images, in both test runs. SRResNet also outperforms ProSR when tested against these metrics, which is to be expected. ESPCN falls behind bicubic scaling in many of these test cases, in both SSIM and PSNR. The majority of results gathered in this test show that error-focused architectures do outperform both traditional scaling methods and perceptual-focused architectures. It is clear when looking at the images PSNR and SSIM alone do not provide the optimal method for judging the visual quality of a super-resolved image (Table 1).

Table 1. The results of the PSNR and SSIM Test on a custom set of 5 images.

Image set	Test	ProSR	SRResNet	LapSRN	ESPCN	Bicubic
Sign	PSNR (dB)	27.674	27.792	28.006	24.283	24.844
-	SSIM	0.888	0.883	0.885	0.696	0.787
Dog	PSNR (dB)	25.325	27.146	25.478	25.738	26.438
-	SSIM	0.753	0.800	0.791	0.741	0.776
Statuette	PSNR (dB)	24.300	26.054	25.007	24.004	27.766
-	SSIM	0.836	0.826	0.821	0.813	0.856
Bluebell	PSNR (dB)	22.604	23.815	22.767	21.235	22.061
-	SSIM	0.746	0.788	0.765	0.689	0.693
View	PSNR (dB)	21.950	23.504	22.735	22.367	21.083
-	SSIM	0.644	0.704	0.697	0.663	0.574

3.2 Perceptual Study

ESPCN was ranked lowest of the tested group, only barely contesting bicubic scaling in most cases. Looking at the images, there is only a minute difference between ESPCN and Bicubic, with ESPCN looking slightly sharper than the Bicubic images. As expected, Bicubic scaling provides the worst quality results and this is reflected in the participants' response. In 3 out of 5 test cases, Bicubic scaling is ranked higher than or equal to ESPCN. Therefore, it can be determined from this that ESPCN provides an alternative to Bicubic scaling, not a true replacement as was expected with the other models (Fig. 3).

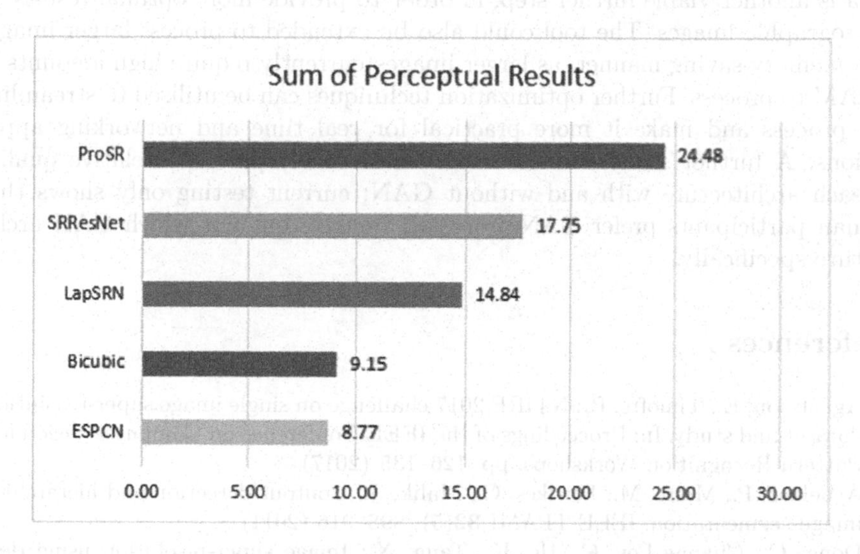

Fig. 3. A bar graph showing the total aggregate results of perceptual testing on the same 5 images.

4 Conclusion

This paper has presented a study and prototype implementation of state-of-the-art techniques for super-resolution within a x64 and Unix compatible application, allowing for any user to upscale a desired image using these techniques without the need for knowledge of programming or deep learning. Through testing, we find that the error-focused architectures (based around PSNR, SSIM, and Mean-Square Error testing) provide some excellent techniques for objective super-resolution, but result in often murky and smudged images. The perceptually-focused architectures, a more recent development making use of adversarial networks, give promising results that better represent true, high-resolution images able to fool the human perception. In the context of applications of these models, perceptual approaches that hallucinate finer detail might be less suited for medical applications or surveillance because the data they produce is technically not present within the original image, giving an advantage to error-focused approaches. Perceptual approaches may therefore be more useful for applications that do not specifically require the content of the images to be accurate (such as personal photos). This gives merit to the suggestion that one path for super-resolution is not necessarily better than another.

5 Further Work

The application can be extended to work on other forms of media with further training, such as drawings or animations. Re-training each network with more data is another viable further step, in order to provide more optimal results on photographic images. The tool could also be extended to process larger images in a memory-saving manner, as larger images currently require high amounts of VRAM to process. Further optimization techniques can be utilised to streamline the process and make it more practical for real time and networking applications. A further study could be conducted to compare the relative quality of each architecture with and without GAN; current testing only shows that human participants prefer GAN-processed images, but not which GAN architecture specifically.

References

1. Agustsson, E., Timofte, R.: NTIRE 2017 challenge on single image super-resolution: dataset and study. In: Proceedings of the IEEE Conference on Computer Vision and Pattern Recognition Workshops, pp. 126–135 (2017)
2. Arbelaez, P., Maire, M., Fowlkes, C., Malik, J.: Contour detection and hierarchical image segmentation. IEEE TPAMI **33**(5), 898–916 (2011)
3. Dong, C., Change Loy, C., He, K., Tang, X.: Image super-resolution using deep convolutional networks. IEEE Trans. Pattern Anal. Mach. Intell. **38**(2), 295–307 (2015)
4. Shi, W., et al.: Real-time single image and video super-resolution using an efficient sub-pixel convolutional neural network. In: 2016 IEEE Conference on Computer Vision and Pattern Recognition (CVPR), Las Vegas, NV, USA, 27–30 June 2016. IEEE (2016)
5. Yue, L., Shen, H., Li, J., Yuan, Q., Zhang, H., Zhang, L.: Image super-resolution: the techniques, applications, and future. Signal Process. **128**, 389–408 (2016)
6. Ledig, C., et al.: Photo-realistic single image super-resolution using a generative adversarial network. In: 2017 IEEE Conference on Computer Vision and Pattern Recognition (CVPR), Honolulu, HI, USA, 16–21 July 2017. IEEE (2017)
7. Lai, W., Huang, J., Ahuja, N., Yang, M.: Fast and accurate image super-resolution with deep Laplacian pyramid networks. IEEE Trans. Pattern Anal. Mach. Intell. (2018)
8. Wang, Y., Perazzi, F., McWilliams, B., Sorkine-Hornung, A., Sorkine-Hornung, O., Schroers, C.: A fully progressive approach to single-image super-resolution. In: 2018 IEEE/CVF Conference on Computer Vision and Pattern Recognition Workshops (CVPRW), Salt Lake City, UT, USA, 18–22 June 2018. IEEE (2018)

Embeddings and Feature Fusion

A Robust Embedding for Attributed Networks with Outliers

Cheng Zhang[1,2], Le Zhang[1,2], Yuanye He[2], and Daren Zha[2(✉)]

[1] School of Cyber Security, University of Chinese Academy of Sciences, Beijing, China
[2] Institute of Information Engineering, Chinese Academy of Sciences, Beijing, China
{zhangcheng,zhangle,heyuanye,zhadaren}@iie.ac.cn

Abstract. Network embedding, as a promising tool, aims to learn low-dimensional embeddings for nodes in a network. Most existing methods work well when the topological structure is closely correlated to node attributes. However, real-world networks often contain outliers that have abnormal attributes. These attributes are quite different from the properties of their neighboring nodes, and they are not consistent with network structure. Thus, outliers exert negative impacts on the learned embeddings. Several methods only obtain unsatisfied results as they don't consider the effects of outliers. Hence, how to eliminate outlier impacts is essential for network embedding. In this paper, we propose a novel method called REANO for learning a Roust Embedding for an Attributed Network with Outliers. An overview of REANO combines residual analysis with attributed network embedding. In detail, residual analysis smooths out the negative impacts from outliers. Meanwhile, network embedding aggregates node attributes with network structure by using deep neural networks. By developing a joint optimization framework, REANO effectively alleviates outlier effects on the learned embeddings, and improve the robustness of node embeddings. Experiments on real-world datasets and manually planted outliers show that REANO learns more robust embeddings of nodes than the state-of-the-art algorithms.

Keywords: Network outliers · Network embedding · Attributed networks

1 Introduction

Attributed networks are ubiquitous in real-world systems such as social networks and academic citation networks. They often contain both network structure and node attributes [15,24]. And these rich properties of nodes reflect the homophily correlation with the topological structure [15,24]. Specifically, nodes usually share similar properties with their structural neighboring nodes, and their proximities are enhanced by considering both of network structure and node attributes [7,23]. Centered around this goal, researchers have proposed

T. Gedeon et al. (Eds.): ICONIP 2019, CCIS 1142, pp. 351–362, 2019.
https://doi.org/10.1007/978-3-030-36808-1_38

attributed network embedding. It aims to embed each node of the network into a low-dimensional vector space so that its proximity in terms of both structure and attribute information are well preserved [3,6,22]. The learned embeddings are directly applied to various applications such as node classification [9,10] and anomaly detection [4,8].

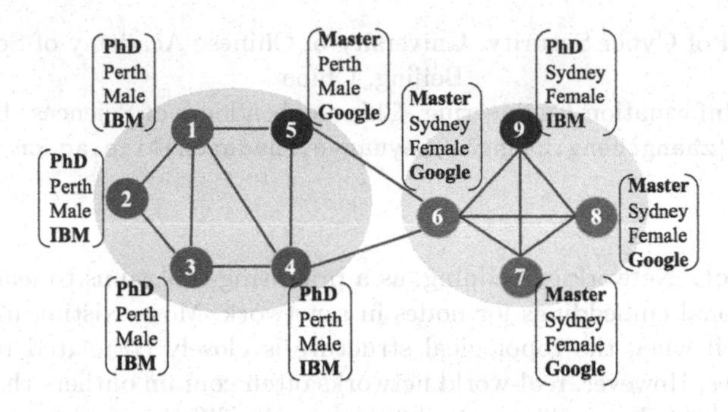

Fig. 1. An illustration of an attributed network with outliers. We highlight the outlier node and its abnormal attributes by red. In this network, nodes denote users having four attributes (*education, country, gender, company*), and edges describe their friendship relations. Node 5 and 9 are two typical outliers. They contain abnormal attributes that are obviously deviated from their normal neighbors. Respectively, the *education* and *company* attributes of node 5 and 9 are very different from their structural neighbors. (Color figure online)

Recently, most existing attributed network embedding models have been proposed [11,22]. They generate desirable node embeddings by exploiting the structure and attribute information jointly. These methods work reasonably well when they run on a normal environment, and assume that all attributes of nodes are closely correlated to network structure [7,23]. However, real-world networks often contain abnormal information, and even face noisy scenarios with anomalous nodes [1,2]. These outliers, which are shown in Fig. 1, have very different attributes from their neighborhoods. Actually, these outliers bring adverse impacts on the embeddings of nodes [3,13]. Hence, it's essential to eliminate the outlier effects and improve the robustness of the learned node embeddings.

It is a nontrivial task to develop a robust network embedding for attributed networks with outliers. There are two main reasons. Firstly, due to the large-scale and high dimensionality of node attributes, there are usually anomalous nodes in real-world datasets [1]. And how to address these outliers is essential for network embedding [3,13]. Besides, although current works on attributed networks could effectively either detect outliers [4,12] or obtain desirable node embeddings [9, 22], the bewildering combination of network embedding and detection outliers makes the robust embedding learning still difficult. If we regard them as two independent steps, there merely produce unsatisfied results [7,23].

To overcome the aforementioned challenges, in this paper, we propose a novel model called REANO that learns a Robust Embedding for Attributed

Networks with Outliers. In detail, REANO uses residual analysis [12,19] to smooth out abnormal information arising from anomalous nodes. Meanwhile, REANO employs deep neural networks [6,13] to incorporate node attributes and network structure into a joint embedding space. Besides, by designing a joint optimization framework, REANO combines residual analysis with network embedding as a whole. Thus, REANO improves the robustness of the learned node embeddings, while effectively alleviating adverse impacts from outliers. And we conduct experimental tasks and manually planted outliers to demonstrate the effectiveness of our proposed model. In summary, we have the following main contributions:

- We explicitly account for negative effects of outliers on the learned node embeddings, and leverage residual analysis for smoothing out these impacts.
- We employ network embedding based on deep neural networks to incorporate node attributes and network structure into a joint embedding space.
- We perform experimental tasks on real-world datasets and manually planted outliers to verify its superior performance than the state-of-the-art baselines.

2 Related Works

Existing network embedding algorithms consist of plain network embedding and attributed network embedding [7,23]. The former only considers network structure, while the latter preserves both structural and attribute information. Specifically, inspired by Skip-Gram [16], DeepWalk [17], LINE [18] and Node2Vec [5] ensure that the embeddings of the two structurally connected nodes are similar. SDNE applies autoencoder to capture the non-linear structure for network embedding [20]. These algorithms only exploit the topological structure for generating embeddings. Researchers have proposed ideas for embedding attributed networks also [7,23]. As the first attempt, TADW improves network representation by injecting texts [21]. LANE jointly leverages topological structure, node attributes and labels [9]. Furthermore, GCN [11] and GraphSage [6] use convolutional neural networks for learning the embeddings on attributed networks. However, all of these methods mentioned above do not directly account for outliers, and hence are often prone to be affected heavily by them [13].

Current efforts attempt to detect network outliers by adopting the idea of embedding [7,23]. For instance, Embed employs embedding for discovering structurally inconsistent nodes and regards them as outliers [8]. APE maps entities into a unified embedding space, then uses Noise-Contrastive Estimation to find abnormal events in this space [4]. Although these algorithms explicitly account for the outlier impacts, they target for spotting outliers, not for network embedding [1]. More recently, SEANO effectively smooths out the effects of outliers on the learned embeddings by predicting the class labels and node context [13].

3 Problem Statement

In this section, we give some notations that will be used in the paper and describe our problem. Given an attributed network $G = (\mathcal{V}, \mathcal{E}, A)$, where $\mathcal{V}$ is the set of $|\mathcal{V}|$

nodes and $\mathcal{E}$ denotes undirected edges of the network. $A \in \mathbb{R}^{|\mathcal{V}| \times m}$ is the attribute matrix that collects all node attributes, and a row $a_i \in \mathbb{R}^m$ in the matrix A represents the node v_i's attributes. Since the given network G has abnormal attributes arising from outliers, our problem is to learn a robust low-dimensional embedding vector $z_i \in \mathbb{R}^d$ for each node $v_i \in \mathcal{V}$, where $d \ll |\mathcal{V}|$. Specifically, the learned embeddings not only effectively eliminate adverse impacts from outliers, but also seamlessly encode the topological structure with node attributes.

4 Methodology

In this section, we firstly present the overall architecture of REANO and its detailed implementation, then introduce a joint framework to optimize this model (Fig. 2).

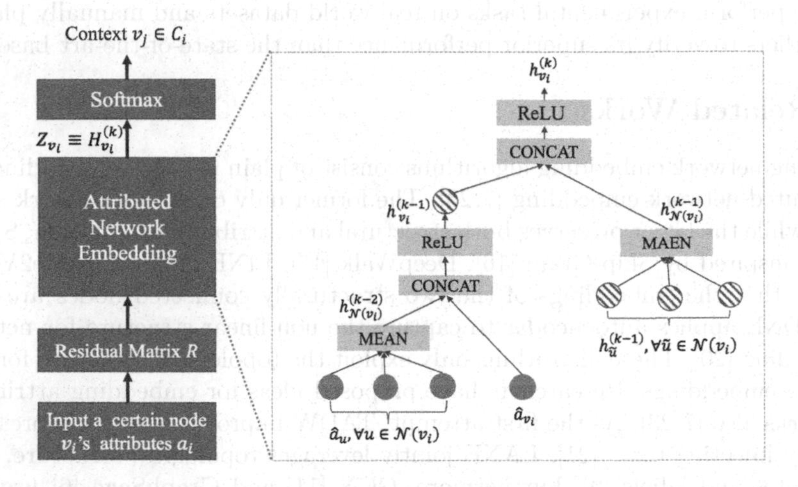

Fig. 2. The overall framework of REANO combines residual analysis with attributed network embedding. Inspired by residual analysis [19], the residual matrix R firstly filters out the abnormal information arising from outliers, then obtains normal attributes $\hat{A}$ of nodes. Meanwhile, network embedding employs multilayer neural networks to map the remaining attributes $\hat{A}$ and structural information $\mathcal{E}$ into a joint embedding space $z_i \in \mathbb{R}^d$, where a node v_i's attributes are aggregated with the properties of its neighbors $\mathcal{N}(v_i)$. Then, the generated embedding z_i of node v_i are fed into the softmax function to predict its structural contexts C_{v_i}. Through iteratively optimizing residual analysis and network embedding, REANO learns robust node embeddings $z_i \equiv h_{v_i}^{(k)}$, $\forall v_i \in \mathcal{V}$, while effectively alleviating adverse impacts of outliers.

The core idea of REANO is that an attributed network exists anomalous nodes that have quite different attributes from its normal neighbors, and these outliers can exert adverse effects on the learned embeddings. In order to improve the robustness of node embeddings, we need to capture and eliminate the abnormalities from node attributes, then get the remaining attributes without outliers.

Simultaneously, we incorporate these normal attributes and network structure into a joint embedding space, in which effectively preserves the proximities of nodes in terms of both structural and attribute information.

4.1 Residual Analysis

We start from the situation that only node attributes are available. As proposed to [12, 19], residual analysis has a good capability of detecting outliers on networks. And it performs detection outliers by studying the residual errors between true data and the estimated one. For instance, if a certain node contains abnormal information, the residual matrix designed by residual analysis could effectively distill the abnormalities from the original data [25]. Therefore, this node is of a large possibility to be anomalous as it has a pretty large error [12]. On the contrary, if the node has any abnormal information, its residual value is quite small, and this node has a small chance to be an anomalous.

Motived by this fact that residual analysis is capable of eliminating abnormal attributes of nodes and measuring the degrees of these abnormalities, similar to [19, 25], we also construct a residual matrix $R \in \mathbb{R}^{|\mathcal{V}| \times m}$ to filer out the abnormal information arising from network outliers. Mathematically, it is formulated as:

$$\tilde{A} = A - R \tag{1}$$

Here the vector $R(i,:)$ denotes the residual value of a node v_i's attributes $A(i,:)$, and $\tilde{A}(i,:)$ represents the remaining attributes of the corresponding node v_i. Obviously, these node attributes involve any abnormalities. As proposed in [19, 25], we also compute the regularization term of $R(i,:)$ to reflect the abnormality degree of the node v_i.

4.2 Attributed Network Embedding

In order to learn robust embeddings of nodes in the network, it's demand for network embedding that not only explicitly eliminates the noisy information from node attributes, but also exploits the normal attribute and structural information jointly. As shown in Sect. 4.1, through smoothing out the abnormalities arising from outliers, REANO gets the remaining attributes $\hat{A}$, which are consistent with the topological structure. According to the homophily theories [15, 24], nodes sharing similar attributes tend to connect with each other when nodes behave as expected. Therefore, jointly learning from the two information of both the remaining node attributes and network structure could draw towards a better embedding. As proposed in [6, 13], REANO also designs k-layer neural networks (where $k \geq 2$) to incorporate node attributes and network structure.

Specifically, at the bottom layer of this neural network architecture, REANO randomly samples a fixed-size set of neighborhoods $\mathcal{N}(v_i)$ from a node $v_i \in \mathcal{V}$, then applies an aggregator to assimilate the aforementioned two heterogeneous information into a joint embedding space as following:

$$h_{\mathcal{N}(v_i)}^{(k-2)} = \text{MEAN}\left(\hat{a}_u, \forall u \in \mathcal{N}(v_i)\right) \tag{2}$$

Here MEAN denotes the aggregator. Similar to [6], it also employs the mean operator to calculate the element-wise mean of the vectors $\hat{a}_u, \forall u \in \mathcal{N}(v_i)$, then gets a single vector $h^{(k-2)}_{\mathcal{N}(v_i)}$. Note that this aggregated vector has the same size as the dimensionality of current node's attributes. Subsequently, to achieve desirable training, we concatenate the vector $h^{(k-2)}_{\mathcal{N}(v_i)}$ with the attribute vector a_i as proposed to [6,13]. Then the concatenation $\text{CONCAT}\left\{\hat{a}_i \cup h^{(k-2)}_{\mathcal{N}(v_i)}\right\}$ is fed through a fully connected layer, where it contains a series of non-linear activation functions $\sigma(\cdot)$. Methodically, it's formulated as:

$$h^{(k-1)}_{v_i} = \sigma\left(\mathbf{W}^{(k-1)} \cdot \text{CONCAT}\left\{\hat{a}_{v_i} \cup h^{(k-2)}_{\mathcal{N}(v_i)}\right\} + b^{(k-1)}\right) \tag{3}$$

where $\mathbf{W}^{(k-1)}$ and $b^{(k-1)}$ are the weights and biases at the bottom layer respectively, and the activation function $\sigma(\cdot)$ is chosen as $\text{ReLU}(x) = \max(0, x)$. And this aggregation step transforms the input information to the node embedding $h^{(k-1)}_{v_i}$, which is regarded as the feature vector used at the next step. Similarly, at the middle layer, REANO also aggregates the features of the nodes in its immediate neighboring nodes $\mathcal{N}(v_i)$ as following:

$$h^{(k-1)}_{\mathcal{N}(v_i)} = \text{MEAN}\left(h^{(k-1)}_u, \forall u \in \mathcal{N}(v_i)\right) \tag{4}$$

$$h^{(k)}_{v_i} = \sigma\left(\mathbf{W}^{(k)} \cdot \text{CONCAT}\left\{h^{(k-1)}_{v_i} \cup h^{(k-1)}_{\mathcal{N}(v_i)}\right\} + b^{(k)}\right) \tag{5}$$

where $W^{(k)}$ and $b^{(k)}$ are the weights and biases at the middle layer separately, and the activation function $\sigma(x)$ is also used as $\text{ReLU}(x)$. Therefore, the middle layer generates the embedding $h^{(k)}_{v_i}$ of a certain node v_i. For simplicity, the learned node embedding is rewritten as $z_i \equiv h^{(k)}_{v_i}, \forall v_i \in \mathcal{V}$.

In order to maximize the correlations between network structure and node attributes, as proposed to [22,24], we encourage a node v_i and its contextual node v_j to have similar embeddings. And this process is written as following:

$$\text{P}(v_j|v_i) = \frac{\exp(z_j \cdot z_i)}{\sum_{v' \in \mathcal{V}} \exp(z_{v'} \cdot z_i)} \tag{6}$$

Here the contexts $C_i = \{v_{i-t}, \cdots, v_{i+t}\} \setminus v_i$ of a certain node v_i are generated by performing random walk on the network as proposed to [5,17]. Specifically, the probability $\text{P}(v_j|v_i)$ in Eq. (6) is computed by the softmax function as follows:

$$\mathcal{L}_{ne} = -\sum_{v_i \in \mathcal{V}} \sum_{v' \in C_i} \log\left(\text{P}(z_{v'}|z_i)\right) \tag{7}$$

Actually, calculating the Eq. (7) directly is rather expensive because it needs to run through all nodes. We therefore adopt the negative sampling strategy [16] to speed up the training process. Accordingly, Eq. (7) is rewritten as follows:

$$\mathcal{L}_{ne} = -\sum_{v_i \in V} \sum_{v_j \in C_i} \left\{\log\sigma\left(z^T_{v_i} \cdot z_{v_j}\right) + \sum_{l=1}^{|\text{neg}|} \mathbb{E}_{v_l \sim P_n(v_i)} \log\left(-z^T_{v_i} \cdot z_{u^l_i}\right)\right\} \tag{8}$$

where $z_{u_i^l}$ is randomly sampled the lth negative node for node v_i. In total, we sample |neg| negatives. And the sampled ratio is $P_n(v_i) \propto r_i^{0.75}$, where r_i is the node v_i's degree. Hence, the loss function in Eq. (8) encourages nearby nodes to have similar embeddings, while making the embeddings of disparate nodes be highly distinct.

4.3 A Joint Optimization Framework

Algorithm 1. The REANO Algorithm

Input: an attributed network $G = (\mathcal{V}, \mathcal{E}, A)$, walk length l, paths per node γ, embedding dimension d, constraint parameters β and α.
Output: the embedding $\Phi(\mathrm{v_i})$ for each node $v_i \in \mathcal{V}$
1: Perform random walk(l,γ) on the network G;
2: Generate the positive and negative contexts for each node $v_i \in \mathcal{V}$;
3: Initialize the residual matrix R with zeros matrix
4: **while** not converged **do**
5: Employ the residual matrix R to filter out origin attributes A of nodes;
6: **for each node** $v_i \in \mathcal{V}$ **do**
7: Aggregate v_i's attributes $\hat{a}_i$ with its nearby attributes $\hat{a}_{\mathcal{N}(v_i)}$ on Eq.(2).
8: Learn the node v_i's features $h_{v_i}^{(k-1)}$ from Eq.(3).
9: Aggregate v_i's features $h_{v_i}^{(k-1)}$ with its neighbors' features $h_{\mathcal{N}(v_i)}^{(k-1)}$ on Eq.(4).
10: Learn the node v_i's embedding $h_{v_i}^{(k)}$ from Eq.(5).
11: **end for**
12: **for each node** $v_i \in \mathcal{V}$ **do**
13: **for each context node** $v_j \in C_i$ **do**
14: Compute the loss function $\mathcal{L}$ on Eq.(9)
15: **end for**
16: **end for**
17: Update weights/biases and the residual matrix R by the gradient $\nabla\mathcal{L}$ and the learning ratio η
18: **end while**
19: Obtain the robust embedding z_{v_i} for each node $v_i \in \mathcal{V}$ in G.

According to [12,25], there always exists a small part of outliers in the network. In order to have better performance, the $\ell_{1,2}$-norm $\left\|R^T\right\|_{2,1}$ should be imposed on the Eq. (8). Similar to [13,25], Eq. (8) should also add the regularization terms $\sum_{k=1}^{K}\left(\left\|\mathbf{W}^{(k-1)}\right\|_F^2 + \left\|b^{(k-1)}\right\|_F^2\right)$ to get better training. And the final loss function is formulated as follows:

$$\mathcal{L} = \mathcal{L}_{ne} + \alpha\left\|R^T\right\|_{2,1} + \frac{\beta}{2}\sum_{k=1}^{K}\left(\left\|\mathbf{W}^{(k-1)}\right\|_F^2 + \left\|b^{(k-1)}\right\|_F^2\right) \tag{9}$$

where the parameter $\beta \geq 0$ controls the column sparsity of matrix R and the weight α is to balance the loss. To minimize the Eq. (8), the learned final embeddings of nodes in the network can seamlessly encode node attributes with network structure, while effectively eliminating negative impacts from outliers.

Noting that the REANO consists of two parts: residual analysis and attributed network embedding. If we treat them as two independent steps rather than a whole, it may result in a suboptimal performance [22,24]. Therefore, we develop a joint optimization framework, which is summarized in Algorithm 1. By using the stochastic gradient algorithm to optimize for Eq. (9), we can achieve the goal of improving the robustness of the node embeddings.

5 Experiments

5.1 Experimental Settings

Datasets. There are four real-world attributed networks used in our experiments. In detail, the former three datasets are used for node classification task, and the later one presents at a case study. These first three datasets including Cora, Citeseer, and Pubmed[1], were used in previous works [6,13], in which nodes denote published papers and edges represent the citation links between them. The attributes of each node are a list of keywords of the corresponding paper. And each node only contains a class label. Besides, the last dataset is Disney[2] collected from Amazon co-purchased network. Nodes in this dataset represent movie products and attributes describe their properties such as ratings, product prices and so on. Then the class label indicates whether the node is an outlier or not. We summary the details of these network datasets in Table 1.

Table 1. A summary of the three real-world network datasets

Dataset	# of Nodes	# of Edges	# of Attributes	# of Labels
Cora	2,708	5,278	1,433	7
Citeseer	3,312	4,660	3,703	6
PubMed	19,717	44,338	500	3
Disney	124	334	28	2

To check the robustness of the methods to outliers, as proposed in [3,13], we manually plant a total of 5% outliers in the first three datasets. This planted process involves: firstly we randomly select 5% of nodes, including both labeled and unlabeled nodes; then we modify their attributes based on the natural perturbation scheme depicted by [13,14]. Specifically, for a planted outlier v_i, we randomly select another $m = \min\left(100, \frac{n}{4}\right)$ nodes and pick the node v_j with the most different attributes from v_i among these selected nodes, *i.e.*, maximizing $\|a_i - a_j\|_2$. Finally, we replace the attributes a_i of node v_i by v_j.

[1] http://linqs.cs.umd.edu/projects/projects/lbc.

[2] http://www.ipd.kit.edu/muellere/consub/.

Baselines. To evaluate the effectiveness of our proposed model, we compare it with the following baseline methods:

- **DeepWalk** [17] performs random walks to get structure information, and learns node embeddings by inputing this information into Skip-Gram [16].
- **LINE** [18] defines the first- and second-order proximity among nodes. After learning from these two definitions, LINE concatenates them together to form the final embeddings. And it considers the structural information only.
- **TADW** [21] is the first attempt to jointly exploit the topological structure and text features for network embedding. In this version, we regard node attributes as text features of nodes.
- **LANE** [9] maps three kinds of information (*i.e.*, network structure, node attributes and labels) into a unified space to generate the embeddings. Here, we only use the version that does not consider label information.
- **GCN** [11] is a neural-network method, and it generates node embeddings by performing convolutional operation on the attributed network.
- **SEANO** [13] obtains a robust network embedding by designing dual inputs and dual outputs to aggregate a node's neighborhood attributes and its labels, while mitigating the adverse impact of outliers in the learning process.

In our baselines, the former two methods consider network structure only, while the latter four algorithms leverage both the structural and attribute information jointly. Besides, among these baselines, SEANO is capable of eliminating the adverse effects of outliers on the learned embeddings, but the rest of these methods don't account the outlier impacts during embedding process.

Parameter settings. For all baselines, we implement them following the original authors, and set default parameters as their report. To be fair comparison, we set the size d of the learned embeddings as 128 for all baseline algorithms. For SEANO, the variant λ is set to 0.5. For REANO, we set window size t as 10, walk length l as 80, walks per node γ as 5, and the sizes of negative samples $|neg|$ as 5, neural network layers k as 3 for all datasets, and the regularizer coefficient α is to 0.5. And we implement REANO using the Pytorch package in Python.

5.2 Node Classification

In this section, we carry out node classification task on the first three real world datasets as well as their outlier version to verify the effectiveness of our proposed model. For these tasks, we get the embeddings of nodes and treat them as the features to train a SVM classifier [17,22]. We split the set of nodes into training set and testing set. The training set size is fixed at 50% of nodes in each data. Then the remaining nodes, which are removed their labels, are used to compare the performance of different algorithms. And we use the popular evaluation criteria, *i.e.*, Accuracy to measure the classification performance. In general, the higher accuracy is, the better classification performance can obtain. We repeat this process 10 times and report the average classification results in Table 2.

Table 2. Classification accuracy of different algorithms on the original datasets and the noisy datasets (mark with *). We use bold to highlight the best results.

Datasets	Cora	Citeseer	Pubmed	Cora*	Citeseer*	Pubmed*
DeepWalk	0.547	0.562	0.682	0.539	0.557	0.671
LINE	0.527	0.532	0.690	0.509	0.519	0.685
TADW	0.606	0.686	0.738	0.464	0.516	0.594
LANE	0.596	0.626	0.711	0.456	0.443	0.572
GCN	0.626	0.681	0.753	0.466	0.531	0.604
SEANO	0.652	0.719	0.801	0.642	0.712	0.793
REANO	**0.678**	**0.741**	**0.821**	**0.662**	**0.729**	**0.809**

On all the datasets, REANO consistently achieves the best classification accuracy both in the presence of original data and outlier version since it effectively eliminates adverse effects of outliers on the learned node embeddings. For Cora and Cora* datasets, REANO respectively gains about 3.6% and 3.1% improvement than SEANO, which is the second best algorithm for classification. It's obvious that TADW, LANE and GCN, which work well on normal datasets, produce unsatisfied results in the presence of just 5% outliers. That's because these three approaches fail to account for the negative impacts from outliers, thus they merely have suboptimal accuracy. Besides, DeepWalk and LINE suffer few impacts from outliers as they only focus on the topological structure. However, they ignore rich attributes of nodes, and gain weak results than the rest of baseline methods on the normal datasets. In addition, SEANO leverages partial label information to alleviate noise impacts from outliers, therefore obtains stable classification performances in terms of both original and outlier datasets.

5.3 A Case Study

In this part, we employ a case study to show the effectiveness of REMAD in detecting outliers. Since the Disney dataset has intrinsic outlier nodes, we should not inject the anomalous into original dataset. REMAD uses the residual matrix R to filter out abnormal attribute information. Therefore, we rank the nodes in order of the higher norm of the matrix R, and choose the top three nodes as outliers. The detection result is shown in Fig. 3.

From the Fig. 3, we observe that there are three outliers like Node N_1, N_2 and N_3 in the Disney dataset. Among these outliers, node N_1 refers to the film *The Many Adventures of Winnie the Pooh*, and node N_2 refers to the film *Buzz Lightyear of Star Command*. They have quit different *rating* attributes from their neighboring products, and are associated with large residual values, thus are typical outliers. Moreover, node N_3 corresponds to the film *The Nightmare Before Christmas/James and the Giant Peach* is an another anomaly. Obviously, this node is a structurally isolated product so that it also has a larger residual error than other products. In a nutshell, our proposed model REMAD can help us discover outliers of different formats.

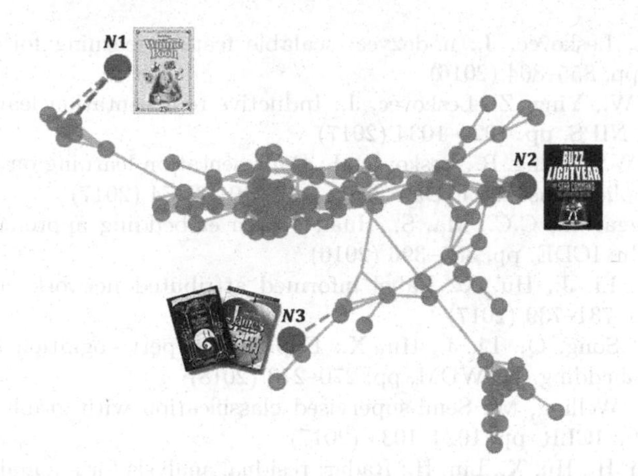

Fig. 3. Visualization of three outliers (N_1, N_2 and N_3) detected by REMAD on Disney dataset. And we use the pink color to highlight outlier nodes. (Color figure online)

6 Conclusion

In this work, we propose a novel embedding method called REANO for learning a Robust Embedding for an Attributed Network while accounting for negative effects of the network outliers. Methodologically, our proposed model combines residual analysis with deep attributed embedding. In detail, REANO filters out the noisy attributes of outliers by residual matrix. Furthermore, network embedding based on deep neural networks aggregates the remaining attribute information and the topological structure into a joint embedding space. Through a joint training framework, REANO obtains the robustness of node embeddings while eliminating the outlier effects. Our experimental results on real-world datasets and manually planted outliers show the effectiveness of our proposed model. In the future, we plan to study how the outliers would bring the impacts on the embedding for a dynamic attributed network.

Acknowledgments. This work is supported by National Key R&D Plan of China (Grant No. 2016QY02D0400), and National Natural Science Foundation of China (Grant Nos. U163620068).

References

1. Aggarwal, C.C.: Outlier analysis. In: Data Mining, pp. 237–263 (2015)
2. Akoglu, L., Tong, H., Koutra, D.: Graph based anomaly detection and description: a survey. DMKD **29**(3), 626–688 (2015)
3. Bandyopadhyay, S., Lokesh, N., Murty, M.: Outlier aware network embedding for attributed networks. In: AAAI
4. Chen, T., Tang, L.-A., Sun, Y., Chen, Z., Zhang, K.: Entity embedding-based anomaly detection for heterogeneous categorical events. In: IJCAI, pp. 1396–1403 (2016)

5. Grover, A., Leskovec, J.: node2vec: scalable feature learning for networks. In: SIGKDD, pp. 855–864 (2016)
6. Hamilton, W., Ying, Z., Leskovec, J.: Inductive representation learning on large graphs. In: NIPS, pp. 1024–1034 (2017)
7. Hamilton, W.L., Ying, R., Leskovec, J.: Representation learning on graphs: methods and applications. IEEE Data Eng. Bull. **40**, 52–74 (2017)
8. Hu, R., Aggarwal, C.C., Ma, S., Huai, J.: An embedding approach to anomaly detection. In: ICDE, pp. 385–396 (2016)
9. Huang, X., Li, J., Hu, X.: Label informed attributed network embedding. In: WSDM, pp. 731–739 (2017)
10. Huang, X., Song, Q., Li, J., Hu, X.: Exploring expert cognition for attributed network embedding. In: WDM, pp. 270–278 (2018)
11. Kipf, T.N., Welling, M.: Semi-supervised classification with graph convolutional networks. In: ICLR, pp. 1024–1034 (2017)
12. Li, J., Dani, H., Hu, X., Liu, H.: Radar: residual analysis for anomaly detection in attributed networks. In: IJCAI, pp. 2152–2158 (2017)
13. Liang, J., Jacobs, P., Sun, J., Parthasarathy, S.: Semi-supervised embedding in attributed networks with outliers. In: SDM, pp. 153–161 (2018)
14. Liang, J., Parthasarathy, S.: Robust contextual outlier detection: where context meets sparsity. In: CIKM, pp. 2167–2172 (2016)
15. McPherson, M., Smith-Lovin, L., Cook, J.M.: Birds of a feather: homophily in social networks. Annu. Rev. Sociol. **27**(1), 415–444 (2001)
16. Mikolov, T., Sutskever, I., Chen, K., Corrado, G.S., Dean, J.: Distributed representations of words and phrases and their compositionality. In: NPIS, pp. 3111–3119 (2013)
17. Perozzi, B., Al-Rfou, R., Skiena, S.: DeepWalk: online learning of social representations. In: SIGKDD, pp. 701–710 (2014)
18. Tang, J., Qu, M., Wang, M., Zhang, M., Yan, J., Mei, Q.: LINE: large-scale information network embedding. In: WWW, pp. 1067–1077 (2015)
19. Tong, H., Lin, C.-Y.: Non-negative residual matrix factorization with application to graph anomaly detection. In: SDM, pp. 143–153 (2011)
20. Wang, D., Cui, P., Zhu, W.: Structural deep network embedding. In: SIGKDD, pp. 1225–1234 (2016)
21. Yang, C., Liu, Z., Zhao, D., Sun, M., Chang, E.Y.: Network representation learning with rich text information. In: IJCAI, pp. 2111–2117 (2015)
22. Zhang, D., Yin, J., Zhu, X., Zhang, C.: User profile preserving social network embedding. In: IJCAI, pp. 3378–3384 (2017)
23. Zhang, D., Yin, J., Zhu, X., Zhang, C.: Network representation learning: a survey. IEEE Trans. Big Data (2018)
24. Zhang, L., Li, X., Shen, J., Wang, X.: Structure, attribute and homophily preserved social network embedding. In: Cheng, L., Leung, A.C.S., Ozawa, S. (eds.) ICONIP 2018. LNCS, vol. 11306, pp. 118–130. Springer, Cham (2018). https://doi.org/10.1007/978-3-030-04224-0_11
25. Zhou, C., Paffenroth, R.C.: Anomaly detection with robust deep autoencoders. In: SIGKDD, pp. 665–674 (2017)

Pay Attention to Deep Feature Fusion in Crowd Density Estimation

Huimin Guo[1,2(✉)], Fujin He[1,2(✉)], Xin Cheng[1,2(✉)], Xinghao Ding[1,2(✉)], and Yue Huang[1,2(✉)]

[1] Key Laboratory of Underwater Acoustic Communication and Marine Information Technology, Ministry of Education, Xiamen University, Xiamen 361005, Fujian, China
871568261@qq.com, 475872590@qq.com, 1500785685@qq.com

[2] School of Informatics, Xiamen University, Xiamen 361005, Fujian, China
{dxh,yhuang2010}@xmu.edu.cn

Abstract. Crowd density estimation has important practical significance for effectively suppressing the occurrence of stampede accidents. However, the crowd counting task can be easily interfered by various factors such as perspective, congestion, occlusion, density, etc., which makes accurate crowd counting a challenging task. To solve these problems, in this paper, we propose an effective hierarchical aggregation module to fuse different scale information in the network. Since the crowd counting task is seriously interfered by the surrounding environment, in this paper we propose to use attention mechanism module to weight the spatial position of the network learned feature map to effectively limit the interference of the background region to the crowd counting task. Finally, a large number of related experiments show that our model in this paper has strong generalization ability while having better performance on several public datasets compared to existing model algorithms.

Keywords: Crowd counting · Effective hierarchical aggregation · Attention mechanism

1 Introduction

In recent years, urban population continues to increase. In this case, the scene of large-scale crowd gathering becomes more frequent and this may lead to severe congestion. Although video surveillance is generally available in these public places, the utilization of monitoring information is very low. Therefore, using computer vision technology to effectively estimate crowd density has received more and more attention. With this technology, people can quickly estimate

This work was supported in part by the National Natural Science Foundation of China under Grants 61571382, 81671766, 61571005, 81671674, 6197136961671309 and U1605252, in part by the Fundamental Research Funds for the Central Universities under Grants 20720160075 and 20720180059, in part by the CCF-Tencent open fund, and the Natural Science Foundation of Fujian Province of China (No. 2017J01126).

T. Gedeon et al. (Eds.): ICONIP 2019, CCIS 1142, pp. 363–370, 2019.
https://doi.org/10.1007/978-3-030-36808-1_39

the number of people in the monitoring scene, and make some abnormal warnings based on the population density distribution information to minimize the possibility of accidents and ensure crowd safety.

The ultimate goal of crowd density estimation is to accurately estimate the total number of people in a given image. In the actual scene, which makes crowding counting task more difficult are the shooting angle, camera focal length, crowd intensity, background interference, etc. Specifically, the existing related research methods can be mainly divided into two major categories: a target-based detection algorithm and a feature-based regression algorithm.

Target-based detection algorithms generally use haar-like wavelet [1], edge, shape and other manual features to extract the whole-body features of pedestrians, and then use classical methods, random forests, etc.) to detect pedestrians. Navnne [2] et al. used the Histograms of Oriented Gradients (HOG) combined with SVM to achieve more accurate pedestrian detection. Zhang [3] et al. used the background difference method to extract the foreground region of the image, and then they used a contour detection algorithm to find the image of the mountain-like shape in the foreground region to achieve simple segmentation of pedestrians. Li [4] et al. proposed a reasonable combination of Mosaic Image Difference (MID) algorithm and HOG feature to achieve accurate counting. Although these improved algorithms solve the performance variation problems caused by overlap and occlusion to a certain extent, but for extremely dense scenes, they are still powerless. And the algorithm based on target detection are generally time consuming, which severely limits their application in real life.

Feature-based regression algorithms are usually function maps for learning image features and the number of people in an image. In [5], Cho et al. proposed a counting algorithm based on feed-forward neural network (FFNN). Lempitsky V first proposed in the [6] to use the density map to indirectly realize the crowd counting method. Zhang et al. [7] proposed to learn two related objective functions of the degree of intensiveness and the number of people in a network. These research methods focus more on the accuracy of the final count than on the clarity and correctness of the density map itself. However, the density map carries a lot of useful information.

In view of the above problems, in this paper, we propose the following two methods to improve the accuracy of crowd counting:

- *Effective hierarchical aggregation module.* In this paper, we explore the importance of scale information for crowd counting, and propose an effective hierarchical aggregation module to effectively fuse different scale information in the network.
- *Attention mechanism.* Considering that the crowd counting task is to estimate the specific targets in the image and these targets are seriously interfered by the surrounding environment. We propose to use the attention mechanism module to weight the network learned feature map in a spatial position to effectively limit the interference of the background region to the crowd counting task.

In addition, through reasonable model design and parameters debugging, the model of this paper achieves state-of-the-art on several public datasets than existing model algorithms.

2 Methods

This paper proposes a crowd counting model based on deep feature fusion. The network structure designed in this paper is given in Fig. 1. The network consists of three parts: scale-aware module, effective hierarchical aggregation module and the attention mechanism module. Below we will detail each module. Consider sharp changes of head size in crowd counting task. Inspired by the literature [8,21], in this paper, we use a simple and efficient scale-aware model. The network adopts an iterative feature transfer method, which not only enhances the transmission of features in the network, but also gradually uses deep features to extract the semantic information of shallow features. So we choose the scale-aware model as our basic model.

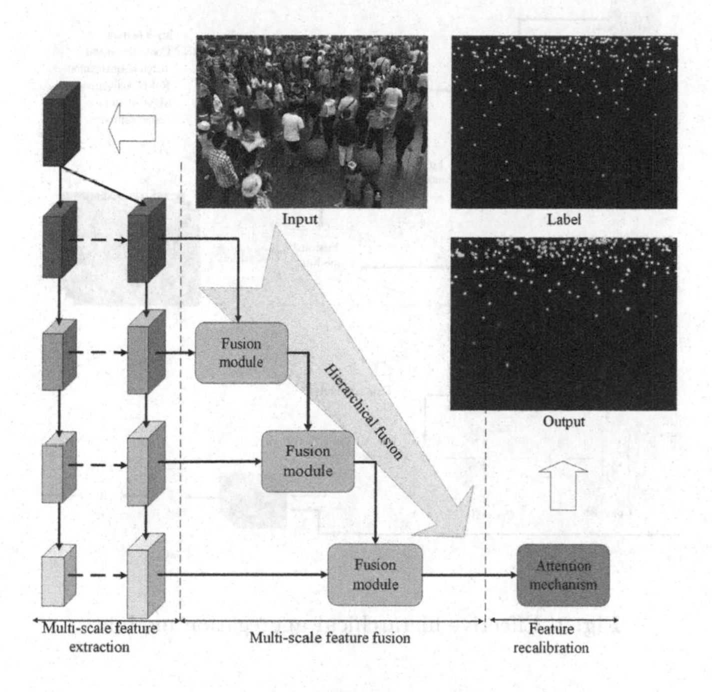

Fig. 1. Network structure diagram

2.1 Effective Hierarchical Aggregation Module

In the network, it can be found that the part feature output of $X(0,1)$ is the feature information of the small-scale target, and $X(0,4)$ is the feature extraction of the large-scale target in the image, which just matches the law of the distribution of human head in the crowd counting task. Therefore, this paper innovatively combines the shallow target feature information with the deep feature information.

The design of this module firstly passes small-scale features through a residual module to extract more scale semantic information, and the residual module is added to deliver more information backwards. The module superimposes the features of the two scales as a rough fusion feature, and then combines a convolution operation for feature fusion to effectively fuse the two scale features together. And the output of current module will directly serve as the input of next fusion module to realize a hierarchical fusion mode. After three fusion operations, the feature information of four scales can be effectively fuse together. Finally, the output characteristics of the final fusion module are used to generate the final estimated density map. The specific hierarchical aggregation module is shown in the Fig. 2.

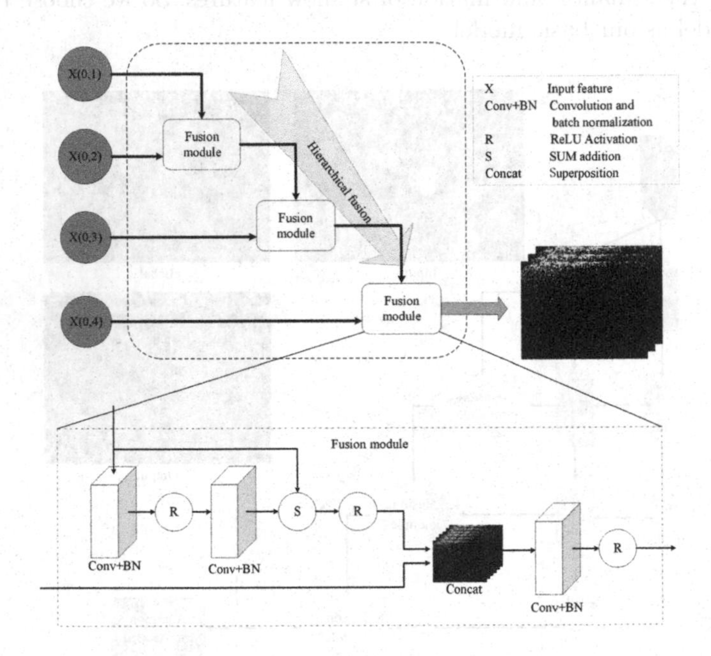

Fig. 2. Effective hierarchical aggregation module

2.2 Attention Mechanism Module

For the output of the effective hierarchical aggregation module, the features of different layers are redundant, which makes it difficult to directly learn the best mapping between the network feature and the density map. Therefore, in this paper, the attention mechanism module is embedded in effective hierarchical aggregation module. This self-supervised attention mechanism brings two benefits. First, it makes it easier for the network to distinguish between background and prospects. Second, the network generates the final output density map in an appropriate ratio.

For attention mechanism module, assuming that the input feature is $X \in \Re^{H \times W \times C}$, where H, W, C denote respectively the height, width, and channel number of the feature map. Firstly, the network learns spatial information through convolution layer to get corresponding mapping $X' = W \times X + B$, where W denotes the weight parameter of convolution kernel, and B denotes bias term, $X' \in \Re^{H \times W \times C}$. Then a sigmoid function is used for nonlinear activation to obtain the weight of each position of the feature. Finally, the $f(W * X + B)$ is multiplied by the original input feature X to recalibrate the input features. This process can be summarized as:

$$Y = X * f(W * X + B), Y \in \Re^{H * W * C} \quad (1)$$

The weighting method based on feature recalibration can be changed according to the change of the data, that is, the attention mechanism module can output the most suitable weight according to the input change, thereby implementing an adaptive weighting.

3 Experiments

This paper use MAE (Mean Absolute Error) and MSE (Mean Square Error) to measure the accuracy and robustness of the model.

3.1 Dataset

In this paper, crowd counting task mainly uses three datasets. Shanghai Tech [9] dataset mainly consists of Part_A and Part_B. The UCF_CC_50 [10] dataset is a dataset with extremely crowded distribution. UCF_QNRF [11] is a dataset of a large dense scene. A comparison of their specific data information is given in Table 1.

Table 1. Information on three datasets used in the experiments. Total images are the sum of train images and test images; Resolution means the resolution of the images in this dataset; Min, Max and Ave mean minimum, maximum and average number of people in a images respectively.

Datasets	Total images	Resolution	Min	Max	Ave	Total count
Shanghai Tech Part_A	482	Difference	33	3139	501	241677
Shanghai Tech Part_B	716	768×1024	9	578	123	88488
UCF_CC_50	50	Difference	94	4543	1279	63974
UCF_QNRF	1535	Difference	49	12865	815	1251642

3.2 Ablation Experiments

In order to explore the effectiveness of the proposed module, we perform an ablation experiment. ModelA is the Scale-aware Model, that is our baseline model. ModelB adds a multi-scale feature Concatenate module (MS_Concat) based on ModelA. ModelC consists of a base model and an effective hierarchical aggregation module (EHA). ModelD introduces attention mechanism (AT) based on ModelC. For the model self-evaluation, all experiments used ShanghaiTech Part_B dataset, and final experimental verification results are shown in Table 2.

Comparing the results of ModelC and ModelB, we can find that ModelC has a significant improvement over ModelB. It can be clearly stated that the effective hierarchical aggregation module proposed in this paper is more effective than the direct superposition method. And the paper finds that ModelC has achieved a great improvement on MAE and MSE compared with ModelA, which shows that the aggregation of hierarchical information can effectively achieve accurate estimation of human heads at different scales. By comparing the ModelD and ModelC, it can be found that attention mechanism improves the performance of the model. This method refines the original features and make the model recalibrate the feature in a dynamic way.

Table 2. Evaluation results of each module

Name	Model	MAE	MSE
ModelA	Baseline	9.2	13.5
ModelB	Baseline+MS_Concat	8.4	13.7
ModelC	Baseline+EHA	8.1	12.4
ModelD	Baseline+EHA+AT	**7.4**	**11.4**

3.3 Comparison with Other Models

In order to verify the comprehensive performance of our final model(ModelD) from various angles, in this chapter, we will verify the performance of our model on each large dataset. Table 3 shows the performance of these models on the Shanghai Tech dataset and the UCF_CC_50 dataset. Table 4 shows the performance of these models on the UCF_QNRF dataset.

Through the above comparison experiments, it can be found that the model of this paper has achieved superior performance on a number of open challenge datasets, and the generalization ability is relatively strong. In particular, for the ShanghaiTech Part_B dataset, our model achieve nearly 50% performance improvement over CP-CNN. The results of this experiment show that the model in this paper has comparative performance on a general dense dataset. Especially for the UCF_CC_50 dataset with very few images, our model still achieves an excellent performance. The MAE of our model for the UCF_QNRF dataset also get good results.

Table 3. Comparison of experimental results of the model on Shanghai Tech and UCF_CC_50

	Part_A		Part_B		UCF_CC_50	
	MAE	MSE	MAE	MSE	MAE	MSE
Crowdnet [12]	–	–	–	–	452	–
Hydra 2s [13]	–	–	–	–	333.7	425.3
Zhang [7]	181.8	277.7	32.0	49.8	467.0	498.5
MCNN [9]	110.2	173.2	26.4	41.3	377.6	509.1
FCN [14]	126.5	173.5	23.8	33.1	338.6	424.5
Switch-CNN [15]	90.4	135.0	21.6	33.4	318.6	439.2
CP-CNN [20]	73.6	106.4	20.1	30.1	295.8	**320.9**
SaCNN [16]	86.8	139.2	16.2	25.8	314.9	424.8
DR-ResNet [17]	86.3	124.2	14.5	21.0	307.4	421.6
ACSCP [18]	75.7	**102.7**	17.2	27.4	291.0	404.6
Ours	**70.6**	117.5	**7.4**	**11.4**	**268.2**	384.4

Table 4. Experiment results on UCF_QNRF

Model	MAE	MSE
Idrees [10]	31.5	508
MCNN [9]	277	–
CMTL [19]	252	514
Switch-CNN [15]	228	445
CL [11]	132	**191**
Ours	**130**	208

4 Conclusion

An effective hierarchical aggregation module proposed in this paper realize the effective extraction of human head information at different scales. The introduction of attention mechanism can achieve dynamic fine-tuning of feature information. Due to the introduction of the above method, our model finally achieves excellent performance on each dataset and has strong generalization ability.

References

1. Lienhart, R., Maydt, J.: An extended set of Haar-like features for rapid object detection. In: International Conference on Image Processing, pp. 900–903 (2002)
2. Dalal, N., Triggs, B.: Histograms of oriented gradients for human detection. In: IEEE Computer Society Conference on Computer Vision and Pattern Recognition, pp. 886–893 (2005)

3. Zhang, E., Feng, C.: A fast and robust people counting method in video surveillance. In: International Conference on Computational Intelligence and Security, pp. 339–343 (2008)
4. Min, L., Zhang, Z., Huang, K., Tan, T.: Estimating the number of people in crowded scenes by mid based foreground segmentation and head-shoulder detection. In: International Conference on Pattern Recognition, pp. 1–4 (2008)
5. Cho, S.Y., Chow, T.S., Leung, C.T.: A neural-based crowd estimation by hybrid global learning algorithm. IEEE Trans. Syst. Man Cybern. B Cybern. **29**(4), 535–541 (1999)
6. Lempitsky, V.S., Zisserman, A.: Learning to count objects in images. In: Neural Information Processing Systems, pp. 3–16 (2017)
7. Cong, Z., Li, H., Wang, X., Yang, X.: Cross-scene crowd counting via deep convolutional neural networks. In: IEEE Conference on Computer Vision and Pattern Recognition, pp. 833–841 (2015)
8. Yu, F., Wang, D., Shelhamer, E., Darrell, T.: Deep layer aggregation. In: Computer Vision and Pattern Recognition, pp. 2403–2412 (2018)
9. Zhang, Y., Zhou, D., Chen, S., Gao, S., Yi, M.: Single-image crowd counting via multi-column convolutional neural network. In: Computer Vision and Pattern Recognition, pp. 589–597 (2016)
10. Idrees, H., Saleemi, I., Seibert, C., Shah, M.: Multi-source multi-scale counting in extremely dense crowd images (2013)
11. Idrees, H., Saleemi, I., Seibert, C., Shah, M.: Multi-source multi-scale counting in extremely dense crowd images. In: 2013 IEEE Conference on Computer Vision and Pattern Recognition, pp. 2547–2554 (2013)
12. Boominathan, L., Kruthiventi, S.S.S., Babu, R.V.: Crowdnet: a deep convolutional network for dense crowd counting. In: ACM Multimedia, pp. 640–644 (2016)
13. Oñoro-Rubio, D., López-Sastre, R.J.: Towards perspective-free object counting with deep learning. In: Leibe, B., Matas, J., Sebe, N., Welling, M. (eds.) ECCV 2016. LNCS, vol. 9911, pp. 615–629. Springer, Cham (2016). https://doi.org/10.1007/978-3-319-46478-7_38
14. Marsden, M., Mcguiness, K., Little, S., O'Connor, N.E.: Fully convolutional crowd counting on highly congested scenes. In: International Conference on Computer Vision Theory and Applications, pp. 27–33 (2017)
15. Sam, D.B., Surya, S., Babu, R.V.: Switching convolutional neural network for crowd counting. In: Computer Vision and Pattern Recognition, pp. 4031–4039 (2017)
16. Lu, Z., Shi, M.: Crowd counting via scale-adaptive convolutional neural network. In: Workshop on Applications of Computer Vision, pp. 1113–1121 (2017)
17. Ding, X., Lin, Z., He, F., Yu, W., Yue, H.: A deeply-recursive convolutional network for crowd counting. In: International Conference on Acoustics, Speech, and Signal Processing, pp. 1942–1946 (2018)
18. Shen, Z., Xu, Y., Ni, B., Wang, M., Hu, J., Yang, X.: Crowd counting via adversarial cross-scale consistency pursuit. In: IEEE Computer Society Conference on Computer Vision and Pattern Recognition, pp. 5245–5254 (2018)
19. Sindagi, V.A., Patel, V.M.: CNN-based cascaded multi-task learning of high-level prior and density estimation for crowd counting. In: 2017 14th IEEE International Conference on Advanced Video and Signal Based Surveillance (AVSS), pp. 1–6 (2017)
20. Sindagi, V.A., Patel, V.M.: Generating high-quality crowd density maps using contextual pyramid CNNs. In: International Conference on Computer Vision, pp. 1879–1888 (2017)
21. He, K., Zhang, X., Ren, S., Sun, J.: Deep residual learning for image recognition. In: Computer Vision and Pattern Recognition, pp. 770–778 (2016)

Knowledge Reuse of Learning Agent Based on Factor Information of Behavioral Rules

Fumiaki Saitoh(✉)

Chiba Institute of Technology, Tsudanuma, Japan
fumiaki.saitoh@p.chibakoudai.jp

Abstract. In this study, we attempt to extract knowledge by collecting results from multiple environments using an autonomous learning agent. A common factor of the environment is extracted by applying non-negative matrix factorization to the set of learning results of the reinforcement learning agent. In transfer learning of knowledge management of agents, as the number of experienced tasks increases, the knowledge database becomes larger and the cost of knowledge selection increases. By the proposed approach, an agent that can adapt to multiple environments can be developed without increasing cost of knowledge selection.

Keywords: Reinforcement learning · Non-negative matrix factorization · Transfer learning · Knowledge reuse · Agent

1 Introduction

In this study, we aim to improve the learning efficiency in a new environment by reusing past learning results. In recent years, the development of AI peripheral technology centering on machine learning is remarkable, and the intelligence of many artifacts is in progress. Behavioral selection by autonomous agents is of great importance, and research has been actively conducted on various approaches such as reinforcement learning for a long time. Adaptation and learning to a single task are not enough to achieve higher-level artificial intelligence. Hence, it is essential to build a learning agent that can adapt to multiple tasks and environments.

When we look at the intellectual activities that we carry out on a day-to-day basis, we can respond differently to various problem settings such as multiple tasks and environments. Even in an inexperienced task, intelligent species such as humans can use similar knowledge experienced in the past, to cope with unknown problems quickly and efficiently, to learn easily. Abilities like this is one of the essential functions required for future autonomous agents to deepen their relationship with life and society.

Despite the fact that such knowledge processing is easy for human beings, the achievement of engineering in general learning systems and their discussions have not been enough, and previous knowledge can be directly identified. It remains within the framework of reusing transfer learning.

Extracting useful information from the database of learning results is important for knowledge management of agents, that have been used in multiple environments. For general transfer of knowledge to reinforcement learning agents, a database of

T. Gedeon et al. (Eds.): ICONIP 2019, CCIS 1142, pp. 371–379, 2019.
https://doi.org/10.1007/978-3-030-36808-1_40

knowledge, acquired from multiple environments have been constructed. Reusable knowledge is chosen for this knowledge database; thus, knowledge that is most suitable for accomplishing the current task, can be selected from past experiences.

In the previous studies on knowledge reuse, behavioral rules in the knowledge database included those that are used directly along with those that reconstruct the database by clustering. Previous approaches to reuse learning results caused problems in the knowledge selection as they directly used past learning results. In this research, we apply unsupervised learning to a database of knowledge composed of multiple learning results, and extract factored information from the agent's behavioral rules, to streamline the reuse of knowledge. Many conventional approaches have the potential to streamline learning with small knowledge databases. However, it is not efficient enough to build an intelligent system for large-scale knowledge, which will be required in recent years to accomplish various tasks.

As the number of tasks experienced in the past increases, a growth in the size of the knowledge database has been observed. Due to this reason, the computational efficiency may deteriorate because the agent has too many options on the task of reusing knowledge. Furthermore, in the transfer learning framework, it is necessary to carry out a large number of useless trials in situations where there is a large amount of non-reusable knowledge. Reusing knowledge in a multi-user agent-like framework is not compatible with large-scale knowledge databases because it uses hierarchical clustering. While the prior studies use knowledge directly, this approach extracts information of partially similar factors from the knowledge database and uses it for knowledge reuse. We aim to streamline knowledge selection in a large-scale knowledge database by reusing factorial information on behavior.

2 Related Works

Importance of transfer learning has increased in recent years because it can reuse model of learning results in situations where data acquisition is limited. And in recent years, it is applied also to transfer learning of the action rule of a learning agent. In general, many approaches of agent's knowledge transfer method construct a database of acquired knowledge for multiple environments and streamline learning by reusing knowledge based on them. Reusable knowledge to this knowledge database, that is, knowledge most suitable for accomplishing the current task is selected from past experiences. The main theme of many previous researches is how to select appropriate knowledge and how to use the selected knowledge, etc., here, the rules of behavior in the knowledge database are often used directly. These have the problem that, as the size of the database increases, the options for knowledge to be reused increase and the system operation becomes inefficient.

There is also an approach to deal with by clustering multiple environments. However, there is a limit to the size of problems that can be dealt with and it is not suitable for large-scale environments. In addition, deep learning approaches have also been studied in recent years, and they have been successful. However, the agent must be trial-and-error massively in advance because data of a sufficiently large size is required in advance to use deep learning models. Because transfer learning requires the construction of a model covering a sufficient environment across tasks, the application of deep learning is limited to situations where massive data acquisition and trial and error are possible. Many of these methods are not factors related to multiple tasks, meaning that they are stored one by one and reused as learning multiple environments. This research proposes a new framework aiming at the efficiency of knowledge reuse by extracting features that are common factors in multiple environments.

3 Proposed Method

To address the problems mentioned above, agents apply unsupervised learning to a knowledge database accumulated through past experiences, and extract factors common to tasks, as concepts acquired, in this paper. Non-negative matrix factorization (NMF) has been selected as a learning model to extract partially common factors as factor information in multiple environments. By using the NMF, it is possible to remove factors attributed to a specific environment such as noise, and it is possible to extract only important factors from the environmental knowledge experienced in the past. Furthermore, by achieving factor extraction in action rules through dimensional reduction of a large-scale knowledge database, it is possible to reduce knowledge options to be reused in an environment, as newly learned.

NMF is a dimension reduction that avoids the orthogonality of the axes occurring in PCA and LSI by providing non-negative constraints. Therefore, if the knowledge database can be expressed as a real value matrix that satisfies the non-negative constraint, extraction of common factors can be achieved by the proposed method.

In tasks where negative rewards exist, it is necessary to convert Q-table into non-negative behavioral selectivity before learning in NMF, such as by Boltzmann selection. The matrix of knowledge set is constructed with multiple Q-tables comprising of knowledge databases like row vector, and NMF is trained with the result. In the proposed method, the agent decides the action based on the factor matrix representing the local feature extracted from the learning of NMF, by reuse of knowledge. In matrix X, which represents the original knowledge database, the number of rows and columns correspond to the number of tasks experienced and the number of Q values, respectively.

In the proposed method, the matrix X is decomposed into two matrices by NMF. First, there is a matrix T, where the number of rows correspond to the number of tasks, and the number of columns correspond to the number of factors. Second is expressed as matrix V, in which the number of rows correspond to the number of factors and the number of columns correspond to the Q value. Matrix V is used for knowledge reuse, and the number of experienced tasks accumulated in the knowledge database is reduced to the value of the base number.

Since it is necessary to extract local knowledge, application of unlearned tasks are examined using matrix V, having elements of base and Q value. Since the element of each row vector of V is a Q value, the agent performs selection, considering this to be a Q-table. After the selection of behavioral factors, it is applied to a new task as knowledge, to reuse Q values of factors for which a reduction in the number of steps has been achieved, compared to the results of completely random trials.

The procedure of the proposed method is as follows:

Step 1: The knowledge database X is constructed based on the action rules for multiple tasks acquired by reinforcement learning. Acquisition of the action rule for each task is learning based on the updated formula of general Q learning.

$$Q(s_{it}, a) \leftarrow Q(s_{it}, a) + \alpha\left[r_{t+1} + \gamma \max_p Q(s_{it+1}, p) - Q(s_{it}, a)\right] \quad (1)$$

where, Q value for the action a in the state s_{it} is expressed as $Q(s_{it}, a)$. α and γ are expressed as learning rates and discount rates, respectively, which takes values > 0 or < 1. Subscript i represents an index that distinguishes between tasks.

Step 2: A knowledge database matrix X is constructed by converting each of the Q-tables collected in Step 1 into a row vector. Here, the elements of each row vector of matrix X corresponds to the Q value of Q-table.

Step 3: Matrix X is calculated by product of matrix T and matrix V through learning of NMF. The NMF update formula for the knowledge database X is as follows.

$$\bar{u}_{ik} \leftarrow u_{ik} \frac{(XV)_{ik}}{(UVV^T)_{ik}} \quad (2)$$

$$\bar{v}_{kj} \leftarrow v_{kj} \frac{(U^T X)_{kj}}{(U^T UV)_{kj}} \quad (3)$$

where, u and v represent components of the matrices U and V, respectively, and subscripts i, j and k are indices representing the components.

Step 4: By creating the corresponding Q-table for each factor, reconstruction of the factor matrix of action law is achieved with knowledge of the factor of matrix V. Here, the elements of the row vector of matrix V is rearranged as Q-table in each factor.

Step 5: The actions are selected based on each Q-table constructed in Step 4. Only the efficient Q-tables are selected for new tasks.

Step 6: For a new task or environment, the behavior is learned with the knowledge selected in Step 5 as the initial value.

For the selection of knowledge in Step 5, knowledge of the factor is applied when the number of steps required to achieve the task is used as a baseline for multiple episodes performed on the unlearned task. Reuse those factors that have been found to have a reduced number of steps in applicable knowledge for unlearned tasks. (Fig. 1)

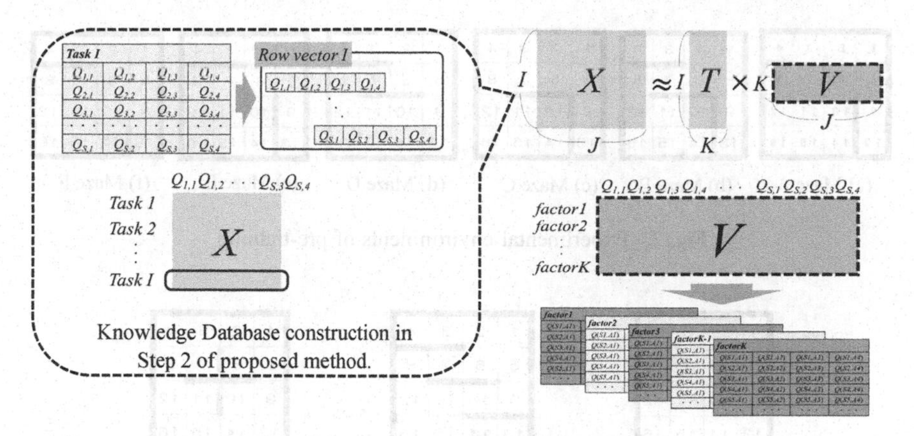

Fig. 1. Structure of NMF in the research

4 Experimental

4.1 Experimental Settings

In this section, we evaluate the behavior of the proposed method based on computer simulations. We adopted the maze problem, the most common experimental setup for learning agent evaluation experiment. To simulate multiple environments in the experimental setting, six types of mazes were prepared as learning environments.

Moreover, to perform different tasks using these environments, different initial states (start) and goal states (goal) were set for each maze, and multiple tasks were constructed. We collected multiple Q-tables as a database of knowledge, which were the learning results for these environments and tasks.

In the proposed method, the superordinate concept is constructed by NMF learning after matrix transformation. To examine the effectiveness of knowledge reuse, three unlearned environments were prepared separately as shown in Fig. 4. The concept acquired by the proposed method confirmed that it can be reused.

The agent's behavior is set such that four actions can be selected from top, bottom, left, and right in the maze, and it returns to the initial position, prior to the selected action that collided with the wall. Boltzmann selection is adopted for action selection in the experiment. The trial from the start to the goal is defined as one episode, and in each setting, the Q-table after 1000 episodes of learning, is used as a sample knowledge for that task.

The parameter setting of NMF used in the experiment was based on the number of bases, where k = 10 and the number of updates is 100. The following values were used for agent behavioral learning, mainly, parameter setting for reinforcement learning. For all the environments and tasks, the learning rate $\alpha = 0.1$, the discount rate $\gamma = 0.9$, the reward r = 1.0, and the initial temperature value T = 5.

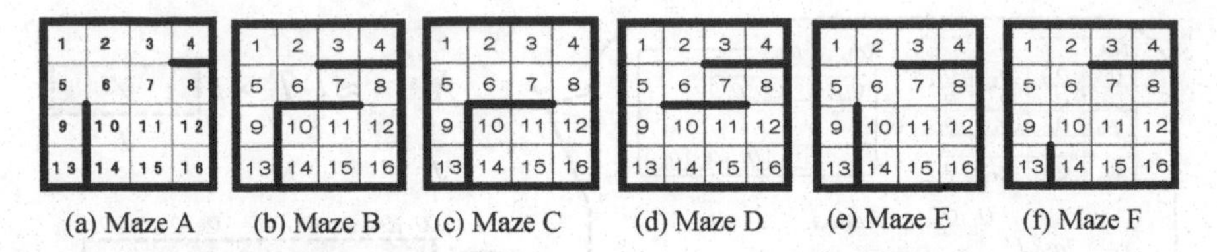

Fig. 2. Experimental environments of pre-training

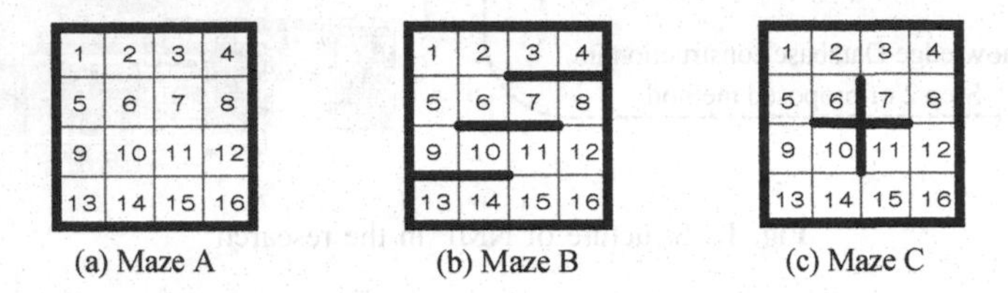

Fig. 3. Experimental environments of post-training

4.2 Experimental Results

Here we describe the results of the experiment. Figure 4 represents a factor space for a knowledge data set. Here, each heat map represents the weight of a matrix, and the magnitude of the weight value of each element is expressed by the corresponding shade of color. Figure 4(a) represents the matrix of the original knowledge database. Figure 4 ((b) and (c)) show factor matrices U and V acquired through NMF learning, respectively. In this experiment, since each task can select 4 types of action for 16 states, hence, it becomes a row vector of 64 (16×4) dimensions.

In this experiment, for each environment in Fig. 4, state No. 1, No. 4, No. 13, No. 16 are set as the target state (goal), while the states other than these target states are in the initial state (start). The agent learns the task, and processes the knowledge database acquired, as matrix X. To confirm the process of the proposed method, we performed the task of adapting to an unknown environment by using factor information (Factor 1–10), representing the obtained upper concept.

For each of the unknown environments shown in Fig. 2, knowledge is provided to the task in which the initial state is set to No. 1 and the target state to No. 16 along with the task in which the initial state is set to No. 16 and target state is No. 1. Table 1 show the results of confirming the reuse status of Table 1 show the experimental results for the mazes A–C in Fig. 3, respectively.

To evaluate the performance that reuses knowledge, the Q value is not updated, but the degree of task achievement in the case where only factor information is used is evaluated. The values in the table represent the average value of the number of steps when each task was performed 1000 times. Random in the first column of each table is the result when the task is performed without learning, and this value is a baseline.

In the table, cells that are the results of factors below this value, that is, the behavioral rules that can accomplish the task in few steps without learning, are shaded.

The best results value for each setting is shown in bold. This shows that it is the factor information of action applicable to an unknown task.

On Comparing the results, it can be concluded that there is a mixture of those that are significantly below the baseline but can be applied to unknown tasks efficiently and those that are worse on the contrary. The tasks on the left side of the table with better values are conversely worse than the tasks on the right side. This indicates that, locally, different knowledges can be properly extracted for each task. Figure 5 shows an example of a learning curve in environment A (initial 1, target 16). The red line shows factor 1 and the blue line shows factor 7 as the initial state of the Q value. Improvement has been noticed henceforth.

4.3 Discussion

By applying the factor matrix acquired by NMF to the tasks in an unknown environment as the initial value of Q-table, the average number of steps decreases compared to randomly acting on the initial value which is the baseline. By extracting factors common to multiple tasks and environments, the result shows that knowledge reuse works effectively in an environment which is matching the factors. It was also confirmed that the average number of steps was higher than the baseline and the corresponding factor was worse.

This means that it is possible to properly perform the feature of the tasks because significant results may be obtained except for tasks whose results have deteriorated. In transfer learning, a learning agent's experience is based on a knowledge database which is further based on the conventional methods, whereas, in an unlearned task, an agent who has experienced many tasks have vast applicable knowledge when selecting available knowledge. Hence, the calculation cost is likely to increase for a learning agent.

On the other hand, it is possible to significantly reduce the applicable knowledge of the candidate by extracting the rule that holds common locally as a factor of the knowledge acquired for multiple tasks. From the result of Fig. 5, it can be confirmed that since unnecessary learning processes can be eliminated by reusing factor information acquired by the proposed method, as an initial value in action learning, it is efficient in aiming at application of reinforcement learning in various tasks.

In this study, while the process of factor selection is simple, efficiency of the process of factor information selection requires improvement. Construction of a selection algorithm of factor information based on behavior, needs to be considered as upcoming task. Furthermore, the selection of parameters suitable for the number of factors and tasks could turn out to be an issue in future.

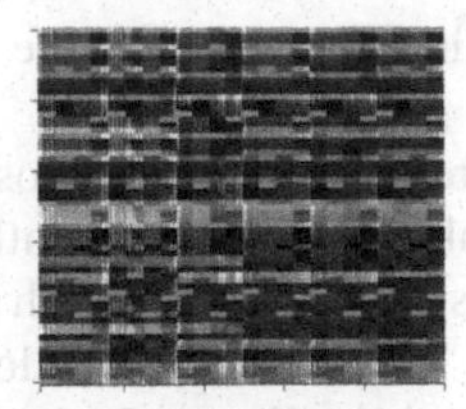

(a) Matrix X that shows knowledge database

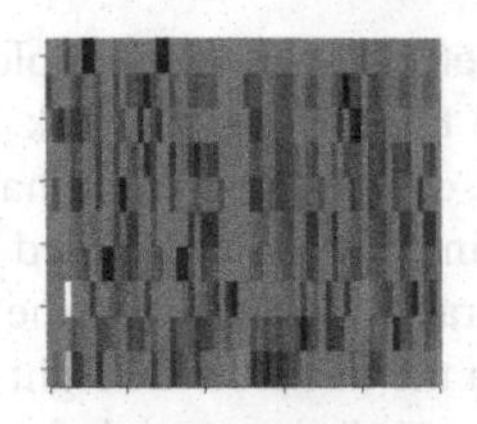

(b) Factor matrix V that represents task

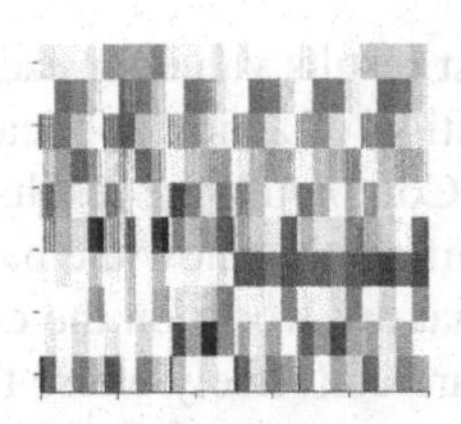

(c) Factor matrix V that represents Q-table

Fig. 4. Extracted knowledge space.

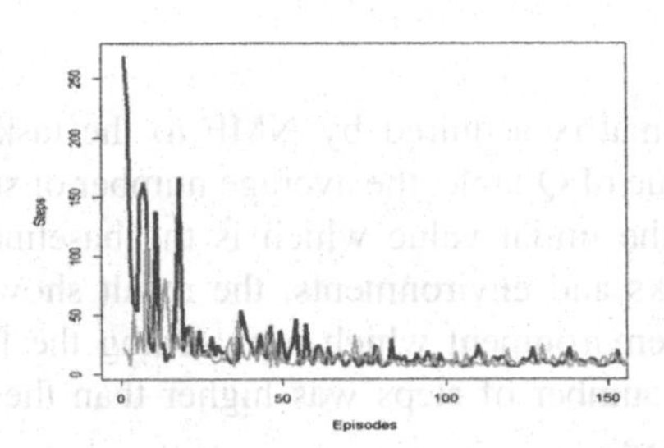

Fig. 5. Learning curves.

Table 1. Experimental results.

	Maze A		Maze B		Maze C	
Start	No. 1	No. 16	No. 1	No. 16	No. 1	No. 16
Goal	No. 16	No. 1	No. 16	No. 1	No. 16	No. 1
Random	71.56	65.19	66.47	61.78	75.85	96.24
Factor 1	100.24	29.04	99.40	29.93	128.66	40.53
Factor 2	96.86	55.40	92.69	52.48	104.22	75.16
Factor 3	38.46	95.17	38.98	88.09	63.87	95.67
Factor 4	73.06	62.83	74.99	63.90	99.01	97.85
Factor 5	31.68	104.65	32.27	105.27	48.80	129.16
Factor 6	99.17	29.24	99.71	30.96	142.75	45.13
Factor 7	29.97	102.73	30.59	94.84	35.78	156.04
Factor 8	69.48	47.47	77.57	47.96	76.30	86.35
Factor 9	52.97	56.18	54.47	55.29	69.31	84.72
Factor 10	55.40	93.30	91.12	90.02	115.94	109.74

5 Conclusion

In this study, we constructed a distributed representation model of learning agent knowledge by extracting common factors from a set of knowledge for multiple learned environments. By applying non-negative matrix factorization with multiple Q-tables representing the action rules as a data set, we could extract common factors from the set of accumulated knowledge. As a result, by combining the factors of knowledge represented in a distributed manner, it is expected to improve the efficiency and flexibility of learning through knowledge reuse, even for tasks intended for unknown environments.

To construct evaluation criteria, in future, it will be considered necessary to establish a framework for automatically discriminating factor information of acquired knowledge that can be applied in unknown environments from those that cannot be applied in the future. Further, extensions to more complex tasks are desired. In that case, transfer learning using deep learning may be effective.

Acknowledgements. This work was supported by JSPS KAKENHI Grant-in-Aid for Young Scientists (B) Numbers 15K16295 and Scientific Research C 19K04887.

References

1. Lee, D.D., Seung, H.S.: Learning the parts of objects by non-negative matrix factorization. Nature **401**(6755), 788–791 (1999)
2. Saito, M., Kobayashi, I.: A study on efficient transfer learning for reinforcement learning using sparse coding. J. Autom. Control Eng. **4**(4), 324–330 (2016)
3. Ohmura, H., Katagami, D., Nitta, K.: Multi user learning agent with clustering. In: Proceeding of 8th International Symposium on Advanced Intelligent Systems (ISIS 2007), pp. 70–72 (2007)
4. Fernández, F., García, J., Veloso, M.: Probabilistic policy reuse for inter-task transfer learning. Robot. Auton. Syst. **58**, 866–871 (2010)
5. Fernandez, F., Veloso, M.: Probabilistic policy reuse in a reinforcement learning agent. In: Proceedings of the 5th International Joint Conference on Autonomous Agents and Multi-Agent Systems, pp. 720–727 (2006)
6. Fachantidis, A., Partalas, I., Tsoumakas, G., Vlahavas, I.: Transferring task models in reinforcement Learning agents. Neurocomputing **107**, 23–32 (2013)
7. Mnih, V., et al.: Human-level control through deep reinforcement learning. Nature **518**, 529–533 (2015)
8. Minato, T., Asada, M.: Environmental change adaptation for mobile robot navigation. J. Robot. Soc. Jpn, **18**(5), 706–712 (2000)
9. Laroche, R., Barlier, M.: Transfer reinfrocement learning with shared dynamics. In: Proceedings of the 31st AAAI Conference on Artificial Intelligence, pp. 2147–2153 (2017)
10. Silva, F.L., Taylor, M.E., Costa, R.A.H.: Autonomously reusing knowledge in multiagent reinforcement learning. In: Proceedings of the Twenty-Seventh International Joint Conference on Artificial Intelligence (IJCAI), pp. 5487–5493 (2018)
11. Zhang, H., et al.: Learning to design games: strategic environments in reinforcement learning. In: Twenty-Seventh International Joint Conference on Artificial Intelligence (IJCAI), pp. 3068–3074 (2018)

Community Based Node Embeddings for Networks

P. Meghashyam[1(✉)] and V. Susheela Devi[2(✉)]

[1] Archeron Group, Bengaluru, India
meghashyam@archerongroup.com
[2] Indian Institute of Science, Bengaluru, India
susheela@iisc.ac.in

Abstract. Network embedding has got enormous attention in recent past for their wide range of applications across different types of networks. This paper mainly includes a simple and novel model which is used for better node embeddings with respect to community detection in social networks. We use existing algorithms (mainly community detection algorithm) and Representation Learning (RL) techniques to find better embeddings that assist in better community detection.

Keywords: Community detection · Node embeddings · Representation learning

1 Introduction

Graphs are found everywhere and have a wide range of applications across different networks. Some of them are social networks, molecular graph structures, biological protein-protein networks, and recommender systems. These applications can be readily modeled as graphs, in which interactions (i.e., edges) can be captured between individual units (i.e., nodes). So there is always a need for an efficient store or access.

RL is used to map or embed the nodes, or entire (sub)graphs, as points in a low-dimensional vector space, R_d. The geometric relationships between the embeddings in learned space should reflect the structure of the original graph. The optimized embeddings then can be used for many subsequent downstream machine learning tasks.

This paper mainly contains the architecture of our proposed model in Sect. 2 followed by experiments and applications conducted by us to validate our model in Sect. 3 followed by **Conclusions and Future Work**.

2 Our Model

The model is simple but efficient in community detection. The steps involved in the model are as follows:

T. Gedeon et al. (Eds.): ICONIP 2019, CCIS 1142, pp. 380–387, 2019.
https://doi.org/10.1007/978-3-030-36808-1_41

- First identify the communities in the graph using some existing algorithm.
- Find the embeddings for all the communities separately.
- Merge the embeddings based on some criteria.

The following sections will give more information about the algorithms used in the model.

2.1 Community Detection

There are many algorithms that find communities in the graph. They use metrics like edge betweenness, fast greedy, infomap, label propagation, leading eigenvector, multilevel, spinglass, walktrap, etc. The algorithm we use in the model is Clauset-Neuman-Moore greedy modularity maximization algorithm [6].

Greedy modularity maximization starts with each node in its own community and joins the pair of communities that most increases modularity until no such pair exists.

2.2 Node Embeddings

Once the communities are detected, the original graph edgelist is partitioned according to the individual communities and one of the RL technique is applied on the individual edgelist and embeddings are learned. Some of the RL techniques are Laplacian Eigenmaps [2], Graph Factorization [1], GraRep [4], HOPE [8], DeepWalk [9], node2vec [7], HARP [5], LINE [11], SDNE [12], etc.

We mostly use the default settings of the parameters values in the publicly available implementations of the respective baseline algorithms such as https://github.com/thunlp/OpenNE. After the node embeddings are learned according to communities, they can be merged using one of the criteria like average, weighted average, etc. Once we get the merged embeddings, any clustering algorithms like spectral clustering, KMeans or DBSCAN can be used to find the communities. We use spectral clustering for detecting the communities from embeddings.

2.3 Evaluation Metric

The major problem in evaluating a community detection algorithm is that there is no shared and universally accepted definition for a community. Anyways, if ground truth is known then evaluation of communities can be done by assigning labels to the nodes in the communities and testing the accuracy with respect to the ground truth. But assigning appropriate labels to detected communities can be done in $n!$ ways which makes it computationally heavy to evaluate, (where n is the number of communities detected by the algorithm). Another variant called Normalized Mutual Information can also be used which is again computationally expensive.

In our work, we take the idea of Normalized F1 - Communities from [10] where they compute an average F1-Score that captures the level of approximation reached by network partitions obtained through community detection algorithms with respect to ground truth ones.

3 Experiments

For simplicity, we used a community detection algorithm that will give disjoint communities. While splitting the nodes according to the communities, inter-community edges are considered in the new edgelists. This indicates for a specific community, edgelist will also contain other community node edges and embeddings will also be calculated to other nodes which all are merged later by taking an average.

The following sections will talk about datasets used in the experiments and results followed by some applications of our model in different scenarios.

3.1 Datasets

Brief description of the datasets are given below:

- **Wiki:** Wiki dataset is a directed graph containing 2405 nodes with 17981 edges and 17 labels.
- **Cora:** Cora dataset is a directed graph containing 2708 nodes with 5429 edges and 7 labels.
- **BlogCatalog:** This dataset is a social relationships network of the bloggers listed on the BlogCatalog website. The network has 10,312 nodes, 333,983 edges and 39 different labels.
- **Karate:** Zachary's karate dataset is a social network of a university karate club. It has 34 nodes with 78 edges and 5 labels.
- **Email:** The email dataset was generated using email data from a large European research institution. It contains 1005 nodes with 25571 edges and 42 labels.

  To make it simple, we have only considered single labels for the datasets. For the case of multi-label instances in dataset, we have considered only first label as true ground truth label.

3.2 Normalized F1 - Communities

The column heads in the tables are appended with '**O**' or '**M**' which indicates

- **Original(O):** The baseline RL algorithm is used, i.e., embeddings are unaltered.
- **Merged(M):** The baseline RL algorithm is used on our model, i.e., embeddings are altered by finding communities followed by merging individual embeddings.

By comparing the two columns in Table 1, '**AMQO** - Average Matching Quality (F1) Original' and '**AMQM** - Average Matching Quality (F1) Merged', we can observe that our model is consistently performing better when compared to that of original baseline algorithm for Cora, Wiki, and Karate datasets. And this is true across different baseline RL algorithms. But when it comes to blog-Catalog dataset, our model is not performing to the expected level compared to that of other datasets. This is because it is a huge dataset, with 39 different labels or communities.

The performance can be easily understood by analysing the column '**GOF** - Gain of F1'.

$$G = \frac{x - y}{y} * 100$$

where G is 'GOF', x is 'AMQM' and y is 'AMQO'.

The important point to note here is ground truth matched values are increased for all the datasets consistently across different RL algorithms which can be seen by comparing '**GTMO** - Ground Truth Matched Original' and '**GTMM** - Ground Truth Matched Merged' columns. The ground truth matched values are increasing even in the blogCatalog which performed poorly in average F1 score. The ground truth matched is improved by at least 15% for the blogCatalog dataset across different RL techniques.

The overall quality of the communities detected in the existing model and our merged model can be compared with '**OQO** - Overall Quality Original' and '**OQM** - Overall Quality Merged' columns from Table 1. Our model is performing better compared to the original model in almost all the datasets across different RL techniques. Cora and Wiki datasets are having higher overall quality gain followed by Karate and blogCatalog.

3.3 Node Classification

Node classification is one of the important applications when the ground truth information of nodes is available, i.e., labels of the nodes are available. Once the embedding representations of the nodes are available by two models (original and merged), the embeddings are considered as features and are used to train a random forest classifier [3] for single-label multi-class classification. For training and testing data, entire embedding data is split into 70% and 30% respectively. The testing accuracies are tabulated according to different algorithms and different baseline RL techniques in Table 2.

Our merged model is performing better than all other RL techniques for almost all the datasets. This can be seen by comparing the '**TAO** - Testing Accuracy Original' and '**TAM** - Testing Accuracy Merged' columns in the Table 2.

3.4 Link Prediction

Link Prediction is one of the important real-life applications in social networks. Given a social network, predicting a new link between nodes will enhance the connectivity of the nodes. For link prediction, we have taken the existing implementation from https://github.com/lucashu1/link-prediction.

In this task, the entire edgelist of the graph is split into 60%, 30% and 10% as training edgelist, testing edgelist and validation edgelist respectively. Giving the train_edges to the original and merged models, we will get the node embeddings.

Table 1. GTMO - Ground Truth Matched Original, GTMM - Ground Truth Matched Merged, AMQO - Average Matching Quality (F1) Original, AMQM - Average Matching Quality (F1) Merged, OQO - Overall Quality Original, OQM - Overall Quality Merged and GOF - Gain of F1 values of different datasets for various baseline RL algorithms.

Dataset	GTMO	GTMM	AMQO	AMQM	OQO	OQM	GOF
node2vec							
BlogCatalog	0.513	0.718	0.021	0.012	0.006	0.006	−42.8571
Cora	0.429	0.571	0.081	0.096	0.015	0.031	18.5185
Wiki	0.471	0.647	0.076	0.219	0.017	0.092	188.1578
Karate	1	1	0.474	0.8	0.474	0.8	68.7763
deepWalk							
BlogCatalog	0.487	0.692	0.018	0.01	0.004	0.005	−44.4444
Cora	0.286	0.286	0.091	0.147	0.007	0.012	61.5384
Wiki	0.529	0.529	0.058	0.091	0.016	0.025	56.8965
Karate	1	1	0.49	0.8	0.49	0.8	63.2653
LINE							
BlogCatalog	0.385	0.538	0.034	0.032	0.005	0.009	−5.8823
Cora	0.143	0.571	0.193	0.501	0.004	0.164	159.5854
Wiki	0.294	0.294	0.071	0.118	0.006	0.01	66.1971
Karate	0.5	0.75	0.27	0.344	0.086	0.152	27.4074
Graph factorization							
BlogCatalog	0.103	0.59	0.046	0.028	0	0.01	−39.1304
Cora	0.143	0.571	0.191	0.501	0.004	0.164	162.3036
Wiki	0.294	0.412	0.124	0.206	0.011	0.035	66.1290
Karate	1	1	0.49	0.8	0.49	0.8	63.2653

Table 2. Node classification results where TAO is Testing Accuracy Original and TAM is Testing Accuracy Merged.

Dataset	node2vec		deepWalk		LINE		GF	
	TAO	TAM	TAO	TAM	TAO	TAM	TAO	TAM
BlogCatalog	0.0845	0.0830	0.0981	0.0930	0.0861	0.0841	0.0977	0.0903
Cora	0.2644	0.8389	0.2614	0.8522	0.2850	0.3781	0.2850	0.5612
Wiki	0.3588	0.7475	0.3853	0.7558	0.3205	0.4784	0.3355	0.6694
Karate	0.5555	0.8888	0.3333	0.7777	0.2222	0.3333	0.3333	0.5555

But our aim is to find the edge embeddings as we are doing link prediction. Edge embeddings can be calculated using node embeddings. Edge embedding for edge $(v1, v2)$ is taken as Hadamard product of the node embeddings of $v1, v2$.

Table 3. VRSO - Validation ROC Score Original, VRSM - Validation ROC Score Merged, VASO - Validation AP Score Original, VASM - Validation AP Score Merged, TRSO - Test ROC Score Original, TRSM - Test ROC Score Merged, TASO - Test AP Score Original, TASM - Test AP Score Merged.

node2vec link prediction								
Dataset	VRSO	VRSM	VASO	VASM	TRSO	TRSM	TASO	TASM
BlogCatalog	0.7189	0.8853	0.6986	0.8637	0.7203	0.8861	0.6989	0.8661
Cora	0.4651	0.6365	0.4823	0.6455	0.4909	0.6083	0.4893	0.6157
Wiki	0.5824	0.7438	0.5691	0.7381	0.5922	0.7473	0.5770	0.7362
Karate	0.4693	0.5510	0.5793	0.6728	0.5198	0.4347	0.5679	0.4998
Email	0.5796	0.7321	0.5692	0.6919	0.5896	0.7295	0.5766	0.6936

Once the edge embeddings are calculated, the train edge embeddings are used to train the logistic regression model. And testing and validation edge embeddings are used to compare different algorithms performance with our merged model. For the evaluation, metrics like the Area Under ROC Curve (AUC_ROC_Score) and Average Precision score (AP) are used.

The scores in Table 3 clearly show that our merged model performs better than the node2vec model. There is a slight decrease in performance for the Karate dataset in the AUC_ROC_Score and AP, which is due to a very less number of edges in the dataset.

Figure 1 clearly shows that the merged model outperforms the deepWalk model for almost all of the datasets. The $2^{nd}, 4^{th}, 6^{th}$ and 8th bars in the chart indicates the merged model scores which are higher than $1^{st}, 3^{rd}, 5^{th}$ and 7th which are original deepWalk model ones.

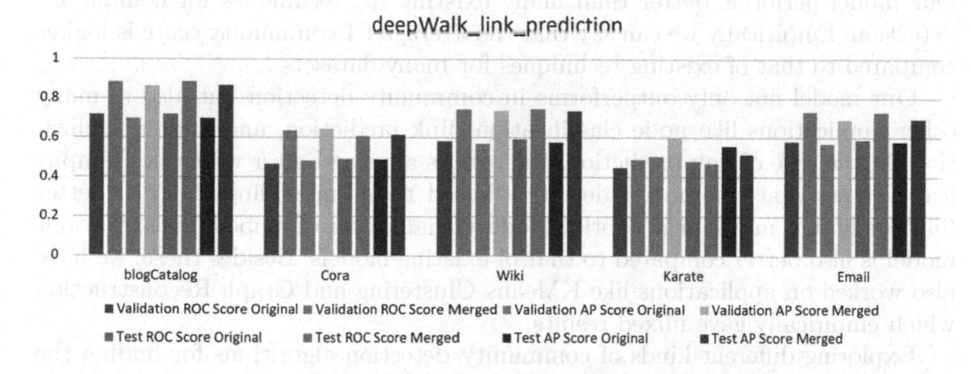

Fig. 1. Link prediction bar graph with deepWalk

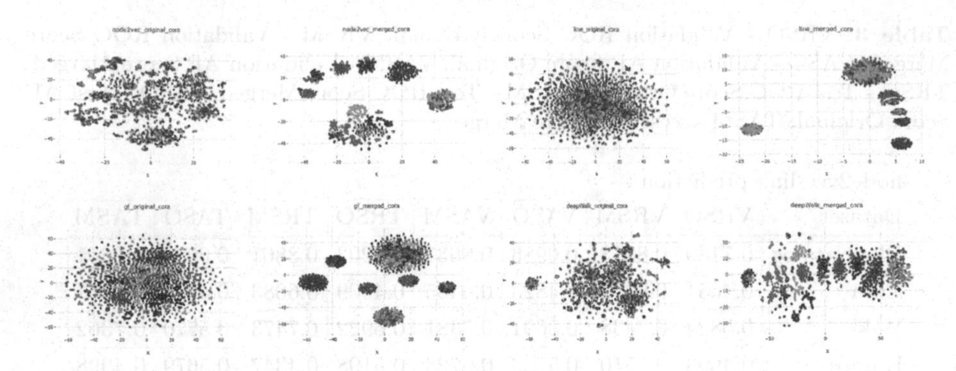

Fig. 2. Node visualization for cora dataset

3.5 Node Visualization

Node visualization is very useful to visualize any patterns or similarities in the data. It helps to reveal data that lies in multiple, different, manifolds, or clusters.

The cora dataset visualization is done using t-SNE for both merged model and original model with different RL techniques and the visuals are given in the Fig. 2. The figure show that in the merged model (2nd, 4th, 6th and 8th plots from top-left to bottom-right), there are clear and almost well-separated communities compared to that of the original model (1st, 3rd, 5th and 7th plots from top-left to bottom-right) respectively. This indicates that our model is performing well with respect to the communities.

4 Conclusions and Future Work

Our model performs better than many existing RL techniques for community detection. Empirically we can say that the average F1 community score is higher compared to that of existing techniques for many datasets.

Our model not only outperforms in community detection but also in many other applications like node classification, link prediction, and node visualization. In the case of link prediction, the results are way better which now empirically shows that community-detection-based node embeddings help in better link prediction in social networks. Node visualization of embeddings from our model is also better compared to that of existing models. Besides these, we have also worked on applications like KMeans Clustering and Graph Reconstruction which empirically gave mixed results.

Exploring different kinds of community detection algorithms for finding the communities from the edgelist is left for future work. As the embeddings are used for further downstream machine learning tasks, we would like to work on how our model embeddings will impact on some of the machine learning and deep learning applications.

References

1. Ahmed, A., Shervashidze, N., Narayanamurthy, S., Josifovski, V., Smola, A.J.: Distributed large-scale natural graph factorization. In: Proceedings of the 22nd International Conference on World Wide Web, pp. 37–48. ACM (2013)
2. Belkin, M., Niyogi, P.: Laplacian eigenmaps and spectral techniques for embedding and clustering. In: Dietterich, T.G., Becker, S., Ghahramani, Z. (eds.) Advances in Neural Information Processing Systems, vol. 14, pp. 585–591. MIT Press (2002). http://papers.nips.cc/paper/1961-laplacian-eigenmaps-and-spectral-techniques-for-embedding-and-clustering.pdf
3. Breiman, L.: Random forests. Mach. Learn. **45**(1), 5–32 (2001). ISSN 1573–0565. https://doi.org/10.1023/A:1010933404324
4. Cao, S., Lu, W., Xu, Q.: GraRep: learning graph representations with global structural information. In: Proceedings of CIKM, pp. 891–900 (2015)
5. Chen, H., Perozzi, B., Hu, Y., Skiena, S.: HARP: hierarchical representation learning for networks. CoRR, abs/1706.07845 (2017). http://arxiv.org/abs/1706.07845
6. Clauset, A., Newman, M.E.J., Moore, C.: Finding community structure in very large networks. Phys. Rev. E **70**, 066111 (2004). https://doi.org/10.1103/PhysRevE.70.066111.
7. Grover, A., Leskovec, J.: node2vec: scalable feature learning for networks. In: Proceedings of KDD, pp. 855–864 (2016)
8. Ou, M., Cui, P., Pei, J., Zhang, Z., Zhu, W.: Asymmetric transitivity preserving graph embedding. In: Proceedings of the 22nd ACM SIGKDD International Conference on Knowledge Discovery and Data Mining, KDD 2016, pp. 1105–1114. ACM, New York (2016). ISBN 978-1-4503-4232-2. https://doi.org/10.1145/2939672.2939751.URL. http://doi.acm.org/10.1145/2939672.2939751
9. Perozzi, B., Al-Rfou, R., Skiena, S.: DeepWalk: online learning of social representations. In: Proceedings of KDD, pp. 701–710 (2014)
10. Rossetti, G., Pappalardo, L., Rinzivillo, S.: A novel approach to evaluate community detection algorithms on ground truth. In: Cherifi, H., Gonçalves, B., Menezes, R., Sinatra, R. (eds.) Complex Networks VII. SCI, vol. 644, pp. 133–144. Springer, Cham (2016). https://doi.org/10.1007/978-3-319-30569-1_10
11. Tang, J., Qu, M., Wang, M., Zhang, M., Yan, J., Mei, Q.: Line: Large-scale information network embedding. In: Proceedings of WWW, pp. 1067–1077 (2015)
12. Wang, D., Cui, P., Zhu, W.: Structural deep network embedding. In: Proceedings of the 22nd ACM SIGKDD International Conference on Knowledge Discovery and Data Mining, pp. 1225–1234. ACM (2016)

Code Generation from Supervised Code Embeddings

Han Hu[1], Qiuyuan Chen[2(✉)], and Zhaoyi Liu[3]

[1] School of Software, Tsinghua University, Beijing, China
hh17@mails.tsinghua.edu.cn

[2] College of Computer Science and Technology, Zhejiang University, Hangzhou, China
chenqiuyuan@zju.edu.cn

[3] School of Shenzhen Graduate, Peking University, Shenzhen 518055, China
1701213615@sz.pku.edu.cn

Abstract. Code generation, which generates source code from natural language, is beneficial for constructing smarter **I**ntegrated **D**evelopment **E**nvironments (IDEs), retrieving code more effectively and so on. Traditional approaches are based on matching similar code snippets, and recently researchers pay more attention to machine learning, especially the encoder-decoder framework. Faced with code generation, most encoder-decoder frameworks suffer from two drawbacks: (a) The length of the code snippet is always much longer than the length of its corresponding natural language, which makes it hard to align them, especially for encoders at word level; (b) Code snippets with the same functionality could be implemented in various ways, even completely different at word level. For drawback (a), we propose a new **S**upervised **C**ode **E**mbedding (SCE) model to promote the alignment between natural language and code. For drawback (b), with the help of **A**bstract **S**yntax **T**ree (AST), we propose a new distributed representation of code snippets which overcomes this drawback. To evaluate our approaches, we build a variant of the encoder-decoder model to generates code with the help of pre-trained code embedding. We perform experiments on several open source datasets. The experiment results indicate that our approaches are effective and outperform the state-of-the-art.

Keywords: Code generation · Code embedding · Supervised learning

1 Introduction

Generating code through natural languages (NL) is considered to be an important future direction of programming. On the one hand, it can lower the threshold of programming and facilitate programming process. On the other hand, it makes programmers more productive, for example, by generating non-core code automatically, which allows programmers to focus more on the core code. So

T. Gedeon et al. (Eds.): ICONIP 2019, CCIS 1142, pp. 388–396, 2019.
https://doi.org/10.1007/978-3-030-36808-1_42

using NL to map complex operations to basic code blocks receives tremendous interest and has shown great benefits.

Traditional code generation approaches are usually based on matching similar code snippets [5–7]. Recently, many researchers pay more attention to generating code by machine learning. Some researchers try to bridge the gap between two corpora by utilizing rich, existing code bases and program contexts [8,16]. Some researchers utilize a standard or a variant encoder-decoder model to map NL to a snippet of executable code directly [1,4,8–10,17].

Most existing approaches regard a code snippet as a simple plain sequence of words, without taking features of itself into account. In this way, current approaches suffer from two drawbacks: (a) the length of code snippets usually differ a lot from that of NL, it seems a tough work for regular encoder-decoder models or attention mechanism to align them. (b) Code snippets usually contain multiple functional processes, and the same functionality could be implemented in various ways. Mapping directly is likely to be disordered because there is no strict bijection between two corpora (i.e., code snippets and the corresponding natural language).

Faced with the drawback (a), we propose a **S**upervised **C**ode **E**mbedding (SCE) model to pre-train distributions of code and NL at the same time, which helps align NL and code better. Faced with (b), only considering the word-level feature of the code is far from enough, for example, Fig. 1(a) and Fig. 1(b) show an example of two Java functions which have the same functionality: counting characters of a string, but implemented in two different ways. Their word-level features are quite different, yet they own the same functionality, which will put a heavy burden on the model training process. So we turn our attention to proposing a new representation of code which not only takes word-level features into account.

Abstract **S**yntax **T**ree (AST) is a tree-structural representation of source code which describes its functionality in a specific programming language. The leaves of the tree usually refer to user-defined values which represent identifiers or variable types in the source code. The non-leaf nodes represent a set of structure in the programming language (such as loops or variable declarations). In Figs. 1(a) and 1(b), the lower subfigure is the visualized AST of the above code snippet, respectively.

As shown in Fig. 1, user-defined identifiers (such as *num*, *str*) and variable types (such as *String*, *int*) are represented as leaves of the tree, and syntactic structure such as judgment statement (*ifStmt*) and loop (*ForStmt*) are represented as non-leaf nodes. In AST, we call a path between two leaves or a leaf and a root an AST path, which are marked red, yellow, blue or green respectively in Fig. 1(a) and 1(b). Intuitively, every path is a functional module in the code snippet. It is clear that although two methods are quite different in token-level representation, their AST paths differ only in two nodes, a *ForeachSt* node instead of a *ForStmt* node and a *Method Call Expr* node instead of a *Unary Expr*, which are circled by red dotted lines. So faced with drawback (b), we propose a new distributed representation of code which combines AST paths features (i.e.,

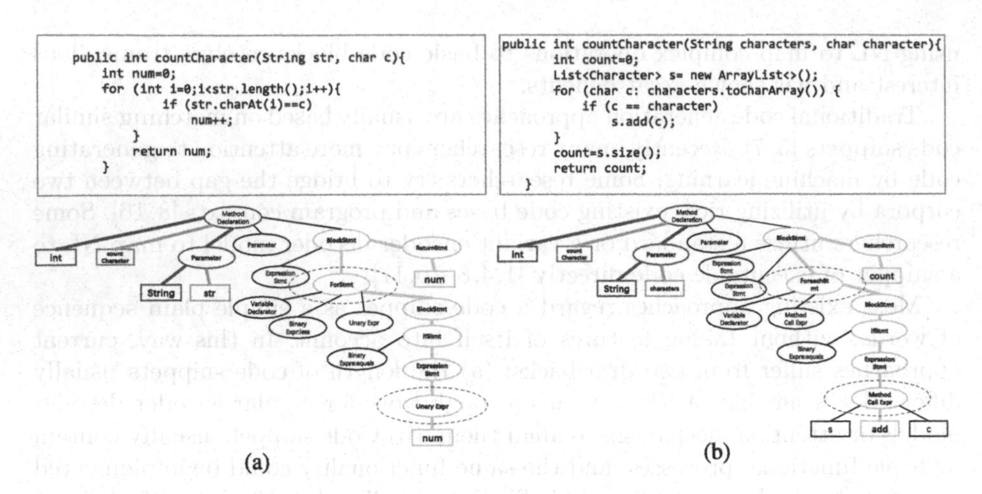

Fig. 1. Two Java functions and their ASTs. Two Java functions have the same functionally while implemented in different ways. Although They differ a lot in token-level representation, considering their AST paths, only differ in two nodes, which are circled by red dotted lines (Color figure online)

syntactic features) and word-level features (i.e., lexical features) of code in SCE model.

In summary, our contributions in this paper are as follows:

- We propose a new distributed representation of code snippets which combines AST paths features and word-level features of code.
- We propose SCE model. The model uses supervised learning to pre-train distributions of code and NL at the same time, aims to promote aligning them. Based on pre-trained code embeddings, we build a variant of the encoder-decoder model to align NL and code.
- We conducted comparative experiments on real-world datasets including code in popular high-rank Github repositories to evaluate our approaches, and the experiment results indicate that our models are effective and outperform mainstream approaches by 10.15% on performance in code generation.

The rest of this paper is organized as follows: our approach is presented in Sect. 2. Section 3 introduces the experiments details. Section 4 introduces research related to this work. Then the conclusion is shown in Sect. 5.

2 Approach

In this section, we introduce the details of our approach. The approach consists of two stages: code embedding and code generation. Figure 2 provides an overview of SCE model for code embedding.

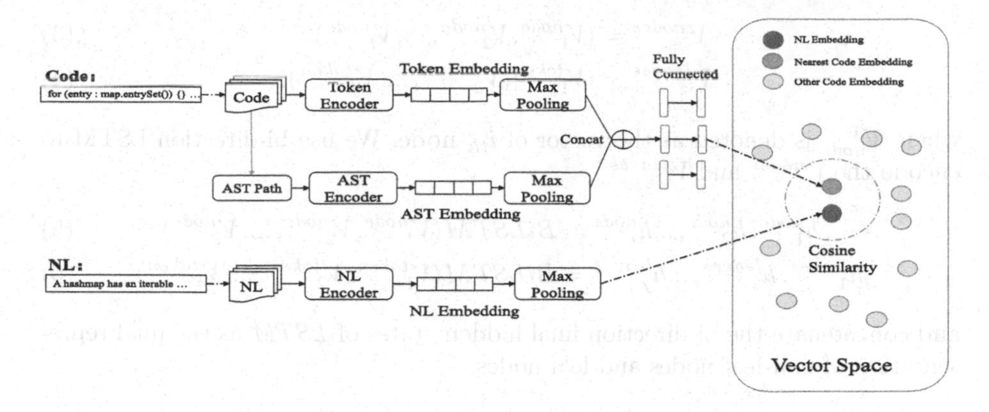

Fig. 2. Overview of Code Embedding Task. Feed NL to NL encoder to get NL embedding E^{nl}, feed code to Token Encoder and code's AST Paths to AST encoder, then combine their output embeddings by a fully connected layer to get E^{code}. Finally, fit the distribution of E^{nl} and E^{code}

A pair of natural language (*NL*) and code snippet (abbreviated as *Code*) is an input for training, which all consist of a sequence of tokenized tokens[1]. After represented as a vector by one-hot, NL would be fed into NL encoder. The NL encoder is a bidirectional transformer, which captures the contextual information for each word of NL, and generates their contextual embedding vectors $E_1^{nl}, E_2^{nl}, ...E_i^{nl}$. Finally, concatenate them to E^{nl}. Let

$$E_1^{nl}, E_2^{nl}, ...E_i^{nl} = BiTransformer(V_1^{nl}, V_2^{nl}, ...V_i^{nl}) \quad (1)$$

$$E^{nl} = [E_1^{nl}; E_2^{nl}; ...E_i^{nl}] \quad (2)$$

where V_{nl}^{i} is denoted as the vector of i_{th} NL token. Then E^{nl} would be fed to a max-pooling layer.

The same as NL representation, we feed code tokens to Token Encoder, which is a bidirectional transformer, the encoder encodes the tokens to Token Embedding vectors and concatenate to E^{tokens}. A max-pooling layer is followed too.

As mentioned in Sect. 1, we not only capture token-level feature but also capture functionality feature of a code snippet by its AST paths. Every AST path consists of two leaves and non-leaf nodes, so every path is seen as a sequence of its non-leaf nodes' embedding vectors and a sequence of two leaves' tokens embedding vectors. Let

[1] More details about tokenization phase, please refer to Sect. 3.2.

$$V^{nodes} = (V_1^{node}, V_2^{node}, ..., V_i^{node}) \tag{3}$$
$$V^{leaves} = (V_1^{token}, V_2^{token}, ..., V_j^{token}) \tag{4}$$

where V_{node}^i is denoted as the vector of i_{th} node. We use bi-direction LSTM to encode the V^{nodes} and V^{leaves}

$$h_1^{node}, h_2^{node}, ...h_i^{node} = BiLSTM(V_1^{node}, V_2^{node}, ..., V_i^{node}) \tag{5}$$
$$h_1^{leaves}, h_2^{leaves}, ...h_j^{leaves} = BiLSTM(V_1^{token}, V_2^{token}, ..., V_j^{token}) \tag{6}$$

and concatenate the bi-direction final hidden states of $LSTM$ as the final representation of non-leaf nodes and leaf nodes.

$$E^{nodes}(V_1^{node}, V_2^{node}, ..., V_i^{node}) = [h_i^{node}; h_1^{node}] \tag{7}$$
$$E^{leaves}(V_1^{token}, V_2^{token}, ..., V_j^{token}) = [h_j^{leaves}; h_1^{leaves}] \tag{8}$$

Suppose a snippet of code has k AST paths, we concatenate E^{nodes} and E^{leaves}, then average the combined vector to E^{AST}:

$$E^{AST} = \frac{1}{k}\sum_{k=1}^{k}[E^{nodes}; E^{leaves}] \tag{9}$$

A max-pooling layer is followed to reduce dimensions of E^{AST}.

To represent the final code embedding E^{code}, we concatenate the AST path representation and token-level representation and apply a fully connected layer to combine them

$$E^{code} = tanh(W \cdot [E^{AST}; E^{tokens}]) \tag{10}$$

where W is a $(d_{AST} + d_{tokens}) \times d_{hidden}$ weight matrix.

we choose cosine similarity as our loss function to describe the distance in distribution between E^{nl} and E^{code},

$$D_{loss} = cos(E^{nl}, E^{code}) = \frac{\sum_{i=1}^{n} E_i^{nl} \times E_i^{code}}{\sqrt{\sum_{i=1}^{n}(E_i^{nl})^2} \times \sqrt{\sum_{i=1}^{n}(E_i^{code})^2}} \tag{11}$$

where E_i^{nl}, and E_i^{code} are the i_{th} dimension of E^{nl} and E^{code}.

When SCE model finishes pre-training, we use pre-trained Token encoder to encode our code dictionary to get dictionary embedding matrix E_{dic}.

To evaluate our model, we build a variant of the encode-decoder model with global attention, and it focuses on the stage of generating code from NL with the support of SCE model. Its NL encoder is transferred from SCE model's. Its decoder is a $LSTM$ with global attention and uses embedding matrix E_{dic} to convert code snippets to continuously distributed vectors when in training stage.

3 Experiments

3.1 Datasets

Our collected dataset[2] consists of four open source datasets: Awesome Java[3], CONCODE [8][4], BigCloneBench[5], and JDK source codes. Table 1 shows statistics of our base datasets.

Table 1. Statistics of datasets

Datasets	Projects	Files	Lines
Awesome java	535	264,284	26,407,592
CONCODE	~33,000	~300,000	~13,000,000
BigCloneBench	10	9,376	2,065,108
JDK	-	7,700	1,009,560

3.2 Data Preprocessing

For every Java function, function's comments are extracted as NL inputs. The function bodies are treated as our target code to be generated. The NL words and code words are lower-cased. The camel-cased and underline identifiers are split into several words, for example, split *checkJavaFile* to three words: check, Java, File, split *get_user_name* to three words: get, user, name. All punctuation marks are removed. We add *[CLS]* at the beginning of every sentence and add *[SEP]* at the end. *[UNK]* is used to represent words outside the dictionary. After these steps, every word is tokenized to token.

Javaparser lib[6] is used to parse Java source codes. ASTParser lib[7] is used to build AST of code. In order to decrease noise and reinforce the learning process, we only use the first sentence of comments since they already summarize the function of methods according to Javadoc guidance[8]. Some redundant comments, such as empty comments, one-word comments, and non-English comments, are filtered.

Finally, we collect 3,950,164 pairs of $(NL, Code)$ for code embedding, and 1,074,963 pairs of $(NL, Code)$ for code generation. Each dataset is split into a training set, and a test set in proportion with 8:2 after shuffling the pairs.

[2] https://drive.google.com/open?id=1nOuZjSS9lUqWfQptUOhfX9kNKd_FeCkn.
[3] https://github.com/akullpp/awesome-java.
[4] https://drive.google.com/drive/folders/1kC6fe7JgOmEHhVFaXjzOmKeatTJy1I1W.
[5] https://github.com/clonebench/BigCloneBench/blob/master/README.md.
[6] https://github.com/javaparser/javaparser.
[7] http://help.eclipse.org/mars/index.jsp.
[8] https://www.oracle.com/technetwork/articles/java/index-137868.html.

3.3 Experiment Setting

When in code embedding, we restrict the maximum length of NL to 20 words, and the length of the code is limited to 100. We use 12 hidden layers to encode NL and code tokens in the transformer, and the hidden size is 768. When in code generation, we set the hidden size of the *LSTM* cells to 512, and all cells are 2-layers. Max-pooling layer is used to reduce computation and align matrix. We use dropout with p = 0.4. The ratio of teacher forcing is 0.5. Adam is used as our optimizer with an initial learning rate of 0.0001 for optimization. We use TensorFlow to implement our models.

3.4 Experiment Results

To evaluate the quality of the output, following recent works in code generation [8,11], we choose the **BLEU**, **Precision**, **Recall** and **F-score** of generated words as our metrics. We compare our model with two baselines:

- **Code Generated Methods Proposed by Iyer et al.** [8]. This method is proposed recently and outperforms the state-of-the-art in code generation. It is abbreviated as *Iyer et al.*.
- **Code Generation Task without Code Embedding.** To evaluate the effect of code embedding, we conducted an experiment that only utilizes code generation task to generate code which is abbreviated as *without Embedding.*

Table 2 illustrates the precision, recall, F-score, and BLEU results of our approach and other baseline methods. Our approach, abbreviated as *ours*, outperforms all baselines in Precision, Recall, F-score and BLEU. Our approach achieves an improvement of 2.08 BLEU points compared with the best results of other approaches, which outperforms current state-of-the-art methods by 10.15%.

Table 2. Precision, Recall, F-score and BLEU of Our Approach and Other Baselines

Approaches	Precision	Recall	F-score	BLEU
Iyer et al	22.67	13.56	16.97	19.35
Without embedding	19.35	11.32	14.28	20.50
Ours	**28.58**	**16.36**	**20.81**	**22.58**

The BLEU scores of *ours* and *without Embedding* show that our SCE model could promote the effect of code generation.

4 Related Work

A number of previous researches have explored mapping NL to code blocks [4,9,17], regular expression [12] and SQL statements [18]. Some researches generate code on a certain context: [13,15] generate code in the environment of database querying; Some specific research [10] generates codes in the field of a card game, conditioned on categorical card attributes. These works are based on a chunk of codes that implement certain business logic. Recent researches propose models and evaluate them on domain-specific dataset (Hearthstone & MTG, [10]; CONCODE [8]), and manual labeled per-line comments (DJANGO [14]). Domain-specific data is organized based on specific business logic. Each functionality contains business knowledge and consists of several basic operations. Manually labeled data (DJANGO) contains programs with short description possibly mapping to categorical data. The values need to be copied onto the resulting code from a single domain. Some researchers use AST to represent code [2,3] too, but they don't fuse word-level features.

In Neural Machine Translation area, neural encoder-decoder has proved to be effective. It also has good performance in mapping NL to programming logic and code generation. Some methods directly generate code blocks or domain-specific programming language using encoder-decoder model [10,14]. Some methods use a customized decoder for capturing code structure and perform generation [16].

5 Conclusion

In the paper, we propose a new representation of code snippets which combine features from lexical level and syntactic level. We propose a novel **S**upervised **C**ode **E**mbedding (SCE) model to learn distributed representation of the code and NL at the same time. We conducted several comparative experiments to prove our approach, and experimental results show that our approach, which outperforms state-of-the-art baselines, is significantly effective and can generate more high-quality codes. Our future work will focus on two aspects: how to better fuse other information, and how to generate executable code directly.

Author contributions. Han Hu,Qiuyuan Chen and Zhaoyi Liu

References

1. Allamanis, M., Tarlow, D., Gordon, A.D., Wei, Y.: Bimodal modelling of source code and natural language. In: ICML (2015)
2. Alon, U., Brody, S., Levy, O., Yahav, E.: code2seq: generating sequences from structured representations of code. In: International Conference on Learning Representations (2019)
3. Alon, U., Zilberstein, M., Levy, O., Yahav, E.: Code2vec: learning distributed representations of code. In: Proceedings ACM Program Language 3(POPL), 40:1–40:29 (2019). https://doi.org/10.1145/3290353, https://doi.acm.org/10.1145/3290353
4. Balog, M., Gaunt, A.L., Brockschmidt, M., Nowozin, S., Tarlow, D.: Deepcoder: learning to write programs. arXiv preprint (2016). arXiv:1611.01989
5. Dieumegard, A., Toom, A., Pantel, M.: Model-based formal specification of a DSL library for a qualified code generator. In: Proceedings of the 12th Workshop on OCL and Textual Modelling, Innsbruck, Austria, September 30, 2012, pp. 61–62 (2012). https://doi.org/10.1145/2428516.2428527

6. Glück, R., Lowry, M.R. (eds.): Generative Programming and Component Engineering, 4th International Conference, GPCE 2005, Tallinn, Estonia, September 29 – October 1, 2005, Proceedings, Lecture Notes in Computer Science, vol. 3676, Springer (2005). https://doi.org/10.1007/11561347
7. Hemel, Z., Kats, L.C.L., Groenewegen, D.M., Visser, E.: Code generation by model transformation: a case study in transformation modularity. Softw. Syst. Model. **9**(3), 375–402 (2010)
8. Iyer, S., Konstas, I., Cheung, A., Zettlemoyer, L.: Mapping language to code in programmatic context. In: Proceedings of the 2018 Conference on Empirical Methods in Natural Language Processing, pp. 1643–1652 (2018)
9. Liang, P., Jordan, M.I., Klein, D.: Learning dependency-based compositional semantics. Comput. Linguist. **39**(2), 389–446 (2013)
10. Ling, W., et al.: Latent predictor networks for code generation. arXiv preprint (2016). arXiv:1603.06744
11. Ling, W., et al.: Latent Predictor Networks for Code Generation (2016)
12. Locascio, N., Narasimhan, K., DeLeon, E., Kushman, N., Barzilay, R.: Neural generation of regular expressions from natural language with minimal domain knowledge. arXiv preprint (2016). arXiv:1608.03000
13. Manshadi, M.H., Gildea, D., Allen, J.F.: Integrating programming by example and natural language programming. In: AAAI (2013)
14. Oda, Y., et al.: Learning to generate pseudo-code from source code using statistical machine translation. In: 2015 30th IEEE/ACM International Conference on Automated Software Engineering (ASE), pp. 574–584. IEEE (2015)
15. Quirk, C., Mooney, R., Galley, M.: Language to code: learning semantic parsers for if-this-then-that recipes. In: Proceedings of the 53rd Annual Meeting of the Association for Computational Linguistics and the 7th International Joint Conference on Natural Language Processing (vol. 1: Long Papers). vol. 1, pp. 878–888 (2015)
16. Yin, P., Neubig, G.: A syntactic neural model for general-purpose code generation. In: Proceedings of the 55th Annual Meeting of the Association for Computational Linguistics (volume 1: Long Papers). vol. 1, pp. 440–450 (2017)
17. Zettlemoyer, L.S., Collins, M.: Learning to map sentences to logical form: structured classification with probabilistic categorial grammars. arXiv preprint (2012). arXiv:1207.1420
18. Zhong, V., Xiong, C., Socher, R.: Seq2sql: generating structured queries from natural language using reinforcement learning. arXiv preprint (2017). arXiv:1709.00103

ComNE: Reinforcing Network Embedding with Community Learning

Ahmed Fathy(✉) and Kan Li

School of Computer Science and Technology,
Beijing Institute of Technology, Beijing 10081, China
{ahmedfathy,likan}@bit.edu.cn

Abstract. Learning network embedding for large-scale networks have been attracting increasing attention due to their importance in supporting numerous network analytic and data mining tasks such as node classification, clustering and visualization. In this paper, we present a novel framework for learning large-scale network embedding incorporating network topology and community structural information. Most existing network embedding methods tend to embed network topology and ignore the partially labeled community structure information that exist in real-world networks and thus are unable to efficiently learn and capture the community structure of real-world networks. Unlike existing works, our framework integrates the network topology and community structure into the learning process. We propose a deep autoencoder model to generate low-dimensional feature representations efficiently through learning network reconstruction and community classification tasks. The experimental results on several real-world networks show that our framework outperforms the state-of-the-art methods.

Keywords: Large-scale network embedding · Network representation learning · Autoencoder · Community prediction

1 Introduction

Large-scale networks are prevalent in our daily lives such as social, collaboration and citations networks. Learning a meaningful network embedding is a critical prerequisite for applying network analytic and data mining tasks such as node classification [7,11], link prediction [4], clustering [2], and visualization [16].

The main challenge in network embedding is to find the most informative data representation that preserves the structural information between the vertices in the network. There has been substantial interest in the work of learning network embedding [3,5,16] which attracted the attention of many experts owing to its comprehensive use in real-world applications.

Traditionally, the dimensionality reduction techniques such as Isomap [15] and Laplacian Eigenmap [1] find a low-dimensional manifold embedded in the high-dimensional data of the network. The main drawback of these methods

T. Gedeon et al. (Eds.): ICONIP 2019, CCIS 1142, pp. 397–405, 2019.
https://doi.org/10.1007/978-3-030-36808-1_43

is having a quadratic time complexity with respect to the number of vertices. Therefore, these techniques are inefficient when applied to large-scale networks.

Recently, several network embedding approaches adopt random walk to exploit the network structure and learn representations using skip-gram model such as DeepWalk [11] and Node2Vec [4]. These methods are proofed to be identical to matrix factorization of networks [17]. However, random walks are inefficient when applied on large-scale networks [13,16]. In addition, these methods adopt shallow model and cannot leverage the representation ability of deep learning [6] which can efficiently learn the non-linear structure of real-world networks.

Another line of works address network embedding problem using deep learning, such as SDNE [16] and DNGR [2]. However, these methods suffer from performance issues, for instance, SDNE requires model pertaining and setup of many hyperparameters, and DNGR applies random surfing model to generate a node probabilistic co-occurrence matrix, which makes these methods unscalable and difficult to apply to large-scale networks. In addition, these methods ignore the community labels that exist in real-world networks and cannot efficiently learn the community structure of the network.

Recently, the network embedding problem was addressed using convolution layers such as [3,5,10]. These methods recursively aggregate the neighborhood information of each vertex in the network. These methods generally suffer from computation and memory issues due to recursive expansion of neighborhoods. The modeling ability of our approach is different from GCN-based methods since we explicitly model the relationship between all vertices using the adjacency matrix A, while these methods learn from vertex features X and adjacency matrix A to community labels Y.

To address these challenges, we propose a novel framework based on semi-supervised stacked sparse autoencoder model which can resolve the above issues and improve the effectiveness and efficiency of network embedding. In summary, the contributions of our paper are as follows:

- We propose a novel network embedding framework based on semi-supervised stacked sparse autoencoder, named ComNE, to jointly learn features for large-scale networks using network topology and community structure using partially labeled vertices.
- We conduct extensive experiments using various real-world network datasets and compare with state-of-the-art methods to validate our approach. Precisely, our framework outperform the baselines on challenging classification, clustering and network visualization tasks.

2 Related Work

Over the past few years, several network embedding techniques have emerged to address the applications of network-structured data such as vertex classification and link prediction. The earlier works for dimensionality reduction methods

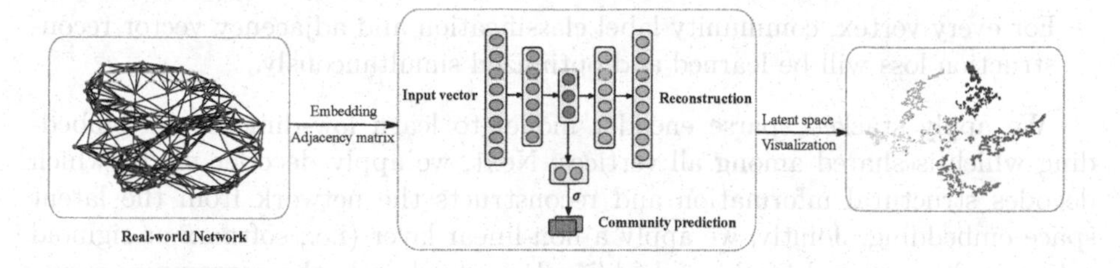

Fig. 1. The framework of ComNE.

have been studied in the literature [1,15]. These methods generally suffer from performance drawbacks and cannot scale to large networks.

Several approaches for learning network embedding have been proposed such as [4,11], which adopt random walk and skip-gram procedure to learn network embedding and generate low-dimensional representations for vertices. Recently, Tang et al. [13] proposed two loss functions to learn the first-order and second-order proximities of the network. Their method learns to embed local and global structure separately and then concatenates the embedding.

Recently, network embedding approaches adapt convolutions and rely on vertex neighborhood to learn representations such as [5,7]. These methods usually provide additional gains on different network analytic tasks. However, these methods have performance issues due to recursive expansion of neighborhoods.

3 Method

3.1 Problem Formulation

Let $G = (V, E, L)$ be a given network, where $V = \{v_1, v_2, ..., v_n\}$ is the set of vertices, $E = \{e_1, e_2, ..., e_m\}$ is the set of edges between a pair of vertices and $L = \{l_1, l_2, ..., l_k\}$ is the set of community labels in the network. Each edge is associated with a weight $w_{i,j}$. A partially labeled network can be defined as $G = (V_L, V_U, E)$ where V_L is the set of labeled vertices, V_U is the set of unlabeled vertices, where $V = V_L \cup V_U$. We assume that each vertex can have multiple labels. The main objective of network embedding is to learn a mapping function $f : V \rightarrow \mathbb{R}^{|V| \times d}$, $d \ll n$.

3.2 ComNE Framework

We propose a novel framework to preserve the network topology and community structure using deep architecture. The fundamental ideas of the proposed framework are as follows:

- Each vertex is mapped to a low-dimensional space that is shared across all network vertices using stacked sparse autoencoder model.
- The linked vertices tend to have similar representations. Similarly, vertices belong to the same community will be embedded closer to each other.

- For every vertex, community label classification and adjacency vector reconstruction loss will be learned and optimized simultaneously.

We apply stacked sparse encoder model to learn low-dimensional embedding which is shared among all vertices. Next, we apply decoder model which decodes structural information and reconstructs the network from the latent space embedding. Jointly, we apply a non-linear layer (i.e., softmax or sigmoid activation) connected to the embedding layer to decode the community structure of the latent space embedding using the labeled vertices, see Fig. 1 for illustration.

The main goal is to optimize the encoder and decoder mappings which can be achieved by minimizing two losses: reconstruction loss (to preserve the network topology) and semi-supervised classification loss (to preserve community structure). The network reconstruction loss is denoted by the mean square error (MSE) given by:

$$\mathcal{L}_1 = \frac{1}{n}\sum_{i=1}^{n}(x_i - \hat{x}_i)^2 \tag{1}$$

where n is the number of vertices, x_i is the input adjacency vector of vertex v_i, $\hat{x}_i$ is the reconstructed vector, and $\mathcal{L}_1$ is the mean of sum of squared distances between input and reconstructed vectors.

Next, the semi-supervised classification loss to preserve the community structure of the network is represented by the cross-entropy loss can be defined as:

$$\mathcal{L}_2 = -\frac{1}{n}\sum_{i=1}^{n}\sum_{j=1}^{k} y_{ij}\log(\hat{y}_{ij}) \tag{2}$$

where k is the number of communities, y_{ij} is the community label for vertex v_i and $\hat{y}_{ij}$ is the community label prediction.

Finally, we impose the l_2 norm regularization prevent model over-fitting:

$$\mathcal{L}_{reg} = \frac{\lambda}{2}\sum_{\ell=1}^{t}\left\|W^{\ell}\right\|_2^2 \tag{3}$$

where λ represents the regularization constant and W^{ℓ} is the weight matrix of the layer ℓ of the deep architecture.

Overall, we minimize the following loss function:

$$\begin{aligned}\mathcal{L} &= \mathcal{L}_1 + \mathcal{L}_2 + \mathcal{L}_{reg}\\ &= \frac{1}{n}\sum_{i=1}^{n}(x_i - \hat{x}_i)^2 - \frac{1}{n}\sum_{i=1}^{n}\sum_{j=1}^{k} y_{ij}\log(\hat{y}_{ij}) + \frac{\lambda}{2}\sum_{\ell=1}^{t}\left\|W^{\ell}\right\|_2^2\\ &= \frac{1}{n}\sum_{i=1}^{n}\left((x_i - \hat{x}_i)^2 - \sum_{j=1}^{k} y_{ij}\log(\hat{y}_{ij})\right) + \frac{\lambda}{2}\sum_{\ell=1}^{t}\left\|W^{\ell}\right\|_2^2\end{aligned} \tag{4}$$

3.3 Complexity Analysis

The computational complexity of ComNE is of order $\mathcal{O}(n \times t \times d \times c_t)$ where n is the number of vertices and t is the number of hidden layers, d is the maximum dimension of the hidden layers, and c_t is the number of iterations till convergence. The parameters t, d and c_t are unrelated with n, but are associated to the deep autoencoder model. Hence, the computational complexity of ComNE framework grows linearly with the number of vertices n.

4 Experiments

4.1 Datasets

An overview of the network datasets we consider in our experiments is given in Table 1. The datasets description are as follows:

- BlogCatalog [14] is an online social network formed by the online users.
- Cora and CiteSeer [9] are research papers datasets.
- Wiki [12] is composed of real-world web pages and hyperlinks.

Table 1. Network datasets used in our experiments and autoencoder layers structure.

Dataset	$\|V\|$	$\|E\|$	$\|Y\|$	Average degree	Category	Layer neurons
BlogCatalog	10,312	333,983	39	64.7756	Social Network	1000-512-128
CiteSeer	3,312	4,732	6	2.8231	Citation Network	512-256-128
Cora	2,708	5,429	7	3.8981	Citation Network	512-256-128
Wiki	2,405	17,981	17	10.6121	Web page Network	512-256-128

4.2 Baseline Algorithms

We evaluate our method against the following baselines:

- **DeepWalk** [11]: It transforms network structure into linear sequences by random walks and employs skip-gram model to learn network embedding.
- **LINE** [13]: It preserves first-order and second-order proximities in the network and uses skip-gram with negative sampling to learn representation.
- **SDNE** [16]: It is a semi-supervised autoencoder method with objective function that exploit the first-order and second-order proximities in the network.
- **Node2Vec** [4]: It improves DeepWalk by adopting biased random walk to explore diverse neighborhoods using depth-first and breath-first sampling.

Table 2. NMI and ARI scores for vertex clustering in Cora, CiteSeer and Wiki datasets.

Algorithm	Cora		CiteSeer		Wiki	
	NMI	ARI	NMI	ARI	NMI	ARI
DeepWalk	0.40	0.30	0.14	0.14	0.36	0.18
LINE	0.13	0.04	0.05	0.01	0.26	0.07
SDNE	0.25	0.17	0.07	0.05	0.30	0.16
Node2Vec	0.44	0.36	0.20	0.16	0.36	0.21
ComNE (ours)	**0.58**	**0.39**	**0.43**	**0.23**	**0.49**	**0.23**

4.3 Experimental Setup

For the unsupervised network embedding baselines, we start with obtaining the latent representations of vertices. Following, a portion of vertices and their labels are randomly sampled from the network and used as training data for a one-vs-rest logistic regression model. The goal is to predict the labels of the remaining vertices. To guarantee that our experiments are reliable, the above process is repeated for 10 times, and we report the average of the accuracy scores.

For our method, we use community labels during training. For fair comparison, we split each dataset into 40%, 10% and 50% train, validation, and test sets respectively (we only use half of each dataset for model training). The structure of autoencoder layers are listed in Table 1. We adapt early stopping procedure with patience of 10 epochs and set a max of 100 epochs. The regularization constant λ is set as 0.001. We adopt Adam optimizer with default learning rate.

For baseline methods, the parameters are tuned to be optimal and fair. For DeepWalk and Node2Vec, the number of random walks to start at each node is set as 10. The length of random walk started at each node is 80 and the window size of skip-gram model is 10. For Node2Vec, the hyperparameters p and q are set as 1.0. For LINE, the starting value of learning rate is 0.025. The number of negative samples is set as 5 and the number of training samples are set as 10,000. For SDNE, the hyperparameters α and β are set as 0.2 and 10 respectively. For fair comparison, we use the same autoencoder structure as shown in Table 1.

4.4 Vertex Clustering with Label Information

In this experiment, we ran each baseline algorithm to obtain the embedding, which is used as a feature representation for clustering and apply k-means algorithm to the learned vertex embedding. Networks communities are used as the ground truth to assess the quality of clustering results. We report the normalized mutual information (NMI) and Adjusted Rand Index (ARI) as the performance metrics. We run k-means algorithm 10 times and report the average in Table 2.

From the results, we can see that ComNE significantly outperforms the baselines. In other words, our model can generate vertex embedding which preserve the structure of the original network topology and community structure.

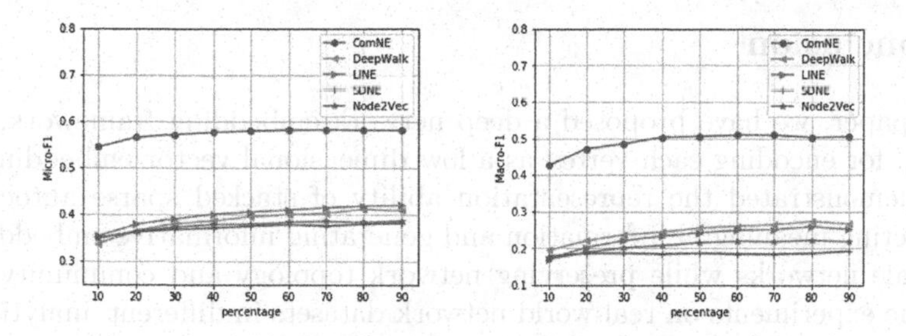

Fig. 2. Average of Micro-F1 and Macro-F1 scores in BlogCatalog dataset.

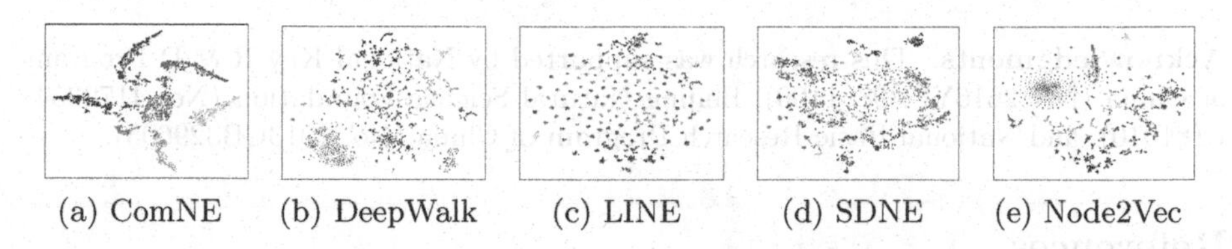

(a) ComNE (b) DeepWalk (c) LINE (d) SDNE (e) Node2Vec

Fig. 3. Visualization of CiteSeer dataset. Each point indicates a research paper and the color of a point represents the category of the paper.

4.5 Multi-label Classification

In multi-label classification, each vertex is assigned multiple communities (i.e., communities overlap). This experiment is conducted by training a classifier on labeled vertices using its low-dimensional embedding as features. Next, the trained classifier is used to classify each vertex into a set of labels. We randomly sample a portion of the labeled vertices as the training set and the remaining as test set to evaluate the performance. We vary the training set from 10% to 90%, and report Micro-F1 and Macro-F1 scores in Fig. 2.

From the results, we observe that our method outperforms the baselines, which illustrates the importance of incorporating the community information.

4.6 Network Visualization

Visualization of the vertex representations assists in validating the quality of learned vertex embedding. We conducted visualization experiment by following [13]. After vertex embedding are learned by different network embedding methods, we apply t-SNE [8], which maps the vertex embedding vectors into a 2D space, where each community is highlighted with a unique color.

Under the same setting, a good visualization is to see the vertices within the same community are expected to be embedded closely. Intuitively, a visualization with clearer boundaries between different color groups indicates better learned representations. We can see from Fig. 3 that the visualization of our method shows clear boundaries and compact clusters.

5 Conclusion

In this paper, we have proposed a deep network embedding framework, called ComNE, for encoding each vertex as a low dimensional vector embedding. Our model demonstrated the representation ability of stacked sparse autoencoder in extracting meaningful information and generating informative embedding for large-scale networks while preserving network topology and community structure. The experiments on real-world network datasets in different analytic tasks showed that the performance of the proposed framework outperformed several state-of-the-art baselines.

Acknowledgments. This research was supported by National Key R & D Program of China (No. 2016YFB0801100), Beijing Natural Science Foundation (No. 4172054, L181010), and National Basic Research Program of China (No. 2013CB329605).

References

1. Belkin, M., Niyogi, P.: Laplacian eigenmaps for dimensionality reduction and data representation. Neural Comput. **15**(6), 1373–1396 (2003)
2. Cao, S., Lu, W., Xu, Q.: Deep neural networks for learning graph representations. In: AAAI, pp. 1145–1152 (2016)
3. Chen, J., Ma, T., Xiao, C.: Fastgcn: fast learning with graph convolutional networks via importance sampling. arXiv preprint (2018). arXiv:1801.10247
4. Grover, A., Leskovec, J.: node2vec: Scalable feature learning for networks. In: Proceedings of the 22nd ACM SIGKDD International Conference on Knowledge Discovery and Data Mining, pp. 855–864. ACM (2016)
5. Hamilton, W., Ying, Z., Leskovec, J.: Inductive representation learning on large graphs. In: Advances in Neural Information Processing Systems, pp. 1024–1034 (2017)
6. Hinton, G.E., Salakhutdinov, R.R.: Reducing the dimensionality of data with neural networks. Science **313**(5786), 504–507 (2006)
7. Kipf, T.N., Welling, M.: Semi-supervised classification with graph convolutional networks. arXiv preprint (2016). arXiv:1609.02907
8. Maaten, L.V.D., Hinton, G.: Visualizing data using t-sne. J. Mach. Learn. Res. **9**, 2579–2605 (2008)
9. McCallum, A.K., Nigam, K., Rennie, J., Seymore, K.: Automating the construction of internet portals with machine learning. Inf, Retrieval **3**(2), 127–163 (2000)
10. Niepert, M., Ahmed, M., Kutzkov, K.: Learning convolutional neural networks for graphs. In: International Conference on Machine Learning, pp. 2014–2023 (2016)
11. Perozzi, B., Al-Rfou, R., Skiena, S.: Deepwalk: online learning of social representations. In: Proceedings of the 20th ACM SIGKDD International Conference on Knowledge Discovery and Data Mining, pp. 701–710. ACM (2014)
12. Sen, P., Namata, G., Bilgic, M., Getoor, L., Galligher, B., Eliassi-Rad, T.: Collective classification in network data. AI magazine **29**(3), 93 (2008)
13. Tang, J., Qu, M., Wang, M., Zhang, M., Yan, J., Mei, Q.: Line: large-scale information network embedding. In: Proceedings of the 24th International Conference on World Wide Web, pp. 1067–1077. International World Wide Web Conferences Steering Committee (2015)

14. Tang, L., Liu, H.: Relational learning via latent social dimensions. In: Proceedings of the 15th ACM SIGKDD International Conference on Knowledge Discovery and Data Mining, pp. 817–826. ACM (2009)
15. Tenenbaum, J.B., De Silva, V., Langford, J.C.: A global geometric framework for nonlinear dimensionality reduction. Science **290**(5500), 2319–2323 (2000)
16. Wang, D., Cui, P., Zhu, W.: Structural deep network embedding. In: Proceedings of the 22nd ACM SIGKDD International Conference on Knowledge Discovery and Data Mining, pp. 1225–1234. ACM (2016)
17. Yang, C., Liu, Z., Zhao, D., Sun, M., Chang, E.Y.: Network representation learning with rich text information. In: IJCAI, pp. 2111–2117 (2015)

D²PLS: A Novel Bilinear Method for Facial Feature Fusion

Yun-Hao Yuan[1,2](✉), Li Zhu[1], Yun Li[1], Jipeng Qiang[1], Bin Li[1], Jianping Gou[3], and Chaofeng Li[4]

[1] School of Information Engineering, Yangzhou University, Yangzhou, China
{yhyuan,liyun}@yzu.edu.cn, 1294691065@qq.com
[2] School of Computer Science and Technology, Fudan University, Shanghai, China
[3] School of Computer Science, Jiangsu University, Zhenjiang, China
[4] Institute of Logistics Engineering, Shanghai Maritime University, Shanghai, China

Abstract. Two-dimensional partial least squares (2DPLS) is an effective two-view data analysis technique. However, conventional 2DPLS only takes into account the column information of two-dimensional images. In this paper, we simultaneously consider the column-wise and row-wise information of two-dimensional face images. We first propose a row-based two-dimensional PLS (r2DPLS) approach and then further present a novel double-directional PLS (D²PLS) method. The proposed D²PLS method can be optimized by two eigenvalue subproblems. Experimental results on the AR, Yale, and AT&T face databases show that our D²PLS method can overall achieve better recognition accuracy than existing related methods.

Keywords: Partial least squares · Feature fusion · Face recognition

1 Introduction

Face recognition (FR) has received significant attention in the past decades. Most of conventional FR methods are based on single facial feature descriptor. However, in real-world face analysis system, a face image can usually be described by multiple different representations (views) due to distinct feature extractors or data sources. Hence, it is necessary to investigate how to recognize multi-view face images via fully using the complementary information of different views.

Partial Least Squares (PLS) is a classical multi-view data analysis technique. It was first proposed by Wold [1] in 1975. PLS is able to be actually regarded

Supported by Undergraduate Education and Teaching Reform Project of Yangzhou University under Grant YZUJX2016-32C; National Natural Science Foundation of China under Grants 61402203, 61703362, and 61611540347; Natural Science Foundation of Jiangsu Province of China under Grants BK20161338 and BK20170513; Yangzhou Science Project Fund of China under Grants YZ2016238 and YZ2017292; Excellent Young Backbone Teacher (Qing Lan) Project and Scientific Innovation Project Fund of Yangzhou University of China under Grant 2017CXJ033.

T. Gedeon et al. (Eds.): ICONIP 2019, CCIS 1142, pp. 406–413, 2019.
https://doi.org/10.1007/978-3-030-36808-1_44

as penalized Canonical Correlation Analysis (CCA) [2] with Principal Component Analysis (PCA) [3]. That is, it combines the merits of both CCA and PCA approaches for two-view data analysis. In recent years, PLS has developed rapidly regardless of theory or applications; see, for example, [4–12]. Since traditional PLS adopts a successive strategy to solve all the directions one by one, it is possible to yield a suboptimal solution in practical calculation. To solve this issue, Chen et al. [6] proposed a manifold optimization method to solve PLS regression model. Xie et al. [7] combined the PCA and PLS models to present a Principal Model Analysis (PMA) method for dimension reduction and classification tasks. Liu et al. [8] proposed a regularized PLS for multi-label learning. In addition, Liquet et al. [12] proposed two PLS extensions referred to as group PLS and sparse group PLS, which can be used to capture the relationship between two sets of data vectors.

The foregoing PLS-related methods are based on sets of data vectors. Different from vector form, a face image is essentially in the form of matrix, thus having obvious spatial structure information. How to utilize such kind of spatial information is an attractive topic. PLS and its preceding variants can not directly handle the two-dimensional face matrices. They are only applicable when all image matrices are transformed into vectors. To directly deal with image matrices, Sun [13] and Yang et al. [14] proposed two-dimensional (2D) PLS methods, where a face image does not need to be transformed into a vector in advance. Later, Zhang et al. [15] presented a 2D Non-negative Sparse PLS (2DNSPLS) for face recognition. Experimental results demonstrate the effectiveness of 2DNSPLS. It should be noted that the above-mentioned 2D improvements of PLS only consider the column-wise information of 2D face images and ignore the row-wise information.

In this paper, we take into account column-wise as well as row-wise information of 2D face features. We first propose a row-based two-dimensional PLS (r2DPLS) approach and then further present a novel 2D approach called Double-Directional PLS or D^{2}PLS. The proposed D^{2}PLS can be optimized by two eigenvalue subproblems. It is applied to 2D facial feature fusion and face recognition. Experimental results on real-world face datasets demonstrate the proposed D^{2}PLS method can achieve better recognition accuracy than existing related approaches.

2 Two-Dimensional PLS

Suppose there are two random matrices $X \in \mathbb{R}^{m\times n}$ and $Y \in \mathbb{R}^{m\times n}$. Let $\tilde{X} = Xw_x^c$ and $\tilde{Y} = Yw_y^c$, where $w_x^c \in \mathbb{R}^n$ and $w_y^c \in \mathbb{R}^n$ are, respectively, the projection axes of X and Y. Two-dimensional PLS (2DPLS) [13] aims to search for pairs of projection axes which maximize the covariance between $\tilde{X}$ and $\tilde{Y}$. Concretely, one pair of directions w_x^c and w_y^c can be found by

$$\max_{w_x^c, w_y^c} \operatorname{cov}(\tilde{X}, \tilde{Y}) \quad s.t.\ (w_x^c)^T w_x^c = 1,\ (w_y^c)^T w_y^c = 1, \tag{1}$$

where $\mathrm{cov}(\cdot)$ denotes the covariance operator and

$$\begin{aligned}\mathrm{cov}(\tilde{X},\tilde{Y}) &= E[Xw_x^c - E(Xw_x^c)]^T[Yw_y^c - E(Yw_y^c)] \\ &= (w_x^c)^T[E(X-EX)^T(Y-EY)]w_y^c \\ &= (w_x^c)^T\Sigma_{xy}^c w_y^c\end{aligned} \tag{2}$$

with $\Sigma_{xy}^c = E(X-EX)^T(Y-EY)$ referred to as *column-wise dispersion matrix* in this paper and $E(\cdot)$ denotes the expectation operator. The optimization problem in (1) can be solved by the following eigenvalue problem [13]:

$$\begin{bmatrix} & \Sigma_{xy}^c \\ (\Sigma_{xy}^c)^T & \end{bmatrix}\begin{bmatrix} w_x^c \\ w_y^c \end{bmatrix} = \lambda \begin{bmatrix} w_x^c \\ w_y^c \end{bmatrix}, \tag{3}$$

where λ is the eigenvalue corresponding to the eigenvector $[(w_x^c)^T\ (w_y^c)^T]^T$.

3 Approach

From Sect. 2, we can clearly find that conventional 2DPLS approach only consider the column information of matrices. In fact, a matrix not only contains columns but also rows. In this section, we employ the two kinds of information and propose a Double-Directional PLS (D^{2}PLS) method for 2D feature fusion.

3.1 Row-Based 2DPLS

Assume two random facial feature matrices are given as $X \in \mathbb{R}^{m\times n}$ and $Y \in \mathbb{R}^{m\times n}$. Let row-based linear transformations of X and Y be $V_x = (w_x^r)^T X$ and $V_y = (w_y^r)^T Y$, respectively, where $w_x^r \in \mathbb{R}^m$ and $w_y^r \in \mathbb{R}^m$. Then, the *row-wise dispersion matrix* of V_x and V_y can be defined by

$$\begin{aligned}\mathrm{cov}(V_x, V_y) &= E[(w_x^r)^T X - E((w_x^r)^T X)][(w_y^r)^T Y - E((w_y^r)^T Y)]^T \\ &= (w_x^r)^T[E(X-EX)(Y-EY)^T]w_y^r \\ &= (w_x^r)^T\Sigma_{xy}^r w_y^r,\end{aligned} \tag{4}$$

where $\Sigma_{xy}^r = E(X-EX)(Y-EY)^T$. With (4), the model of our row-based 2DPLS (r2DPLS) can be formulated as

$$\max_{w_x^r, w_y^r} (w_x^r)^T\Sigma_{xy}^r w_y^r \quad s.t.\ (w_x^r)^T w_x^r = 1,\ (w_y^r)^T w_y^r = 1. \tag{5}$$

With the Lagrange multipliers, the optimization problem in (5) can be solved by the following eigenvalue problem:

$$\begin{bmatrix} & \Sigma_{xy}^r \\ (\Sigma_{xy}^r)^T & \end{bmatrix}\begin{bmatrix} w_x^r \\ w_y^r \end{bmatrix} = \eta \begin{bmatrix} w_x^r \\ w_y^r \end{bmatrix}, \tag{6}$$

where η is the eigenvalue corresponding to the eigenvector $[(w_x^r)^T\ (w_y^r)^T]^T$.

3.2 Double-Directional PLS

Now, combining the foregoing two objectives in (1) and (5) leads to our double directional PLS method as follows.

$$\max_{w_x^c, w_y^c, w_x^r, w_y^r} (1-\alpha)(w_x^c)^T \Sigma_{xy}^c w_y^c + \alpha (w_x^r)^T \Sigma_{xy}^r w_y^r$$
$$s.t. \begin{cases} (w_x^c)^T w_x^c + (w_y^c)^T w_y^c = 1, \\ (w_x^r)^T w_x^r + (w_y^r)^T w_y^r = 1, \end{cases} \tag{7}$$

where α is a balance parameter satisfying $0 \leq \alpha \leq 1$. It is obvious that when $\alpha = 0$, our proposed D^{2}PLS method reduces to 2DPLS as described in Sect. 2; when $\alpha = 1$, D^{2}PLS reduces to r2DPLS. Thus, the proposed D^{2}PLS integrates the column as well as row information of 2D face images.

The Lagrangian of the problem in (7) is

$$\begin{aligned} \mathcal{L} =& (1-\alpha)(w_x^c)^T \Sigma_{xy}^c w_y^c + \alpha (w_x^r)^T \Sigma_{xy}^r w_y^r \\ & - \frac{\eta_1}{2}[(w_x^c)^T w_x^c + (w_y^c)^T w_y^c - 1] - \frac{\eta_2}{2}[(w_x^r)^T w_x^r + (w_y^r)^T w_y^r - 1], \end{aligned} \tag{8}$$

where η_1 and η_2 are the Lagrange multipliers. Let $\partial \mathcal{L}/w_x^c = 0$, $\partial \mathcal{L}/w_y^c = 0$, $\partial \mathcal{L}/w_x^r = 0$, and $\partial \mathcal{L}/w_y^r = 0$. Then, we obtain

$$\begin{cases} \partial \mathcal{L}/w_x^c = (1-\alpha)\Sigma_{xy}^c w_y^c - \eta_1 w_x^c = 0, \\ \partial \mathcal{L}/w_y^c = (1-\alpha)(\Sigma_{xy}^c)^T w_x^c - \eta_1 w_y^c = 0, \\ \partial \mathcal{L}/w_x^r = \alpha \Sigma_{xy}^r w_y^r - \eta_2 w_x^r = 0, \\ \partial \mathcal{L}/w_y^r = \alpha (\Sigma_{xy}^r)^T w_x^r - \eta_2 w_y^r = 0. \end{cases} \tag{9}$$

It follows that

$$\begin{bmatrix} & (1-\alpha)\Sigma_{xy}^c & & \\ (1-\alpha)(\Sigma_{xy}^c)^T & & & \\ & & & \alpha \Sigma_{xy}^r \\ & & \alpha (\Sigma_{xy}^r)^T & \end{bmatrix} \begin{bmatrix} w_x^c \\ w_y^c \\ w_x^r \\ w_y^r \end{bmatrix} = \begin{bmatrix} \eta_1 w_x^c \\ \eta_1 w_y^c \\ \eta_2 w_x^r \\ \eta_2 w_y^r \end{bmatrix}. \tag{10}$$

Clearly, (10) is not an usual eigenvalue problem. Actually, it is referred to as multivariate eigenvalue problem (MEP) [16], which has no closed-form solution except some special cases. Thus, we relax the MEP in (10) into the following two eigenvalue subproblems.

$$\begin{bmatrix} & \Sigma_{xy}^c \\ (\Sigma_{xy}^c)^T & \end{bmatrix} \begin{bmatrix} w_x^c \\ w_y^c \end{bmatrix} = \frac{\eta_1}{(1-\alpha)} \begin{bmatrix} w_x^c \\ w_y^c \end{bmatrix}, \tag{11}$$

and

$$\begin{bmatrix} & \Sigma_{xy}^r \\ (\Sigma_{xy}^r)^T & \end{bmatrix} \begin{bmatrix} w_x^r \\ w_y^r \end{bmatrix} = \frac{\eta_2}{\alpha} \begin{bmatrix} w_x^r \\ w_y^r \end{bmatrix}. \tag{12}$$

Note that when $\alpha = 0$ or $\alpha = 1$, we only compute (11) or (12). We separately select the d eigenvectors of (11) and (12) corresponding to the first d largest eigenvalues to generate the projection matrices P_x^c, P_y^c, P_x^r, and P_y^r.

Table 1. Average recognition accuracy (%) on different face databases.

Method	AR	Yale	AT& T
D^{2}PLS-FFS1	86.45	75.33	94.35
D^{2}PLS-FFS2	82.03	74.00	93.55
r2DPLS	83.49	74.13	91.90
2DPLS	82.39	73.60	93.15
2DCCA	66.51	70.40	93.05

Feature Fusion. For 2D facial features X and Y, we are able to obtain 2D-projection features XP_x^c, YP_y^c, $(P_x^r)^T X$, and $(P_y^r)^T Y$. Borrowing the idea in [13], we use the following strategies to fuse them.

$$[XP_x^c; YP_y^c; X^T P_x^r; Y^T P_y^r], \tag{13}$$

$$[XP_x^c + YP_y^c; X^T P_x^r + Y^T P_y^r]. \tag{14}$$

We call (13) Feature Fusion Strategy 1 (FFS1) and (14) Feature Fusion Strategy 2 (FFS2).

4 Experiment

In this section, we perform several experiments on the AR, Yale, and AT&T face databases to test the performance of r2DPLS and D^{2}PLS, and compare them with 2DPLS and two-dimensional CCA (2DCCA) [13].

4.1 Data Preparation

Data Sets. The AR database includes more than 4,000 color images of 126 people. These images are frontal view of face with different expressions, lighting conditions, and occlusions. In this experiment, we select 120 people and each person has 14 images with 50×40 pixels. The Yale database contains 165 images of 15 individuals. Each person has 11 different images under various expressions and lighting conditions. Each image is resized to 80×80. The AT&T database contains 400 images from 40 people. Each person has 10 grayscale images with size as 112×92 under different expressions, lighting conditions, facial details and at different times.

Settings. To yield two sets of 2D facial features, we employ original 2D face images as X and 2D wavelet-transform images as Y. The nearest neighbor classifier is used for the classification performance test. On AR database, we randomly select 8 images per people for training and the rest for testing. On Yale database, we randomly choose 6 images for training and the remaining 5 images for testing. On AT&T database, we randomly choose 5 images for training and the rest for testing. On all the three databases, 10 classification tests are carried out independently and the average recognition results are computed.

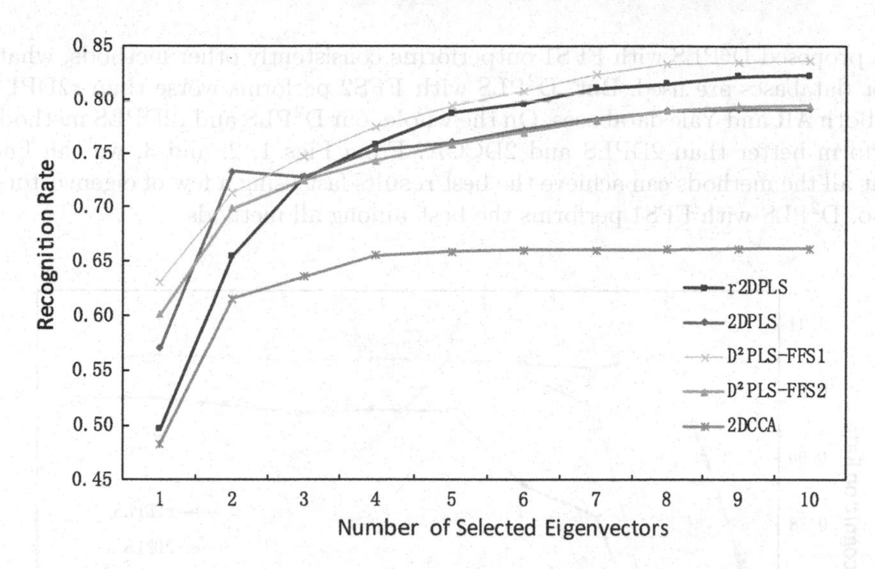

Fig. 1. Average recognition accuracy of different methods versus the number of selected eigenvectors on the AR database.

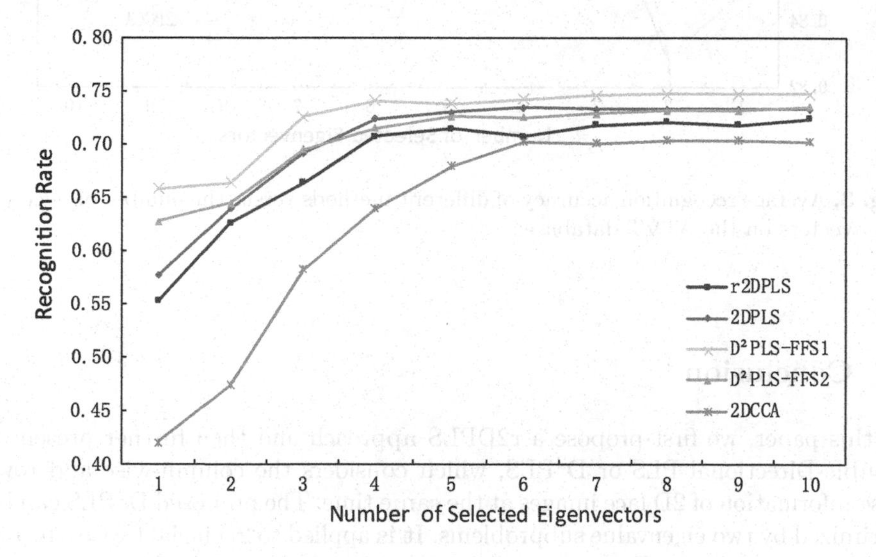

Fig. 2. Average recognition accuracy of different methods versus the number of selected eigenvectors on the Yale database.

4.2 Result

Table 1 shows the maximal average recognition results of 2DCCA, 2DPLS, r2DPLS, and D^{2}PLS across ten runs on all possible dimensions. Figures 1, 2, and 3 show the average recognition results of different methods versus the top 10 eigenvectors on different databases, respectively. As we can see from Table 1, the proposed D^{2}PLS with FFS1 outperforms consistently other methods, whatever databases are used. But, D^{2}PLS with FFS2 performs worse than r2DPLS on both AR and Yale databases. On the whole, our D^{2}PLS and r2DPLS methods perform better than 2DPLS and 2DCCA. From Figs. 1, 2, and 3, we can find that all the methods can achieve the best results fast using a few of eigenvectors. Also, D^{2}PLS with FFS1 performs the best among all methods.

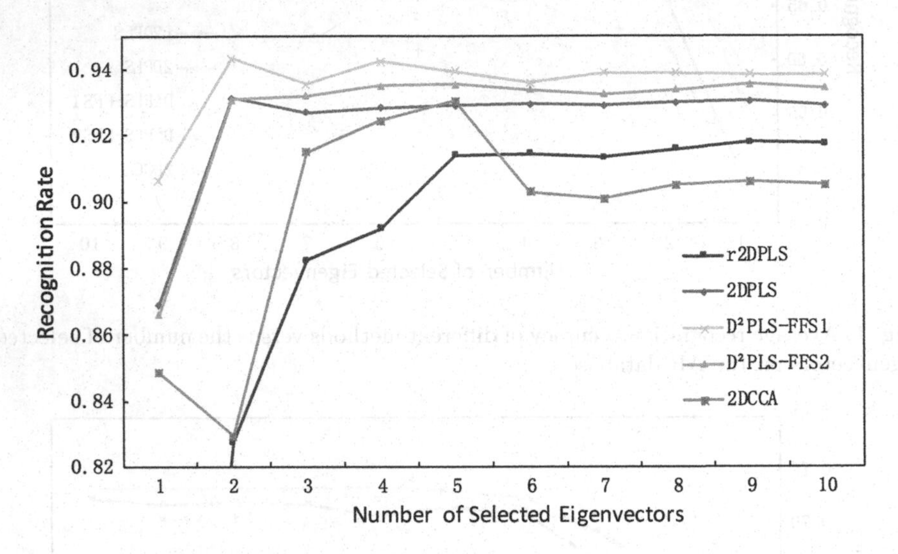

Fig. 3. Average recognition accuracy of different methods versus the number of selected eigenvectors on the AT&T database.

5 Conclusion

In this paper, we first propose a r2DPLS approach and then further present a Double-Directional PLS or D^{2}PLS, which considers the column-wise and row-wise information of 2D face images at the same time. The proposed D^{2}PLS can be optimized by two eigenvalue subproblems. It is applied to 2D facial feature fusion and face recognition. Experimental results on real-world face databases show D^{2}PLS can achieve better recognition accuracy than existing related approaches.

References

1. Wold, H.: Soft modelling by latent variables: the non-linear iterative partial least squares (NIPALS) approach. Perspect. Probab. Stat. **12**(S1), 117–142 (1975)
2. Hotelling, H.: Relations between two sets of variates. Biometrika **28**, 321–377 (1936)

3. Aleix, M.M., Avinash, C.K.: PCA versus LDA. IEEE Trans. Pattern Anal. Mach. Intell. **23**(2), 228–233 (2001)
4. Hubert, M., Branden, K.V.: Robust methods for partial least squares regression. J. Chemometr. **17**(10), 537–549 (2003)
5. Kondylis, A., Hadi, A.S.: Derived components regression using the bacon algorithm. Comput. Stat. Data Anal. **51**(2), 556–569 (2006)
6. Chen, H., Sun, Y., Gao, J., Hu, Y., Yin, B.: Solving partial least squares regression via manifold optimization approaches. IEEE Trans. Neural Netw. Learn. Syst. **30**(2), 588–600 (2019)
7. Xie, Q., Tang, L., Li, W., John, V., Hu, Y.: Principal model analysis based on partial least squares (2019). arXiv: 1902.02422
8. Liu, H., Ma, Z., Han, J., Chen, Z., Zheng, Z.: Regularized partial least squares for multi-label learning. Int. J. Mach. Learn. Cybern. **9**(2), 335–346 (2018)
9. Talukdar, U., Hazarika, S.M., Gan, J.Q.: A kernel partial least square based feature selection method. Pattern Recogn. **83**, 91–106 (2018)
10. Tao, J.-L., Zhang, J.-M., Wang, L.-J., Shen, X.-J., Zha, Z.-J.: Near-duplicate video retrieval through toeplitz kernel partial least squares. In: Kompatsiaris, I., Huet, B., Mezaris, V., Gurrin, C., Cheng, W.-H., Vrochidis, S. (eds.) MMM 2019. LNCS, vol. 11296, pp. 352–364. Springer, Cham (2019). https://doi.org/10.1007/978-3-030-05716-9_29
11. Qiao, J., Wang, G., Li, W., Li, X.: A deep belief network with PLSR for nonlinear system modeling. Neural Netw. **104**, 68–79 (2018)
12. Liquet, B., de Micheaux, P.L., Hejblum, B.P., Thiébautt, R.: Group and sparse group partial least square approaches applied in genomics context. Bioinformatics **32**(1), 35–42 (2016)
13. Sun, Q.S.: Research on feature extraction and image recognition based on correlation projection analysis. Ph.D. dissertation. Nanjing University of Science and Technology, Nanjing (2006)
14. Yang, M.-L., Sun, Q.-S., Xia, D.-S.: Two-dimensional partial least squares and its application in image recognition. In: Huang, D.-S., Wunsch, D.C., Levine, D.S., Jo, K.-H. (eds.) ICIC 2008. CCIS, vol. 15, pp. 208–215. Springer, Heidelberg (2008). https://doi.org/10.1007/978-3-540-85930-7_28
15. Zhang, Y.X., Huang, S., Feng, X., Zhang, J.H., Bu, W.B., Yang, D.: Two dimensional non-negative sparse partial least squares for face recognition. In: ICMEW, pp. 1–6, IEEE, Chengdu (2014)
16. Chu, M.T., Watterson, J.L.: On a multivariate eigenvalue problem: I. Algebraic theory and a power method. SIAM J. Sci. Comput. **14**(5), 1089–1106 (1993)

Learning Network Representation via Ego-Network-Level Relationship

Bencheng Yan[1(✉)] and Shenglei Huang[2]

[1] School of Software, Tsinghua University, Beijing 100084, People's Republic of China
ybc17@mails.tsinghua.edu.cn

[2] Department of Computer Science and Engineering,
Shanghai Jiao Tong University, Shanghai, China
shengleihuang@sjtu.edu.cn

Abstract. Network representation, aiming to map each node of a network into a low-dimensional space, is a fundamental problem in the network analysis. Most existing works focus on the self-level or pairwise-level relationship among nodes to capture network structure. However, it is too simple to characterize the complex dependencies in the network. In this paper, we introduce the theory of the ego network and present an ego-network-level relationship. Then a deep recurrent auto-encoder model is proposed to preserve the complex dependencies in each ego network. In addition, we present two strategies to solve the sparsity problem. Finally, we conduct extensive experiments on three real datasets. The experimental results demonstrate that the proposed model can well preserve network structure and learn a good network representation.

Keywords: Network representation · Ego-network-level relationship · Recurrent auto-encoder

1 Introduction

Graph data have been widely analyzed [6,8–10]. In recent years, network representation has been proposed and aroused considerable research interest. The goal is to learn a low-dimensional vector for each node as its representation. One basic requirement of network representation is to preserve the inherent network structure. Recently, a random walk strategy is adopted to capture network structure. For example, shown in Fig. 1(b), DeepWalk [7] and Node2vec [5] exploit the node pairwise-level relationship in the truncated random walks over a network. Actually, these methods model the probabilities $p(v_j|v_i)$ and $p(v_i|v_j)$ for each co-occurring node pair v_i and v_j. In addition, deep learning is introduced in the network representation. As shown in Fig. 1(a), the basic idea of these deep learning methods is characterizing the self-level relationship by modeling the probability $p_{autoencoder}(\hat{x}_i|x_i)$ (i.e., $p_{encoder}(h|x_i)$ and $p_{decoder}(\hat{x}_i|h)$), where x_i and $\hat{x}_i$ refer to the raw and reconstruction feature of node v_i, and h refers to the output of the encoder.

T. Gedeon et al. (Eds.): ICONIP 2019, CCIS 1142, pp. 414–422, 2019.
https://doi.org/10.1007/978-3-030-36808-1_45

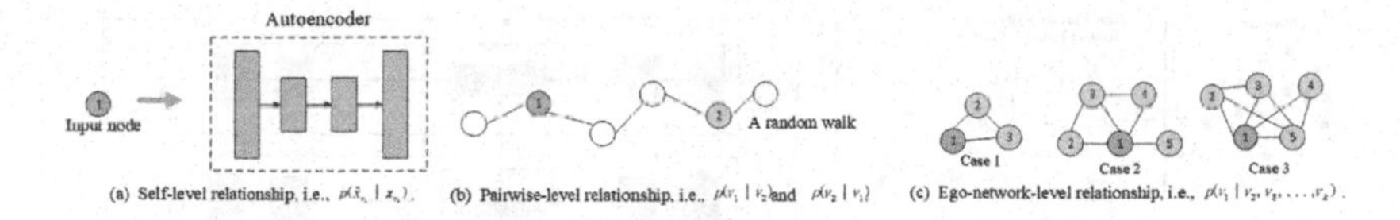

Fig. 1. Different relationship considered in existing works and our paper.

However, we argue that the pairwise-level or self-level relationship is too simple to capture the complex network. For example, as shown in Fig. 1, because these two methods only consider the single or two-tuples dependencies in the network representation, a multi-tuple relationship cannot be expressed directly, even if it is just a simple triangle. Thus, the pairwise-level or self-level relationship is not enough to characterize such a complex relationship in the network, and a more complex dependence should be considered to have a better embedding performance.

In this paper, to characterize the complex relationship in the network, we introduce an ego-network-level relationship in the network representation, and the difference between ego-network-level representation and others is shown in Fig. 1. Ego networks have been studied in social networks to understand how the node interacts with others [2,4]. It consists of a focal node (i.e., ego) and the nodes to whom ego is directly connected to (these are called alters) plus the connections, if any, among the alters. Of course, each alter in an ego network has its own ego network, and all ego networks interlock to form a network. As discussed above, the connection is complex in the network. We can hardly cover all possible dependencies. Therefore, in this paper, we mainly care about the surroundings in an ego network. The reason is that the connection from an ego to an alter is more likely influenced by alters rather than other nodes in a network. Then, to characterize the ego-network-level relationship, we sample several ego-node-sequences over each ego network, and propose a deep recurrent auto-encoder to capture each sequence. In this way, we can flexibly preserve the ego-network-level relationship.

In summary, our main contributions are listed as follows: (1) We introduce the ego network into network representation and propose an ego-network-level relationship to preserve the complex network structure. (2) We design an ego-node-sequences sampling, and propose a Deep Recurrent Auto-Encoder model called DRAE, to characterize each ego-node-sequence, and to learn a useful representation of each node. (3) In order to exhaustively evaluate the proposed model, we conduct extensive experiments on three real datasets. These results demonstrate that DRAE can effectively uncover more complex hidden features than baselines.

2 Related Work

Recently, representation learning has become a very important task in the network research. In this section, we mainly introduce some related methods

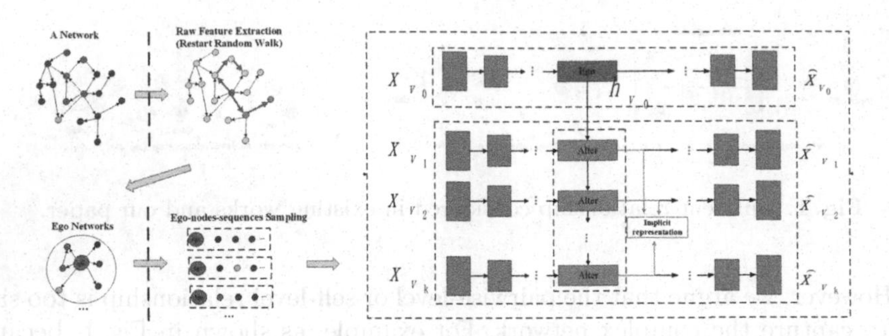

Fig. 2. The framework of DRAE

including the self-level and pairwise-level methods. (1) **Pairwise-level Methods**. Many recent successful methods such as DeepWalk [7] and Node2vec [5] learn the node representation based on random walk statistics. Their key innovation is that if two nodes can be reached by each other on short random walks over the graph, they are expected to have a similar representation. In this way, these methods can naturally capture the pairwise-level relationship over the network. (2) **Self-level Methods**. To extract complex structural features and learn deep, highly non-linear node representations, the deep learning techniques are also applied to network representation. Most of them construct a similarity matrix and make use of deep auto-encoder to capture the network information. By reconstructing each node feature, these methods can characterize the self-level relationship. GraphEncoder [11] takes the adjacency matrix as input. DNGR [3] constructs a high-order proximity matrix to capture global information. SDNE [12] designs the second-order proximity loss to learn the global information.

3 DRAE

In this section, we introduce the proposed Deep Recurrent Auto-Encoder model (DRAE). The overall framework of DRAE is shown in Fig. 2. Specifically, given a network $G = (V, E)$, we firstly adopt a restart random walk to extract raw feature of each node. Then for each node, we focus on the ego network, and an ego-node-sequences sampling is proposed to generate several ego-node-sequences over the ego network. Then we feed the raw feature sequence corresponding to the ego-node-sequence into the proposed model. Finally, with the help of the deep model, we obtain the representation of each node.

3.1 Raw Feature Extraction

Inspired by the random walk, we adopt a restart random walk strategy to extract raw network feature. Specifically given a network $G = (V, E)$, we define $d_i = \sum_j a_{i,j}$, and let $D = diag\{d_1, d_2, ..., d_n\}$ be a degree matrix. Then $P = D^{-1}A$ is the random walk probability between two nodes in 1-step. Consider one step of a restart random walk from node v_i, the random walker randomly jumps to one of the neighbors of current node with probability $1-\alpha$, and jumps back to v_i with the probability α. Then after $w-1$ jumps, the next jump can be expressed as: $s_i^w = (1-\alpha)s_i^{w-1}P + \alpha s_i^0$,where s_i^w is the i-th row of S^w, and S^0 is the identity matrix.

Then the raw feature of node v_i can be obtained by summarizing $s_i^1, s_i^2, ..., s_i^W$, i.e., $x_i = \sum_{w=1}^{W} s_i^w$, where x_i characterizes the relationship between node v_i and any other nodes in G. Finally, we adopt a normalization strategy with point-wise mutual information as suggested in [3]: $x_{i,j}^* = log((x_{i,j}Q)/(\sum_{t=1}^{n} x_{i,t} \sum_{t=1}^{n} x_{t,j}))$,where $Q = \sum_{i=1}^{n} \sum_{j=1}^{n} x_{i,j}$. Then the negative value is set to 0. After the restart random walk and normalization, the raw feature of node v_i can be formed as x_i^*, and for simplify, in rest of the paper, the symbol x_i is regarded as the vector after normalization.

Ego-Node-Sequences Sampling. Although we simplify the surroundings of each connection from the whole network into an ego network, there still exists an amount of possible dependencies. It is impossible to consider all of these dependencies. Thus, we design a novel sampling strategy to characterize the ego network, and then apply the dependencies in each sample. Specifically, given a network $G = (V, E)$, for any node v_0, we can get its ego network. We denote N_{v_0} as the alters set (i.e., the neighbor nodes set) of the ego v_0. Then, we randomly select k alters from N_{v_0} and generate a $k+1$ length ego-node-sequence $S_{N_{v_0}} = (v_0, v_1, ..., v_k)$, where $v_i \in N_{v_0}$ $(i = 1, 2, ..., k)$. Of course, using only one ego-node-sequence cannot fully characterize an ego network. Thus, for each ego network, we randomly generate q ego-node-sequences.

3.2 Deep Recurrent Auto-Encoder

In this section, we first introduce a model only considering the self-level relationship, and then we extend it into the ego-network-level representation with the help of the ego-node-sequences.

Self-Level Representation. Here we mainly apply the auto-encoder to model the self-level relationship. Specifically, given a network $G = (V, E)$, we can get its raw feature matrix X by the restart random walks. For each node v, we feed x_v into the auto-encoder network, and then the hidden layer L_i can be expressed as: $y_i = f(W_i \cdot y_{i-1} + b_i)$, where W_i and b_i are the weight and bias of the hidden layer L_i respectively, y_{i-1} is the output of the hidden layer L_{i-1}, and f is the activation function. Then the loss function can be expressed as: $l(x_v; \hat{x}_v) = \|x_v - \hat{x}_v\|_2^2$, where $\hat{x}_v$ is the output of the auto-encoder. Finally, we take $h_v = y_{mid} \in \Re^d$ as the self-level representation, where y_{mid} is the output of the encoder (i.e., the middle layer L_{mid}).

Ego-Network-Level Representation. Although, auto-encoder is a powerful model to learn the self-level representation, it is not suitable to model the ego-network-level relationship. Therefore, based on auto-encoder, we design a deep recurrent auto-encoder, and take place the single input with a sequence input. Specifically, for each node v_0, we sample q ego-node-sequences in the corresponding ego network. For each sequence $S_{N_{v_0}} = (v_0, v_1, ..., v_k)$, we feed $(x_{v_0}, x_{v_1}, ..., x_{v_k})$ into the deep recurrent auto-encoder. In other words, for the each time point t, the middle hidden layer L^t_{mid} not only receives the output of hidden layer L^t_{mid-1} produced by x_{v_t} at the time point t, but also receives the output of layer L^{t-1}_{mid} produced by $(x_{v_0}, x_{v_1}, ..., x_{v_{t-1}})$ at the time point $t-1$. So for each time point t, we can get the output expression of its middle hidden layer L^t_{mid}:

$$y^t_{mid} = f(W_{mid} \cdot y^t_{mid-1} + U \cdot y^{t-1}_{mid} + b_{mid}) \tag{1}$$

where W_{mid} and b_{mid} are the weight and bias of the hidden layer L_{mid}, U is the weight of the information transmission between the previous time point $t-1$ and the current point t, y^t_{mid-1} is the output of hidden layer L_{mid-1} at time point t, y^{t-1}_{mid} is the output of hidden layer L_{mid} at time point $t-1$, and f is the activation function. Then the loss function can be expressed as follow: $Loss = \sum_{t=0}^{k} Loss_t = \sum_{t=0}^{k} \|x_{v_t} - \hat{x}_{v_t}\|_2^2$.

Intuitively, at time point t_0, DRAE models the self-level relationship (i.e., $p(\hat{x}_{v_0}|x_{v_0})$). For each time point t, DRAE models the probability $p(\hat{x}_{v_t}|x_{v_0}, x_{v_1}, ..., x_{v_{t-1}}, x_{v_t})$. In this way, the connection of the ego v_0 and the alter v_t can be not only influenced by these two nodes, but also by the surrounding alters. Furthermore, multi-sampling ego-node-sequences in this ego network can be allowed to capture as much dependencies as possible. Finally, for each ego v_0, we take the mean of y^0_{mid} over the q ego-node-sequences as the learned representation h_{v_0}.

3.3 Sparsity Problem

In real life, a network is often huge and contains a large number of nodes, such as Youtube and Wiki. However few of nodes are connected. Although, the restart random walk enriches the reachability, there still are a lot of zero elements in the matrix X which may bring a difficulty to learn a useful representation. To address this problem, we introduce two strategies including penalty strategy and implicit representation.

Penalty Strategy. Inspired by SDNE [12], we add a penalty weight E to the reconstruction error. Thus given the input $(x_{v_0}, x_{v_1}, ..., x_{v_k})$ for an ego-node-sequence $S_{N_{v_0}} = (v_0, v_1, ..., v_k)$, the loss function in Equation (2) can be rewritten as $Loss = \sum_{t=0}^{k} Loss_t = \sum_{t=0}^{k} \|(x_{v_t} - \hat{x}_{v_t}) \odot r_{v_t}\|_2^2$, where $\odot$ represents the element wise product, $r_{v_i} = (r_{v_i,v_j})_{j=1}^{n}$ and $R = (r_{v_i,v_j})_{n \times n}$. If $x_{v_i,v_j} > 0$, then $r_{v_i,v_j} = \epsilon > 1$, else $r_{v_i,v_j} = 1$. In this way, DRAE can focus on the reachability features, rather than zeros.

Implicit Representation. Actually, the representations of alters in each time point $t(t = 1, 2, ..., k)$ can also be helpful to the representation of the ego

v_0 and alleviate the sparsity problem. Here we take these representations as the implicit representations. Specifically, for each ego network, we take the mean of $h_{v_i}(i = 1, ..., k)$ over the q ego-node-sequences as implicit representation (i.e., $h_{implicit} = \sum_{q\_sequences} \sum_{i=1}^{k} h_{v_i}$). Finally, by concatenating the two vectors h_{v_0} and $h_{implicit}$, we obtain the final representation of the ego v_0.

4 Experiments

4.1 Setting

Data. Here, we introduce three real datasets used in this paper. **BlogCatalog** is a social network and represents the friendship between bloggers. **Wiki** contains the links between documents in Wikipedia. **Email-Eu-core** represents the relationship of mail exchanging between members of European research institutions. To sum up, the detailed statistics of these datasets are presented in Table 1.

Table 1. Datasets Information. Task 'c' denotes node classification, 's' denotes sparsity strategies analysis, 'd' denotes different level relationships comparison

Dataset	#Nodes	#Edges	#Class	Task
BlogCatalog	10312	333983	39	c
Wiki	2,405	17,981	19	c
Email-Eu-core	1,005	25,571	42	c | s | d

Baselines. In this paper, we take six different methods which focus on network structure as baselines, including the self-level and pairwise-level network representation methods. **Self-level Method** (1) Graph Factorization (GF) [1] factorizes the adjacency matrix of a network by singular value decomposition (SVD). (2) GraphEncoder [11] makes use of sparse auto-encoder, and designs a stack model. (3) SDNE [12] considers the first and second order proximities. (4) DNGR [3] constructs a high order matrix and learns the network structure by denoising auto-encoder. **Pairwise-level Method** (1) DeepWalk [7] first generates a path by random walk, and then takes this path as the input data to the word2vec. (2) Node2vec [5] considers a high order information of nodes and proposes a random walk strategy.

Parameters Settings. In the baseline system, for DeepWalk, we set $t = 40$, $\gamma = 80, w = 10$ as suggested in [7] for the datasets BlogCatalog, Wiki and Email-Eu-core. For Node2vec, the walk length, per walk and windows size are set the same as DeepWalk, and we set $p = 0.25, q = 0.25$ for BlogCatalog, and $p = 4, q = 0.25$ for Wiki and Email-Eu-core. For SDNE, we set $\alpha = 0.2, \beta = 10$ as suggested in [12] for all datasets. For DNGR, we set $\alpha = 0.02$ as suggested in [3] for all datasets. For our method, to fully evaluate the performance of DRAE,

Table 2. Node Classification results

Datasets	BlogCatalog						Wiki						Email-Eu-core					
	Micro-F1(%)			Macro-F1(%)			Micro-F1(%)			Macro-F1(%)			Micro-F1(%)			Macro-F1(%)		
Method	10%	50%	90%	10%	50%	90%	10%	50%	90%	10%	50%	90%	10%	50%	90%	10%	50%	90%
DRAE	33.28	39.03	40.93	21.05	26.43	28.35	57.88	66.81	71.12	41.10	48.85	51.50	64.57	75.77	80.00	37.85	55.16	52.41
Node2vec	34.70	37.90	38.91	16.47	21.22	21.66	58.01	65.24	67.43	37.35	45.43	47.23	61.10	72.54	75.25	27.41	45.62	42.08
DeepWalk	32.15	35.00	36.09	14.80	18.73	19.86	58.06	66.32	67.59	38.84	48.23	48.34	59.51	71.33	72.18	26.78	41.53	40.58
SDNE	31.67	36.68	38.57	13.99	20.57	23.45	56.88	65.75	68.80	35.54	44.79	45.89	48.35	67.00	73.47	23.81	42.08	42.67
DNGR	33.32	36.79	38.02	19.63	21.95	22.48	36.35	45.95	47.65	36.35	45.95	47.65	62.32	73.04	78.81	35.29	51.75	48.78
GraphEncoder	29.70	33.78	35.05	12.72	16.49	18.09	47.80	59.39	65.23	30.86	41.43	45.07	41.26	57.14	63.17	24.47	41.65	39.35
GF	27.77	31.99	32.76	13.10	16.32	16.62	49.39	60.68	62.16	32.08	42.29	44.22	48.65	60.20	66.83	30.46	44.48	41.82

we just set the same parameters for all datasets rather by experimental tuning. Specifically, we set the restart parameter $\alpha = 0.02$, the walk length $W = 6$, the sampling times $p = 8$, the ego-node-sequence length $k = 8$, and the penalty parameter $\epsilon = 10$.

4.2 Classification Task

In order to compare the performance between DRAE and baselines, we take these node representation vectors as features and apply them to the classification task. We randomly select a part (10%, 50% and 90%) of the dataset with the labeled nodes as the training data and take the rest as the test data. Then we use the training data to train a one-vs-rest logistic regression classifier. We repeat this process 10 times and report the average Macro-F1 and Micro-F1. The results are shown in Table 2. We see that DRAE outperforms the baselines at most of the time. For example, DRAE achieves gains of 2.02% to 8.71% and 4.9% to 11.73% on Micro-F1 and Macro-F1 on BlogCatalog(90%). It indicates DRAE, compared with baselines, can capture a better structure of the network.

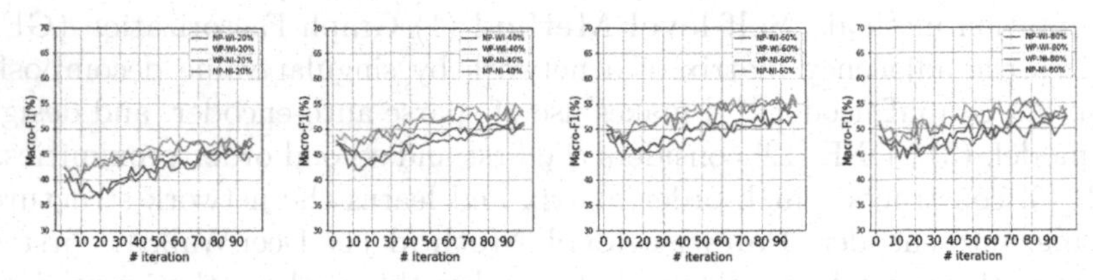

(a) Macro-F1 on 20% training data (b) Macro-F1 on 40% training data (c) Macro-F1 on 60% training data (d) Macro-F1 on 80% training data

Fig. 3. Sparsity strategy analysis on Email-Eu-core

4.3 Sparsity Strategies Analysis

As we know, in the network representation, it is a big challenge to deal with the sparsity problem. In this paper, we propose two strategies to solve this

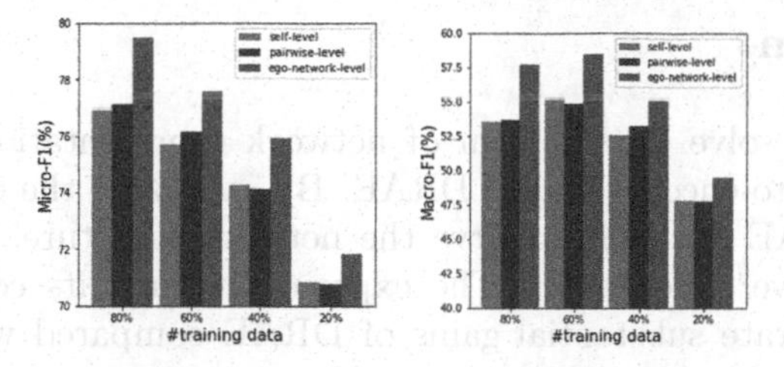

Fig. 4. Different level representations comparison on Email-Eu-core

problem. Here, we conduct some experiments to further analyze these two strategies. Specifically, we test four variants of DRAE (denoted as WP-WI, WP-NI, NP-WI and NP-NI). WP-NI refers to the method adopting these two strategies. WP-NI only adopts penalty strategy. NP-WI only adopts implicit representation, and NP-NI drops all strategies. All of these variants are set in the same parameters. Then similar to the node classification task, we take the learned representation from different variants and test them on Email-Eu-core with training data from 10%–90%. Because of the limited space, we only report the Micro-F1 and Macro-F1 results of 20%, 40%, 60% and 80% training data in different iteration (Similar conclusion can be found in the rest results). The WP-NI-80% in Fig. 3 refers to the result of model WP-NI trained on 80% training data, and the others have similar meanings.

From Fig. 3, the baseline NP-NI without any sparsity strategies performs very poorly in all cases. It demonstrates that it is very necessary to take the sparsity problem into consideration. Besides, compared with the implicit representation (NP-WI), the penalty strategy (WP-NI) gains more improvement. The reason is that penalty strategy is a more direct strategy to alleviate the sparsity problem. Furthermore, whenever any sparsity strategy is considered, the model gains a significant improvement. Especially, DRAE (i.e., WP-WI) which considers all of these two strategies obtains the best performance. It demonstrates that the proposed sparsity strategies can well solve the sparsity problem.

4.4 Different Level Relationships Comparison

As discussed above, our model can be easily modified as a self-level or pairwise-level model. In this part, we conduct experiments to give a comparison of these three different level relationships models. Specifically, the self-level model of DRAE refers that we directly take place the ego-node-sequence with a single node, and the pairwise-level model refers to DRAE with $k = 1$. Similar to the node classification task, we report the best Micro-F1 and Macro-F1 results on Email-Eu-core in 100 iterations. From Fig. 4, we can see that ego-network-level relationship makes a significant contribution to the results. It demonstrates that it is necessary to introduce the complex dependencies in the network representation, and the ego-network-level relationship can better characterize such dependencies than the other two kinds of relation.

5 Conclusion

In this paper, to solve the problem of network representation, we propose a deep recurrent auto-encoder, called DRAE. By analyzing the ego-network-level relationship, DRAE can well preserve the network structure. We evaluate our model by extensive experiments. The experimental results conducted on real datasets demonstrate substantial gains of DRAE compared with the state-of-the-art.

References

1. Ahmed, A., Shervashidze, N., Narayanamurthy, S., Josifovski, V., Smola, A.J.: Distributed large-scale natural graph factorization. In: WWW, pp. 37–48. ACM (2013)
2. Burt, R.S.: Models of network structure. Ann. Rev. Sociol. **6**(1), 79–141 (1980)
3. Cao, S., Lu, W., Xu, Q.: Deep neural networks for learning graph representations. In: AAAI, pp. 1145–1152 (2016)
4. Everett, M., Borgatti, S.P.: Ego network betweenness. Soc. Netw. **27**(1), 31–38 (2005)
5. Grover, A., Leskovec, J.: node2vec: scalable feature learning for networks. In: SIGKDD, pp. 855–864. ACM (2016)
6. Linderman, S., Adams, R.: Discovering latent network structure in point process data. In: ICML, pp. 1413–1421 (2014)
7. Perozzi, B., Al-Rfou, R., Skiena, S.: Deepwalk: online learning of social representations. In: SIGKDD, pp. 701–710. ACM (2014)
8. She, Q., Chen, G., Chan, R.H.: Evaluating the small-world-ness of a sampled network: functional connectivity of entorhinal-hippocampal circuitry. Sci. Rep. **6**, 21468 (2016)
9. She, Q., So, W.K., Chan, R.H.: Reconstruction of neural network topology using spike train data: small-world features of hippocampal network. In: EMBC, pp. 2506–2509. IEEE (2015)
10. She, Q., So, W.K., Chan, R.H.: Effective connectivity matrix for neural ensembles. In: EMBC, pp. 1612–1615. IEEE (2016)
11. Tian, F., Gao, B., Cui, Q., Chen, E., Liu, T.Y.: Learning deep representations for graph clustering. In: AAAI, pp. 1293–1299 (2014)
12. Wang, D., Cui, P., Zhu, W.: Structural deep network embedding. In: SIGKDD, pp. 1225–1234. ACM (2016)

Human Centred Computing

DMCM: A Deep Multi-Channel Model for Dynamic Movie Recommendation

Xinyi Wang[1,2], Min Gao[1,2(✉)], Zhenni Lu[1,2], Zongwei Wang[1,2], Junwei Zhang[1,2], and Yi Zhang[1,2]

[1] School of Big Data and Software Engineering, Chongqing University, Chongqing 400044, China

[2] Key Laboratory of Dependable Service Computing in Cyber Physical Society, Chongqing University, Ministry of Education, Chongqing 400044, China
{xywang,gaomin,jennylu,zongwei,jw.zhang,cquzhangyi}@cqu.edu.cn

Abstract. Online movie recommender systems aim to address information overload problem in movie perspective. Recently, incorporating knowledge graph into recommender systems as auxiliary information has attracted much attention due to its rich semantic content. In this paper, we propose a deep multi-channel model for dynamic movie recommendation (DMCM), which makes full use of user-item interaction and knowledge graph. First, we learn item embedding, entity embedding and genre embedding from interaction matrix and knowledge graph. Then we design a CNN-based network which can fuse the learnt embeddings and acquire the final movie representation, among which an attention module is applied to better represent the user. Finally, the click-through rate for the user-movie pair is calculated utilizing the obtained user and movie representation. Results of extensive experiments on a real-world dataset show that the proposed DMCM outperforms state-of-art baselines.

Keywords: Movie recommendation · Knowledge graph · Attention module

1 Introduction

Due to the ever-growing volume of online movies, movie recommender system is essential to address the information overload problem and guides users in a personalized way. Among all kinds of recommendation strategies, collaborative filtering has achieved significant success due to its efficiency. Nevertheless, the performance of collaborative filtering based recommender system suffers severely from data sparsity and cold start problems. To alleviate the above-mentioned problems, auxiliary information such as social networks [11], images [12] and texts [8] were incorporated in order to better comprehend user's taste and boost the performance of recommender system. Among various side information, knowledge graph has drew much attention in recent years.

Knowledge graph is a centralized repository for heterogeneous information which makes an excellent auxiliary information for movie recommendation,

T. Gedeon et al. (Eds.): ICONIP 2019, CCIS 1142, pp. 425–432, 2019.
https://doi.org/10.1007/978-3-030-36808-1_46

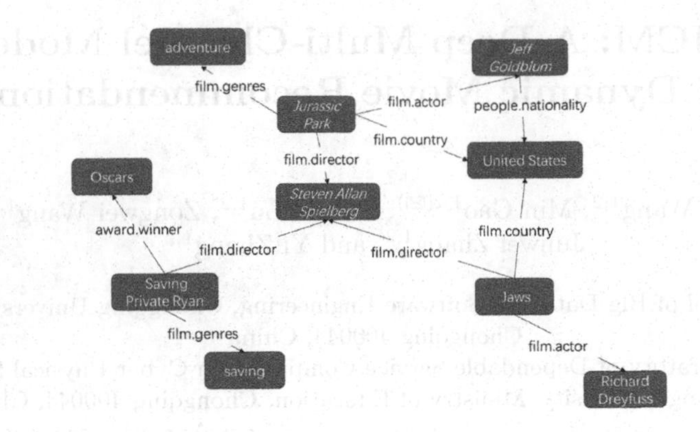

Fig. 1. Illustration of knowledge graph providing supplementary information for movies.

as is illustrated in Fig. 1; There already exist some methods that incorporate knowledge graph into movie recommendation. For example, CKE [12] combines an item's knowledge embedding, textual embedding and image embedding which are extracted respectively through a CF module in a unified Bayesian framework.

Although existing methods utilizing knowledge graphs have proved their effectiveness, most of them did not take full advantage of user-item interaction. To address this issue, we propose DMCM, a novel framework which leverages knowledge graph and user-item interaction and incorporates them into movie recommendation. DMCM is a hybrid deep recommendation model for click-through rate prediction, which takes a user's click history and a movie as input, and outputs the possibility that the user will click the movie.

Empirically, our studies can be mainly divided into two parts. First, we conduct extensive experiments to evaluate the performance of our model. Next, we evaluate the effectiveness of our framework compared with several up-to-date baselines. The experiment results show that DMCM gains significant improvements over comparison methods.

The key contributions of this paper are summarized as follows:

- We simultaneously leverage rich heterogeneous information from knowledge graph and content learned from user-item interaction. The representations extracted are fused through a multi-channel module.
- We propose DMCM, a novel framework that incorporates knowledge graph with interaction matrix for movie recommendation.
- We conduct extensive experiments on a real-world data set and evaluate the effectiveness of our framework.

2 Preliminaries

In this section, we will give a brief introduction to the concepts and terminologies which we shall use in the following parts.

2.1 User-Item Interaction Matrix

Assume there are m users and n items in total, we can define the interaction matrix $R \in \mathbb{R}^{m \times n}$, for each entry in the interaction matrix:

$$R_{ij} = \begin{cases} 1, \text{if an interaction between user } i \text{ and item } j \text{ is observed;} \\ 0, \text{otherwise.} \end{cases} \tag{1}$$

2.2 Knowledge Graph Embedding

Performance of recommender system can be significantly enhanced if we take full advantage of knowledge graphs. By applying knowledge graph embedding methods we can obtain low-dimensional representation vectors for each entity and relation that preserves the original structure and semantic relationship of the knowledge graph. In this paper some typical translation-based methods, namely TransE [1], TransH [13], TransR [4] and TransD [3] are adopted to acquire the embedding of the entities.

3 DMCM Framework

In this section, we present in detail how our framework leverages information from interaction matrix and knowledge graph respectively and utilizes them for recommendation afterwards. Our framework is illustrated in Fig. 2.

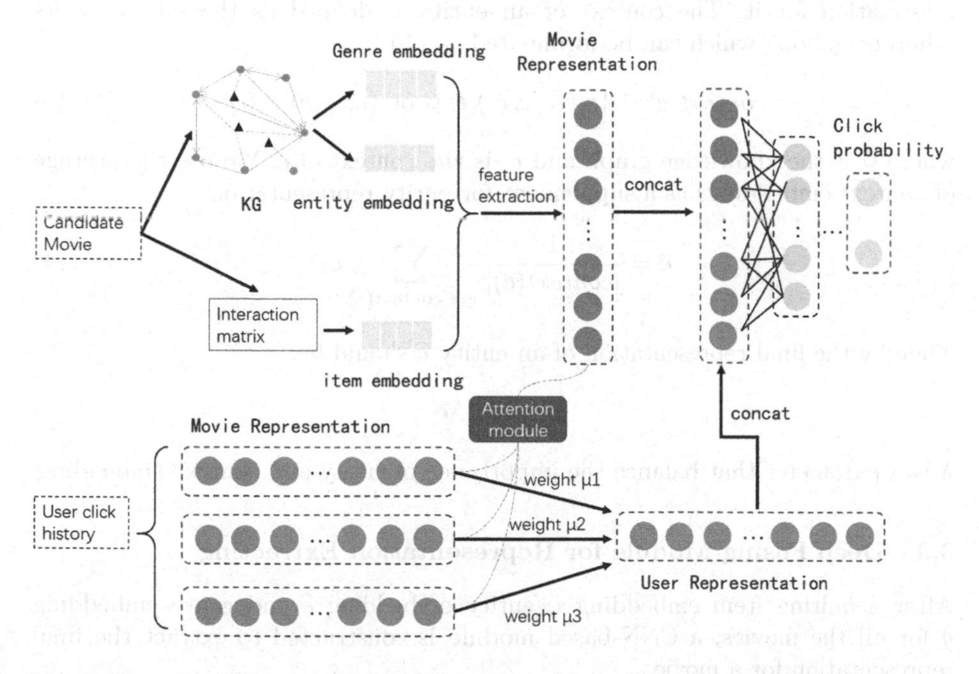

Fig. 2. The overall framework of DMCM, which takes a piece of movie and user history as input and outputs a user's CTR for the movie.

3.1 Item Representation Learning

We apply Bayesian Personalized Ranking [6] (BPR) to acquire the item latent matrix $\widehat{i}$, in which each row can be regarded as the item embedding of the corresponding movie.

3.2 Knowledge Representation Learning

We adopt a large-scale public knowledge base Freebase as our linked knowledge graph. To begin with, we have to apply entity linking [5] to match movies in our dataset with entities in knowledge graph, then we could extract the needed sub-graph. Finally, knowledge representation learning methods can be utilized to acquire embeddings of entities.

Through the above-mentioned procedure we could obtain the embeddings for movies and all their corresponding genres which are denoted as e and g respectively. A movie's genre can reflect its content to some extent. We formulate final genre embedding $\widehat{g}$ of a movie as Formula (9) :

$$\widehat{g} = \frac{1}{|g(m)|} \sum_{g_i \in \mathrm{g}(m)} \mathbf{g}_i \tag{2}$$

where $g(m)$ contains all the genres of a movie m.

Intuitively, a node's direct neighbor should provide some supplementary information for it. The context of an entity is defined as the set of all its 1-hop neighbors, which can be formulated as (10) :

$$context(e) = \{e_i | \, (e, r, e_i) \in \mathcal{G} \text{ or } \, (e_i, r, e) \in \mathcal{G}\} \tag{3}$$

where $\mathcal{G}$ is the knowledge graph and e_i is the context of e. We use the average of context embeddings as a supplement for entity representation:

$$\overline{\mathbf{e}} = \frac{1}{|context(e)|} \sum_{e_i \in \mathrm{context}(e)} \mathbf{e}_i \tag{4}$$

Thereby the final representation of an entity $\widehat{e}$ should be:

$$\widehat{e} = e + \lambda \overline{\mathbf{e}} \tag{5}$$

λ is a parameter that balance the importance of entity and context embedding

3.3 Deep Fusing Module for Representation Extracting

After acquiring item embedding $\widehat{i}$, entity embedding $\widehat{e}$ and genre embedding $\widehat{g}$ for all the movies, a CNN-based module is constructed to extract the final representation for a movie.

It should be noted that entity embedding and item embedding are learnt in different space, hence a projection procedure that maps embeddings from entity space to item space is proceeded, which can be formulated as (13) :

$$f(\mathbf{e}) = \tanh(\mathbf{M}\mathbf{e} + \mathbf{b}) \tag{6}$$

In which $\mathbf{e}$ is the embedding to be mapped. Then we could treat the embedding matrices as different channels of a movie. The matrices are aligned and stacked together and then fed into a CNN to extract features for movies.

3.4 Dynamic User Modeling

With the final movie representation $\mathbf{v}(t)$ obtained above. Based on the intuition that different movie should have impact on user's interest with varying degrees, we apply an attention module [9] to dynamically model a user's interest. Concretely, for a user i's clicked movie t_k^i and a candidate movie t_j, we can calculate the attention weight through:

$$s_{t_k^i, t_j} = \text{softmax}\left(\text{v}\left(t_k^i\right) \cdot \text{v}\left(t_j\right)\right) = \frac{\exp\left(\text{v}\left(t_k^i\right) \cdot \text{v}\left(t_j\right)\right)}{\sum_{k=1}^{N_i} \exp\left(\text{v}\left(t_k^i\right) \cdot \text{v}\left(t_j\right)\right)} \tag{7}$$

After acquiring the attention weight, we can dynamically model a user with respect to candidate movie t_j by:

$$\mathbf{v}(i) = \sum_{k=1}^{N_i} s_{t_k^i, t_j} \mathbf{v}\left(t_k^i\right) \tag{8}$$

Finally, user and candidate movie's representation are concatenated and fed into a DNN D to calculate the final click-through-rate.

4 Experiments

4.1 Dataset Introduction

We adopt a real-world dataset MovieLens-1M [2] to demonstrate the effectiveness of our proposed method. MovieLens-1M is a widely used dataset which consists of 6,040 users and their ratings on more than 3,900 movies. Since the dataset consists of explicit feedbacks while we need implicit ones, the labels in the dataset are transformed into 1 which indicate an existing interaction between user and items.

Freebase is a knowledge base which contains more than 570 million entities and 19 billion triples. We utilize Freebase to construct the knowledge graph for our dataset. Movies which failed to link to knowledge graph were removed from the MovieLens-1M dataset. The statistics of the final datasets are showed as follow (Table 1).

Table 1. Dataset statistics

Datasets		
Movielens 1M	#Users	6,040
	#Items	3,689
	#Ratings	998,141
	Sparsity	95.52%
Knowledge Graph	#Entities	984,583
	#Link Types	542
	#Triples	1,684,901

4.2 Experiment Setup

We choose TransD [3] as knowledge representation learning algorithm. The dimension of entity embedding is chosen among 25, 50, 75, 100, 125 and 150, while the dimension item embedding is set to 50, which is its optimum value. A grid search is applied in 64, 128, 192 and 256 to find the optimal number of filters. We utilize Adam to optimize the train lost. We choose *AUC* and *Recall* as evaluation metrics to compare our proposed framework with baselines. For baselines, dimension of entity embedding is set to 100, other parameters are set as default value. We consider five methods for comparison, namely **BPRMF** [6], **CKE** [12], **SHINE** [7], **DKN** [8] and **PER** [10]. BPR ignores structural knowledge, DKN is fed with movie names, other methods take interaction matrix and knowledge graph as input.

4.3 Result

In this subsection performance comparison between models is presented, and we analysis the variants of DMCM.

Table 2. Model comparison

Model	AUC	Recall@1	Recall@2	Recall@5	Recall@10	Recall@20
BPRMF	0.842	0.33%	0.64%	1.53%	2.93%	5.51%
CKE	0.803	0.49%	1.51%	3.14%	4.25%	6.62%
SHINE	0.781	0.36%	1.41%	3.32%	4.11%	5.67%
DKN	0.667	0.24%	0.53%	1.11%	1.62%	3.11%
PER	0.697	3.25%	5.92%	**8.85%**	13.95%	19.88%
DMCM	**0.856**	**3.66%**	**6.12%**	8.76%	**14.54%**	**20.13%**

Performance Comparison. In Table 2 we presented the experiment results of different methods. We can conclude:

- **BPRMF** acquires a rather good result on AUC, which mainly owes to the low sparsity of the dataset. Nevertheless, BPRMF has a comparatively low Recall value compared with other knowledge-based methods.
- **CKE** performs relatively poor, probably owing to that we only utilize structural information since we have no access to textual and visual input.
- **SHINE** performs badly, which is probably because of the fact that knowledge representation learning methods are more efficient than autoencoders when leveraging information in knowledge graph.
- **DKN** performs worst among all methods, because news title contains abundant entities and semantic information while movie names do not.
- **PER** performs rather well, indicating that well designed meta-path can preserve structural information and enhance recommend result.
- **DMCM** achieves significant improvement over baselines which can be shown in Table 2, demonstrating our proposed framework can take full advantage of information in interaction matrix and knowledge graph.

Study of DMCM. In this subsection we analyze the structure of our framework and influence of different module on the final result. The results are presented in Table 3, which suggest: (1) entity, context, genre and item embeddings have positive impact on AUC; (2) TransD, which is the most complicated representation learning model, can make the most of knowledge graph; (3) attention module can capture different users' taste and model user dynamically, thereby bringing a rise in AUC.

Table 3. Study of proposed framework

Variants	AUC
genre+entity(context)+item	**0.852**
genre+entity(without context)+item	0.850
entity(context)+item	0.847
genre+item	0.839
entity+genre	0.842
DMCM+TransE	0.839
DMCM+TransH	0.837
DMCM+TransR	0.844
DMCM+TransD	**0.851**
with attention module	**0.852**
without attention module	0.841

5 Conclusion

In this paper, we propose DMCM, a novel framework that leverages rich heterogeneous information from knowledge graph and incorporates it with content learned from interaction matrix. DMCM fuses the learned representations in a common vector space and extracts movie representations automatically. An attention module is also adopted to capture user's interest and model user dynamically. Extensive experiments on real-world datasets demonstrated the efficiency of the proposed model.

Acknowledgements. This research is supported by graduate research and innovation foundation of Chongqing,China(Grant No. CVS19052), the Fundamental Research Funds for the Central Universities (No. 2019CDXYRJ0011), the National Key Research and Development Program of China (No. 2018YFF0214706), Chongqing Research Program of Basic Research and Frontier Technology (No. cstc2017jcyjBX0025) and Science and Technology Major Special Project of Guangxi (GKAA17129002).

References

1. Bordes, A., Usunier, N., García-Durán, A., Weston, J., Yakhnenko, O.: Translating embeddings for modeling multi-relational data. In: Advances in Neural Information Processing Systems 26: 27th Annual Conference on Neural Information Processing Systems 2013 (2013)
2. Harper, F.M., Konstan, J.A.: The MovieLens Datasets: History and Context (2015)
3. Ji, G., He, S., Xu, L., Kang, L., Zhao, J.: Knowledge graph embedding via dynamic mapping matrix. In: Meeting of the Association for Computational Linguistics & The International Joint Conference on Natural Language Processing (2015)
4. Lin, Y., Liu, Z., Sun, M., Liu, Y., Zhu, X.: Learning entity and relation embeddings for knowledge graph completion. In: Proceedings of the Twenty-Ninth AAAI Conference on Artificial Intelligence (2015)
5. Milne, D., Witten, I.H.: Learning to link with wikipedia (2008)
6. Rendle, S., Freudenthaler, C., Gantner, Z., Schmidt-Thieme, L.: Bpr: bayesian personalized ranking from implicit feedback. In: Conference on Uncertainty in Artificial Intelligence (2009)
7. Wang, H., Zhang, F., Min, H., Xing, X., Guo, M., Qi, L.: Shine: signed heterogeneous information network embedding for sentiment link prediction (2017)
8. Wang, H., Zhang, F., Xing, X., Guo, M.: Dkn: Deep knowledge-aware network for news recommendation (2018)
9. Wang, X., et al.: Dynamic attention deep model for article recommendation by learning human editors' demonstration. In: ACM SIGKDD International Conference on Knowledge Discovery & Data Mining (2017)
10. Xiao, Y., Xiang, R., Sun, Y., Gu, Q., Han, J.: Personalized entity recommendation: a heterogeneous information network approach (2014)
11. Yu, J., Min, G., Li, J., Yin, H., Liu, H.: Adaptive implicit friends identification over heterogeneous network for social recommendation. pp. 357–366 (2018)
12. Zhang, F., Yuan, N.J., Lian, D., Xie, X., Ma, W.Y.: Collaborative knowledge base embedding for recommender systems. In: ACM SIGKDD International Conference on Knowledge Discovery & Data Mining (2016)
13. Zhen, W., Zhang, J., Feng, J., Zheng, C.: Knowledge graph embedding by translating on hyperplanes. In: Twenty-Eighth AAAI Conference on Artificial Intelligence (2014)

Dance to Music Expressively: A Brain-Inspired System Based on Audio-Semantic Model for Cognitive Development of Robots

Dengju Li[1], Rui Yan[1(✉)], Xiaoliang Xu[2], and Huajin Tang[1]

[1] Neuromorphic Computing Research Center, College of Computer Science, Sichuan University, Chengdu, China
kevinleeex@gmail.com, huajin.tang@gmail.com, ryan@scu.edu.cn
[2] School of Computer Science and Technology, Hangzhou Dianzi University, Hangzhou 310018, China
xxl@hdu.edu.cn

Abstract. Cognitive development is one of the most challenging and promising research fields in robotics, in which emotion and memory play an important role. In this paper, an audio-semantic (AS) model combining deep convolutional neural network and recurrent attractor network is proposed to associate music to its semantic mapping. Using the proposed model, we design the system inspired by the functional structure of the limbic system in our brain for the cognitive development of robots. The system allows the robot to make different dance decisions based on the corresponding semantic features obtained from music. The proposed model borrows some mechanisms from the human brain, using the distributed attractor network to activate multiple semantic tags of music, and the results meet the expectations. In the experiment, we show the effectiveness of the model and apply the system on the NAO robot.

Keywords: Cognitive robot · Brain-inspired system · Emotional model · Semantic representation

1 Introduction

With the development of robotics, a growing number of social robots have entered people's lives. Many robots play the role of human beings, such as caring for the elderly, teaching, assisting the treatment of autistic children [3,4]. Although the intelligence level of the robot is gradually improving, in the field of cognitive development, how to get robots to have compatible cognitive abilities

This work was supported by the National Natural Science Foundation of China under grant number 61773271 and the National Key R&D Program of China under grant number 2017YFB1300201.

T. Gedeon et al. (Eds.): ICONIP 2019, CCIS 1142, pp. 433–440, 2019.
https://doi.org/10.1007/978-3-030-36808-1_47

as humans, to interact naturally with humans, or to respond quickly in changing environments, still face significant challenges and difficulties [2,13]. In recent years, the research on the cognitive development of robots has attracted wide attention from scholars [1,5,9,12], and these studies have demonstrated that emotion, memory, and biological plausibility play essential roles in the cognitive development of robots.

Music processing is a whole-brain phenomenon [15], while the Limbic system plays a vital role for associating the auditory perception with meaning and memory, and guide behavioral responses to music, which is consists of the hippocampus, amygdala, cingulate cortex, and hypothalamus. The hippocampus remembers songs and related experiences and contexts. The amygdala is mainly responsible for emotional responses to music, while the prefrontal cortex and cingulate cortex participate in behavioral decision evoked by music. The coordination of our brain regions allows us to dance to music and to feel and express our emotions. To enable cognitive robots to perform similar functions, we design a simple brain-inspired system based on the proposed audio-semantic model. In the aspect of obtaining the labels of music, our work's idea is different from the methods like [11,14] based on the multi-label classification. Ours draws on some mechanisms of the human brain, using the distributed attractor network to activate multiple semantic tags.

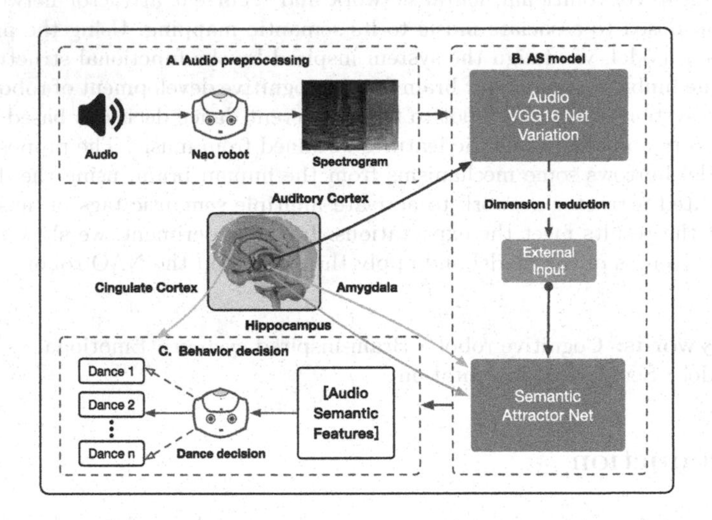

Fig. 1. System architecture.

2 System Design

This section shows the overall architectural design of the system, as shown in Fig. 1. In this paper, we apply the proposed system on the NAO robot, which is a widely used programmable humanoid robot designed by Aldebaran Robotics.

Part A is the preprocessing module of audio. For input music, the 6 s time window is used to intercept music into segments, and we apply the short-time Fourier transform (STFT) to each music segment to obtain its mel-scale spectrogram, and regard it as the input of the audio-semantic model. Part B is the proposed audio-semantic model for mapping the high-level auditory pattern of music to the corresponding semantic features, which will be described in detail in Sect. 4. Part C is the behavioral decision module of the system, roughly corresponding to the functional role of prefrontal cortex and cingulate cortex. It is used to make related dance decisions after receiving the semantic features of music. We make decisions by comparing the cosine similarity with the semantic features of different dance types, and then randomly select a dance for display in that type.

3 Data Acquisition

GTZAN music genre dataset is utilized as the input source of music, which contains blues, classical, country, disco, hiphop, jazz, metal, pop, reggae, and rock in 10 genres, each of which contains 100 music clips with a length of 30 s. We use 6 s time window, 1.5 s offset to intercept each song, and obtained a total of 17,000 samples of 6 s in length. Each sample gets a mel-scale frequency spectrogram through STFT. Because we use CNN network to process audio like images, we copy the transformed data into three channels, and then divide the training set and validation set by 7:3 to train the audio network. In order to complete the designed experiment and extract semantic commonness from the original features of music, it is necessary to tag music with corresponding semantic labels consists of emotions, characteristics, and contexts (ECC). Thus a song can be represented in semantic vector space, as shown in Fig. 2.

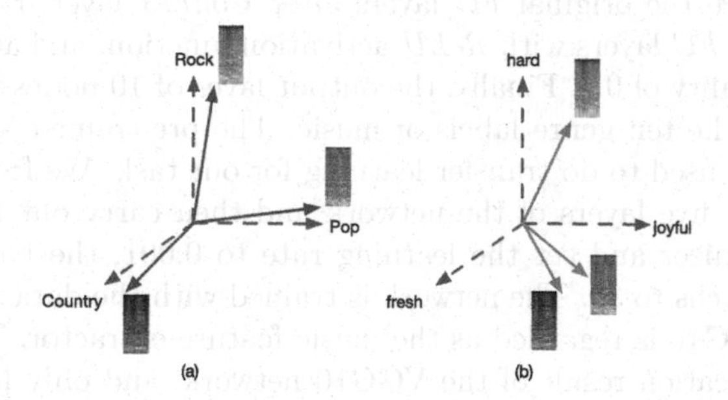

Fig. 2. Representation space for music semantics. (a) shows the representation space of music genre labels. The vectors of music with different genres in this space are orthogonal relations. (b) shows representation space of music ECC features. The vectors of music, which express similar emotions, are closer in this space. (Color figure online)

We develop a multi-user online tagging system. For the songs that need to be labeled, we do not show any visible features to the participates, who are required to label the songs only by listening. In order to reduce the extra workload, we only randomly select 10 songs of each category, 100 songs in total, and label the content of the first 12 s for each song, the songs are intercepted to 6 s with 2 s offset as training data, in a total of 400. Moreover, 12 to 18 s of each song as testing data with the same tags, in a total of 100. We provide 50 tags in three categories: Emotion (such as *fresh, joyful, sad*), Characteristic (such as *fast-pace, guitar, piano*), and Context (such as *dinner, morning, working*), named ECC[1] tags.

Participants are three males and three females, a total of 6 non-music professionals aged 20 to 28 years old. The language of tags is the native language of each participant, which is later uniformly translated into English. Each annotator labels the same 100 song segments with a length of 12 s in random order. Finally, for each labeled song, the tag with term frequency equal or greater than three will be included as the ECC semantic feature of the song.

4 Model and Method

Audio-semantic model is the core of the system. The model consists of two parts, which are responsible for mapping high-level auditory perceptual patterns to the activation pattern representing semantic features of music. The structure is shown in Fig. 3.

4.1 VGG16 Network for Audio Processing

In order to process music, we construct a deep convolutional network based on the VGG16 as shown in Fig. 3 part A. The original network structure consists of 16 layers including the convolution (*Conv*) layer and the full connection (*FC*) layer. We remove the original *FC* layers after *Conv13* layer, replace with two lower dimension *FC* layers with *ReLU* activation function, and add the Dropout with the probability of 0.5. Finally, the output layer of 10 nodes is added, which corresponds to the ten genre labels of music. The pre-trained VGG16 on ImageNet dataset is used to do transfer learning for our task. We freeze the parameters of the first five layers of the network and then carry out fine-tuning. We use Adam optimizer and set the learning rate to 0.001, the batch size to 128 and training epochs to 30. The network is trained with the data set described in Sect. 2. The VGG16 is regarded as the music feature extractor. Thus we do not need the classification result of the VGG16 network, and only keep the output value of the penultimate layer (FC15) without activation function. In order to get data used for driving the semantic attractor network in part B, we feed the labeled 100 pieces of the 6-second song into the trained audio network. For 100 samples' FC15 output, PCA is used to reduce original 512 dimensionalities to 60, to reduce computation and aggregate effective features. The dimensionality-reduced data is used as the audio input of the semantic attractor network for training.

[1] The full ECC tags can be obtained via https://github.com/kevinleeex/DTME.

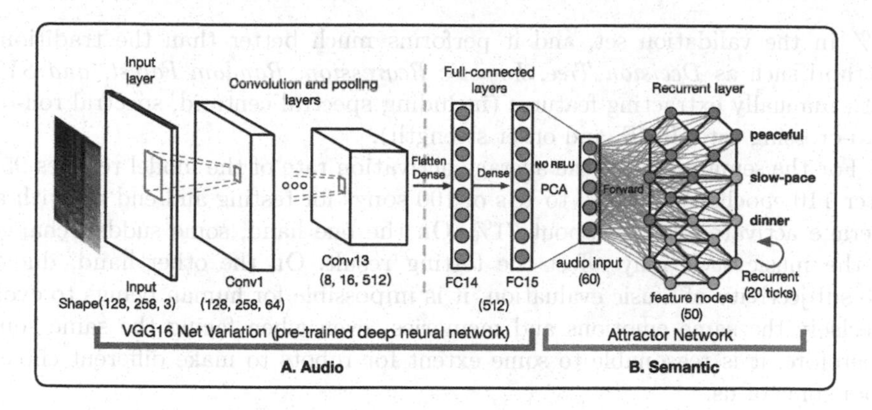

Fig. 3. Illustration of audio-semantic model structure.

4.2 Attractor Network for Semantics

Music can activate brain mechanisms related to semantic processing, as language does [8]. Research on concept processing using a feature-based distributed semantic model shows that statistical structural similarity of semantic and conceptual features between objects can explain a series of behavioral and neuroimaging data [6,10]. In this paper, we use attractor networks to obtain the semantic features corresponding to music. Attractor network [7] is a dynamic recurrent network which evolves into the stable pattern over time. We use the given ECC feature tag set to construct attractor points, in a total of 50. Emotional features correspond to the amygdala, while characteristic and context features correspond to the hippocampus. The audio input nodes are fully forward connected to the attractor network, and the internal nodes of the attractor network are connected with each other. We train the attractor network to learn the corresponding binary patterns from the input music, where '1' represents the presence of this feature, and '0' represents the absence. Moreover, we use the cross-entropy of the desired activation pattern and actual activation pattern as the loss function and use back-propagation through time (BPTT) with 20 time-ticks iteration for training. The neuronal computation process is described in [6]. We use the first 12 s with 6 s offset labeled data of music clips to train the attractor network, and training will stop until 95% nodes' activation value reaches more than 0.7. The network uses AdaGrad optimization method with learning rate $\eta = 0.02$, which can dynamically adjust the learning rate and is more suitable for sparse pattern learning. The input of each epoch is a random sample sequence, and weight will start adaptation after five time-ticks.

5 Experimental Results

5.1 Model Results

For the audio part, under the task of the music genre classification, the VGG16 variant network achieves accuracy of 95% in the training set and accuracy of

92% in the validation set, and it performs much better than the traditional method such as *Decision Tree, Logistic Regression, Random Forest, and SVM* with manually extracting features (including spectral centroid, spectral roll-off, zero-crossing rate, RMS, and onset strength).

For the semantic part, the average activation rate of the model reaches 95% after 110 epochs. We use 12 to 18 s of 100 songs for testing and end up with an average activated rate of about 71%. On the one hand, some sudden changes in the music style may affect the testing result. On the other hand, due to the subjectivity of music evaluation, it is impossible for human beings to evoke precisely the same emotions and memories, even when facing the same song. Therefore, it is reasonable to some extent for robots to make different choices from some of us.

5.2 System Results

We integrate the trained AS model into the system and deploy the system on the NAO robot. We program some dance clips for the robot with its developer kit. Music and dance often need to express the same emotions. In this paper, dance clips are classified into four types, and corresponding semantic features are tagged with ECC feature set, see Fig. 4. Then, we select four types of songs from the GTZAN dataset, and randomly select a song in each type and ensure that the song did not participate in the training of semantic attractor network. Each song captures the first 6-second segment, and then the segments are spliced into a 24-second testing clip[2], as shown in Fig. 5(a).

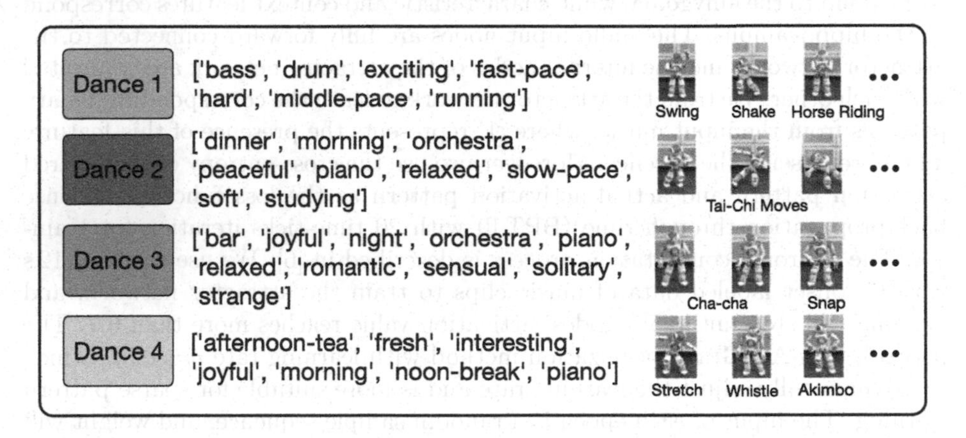

Fig. 4. Illustration of robot dance clips in different styles with semantic features.

Scanning the testing clip with the 6 s time window and 2 s offset to get the corresponding semantic features and make behavioral responses. Each test segment

[2] Readers can download a copy for listening via https://github.com/kevinleeex/DTME/blob/master/assets/testing_clip.wav.

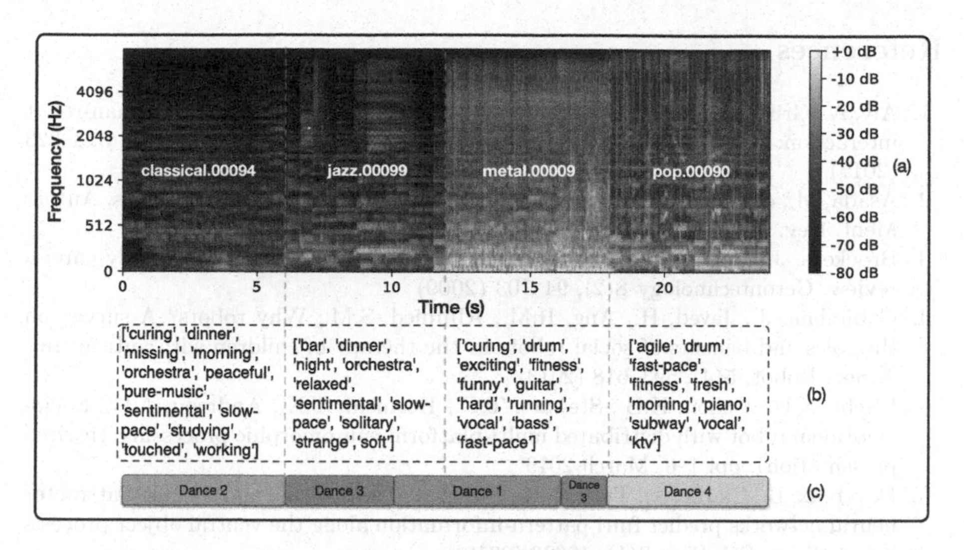

Fig. 5. Illustration of the results on testing clips. (a) shows the mel-spectrogram of the testing music clips. (b) denotes the associated semantic features related to the music. (c) shows the dance decisions made by the robot. (Color figure online)

is fed to the system to obtain the corresponding output, as shown in Fig. 5(b) and (c). Figure 5(b) shows the semantic output corresponding to the four segments with an interval of 6 s. The tags marked in black and red are the output value of the model, the reds are the wrong output tags, the greens are the correct tags but do not appear, and the blacks are the consistent output with the provided by the annotators. (c) shows the dance decisions made by comparing the cosine similarity with the emotion and memory features evoked by music, and the dance with the highest similarity score will be performed.

6 Conclusion

This work inspired by the functional structure of Limbic system of human brain constructs a system for cognitive development of robots, which based on the proposed model of audio mapping to semantics, to realize that robots can evoke emotions and related memories from music. Moreover, behavioral decisions are made through the similarity comparison between music semantics and dance semantics in their bag-of-words vector representation, and then the robot will dance to the music as feedback. The working principle of the model is described in detail, and the experimental results show the effectiveness of the model. However, the audio part of the proposed model lacks some biological plausibility, and we can consider using the spiking neural network (SNN) to enhance the biological plausibility of the AS model, while it is suitable to capture the spatial-temporal patterns and is advantaged in dealing with sound coding and learning [16]. Furthermore, the decision module can be constructed more sophisticated according to specific application scenarios.

References

1. Aly, A., Griffiths, S.S., Stramandinoli, F.: Metrics and benchmarks in human-robot interaction: recent advances in cognitive robotics. Cogn. Syst. Res. **43**, 313–323 (2017)
2. Asada, M., et al.: Cognitive developmental robotics: a survey. IEEE Trans. Auton. Ment. Dev. **1**(1), 12–34 (2009)
3. Broekens, J., Heerink, M., Rosendal, H.: Assistive social robots in elderly care: a review. Gerontechnology **8**(2), 94–103 (2009)
4. Cabibihan, J., Javed, H., Ang, H.M., Aljunied, S.M.: Why robots? A survey on the roles and benefits of social robots in the therapy of children with autism. Int. J. Soc. Robot. **5**(4), 593–618 (2013)
5. Fischl, K.D., Cellon, K.B., Stewart, T.C., Horiuchi, T.K., Andreou, A.G.: Socio-emotional robot with distributed multi-platform neuromorphic processing: (invited presentation), pp. 1–6, March 2019
6. Devereux, B., Clarke, A., Tyler, L.K.: Integrated deep visual and semantic attractor neural networks predict fmri pattern-information along the ventral object processing pathway. Sci. Rep. **8**(1), 10636 (2018)
7. Hinton, G.E., Shallice, T.: Lesioning an attractor network: investigations of acquired dyslexia. Psychol. Rev. **98**(1), 74 (1991)
8. Koelsch, S., Kasper, E., Sammler, D., Schulze, K., Gunter, T., Friederici, A.D.: Music, language and meaning: brain signatures of semantic processing. Nat. Neurosci. **7**(3), 302–307 (2004)
9. Masuyama, N., Islam, M.N., Seera, M., Loo, C.K.: Application of emotion affected associative memory based on mood congruency effects for a humanoid. Neural Comput. Appl. **28**(4), 737–752 (2017)
10. Nishida, S., Nishimoto, S.: Decoding naturalistic experiences from human brain activity via distributed representations of words. NeuroImage **180**, 232–242 (2017)
11. Oramas, S., Nieto, O., Barbieri, F., Serra, X.: Multi-label music genre classification from audio, text, and images using deep features. arXiv preprint: arXiv:1707.04916 (2017)
12. Tang, H., Huang, W., Narayanamoorthy, A., Yan, R.: Cognitive memory and mapping in a brain-like system for robotic navigation. Neural Netw. **87**, 27–37 (2017)
13. Tikhanoff, V., Cangelosi, A., Metta, G.: Integration of speech and action in humanoid robots: iCub simulation experiments. IEEE Trans. Auton. Ment. Dev. **3**(1), 17–29 (2011)
14. Trohidis, K., Tsoumakas, G., Kalliris, G., Vlahavas, I.: Multi-label classification of music by emotion. EURASIP J. Audio Speech Music Process. **2011**(1), 4 (2011)
15. Warren, J.D.: How does the brain process music. Clin. Med. **8**(1), 32–36 (2008)
16. Xiao, R., Yan, R., Tang, H., Tan, K.C.: A spiking neural network model for sound recognition. In: Sun, F., Liu, H., Hu, D. (eds.) ICCSIP 2016. CCIS, vol. 710, pp. 584–594. Springer, Singapore (2017). https://doi.org/10.1007/978-981-10-5230-9_57

Identifying EEG Responses Modulated by Working Memory Loads from Weighted Phase Lag Index Based Functional Connectivity Microstates

Li Zhang[1], Bo Shi[1], Mingna Cao[1], Sai Zhang[1], Yiming Dai[1], and Yanmei Zhu[2](✉)

[1] School of Medical Imaging, Bengbu Medical College, Bengbu 233030, Anhui, China
li_zhang@seu.edu.cn,
{shibo,mingna_mit,xmxu,daiyiming}@bbmc.edu.cn
[2] School of Biological Science and Medical Engineering, Southeast University, Nanjing 210096, Jiangsu, China
zhuyanmei@seu.edu.cn

Abstract. Working-memory training has been viewed as an important intervention way to improve the working memory capacity of children's brain. However, effective electroencephalogram (EEG) features and channel sites correlated with working memory loads still need to be identified for future application to brain-computer interface (BCI) system. In this experiment, 21 adolescent subjects' EEG was recorded while they performed an n-back working-memory task with adjustable loads (n = 1, 2, 3). Based on global neuronal workspace (GNW) theory, α-band (4–8 Hz) weighted phase lag index (wPLI) between signals was computed in consecutive 200-ms time windows of each trial to construct continuously evolving functional connectivity microstates. Statistical analysis reveals that, in post-stimulus 200–400 ms and 400–600 time intervals, working-memory loads significant modulate functional integration of global network, showing increasing connectivity density and decreasing characteristic path length with the increase of memory loads. Classifications between single-trail samples from high- and low-loads were conducted for local nodal connection strength. Analytical results indicate that network vertices in right-lateral prefrontal cortex, right inferior frontal gyrus and pre-central cortices are highly involved in identifiable brain responses modulated by working-memory loads, suggesting feasible EEG reference locations and novel features for future BCI study on the development of children/adolescents' working memory resource.

Keywords: Working memory loads · Weighted phase lag index · EEG functional connectivity microstates · Right-lateral frontal cortices

T. Gedeon et al. (Eds.): ICONIP 2019, CCIS 1142, pp. 441–449, 2019.
https://doi.org/10.1007/978-3-030-36808-1_48

1 Introduction

Working memory can be linked to IQ, ageing and mental health, and is viewed as a central intellectual faculty of the brain [1]. In the field of educational neuroscience, how to improve the capacity of children/adolescents' working memory system is always a hot topic, which has been proven to be beneficial for the intervention of attention deficit disorder, hyperactivity disorder, dyscalculia etc. [1, 2]. In this case, identification and extraction of neural features correlated with working memory loads is crucial to an effective application to brain-computer interface (BCI) system.

Previous studies on working memory training mainly focused on event-related changes of EEG power in low-frequency neuronal oscillations [3]. Global neuronal workspace (GNW) theory has pointed out that, a global "workspace" that potentially interconnects multiple distributed and specialized brain areas can be driven by individual's cognitive effort, which is usually positively correlated with workloads of cognitive tasks [4, 5]. Recent dynamic network research reveals that different workloads can modulate long-distance phase synchronizations among discrete brain areas, which can lead to changes of functional connection status due to spatial redistribution of links in a network [5]. In order to extract effective load-dependent EEG features from phase-synchronized networks, in this study, we collected 21 adolescents' EEG data by using an n-back working memory task with gradually increasing loads (n = 1, 2, 3). According to the GNW model, a weighted phase lag index (wPLI) method was employed to construct consecutive functional networks in every 200-ms time window, *i.e.*, functional connectivity microstates, over the cognitive task [6], which has been proven its effectiveness in capturing rapid reconfiguration of network topologies in our previous study [7]. Furthermore, EEG features in individual nodes of the wPLI networks were extracted and classifications were performed to find out the most distinguishable EEG channel sites involved in the modulation of working memory loads.

2 Materials and Methods

2.1 EEG Experiment and Data Preprocess

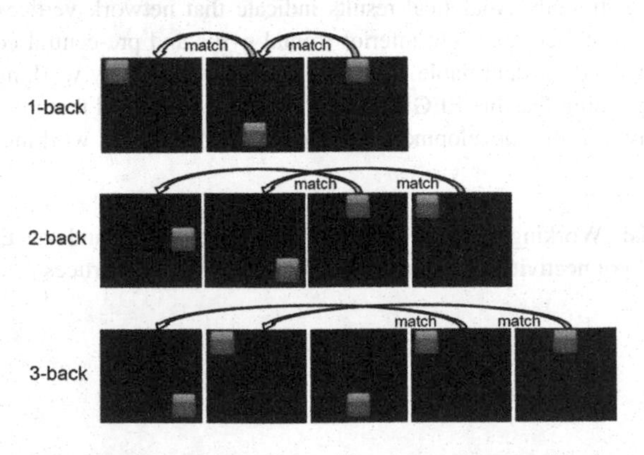

Fig. 1. Examples of an n-back (n = 1, 2, 3) working memory task with the type of visuospatial graph matching

The electroencephalogram (EEG) experiment was approved by the Academic Committee of the Research Center for Learning Science, Southeast University, China. EEG data were recorded by a 60-channel Neuroscan international 10–20 system with sampling rate at 1000 Hz. In this experiment, 21 adolescents composed of 10 males and 11 females aged 17.4 ± 3.3 (mean ± SD) performed an n-back (n = 1, 2, 3) working memory task with the type of visuospatial graph matching (Fig. 1).

The raw EEG signals were preprocessed by the Scan 4.3 software. After extracting the trials with the epoch of 1200 ms (200-ms pre-stimulus and 1000-ms post-stimulus intervals), baseline correction, artifact rejection and low-pass filtering (1–60 Hz) were performed subsequently for each subject. As a result, 18–45 trials were retained for each subject under 1-back, 2-back and 3-back task conditions.

2.2 Weighted Phase Lag Index Based Functional Connectivity Microstates

According to the GNW model, the intense mobilization/driving of workspace configuration of functional network is correlated with individual cognitive effort, which exerts cognitive loads on the working memory system [4, 5]. In terms of functional connectivity of brain network, the workspace formation can be quantified by a topological transition from locally synchronized and modular subsystems to a highly integrated global configuration pattern [5]. Therefore, in this study, continuously evolving functional connectivity microstates were constructed over task course, in order to find out load-modulated topological reconfigurations and specific time windows.

In sensor-level functional network construction, volume conduction of the brain is a considerable factor that can cause spurious increase of connectivity among distributed brain areas. Through measuring the asymmetry of the distribution of phase differences around zero, phase lag index (PLI) quantifies the time-lagged interdependence between two time series, which is defined as

$$PLI = |\langle sign(\Delta\varnothing_{rel}(t))\rangle| = \left|\frac{1}{X}\sum\nolimits_{x=1}^{X} sign(\Delta\varnothing_{rel}(t_x))\right| \tag{1}$$

where $\Delta\varnothing_{\mathrm{rel}}$ refers to phase difference at time-point x between two signals, *sign* is used to stand for signum function, $|\;|$ refers to the absolute value, and $<>$ indicates the operation of mean value. Instantaneous phases were produced from the Hilbert transformation. Although PLI exhibits robustness against the presence of common sources in estimating functional connectivity, it is biased and lacks ability in detecting some changes in phase synchronization caused by noisy signals for weak coupling [6]. WPLI is a weighted version of PLI to tackle the problems of small-magnitude synchronization effect, by weighting each phase difference based on the magnitude of the imaginary component of the cross-spectrum. In this study, wPLI was computed between pairwise signals from EEG sensors to quantify phase synchrony [6].

After further filtering EEG time series into α frequency band (4–8 Hz), wPLI was calculated within each 200-ms time window of each trial, because α-band oscillations can be associated with cognitive and memory performance [3, 8]. For each trial composed of 200-ms pre-stimulus and 1000-ms post-stimulus intervals, sequential

60 × 60 association matrices were created over the task course. For each association matrix, an adjacent matrix can be acquired after setting a threshold according the following steps: Firstly, a fixed connection density p was set for the association matrix from the pre-stimulus period, abiding by the Erdös-Rényi model, where $p = 2lnn/n$ (n is the number of the EEG channels), which produced a no-task wPLI adjacency matrix. Then the minimum wPLIs of all non-task adjacency matrices were averaged. The mean was used as a fixed threshold that was applied to all association matrices within the time windows in post-stimulus period, through which we got a series of task-related adjacency matrices. For these adjacency matrices, time-sequential undirected graphs can be constructed, which represent sequential functional connectivity microstates of brain network in working memory information processing.

2.3 Extracting Graph Features Modulated by Working Memory Loads

To reveal the modulation effect of working memory loads on the functional connectivity microstates, we measured the global structure and local node characteristic according to graph theory [9]. In the definition of network topology from graph theory, N represents the set of all the nodes in a network and (i, j) indicates the edge between nodes i and j $(i, j \in N; i \neq j)$. In the case that there is connection status between nodes i and j, $a_{ij} = 1$; otherwise, $a_{ij} = 0$.

Here, connection density and characteristic path length of a functional connectivity microstate were estimated to reflect global integration of functional network. Connection density refers to the number of edges in a graph with n nodes divided by the maximum number of possible edges $[(n^2 - n)/2]$. Characteristic path length indicates the average number of edges in the shortest paths between all nodes:

$$L = \frac{1}{n}\sum_{i \in N} L_i = \frac{1}{n}\sum_{i \in N} \frac{\sum_{j \in N, j \neq i} d_{ij}}{n - 1} \tag{2}$$

where L_i refers to the average distance between node i and other nodes, and $d_{ij} = \sum_{a_{uv} \in g_{i \to j}} a_{uv}$ represents the shortest path length between i and j ($g_{i \to j}$ is the shortest geodesic path). If node pairs i and j are disconnected, $d_{ij} = \infty$. For the two global measurements of functional networks, one-way analysis of variance (ANOVA) was performed between every two task conditions, using all trials as the testing samples, in order to discover differences in network configuration caused by changed working memory loads.

On the other hand, connection strength of individual nodes was extracted from each functional microstate network. Nodal connection strength refers to the sum of weights attached to ties that belong to a node in a weighted network. Here connection strength can be given by the sum of the wPLIs of the adjacent edges connected to node i, *i.e.*,

$$\mathrm{s_i} = \sum_{\mathrm{j \in N}} \mathrm{a_{ij} w_{ij}} \tag{3}$$

where w_{ij} represents wPLI between node i and node j.

For each node, connection strength was extracted in the time windows with statistically significant modulation effect by the working memory loads, which constitutes multiple-dimensional (*i.e.*, nodal connection strength from multiple time windows) input features for this channel used for further discriminant analysis between task conditions. For each subject, the single-trial feature samples of a channel were

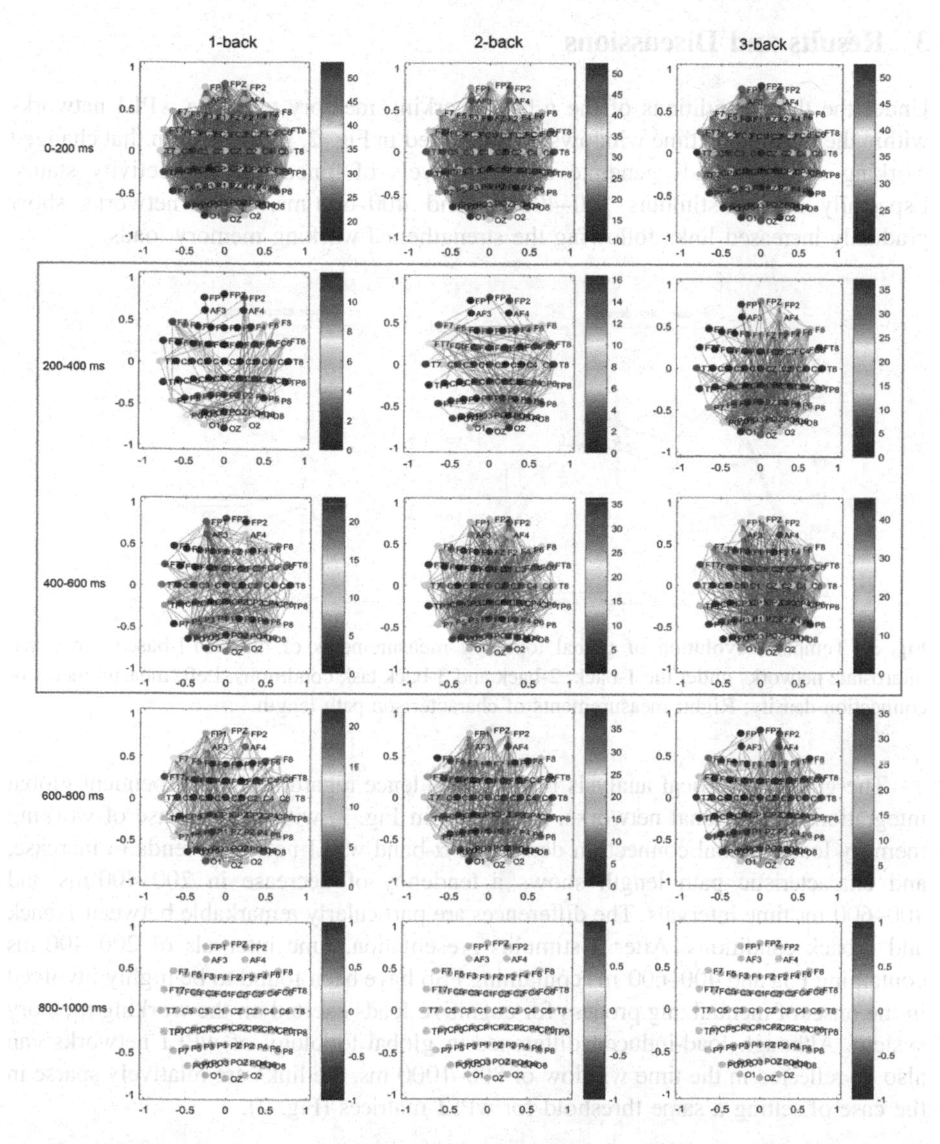

Fig. 2. WPLI-based functional microstate networks during continuous five 200-ms time windows under 1-back, 2-back and 3-back conditions of the n-back working memory task. The functional networks are constructed by setting a fixed threshold for association matrices of wPLIs. The color marked in an EEG electrode represents the number of connections of the node in a network, with its value indicated in the corresponding color bar.

recognized by linear discriminant analysis (LDA), support vector machine (SVM) and Naive Bayes methods, respectively, in order to reveal the distinguishable channel sites for isolating the load-modulated network topologies. Finally, for the total subjects, the classification results were statistically analyzed in terms of mean value and standard deviation.

3 Results and Discussions

Under the three conditions of the n-back working memory task, the wPLI networks within the continuous time windows are presented in Fig. 2. It can be seen that changed working memory loads generate different levels of functional connectivity status. Especially in post-stimulus 200–400 ms and 400–600 ms, wPLI networks show gradually increased links following the strengthened working memory loads.

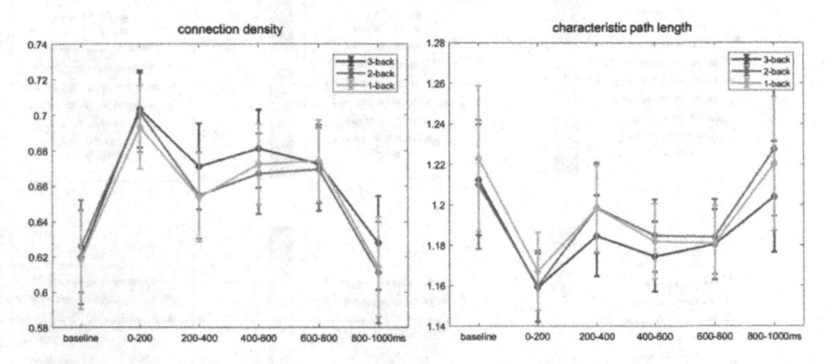

Fig. 3. Temporal evolution of global topology measurements of the wPLI-based functional microstate networks under the 1-back, 2-back and 3-back task conditions. Left: measurements of connection density; Right: measurements of characteristic path length.

The graph-theoretical analysis provides evidence regarding load-dependent global integration of functional networks. As shown in Fig. 3, with the increase of working memory loads, global connection density of α-band wPLI networks tends to increase, and characteristic path length shows a tendency of decrease in 200–400 ms and 400–600 ms time intervals. The differences are particularly remarkable between 1-back and 3-back conditions. After a stimulus presentation, time intervals of 200–400 ms containing P3a and 400–600 ms containing P3b have been found to be highly involved in the onset of mentalizing process for cognitive loads exerted on the working memory system. Although load-induced difference in global topology of wPLI networks can also be reflected in the time window of 800–1000 ms, the links are relatively sparse in the case of setting a same threshold for wPLI matrices (Fig. 2).

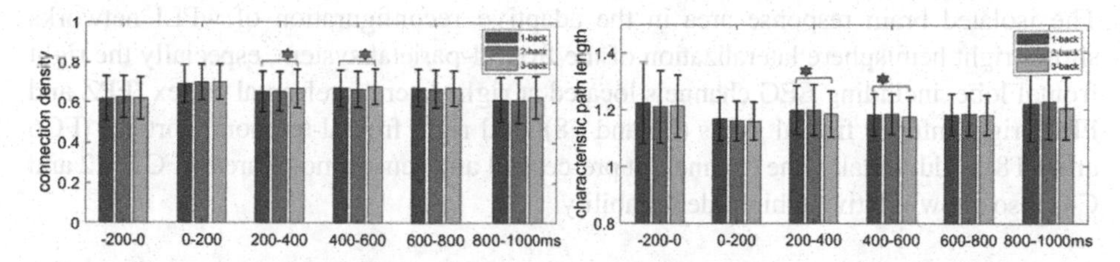

Fig. 4. Statistical bars of connection density and characteristic path length of wPLI networks formed in pre-stimulus −200–0 ms time window and five post-stimulus 200-ms time intervals, for comparisons between different working memory loads (1-back, 2-back and 3-back) (* indicates significance level $p < 0.05$ in the ANOVA).

Further ANOVA for the single-trial samples demonstrates the modulation effect of the working memory loads on global network topology in post-stimulus 200–400 ms and 400–600 ms time intervals. Especially compared to 1-back task condition, there are significant higher connection density and shorter characteristic path length of wPLI networks formed under 3-back task condition (Fig. 4), indicating that higher cognitive effort invested in high-load working memory task significantly induces strengthened phase synchronizations and global integration of functional network.

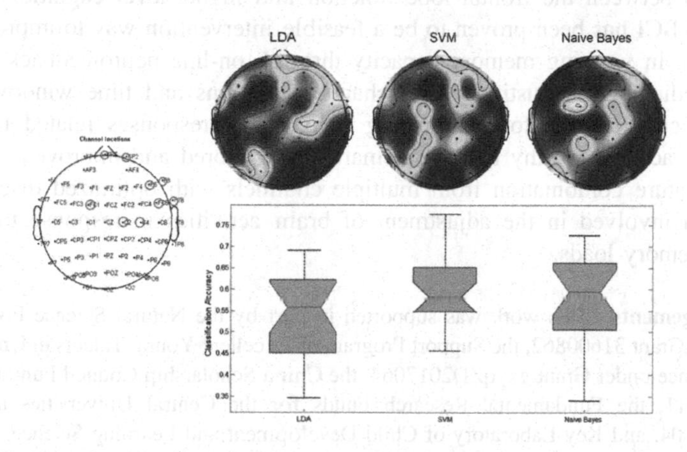

Fig. 5. EEG channels with relatively high classification accuracy in terms of nodal connection strength of wPLI networks between 1-back and 3-back task conditions (two-dimensional input features constructed by nodal connection strength in 200–400 ms and 400–600 ms intervals). Left: EEG channel locations; Right: Brain mapping of mean classification accuracy of individual channels, and statistical boxplots of classification accuracy for all subjects through LDA, SVM and Naive Bayes methods.

For each node, connection strengths were extracted in time windows of 200–400 ms and 400–600 ms to constitute two-dimensional input features for the discriminant analysis between 1-back and 3-back conditions. The EEG channel sites with relatively high accuracy through LDA, SVM and Naive Bayes methods can be seen in Fig. 5.

The isolated brain response area in the adaptive reconfiguration of wPLI networks shows right hemisphere lateralization of the frontal-parietal system, especially the right frontal lobe, including EEG channels located at right-lateral prefrontal cortex (FPZ and FP2), right inferior frontal gyrus (F6 and F8), and right frontal-temporal cortices (FC6 and FT8). Additionally, the channels at pre-central and sensorimotor areas (FC1, C2 and C4) also show relatively high identifiability.

4 Conclusions

By constructing α-band wPLI networks affected by different working memory loads, our EEG study confirms different levels of global integration of functional networks. Specifically, significant modulation effect of working memory loads can be found in post-stimulus 200–400 ms and 400–600 ms time windows, within which individual nodes at right-lateral frontal-temporal-precentral areas show relatively high identifiability in isolating brain responses in changing wPLI network topologies.

In the brain locations involved in working memory loads, the right frontal regions, including right-lateral prefrontal cortex and inferior frontal grus, have great potentials in developing the cortical resource of children and early adolescents, since previous studies have found that the brain maturation at this age stage mainly focuses on the relationship between the frontal lobe function and higher-level cognitions [10, 11]. EEG-based BCI has been proven to be a feasible intervention way to improve child's performance in working memory capacity through on-line neurofeedback manipulation. According to the justified EEG channel locations and time windows, further single-trial classification for recognizing individual's responses related to working memory capacity is worthy to be systematically explored and improved, such as an effective feature combination from multiple channels with optimized discriminatory information involved in the adjustment of brain activities in response to changing working memory loads.

Acknowledgements. This work was supported in part by the Natural Science Foundation of China under Grant 31600862, the Support Program of Excellent Young Talents in Universities of Anhui Province under Grant gxyqZD2017064, the China Scholarship Council Fund under Grant 201808340011, the Fundamental Research Funds for the Central Universities under Grant CDLS-2018-04, and Key Laboratory of Child Development and Learning Science.

References

1. Melby-Lervåg, M., Hulme, C.: Is working memory training effective? Dev. Psychol. **49**, 270 (2013)
2. Klingberg, T.: Training and plasticity of working memory. Trends Cogn. Sci. **14**, 317–324 (2010)
3. Fuster, J.M., Bressler, S.L.: Cognit activation: a mechanism enabling temporal integration in working memory. Trends Cogni. Sci. **16**, 207–218 (2012)
4. Dehaene, S., Naccache, L.: Towards a cognitive neuroscience of consciousness: basic evidence and a workspace framework. Cognition **79**, 1–37 (2001)

5. Kitzbichler, M.G., Henson, R.N., Smith, M.L., Nathan, P.J., Bullmore, E.T.: Cognitive effort drives workspace configuration of human brain functional networks. J. Neurosci. **31**, 8259–8270 (2011)
6. Vinck, M., Oostenveld, R., Wingerden, M.V., Battaglia, F., Pennartz, C.M.A.: An improved index of phase-synchronization for electrophysiological data in the presence of volume-conduction, noise and sample-size bias. Neuroimage **55**, 1548–1565 (2011)
7. Zhang, L., Gan, J.Q., Zheng, W., Wang, H.: Spatiotemporal phase synchronization in adaptive reconfiguration from action observation network to mentalizing network for understanding other's action intention. Brain Topogr. **31**, 447–467 (2018)
8. Hsueh, J.J., Chen, T.S., Chen, J.J., Shaw, F.Z.: Neurofeedback training of EEG alpha rhythm enhances episodic and working memory. Hum. Brain Mapp. **37**, 2662–2675 (2016)
9. Sporns, O.: Contributions and challenges for network models in cognitive neuroscience. Nat. Neurosci. **17**, 652–660 (2014)
10. Zhang, L., Gan, J.Q., Wang, H.: Neurocognitive mechanisms of mathematical giftedness: a literature review. Appl. Neuropsychol. Child **6**, 79–94 (2017)
11. Zhang, L., Gan, J.Q., Wang, H.: Localization of neural efficiency of the mathematically gifted brain through a feature subset selection method. Cogn. Neurodyn. **9**, 495–508 (2015)

Combining Fisheye Camera with Odometer for Autonomous Parking

Donglin Bai and Jianbo Su(✉)

School of Electronic Information and Electrical Engineering,
Shanghai Jiao Tong University, Shanghai 200240, China
{baidonglin,jbsu}@sjtu.edu.cn

Abstract. Simultaneous Localization and Mapping (SLAM) is one of the key technologies for autonomous driving. This paper focuses on the autonomous parking problem. A fisheye camera with a very large field of view is combined with the odometer inside the car to provide the localization information in an underground garage. The odometer provides an initial estimation of pose increment, and then the odometer and camera measurements are jointly optimized by graph optimization. The proposed strategy is evaluated on an autonomous driving platform, and high accuracy is achieved for the trajectory estimation with real scale.

Keywords: Autonomous driving · Visual SLAM · Sensor fusion

1 Introduction

Autonomous parking is the task to have a car park into a garage by itself without human entering the garage. Visual SLAM is a critical technique in this process which provides a precise localization information, especially in an underground garage where the GPS signal is unavailable. Many excellent visual SLAM frameworks have been proposed such as ORB-SLAM [9]. However, monocular visual SLAM cannot get the real metric scale, and traditional pin-hole cameras have very limited Field of View (FOV), and are prone to tracking lost, which may be very dangerous for autonomous driving. The issue of FOV can be amended by using large-FOV cameras such as fisheye camera, while the resolution of undetermined metric scale requires either using stereo vision or combining vision with other types of sensor, such as Lidar [5], IMU [8] and Odometer [2,7,12]. However, since wheel odometer is the most common sensor for autonomous driving, the combination of monocular vision with odometer is a good choice. [7] incorporates odometer into the SVO [3] to improve localization accuracy and prevent tracking lost, and [2] does a similar work based on ORB-SLAM [9]. [12] extendes the functionality with map re-creation and merging. While having impressive performance, none of those works above is for the industrial autonomous driving situation, and all those approaches use a traditional camera.

T. Gedeon et al. (Eds.): ICONIP 2019, CCIS 1142, pp. 450–457, 2019.
https://doi.org/10.1007/978-3-030-36808-1_49

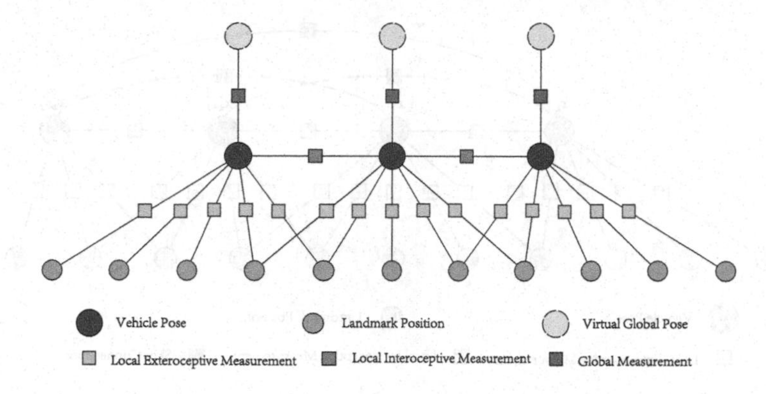

Fig. 1. Illustration of graph based sensor fusion strategy. Vertices are drawn with circles and represent state variables. Sensor measurements are drawn with squares on the edges. Each sensor measurement serves as one constraint on the state variables connected to the edge.

Based on the well-known ORB-SLAM framework, this paper combines the odometer measurements with fisheye camera for visual SLAM. The measurements of odometer first provide an initial estimation of vehicle pose increment between consecutive frames, and then jointly optimized in a local map with visual measurements via bundle adjustment, which achieves accurate estimation.

In the remainder of this paper, we first describe the overall sensor fusion scheme in Sect. 2. Then the tightly fusion is formulated in Sect. 3. In Sect. 4 the experimental evaluation of fusion strategy is shown and the Sect. 5 concludes this paper.

2 Sensor Fusion Scheme

The sensor fusion scheme is based on the nonlinear optimization, especially the graph-optimization [6], formulated as a large scale bundle adjustment problem, which is a special nonlinear least squares optimization. State variables, i.e. vehicle poses at different time steps and landmark positions, are represented as vertices in the graph, and the observation errors are represented as edges between corresponding vertices. Once the formulation is done, the problem is then solved with graph optimization tools like g2o [6]. We take a similar strategy for sensor fusion to VINS-Fusion [10,11].

2.1 General Sensor Fusion Strategy

The graph structure of optimization is illustrated in Fig. 1. Sensors on a vehicle can be classified into three types, i.e. local interoceptive sensor (Odometer, IMU), local exteroceptive sensor (Camera), and global sensor (GPS) [1]. Each measurement contributes one single term to the overall least squares formulation, and appears in the graph at a specific place depending on its type. All of

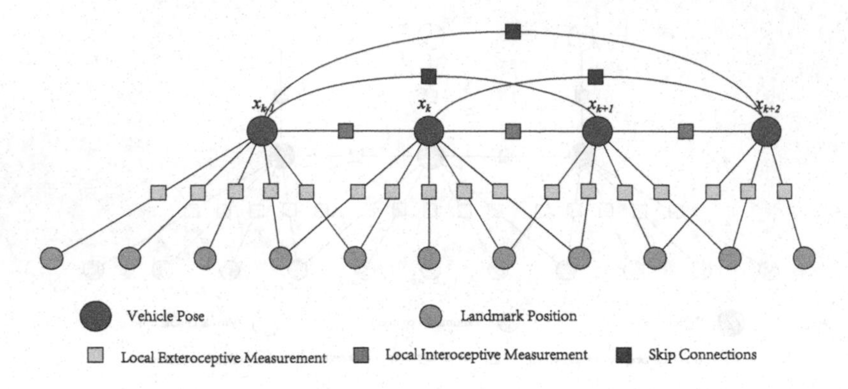

Fig. 2. Graph structure of vision and odometer fusion. The constraint obtained from odometer measurements can not only be applied to directly adjacent poses, but also to indirectly nearby poses, forming arbitrarily complex graphs.

the sensors are associated by extrinsic parameters with respect to a reference body frame, which are known during the offline calibration.

2.2 Odometer Model

The Ackermann model is used, and it is assumed that the odometer measures the forward translational velocity v and rotational velocity of yaw angle ω_z. The state of odometer is represented by $\mathbf{x} = [x, y, \phi]$. Each time the velocity is sampled, the state of odometer is updated as following

$$\begin{cases} x_{r+1} = x_r + v\cos(\phi_r)\Delta t \\ y_{r+1} = y_r + v\sin(\phi_r)\Delta t \\ \phi_{r+1} = \phi_r + \omega_z \Delta t, \end{cases} \tag{1}$$

where Δt is the sampling period and r represents time step index.

2.3 Odometer Pre-integration

Local bundle adjustment modifies state estimations of each pose several times. Since the odometer measurements are accumulated with respect to time, it is necessary to pre-integrate the odometer measurements to avoid repropagating pose estimations. Following a similar idea of IMU preintegration [4], the accumulated measurement values are isolated from the pose estimations of keyframes. The idea of preintegration is illustrated in Fig. 3. Considering two states at keyframe i and j, with the state at keyframe j accumulated based on the state at keyframe i, the difference between the two states is

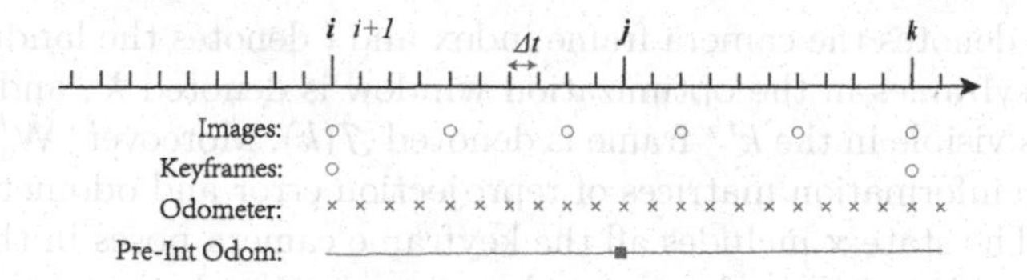

Fig. 3. Odometer measurement pre-integration.

$$
\begin{cases}
\Delta\phi_{ij} = \phi_j - \phi_i = \sum_{k=i}^{j-1}(\omega_k - n_{\omega_k})\Delta t \\
\alpha_{ij} = \sum_{k=i}^{j-1}(v_k - n_{v_k})\cos(\Delta\phi_{ik})\Delta t \\
\beta_{ij} = \sum_{k=i}^{j-1}(v_k - n_{v_k})\sin(\Delta\phi_{ik})\Delta t \\
\Delta x_{ij} = x_j - x_i = \alpha_{ij}\cos(\phi_i) - \beta_{ij}\sin(\phi_i) \\
\Delta y_{ij} = y_j - y_i = \alpha_{ij}\sin(\phi_i) + \beta_{ij}\cos(\phi_i),
\end{cases}
\tag{2}
$$

where n_{v_k} and n_{ω_k} are measurement noises. Notice that $\Delta\phi_{ij}$, α_{ij}, β_{ij} are independent of the state at i, thus can be treated as pre-integration terms. This form coincides with simple planar Euclidean transformation, and the pre-integrated measurements can be easily composed or decomposed. When to be used, the pre-integrated measurements are converted to the 3D transformation as

$$
\Delta T_{ij}^{b} = \begin{bmatrix} \cos(\Delta\phi_{ij}) & -\sin(\Delta\phi_{ij}) & 0 & \alpha_{ij} \\ \sin(\Delta\phi_{ij}) & \cos(\Delta\phi_{ij}) & 0 & \beta_{ij} \\ 0 & 0 & 1 & 0 \\ 0 & 0 & 0 & 1 \end{bmatrix},
\tag{3}
$$

which can be further transformed to camera frame via extrinsics

$$
\Delta T_{ij}^{c} = T_{cb}\Delta T_{ij}^{b} T_{cb}^{-1}.
\tag{4}
$$

3 Tightly Fusion Formulation

The least squares formulation which jointly optimizes the measurements of the two sensors consists of two terms, reprojection errors of landmarks and odometer measurement constraints. The formulation is adapted from OKVIS [8].

$$
J(\mathbf{x}) = \sum_{k\in\mathcal{K}} \sum_{l\in\mathcal{J}(k)} \mathbf{e}_r^{k,l^T} \mathbf{W}_r^{k,l} \mathbf{e}_r^{k,l} + \sum_{(i,j)\in\mathcal{K}} \mathbf{e}_b^{i,j^T} \mathbf{W}_b^{i,j} \mathbf{e}_b^{i,j},
\tag{5}
$$

where i, j, k denotes the camera frame index and l denotes the landmark index. The set of keyframes in the optimization window is denoted $\mathcal{K}$, and the indices of landmarks visible in the kth frame is denoted $\mathcal{J}(k)$. Moreover, $\mathbf{W}_r^{k,l}$ and $\mathbf{W}_b^{i,j}$ represent the information matrices of reprojection error and odometer measurement error. The state $\mathbf{x}$ includes all the keyframe camera poses in the optimization window and positions of associated landmarks. Landmarks are represented by 3D vector and camera poses are parameterized with both transformation matrix and Lie algebra for $SE(3)$, i.e.

$$\mathbf{x}_{C_k} = T_{cw}^k = \exp(\xi_k^\wedge). \tag{6}$$

3.1 Reprojection Error Formulation

The formulation of reprojection error is straightforward

$$\mathbf{e}_r^{k,l} = \mathbf{z}^{k,l} - \pi(T_{cw}^k \, {}^w P_l), \tag{7}$$

where $z^{k,l}$ is the measurement of lth landmark in k^{th} keyframe, $\pi(\cdot)$ is the camera projection function, ${}^w P_l$ is l^{th} landmark in the world frame.

3.2 Odometer Measurement Constraint Formulation

The odometer measurement constraint is defined as the error of predicted and pre-integrated pose difference between two keyframes,

$$\begin{aligned} \mathbf{e}_b^{i,j} &= \log(\Delta T_{ij}^{c\,-1} T_{cw}^i T_{cw}^{j\;-1})^\vee \\ &= \log(\exp(-\xi_{ij}) \exp(\xi_i) \exp(-\xi_j))^\vee. \end{aligned} \tag{8}$$

In practice, the pose-wise constraint may not only be applied to directly adjacent keyframes, but also be applied between arbitrary keyframes, as shown in Fig. 2. However, as the measurement error of odometer accumulates with time, this constraint should not be applied to far keyframe pairs.

4 Experiment Results

We evaluate the proposed fusion scheme on the autonomous driving platform shown in Fig. 4. A car with four fisheye cameras is driven in an underground garage, but only the front camera is used for the evaluation. The car starts from the entrance of the garage and travels alongside the main road, and after a full cycle it returns to the starting place with a loop closure. The SLAM algorithm is run on a laptop with Ubuntu 18.04 operation system and Intel Core i7-7700HQ processor. Since the fisheye camera has large distortion and there are more outliers, we set a large value of the number of ORB features to be extracted and a large number of RANSAC iterations. An ultra-high precision IMU is used to get a highly accurate estimation of trajectory for comparation as groundtruth.

Fig. 4. Experiment platform.

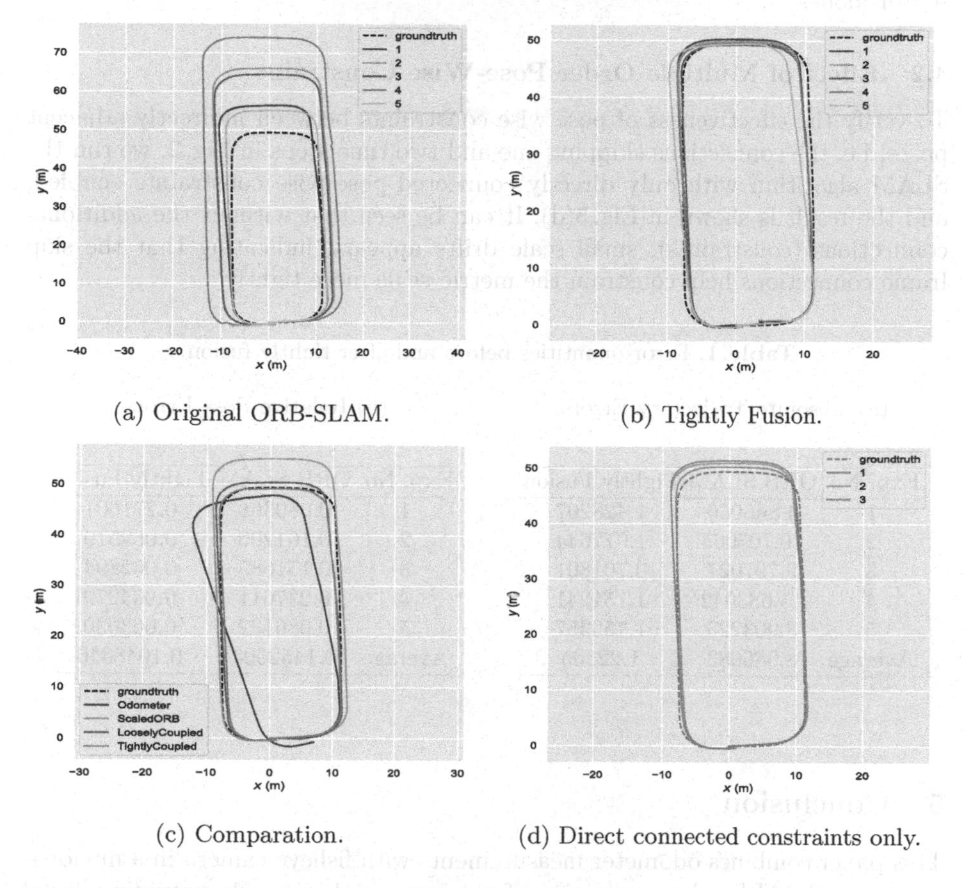

Fig. 5. Experiment results.

4.1 Test of Overall Fusion Strategy

The tightly fusion scheme is run several times and the resulting trajectories are shown in Fig. 5(b). For comparation purpose, the original ORB-SLAM framework with the fisheye camera model is also run several times and the resulting trajectories are shown in Fig. 5(a), in which the initial estimation of pose and landmarks are scaled to fit the real metric scale. The evaluation of absolute trajectory error (ATE) and relative pose error (RPE) are shown in Table 1a and b respectively. It can be seen that the resulting trajectories of scaled ORB-SLAM varies greatly, while all the trajectories after fusion are very close to the groundtruth and has a constant metric scale, and low ATE and RPE are achieved. Figure 5(c) shows a comparation of trajectories obtained by multiple approaches, including the tightly fusion method, scaled ORB-SLAM and from accumulating odometer alone. The result indicates that cooperating different types of sensors achieves better accuracy than any of the individual sensor alone.

4.2 Effect of Multiple Order Pose-Wise Constraint

To verify the effectiveness of pose wise constraints between indirectly adjacent poses, i.e. the connections skipping one and two time steps in Fig. 2, we run the SLAM algorithm with only directly connected pose wise constraints enabled, and the result is shown in Fig. 5(d). It can be seen that without the additional connections (constraints), small scale drifts appears, indicating that the skip frame connections help constrain the metric scale more tightly.

Table 1. Error quantities before and after tightly fusion.

(a) Absolute Trajectory Errors.

Exp. No.	ORB-SLAM	Tightly Fusion
1	4.845056	1.428207
2	10.705065	1.077644
3	9.797027	0.701801
4	15.683042	1.151241
5	3.903225	1.759357
Average	8.986683	1.22365

(b) Relative Pose Errors.

Exp. No.	ORB-SLAM	Tightly Fusion
1	0.089368	0.271601
2	0.161206	0.062519
3	0.151685	0.062894
4	0.237611	0.064279
5	0.086132	0.063030
Average	0.1452004	0.1048646

5 Conclusion

This paper combines odometer measurements with fisheye camera in a monocular visual SLAM for the application of autonomous driving. By providing visual SLAM an initial estimation of pose increment acquired from odometer between consecutive frames and jointly optimizing the measurements from both sensors, the metric scale of translation is successfully recovered and a better accuracy of estimation is achieved. Future work will be to utilize all of the four cameras installed on the car, which is expected to be more robust to featureless environment and more resistant to tracking lost.

Acknowledgement. This work was partially financially sponsored by the National Natural Science Foundation of China under grants 61533012, 91748120, and project of Shanghai Automotive Industry Science and Technology Development Foundation.

References

1. Barfoot, T.D.: State estimation for robotics (2017). https://doi.org/10.1017/9781316671528, ISBN 9781316671528
2. Caldato, B.A.C., Filho, R.A., Castanho, J.E.C.: ORB-ODOM: stereo and odometer sensor fusion for simultaneous localization and mapping. In: 2017 Latin American Robotics Symposium (LARS) and 2017 Brazilian Symposium on Robotics (SBR), pp. 1–5, November 2017
3. Forster, C., et al.: SVO: semidirect visual odometry for monocular and multicamera systems. IEEE Trans. Robot. **33**(2), 249–265 (2017). https://doi.org/10.1109/TRO.2016.2623335. ISSN: 1552-3098
4. Forster, C., et al.: On-manifold preintegration for real-time visual-inertial odometry. IEEE Trans. Robot. (2017). https://doi.org/10.1109/TRO.2016.2597321. ISSN: 1552-3098
5. Graeter, J., Wilczynski, A., Lauer, M.: Limo: lidar-monocular visual odometry. In: 2018 IEEE/RSJ International Conference on Intelligent Robots and Systems (IROS), pp. 7872–7879 (2018)
6. Grisetti, G., et al.: g2o: a general framework for (hyper) graph optimization. Technical report (2011)
7. He, Y., et al.: Camera-odometer calibration and fusion using graph based optimization. In: 2017 IEEE International Conference on Robotics and Biomimetics (ROBIO), pp. 1624–1629, December 2017
8. Leutenegger, S., et al.: Keyframe-based visual-inertial odometry using nonlinear optimization. Int. J. Robot. Res. **34**(3), 314–334 (2015). ISSN: 1741-3176
9. Mur-Artal, R., Tardos, J.D.: ORB-SLAM2: an open-source SLAM system for monocular, stereo, and RGB-D cameras. IEEE Trans. Robot. **33**, 1255–1262 (2017). ISSN: 1552-3098
10. Qin, T., et al.: A general optimization-based framework for global pose estimation with multiple sensors. eprint: arXiv:1901.03642 (2019)
11. Qin, T., et al.: A general optimization-based framework for local odometry estimation with multiple sensors. eprint: arXiv:1901.03638 (2019)
12. Wang, J., Shi, Z., Zhong, Y.: Visual SLAM incorporating wheel odometer for indoor robots. In: 2017 36th Chinese Control Conference (CCC), pp. 5167–5172, July 2017. https://doi.org/10.23919/ChiCC.2017.8028171

Deep Learning and Statistical Models for Time-Critical Pedestrian Behaviour Prediction

Joel Janek Dabrowski[1(✉)], Johan Pieter de Villiers[2,3], Ashfaqur Rahman[1], and Conrad Beyers[2]

[1] Data61, CSIRO, St Lucia, Australia
{joel.dabrowski,ashfaqur.rahman}@data61.csiro.au
[2] University of Pretoria, Pretoria, South Africa
[3] Council for Scientific and Industrial Research (CSIR), Pretoria, South Africa
{pieter.devilliers,conrad.beyers}@up.ac.za

Abstract. The time it takes for a classifier to make an accurate prediction can be crucial in many behaviour recognition problems. For example, an autonomous vehicle should detect hazardous pedestrian behaviour early enough for it to take appropriate measures. In this context, we compare the switching linear dynamical system (SLDS) and a deep long short-term memory (LSTM) neural network, which are applied to infer pedestrian behaviour from motion tracks. We show that, though the neural network model achieves an accuracy of 80%, it requires long sequences to achieve this (100 samples or more). The SLDS, has a lower accuracy of 74%, but it achieves this result with short sequences (10 samples). To our knowledge, such a comparison on sequence length has not been considered in the literature before. The results provide a key intuition of the suitability of the models in time-critical problems.

1 Introduction

Research in pedestrian behaviour analysis and detection has made several significant advances in the past decade. A large portion of the literature has been devoted to tracking and path prediction [18]. Many studies use the Switching Linear Dynamic System (SLDS) as a framework [12–14,21]. Recently, the recurrent neural network (RNN) has shown to be a promising approach [2,10,19,20]. Owing to the significant advancement of the state-of-the-art in computer vision-based pedestrian detection systems (see [15] for example), the trajectories of the pedestrians are assumed to be known in this study. Given the pedestrian trajectories, the behavioural class is predicted.

Several studies exist where the problem of pedestrian behaviour prediction is considered. Probabilistic models such as the latent dynamic conditional random field [22] and balanced Gaussian process dynamical models [16] have been applied. Various forms of the RNN have been also been considered [9,23]. Although both statistical and machine learning models have been applied to the

T. Gedeon et al. (Eds.): ICONIP 2019, CCIS 1142, pp. 458–465, 2019.
https://doi.org/10.1007/978-3-030-36808-1_50

problem, to our knowledge no specific analyses between these two model types in terms of time-to-detection have been considered in the literature.

Our contribution is a comparison between a SLDS and a multi-layered bi-directional LSTM neural network in the context of time-to-detection. This is performed by applying the models to classify various pedestrian behaviours from the raw motion tracks under varying sequence lengths. Varying the sequence length provides a means to measure detection time as the number of sequential samples the model requires to make an accurate prediction. Through the comparison, we gain novel insight into a key difference between the models: though the neural network is more accurate than the SLDS overall, it requires 10 times as many sequential samples to achieve this accuracy. The SLDS is able to provide its most accurate classification within the first few samples of the sequence. This result is important in situations where early detection is imperative.

2 Switching Linear Dynamical System

The SLDS models a system that switches between various dynamical models. It comprises a switching state variable s_t, a hidden or latent variable h_t, and a visible or observable variable v_t at time t. The latent variables $h_{1:T}$ and observable variables $v_{1:T}$ form a Linear Dynamic System (LDS) (where the subscript $1{:}T$ describes the sequence over all discrete time instances from 1 to T). For each switching state s_t, a LDS is defined. The model transitions between the LDSs according to the switching state. The continuous dynamics of the LDS are represented by a linear-Gaussian state space model. The SLDS is thus described by following equations [1]

$$h_t = A_t(s_t)h_{t-1} + \eta_t^h(s_t), \tag{1}$$

$$v_t = B_t(s_t)h_t + \eta_t^v(s_t). \tag{2}$$

Equation 1 describes the transition model and (2) describes the emission model. The matrix $A_t(s_t)$ is the state matrix and $B_t(s_t)$ is the measurement matrix. The noise components are modelled as white noise such that $\eta_t^h(s_t) \sim \mathcal{N}(0, \Sigma_H(s_t))$ and $\eta_t^v(s_t) \sim \mathcal{N}(0, \Sigma_V(s_t))$. All the LDS model parameters are conditionally dependent on s_t at time t. Through this conditioning, the different dynamic models are defined for each switching state.

The joint distribution describing the SLDS is given by:

$$p(s_{1:T}, h_{1:T}, v_{1:T}) = p(s_1)p(h_1)\prod_{t=2}^{T} p(s_t|s_{t-1})p(h_t|h_{t-1}, s_t)\prod_{t=1}^{T} p(v_t|h_t). \tag{3}$$

The switching state transition probability $p(s_t|s_{t-1})$ describes how the model switches between various states. The state transition distribution $p(h_t|h_{t-1}, s_t)$ and emission distribution $p(v_t|h_t)$ are assumed to be Gaussian. These describe the dynamics of the system through the linear state space equations.

Inference in the SLDS involves inferring the latent variables s_t and h_t given the observations $v_{1:t}$. This is typically performed using filtering and smoothing methods. Filtering computes $p(s_t, h_t|v_{1:t})$ and smoothing computes $p(s_t, h_t|v_{1:T})$. Exact inference in the SLDS is however intractable [1]. An approximate inference algorithm such as the Generalised Pseudo Bayesian (GPB) algorithm [17] is required. Finally, the model parameters can be computed using the Expectation Maximisation (EM) algorithm [17].

3 Multi-layered Bidirectional LSTM

A three-layered bi-directional RNN with LSTM [8] cells is constructed. Each LSTM layer comprises two sequences of LSTM cells propagating in opposite directions. Together, these cell sequences form a bi-directional LSTM (BiLSTM). The bi-directional structure provides a means to make a prediction at time t according to the full sequence from 1 to T. Three BiLSTMs are stacked to form three distinct layers. Multiple layers provide a deep structure which promotes higher level feature extraction. Data samples are provided to the inputs of the first BiLSTM layer. For each sequence step, the outputs of the third BiLSTM layer are passed through a softmax layer which predicts the class associated with each data sample. In the remainder of this text, this model is referred to as the RNN.

4 Experiments

The well-known Daimler Pedestrian Path Prediction Benchmark Dataset [21] is used in this study. The dataset comprises a collection of 68 pedestrian sequences with 4 different pedestrian behaviour types: BendingIn, Crossing, Starting, and Stopping. For each pedestrian sequence, bounding boxes, disparity, X, and Z coordinates are provided. The Z-coordinate represents depth and the X-coordinate represents the horizontal axis relative to the dataset video images.

The model parameters are estimated using the predefined training dataset and the presented results are computed with the predefined test set. To measure the performance of the models, accuracy, precision and recall are used.

The models are tested on sequences of varying length. This is achieved by truncating the sequences in increments of 10 samples. That is, the models are tested on the first $10, 20, 30, \ldots, N$ samples of a sequence in the test set, where N is the length of the particular sequence. Limiting the number of timesteps provides an indication of how well the method is able to predict a behaviour class in a short period of time. Furthermore, it provides some form of consistency over the varying sequence lengths in the dataset.

The SLDS motion model is configured as a constant acceleration model. The tracked X and Z coordinates are provided as observations to the SLDS. The model parameters are learned using the EM algorithm. The switching state is defined according to the 4 behaviour classes. The switching state transition distribution is configured with a 0.97 probability of remaining in the current switching state and a 0.01 probability of transitioning to one of the other three switching states. The prior switching state probability distribution is set to the uniform distribution.

The RNN is configured with 32 hidden units in each LSTM cell. The ADAM algorithm [11] is used to minimise the cross entropy of the softmax outputs. Hyperparameters are optimised through a grid search. The model is trained over 110 epochs with a learning rate of 0.0001. The remaining ADAM parameters are set as recommended in [11]. The model inputs include the X and Z coordinates, the disparity, and a timestamp index.

5 Results and Discussion

The accuracy over the set of truncated sequences is presented in Fig. 1. The RNN increases in accuracy with increasing sequence length, whereas the SLDS decreases in accuracy with increasing sequence length. The SLDS has the highest accuracy with a sequence of 10 samples. This implies that within the first 10 samples, the SLDS is able to classify the sequence. The RNN's accuracy curve saturates at 100 samples. This indicates that the RNN requires a sequence of at least 100 samples to achieve its highest accuracy.

The SLDS performs better when provided with the shorter sequences as it assumes a first order Markov model [1]. A first order Markov model assumes that the current state is conditionally dependent *only* on the previous state. The result is that the SLDS is not designed to model long-term dependencies. Furthermore, the SLDS performance decreases with sequence length as it is designed to switch between dynamics. It is more likely to switch behaviour class in a longer sequence. The LSTM cell is specifically designed to model longer-term dependencies in the data [8]. The result is that the RNN model performs best with longer sequences. Another relevant difference between the models is that the SLDS is a structured model where the dynamics have been predefined. The RNN is a black-box model, which often requires more data for training.

The precision and recall over the set of truncated sequences are presented in Fig. 2. Confusion matrices for the 10-sample-length and complete sequences are presented in Table 1. As for accuracy, the RNN precision and recall values saturate at 100 samples and the SLDS values are highest at 10 samples.

The RNN generally has a higher precision and recall than the SLDS. The RNN however struggles to correctly predict the Starting behaviour class. The majority of Starting samples are incorrectly associated with the BendingIn class

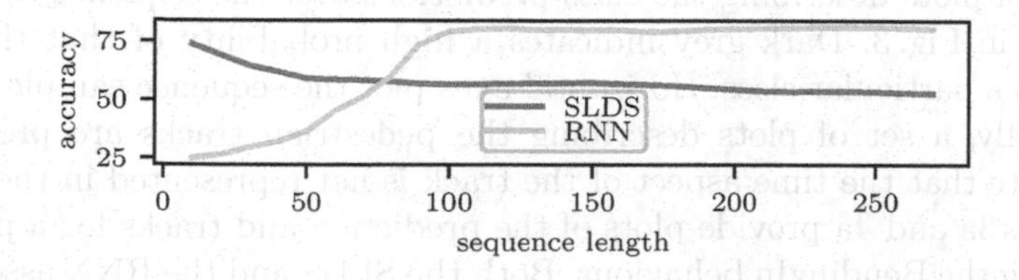

Fig. 1. Accuracy (%) over the set of truncated sequences.

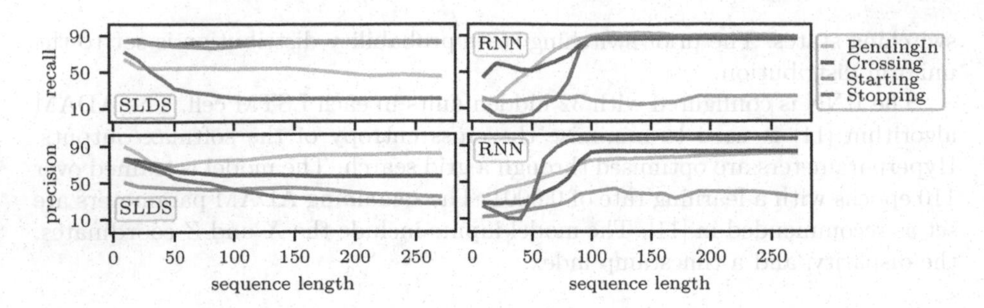

Fig. 2. Precision (%) and recall (%) over the set of truncated sequences.

Table 1. Confusion matrices for the SLDS and RNN for the 10-sample-length and complete sequence predictions. Rows and columns are ordered: BendingIn, Crossing, Starting, and Stopping. The matrices are normalised over the rows to indicate a form of recall.

SLDS 10 samples	SLDS Complete sequence	RNN 10 samples	RNN Complete sequence
$\begin{bmatrix} 0.63 & 0.02 & 0.16 & 0.00 \\ 0.00 & 0.84 & 0.00 & 0.29 \\ 0.30 & 0.00 & 0.84 & 0.00 \\ 0.07 & 0.13 & 0.00 & 0.71 \end{bmatrix}$	$\begin{bmatrix} 0.46 & 0.04 & 0.04 & 0.23 \\ 0.02 & 0.83 & 0.12 & 0.51 \\ 0.37 & 0.01 & 0.84 & 0.06 \\ 0.15 & 0.12 & 0.00 & 0.21 \end{bmatrix}$	$\begin{bmatrix} 0.19 & 0.19 & 0.19 & 0.19 \\ 0.46 & 0.45 & 0.47 & 0.47 \\ 0.21 & 0.22 & 0.21 & 0.22 \\ 0.13 & 0.14 & 0.13 & 0.12 \end{bmatrix}$	$\begin{bmatrix} 0.89 & 0.10 & 0.67 & 0.00 \\ 0.00 & 0.87 & 0.09 & 0.10 \\ 0.11 & 0.02 & 0.23 & 0.00 \\ 0.00 & 0.02 & 0.00 & 0.89 \end{bmatrix}$

as indicated in the complete sequence confusion matrix in Table 1. This could be due to the short length of the Starting sequences. The poor results for the Starting class lowers the overall accuracy of the RNN.

The lowest recall for the SLDS model is the BendingIn class, with a value of 63%. Considering the confusion matrix, many samples were misclassified as Starting behaviour. The model however performs well on the Crossing and Starting classes. As also indicated in Fig. 2, the recall for the Crossing and Starting classes remain fairly constant. For longer sequences, the precision and recall for the Stopping class decreases significantly.

For the RNN with 10-sample sequences, 46% of the BendingIn samples were incorrectly associated with the Crossing class as indicated in Table 1. When provided with the complete sequence, this reduces to 0%. Similarly, most of the Starting samples are incorrectly associated with the Crossing class with short sequences. When provided with the complete sequence, the incorrect classifications shift to the BendingIn class.

A set of plots describing the class predictions over the sequence samples are presented in Fig. 3. Dark grey indicates a high probability of that the sample belongs to a particular class. Horizontal axes plot the sequence sample numbers. Additionally, a set of plots describing the pedestrian tracks are presented in Fig. 4. Note that the time aspect of the track is not represented in these plots.

Figures 3a and 4a provide plots of the prediction and tracks for a pedestrian performing the BendingIn behaviour. Both the SLDS and the RNN associate the behaviour with the BendingIn class for the first 160 time steps. The predictions

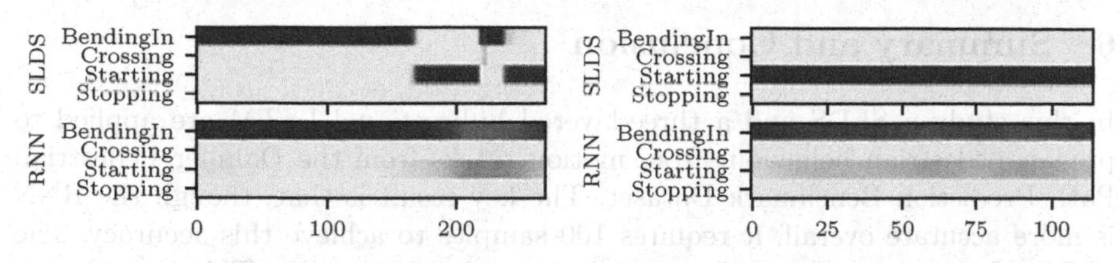

(a) Test sequence 0, true class: BendingIn. (b) Test sequence 11, true class: Starting.

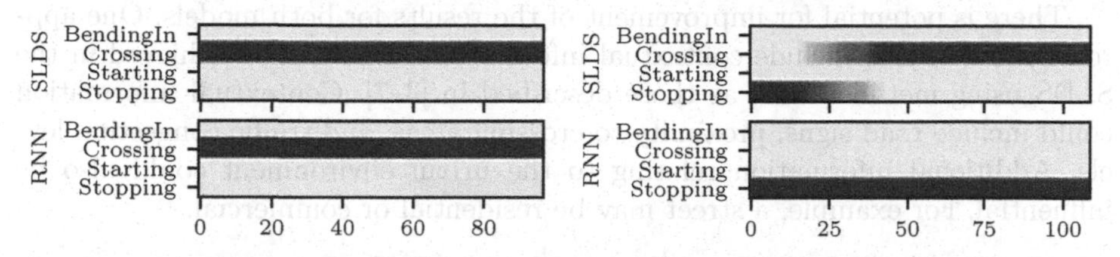

(c) Test sequence 8, true class: Crossing. (d) Test sequence 9, true class: Stopping.

Fig. 3. Behaviour prediction for various test sequences and classes.

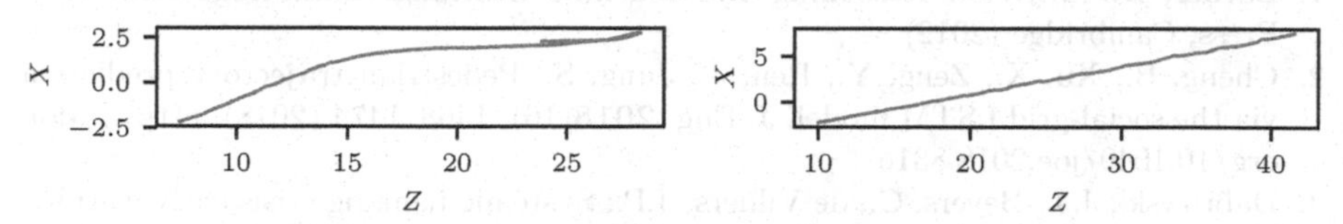

(a) Test sequence 0, true class: BendingIn. (b) Test sequence 8, true class: Crossing.

Fig. 4. Pedestrian tracks in the $X - Z$ plane for various test sequences and classes.

subsequently transition to the Starting class. This may be explained by the fact that the pedestrian seems to back-track as illustrated in Fig. 4a.

A Starting behaviour class prediction is illustrated in Fig. 3b. The SLDS predicts the correct class for the entire sequence. The RNN incorrectly predicts the BendingIn class, but with some probability associated with the Starting class. This result corresponds to the confusion matrix in Table 1.

Figure 3c illustrates results for the Crossing behaviour class. The pedestrian track corresponding to this prediction is approximately linear over the space as illustrated in Fig. 4b. With such behaviour, both models generally perform well.

An example of the Stopping behaviour class is presented in Fig. 3d. The SLDS correctly begins by classifying the stopping behaviour class and then transitions to the crossing class. This result corresponds to the complete sequence confusion matrix presented in Table 1. The RNN correctly classifies the stopping class for the entire sequence, which corresponds to the high recall for this class.

6 Summary and Conclusion

In this study a SLDS and a three-layered bidirectional LSTM are applied to predict pedestrian behaviour from motion tracks from the Daimler Pedestrian Path Prediction Benchmark Dataset. The key result is that, though the RNN is more accurate overall, it requires 100 samples to achieve this accuracy. The SLDS achieves its most accurate predictions with 10 samples. This suggests the SLDS may be the preferred model when quick detections are required.

There is potential for improvement of the results for both models. One approach would be to include contextual information. This can be achieved in the SLDS using methods such as those described in [3–7]. Contextual information could include road signs, proximity to crossing areas, and traffic congestion levels. Additional information relating to the urban environment could also be influential. For example, a street may be residential or commercial.

References

1. Barber, D.: Bayesian Reasoning and Machine Learning. Cambridge University Press, Cambridge (2012)
2. Cheng, B., Xu, X., Zeng, Y., Ren, J., Jung, S.: Pedestrian trajectory prediction via the social-grid LSTM model. J. Eng. **2018**(16), 1468–1474 (2018). https://doi.org/10.1049/joe.2018.8316
3. Dabrowski, J.J., Beyers, C., de Villiers, J.P.: Systemic banking crisis early warning systems using dynamic Bayesian networks. Expert Syst. Appl. **62**, 225–242 (2016). https://doi.org/10.1016/j.eswa.2016.06.024
4. Dabrowski, J.J., de Villiers, J.P.: Maritime piracy situation modelling with dynamic Bayesian networks. Inf. Fusion **23**, 116–130 (2015). https://doi.org/10.1016/j.inffus.2014.07.001
5. Dabrowski, J.J., de Villiers, J.P.: A unified model for context-based behavioural modelling and classification. Expert Syst. Appl. **42**(19), 6738–6757 (2015). https://doi.org/10.1016/j.eswa.2015.04.061
6. Dabrowski, J.J., de Villiers, J.P., Beyers, C.: Context-based behaviour modelling and classification of marine vessels in an abalone poaching situation. Eng. Appl. Artif. Intell. **64**, 95–111 (2017). https://doi.org/10.1016/j.engappai.2017.06.005
7. Dabrowski, J.J., de Villiers, J.P., Beyers, C.: Naive Bayes switching linear dynamical system: a model for dynamic system modelling, classification, and information fusion. Inf. Fusion **42**, 75–101 (2018). https://doi.org/10.1016/j.inffus.2017.10.002
8. Hochreiter, S., Schmidhuber, J.: Long short-term memory. Neural Comput. **9**(8), 1735–1780 (1997). https://doi.org/10.1162/neco.1997.9.8.1735
9. Hoy, M., Tu, Z., Dang, K., Dauwels, J.: Learning to predict pedestrian intention via variational tracking networks. In: 2018 21st International Conference on Intelligent Transportation Systems (ITSC), pp. 3132–3137, November 2018. https://doi.org/10.1109/ITSC.2018.8569641
10. Hug, R., Becker, S., Hbner, W., Arens, M.: Particle-based pedestrian path prediction using LSTM-MDL models. In: 2018 21st International Conference on Intelligent Transportation Systems (ITSC), pp. 2684–2691, November 2018. https://doi.org/10.1109/ITSC.2018.8569478

11. Kingma, D.P., Ba, J.: Adam: a method for stochastic optimization. arXiv preprint: arXiv:1412.6980 (2014)
12. Kooij, J.F.P., Englebienne, G., Gavrila, D.M.: Mixture of switching linear dynamics to discover behavior patterns in object tracks. IEEE Trans. Pattern Anal. Mach. Intell. **38**(2), 322–334 (2016). https://doi.org/10.1109/TPAMI.2015.2443801
13. Kooij, J.F.P., Schneider, N., Gavrila, D.M.: Analysis of pedestrian dynamics from a vehicle perspective. In: 2014 IEEE Intelligent Vehicles Symposium Proceedings, pp. 1445–1450, June 2014. https://doi.org/10.1109/IVS.2014.6856505
14. Kooij, J.F.P., Schneider, N., Flohr, F., Gavrila, D.M.: Context-based pedestrian path prediction. In: Fleet, D., Pajdla, T., Schiele, B., Tuytelaars, T. (eds.) ECCV 2014, Part VI. LNCS, vol. 8694, pp. 618–633. Springer, Cham (2014). https://doi.org/10.1007/978-3-319-10599-4_40
15. Li, J., Liang, X., Shen, S., Xu, T., Feng, J., Yan, S.: Scale-aware fast R-CNN for pedestrian detection. IEEE Trans. Multimed. **20**(4), 985–996 (2018). https://doi.org/10.1109/TMM.2017.2759508
16. Minguez, R.Q., Alonso, I.P., Fernandez-Llorca, D., Sotelo, M.A.: Pedestrian path, pose, and intention prediction through Gaussian process dynamical models and pedestrian activity recognition. IEEE Trans. Intell. Transp. Syst., 1–12 (2018). https://doi.org/10.1109/TITS.2018.2836305
17. Murphy, K.P.: Switching Kalman filters. Technical report, Department of Computer Science, UC Berkeley (1998)
18. Ridel, D., Rehder, E., Lauer, M., Stiller, C., Wolf, D.: A literature review on the prediction of pedestrian behavior in urban scenarios. In: 2018 21st International Conference on Intelligent Transportation Systems (ITSC), pp. 3105–3112, November 2018. https://doi.org/10.1109/ITSC.2018.8569415
19. Saleh, K., Hossny, M., Nahavandi, S.: Intent prediction of pedestrians via motion trajectories using stacked recurrent neural networks. IEEE Trans. Intell. Veh. **3**(4), 414–424 (2018). https://doi.org/10.1109/TIV.2018.2873901
20. Saleh, K., Hossny, M., Nahavandi, S.: Long-term recurrent predictive model for intent prediction of pedestrians via inverse reinforcement learning. In: 2018 Digital Image Computing: Techniques and Applications (DICTA), pp. 1–8, December 2018. https://doi.org/10.1109/DICTA.2018.8615854
21. Schneider, N., Gavrila, D.M.: Pedestrian path prediction with recursive Bayesian filters: a comparative study. In: Weickert, J., Hein, M., Schiele, B. (eds.) GCPR 2013. LNCS, vol. 8142, pp. 174–183. Springer, Heidelberg (2013). https://doi.org/10.1007/978-3-642-40602-7_18
22. Schulz, A.T., Stiefelhagen, R.: Pedestrian intention recognition using latent-dynamic conditional random fields. In: 2015 IEEE Intelligent Vehicles Symposium (IV), pp. 622–627, June 2015. https://doi.org/10.1109/IVS.2015.7225754
23. Volz, B., Behrendt, K., Mielenz, H., Gilitschenski, I., Siegwart, R., Nieto, J.: A data-driven approach for pedestrian intention estimation. In: 2016 IEEE 19th International Conference on Intelligent Transportation Systems (ITSC), pp. 2607–2612, November 2016. https://doi.org/10.1109/ITSC.2016.7795975

Multi-class Human Body Parsing with Edge-Enhancement Network

Xi Huang, Keyu Wu, Gang Hu, and Jie Shao(✉)

Center for Future Media, School of Computer Science and Engineering, University of Electronic Science and Technology of China, Chengdu 611731, China
{xihuang,wukeyu,hugang}@std.uestc.edu.cn, shaojie@uestc.edu.cn

Abstract. Single human parsing aims at partitioning an image into semantically consistent regions belonging to the body parts or clothing items, which has gained remarkable improvement owing to a wide range of proposed methodologies. From the perspective of the loss design, besides the parsing loss of the final output, most existing studies target on exploiting multiple other losses to enhance parsing results, which is hard to make the model reach balanced condition by adjusting their ratios and may weaken the potential of some losses. In this work, we propose an edge enhancement module to emphasize the potential of edge loss and boundary information. At the same time, local and global information will be explored for complex multi-class human body parsing problem by densely connected atrous spatial pyramid pooling. This scheme results in a simple yet powerful Edge-Enhancement Network (EEN). Extensive experiments demonstrate that EEN achieves 56.55% mIoU on LIP dataset and 62.60% mIoU on CIHP dataset, which outperform the state-of-the-arts by 3.45% and 4.02%, respectively. The code of EEN is available at https://github.com/huangxi6/EEN.

Keywords: Human body parsing · Multi-class · Edge-enhancement network

1 Introduction

Human parsing is a sub-task of semantic segmentation, aiming at partitioning a human body into multiple semantic parts on pixel level. There are multiple research problems derived from this task owing to concentration on human-centric study, such as human pose estimation [24] and fashion synthesis [29].

As for human parsing task, a number of researchers are exploring how to design an accurate and reliable human parsing architecture upon various methods. From the perspective of the loss design, besides the parsing loss of the final output, most of existing studies target on exploiting multiple other losses (which denotes that the objective loss function formula contains two or more loss terms on the basis of the initial parsing loss) to optimize parsing results, such as adversarial loss [22], joint structure loss [12,16], edge loss [11,20], and the other

T. Gedeon et al. (Eds.): ICONIP 2019, CCIS 1142, pp. 466–477, 2019.
https://doi.org/10.1007/978-3-030-36808-1_51

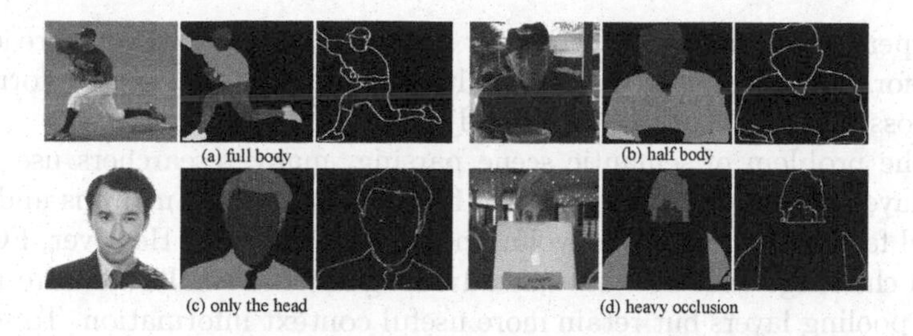

Fig. 1. Several parsing images and edges images from LIP dataset [12] in four scenarios.

Table 1. The percentages of edge pixels in parsing pixels and all pixels on training set.

	Dataset	1000	2000	5000	10000	20000	30000	All
In parsing pixels	LIP [12]	10.27	10.09	10.21	10.01	9.96	9.93	9.93
	CIHP [11]	12.83	12.89	12.83	12.77	12.84	–	12.75
In all pixels	LIP [12]	4.89	4.79	4.83	4.74	4.73	4.72	4.72
	CIHP [11]	5.38	5.35	5.36	5.35	5.17	–	5.07

parsing loss of previous layer output [11,20]. Fortunately, these studies manifest that other losses can encourage their models to generate more reliable results. However, previous approaches only simply add all losses up or multiply out, which is hard to express the influence of each loss to some extent (e.g., CE2P [20] adds up three losses with a ratio of 1:1:1). Multiple losses from different views are hard to make the model reach balanced condition by adjusting their ratios, and may impair the parsing performance. In addition, such strategy may weaken the potential of some losses applied to the loss function formula.

Different from previous works, one novelty of this work lies in we emphasize edge loss, which is an extremely essential constraint for multi-class human body parsing. Noticing the 9 joints defined in [12], we observe that edge information of each class is more critical than joints information on pixel level. In Fig. 1, we show several parsing images and the edge images in four kinds of scenes, (i.e., from full-body, half-body, only the head, to heavy occlusion). In the absence of entire body information (Fig. 1(d)), there may be only one joint (head), but for edge image there are still multiple keypoints. In Table 1, we compute the ratios of edge pixels in human body parsing pixels and all pixels on the training set of LIP dataset [12] and CIHP dataset [11], respectively. There are different ratios of edge pixels in each image, while they seem to fluctuate within a small range on each dataset. However, the 9 joints only occupy a small percentage of pixels in an image. Based on the observations above, we design an edge enhancement module to refine the edge information and optimize our network with edge loss. Especially, Ruan et al. [20] identified the importance of edge details and achieved outstanding results according to their experiments about CE2P. Different from the edge perceiving module of CE2P, our module is able to extract more local and global information by incorporating multiple scale features, and we formulate a simpler loss function that is easier to adjust parameter.

For the problem of semantic scene parsing, many researchers use multiple pooling layers to increase the receptive field size of output neurons and extract high-level features with fully convolutional network (FCN). However, FCN faces a serious challenge that how to take advantage of the smaller feature maps by multiple pooling layers but retain more useful context information. To solve this problem, various solutions are proposed to capture wealthy context information, such as Atrous Spatial Pyramid Pooling (ASPP) [3], pyramid pooling module (PSP) [27], and Densely connected Atrous Spatial Pyramid Pooling (DenseASPP) [26]. ASPP concatenates feature maps generated by atrous convolution with four different dilation rates. Pyramid pooling module incorporates suitable global features by a pyramid parsing module for more reliable scene parsing results. Yang et al. [26] considered that ASPP suffered from the limitation of receptive field in the high-resolution scene images.

Compared with scene parsing, human body parsing seems to be effortless because the human body is more structural than the scene. Actually otherwise, there are rich categories and complex conjunctions between categories in human parsing datasets. This motivates us to design edge enhancement module and human parsing module which both utilize the multiple scale features generated by densely connected atrous spatial pyramid pooling to solve multi-class human body parsing. To this end, we present a simple yet powerful Edge-Enhancement Network (EEN), which consists of three key components, i.e., backbone network, human parsing module and edge enhancement module. We evaluate our model on benchmark datasets and it outperforms state-of-the-art methods. To summarize, this paper makes three following contributions:

- We further tap the potential of edge information for complex multi-class human body parsing, where its particularity is discovered by us.
- We formulate a powerful architecture by embedding edge enhancement module and human parsing module together.
- The proposed EEN model surpasses previous approaches on two large benchmarks, LIP and CIHP.

2 Related Work

2.1 Human Parsing

Human body parsing is a nontrivial task that has been approached in varied schemes. Early researches [17,19,25] faced a particular challenge that there were no large datasets, until Gong et al. [11,12] filled this gap. Recently, many solutions have been developed from different views to explore this task based on abundant human parsing datasets. For instance, Luo et al. [22] proposed Macro-Micro Adversarial Network (MMAN) architecture, using two discriminators to enhance local and global body parsing and avoid the poor convergence problems of adversarial networks when dealing with high resolution images. To incorporate the context and detailed information into human parsing, Ruan et al. [20] designed a CE2P framework which leveraged three key properties, including feature resolution, global context information and edge details. With the motivation of transfer learning, the presented Graphonomy [10] was a novel model that can spread graph representation among the labels within one dataset and transfer semantic information across multiple datasets.

2.2 Edge Detection

Edge detection, which is a fundamental and critical task, has far-reaching applications in different domains. During the previous years, there were many early works [2,8,23] to encourage the development of edge detection, but these methods were less accurate and less suitable for diverse modern applications. Recently, most of researchers primarily focused on deep learning, greatly improving edge detection performance. For example, Hou et al. [13] explored a universal architecture for three tasks, i.e., salient object segmentation, edge detection and skeleton extraction, which indicates some similar tasks can use a model to perform well. SE2Net [28] incorporated the edge detection and object detection tasks into a siamese network by parallelly estimating the salient maps of edges and regions. Besides, many research efforts [14,21] were devoted into the edge detection. Remarkably, most of these works used ResNet-101 as the backbone network.

3 Our Approach

In this section, we first introduce the architecture of Edge-Enhancement Network (EEN), and then describe it in detail. Figure 2 depicts the overall framework of EEN. Being constructed based on the FCN architecture, EEN contains three main components to learn for single human parsing in an end-to-end manner, including backbone network, human parsing module and edge enhancement module. Firstly, backbone network produces common features for human parsing module and edge enhancement module. Secondly, since the edge pixels make up about 10% of the parsing pixels which motivates us to explore the correlation of human parsing task and edge detection task, we design an edge enhancement module to generate edge score maps and transmit edge feature maps to human parsing module. Thirdly, the human parsing module is used for incorporating useful edge information from edge enhancement module to improve the final outputs. In both the training and testing phases, the input of EEN is RGB human images. The final outputs are edge score maps and human body parsing maps. During training, we use the edge labels for edge enhancement module and the parsing labels for human parsing module.

Backbone Network: ResNet-101 is used as the feature extraction backbone of our network because of its high computational effectiveness. As mentioned in Sect. 2, many researches have demonstrated the robustness of ResNet-101 for human parsing task and edge detection task. To evaluate the effectiveness of

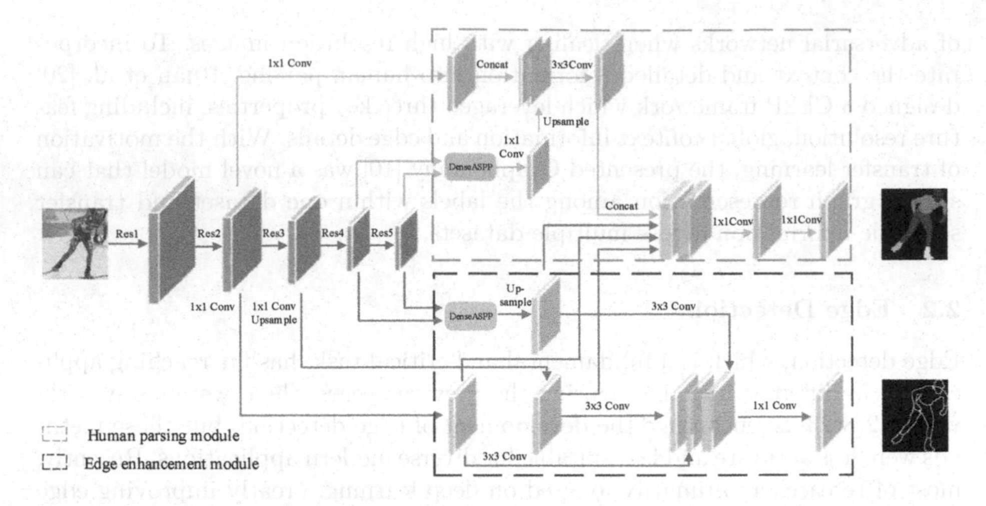

Fig. 2. Illustration of Edge-Enhancement Network (EEN).

human parsing module and edge enhancement module and obtain better representation, we employ ResNet-101 as our backbone to further investigate the activation of the final four feature maps.

Human Parsing Module: There are some techniques for scene parsing to improve the performance, such as inputting the multi-scale images and using dilated convolution layers. Nevertheless, to avoid consuming enormous memory because of multi-scale inputs, we apply DenseASPP [26] to assemble global rich context information to deal with the complex multi-class human body images. DenseASPP can be described as a variant of ASPP, which connects a set of atrous convolutional layers in a dense way. The employed DenseASPP consists of five 1×1 convolutions and five 3×3 astrous convolutions with dilation rates of 3, 6, 12, 18 and 24. The outputs of each convolutional layer are concatenated with input feature maps. Then, the outputs of each convolutional layer are fed into the next dilated layer. Compared with the original DenseASPP module, the difference in our work is that each feature map undergoes In-Place Activated BatchNorm (InPlace-ABN) [1] after convolution in each block with different rates, whose goal is to adapt to the small batch size. Moreover, motivated by the variant "U-Net" architecture [15], we add skip connections between the early layer and the latter layer to retain applicable features. In the final classification layers for human parsing, we just apply two 1×1 convolutional layers for K channels parsing outputs to obtain parsing maps, where K is the numbers of categories.

Edge Enhancement Module: This module targets on benefitting the human parsing module with edge loss and strengthened features. To generate reliable edge detection maps, edge enhancement module also embeds DenseASPP module in the same way as human parsing module. We extract spatial features from the early blocks to retain more fine-grained spatial details by 1×1 convolutions. On the one hand, we feed the upsampled feature maps from DenseASPP module and the early blocks into 3×3 convolutions. On the other hand, these maps are concatenated with the output feature maps of decoder operation as the input of next layers in human parsing module. Finally, these concatenated features after 3×3 convolutions are further reduced by 1×1 convolution.

In summary, the outputs of EEN consist of edge prediction maps and human parsing maps. Hence, the whole training objective of EEN can be simply formulated as:

$$\mathcal{L} = \mathcal{L}_{parsing} + \lambda \mathcal{L}_{edge}, \quad (1)$$

where $\mathcal{L}_{parsing}$ denotes the weighted cross entropy loss function between the human parsing map and the human parsing label map, and $\mathcal{L}_{edge}$ denotes the weighted cross entropy loss function between the edge map and the edge label map. In addition, λ is used to tradeoff the importance between these two losses.

Due to the limitation of memory on GPUs, our network is trained with a small batch size. To reduce its impact, InPlace-ABN [1] is applied following each convolution layer except for the final classification layers.

4 Experiments

4.1 Datasets and Evaluation Metric

Three benchmark datasets are used in our experiments.

- **LIP** dataset [12] is the largest benchmark for human parsing, which contains images in special scenarios, such as heavy occlusions, multiple person and scene complexity. LIP contains 50462 images in total, including 30462 images for training. 19 human part labels and a background class are defined in LIP.
- **CIHP** dataset [11] is collected from the real-world scenarios for single human parsing and instance-level human parsing with multiple persons. CIHP contains 38280 images, including 28280 images for training. It also contains 20 classes like LIP.
- **Pascal-Person-Part** dataset [7] is a subset of Pascal-VOC 2010. In this dataset, 1716 images are annotated for training and 1817 images are annotated for testing. There are 6 human part labels and a background class.

The mean Intersection over Union (mIoU) criterion is commonly applied to measure the accuracy of human parsing models. mIoU is computed by averaging the IoU values across all classes corresponding to the classification of each benchmark. We adopt IoU for each class and mIoU for each dataset to appraise the competence of human parsing model.

4.2 Implementation Details

Our method is implemented by extending the PyTorch framework. We use pretrained resenet-101 on Imagenet as the backbone network. All models are trained on two NVIDIA TITAN RTX 2080 Ti GPUs.

Table 2. Performance comparison on the validation set of LIP.

Method	Overall accuracy	Mean accuracy	mean IoU
Attention [5]	83.43	54.39	42.92
SS-JPPNet [12]	84.36	54.94	44.73
MMAN [22]	–	–	46.81
JPPNet [16]	86.39	62.32	51.37
CE2P [20]	87.37	63.20	53.10
EEN	**88.16**	**67.86**	**56.55**

Due to some diversities in these datasets, we aptly use different input size for each dataset. For LIP, the input size of image is 473 × 473 during training and testing. For CIHP and Pascal-Person-Part, the input size of image is 512 × 512. For data augmentation, we apply the random scaling (from 0.5 to 1.5), cropping and left-right flipping for all datasets. In our method, the edge labels are generated automatically from the annotated parsing images by computing the correlation of adjacent pixels.

We apply a "poly" learning rate policy following [4]. The initial learning rate is 0.007 during all training processes. During the experiments, we adjust the edge loss weight λ to adapt to different datasets. We set $\lambda = 2$ for LIP dataset, $\lambda = 3$ for CIHP dataset, and $\lambda = 0.1$ for Pascal-Person-Part dataset. The batch size in all models is 8 and the momentum is 0.9. For fair comparisons, the models are trained with 150 epochs on LIP dataset, 80 epochs on CIHP dataset, and 300 epochs on Pascal-Person-Part dataset. Source code is available at https://github.com/huangxi6/EEN.

4.3 Comparison with Other Methods

In this section, we evaluate the performance of our method EEN, and compare it with the state-of-the-art methods on three datasets. On LIP dataset, our model yields an mIoU of 56.55%, and the details are reported in Tables 2 and 3. EEN outperforms these five state-of-the-art methods. Especially, EEN achieves 3.45% improvement in terms of mIoU with the best competitor CE2P and obviously outperform SS-JPPNet [12] using joints information. Thanks to the extracted local and global information by densely connected atrous spatial pyramid pooling, our model with edge enhancement module can be more effective to analyze multi-class human body, particularly in the case that some part classes just have lower ratio in the entire human body, such as right shoes (see r-sh class in Table 3).

To further evaluate the performance of our model in multi-class human parsing task, we conduct experiments on CIHP dataset. The results is shown in Table 4. As Table 4 shows, the previous work [10] achieved a most capable performance with 58.58% mIoU on CIHP dataset. While our EEN surpasses their

Table 3. Method comparison of per-class IoU and mIoU on LIP validation set. "CE2P-o" means the original CE2P without edge perceiving module. "Baseline" means baseline backbone network. "H" and "E" denote human parsing module and edge enhancement module, respectively.

Method	hat	hair	glov	glas	u-cl	dress	coat	sock	pant	suit	scarf	skirt	face	l-ar	r-ar	l-leg	r-leg	l-sh	r-sh	bkg	mIoU
Attention [5]	58.57	66.78	23.32	19.48	63.20	29.63	49.70	35.23	66.04	24.73	12.84	20.41	70.58	50.17	54.03	38.35	37.70	26.20	27.09	84.00	42.92
SS-JPP [12]	59.75	67.25	28.95	21.57	65.30	29.49	51.92	38.52	68.02	24.48	14.92	24.32	71.01	52.64	55.79	40.23	38.80	28.08	29.03	84.56	44.73
MMAN [22]	57.66	65.63	30.07	20.02	64.15	28.39	51.98	41.46	71.03	23.61	9.65	23.20	69.54	55.30	58.13	51.90	52.17	38.58	39.05	84.75	46.81
JPPNet [16]	63.55	70.20	36.16	23.48	68.15	31.42	55.65	44.56	72.19	28.39	18.76	25.14	73.36	61.97	63.88	58.21	57.99	44.02	44.09	86.26	51.37
CE2P-o [20]	87.22	65.34	72.13	36.18	31.97	68.86	31.02	55.81	47.35	73.23	26.91	12.28	20.58	74.49	62.95	65.18	56.31	55.59	43.49	43.80	51.54
CE2P [20]	65.29	72.54	39.09	32.73	69.46	32.52	56.28	49.67	74.11	27.23	14.19	22.51	75.50	65.14	66.59	60.10	58.59	46.63	46.12	87.67	53.10
Baseline	63.24	70.14	33.75	29.45	66.15	24.10	51.79	45.70	70.75	21.52	13.56	20.36	73.04	58.24	60.88	49.70	48.06	36.50	36.40	86.08	47.97
EEN (H)	65.92	72.09	44.61	32.54	69.68	35.12	56.45	49.33	75.06	30.42	16.83	28.07	75.39	66.43	68.91	61.69	61.20	47.52	48.42	87.89	54.68
EEN (H+E)	68.41	73.13	44.80	35.47	70.49	38.87	57.22	50.67	75.90	32.12	18.54	28.49	76.23	68.30	70.15	65.35	64.45	51.60	52.36	88.47	56.55

Table 4. Performance comparison on the validation set of CIHP.

Method	Mean acc.	mIoU
PGN [11]	64.22	55.80
DeepLab v3+ [6]	65.06	57.13
Graphonomy [10]	66.65	58.58
EEN (H)	71.58	60.81
EEN (H+E)	**73.19**	**62.60**

Table 5. Performance comparison in terms of per-class IoU with six state-of-the-art methods on the PASCAL-Person-Part test set.

Method	head	torso	u-arms	l-arms	u-legs	l-legs	bkg	mIoU
SS-JPPNet [12]	83.26	62.40	47.80	45.58	42.32	39.48	94.68	59.36
MMAN [22]	82.58	62.83	48.49	47.37	42.80	40.40	94.92	59.91
Fang et al. [9]	87.15	72.28	57.07	56.21	52.43	50.36	97.72	67.60
PGN [11]	90.89	75.12	55.83	64.61	55.42	41.57	95.33	68.40
Refinenet [18]	–	–	–	–	–	–	–	68.60
Graphonomy [10]	–	–	–	–	–	–	–	71.14
EEN	86.96	70.95	61.08	60.94	53.18	50.63	95.80	68.51

result by improving the result up to 62.60%, and the improvement over Graphonomy [10] is +4.02%. In view of this experimental result, we suggest that edge information is more critical in human images containing multiple instances.

To observe whether our model is suitable for human body parsing with smaller number of classes, we conduct experiments on Pascal-Person-Part, which contains only 7 classes. The comparison results are shown in Table 5. As Table 5 shows, 68.51% mIoU of EEN is not the best performer. However, it still performs the competency with current state-of-the-art approaches. In particular, our model shows the superiority in u-arms class and background class.

4.4 Ablation Study

This section presents ablation studies of our approach. In particular, we evaluate the effect of edge enhancement module.

In Tables 3 and 4, we report the results when removing edge enhancement module on LIP dataset and CIHP dataset, respectively. On LIP dataset, when removing the edge enhancement module from the full architecture, mIoU will drop by 1.87% compared with the full EEN. According to CE2P-o [20] without edge perceiving module in Table 3, we can observe that our edge enhancement module is better than their module used for extracting edge information. Meanwhile, EEN (H) shows the effectiveness of human parsing module according to 54.68% mIoU compared with the baseline. On CIHP dataset, we only get the mIoU of 60.81% when removing edge enhancement module, leading to a decrease of 1.79%. These results indicate that our proposed network incorporating the edge enhancement module can be generalized and work well in the case of multi-class human body parsing.

In Fig. 3, we provide some parsing examples obtained by EEN. We evaluate the effectiveness of edge enhancement module on LIP and CIHP datasets by showing the ground-truth and the predictions generated by EEN without or with the edge enhancement module.

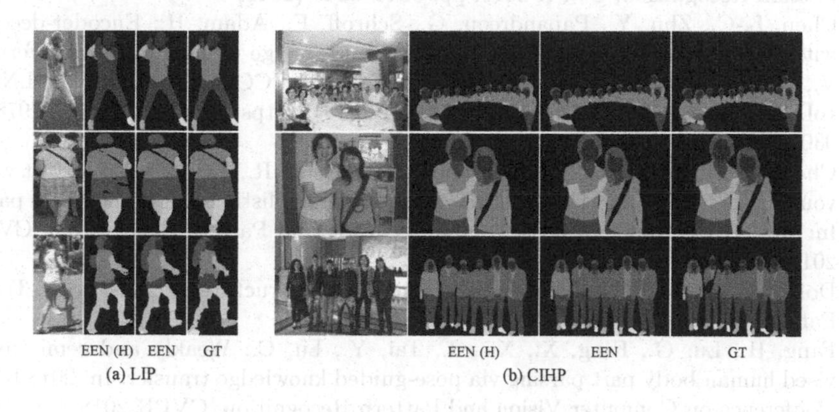

Fig. 3. Visualized results on LIP and CIHP datasets.

5 Conclusion

In this paper, we explore the potential of edge information to solve the problem of multi-class human body parsing. Based on the similarity between scene parsing and human parsing tasks, we propose to use densely connected atrous spatial pyramid pooling to capture wealthy context information. More importantly, we present a simple yet effective EEN including edge enhancement module to solve multi-class complex human body parsing problem. Extensive experimental results demonstrate its superiority on LIP and CIHP datasets, and the results on PASCAL-Person-Part dataset shows its universality.

Acknowledgments. This work is supported by National Natural Science Foundation of China (grants No. 61672133 and No. 61832001).

References

1. Bulò, S.R., Porzi, L., Kontschieder, P.: In-place activated BatchNorm for memory-optimized training of DNNs. In: 2018 IEEE Conference on Computer Vision and Pattern Recognition, CVPR 2018, pp. 5639–5647 (2018)

2. Canny, J.F.: A computational approach to edge detection. IEEE Trans. Pattern Anal. Mach. Intell. **8**(6), 679–698 (1986)
3. Chen, L., Papandreou, G., Kokkinos, I., Murphy, K., Yuille, A.L.: DeepLab: semantic image segmentation with deep convolutional nets, atrous convolution, and fully connected CRFs. IEEE Trans. Pattern Anal. Mach. Intell. **40**(4), 834–848 (2018)
4. Chen, L., Papandreou, G., Schroff, F., Adam, H.: Rethinking Atrous convolution for semantic image segmentation. CoRR abs/1706.05587 (2017)
5. Chen, L., Yang, Y., Wang, J., Xu, W., Yuille, A.L.: Attention to scale: scale-aware semantic image segmentation. In: 2016 IEEE Conference on Computer Vision and Pattern Recognition, CVPR 2016, pp. 3640–3649 (2016)
6. Chen, L.-C., Zhu, Y., Papandreou, G., Schroff, F., Adam, H.: Encoder-decoder with Atrous separable convolution for semantic image segmentation. In: Ferrari, V., Hebert, M., Sminchisescu, C., Weiss, Y. (eds.) ECCV 2018, Part VII. LNCS, vol. 11211, pp. 833–851. Springer, Cham (2018). https://doi.org/10.1007/978-3-030-01234-2_49
7. Chen, X., Mottaghi, R., Liu, X., Fidler, S., Urtasun, R., Yuille, A.L.: Detect what you can: detecting and representing objects using holistic models and body parts. In: 2014 IEEE Conference on Computer Vision and Pattern Recognition, CVPR 2014, pp. 1979–1986 (2014)
8. Dollár, P., Zitnick, C.L.: Fast edge detection using structured forests. IEEE Trans. Pattern Anal. Mach. Intell. **37**(8), 1558–1570 (2015)
9. Fang, H., Lu, G., Fang, X., Xie, J., Tai, Y., Lu, C.: Weakly and semi supervised human body part parsing via pose-guided knowledge transfer. In: 2018 IEEE Conference on Computer Vision and Pattern Recognition, CVPR 2018, pp. 70–78 (2018)
10. Gong, K., Gao, Y., Liang, X., Shen, X., Wang, M., Lin, L.: Graphonomy: universal human parsing via graph transfer learning. CoRR abs/1904.04536 (2019)
11. Gong, K., Liang, X., Li, Y., Chen, Y., Yang, M., Lin, L.: Instance-level human parsing via part grouping network. In: Ferrari, V., Hebert, M., Sminchisescu, C., Weiss, Y. (eds.) ECCV 2018, Part IV. LNCS, vol. 11208, pp. 805–822. Springer, Cham (2018). https://doi.org/10.1007/978-3-030-01225-0_47
12. Gong, K., Liang, X., Zhang, D., Shen, X., Lin, L.: Look into person: self-supervised structure-sensitive learning and a new benchmark for human parsing. In: 2017 IEEE Conference on Computer Vision and Pattern Recognition, CVPR 2017, pp. 6757–6765 (2017)
13. Hou, Q., Liu, J., Cheng, M., Borji, A., Torr, P.H.S.: Three birds one stone: a unified framework for salient object segmentation, edge detection and skeleton extraction. CoRR abs/1803.09860 (2018)
14. Hu, Y., Chen, Y., Li, X., Feng, J.: Dynamic feature fusion for semantic edge detection. CoRR abs/1902.09104 (2019)
15. Isola, P., Zhu, J., Zhou, T., Efros, A.A.: Image-to-image translation with conditional adversarial networks. In: 2017 IEEE Conference on Computer Vision and Pattern Recognition, CVPR 2017, pp. 5967–5976 (2017)
16. Liang, X., Gong, K., Shen, X., Lin, L.: Look into person: joint body parsing & pose estimation network and a new benchmark. IEEE Trans. Pattern Anal. Mach. Intell. **41**(4), 871–885 (2019)
17. Liang, X., et al.: Human parsing with contextualized convolutional neural network. In: 2015 IEEE International Conference on Computer Vision, ICCV 2015, pp. 1386–1394 (2015)

18. Lin, G., Milan, A., Shen, C., Reid, I.D.: RefineNet: multi-path refinement networks for high-resolution semantic segmentation. In: 2017 IEEE Conference on Computer Vision and Pattern Recognition, CVPR 2017, pp. 5168–5177 (2017)
19. Liu, S., et al.: Matching-CNN meets KNN: quasi-parametric human parsing. In: IEEE Conference on Computer Vision and Pattern Recognition, CVPR 2015, pp. 1419–1427 (2015)
20. Liu, T., et al.: Devil in the details: towards accurate single and multiple human parsing. CoRR abs/1809.05996 (2018)
21. Lu, R., Zhou, M., Ming, A., Zhou, Y.: Context-constrained accurate contour extraction for occlusion edge detection. CoRR abs/1903.08890 (2019)
22. Luo, Y., Zheng, Z., Zheng, L., Guan, T., Yu, J., Yang, Y.: Macro-micro adversarial network for human parsing. In: Ferrari, V., Hebert, M., Sminchisescu, C., Weiss, Y. (eds.) ECCV 2018, Part IX. LNCS, vol. 11213, pp. 424–440. Springer, Cham (2018). https://doi.org/10.1007/978-3-030-01240-3_26
23. Martin, D.R., Fowlkes, C.C., Malik, J.: Learning to detect natural image boundaries using local brightness, color, and texture cues. IEEE Trans. Pattern Anal. Mach. Intell. **26**(5), 530–549 (2004)
24. Nie, X., Feng, J., Zuo, Y., Yan, S.: Human pose estimation with parsing induced learner. In: 2018 IEEE Conference on Computer Vision and Pattern Recognition, CVPR 2018, pp. 2100–2108 (2018)
25. Simo-Serra, E., Fidler, S., Moreno-Noguer, F., Urtasun, R.: A high performance CRF model for clothes parsing. In: Cremers, D., Reid, I., Saito, H., Yang, M.-H. (eds.) ACCV 2014, Part III. LNCS, vol. 9005, pp. 64–81. Springer, Cham (2015). https://doi.org/10.1007/978-3-319-16811-1_5
26. Yang, M., Yu, K., Zhang, C., Li, Z., Yang, K.: DenseASPP for semantic segmentation in street scenes. In: 2018 IEEE Conference on Computer Vision and Pattern Recognition, CVPR 2018, pp. 3684–3692 (2018)
27. Zhao, H., Shi, J., Qi, X., Wang, X., Jia, J.: Pyramid scene parsing network. In: 2017 IEEE Conference on Computer Vision and Pattern Recognition, CVPR 2017, pp. 6230–6239 (2017)
28. Zhou, S., Wang, J., Wang, F., Huang, D.: SE2Net: Siamese edge-enhancement network for salient object detection. CoRR abs/1904.00048 (2019)
29. Zhu, S., Fidler, S., Urtasun, R., Lin, D., Loy, C.C.: Be your own Prada: fashion synthesis with structural coherence. In: IEEE International Conference on Computer Vision, ICCV 2017, pp. 1689–1697 (2017)

vUBM: A Variational Universal Background Model for EEG-Based Person Authentication

Huyen Tran[1], Dat Tran[1(✉)], Wanli Ma[1], and Phuoc Nguyen[2]

[1] Faculty of Science and Technology, University of Canberra, Canberra, Australia
{huyen.tran,dat.tran,wanli.ma}@canberra.edu.au
[2] A2I2, Deakin University, Geelong, Australia
phuoc.nguyen@deakin.edu.au

Abstract. EEG-based person authentication is an important means for modern biometrics. However EEG signals are well-known for small signal-to-noise ratio and have many factors of variation. These variations are caused by intrinsic factors, e.g. mental activity, mood, and health conditions, as well as extrinsic factors, e.g. sensor errors, electrode displacements, and user movements. These create complex variations of source signals going from inside our brain to the recording devices. We propose vUBM, a variational inference framework to learn a simple latent representation for complex data, facilitating authentication algorithms in the latent space. A variational universal background model is created for normalizing scores to further improve the performance. Extensive experiments show the advantages of our proposed framework.

1 Introduction

EEG signals are live electrical brainwave signals emitted at the scalp. They are known for containing a rich amount of information about brain activity. For instance, sleep EEGs contain slow waves and awake EEGs show much higher frequencies. Different mental activities and moods also trigger different neuronal patterns. EEGs are also known for carrying physiology characteristics of individuals. The neuronal excitation waves, propagating through our brain, penetrating the skull, then reaching the scalp, are thought to carry the signature of carrying mediums. Because of the skull thickness, however, the EEG signals are weak and noisy aggregations of the sources. On the other hand, there are also noises due to recording protocols, such as sensor imperfection, variations in the device placements, as well as other session variability factors. All these added variations of a source signal inside our brain introduce much difficulty in modelling EEG signals. Therefore modelling methods should have the capability to account for these variation.

Recent literature uses support vector machine (SVM) methods [8,9] due to its generalization property and ability to handle high dimensional input. The combination of SVM and UBM was introduced in [8]. However, SVM does not

T. Gedeon et al. (Eds.): ICONIP 2019, CCIS 1142, pp. 478–485, 2019.
https://doi.org/10.1007/978-3-030-36808-1_52

scale well to big data, due to the quadratic or cubic complexity in the data size with nonlinear kernel. As a result, some recent deep learning methods have been introduced to modelling EEG signals [2,6,11]. Some recent works also used generative models for augmenting EEG datasets [1] or for classification [4]. These methods scale linearly to the size of dataset, therefore are promising to modelling an increasing amount of EEG data, thanks to the popularity of consumer EEG headsets. These frameworks model directly the feature vectors as opposed to ours, which used a latent variational representation for modelling the complex variation in the observed data, thanks to the powerful function approximation ability by neural networks. Variational modelling in EEG signals is new and open.

We focused on EEG-based person authentification [3] due to its attractive biologically live characteristic. We used imagined speech EEG signals for authentication. Compared to other methods, this seems more natural, convenient, and familiar to users as they would imagine speaking their password instead of typing it. However, unlike passwords which are fixed texts, there are a lot of variations involving the imagined speaking tasks, e.g. speed, intensity, mood, focus, and possible rhythms. These hidden factors are not easy to be captured. We introduced variational latent variable methods for EEG-based person authentification. Particularly a low dimensional latent representation to account for the possible variations in the observed user data. This variational latent representation opens up opportunities to model diverse variabilities inherent in the EEG signals, allowing models to capture hidden factors of variation. We based on the variational autoencoder framework [5] to develop a Variational Universal Background Model, which we termed vUBM. Universal background model provides the basic for the statistical hypothesis testing framework for speaker verification [10].

We use a latent Gaussian Mixture (GM) to model each user latent vector and pool all the user GMs together to create a universal background model for score nomalization purposes. With the introduction of the latent GM, however, the minimization objective becomes difficult. Therefore, we derived an approximation objective which is an upper bound to the target objective. We carried extensive experiments on two datasets: (1) the MNIST dataset for testing our model; and (2) the imagined speech dataset which is the main application of our method.

2 Methods

2.1 Variational Auto Encoder (VAE)

VAE consists of two processes, the inference process maps a input data x to a latent distribution z, and the reconstruction process that maps the laten z to x. Suppose the distribution of x is Bernoulli, and the prior z can be assumed a simple normal distribution $\mathcal{N}(0, I)$. We will find out which latent z is mapped to a given data point x. By using Bayes rule, we can compute the posterior of z given x, $p(z|x;\theta)$ as follow:

$$p(z|x) = \frac{p(x|z;\theta)p(z)}{p(x)} \tag{1}$$

However the normalization constant $p(x) = \int p(x|z;\theta)p(z)dz$ is intractable to compute as it requires summing over all z. Therefore, we use an alternative tractable distribution $q(z|x;\phi)$, such as $\mathcal{N}(\mu, \sigma^2 I)$ for approximating the posterior.

The training objective is to minimize the Kullback-Leibler distance between the true posterior $p(z|x)$ and the variational posterior $q(z|x)$.

$$\min_{(\theta,\phi)} D_{KL}\left(q(z|x;\phi)\|p(z|x;\theta)\right) \tag{2}$$

where we have added the parameters (θ, ϕ) to show the dependence of the distributions on these parameters. The parameters (θ, ϕ) belongs to the decoder and the encoder networks and $p(z|x;\theta)$ is calculated as in Eq. 1. However in practice $p(z|x;\theta)$ is difficult to evaluate (due to the normalizing constant $p(x)$), therefore we use an unnormalized version $p(x, z;\theta) = p(z|x;\theta)p(x)$ instead. The objective becomes:

$$\min_{(\theta,\phi)} D_{KL}\left(q(z|x;\phi)\|p(x,z;\theta)\right)$$

or equivalently as

$$\max_{(\theta,\phi)} -D_{KL}\left(q(z|x;\phi)\|p(x,z;\theta)\right) \tag{3}$$

This objective is called ELBO since it can be shown equivalent to the Evidence ($\log p(x)$) Lower BOund. It can be written as follows for the optimization objective:

$$\begin{aligned}
\mathrm{ELBO}(\theta,\phi,x) &= -\, D_{KL}\left(q(z|x;\phi)\|p(x,z;\theta)\right) \\
&= -\, \mathbb{E}_{q(z|x;\phi)} \log \frac{q(z|x;\phi)}{p(x,z;\theta)} \qquad (4)\\
&= -\, \mathbb{E}_{q(z|x;\phi)} \log \frac{q(z|x;\phi)}{p(z)p(x|z;\theta)} \\
&= -\, D_{KL}(q(z|x;\phi)\|p(z)) + \mathbb{E}_{q(z|x;\phi)} p(x|z;\theta) \qquad (5)
\end{aligned}$$

This framework is called variational auto encoder (VAE) [5]. We use the objective 5 together with the reparameterization trick [5] for training this VAE.

2.2 Mixture of Gaussians Prior for User Modelling

Because of the variation and complexity of user data, we use a Variational Gaussian Mixture (VGM) prior for modelling each user in the z space, instead of using a single distribution. A Gaussian mixture prior for each user u is defined as follows, it is initialized randomly and learnt together with the VAE:

$$p_u(z;\gamma_u) = \sum_{k=1}^{K} \alpha_k \mathcal{N}(z|\mu_k, \sigma_k^2 I) \tag{6}$$

where K is the number of components, $\gamma_u = (\alpha_k, \mu_k, \sigma_k^2)$ is the parameter for the mixing weight, mean and variance of each component Gaussian for the user u.

For simplicity, we use a single Gaussian for modelling the posterior. Training would update these parameters for each user separately. The training objective for each user u is similar to Eq. 5 but with the additional parameter γ_u since the prior is no longer $\mathcal{N}(0, I)$:

$$\text{ELBO}_u(\theta, \phi, \gamma_u, x) = -D_{KL}(q(z|x;\phi)\|p(z;\gamma_u)) + \mathbb{E}_{q(z|x;\phi)} p(x|z;\theta) \tag{7}$$

However the KL distance between the prior mixture distribution $p_u(z;\gamma_u)$ and the posterior single Gaussian distribution need an approximation, which is presented in the following proposition.

Proposition 1. *The variational upperbound of the Kullback-Leibler divergence between a unimodal distribution $f(x)$ and a mixture model $g(x) = \sum_k \alpha_k g_k(x)$ is:*

$$D_{KL}(f(x)\|g(x)) \leq D_{var}(f(x)\|g(x)) \stackrel{\text{def}}{=} -\log \sum_k \alpha_k \exp\left(-D_{KL}(f(x)\|g_k(x))\right). \square \tag{8}$$

2.3 Universal Background Model by Pooling

We approximate the background model of the null hypothesis by a large mixture model constructed by pooling every user's mixture models, except for the target user's model, and renormalizing the component weights, $p(z_x|\theta_{\text{bg}}) = \frac{1}{B}\sum_{b=1}^{B} p(z_x|\theta_b)$. We use the following three versions of the score:

$$\text{score1}(x, u) = \log p(z_x|\theta_u) \tag{9}$$

$$\text{score2}(x, u) = \log p(z_x|\theta_u) - \log \frac{1}{B}\sum_{b=1}^{B} p(z_x|\theta_b) \tag{10}$$

$$\text{score3}(x, u) = \log p(z_x|\theta_u) - \log \max_b p(z_x|\theta_b) \tag{11}$$

where Eq. 9 refers to no background nomalization, while Eqs. 10 and 11 refer to background nomalization applied, B is the number of background models.

3 Experiments

3.1 Datasets

We use two datasets: (1) the popular MNIST dataset for testing our proposed methods; and (2) an EEG imagined speech dataset to demonstrate the main application of our methods.

The MNIST dataset has 70000 handwritten digits of ten classes ranging from 0 to 9. There are 60000 digits for training and the remaining 10000 for testing.

The imagined speech dataset [7] includes EEG signals of 15 healthy subjects imagining to pronounce different sounds. There are four tasks of silent speech corresponding to: (1) pronouncing vowels /a/, /i/ and /u/; (2) short words 'in', 'out' and 'up'; (3) long words 'cooperate' and 'independent'; and (4) short versus long words 'in' and 'cooporate'. There are 100 independent trials for each of the sound, and each lasts for 5 s. Since there is no separate training and test sets for the imagined speech dataset, we randomly split 4/5 trials for training and 1/5 trials for testing. The data were preprocessed to remove artifacts and noises and downsampled to 128 Hz. We extracted common EEG frequency bands as follows. First, the power spectral densities from all 64 channels are computed. Next, the powers of five EEG frequency bands delta, theta, alpha, beta, and gamma are calculated. This makes a feature vector of size $5 \times 64 = 320$ features for each trial. The features are scaled into range $[0, 1]$.

3.2 Models

We compare two baseline methods Gaussian and Gaussian mixture model (GMM) to our proposed methods VAE, VGM with scoring Eq. 9, and their UBM variants with scoring Eq. 10 or 11. Each model was trained on the training data of each individual. Then background model for the null hypothesis is constructed by pooling the models of all other users. The background model score is computed as either the average score over all individual models, Eq. 10, or the maximum, Eq. 11, whichever is better.

Model Parameters. We use a similar setup for both datasets. The Gaussian and GMM models have the mean and covariance parameters for each Gaussian, while the VAE and VGM models have their latent means and covariances parametrized by neural networks. Specifically, the encoder and decoder are multilayer perceptrons with 3 hidden layers, each of size 500. The output layer of the encoder parameterizes the mean and variance of each Gaussian component, while the output layer of the decoder parameterizes the mean of the Bernoulli distribution. Neural network training is done by stochastic gradient descent with batchsize 20, optimizer Adam with learning rate 0.001, and the training is run until convergence. Different latent sizes and number of mixture components were compared in the next section.

3.3 Results

Performance Measures. We use Area Under Curve (AUC) and Equal Error Rate (EER) for performance measures. The following results were reported for the MNIST dataset and imagined speech dataset on the test sets.

MNIST Dataset. Table 1 shows the average authentication performance of different methods on the MNIST dataset. Gaussian method has lowest AUC score and highest EER, GMM method performs better. The variational methods VAE and VGM have better score. With the introduction of UBM model for score normalization all methods improve, by 1.4 to 4.7% on average, with highest AUC score at 98.6% by VGM method. It shows that the variational latent variable model with Gaussian mixture greatly improve the performance. This suggests that the variational latent variable helps with modelling the data variation well.

Table 1. MNIST dataset test AUC and EER of different methods.

	Without UBM	EER	With UBM	EER
Gaussian	61.9 ± 7.4	38.1 ± 7.4	62.6 ± 7.8	37.5 ± 7.7
GMM	66.8 ± 10.1	33.3 ± 10.1	71.5 ± 13.6	28.9 ± 13.6
VAE	76.9 ± 7.7	28.7 ± 6.3	80.1 ± 8.4	26.7 ± 6.8
VGM	97.2 ± 2.1	7.9 ± 4.0	98.6 ± 1.3	5.3 ± 2.9

Table 2 compares the effects of different number of Z dimensions on the MNIST dataset for VAE and VGM methods. It can be seen that increasing the number of latent dimensions help improve the performance. However for VGM model, 100 dimensions are marginally better than 40 latent dimensions, suggesting that 40 dimensions are enough due to the flexibility of the Gaussian mixture. Again, with the introduction of UBM model for score normalization both methods improve.

Table 2. MNIST dataset test AUC scores of VAE and VGM at different number of Z dimensions.

	Z dimensions	Without UBM	EER	With UBM	EER
VAE	20	75.3 ± 11.7	29.7 ± 8.9	77.9 ± 13.0	28.5 ± 9.9
	40	75.5 ± 10.4	24.9 ± 12.4	77.8 ± 11.8	28.2 ± 9.2
	100	76.9 ± 7.7	28.7 ± 6.3	80.1 ± 8.4	26.7 ± 6.8
VGM	20	91.1 ± 7.0	14.8 ± 7.9	93.1 ± 6.8	12.8 ± 7.6
	40	96.7 ± 2.0	8.5 ± 3.8	98.3 ± 1.1	5.8 ± 2.6
	100	97.2 ± 2.1	7.9 ± 4.0	98.6 ± 1.3	5.3 ± 2.9

After confirming the methods work well on the MNIST test dataset, we carried out authentication experiments on the imagined speech dataset, which is the main application of our models.

Imagined Speech Dataset. Table 3 shows the average authentication performance of all methods on the imagined speech dataset. The VAE and VGM models have similar performances and are better than Gaussian and GMM. It is interesting that with UBM score normalization, the improvement jumps up to more than 91% for all methods. The VGM has highest AUC score, at 95.3% on average.

Table 3. Imagined speech dataset test AUC and EER of different methods.

	Without UBM	EER	With UBM	EER
Gaussian	60.2 ± 22.4	36.0 ± 20.6	91.8 ± 18.3	7.9 ± 18.0
GMM	78.0 ± 26.0	19.7 ± 23.7	91.7 ± 18.2	7.9 ± 17.9
VAE	84.6 ± 29.6	10.6 ± 19.6	93.5 ± 13.8	5.5 ± 13.4
VGM	84.5 ± 29.8	8.9 ± 15.9	95.3 ± 14.3	5.7 ± 14.5

We compare different numbers of Z dimensions in Table 4. It can be seen that the best number of latent dimensions for VAE is 100 while it is 40 for VGM. This effect demonstrates that the latent mixture model is flexible enough and need only 40 latent dimensions to represent the variations in user's EEG signals. Using a higher number of dimensions would overfit and make the model perform worse at test time.

Table 4. Imagined speech dataset test AUC scores of VAE and VGM at different Z dimensions.

	Z dimension	Without UBM	EER	With UBM	EER
VAE	20	85.2 ± 25.3	11.6 ± 19.5	92.9 ± 14.5	6.0 ± 13.9
	40	84.1 ± 29.8	10.8 ± 19.6	91.9 ± 16.5	7.1 ± 16.0
	100	84.6 ± 29.6	10.6 ± 19.6	93.5 ± 13.8	5.5 ± 13.4
VGM	20	75.9 ± 33.4	13.3 ± 17.5	93.8 ± 15.3	6.5 ± 15.7
	40	84.5 ± 29.8	8.9 ± 15.9	95.3 ± 14.3	5.7 ± 14.5
	100	84.4 ± 27.7	9.0 ± 16.1	93.8 ± 15.1	6.3 ± 15.3

4 Conclusion

We have developed vUBM, a Variational Universal Background Model framework to model the complex data distribution by a latent variational mixture model for each user. A universal background model pooling all user's models was created for score nomalization in the hypothesis test. We derived a lower bound objective for the optimization problem due to the complexity involved when introducing of the latent GM. Our framework was trained end-to-end. We carried extensive experiments on two datasets, the MNIST and the imagined speech dataset. Experimental results showed that our methods have high performance and are applicable for the imagined speech authentication task. Future research direction would be applying our methods to model the diverse variations in other EEG datasets.

References

1. Abdelfattah, S.M., Abdelrahman, G.M., Wang, M.: Augmenting the size of EEG datasets using generative adversarial networks. In: 2018 International Joint Conference on Neural Networks (IJCNN), pp. 1–6. IEEE (2018)
2. Bashivan, P., Rish, I., Yeasin, M., Codella, N.: Learning representations from EEG with deep recurrent-convolutional neural networks. arXiv preprint: arXiv:1511.06448 (2015)
3. Brigham, K., Vijaya Kumar, B.V.K.: Subject identification from electroencephalogram (EEG) signals during imagined speech. In: 2010 Fourth IEEE International Conference on Biometrics: Theory Applications and Systems (BTAS), pp. 1–8. IEEE (2010)
4. Dai, M., Zheng, D., Na, R., Wang, S., Zhang, S.: EEG classification of motor imagery using a novel deep learning framework. Sensors **19**(3), 551 (2019)
5. Kingma, D.P., Welling, M.: Auto-encoding variational Bayes. arXiv preprint: arXiv:1312.6114 (2013)
6. Lawhern, V.J., Solon, A.J., Waytowich, N.R., Gordon, S.M., Hung, C.P., Lance, B.J.: EEGNet: a compact convolutional neural network for EEG-based brain-computer interfaces. J. Neural Eng. **15**(5), 056013 (2018)
7. Nguyen, C.H., Karavas, G.K., Artemiadis, P.: Inferring imagined speech using EEG signals: a new approach using Riemannian manifold features. J. Neural Eng. **15**(1), 016002 (2017)
8. Nguyen, P., Tran, D., Le, T., Huang, X., Ma, W.: EEG-based person verification using multi-sphere SVDD and UBM. In: Pei, J., Tseng, V.S., Cao, L., Motoda, H., Xu, G. (eds.) PAKDD 2013, Part I. LNCS (LNAI), vol. 7818, pp. 289–300. Springer, Heidelberg (2013). https://doi.org/10.1007/978-3-642-37453-1_24
9. Pham, T., Ma, W., Tran, D., Nguyen, P., Phung, D.: Multi-factor EEG-based user authentication. In: 2014 International Joint Conference on Neural Networks (IJCNN), pp. 4029–4034. IEEE (2014)
10. Reynolds, D.A.: Comparison of background normalization methods for text-independent speaker verification. In: Fifth European Conference on Speech Communication and Technology (1997)
11. Schirrmeister, R.T., et al.: Deep learning with convolutional neural networks for EEG decoding and visualization. Hum. Brain Mapp. **38**(11), 5391–5420 (2017)

Passenger Demographic Attributes Prediction for Human-Centered Public Transport

Can Li[1(✉)], Lei Bai[1], Wei Liu[1,2], Lina Yao[1], and S. Travis Waller[2]

[1] Computer Science and Engineering, University of New South Wales, Sydney, Australia
{can.li4,lei.bai}@student.unsw.edu.au, {wei.liu,lina.yao}@unsw.edu.au
[2] Civil and Environmental Engineering, University of New South Wales, Sydney, Australia
s.waller@unsw.edu.au

Abstract. This study examines the potential of the smart card data in public transit systems to infer passengers' demographic attributes, thereby enabling a human-centered public transport service design while reducing the use of expensive and time-consuming travel surveys. This is challenging since travel behaviors vary significantly over the population, space and time and developing meaningful links between them and passengers' demographic attributes are not trivial. To achieve this, we conduct an extensive analysis of spatio-temporal travel behavior patterns using smart card data from the Greater Sydney area, based on which we develop an end-to-end Hybrid Spatial-Temporal Neural Network. In particular, we first empirically analyze passenger movement and mobility travel patterns from both spatial and temporal perspectives and design a set of discriminative features to characterizing the patterns. We then propose a novel Product-based Spatial-Temporal module which encodes the relationships across a variety of features and harnesses them collectively under an Auto-Encoder Compression module, in order to predict passengers' demographic information. The experiments are conducted using a large-scale real-world public transportation dataset covering 171.77 million users. The experimental results demonstrate the effectiveness of the proposed method against a number of established tools in the literature.

Keywords: Passenger attribute classification · Public transport system · Deep neural networks

1 Introduction

Urban public transportation systems serve a large number of passengers on a daily basis and plays an important role in metropolitan areas. However, current public transportation systems' designs are often capacity-maximizing while individual preference is considered to a limited extent. In fact, different passenger

T. Gedeon et al. (Eds.): ICONIP 2019, CCIS 1142, pp. 486–494, 2019.
https://doi.org/10.1007/978-3-030-36808-1_53

groups normally have totally different requirements. For instance, elders may prefer cheaper but may be slightly slower public transport without noise and chaos. They may be much less demanding for the length of time. On the contrary, young commuters tend to choose fast and in straight lines since they have to save time on the road. There is a growing trend to allow a more human-centered public transport system, which better accommodates, e.g., different age groups and the disabled. A systemic planning of such a human-centered public transit system requires extensive inputs regarding passenger attributes. These inputs may be obtained through travel surveys, which can be expensive, time-consuming and biased. This study develops methods to infer passenger demographic attributes (e.g. age groups) for a human-centered public transit system without surveys.

Human-centered transport design is less emphasized in traditional public transport works, only a few studies are related to passenger attributes classification. Shiftan et al. [1] proposed to categorize travelers into demographic groups based on surveys, which is costly, time-consuming and highly-biased. Electronic smart card, as a widely used tool for accessing public transport services, provides ready-to-use passenger transit data and a potentially more efficient way to automatically classify passenger attributes by mining their travel patterns. Along this line, Mohamed et al. [2] proposed to cluster citizens into several groups with smart card data. They clustered passengers with similar boarding times as one class by constructing temporal passenger profiles based on the Expectation Maximization (EM) algorithm. Hagenauer et al. [3] empirically studied a range of machine learning methods for categorizing passengers' travel modes and analyze the most influential factors affecting people's travel choices. However, while most existing studies using smart card data analyze travel patterns of users, they often stopped at clustering users into groups based on similar patterns observed but did not further infer attributes of users. More critically, inferring passengers' demographic attributes has rarely been considered. The critical features of travel patterns associated with demographic attributes and the complex spatio-temporal inter-correlation among features have not be uncovered.

In this work, we propose to classify and infer the passenger demographic attributes in public transportation systems with the help of large-scale smart card usage data and land use data. Specially, we focus on identifying the passengers as three age groups, i.e., adults, seniors, and children, in this work. These three groups generally have different preferences or needs for a public transport system. We first briefly introduce the dataset used in this work. Then, we present the powerful features including both the spatial and temporal information together with a deep analysis of their relationships with the passenger age groups. Based on the extracted features and analysis, a new hybrid Spatial-Temporal correlation model based on deep neural networks is developed for passenger age groups classification by integrating different types of features and transit stops sparse matrix. Specifically, Product based Spatial-Temporal Module (PSTM) is developed to capture the pairwise latent relations among temporal and spatial features while Auto-Encoder-based Compression module (AECM) is utilized to learn the embedding vectors of transit stops matrix. Our main contributions are:

(1) To the best of our knowledge, this is the first work to classify passenger demographic attributes based on smart card data with deep neural networks. In this context, we propose a hybrid Spatial-Temporal correlation Neural Network to combine PSTM and AECM for classification.
(2) We uncover representative spatial and temporal passenger behavior patterns from the raw data and analyze their correlations with passenger age groups. This provides critical insights regarding mobility associated with age groups.
(3) We evaluate the developed method on a large-scale real-world dataset collected in the largest metropolitan area in Australia (Greater Sydney area) and demonstrate the effectiveness of the method against several baselines.

2 Data Description and Behavioural Features

2.1 Dataset Description

Smart Card Dataset is collected from Opal[1], the electronic smart card ticket system in Sydney covering main public transportation services (buses, trains, ferries, and light rails). The dataset is collected from 01/Apr/2017 to 30/Jun/2017 and records 171.77 million journey transactions covering 6.37 million users. The data does not involve personal information for protecting the privacy of users.

PoI Dataset is collected with the consideration that PoI information is close related to a region's function [4] and travel patterns do not only rely on the distances between two places [5]. Thus, we may infer the passengers' trip purposes to reflect passengers' attributes, through analyzing the PoI information of frequently visited places. In practice, we map PoI data of six categories (shopping mall, church, school, hospital, club, and gym) to related transit stops.

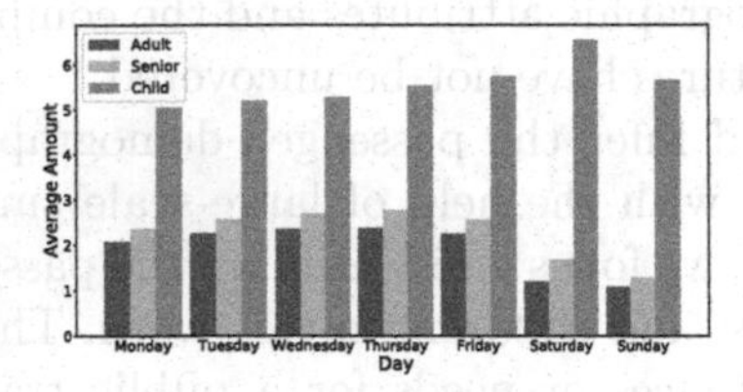

Fig. 1. Journey Transaction account in a week on different age groups

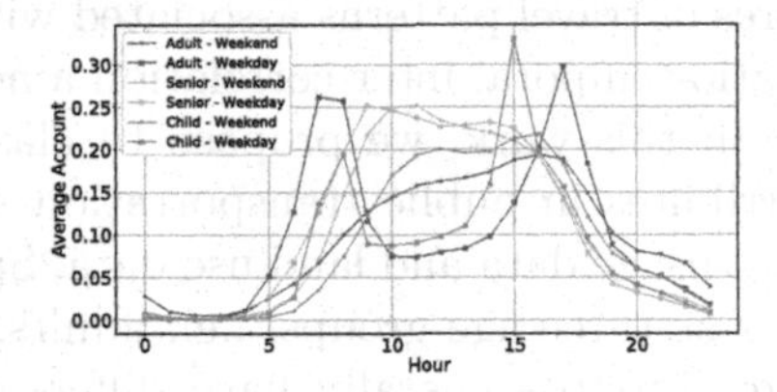

Fig. 2. Journey Transaction account in a day on different age groups

2.2 Feature Analysis

Temporal Distribution consists of the average transaction amount in a week and a day. Figure 1 shows that the average travel amount in a day of children is higher than the other two groups. Children can only arrive at the destinations by public transport without the lead of adults since they are not able to drive.

[1] https://www.opal.com.au/en/about-opal/.

Figure 2 shows that peaks exist around 8 am and 5 pm for adults while peaks exist around 8 am and 3 pm for children during weekdays since they have to work or attend class at a fixed time. The elderly are not under pressure to study or work, so the travel time is relatively flexible without any sharp peak or trough.

Spatial Distribution considers travel distance and PoI categories of the destinations which are shown in Figs. 3 and 4. The percentage of travel distances within 10 miles for children is 90.66% which is the largest among the three groups since it is not safe for children to go too far. The statistical analysis on the six categories collected from PoI of the destinations are performed to infer the possible trip purposes. 30.93% of the places where children go most often are schools and the ratio is higher than that of adults and seniors. Adults hold 0.19% to go to the clubs while the other two groups hold almost zero. Old people have the highest probability to church, 6.82%.

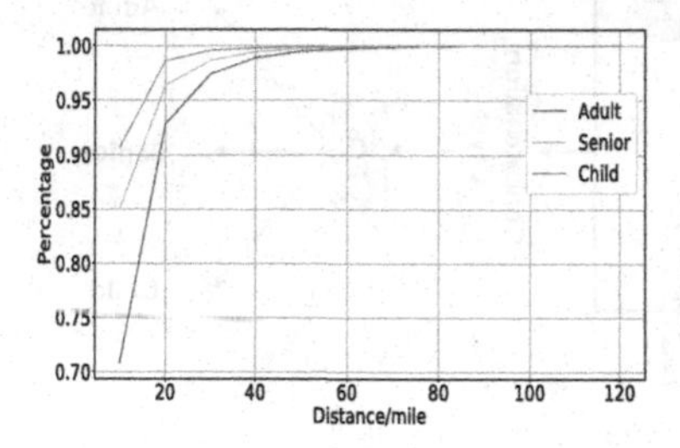

Fig. 3. Journey distance distribution

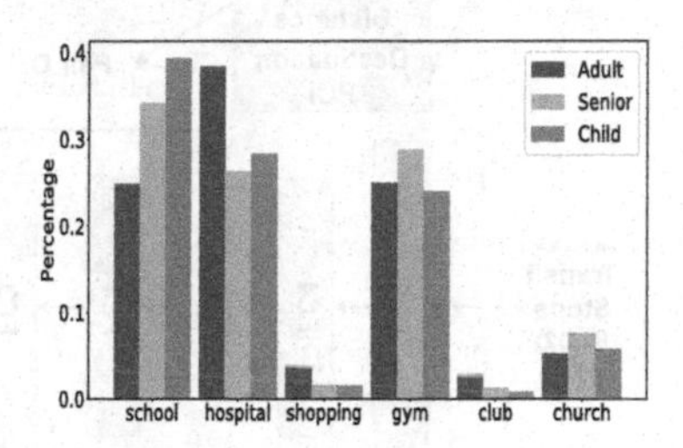

Fig. 4. PoI distribution

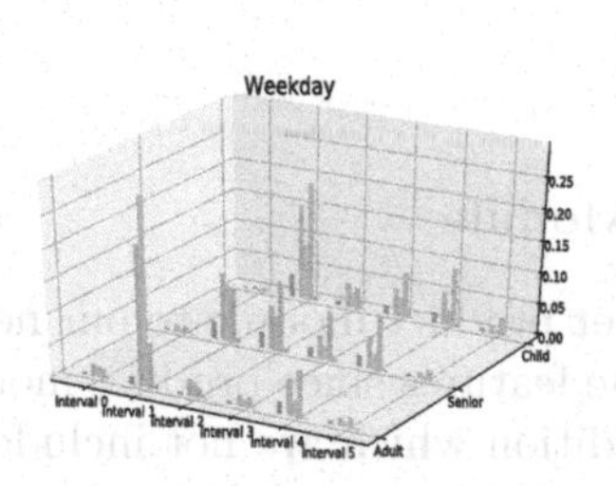

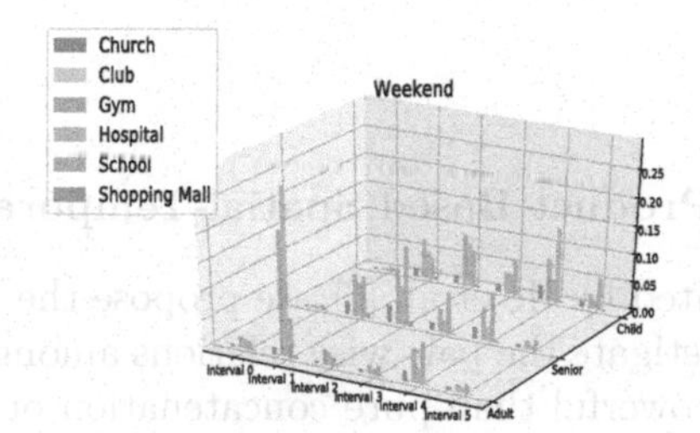

Fig. 5. Relationship between PoI and temporal distribution

Structural Spatio-Temporal Associations will be analyzed in this subsection. Figure 5 shows the proportional distribution of the six categories from destination PoI based on arrival time. Interval 0 is the period of time for the adults to the clubs while kids and seniors do not have this travel pattern. In Interval 1 on weekdays, children choose to go to school while the other two groups hold less probability to school. In Interval 2, the elderly prefer to go to the church and the ratio is higher than the other two groups.

3 Methodology

We now present our framework of Hybrid Spatial-Temporal correlation neural network. The architecture of our model is illustrated in Fig. 6. Throughout the paper, the matrix is shown in the uppercase letter while a vector is represented as a bold lowercase letter. Our model consists of two parallel sub-networks, Product-based Spatial and Temporal Module and Auto-Encoder based Compression Module. The concatenation of them is sent for classification.

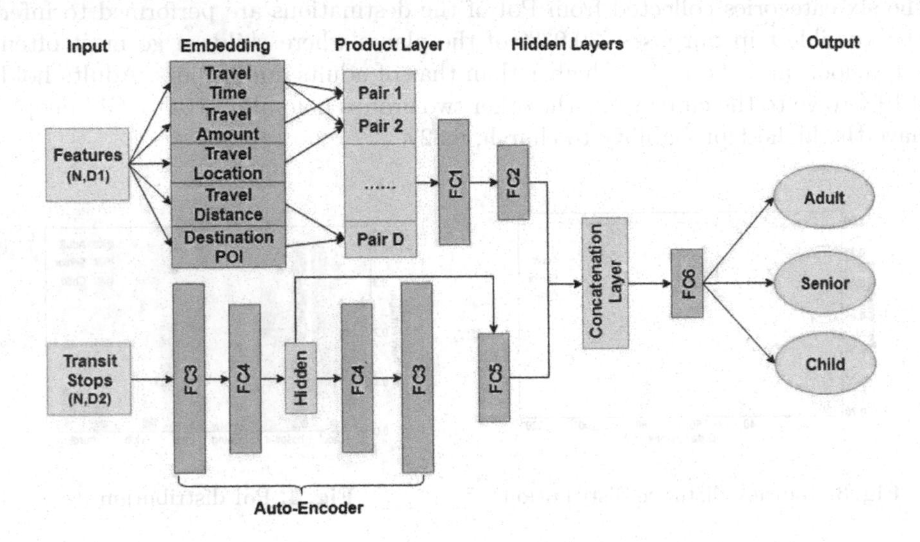

Fig. 6. Neural Network Architecture

3.1 Product Based Spatial-Temporal Module

Motivated by He et al. [6], we propose the inner product module to our network to investigate the pair-wise relations among the features since product module is more powerful than pure concatenation or addition which are not included any correlation among features. Moreover, the inner product module combined with the deep neural network is able to capture non-linear latent patterns.

The feature matrix $P_1 \in \mathbb{R}^{N \times D_1}$ is embedded into five fields, where N and D_1 denote the number of test samples and dimension of features, respectively. Each field represents one type of feature. $\boldsymbol{l}_{0-1} = (\boldsymbol{v}_1, \boldsymbol{v}_2, \cdots, \boldsymbol{v}_m, \cdots, \boldsymbol{v}_I)$ is the output of embedding layer which are sent into inner product layer to find the pairwise connection where I is the number of fields.

The definition of inner product between two vectors is $\boldsymbol{a} \cdot \boldsymbol{b} = \boldsymbol{a}^T \boldsymbol{b}$ where T means transpose. In the geometric sense, we can see the proximity of two vectors in the direction from the inner product values. Therefore, we expand the inner product to two matrices to find the relation between them. The inner product in the layer is defined as: $A \bullet B = \sum_{m,n} A_{m,n} B_{m,n}$. Then let $\boldsymbol{v_m} = (v_{m1}, v_{m2}, \cdots, v_{mn}, \cdots, v_{mN_m})$ as the m_{th} field vector from the embedding layer where $m = (1, 2, \cdots, I)$. The inner product of two fields is $< v_m, v_n >= W_0^m v_m \cdot W_0^n v_n$ where $W_0^i \in \mathbb{R}^{M \times N_i}$. Then we define W_p^i as the i_{th} vector weight of the product layer and the dimension is depended on the embedding vector. The output of product layer is $\boldsymbol{l_{1-1}} = (l_1, l_2, \cdots, l_i, \cdots, l_{D_{PW}})$ where $\boldsymbol{l_{1-1}} \in \mathbb{R}^{D_{PW}}$ and D_{PW} is the number of pairs. l_i is represented as

$$l_i = \sum_{m=1,n=1}^{I,M} (W_p^i)_{m,n} < v_m, v_n > \tag{1}$$

$\boldsymbol{l_{1-1}}$ is then fed into a fully connected layer and get the ouput $\boldsymbol{l_{2-1}} \in \mathbb{R}^{D_{2-1}}$.

3.2 Auto-Encoder Based Compression Module

The transit stop matrix P_2 is sparse with redundant information so it needs to be compressed. Auto-Encoder is used for dimensional reduction which is able to retain most of the original data information.

The AECM is composed of an Auto-Encoder and a fully connected layer. P_2 is fed into an Auto-Encoder to fuse features from different domains together while keeping most of the useful information. The encoding and decoding processes are employed with two-layer fully-connected networks and the transformation can be described as: $H_t(r_i) = encoder(P_2)$ and $\hat{P_2} = decoder(H_t(r_i))$, where $encoder(\cdot)$ and $decoder(\cdot)$ represent the transformation of encoder part and decoder part respectively. $H_t(r_i)$ is the hidden representation of P_2, $\hat{P_2}$ is the output. The cost function of auto-encoder is MSE (mean squared error) of $P_2 - \hat{P_2}$ in order to make sure that $\hat{P_2}$ and $P2$ are as similar as possible. The hidden representation $H_t(r_i)$ is then fed into one fully connected layer for concatenation. And the result of this module is $\boldsymbol{l_{2-2}}$.

3.3 Combination and Classification

To fuse the spatio-temporal relevance information and transit stops information, we concatenate $\boldsymbol{l_{2-1}}$ with $\boldsymbol{l_{2-2}}$ together to form $\boldsymbol{l_2}$. At last, $\boldsymbol{l_2}$ is fed into one fully connected layer to get the final classification result $\hat{y}$. The objective function of the proposed network consists of two parts: constraint of auto-encoder in the second part L_1 and the loss of final classification in the concatenation part L_2.

$$\begin{aligned} L_1 &= MSE(P_2, \hat{P_2}) \\ L_2 &= Softmax\_cross\_entropy(y, \hat{y}) \end{aligned} \tag{2}$$

where y is the true label of the input samples, MSE is the mean square error, and $Softmax\_cross\_entropy$ is the cross entropy loss for softmax function. The overall loss is $L(\theta) = \lambda \times L_1 + (1 - \lambda) \times L_2$ where θ represents all learnable parameters in the network. It is obtained via Gradient Descent optimizer.

4 Experiments

4.1 Overall Comparison

We first compare our model with several widely used classification algorithms: Linear Discriminant Analysis (LDA), Quadratic Discriminant Analysis (QDA), Support Vector Machine (SVM), Adaptive Boosting (Ada), Decision Tree (DT), XGBoost [7], and Multilayer Perceptron (MLP). Jahangiri et al. [8] found that SVM produced the best performance to classify travel mode. Table 1 summarize the results. Our model based on deep neural networks achieves better accuracy since non-linear relations exist among features. The auto-encoder carries out sparse matrix analysis that SVM cannot solve. Our approach significantly outperforms all other listed approaches, indicating that our model can be used to capture the implicit relevance among spatial and temporal characteristics and compress sparse matrices retaining the original data information.

Table 1. Overall comparison

Model	Accuracy	Recall			Precision		
		Adult	Senior	Child	Adult	Senior	Child
LDA	0.6072	0.6859	0.5765	0.5590	0.6787	0.5799	0.5617
QDA	0.4265	0.3741	0.8486	0.0570	0.6345	0.3802	0.3190
Jahangiri et al. [8]	0.5151	0.7720	0.0040	0.7692	0.5955	0.6430	0.4532
Ada	0.6370	0.7518	0.5474	0.5114	0.8729	0.5592	0.5269
DT	0.7613	0.9262	0.7545	0.6027	0.8585	0.7006	0.7143
Chen et al. [7]	0.6879	0.8312	0.5982	0.6341	0.8299	0.6112	0.6218
MLP	0.7849	0.8135	0.7512	0.7867	0.8747	0.7226	0.7576
Our	**0.9237**	**0.8664**	**0.9068**	**0.9989**	**0.9027**	**0.8854**	**0.9831**

4.2 Ablation Study

The results of the ablation study on the architecture of the network are listed in Table 2. We compare Fully Connected Layer (FCL), Auto-Encoder, Inner Product-based Module, Outer Product-based Module, and their combinations.

The outer product-based network proposed by He et al. [6] is used to explicitly model the pairwise correlations among features for the recommendation. We have similar input data structure as theirs consisting of several types of features, the data used here is not one-hot encoded sparse matrix. Also, inner product judges angle while outer product judges direction so the first one will perform better on determining similarity. Thereby, the inner product-based layer achieves a higher accuracy than outer product-based layer. Moreover, FCL and AE only analyze the correlations among transit stops and temporal features were not included.

In the compression process of FCL, important information may be discarded while AE retains most information and achieves a better result. Consequently, the combination of Inner Product and Auto-Encoder takes full advantage of spatial-temporal information, which helps to produce a better result.

Table 2. Performance with different components

Model	Accuracy	Recall			Precision		
		Adult	Senior	Child	Adult	Senior	Child
FCL	0.8051	0.7936	0.7354	0.8864	0.8105	0.7952	0.8086
AE	0.8194	0.6226	0.8518	0.9849	0.8404	0.7275	0.9046
Inner-PNN	0.8624	0.8874	0.8304	0.8694	0.8575	0.8238	0.9087
Outer-PNN	0.8009	0.8682	0.7254	0.8090	0.8130	0.7868	0.8009
Outer-PNN + FCL	0.8792	0.7711	0.8774	0.9897	0.8754	0.8193	0.9441
Inner-PNN + FCL	0.9053	0.8352	0.8896	0.9719	0.8807	0.8615	0.9719
Outer-PNN + AE	0.9023	0.8295	0.8890	0.9889	0.8792	0.8552	0.9723
Inner-PNN + AE	**0.9237**	**0.8664**	**0.9068**	**0.9989**	**0.9027**	**0.8854**	**0.9831**

5 Conclusion

This paper proposes a new neural network to classify passengers based on demographic attributes by exploring the relevance among temporal-spatial information of transit data with the Inner Product based strategy and Auto-Encoder based method. We evaluate our approach by classifying three age groups from real-world collected data and achieve an accuracy of 92.37% which outperforms other classification methods. In the future, the proposed model will be further adapted for more domains. We will maximize the use of other attributes and develop frameworks for inferring demographic attributes of passengers, which can further help operation of a human-centered public transport system.

References

1. Shiftan, Y., Outwater, M.L., Zhou, Y.: Transit market research using structural equation modeling and attitudinal market segmentation. Transp. Policy **15**(3), 186–195 (2008)
2. Mohamed, K., et al.: Clustering smart card data for urban mobility analysis. IEEE Trans. Intell. Transp. Syst. **18**(3), 712–728 (2016)
3. Hagenauer, J., et al.: A comparative study of machine learning classifiers for modeling travel mode choice. Expert Syst. Appl. **78**, 273–282 (2017)

4. Bai, L., Yao, L., Kanhere, S.S., Yang, Z., Chu, J., Wang, X.: Passenger demand forecasting with multi-task convolutional recurrent neural networks. In: Yang, Q., Zhou, Z.-H., Gong, Z., Zhang, M.-L., Huang, S.-J. (eds.) PAKDD 2019. LNCS (LNAI), vol. 11440, pp. 29–42. Springer, Cham (2019). https://doi.org/10.1007/978-3-030-16145-3_3
5. Bai, L., et al.: Stg2seq: Spatial-temporal graph to sequence model for multi-step passenger demand forecasting. In: IJCAI (2019)
6. He, X., et al.: Outer product-based neural collaborative filtering. In: IJCAI, pp. 2227–2233. AAAI Press (2018)
7. Chen, T., et al.: Xgboost: a scalable tree boosting system. In: SIGKDD, pp. 785–794. ACM (2016)
8. Jahangiri, A., Rakha, H.A.: Applying machine learning techniques to transportation mode recognition using mobile phone sensor data. IEEE Trans. Intell. Transp. Syst. **16**(5), 2406–2417 (2015)

Spontaneous EEG Classification Using Complex Valued Neural Network

Akira Ikeda and Yoshikazu Washizawa(✉)

The University of Electro-Communications, Tokyo, Japan
{akira.ikeda,washizawa}@uec.ac.jp

Abstract. Identification of spontaneous brain activity using the electroencephalography (EEG) requires information of the frequency spectrum and the spatial distribution. The complex valued neural network (CVNN) which uses complex weights and inputs has been shown higher performance for periodic data analysis, since spectrum information is represented by complex numbers. In spontaneous EEG analysis, the phase information depends on the onset of the recording, thus it is not informative. However, the conventional CVNN is not able to remove the phase information and extract amplitude spectrum efficiently. In this paper, we introduce two activation functions for CVNN to extract the amplitude spectrum directly, and classify spontaneous EEG. Our experimental results showed that the proposed method is higher classification performance than the conventional CVNN, and comparable to the convolutional neural network (CNN). Furthermore, the proposed method showed high performance when the number of hidden units is small.

Keywords: Complex-valued neural network (CVNN) · Spontaneous EEG · Frequency classification

1 Introduction

Spontaneous EEG analysis is applied for estimation of human states (such as relax, stress, and attentive), emotions (such as anger, happiness, and (un)pleasant), classification of sleep stage, and application for neuro-marketing, bio-feedback, and brain computer interfaces (BCI) [12,15,16,18]. Spontaneous EEG is divided into five frequency bands, delta (0.5–4 Hz), theta (4–8 Hz), alpha (8–13 Hz), beta (13–30 Hz), and gamma (30-Hz) bands. The power and distribution of these frequency bands are related to drowsiness and arousal level, or attentive and relax states. For example, the alpha and beta bands are related to the degree of relaxation and tension, and alpha, delta, and theta bands are associated with the sleep stage. In BCI, the steady state visual evoked potential (SSVEP) or the desynchronization of mu-rhythm in the specific motor area is used [14]. In SSVEP-based BCI, a subject gazes at one of several blinking patterns of different frequencies, and BCI determines the target command by detecting the frequency the subject gazes from EEG [10]. For spontaneous EEG

T. Gedeon et al. (Eds.): ICONIP 2019, CCIS 1142, pp. 495–503, 2019.
https://doi.org/10.1007/978-3-030-36808-1_54

analysis, not only the power spectrum feature, but its spatial distribution is important. For example, the asymmetry index (ASM) is used for estimation of emotion states [9]. Thus, in order to classify these spontaneous EEG, the amplitude spectrum, its spacial distribution, time-varying information are utilized.

The complex valued neural network (CVNN), which has complex valued weight, input, and output, has shown better performance than the real valued neural network (RVNN) in various fields [1,8,16,17]. Complex feature vectors represent periodic/cyclic data, such as oscillation, and wave, in particular, electromagnetics, electric circuits, acoustic/biomedical signals, and imaging radar in nature. Therefore, complex data analysis by CVNN is compatible with such periodic/cyclic data. The activation functions of CVNN are divided into the split type $f(z) = \phi_1(\Re(z)) + j\phi_1(\Im(z))$ and the amplitude-phase type $f(z) = \phi_2(|z|)\exp(j\arg z)$. For the split type, the split-ReLU ($\phi_1 = \text{ReLU}$) and the split-tanh ($\phi_1 = \tanh$) are often used. For the amplitude-phase type function, $\tanh(|z|)\exp(j\arg z)$ is used [6,13]. The amplitude-phase type is used for wave phenomenon analysis [5]. In some applications such as EEG analysis, the phase information depends on the onset of data, and it is not informative. The activation functions listed above keep the phase information, thus they are not suitable for spontaneous EEG analysis.

In this paper, we propose a new activation function to extract features from spontaneous EEG. The discrete Fourier transform (DFT) is computed by the inner product of an input time series data and the complex sinusoidal basis. Then its amplitude is extracted by the complex absolute function. Therefore, we introduce the complex absolute activation function and the complex absolute split ReLU function for CVNN. They include the feature extraction using DFT and the complex absolute function in nature. Recently, the convolutional neural network (CNN) shows very high performance in various research area [4,10]. CNN can extract frequency information and vanish phase information by filtering and the pooling. However, the proposed CVNN is more direct solution, and has compact structure. We show our experimental results on two datasets to demonstrate the proposed method and compare with CNN and the conventional CVNN using split-tanh, split-sigmoid, and split-ReLU.

2 Complex Valued Neural Network

The forward propagation of CVNN is calculated by the following equations,

$$\boldsymbol{u}^{(l)} = \boldsymbol{W}^{(l)}\boldsymbol{o}^{(l-1)} \tag{1}$$

$$\boldsymbol{o}^{(l)} = f^{(l)}(\boldsymbol{u}^{(l)}), \quad l = 2, \ldots, L, \tag{2}$$

where $\boldsymbol{o}^{(1)}$ is the complex valued input to the network, $\boldsymbol{o}^{(l)}$ and $f^{(l)}(\cdot)$ are the complex valued output and the activation function in the lth layer, respectively, and $\boldsymbol{W}^{(l)}$ is the complex weight connecting from the $(l-1)$th to the lth layer.

The weight connection $\boldsymbol{W}^{(l)}$ is optimized to minimize a loss function E. The square error loss or the logistic loss is often used for classification problems.

Let $\boldsymbol{d}_n$ and $\boldsymbol{o}_n^{(L)}$ be the target vector and output loss for the nth sample respectively, then the square error E_n is

$$E_n = \frac{1}{2}\|\boldsymbol{o}_n^{(L)} - \boldsymbol{d}_n\|^2. \tag{3}$$

The weight connection $\boldsymbol{W}^{(l)}$ is iteratively updated,

$$\boldsymbol{W}^{(l)} \leftarrow \boldsymbol{W}^{(l)} + \eta \Delta \boldsymbol{W}^{(l)}, \quad l = 2, \ldots, L, \tag{4}$$

where $\eta > 0$ is the learning rate. The back-propagation (BP) for CVNN is used to optimize $\boldsymbol{W}^{(l)}$.

Suppose that we use the split-type activation function, and let $u_r^{(l)}$ be the rth unit value of $\boldsymbol{u}^{(l)}$ and $\delta_r^{(l)} = \frac{\partial E}{\partial \Re(u_r^{(l)})} + j\frac{\partial E}{\partial \Im(u_r^{(l)})}$. Then the partial derivatives of $\Re(\boldsymbol{W}^{(l)})$ and $\Im(\boldsymbol{W}^{(l)})$ are given by

$$\frac{\partial E}{\partial w_{rp}^{(l)}} = \delta_r^{(l)} \overline{o_p^{(l-1)}}, \quad l = 2, \ldots, L, \tag{5}$$

where $\bar{\cdot}$ stands for the complex conjugate, $o_p^{(l-1)}$ is the pth element of $\boldsymbol{o}^{(l-1)}$, and $w_{rp}^{(l)}$ is the (r,p) element of $\boldsymbol{W}^{(l)}$. We, hereafter, omit the sample index n. $\delta_r^{(l)}$ was computed by the chain rule,

$$\Re(\delta_r^{(l)}) = \Re(\sum_q \delta_q^{(l+1)} \overline{w_{qr}^{(l+1)}}) \frac{\partial \Re(f^{(l)}(u_r^{(l)}))}{\partial \Re(u_r^{(l)})}, \tag{6}$$

$$\Im(\delta_r^{(l)}) = \Im(\sum_q \delta_q^{(l+1)} \overline{w_{qr}^{(l+1)}}) \frac{\partial \Im(f^{(l)}(u_r^{(l)}))}{\partial \Im(u_r^{(l)})}. \tag{7}$$

When the amplitude-phase type $f(z) = \tanh(|z|)\exp(j \arg z)$ is used as the activation function, $|\boldsymbol{W}|$ and $\arg(\boldsymbol{W})$ are independently updated by the following rule [5,12],

$$|w_{rp}^{(l)}| \leftarrow |w_{rp}^{(l)}| = \eta |o_r^{(l)}||d_r^{(l)}| \sin(\arg o_r^{(l)} - \arg d_r^{(l)}) \frac{o_p^{(l-1)}}{u_r^{(l)}} \sin(\theta_{rp}^{(l)})$$
$$- \eta \left(1 - |(o_r^{(l)})^2|\right) \left(|o_r^{(l)}| - |d_r^{(l)}| \cos(\arg o_r^{(l)} - \arg d_r^{(l)})\right) |o_p^{(l-1)}| \cos(\theta_{rp}^{(l)}),$$

$$\arg(w_{rp}^{(l)}) \leftarrow \arg(w_{rp}^{(l)}) = -\eta |o_r^{(l)}||d_r^{(l)}| \sin(\arg o_r^{(l)} - \arg d_r^{(l)}) \frac{o_p^{(l-1)}}{u_r^{(l)}} \cos(\theta_{rp}^{(l)})$$
$$- \eta \left(1 - |(o_r^{(l)})^2|\right) \left(|o_r^{(l)}| - |d_r^{(l)}| \cos(\arg o_r^{(l)} - \arg d_r^{(l)})\right) |o_p^{(l-1)}| \sin(\theta_{rp}^{(l)}),$$

where $\theta_{rp}^{(l)} = \arg(o_r^{(l)}) - \arg(o_p^{(l-1)}) - \arg(w_{rp}^{(l)})$, $\boldsymbol{d}^{(l-1)} = \left(f^{(l)}\left(\boldsymbol{d}^{(l)*}\boldsymbol{W}^{(l)}\right)\right)^*$ and $*$ is the complex conjugate transpose.

3 Proposed Methods

Complex valued information used in CVNN contains information of signal amplitude and phase. In the analysis of spontaneous brain activity, the phase information is determined by the onset time of the measurement start time. Although the phase difference between channels may be informative, the absolute phase information should be removed. Therefore, in this study, we introduce the absolute activation function $f(z) = |z|$ and the absolute split-ReLU function $f(z) = |\mathrm{ReLU}(\Re(z)) + j\mathrm{ReLU}(\Im(z))|$ for CVNN. Since the output of the activation function is real-valued, we consider the standard RVNN structure for latter layers. We derive BP based updating rule for the proposed CVNN. Suppose that two layer network, the activation function of the input layer $f^{(2)}(\cdot)$ in Eq. (2) is the absolute or absolute split-ReLU, and the activation function of the output layer is the soft-max. When we employ the cross entropy function for the error function, the gradient is calculated using the chain rule. The gradient of E with respect to $w_{pq}^{(3)}$ is obtained in the same way as RVNN. We derive the gradient of E with respect to $w_{qr}^{(2)}$.

$$\frac{\partial E}{\partial w_{qr}^{(2)}} = \frac{\partial E}{\partial \Re(w_{qr}^{(2)})} + j\frac{\partial E}{\partial \Im(w_{qr}^{(2)})} = o_r^{(1)}\left(\frac{\partial E}{\partial \Re(u_q^{(2)})} + j\frac{\partial E}{\partial \Im(u_q^{(2)})}\right) \tag{8}$$

$$\frac{\partial E}{\partial \Re(u_q^{(2)})} = \sum_p \frac{\partial E}{\partial u_p^{(3)}}\frac{\partial u_p^{(3)}}{\partial \Re(u_q^{(2)})} = \sum_p \frac{\partial E}{\partial u_p^{(3)}} w_{pq}^{(3)} \frac{\partial o_q^{(2)}}{\partial \Re(u_q^{(2)})} \tag{9}$$

Therefore, when the imaginary part is also calculated and let $f_{\mathrm{sReLU}}(x) = \mathrm{ReLU}(\Re[x]) + j\mathrm{ReLU}(\Im[x])$., the gradients of the proposed networks are

$$\frac{\partial E}{\partial w_{pq}^{(3)}} = \frac{\partial E}{\partial u_p^{(3)}}\frac{\partial u_p^{(3)}}{\partial w_{pq}^{(3)}} = (o_p^{(3)} - d_p)o_q^{(2)} \tag{10}$$

$$\frac{\partial E}{\partial w_{qr}^{(2)}} = \begin{cases} \sum_p (o_p^{(3)} - d_p) w_{pq}^{(3)} \frac{u_q^{(2)}}{o_q^{(2)}} o_r^{(1)} & \text{absolute} \\ \sum_p (o_p^{(3)} - d_p) w_{pq}^{(3)} \frac{f_{\mathrm{sReLU}}(u_q^{(2)})}{o_q^{(2)}} o_r^{(1)} & \text{absolute of split} - \text{ReLU}, \end{cases} \tag{11}$$

The network is learned by the stochastic gradient algorithms (SGD).

4 Experiment

4.1 Dataset

We used two datasets. The first dataset is the following artificial dataset for four class classification problem,

$$x[n] = \sum_{c=1}^{4} k_c \sin(2\pi \frac{f_c}{f_s} n + \theta_c) + \epsilon, \tag{12}$$

where the target frequencies are $(f_1, f_2, f_3, f_4) = (3, 7, 12, 14)$ [Hz], $\epsilon \sim \mathcal{N}(0, 4)$, $\mathcal{N}(\mu, \sigma^2)$ is the normal distribution of average μ and variance σ^2, the sampling frequency is $f_s = 100$[Hz]. When f_c is the target frequency, $k_c = 1$, otherwise k_c is chosen from the uniform distribution in the range of $[0, 1)$. θ_c is chosen from the uniform distribution in the range of $[0, 2\pi)$. We generated 4000 samples of the same length as the input dimension of the network, and conducted 10-fold cross validation.

The second dataset is EEG data for open and closed eyes. This dataset was prepared and provided by the developer of BCI2000 [2,11]. We used three Pz, O1, and O2 channels of subject one among them. The sampling frequency of this dataset is 160 Hz. We clipped out 4000 samples of the same length as the input dimension of the network randomly in both case. As preprocessing of all datasets, we normalized training and test data separately. The target vector of the proposed method was 1 or 0 (one-hot label), that of CVNN using split-ReLU, split-sigmoid was $1 + j$ or 0, and that of CVNN using split-tanh was $1 + j$ or $-1 - j$. We conducted 10-fold cross validation.

4.2 Networks

The structure of our networks used in the experiment is listed in Table 1. We used two layer network for the proposed CVNN network, and CNN has single 1D convolution layer, the max pooling layer, and fully-connected output layer. In the experiment on artificial data, the input dimension was 64, 128, or 256, and the dimensions of the hidden layer were set to the values between the output dimensions and the input dimensions. The number of filters and filter size of CNN were set to the values between $[5, 30]$, $[8, \text{inputdim}/2]$. In the experiments on three channels EEG data, the input dimension was 192, 384, or 768. The dimensions of the hidden layer, the number of filters and filter size were set from the same value range as the experiment on the artificial data. The mini batch size was fixed to 200, and the number of epochs was fixed at 400. He's initialization was used to initialize the weight for CNN [3]. The weight for the others was initialized with a normal distribution of standard deviation 0.01 and mean 0. The max pooling of 1×2 with the stride step two was used. The learning rate in proposed networks, split-tanh, split-sigmoid, split-relu, and CNN were chosen from $\{0.05, 0.01, 0.005\}$, $\{0.005, 0.001, 0.0005\}$, $\{0.01, 0.005, 0.001\}$, $\{0.00005, 0.00001, 0.000005\}$, and $\{0.05, 0.01, 0.005\}$, respectively. We selected the learning rate that maximizes the classification accuracy among these values at (the number of filters, filter size)$= (15, \text{input dim}/2)$ for CNN. We selected the learning rate that maximizes the classification accuracy among these values at the hidden layer 50 for other networks. Table 1 shows the learning rate determined in this way. We implemented CVNN by Python 3.6.8 and Numpy 1.15.2, and used Keras 2.2.4, Tensorflow 1.13.1 as well for CNN.

Table 1. Networks used in experiments.

Network	Hidden activation	Output activation	Error function	Learning rate for three channels EEG	Learning rate for the other
Proposed	abs	soft-max	cross entropy	0.01	0.01
Proposed	abs-split-ReLU	soft-max	cross entropy	0.01	0.01
CVNN	split-tanh [6]	split-tanh	squared	0.0005	0.001
CVNN	split-sigmoid [7]	split-sigmoid	squared	0.01	0.005
CVNN	split-ReLU [13]	split-relu	squared	0.00001	0.00001
CNN	ReLU	soft-max	cross entropy	0.05	0.05

5 Results

Figures 1 and 2 show the relation between the classification accuracy and the number of hidden units for the artificial dataset and EEG dataset. The horizontal axis is log scale and the results of the inputs 256, 128, and 64 are represented in order from the left. Figure 1 shows that the proposed CVNN using the absolute activation exhibited the best classification performance. The proposed method also shows the best classification performance for almost all range of the number of hidden units in Fig. 2. From the figures, the proposed absolute function keeps higher performance when the number of hidden units is small. In other words, the proposed absolute activation function efficiently extracts information by smaller number of units. Figure 3 compares the performance with different input length, kernel size, and the number of kernels, where these values are chosen to have the best test accuracy.

The proposed method shows better performance than CNN for the artificial dataset, and comparable performance to CNN for the EEG dataset.

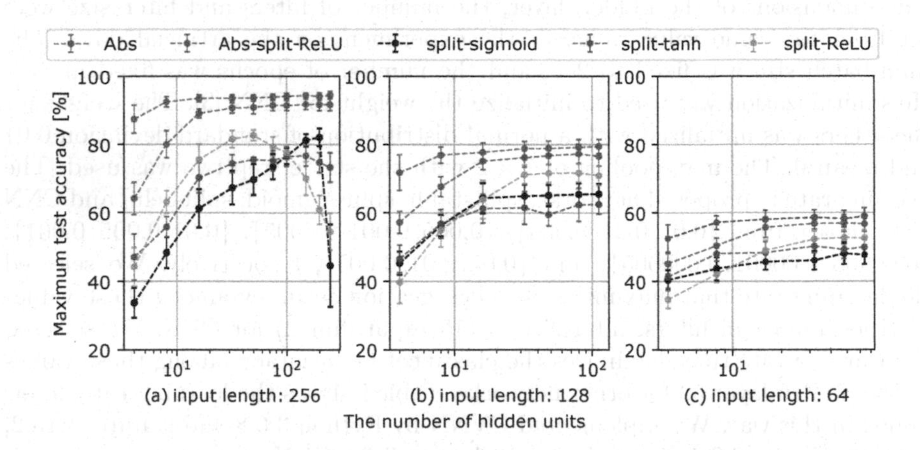

Fig. 1. Test accuracies of artificial data.

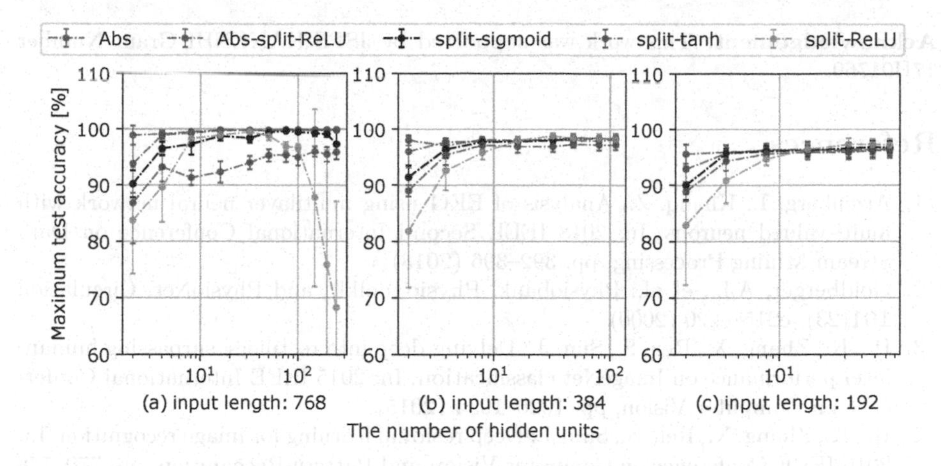

Fig. 2. Test accuracies of three channels EEG data.

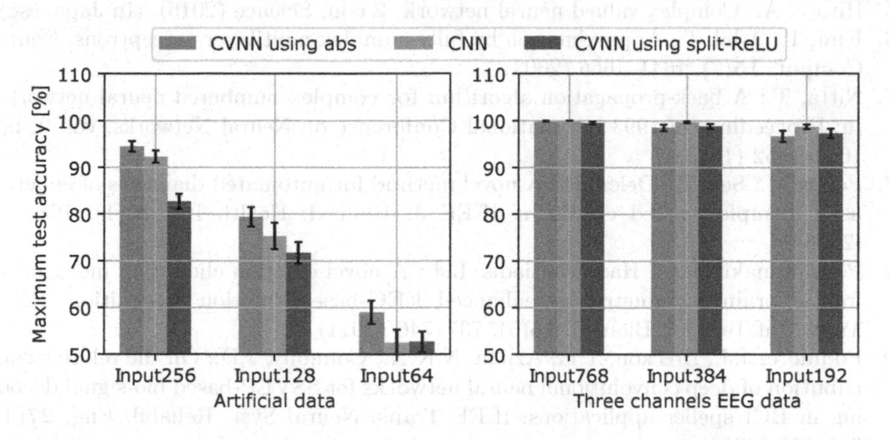

Fig. 3. Comparison of test accuracies of CVNN with absolute, split-ReLU activation and CNN.

6 Conclusion

We proposed CVNN using the absolute activation function and the absolute of split-ReLU function for spontaneous EEG analysis. We derived updating rules for these activation functions. In our experiment, the proposed CVNN outperformed conventional CVNNs, showed better performance than CNN in the artificial dataset. The proposed method showed higher performance even when the number of hidden units is small. That is to say, the proposed absolute activation efficiently extracts the feature from smaller number of units. For future tasks, we will investigate the performance for multi-layered deep network structure of the proposed CVNN.

Acknowledgement. This work was supported by JSPS KAKENHI Grant Number 17H01760.

References

1. Aizenberg, I., Khaliq, Z.: Analysis of EEG using multilayer neural network with multi-valued neurons. In: 2018 IEEE Second International Conference on Data Stream Mining Processing, pp. 392–396 (2018)
2. Goldberger, A.L., et al.: Physiobank, PhysioToolkit, and PhysioNet. Circulation **101**(23), e215–e220 (2000)
3. He, K., Zhang, X., Ren, S., Sun, J.: Delving deep into rectifiers: surpassing human-level performance on ImageNet classification. In: 2015 IEEE International Conference on Computer Vision, pp. 1026–1034 (2015)
4. He, K., Zhang, X., Ren, S., Sun, J.: Deep residual learning for image recognition. In: 2016 IEEE Conference on Computer Vision and Pattern Recognition, pp. 770–778 (2016)
5. Hirose, A.: Complex valued neural network, 2 edn. Science (2016). (In Japanese)
6. Kim, T., Adal, T.: Approximation by fully complex multilayer perceptrons. Neural Comput. **15**(7), 1641–1666 (2003)
7. Nitta, T.: A back-propagation algorithm for complex numbered neural networks. In: Proceedings of 1993 International Conference on Neural Networks, vol. 2, pp. 1649–1652 (1993)
8. Peker, M., Sen, B., Delen, D.: A novel method for automated diagnosis of epilepsy using complex-valued classifiers. IEEE J. Biomed. Health Inf. **20**(1), 108–118 (2016)
9. Petrantonakis, P.C., Hadjileontiadis, L.J.: A novel emotion elicitation index using frontal brain asymmetry for enhanced EEG-based emotion recognition. IEEE Trans. Inf Technol. Biomed. **15**(5), 737–746 (2011)
10. Podmore, J.J., Breckon, T.P., Aznan, N.K.N., Connolly, J.D.: On the relative contribution of deep convolutional neural networks for SSVEP-based bio-signal decoding in BCI speller applications. IEEE Trans. Neural Syst. Rehabil. Eng. **27**(4), 611–618 (2019)
11. Schalk, G., McFarland, D.J., Hinterberger, T., Birbaumer, N., Wolpaw, J.R.: BCI2000: a general-purpose brain-computer interface (BCI) system. IEEE Trans. Biomed. Eng. **51**(6), 1034–1043 (2004)
12. Sunaga, Y., Natsuaki, R., Hirose, A.: Proposal of complex-valued convolutional neural networks for similar land-shape discovery in interferometric synthetic aperture radar. In: Cheng, L., Leung, A.C.S., Ozawa, S. (eds.) ICONIP 2018. LNCS, vol. 11301, pp. 340–349. Springer, Cham (2018). https://doi.org/10.1007/978-3-030-04167-0_31
13. Trabelsi, C., et al.: Deep complex networks. In: International Conference on Learning Representations (2018)
14. Wolpaw, J., Wolpaw, E.W. (eds.): Brain-Computer Interfaces: Principles and Practice (English Edition), 1st edn. Oxford University Press, Oxford (2012)
15. Wu, R., Huang, T.: Learning of phase-amplitude-type complex-valued neural networks with application to signal coherence. In: Liu, D., Xie, S., Li, Y., Zhao, D., El-Alfy, E.S. (eds.) ICONIP 2017. LNCS, vol. 10634, pp. 91–99. Springer, Cham (2017). https://doi.org/10.1007/978-3-319-70087-8_10
16. Zhang, J., Wu, Y.: A new method for automatic sleep stage classification. IEEE Trans. Biomed. Circuits Syst. **11**(5), 1097–1110 (2017)

17. Zhang, Z., Wang, H., Xu, F., Jin, Y.: Complex-valued convolutional neural network and its application in polarimetric SAR image classification. IEEE Trans. Geosci. Remote Sens. **55**(12), 7177–7188 (2017)
18. Zheng, W.: Multichannel eeg-based emotion recognition via group sparse canonical correlation analysis. IEEE Trans. Cogn. Dev. Syst. **9**(3), 281–290 (2017)

Tongue Coating Classification Based on Multiple-Instance Learning and Deep Features

Xiaoqiang Li[1,2](✉), Yonghui Tang[1], and Yue Sun[1]

[1] School of Computer Engineering and Science, Shanghai University, Shanghai, China
xqli@shu.edu.cn

[2] Shanghai Institute for Advanced Communication and Data Science, Shanghai University, Shanghai, China

Abstract. Tongue coating classification has long been a challenging task in Traditional Chinese Medicine (TCM) due to the fact that tongue coatings are multiform. Most existing methods make use of fixed location and handcrafted features, which may lead to inconstant performance when the size or location of the coating region varies. To solve this problem, our paper proposes a new tongue coating classification method. This method is mainly improved from two aspects: feature extraction and classification method. Complex tongue coating features extracted by Convolutional Neural Network (CNN) is used instead of handcrafted features, and a multiple-instance Support Vector Machine (MI-SVM) is applied to solve the uncertain location problem. Experimental results prove that our method shows significant improvements over state-of-the-art tougue coating classification methods.

Keywords: Tongue coating classification · Multiple-instance learning · Deep features

1 Introduction

According to TCM, tongues are closely related to people's health. This paper mainly concentrates on how to distinguish rotten-greasy tongue coating from normal tongue coating. Rotten-greasy tongue coating is thick and loose, looks like residues of bean curd and always exists in the middle and root of the tongue body [1]. Normal tongue coating is usually thin and white. Figure 1 shows normal and different rotten-greasy tongue coatings. The classification of tongue coatings can be viewed as a fine-grained [2] classification problem since normal and rotten-greasy tongue coating are only different symptom of the floating layer of the tongue. It is a challenging task for there lack further information (such as the location or size of the tongue coating patch) if a tongue image is labeled as normal coating tongue or rotten-greasy coating tongue.

Thanks to Shanghai Daosheng Medical Technology Co., Ltd.

T. Gedeon et al. (Eds.): ICONIP 2019, CCIS 1142, pp. 504–511, 2019.
https://doi.org/10.1007/978-3-030-36808-1_55

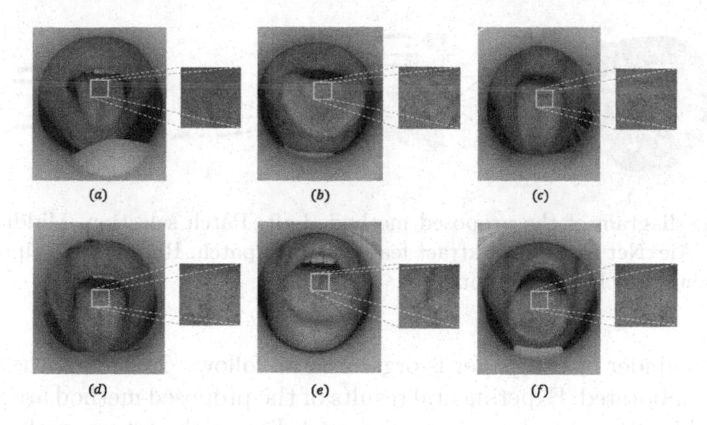

Fig. 1. Different tongue coatings. (a) Normal tongue coating. (b-f) Typical rotten-greasy tongue coatings.

Recently, some works have been conducted on tongue coating classification. Li et al. [3] extracted the center patch of a tongue body and classified tongue coating using Gabor [4] and Tamura [5] features of the patch. Qu et al. [6] proposed a Gabor wavelet transformation-based tongue coating classification method. Fu et al. [7] computerized tongue coating feature using deep neural networks. The methods mentioned above, however, have some drawbacks. Firstly, handcrafted features used in the methods of Li and Qu cannot describe the salient characteristic of tongue coating. Secondly, although the method of Fu based on deep neural networks can extract deep features, it focuses on global information rather than local information, which may capture more irrelevant information and deteriorate the classification.

In this paper, we try to solve these problems by multiple-instance learning (MIL) [8] and deep learning. MIL is first proposed by Dietterich et al. [8]. The classification of tongue coatings is naturally a multiple-instance problem since it shares similar assumptions with the multiple-instance binary classification that a tongue is considered as a rotten-greasy coating tongue if there exist one rotten-greasy coating patch on the tongue. So, we propose a multiple-instance representation of the tongue, in which a rotten-greasy coating tongue containing at least one rotten-greasy coating patch (positive instance) is treated as a positive bag, and a normal coating tongue containing only normal coating patches (negative instances) is treated as a negative bag. The classification task for only coarsely labeled images need to train a MI-SVM.

Since its successful usage in 2012 ImageNet competition, CNN has significantly improved the performance of many computer vision tasks. [9] shows that features extracted from CNN can perform well. Motivated by the success of CNN we use a method of fine-tuned CNN instead of handcrafted feature extraction to extract deep features of the tongue coating patches.

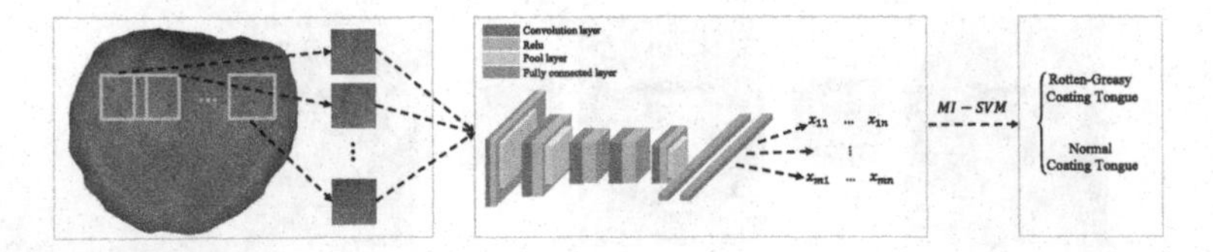

Fig. 2. The diagram of the proposed method. Left: Patch selection. Middle: Feature extraction. AlexNet is used to extract features from patch. Right: A multiple-instance SVM is trained to classify the tongues.

The remainder of this paper is organized as follows. In Sect. 2, the proposed method is elaborated. Experimental results of the proposed method are presented in Sect. 3. Finally, we make a conclusion and discuss the future work in Sect. 4.

2 Method

As shown in Fig. 2, the proposed method contains three stages. First, it utilizes rotten-greasy tougue coating information to select suspected tongue coating patches. Then, a CNN is used to extract fixed-length feature vectors for each tongue coating patch. At last, feature vectors are grouped into bags and a MI-SVM is used to do the classification.

2.1 Obtaining Convinced Rotten-Greasy Coating Patches

CNN can be powerful feature extractors. We hope that CNN can effectively extract features combining color, shape and texture information to describe tongue coating patches. Therefore when training a CNN, we manually obtain patches with salient features in each tongue image as input.

The method of obtaining convinced rotten-greasy coating patches is described as follows. For rotten-greasy tongue coating images, patches are chosen in the area of tongue body with rotten-greasy coating characteristics. For normal tougue coating images, patches are chosen in the area of normal coating characteristics. And for each tongue image, 10–15 patches are obtained and each patch is about 180–300 pixels wide and 240–400 pixels high.

2.2 Obtaining Suspected Rotten-Greasy Coating Patches

The goal of this stage is to find as many rotten-greasy coating patches as possible, and at least one definite rotten-greasy coating patch should be included. According to the theory of TCM and our observation, the rotten-greasy coating always appear in the middle and root of a tongue body, while the rest of the tongue can be ignored.

Patches obtained from a rotten-greasy coating tongue satisfies the assumption of multiple-instance binary classification that there exist at least one positive instance in a positive bag. Patches obtained from a healthy coating tongue include only healthy ones.

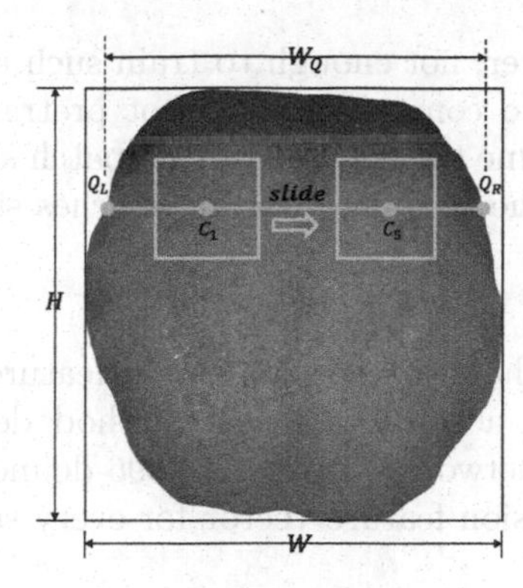

Fig. 3. The diagram of patch selection method.

The steps of obtaining the patches are as follows.

Step1: The circumscribed rectangle of the tongue is denoted as R. The height and the width of the rectangle is denoted as H and W respectively.

Step2: Draw a horizontal line $\frac{H}{3}$ away from the top of the tongue, denoted as Q. Use Q_L, Q_R to denote its left and right intersection point with the edge of the tongue. The width of the intersection line is denoted by W_Q.

Step3: Find C_1 on line Q. C_1 locates $\frac{W_Q}{3}$ to the right of Q_L. Take C_1 as the starting point and $\frac{W_Q}{3n}$ as the step length, find C_i $(i = 2, 3, \ldots, n+1)$ rightwards. For each C_i, draw square with C_i as its center and $\frac{W_Q}{6}$ as its side length. The squares represent the selected patches.

As shown in Fig. 3, by changing the side length and the step length, we can obtain tongue patches of different sizes and numbers.

2.3 Feature Extraction

In this stage, we use a CNN to extract fixed-length feature vectors of the rotten-greasy coating patches instead of the whole tongue image.

Architecture: We use the pretrained AlexNet described in [10]. It has 8 weight layers, 5 of which are convolutional layers and the rest 3 are fully connected layers. There are 4096 units in the second fully connected layer and the outputs of this layer are used as features. Thus, we can extract a 4096-dimension feature vector for each patch. We drop the last 1000-way fully connected layer and replace it with a 2-way fully connected layer during the network training.

Training: The network is first pretrained on ILSVRC2012 [11] dataset and then followed by fine-tuning on tongue coating patches. All tongue coating patches are obtained using the method described in Sect. 2.1. These patches are only used for fine-tuning the network. There are about 3000 rotten-greasy coating patches in total, which are, however, not enough to train such a high-capacity network. The network would fail to converge if it is not pretrained. We use stochastic gradient descent to fine-tune the network with a batch size of 128 and a learning rate of 0.0001. We stop the training after 20 epoches since the accuracy ceases increasing.

Testing: In this stage, the network serves as a feature extractor. The tongue coating patches obtained according to the method described in Sect. 2.2 are used as input, and the network outputs a 4096-deimension vector. Thus, we can extract a 4096-dimension feature vector for every suspected tongue coating patch.

2.4 Classifiction

In this stage, we train a MI-SVM to classify the tongue images. In the MIL task we can learn a classifier based on a training set of bags, where each bag contains multiple feature vectors [12]. The main idea of MI-SVM is to maximize bag margin which serves as an extension of the instance margin of standard SVM, and the details are well introduced in [13]. The input of a MI-SVM is a bag B_I which in our case represents tongue image I. And the instances in the bag are the features $\{x_i : i \in I\}$ we extracted from the tongue coating patches. Instead of explicitly associating a label y_i to each instance, we associate a label Y_I to a bag B_I. If $Y_I = -1$, then $y_i = -1$ for all $i \in I$. If $Y_I = +1$, then at least one instance $x_i \in B_I$ is a positive instance. In MI-SVM, the function margin of a bag is defined as:

$$\gamma_I = Y_I \max_{i \in I} \left(\langle \omega, x_i \rangle + b\right) \tag{1}$$

The MI-SVM aims at maximizing the bag margin, which is defined as follows:

$$\begin{gathered} \min_{\omega, b, \xi} \frac{1}{2} \|\omega\|^2 + C \sum_I \xi_I \\ s.t. \ \forall I : Y_I \max_{i \in I} (\langle \omega, x_i \rangle + b) \geq 1 - \xi_I, \ \xi_I \geq 0 \end{gathered} \tag{2}$$

In MI-SVM, the bag margin is determined by only one of its instance. For a positive bag, the margin is decided by the most positive instance, while the margin of a negative bag is decided by the least negative instance [14]. The label of the bag is then the label of the image.

3 Experiment

Experimental results are evaluated by the following three metrics: (1) accuracy (ACC); (2) ture positive rates (TPR); (3) true negative rate (TNR). True Positive (TP) and False Negative (FN) are samples which are positive and predicted to be positive or negative. False Positive (FP) and True Negative (TN) are samples which are negative and predicted to be positive or negative.

$$ACC = \frac{TP + TN}{TP + FP + FN + TN} \tag{3}$$

$$TPR = \frac{TP}{TP + FN} \tag{4}$$

$$TNR = \frac{TN}{TN + FP} \tag{5}$$

The tongue image dataset used in this paper is provided by Shanghai Daosheng Medical Technology Co., Ltd. It is a dataset including 274 tongue images, 186 of them are normal tongue coating images and 86 of them are rotten-greasy ones. The label of a tongue image is voted by multiple TCM practitioners. It should be noticed that the samples of ConvNet training are convinced tongue coating patches selected using the method described in Sect. 2.1. On the other hand, the samples of MI-SVM training are suspected tongue coating patches selected using the method described in Sect. 2.2.

Table 1. Comparison between using MI-SVM with CNN and using CNN directly.

Method	Accuracy	TPR	TNR
Alexnet [10]	72.2%	77.7%	69.1%
Alexnet+MI-SVM (Ours)	**82.1%**	**82.0%**	**82.2%**

Table 2. Comparison of different classifiers.

Classifier	Accuracy
Decision tree [15]	51.9%
KNN [16]	61.1%
EMDD [17]	67.3%
MI-SVM[13]	**82.1%**

We present three different experiments of the proposed method. The first is the comparison of different classifiers with the same feature extractor. Different classifiers are evaluated for tougue coating classification using the features extracted by a fine-tuned CNN. As shown in Table 1, the results demonstrate that our method achieves an accuracy of 82.1% and a recall rate (TPR) of 82.0% which is 10% and 4% higher respectively than that of using CNN directly.

The performance of other classifiers with the same features extracted using AlexNet model is shown in Table 2. It can be seen that the accuracy of the proposed method is superior to that of Decision Tree [15], KNN [16] and EMDD [17].

The second is the comparison of different feature extractors with the same classifier. SVM is used to test the performance of different feature extraction methods. Experimental results in Table 3 show that deep features perform better than handcrafted features such as the features extracted by GLDM [18], Tamura [5] and Gabor [4].

Table 3. Comparison between different features.

Feature extractor	Accuracy
GLDM [18]	54.0%
Tamura [5]	60.2%
Gabor [4]	68.5%
Alexnet [10]	**82.1%**

The third is the comparison with other works. The three methods are: Li's work [3], Qu's work [6] and Fu's work [7]. The results of the above experiments are listed in Table 4. It can be observed from the table that our method has the highest accuracy.

Table 4. Comparison with otcher methods.

Method	Accuracy
Li's [3]	75.6%
Qu's [6]	67.9%
Fu's [7]	58.3%
AlexNet+MI-SVM (Ours)	**82.1%**

4 Conclusions

In this paper, we have presented a new method for tongue coating classification using MIL and deep features. The method is divided into three stages. First, tongue coating patches are selected. Then, a deep CNN is used to extract the feature of each patch. At last, tongue coating is represented by a bag consisting of multiple feature vectors and MI-SVM is used to make the final classification. Experiment results show that the proposed method outperforms previous methods. Future work includes two aspects: (1) Collecting more tongue samples. Since we use a deep CNN as feature extractor, the proposed model always benefits from a larger dataset. (2) Adopting more advanced network architecture to further improve the accuracy.

References

1. Kirschbaum, B.: Atlas of Chinese Tongue Diagnosis. Eastland Press, Vista (2010)
2. Yao, B., Bradski, G., Fei-Fei, L.: A codebook-free and annotation-free approach for fine-grained image categorization. In: 2012 IEEE Conference on Computer Vision and Pattern Recognition, pp. 3466–3473. IEEE (2012)
3. Li, X., Shao, Q., Wang, J.: Classification of tongue coating using gabor and tamura features on unbalanced data set. In: IEEE International Conference on Bioinformatics and Biomedicine, pp. 108–109 (2013)
4. Lyons, M., Akamatsu, S., Kamachi, M., Gyoba, J.: Coding facial expressions with gabor wavelets. In: IEEE International Conference on Automatic Face and Gesture Recognition, 1998, Proceedings, pp. 200–205 (2002)
5. Tamura, H., Mori, S., Yamawaki, T.: Textural features corresponding to visual perception. IEEE Trans. Syst. Man Cybern. **8**(6), 460–473 (1978)
6. Qu, T.T., Xia, C.M., Wang, Y.Q., Zhu, M.L.M.: Recognition of greasy or curdy tongue coating based of wavelet transformation. Comput. Appl. Software **33**(10), 162–166 (2016)
7. Fu, S., Zheng, H., Yang, Z., Yan, B., Su, H., Liu, Y.: Computerized tongue coating nature diagnosis using convolutional neural network. In: IEEE International Conference on Big Data Analysis, pp. 730–734 (2017)
8. Dietterich, T.G., Lathrop, R.H., Lozano-Pérez, T.: Solving the multiple instance problem with axis-parallel rectangles. Artif. Intell. **89**(1–2), 31–71 (1997)
9. Sharif Razavian, A., Azizpour, H., Sullivan, J., Carlsson, S.: CNN features off-the-shelf: an astounding baseline for recognition. In: Proceedings of the IEEE Conference on Computer Vision and Pattern Recognition Workshops, pp. 806–813 (2014)
10. Krizhevsky, A., Sutskever, I., Hinton, G.E.: Imagenet classification with deep convolutional neural networks. In: International Conference on Neural Information Processing Systems, pp. 1097–1105 (2012)
11. Russakovsky, O., et al.: Imagenet large scale visual recognition challenge. Int. J. Comput. Vision (IJCV) **115**(3), 211–252 (2015)
12. Amores, J.: Multiple instance classification: review, taxonomy and comparative study. Artif. Intell. **201**(4), 81–105 (2013)
13. Andrews, S., Tsochantaridis, I., Hofmann, T.: Support vector machines for multiple-instance learning. Adv. Neural Inf. Process. Syst. **15**(2), 561–568 (2002)
14. Manivannan, S., Cobb, C., Burgess, S., Trucco, E.: Sub-category classifiers for multiple-instance learning and its application to retinal nerve fiber layer visibility classification. In: International Conference on Medical Image Computing and Computer-Assisted Intervention, pp. 308–316 (2016)
15. Quinlan, J.R.: Induction of decision trees. Mach. Learn. **1**(1), 81–106 (1986)
16. Cover, T.M., Hart, P.E., et al.: Nearest neighbor pattern classification. IEEE Tran. Inf. Theory **13**(1), 21–27 (1967)
17. Zhang, Q., Goldman, S.A.: Em-dd: an improved multiple-instance learning technique. In: Advances in Neural Information Processing Systems, pp. 1073–1080 (2002)
18. Gadelmawla, E.S.: A vision system for surface roughness characterization using the gray level co-occurrence matrix. NDT & e Int. **37**(7), 577–588 (2004)

Machine Learning Based Trust Model for Misbehaviour Detection in Internet-of-Vehicles

Sarah Ali Siddiqui[1(✉)], Adnan Mahmood[1(✉)], Wei Emma Zhang[1,2], and Quan Z. Sheng[1]

[1] Department of Computing, Macquarie University, Sydney, NSW 2109, Australia
sarah-ali.siddiqui@hdr.mq.edu.au, adnan.mahmood@mq.edu.au
[2] School of Computer Science, The University of Adelaide, Adelaide, SA 5005, Australia

Abstract. The recent state-of-the-art advancements in vehicular ad hoc networks (VANETs) have led to the emergence and rapid proliferation of the promising notion of the Internet-of-Vehicles (IoV), wherein vehicles exchange safety-critical messages with one another to ensure safe, convenient, and highly efficient traffic flows. Nevertheless, such inter-vehicular communication could not be realized until the network is completely secured as the dissemination of even a *single* malicious message may jeopardize the entire network. Accordingly, numerous trust models have been proposed in the research literature to ensure the identification and elimination of malicious vehicles from a network. These trust models primarily depend on the aggregation of both direct and indirect observations, and which themselves are computed depending on the diverse influential parameters pertinent to dynamic and distributed networking environments. Still, optimum weights need to be allocated to these parameters for generating accurate and intuitive trust values. Furthermore, once the trust for a target vehicle has been computed, a specific threshold value equal to the minimum acceptable trust score has been selected for identifying the malicious vehicles. Quantification of these weights and selecting of an optimal threshold poses a significant challenge in VANETs. Accordingly, this paper focuses on employing machine learning techniques as to cope with the said problems in VANETs. It thus utilizes a real IoT data set by transforming it into an IoV format and computes the feature matrix for three parameters, i.e., *similarity*, *familiarity*, and *packet delivery ratio*, in two different ways, (a) all of the stated parameters computed by each trustor for a trustee are treated as individual features, and (b) the mean of each single parameter computed by all of the trustors for a trustee is regarded as a collective feature. Different machine learning algorithms were employed for classifying vehicles as *trustworthy* and *untrustworthy*. Simulation results revealed that the

The corresponding authors in seriatim acknowledges the kind support of the *Macquarie University's Research Excellence Program (Allocation No. 2018360)* and the *Government of the Commonwealth of Australia's International Research Training Program (Allocation No. 2017560)* respectively for funding the research at hand.

T. Gedeon et al. (Eds.): ICONIP 2019, CCIS 1142, pp. 512–520, 2019.
https://doi.org/10.1007/978-3-030-36808-1_56

classification via the mean parametric scores yielded much more accurate results in contrast to the one which takes into account the parametric score of each trustor for a trustee on an individual basis.

1 Introduction

Over the past few decades, the state-of-the-art technological breakthroughs in VANETs have played a significant role in the advancement of Intelligent Transportation Systems, which is an indispensable constituent of the emerging and promising paradigm of smart cities [1]. Today, the smart connected vehicles employ the notion of *vehicle-to-everything communication* in order to exchange safety-critical messages with the other vehicles on the roads, with the supporting roadside infrastructure and/or backbone networks, and with the vulnerable pedestrians in a bid to guarantee safe, secure, and efficacious traffic flows. Nevertheless, this could only be possible if the messages disseminated and/or exchanged by the vehicles are legitimate and are not altered or counterfeited, or else, this may become a potential source of threat, thereby, resulting in severe injuries and loss of precious human lives on the roads [2,3]. Malicious vehicles are competent of altering or counterfeiting safety messages, could restrict trusted vehicles from taking a part in network operations, and may exhaust network resources subsequently causing serious damage to both local and geographical networks [3]. It is, therefore, of paramount importance to guarantee the integrity of the disseminated and/or exchanged information so as to ensure that its sender is *trustworthy*.

In a trust-based model, a vehicle is evaluated by other vehicles in a vehicular cluster depending on several parameters, i.e., *the interactions between the vehicles*, *how similar their interests are*, and *how familiar/acquainted they are with one another*, among many others. The vehicle evaluating and assigning the trust scores to other vehicles is known as the *trustor*, whereas, the one being evaluated is referred to as the *trustee*. In general, this evaluation is an amalgamation of both the *direct trust* and an *indirect trust* for each vehicle. It is extremely indispensable to allocate weights to these parameters in order to ascertain accurate and intuitive trust values. The resulting trust score highly depends on the assigned weights and quantification of these weights further poses a considerable challenge. This essentially necessitates an in-depth knowledge of the effects of each of these individualized influential parameters on the trust evaluation (i.e., corresponding to the divergent traffic scenarios and vehicular applications) and is a complex analysis problem in its own essence. Furthermore, once the trust for a targeted vehicle has been computed, the malicious vehicles are identified by opting for a specific threshold value, i.e., *equal to the minimum acceptable trust score*, and the vehicles having a trust score below the specified threshold are considered untrustworthy. Thus, an optimal threshold selection is of huge significance, as if the threshold is extremely low, the system would not be able to filter out all the misbehaving nodes, whereas, if the said threshold is set too high, the trustworthy nodes might also get evicted from the network.

Accordingly, this paper primarily focuses on exploiting machine learning techniques to cope with the problems of optimal weights and threshold selection within VANETs. It thus employs a real IoT data set by transforming it into an IoV format and subsequently computes the feature matrix for three parameters, i.e., *similarity* – manifesting as how similar are the interests of the trustor and the trustee, *familiarity* – depicting how good the trustor knows the trustee, and the *packet delivery ratio* – delineating the throughput between the trustor and the trustee. The said feature matrix has been computed in two different ways, (a) all of the stated parameters computed by each trustor for a trustee are treated as individual features, and (b) the mean of each single parameter computed by all of the trustors for a trustee is considered as a collective feature. Subsequent to the feature extraction and labelling process, different machine learning algorithms, i.e., support vector machine (SVM), k-nearest neighbors (KNN), ensemble subspace KNN, and subspace discriminant, etc. have been employed to classify vehicles into two classes, i.e., *trustworthy* and *untrustworthy*. Simulation results revealed that the classification via mean parametric scores yielded more accurate results in contrast to the one which takes into account the parametric score of each trustor for a trustee on an individual basis.

2 Related Work

A brief glimpse of the literature reveals a number of research studies envisaging various *trust management models* and *intrusion detection frameworks* for identifying malicious vehicles and subsequently eliminating them from within the network. Accordingly, in [3], a trust management heuristic based on job marketing signaling scheme has been proposed in order to promote cooperative behavior amongst different vehicles in a network. A credit is allocated to each individual node within the network, and every time a node behaves maliciously, an amount depending on the *cost of the attack* is deducted from the originally allocated credit so as to discourage the malicious vehicles. Similarly, once a node manifests a positive participation, the credit is subsequently increased to encourage the node's participation and its cooperation with the other nodes in the network. In [4], the authors proposed a fuzzy logic-based decentralized trust management framework that flags the unintentional misbehavior of a target vehicle by amalgamating the trustor's own experience and the suggested evaluation of it's neighbors. Moreover, indirect trust was also evaluated for trustees which were not directly connected to the trustor by utilizing the notion of reinforcement learning.

A blockchain-based privacy preserving distributed trust management scheme has been proposed in [5] which breaks the linkability between the public key and vehicle's real identity to achieve the anonymity when the certification authority issues or revokes the respective certificates. All the messages were recorded in the blockchain and trust scores were assigned to each individual vehicle by evaluating the data transmitted by them, thereby, discouraging misconduct. To mitigate the adversarial effects of malicious attacks and misbehaving vehicles in

VANETs, a noteworthy solution is to introduce an intrusion detection system (IDS) which utilizes signature- and anomaly-based detection schemes for the said purpose. Hence, a decentralized cooperative IDS has been proposed in [6] which employed the privacy-preserving distributed machine learning for ensuring a private collaboration. The collaborative nature of the proposed scheme encourages all the vehicles within the network to share their trained data along with the ground truth to provide a scalable, cost-efficacious, and higher quality mechanism. Moreover, a distributed classification solution has also been achieved using ADMM (i.e., alternating direction method of multipliers) algorithm. The IDS suggested in [7] inspects the traffic, employs a deep belief network for simplifying the data dimensionality, and distinguishes the genuine service requests from the counterfeited ones. Furthermore, it implements a service-specific clustering to ensure that the cloud services are available continuously, thereby, guaranteeing both the quality-of-service and quality-of-experience. In [8], the IDS amalgamated support vector machine and the promiscuous mode in order to build the trust table for the identification and the prevention of attacks, wherein every vehicle monitors its neighbor for the misconduct. Similarly, authors in [9] introduced multiple types of attacks in their proposed scheme by altering the safety messages exchanged by vehicles and subsequently classified different malicious (active) attacks by extracting distinguishing features and via utilizing machine learning techniques.

Whilst the existing literature has already demonstrated some significant contributions by applying numerous machine learning techniques, nevertheless, they still lack the potential of being a generic algorithm that could be commonly applied to any of the service domains and across diverse parameters. Moreover, the existing research studies merely rely on the conventional factors in the trust assessment process and the impact of the influential parameters *(i.e., similarity, familiarity, and packet delivery ratio)* on the trust assessment and aggregation process has been completely ignored.

3 System Model and Simulation Results

We envisage a machine learning-based trust management scheme to identify malicious (dishonest) vehicles for eradicating them from the network in a bid to restrict them from causing any further harm and to conserve the precious network resources. The proposed system model comprises of two main steps. The first step utilizes unsupervised learning algorithms to cluster and label the data, whereas, the second step relies on supervised learning algorithms for classifying the vehicles into two groups, i.e., *untrustworthy* and *trustworthy*.

The simulations are performed for a vehicular network (i.e., cluster) comprising of n vehicles as x_i, where $i = 1, \ldots, n$. Every vehicle x_i has j one-hop neighbors, where $j = 1, \ldots, (n-1)$ and $(i \neq j)$, and is evaluated by them, i.e., x_i is the trustee and x_j is the trustor. The evaluation transpires on the basis of three parameters, i.e., similarity ($SMR_{i,j}$), familiarity ($FMR_{i,j}$) and the packet delivery ratio ($PDR_{i,j}$). The parameter values vary in the range of *0* and *1*, wherein *0* represents the lowest correlation between a pair of a trustor and a trustee, whereas, *1* signifies the highest correlation of the said pair.

3.1 Data Set and Feature Extraction

For the envisaged system model, we have employed an IoT data set from CRAWDAD[1] by suitably transforming it into an IoV format. The proposed trust management model has been evaluated via MATLAB simulations for 20 vehicles. We defined three scoring parameters, i.e., *similarity*, *familiarity*, and the *packet delivery ratio* for evaluating each node in the network as follows:

Similarity (SMR) – In a vehicular network, the similarity ($0 \leq SMR \leq 1$) relates to the degree of similar content and services amongst any two vehicles. The similarity is computed as, $SMR_{i,j} = \frac{S_{i,j}}{S_j}$, where $S_{i,j}$ is the number of common content or services accessed by both the trustor and the trustee, and S_j is the total number of content or services accessed by the trustor.

Familiarity (FMR) – Familiarity ($0 \leq FMR \leq 1$) suggests the degree of how well a trustor is acquainted with the trustee. The familiarity is computed as, $FMR_{i,j} = \frac{F_{i,j}}{F_j}$, where $F_{i,j}$ is the number of common friends between both the trustor and the trustee, and F_j is the total number of the trustor's friends.

Packet Delivery Ratio (PDR) – The packet delivery ratio ($0 \leq PDR \leq 1$) depicts the throughput between the trustor and the trustee. The packet delivery ratio is computed as, $PDR_{i,j} = \frac{P_{i,j}}{P_j}$, where $P_{i,j}$ manifests the number of messages disseminated by the trustee i that were successfully received by the trustor j, and P_j is the aggregate number of messages sent to j by i.

These three parameters are calculated for each pair of a trustor and a trustee that exists in the vehicular network and the scores are thus recorded in two different feature matrices. In the first feature matrix, the rows in fact represent the trustees (there are n = *20* number of rows) and the columns represent the said parameters (SMR, FMR, and PDR) ascertained by each trustor for each trustee on an individual basis, i.e., there are $3n - 3$ number of columns. This feature matrix (see, Eq. 1) is formed with an intent to inspect the impact of each trustor for a trustee against each parameter in the final classification.

$$FM_1 = \begin{bmatrix} SMR_{11} \ldots SMR_{1n-1} & FMR_{11} \ldots FMR_{1n-1} & PDR_{11} \ldots PDR_{1n-1} \\ \vdots \quad \ddots \quad \vdots & \vdots \quad \ddots \quad \vdots & \vdots \quad \ddots \quad \vdots \\ SMR_{n1} \ldots SMR_{nn-1} & FMR_{n1} \ldots FMR_{nn-1} & PDR_{n1} \ldots PDR_{nn-1} \end{bmatrix} \tag{1}$$

On the contrary, in the second feature matrix, the rows represent the trustees (there are n = *20* number of rows) and the columns signify the mean of each

[1] The data set can be accessed at: https://crawdad.org/thlab/sigcomm2009/20120715/.

of the parameter (SMR, FMR, and PDR) computed for each trustee by all the trustors, i.e., there are 3 columns in total. This feature matrix (see, Eq. 2) is formulated with an aim to classify the vehicles on the basis of their mean parametric scores.

$$FM_2 = \begin{bmatrix} SMR_{avg_1} & FMR_{avg_1} & PDR_{avg_1} \\ \vdots & \vdots & \vdots \\ SMR_{avg_n} & FMR_{avg_n} & PDR_{avg_n} \end{bmatrix} \quad (2)$$

3.2 Clustering and Labelling

The computed score for each parameter is used to classify the vehicles into two clusters, i.e., *trustworthy* and *untrustworthy*. The said clusters are ascertained by employing the algorithms that we have envisaged on the basis of the unsupervised learning algorithms, i.e., *k-means*, *fuzzy c-means*, *hierarchical clustering*, and *gaussian mixture*, to label the feature matrices obtained in the previous subsection. The key rationale for employing all of these four unsupervised learning algorithms is to ensure a credible, reliable, and persistent ground truth. The cluster closer to the origin is categorized as the *malicious*, whereas, the other is regarded as the *trusted* one. In other words, vehicles with a higher parametric score are more credible in contrast to the ones having a lower parametric score. These obtained labels are subsequently incorporated into the feature matrices. It is pertinent to mention that both of these feature matrices would have different labels as the data points inside the said matrices are in contrast with one another.

Owing to space constraints, only the clustering of data points (i.e., vehicles) from FM_2 into two clusters is illustrated in Fig. 1. To facilitate the visuality, the clustering for each pair of features is depicted.

3.3 Classification Model

Subsequent to both the clustering and the labelling mechanism, the supervised learning classifiers have been employed to the resulting feature matrices for training with a *5-fold cross validation* so as to identify malicious vehicles by obtaining the decision boundary due to their distinct characteristics. A variety of the machine learning techniques based on *k-nearest neighbors*, *support vector machine*, and the *ensemble classification models* have been utilized.

The overall accuracy, malicious nodes' classification accuracy, precision, recall, F1-score, and the decision boundary for each classifier have been computed for performance evaluation purposes. Simulation results revealed that the classification via mean parametric scores yielded comparatively more accurate results in contrast to the one which takes into account the parametric score of each trustor for a trustee on an individual basis, as depicted in Fig. 2. It could be observed that the minimum overall classification accuracy while taking into account the mean parametric score is provided via Cubic KNN and Medium

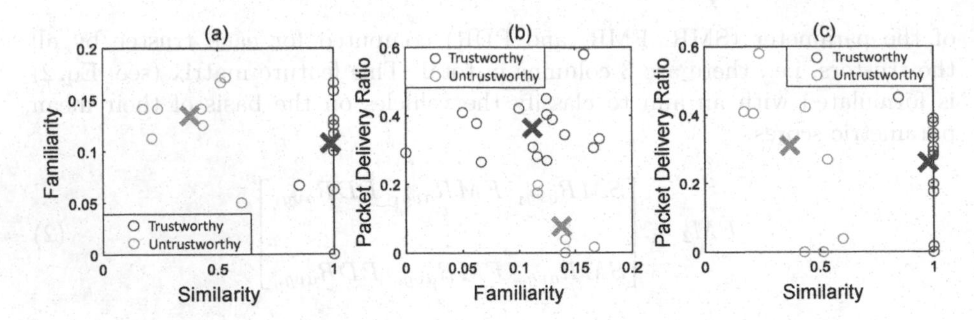

Fig. 1. Clustering of labels using unsupervised learning (a) similarity vs. familiarity, (b) familiarity vs. packet delivery ratio, and (c) similarity vs. packet delivery ratio

KNN as 90%, whereas, while using the individual parametric scores, the minimum overall classification accuracy is yielded via Linear SVM and Subspace KNN and is found to be 65%. It is pertinent to highlight that the best malicious vehicles' classification result of the proposed trust management model is yielded by taking the mean parametric scores and via Subspace KNN.

Figure 3 depicts the performance evaluation of the envisaged trust model with respect to malicious nodes' classification in terms of precision, recall, and F1-score. Precision is defined as the accuracy of the model to classify malicious nodes as malicious, whereas, recall is the proportion of the malicious nodes that have been correctly identified. F1-score represents the weighted mean of the two. All of the three performance evaluation metrics mentioned above ranges from *0* to *1*, i.e., *0* represents the *worst* and *1* manifests the *best* performing model. It could be observed that Subspace KNN yields the perfect precision, recall, and the F1-score equal to 1. Accordingly, Fig. 4 depicts the pair-wise decision boundary between the *trustworthy* and *untrustworthy* vehicles using Subspace KNN classifier.

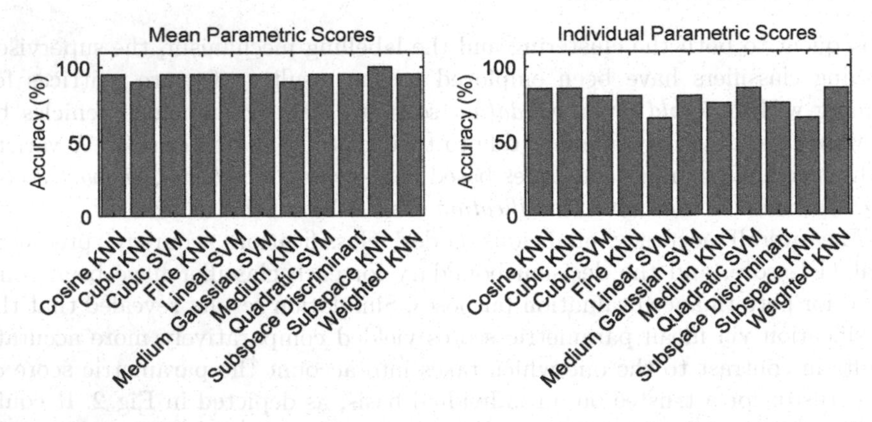

Fig. 2. Malicious vehicles' classification accuracy via different machine learning classifiers

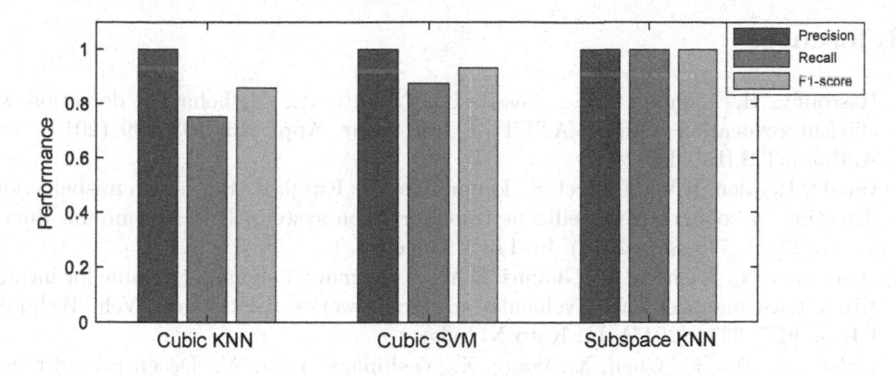

Fig. 3. Performance evaluation for malicious vehicles' classification (Precision, Recall, and F1-score)

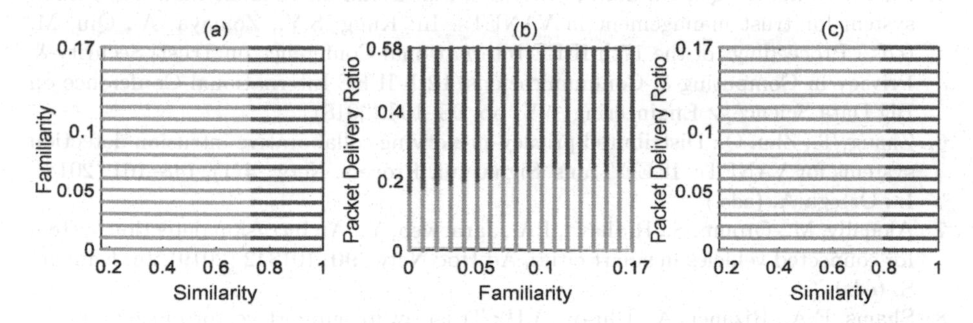

Fig. 4. Decision boundaries (a) similarity vs. familiarity, (b) familiarity vs. packet delivery ratio, and (c) similarity vs. packet delivery ratio (boundary for *untrustworthy vehicles* is depicted in red, whereas, blue manifests the *trustworthy* vehicles' region) (Color figure online)

4 Conclusion and Future Directions

In this paper, we have proposed a distributed trust management scheme that considers the notion of *similarity*, *familiarity*, and *packet delivery ratio* amongst the vehicles and employs supervised learning to identify and subsequently eradicate multiple malicious vehicles in real-time by ascertaining an optimal trust threshold. Our simulation results demonstrate the significance of these selected feature parameters in the classification of the dishonest vehicles. In the near future, the authors intend to apply the paradigm of online machine learning for trust management in vehicular networks to ensure a precise and resource efficient mechanism for the early eradication of misbehaving vehicles before they could disrupt the network performance.

References

1. Hasrouny, H., Samhat, A.E., Bassil, C., Laouiti, A.: Misbehavior detection & efficient revocation within VANET. J. Inf. Secur. Appl. **46**, 193–209 (2019). In: Anthony T.S.H. (eds)
2. van der Heijden, R.W., Dietzel, S., Leinmuller, T., Kargl, F.: Survey on misbehavior detection in cooperative intelligent transportation system. IEEE Commun. Surv. Tutor. **21**(1), 779–811 (2019). In: Lin Y.D. (eds)
3. Haddadou, N., Rachedi, A., Ghamri, D.Y.: A job market signaling scheme for incentive & trust management in vehicular ad hoc networks. IEEE Trans. Veh. Technol. **64**(8), 3657–3674 (2015). In: Kato N. (eds)
4. Guleng, S., Wu, C., Chen, X., Wang, X., Yoshinaga, T., Ji, Y.: Decentralized trust evaluation in vehicular IoT. IEEE Access **7**, 15980–15988 (2019). In: Abbott D. (eds)
5. Lu, Z., Wang, Q., Qu, G., Liu, Z.: BARS: a blockchain-based anonymous reputation system for trust management in VANETs. In: Kung, S.Y., Zomaya, A., Qiu, M. (eds.) Proceedings of the 17th IEEE International Conference on Trust, Security & Privacy in Computing & Communications/12th IEEE International Conference on Big Data Science & Engineering, NY, pp. 98–103 (2018)
6. Zhang, T., Zhu, Q.: Distributed privacy-preserving collaborative intrusion detection systems for VANETs. IEEE Trans. Signal Inf. Process. Netw. **4**(1), 148–161 (2018). In: Ortega A. (eds.)
7. Aloqaily, M., Otoum, S., Ridhawi, I.A., Jararweh, Y.: An intrusion detection system for connected vehicles in smart cities. Ad Hoc Netw. **90**, 101842 (2019). In: Kanhere S. (eds)
8. Shams, E.A., Rizaner, A., Ulusoy, A.H.: Trust aware support vector machine intrusion detection & prevention system in vehicular ad hoc networks. Comput. Secur. **78**, 245–254 (2018). In: Eugene H.S. (eds)
9. Grover, J., Prajapati, N.K., Laxmi, V., Gaur, M.S.: Machine learning approach for multiple misbehavior detection in VANET. In: Abraham, A., Mauri, J.L., Buford, J.F., Suzuki, J., Thampi, S.M. (eds.) ACC 2011. CCIS, vol. 192, pp. 644–653. Springer, Heidelberg (2011). https://doi.org/10.1007/978-3-642-22720-2_68

Toward the Ontology-Based Security Verification and Validation Model for the Vehicular Domain

Abdelkader Magdy Shaaban[1(✉)], Christoph Schmittner[1], Gerald Quirchmayr[2], A. Baith Mohamed[2], Thomas Gruber[1], and Erich Schikuta[2]

[1] Center for Digital Safety and Security, Austrian Institute of Technology, Vienna, Austria
{abdelkader.shaaban,christoph.schmittner,Thomas.Gruber}@ait.ac.at
[2] Faculty of Computer Science, University of Vienna, Vienna, Austria
{gerald.quirchmayr,abdel.baes.mohamed,erich.schikuta}@univie.ac.at
https://www.ait.ac.at/en/
https://www.univie.ac.at/en/

Abstract. Security verification and validation is an essential part of the development phase in current and future vehicles. It is essential to ensure that a sufficient level of security is achieved. This process determines whether or not all security issues are covered and confirms that security requirements and implemented measures meet the security needs. This work proposes a novel ontology-based security verification and validation model in the vehicular area. Ontologies allow creating a comprehensive view of threats and security requirements. The proposed model performs a series of queries and inference rules to the comprehensive view to ensure the compliance of vehicle components with security requirements.

Keywords: Ontology · Verification and validation · Potential threats · Security requirements

1 Motivational Background

Modern vehicles are part of a substantial ecosystem, including communication with stakeholders, infrastructures, customers, and authorities. The increase of connected units in vehicles leads to a considerable number of attack surfaces, which possibly leads to an increasing amount of security incidents. A vehicle might perform correctly according to the functional requirements; however, it can make other unintended tasks in the process. Furthermore, verification and validation (V&V) procedures can miss simply some of the hidden security defects, which lead to threatening the whole vehicle. Accordingly, the vehicular security requirements must be fulfilled [7]. One way to manage the structure of security requirements is to define them in groups called protection profiles. A Protection Profile (PP) is a document that describes the security considerations and resulting requirements for a Target of Evaluation (ToE) according to Common Criteria

T. Gedeon et al. (Eds.): ICONIP 2019, CCIS 1142, pp. 521–529, 2019.
https://doi.org/10.1007/978-3-030-36808-1_57

(CC) [5]. The ToE is an abstract description of a system or a system unit for specific usage. Besides, the PP identifies Security Target (ST) or security properties of ToE(s). It is essential to ensure the compliance of one or more PP(s) with identified ToE(s) to develop secure vehicles. This is especially important since systems designed for vehicular usage are often reused in a different context. Assuring that such a system complies with the PP for this context ensures that it is security needs are covered.

This work introduces a novel ontology-based security V&V model for the vehicular industry. The model creates a comprehensive ontological representation in terms of classes, subclasses, individuals, annotations, properties, and datatypes of vehicular ToE(s), threats, vulnerabilities, and security requirements (according to CC). A series of inference rules are applied to the ontology to determine whether or not the selected security requirements cover the security gaps, and confirms if security requirements meet the actual security condition. If this is not the case, it uses a Knowledge Base (KB) of several PPs to select additional security requirements. These additional requirements are applied to handle existing security weaknesses and assure the compliance with protection profiles to meet the ST of ToE. The ontologies assist in validating and verifying the operational and the performance of the security requirements against the vehicular security gaps. The paper is organized as follows; the related work on automotive cybersecurity is discussed in Sect. 2. The main contribution of this work is presented in Sect. 3. A description of threats and relevant security requirements of some interconnected units in a modern vehicle is described in Sect. 4. Section 5 demonstrates that the importance of ontologies in the V&V process to manage a massive amount of security requirements. Then, the paper ends with a summary, conclusion, and presents future work.

2 Related Work

In 2010 cybersecurity began to take more attention in the automotive industry [11]. The vehicles could have physical changes if malicious messages could be injected into internal parts of a vehicle such as the Controller Area Network (CAN bus) [8]. Nevertheless, the attack surface against vehicles not only by physical access but also there are several remote approaches. Ref. [1] defines four different methods for remote vehicle attacks. In modern vehicles, the diversity in communication protocols and heterogeneity between connected units lead to a potential increase in the number of security vulnerabilities. Furthermore, cybersecurity requires to be considered in all of the vehicular development phases. The development of vehicles is a distributed effort, regarding different organizations which use various methods. The majority of current security requirements verification processes are performed in the late phase of the development process since it needs the System Under Test (SUT) to be implemented, where both budget and time are very limiting circumstances [7].

The ontology approach has been proposed in several works in the cybersecurity domain [9]. Ref. [13] proposed a reference ontology to help in finding

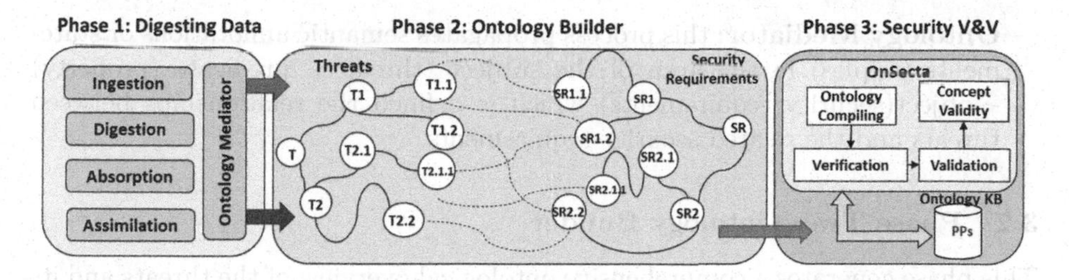

Fig. 1. The proposed ontology-based V&V model

security solutions to the Internet of Things (IoT) environment. The proposed reference ontology is based on the modeling process to unify concepts and explain relationships among the main components of risk analysis of information security. Ref. [12] introduced a technical framework to monitor business process and technology assets using an ontology and knowledge reasoning for IoT security.

3 Ontology-Based Security V&V Model

The proposed model uses ontologies to describe a set of representational primitives of classes, individuals, and annotations of security properties of vehicles. The ontology generates new machine-processable meta-data for the vehicle security information, and then the model creates a domain knowledge. The domain knowledge is essential for identifying the relationships between threats and security requirements to verify and validate these security requirements according to CC in one or multiple PP(s). This Section describes the structure of the proposed model, as shown in Fig. 1. The model consists of three main phases.

3.1 Phase One: Digesting Data

This phase receives data of ToE(s) with all related threats and security requirements. These data are processed by multiple sub-phases to extract the required information [6].

- **Ingestion:** collects the data are as follows:
  - list of identified assets with all related information,
  - all the detected threats with all related information details (i.e., name, id, type, description, and risk severity),
  - list of the security requirements according to the selected PP(s).
- **Digestion:** processes the raw data into a standard form that can facilitate to extract specific values from the original format.
- **Absorption:** extracts all data values which are needed to create an ontological representation from the input.
- **Assimilation:** acts as a filter to get rid of all unnecessary data. For example, the threats with low severity risk are not considered as significant security issues to threaten a vehicle.

- **Ontology Mediator:** this process propagates semantic annotations or statements (triples) in the form of the subject (threat) – predicate (property) – object(security requirements) which is defined the relationships between threats and the related security requirements.

3.2 Phase Two: Ontology Builder

This phase generates a comprehensive ontological overview of the threats and its relationships with security requirements. This overview has two main hierarchies:

- **Threats Hierarchy:** this is a hierarchical representation of a typical construction of vehicle threats.
- **Security Requirements Hierarchy:** it is a semantic representation of security requirements that are related to a specific PP for addressing potential vehicle threats.

Afterward, this phase creates an ontology linking between the threats hierarchical nodes and the security requirements nodes. This process defines links between these two hierarchical ontologies, which represent that the selected security requirements can handle one or more potential threat(s). The output of this phase is called "Ontology Outlook" as is illustrated in Fig. 1 phase two. The left side of the ontology outlook represents the threats, whereas the security requirements are illustrated on the right side.

3.3 Phase Three: Security Verification and Validation

This phase is the core of the proposed model, which consists of two main parts:

Ontology Knowledge Base: this is a set of specific instances of PPs with all included security requirements and common criteria in an ontology representation format.

Ontology Security Testing Algorithm (OnSecta): is an ontology reasoner uses the Ontology Outlook to perform security V&V procedure:

- **Ontology Compiling:** this process compiles the contents of the Ontology Outlook (i.e., classes, subclasses, terms, annotations, and properties); this allows understanding the ontology linking between threats, and security requirements.
- **Verification:** performs a set of queries for ensuring that the vehicular ToEs are developed regarding CC according to specific PP.
- **Validation:** it assures the compliance of ToEs with PP to meet the actual ST. If that is not specified, OnSecta performs a series of inference rules to select new security requirements from other PPs in the KB to reach the actual ST.
- **Concept Validity:** this activity checks the content of the ontology KB to find new security requirements from other PPs.

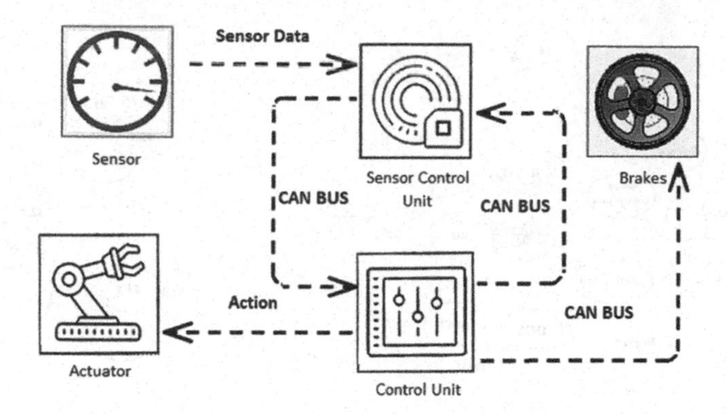

Fig. 2. Data flow between internal components in a modern vehicle

4 Case-Study: Modern Vehicles in Smart Farming

Future farming needs combination with innovative technologies to adapt and improve the production process. Smart farming applies and combines smart things with approaches from industry4.0 and intelligent mobility to address the challenges and improve a holistic system [9]. Integrating modern vehicles with current and future farming applications makes the farming process easier. The case-study shows a simple example of interconnected units in a modern vehicle as is depicted in Fig. 2. The Figure contains a "Sensor" unit that collects data from the external environment. The sensor data are sent to "Sensor Control Unit" to process these inputs. Then the "Control Unit" manipulate the data to deliver the appropriate action to the "Actuator" unit for different action scenarios (i.e., drilling, fetching, cutting, etc.). Besides, the "Control Unit" controls the tracking of the vehicle according to different situations, such as controlling the vehicular "Brakes."

A secure vehicle can be developed only if the exact security requirements are fulfilled against potential threats. In the course of the authors' research, they developed the Threat Management Tool (ThreatGet). ThreatGet identifies, detects, and understands potential threats in the vehicular sector. It integrates the initial steps of the developing vehicular process to guarantee the security-by-design [3]. In addition, the authors created a security requirement tool is called Model-Based Security Requirement Management Tool (MORETO) [16]. MORETO aims to manage a vast number of PPs with all related security requirements according to CC. The ThreatGet and the MORETO tools are applied to this example to define potential threats, manage, and select security requirements. Afterward, the ontology-based model is applied to validate and verify the selected security requirements. The model generates multiple classes, subclasses, individuals, properties, and annotations of all detected potential threats and selected security requirements. Then it generates the Ontology Outlook to depict a comprehensive overview of all threats and security requirements as discussed in Sect. 3.2. Figure 3 shows the structure of the Ontology Outlook; this structure consists of three main parts:

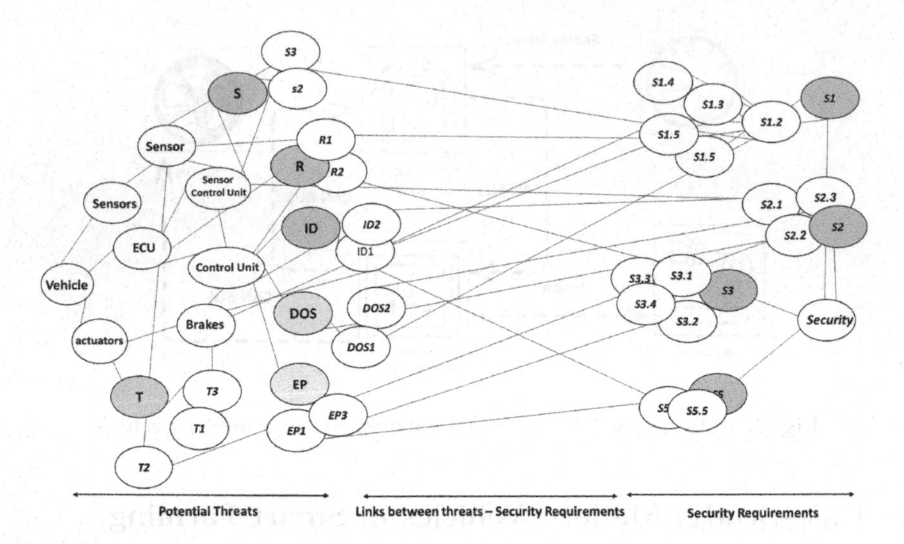

Fig. 3. Ontology outlook: ontology hierarchy between threats (left) and security requirements (right)

- **Potential Threats (left-side):** this hierarchy has all the vehicular units, which are defined in this case study. The colored nodes define the threat categories regarding the STRIDE model (i.e., Spoofing (S), Tampering (T), Repudiation (R), Information Disclosure (ID), Denial of Service (DoS), and Elevation of Privilege (EP)) [17]. The leaf nodes represent the actual detected potential threats.
- **The Security-Requirements (right-side):** this hierarchy represents CC are used to handle the potential threats. The colored nodes represent the category of security requirements (i.e., access control, communication port access, use control, data confidentiality, and so on). The leaf units represent the exact security requirements.
- **The Links Between the Two Ontologies:** the links between these two hierarchies can be defined not only between leaves of the hierarchies but also between internal nodes. Accordingly, a node specifying a more general threat type in the threat ontology can link to a subtree in the security requirements hierarchy identifying a set of similar security requirements can fit for handling related security issues [15].

OnSecta uses SPARQL language to perform queries across diverse data sources (threats and security requirements). These queries are applied to ensure that a vehicle is being developed based on standard security requirements, according to CC. Besides, to assures, the compliance of ToEs with PP meet the actual ST. OnSecta applies a series of rules to specify new PPs and selects additional security requirements. The rules are based on Semantic Web Rule Engine (SWRL), SWRL builds up a Horn clause representing the simple if-then conditional statement likewise formally from the Ontology KB to select proper security requirements [4].

5 Model Evaluation

Ontologies are considered a powerful method that uses regular specifications for knowledge representation such as vocabularies, taxonomies, classes, individuals, and annotations. Ontologies function acts like the human brain. They work and reason with concepts and relationships among multiple entities. That is considered the same way as humans perceive interlinked thoughts [14]. Furthermore, ontologies are integrated with this proposed model to perform security verification and validation in the vehicular domain. The vehicle development process requires to merge a significant number of security requirements according to multiple PPs. For instance, the requirements that relate to the Security Development Lifecycle (SDL) are appropriate to all industrial application such as vehicle development [10]. Managing hundreds or thousands of security requirements is considered a challenging task because it is time-consuming and complex work. The structure of the ontologies has a significant role in reducing the query complexity [2]. Furthermore, OnSecta manages ontologies by applying queries and rules over a massive number of ontology entities and define relationships and concept matching new security requirements to achieve a particular ST. Especially in the automotive domain the basic hardware of ECUs is often used for multiple vehicle types and even roles in the same vehicle where an adaption to new roles is done only by software and configuration. Giving guidance on the necessary security requirements for a specific role will ease the re-usability and adaptability of ECUs.

6 Conclusion and Future Work

To conclude this contribution, security verification and validation in the vehicular domain is one of the most critical challenges in the vehicular industry. On the first hand, it is quite a time, and effort consuming process to manage hundreds of interconnected units with thousands of threats. On the second hand, multiples of security requirements address potential threats according to CC. This work introduced an ontology-based security V&V model for current and modern vehicles. Ontologies are used to define domain knowledge representation of potential threats in vehicles and security requirements in multiples of PPs. The core of this model is OnSecta, which applies queries and a series of inference rules to perform verification and validation process to ensure the compliance of vehicle components with PP to meet a required ST. Future work will include the following points:

- **Protection Profiles:** create ontological representations of the most common security requirements in the vehicular domain.
- **OnSecta Implementation:** OnSecta is still in the developing stage; the authors work on developing the different building blocks of OnSecta.
- **Comparative Study:** the future work will include a comparative study between the proposed method with other kinds of typical techniques in the related domain to validate the superiority of the proposed method.

Acknowledgment. This work has received funding from the "AFarCloud" project, under grant agreement #783221. The project is partially funded by the EC Horizon 2020 Programme, ECSEL JU, and the partner National Funding Authorities (for Austria these are bmvit and FFG).

References

1. Checkoway, S., et al.: Comprehensive experimental analyses of automotive attack surfaces, San Francisco (2011)
2. Choksuchat, C., Chantrapornchai, C.: Benchmarking query complexity between RDB and OWL. In: Kim, T., Lee, Y., Kang, B.-H., Ślęzak, D. (eds.) FGIT 2010. LNCS, vol. 6485, pp. 352–364. Springer, Heidelberg (2010). https://doi.org/10.1007/978-3-642-17569-5_35
3. El Sadany, M., Schmittner, C., Kastner, W.: Assuring compliance with protection profiles with threatget. In: Romanovsky, A., Troubitsyna, E., Gashi, I., Schoitsch, E., Bitsch, F. (eds.) SAFECOMP 2019. LNCS, vol. 11699, pp. 62–73. Springer, Cham (2019). https://doi.org/10.1007/978-3-030-26250-1_5
4. Hebeler, J., Fisher, M., Blace, R., Perez-Lopez, A.: Semantic Web Programming. Wiley, Hoboken (2011)
5. IEC: ISO/IEC 15408-1:2009-information technology-security techniques-evaluation criteria for it security-part 1: Introduction and general model (2009). standard
6. Josverwoerd: Digesting big data. https://blog.bigml.com/2012/11/12/digesting-big-data/. Accessed 24 Sept 2019
7. Kastebo, M., Nordh, V.: Model-based Security Testing in Automotive Industry. Master's thesis, Department of Computer Science and Engineering - University of Gothenburg, Gothenburg, Sweden (2017)
8. Koscher, K., et al.: Experimental security analysis of a modern automobile. In: 2010 IEEE Symposium on Security and Privacy (2010)
9. Magdy, A., Schmittner, C., Gruber, T., Mohamed, A.B., Quirchmayr, G., Schikuta, E.: CloudWoT-a reference model for knowledge-based IoT solutions. In: iiWAS (2018)
10. McAfee: Automotive security best practices. Technical report, McAfee (2016)
11. Miller, C., Valasek, C.: Remote exploitation of an unaltered passenger vehicle. Black Hat USA (2015)
12. Mozzaquatro, B., Agostinho, C., Goncalves, D., Martins, J., Jardim-Goncalves, R.: An ontology-based cybersecurity framework for the internet of things. Sensors **18**, 3053 (2018)
13. Mozzaquatro, B.A., Jardim-Goncalves, R., Agostinho, C.: Towards a reference ontology for security in the internet of things. In: IEEE International Workshop on Measurements & Networking (M&N) (2015)
14. Ontotext: What are ontologies? https://ontotext.com/knowledgehub/fundamentals/what-are-ontologies/. Accessed 22 Sept 2019
15. Schikuta, E., Magdy, A., Mohamed, A.B.: A framework for ontology based management of neural network as a service. In: Hirose, A., Ozawa, S., Doya, K., Ikeda, K., Lee, M., Liu, D. (eds.) ICONIP 2016. LNCS, vol. 9950, pp. 236–243. Springer, Cham (2016). https://doi.org/10.1007/978-3-319-46681-1_29

16. Shaaban, A.M., Kristen, E., Schmittner, C.: Application of IEC 62443 for IoT components. In: Gallina, B., Skavhaug, A., Schoitsch, E., Bitsch, F. (eds.) SAFECOMP 2018. LNCS, vol. 11094, pp. 214–223. Springer, Cham (2018). https://doi.org/10.1007/978-3-319-99229-7_19
17. Shostack, A.: Experiences threat modeling at microsoft, vol. 413 (2008)

Encephalographic Assessment of Situation Awareness in Teleoperation of Human-Swarm Teaming

Raul Fernandez Rojas[1(✉)], Essam Debie[1], Justin Fidock[2], Michael Barlow[1], Kathryn Kasmarik[1], Sreenatha Anavatti[1], Matthew Garratt[1], and Hussein Abbass[1]

[1] School of Engineering and IT, University of New South Wales, Canberra, Australia
r.fernandezrojas@adfa.edu.au
[2] Defence Science and Technology Organisation, Adelaide, Australia

Abstract. An important factor in the operational success of any teleoperated human-swarm system is situation awareness (SA). A loss of SA has been associated with poor human performance, which can lead to misjudgement, errors, and life-threatening situations. One of the major factors that causes loss of SA is the degradation of data transmission. It is imperative to assess the SA of an operator before the performance of a teleoperated system has declined, in particular in situations of delayed relay and/or loss of critical information. We use electroencephalography (EEG) to predict different levels of SA. A human-swarm simulation was used to obtain subjective scores from participants. Quality of information significantly affected the perception of SA of the participants. EEG data provided objective confirmation of the resultant SA level. Theta, Alpha, and Beta band exhibited an increase during loss of SA. Frontal and occipital areas were identified to reflect changes in SA. These preliminary results offer evidence for the potential use of EEG to offer real-time indicators for the objective assessment of SA.

Keywords: Cognitive assessment · Human performance · EEG · Teleoperation

1 Introduction

Teleoperated systems provide humans the ability to perform difficult or dangerous tasks that otherwise cannot be achieved by having humans in-situ. Teleoperation can be defined as doing work at a distance. Typical examples of applications of teleoperated systems are: handling of nuclear materials, assisting in rescue missions underwater or after natural disasters (e.g., earthquakes), performing surgery on unreachable areas in the human body by a surgeon, and conducting surveillance in war and high-risk regions. In these contexts, one or more humans use an interface to control one or several robots from a command room, which might be in a different geographical location from the area of operations.

T. Gedeon et al. (Eds.): ICONIP 2019, CCIS 1142, pp. 530–539, 2019.
https://doi.org/10.1007/978-3-030-36808-1_58

An important factor in the operational and mission success of any teleoperated system is situation awareness (SA); a term that is normally used to describe the recognition, understanding, and future projection of the elements in a situation's context. In her seminal work, Mica Endsley [6] provided a SA framework that decomposes it into three levels: (1) Perception of elements in the current situation, to gather all the information that is currently available; (2) comprehension of current situation, to synthesize raw information into meaningful patterns/clues to understand the current situation; and (3) projection of future states, where the cumulative understanding of the elements in the surrounding environment is used to predict the status and dynamics of these elements in the future.

One of the major factors that causes loss of SA is degradation of data [6]. Failures in data transmission exhibit a common set of problems including information latency (timing is too late to be effective), and information loss (contents are completely lost, not consistently complete or inaccurate) [8]. In particular, information loss, has been shown to have a degrading impact on SA [6]. In aviation, for instance, Thornton [11] has shown that the number of communications requesting clarification made by aircraft crew members was positively correlated with the number of committed errors. Therefore, it is imperative to predict the loss of situation awareness before performance declines.

A method that has often been used to measure SA is the situational awareness global assessment technique (SAGAT), which was developed by Endsley [6] to measure individual SA. In human-machine systems, it is used to evaluate system design to ensure that the system in question supports the operator's SA requirements. SAGAT is used during a simulation of the system, it works by freezing the simulation at randomly selected times and operators answer questions about their perception of the situation at that time. The questions examine the three levels of SA (perception, comprehension, and projection). However, this method is not well suited for contexts where an uninterrupted and objective assessment is desirable.

Previous attempts to measure SA objectively are limited and have shown diverse results. Catherwood et al. [2], used electroencephalography (EEG) to map brain activity during loss of situation awareness in identification of target patterns and threats in urban scenes; their results showed that loss of situation awareness activated cortical areas associated with cognition, such as, prefrontal, anterior cingulate, parietal, and visual regions. Yeo et al. [14] used EEG to monitor SA in an air traffic controller (ATC) task, their model predicted the response latency of the ATC operators with a 10% error. In another study, Berka et al. [1] used EEG to determine SA in a naval command task; the study was based on EEG-engagement and EEG-workload metrics and their results showed that engagement and workload decreased as participants gained experience in the simulation task, which represented better SA.

The effect on quality of information in the perceived SA in teleoperated systems has not been explored in any of the current literature. We close this gap in this paper by using electroencephalography (EEG) to predict different levels of SA in a human-swarm interaction task. The hypothesis of this study is that the latency and loss of information will impact operators' SA in our teleoperation system and that EEG can be used to identify these changes in the perceived SA by the operators.

2 Methods

2.1 Task and Scenario Design

The experimental task is undertaken using the Virtual Battlespace 3 (VBS3) (Bohemia Interactive Simulations, Orlando, FL, USA) simulation environment. Subjects teleoperate an Uninhabited Aerial Vehicle (UAV) to guide a swarm formation of autonomous unmanned ground vehicles (UGVs). Only the UAV remote-operator knows the destination defined by the mission profile. The UGVs consist of a group of 4 vehicles with capabilities to self-organize to autonomously maintain a formation during the mission. The operator's graphical user interface displays sufficient information to successfully guide the UAV.

The interface (Fig. 1) has two main displays located side by side on the top. On the left side, there is a lateral view of the UAV and UGVs' positions on a map. The UAV is presented by a green rectangle and the UGVs are visualized as blue rectangles. A blue star marks the UGVs' destination on the map. On the right side, real-time video streamed from the UAV camera is provided to the operator. At the bottom of the interface, detailed information on the UAV and UGVs' status including their positions, headings and speeds are provided. In the middle of the interface, a panel lists all possible UGV formation options, however, for this study we limit the formation to a boxing formation alone.

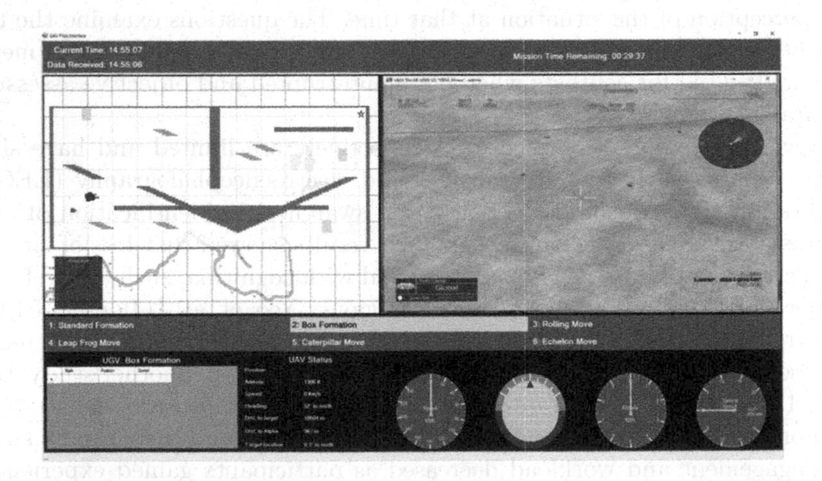

Fig. 1. UAV pilot interface.

Each experiment runs a simulation that combines four scenarios of different levels of information quality. Each scenario lasts 4 min and is repeated twice.

Three SAGAT questions are given to the participants in each scenario, in which the simulation is frozen at randomly-chosen times for 20 s per each question.

2.2 Experimental Design

A within-subject design with four different conditions determined by different levels of quality of information was used in this study. The four conditions (scenarios) are: (1) low latency/delay and low dropout; (2) low delay and high dropout; (3) high delay and low dropout; and (4) high delay and high dropout. The experiment is counterbalanced by using the composite 3×3 Latin Square design to avoid confounding due to order effects. In our experiment, information latency is the amount of time a video frame from the UAV camera and the status of all vehicles to traverse in the camera's field of view are delayed to the interface; while, information loss is the time in which video frames and data about the status of vehicles is lost during data transmission.

However, to study the effect of these two variables on the perceived SA, artificial information latency and information loss are injected into the system. These two variables are modelled using two parameters, d for the delay time (Low $d = 1$ s, High $d = 9$ s) of information transmission, and lf for the number of video frames lost per second (Low $lf = 1$ s, High $lf = 9$ s) in transmission.

2.3 Participants

Ten participants (mean age 31 ± 5.9 std.) were recruited for the study. The experiment was approved by the Faculty Research Ethics Committee (approval number: HC180554) and all participants provided written informed consent to participate in the study. An introduction to the experimental procedure and practice session were provided to the participants before the start of the study. The participants were instructed to start the experiment after a 2-minute break, the complete session lasted approximately 50 min.

2.4 Electroencephalography (EEG)

A wireless EEG acquisition system (Emotiv EPOC) was used to record neural activity. This device has a resolution of 14 channels (plus 2 reference channels) with a sampling frequency of 128 samples per second. Figure 2 presents the headset and the channel positions based on the international 10–20 EEG system of electrode placement. Channel locations correspond to: AF3, F7, F3, FC5, T7, P7, O1, O2, P8, T8, FC6, F4, F8, AF4, M1, and M2. M1 is used as the ground reference channel for measuring the voltage of the other channels, while M2 is used as a feed-forward reference point to reduce external electrical interference.

EEG pre-processing was performed in Matlab (version 2018b, The MathWorks Inc.) by using custom software. Baseline correction was performed by subtracting the corresponding mean from a pretrial (200 ms) period from each channel. Then, EEG signals were band-pass filtered between 2 and 43 Hz using a

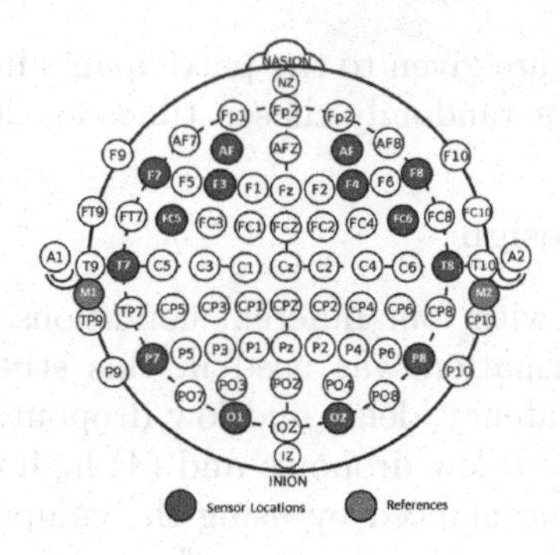

Fig. 2. Channel location of the 14-channel Emotiv headset.

FIR filter, which helps remove high-frequency artefacts and low-frequency drifts. Electrode movement artefacts were manually removed from the data. Artefacts from eye blinks were corrected using the multiple artefact rejection algorithm (MARA) [13].

2.5 Feature Extraction and Classification

Feature extraction was carried out using spectral analysis. First, the power distribution from each channel was studied by transforming the EEG into power spectral density (PSD) using a fast-Fourier transform (FFT) and using 10-s windows with 50% overlapping windows multiplied by the Hamming function to reduce spectral leakage. Second, from each window, the EEG channels were decomposed into sub-bands: delta (2–4 Hz), theta (4–8 Hz), alpha (8–12 Hz), beta (12–30 Hz), and gamma (30–40 Hz). Third, the PSD results of each frequency band were normalized (1/f) to obtain the relative PSD of each band to the baseline time period. This normalization helps to make quantitative comparisons of power across frequency bands. Finally, the resulting PSD values in each band were averaged to obtain the power spectral features used for classification.

The classification task is to determine the level of SA based on the recorded EEG signals from each participant. To identify the four levels of SA, we used the linear discriminant analysis (LDA) algorithm. The reason to choose LDA is because it is the most popular classifier in BCI research due to its good performance and low computational cost, attributes needed for the development of an on-line assessment of SA in our future work. To measure the classifier's performance, the data was divided into two parts with 70% for training and the remaining 30% used for testing and to report generalisation performance. k-fold cross validation ($k = 10$) was performed on the training set; the training set was randomly divided into k partitions. Then, k-1 partitions are used to fit the learning model and the remaining partition used to validate the model, this process is repeated k times, and each time using a different partition to validate the model. The final generalisation results are presented as the average and standard deviation on the 30% untouched test set.

3 Results

3.1 Situation Awareness

SAGAT was used with an hypothesis that in conditions with low drop out and low delay (e.g., Scenario 1), the rate of correct responses to the questionnaire will be higher than the rate of correct responses in conditions with high drop out and high delay (e.g., Scenario 4). Figure 3 presents the SAGAT response obtained from the subjects. The recorded response to each question is stored as correct or incorrect answer for each condition and the three responses in each condition are averaged. The overall trend of subjects' response to SAGAT questionnaire shows higher SA ($mean = 0.83$) in good-quality communication scenarios (Scenario 1) and low SA ($mean = 0.48$) in bad-quality communication scenarios (Scenario 4); these findings are in line with our hypothesis. The SAGAT response was then used for statistical analysis.

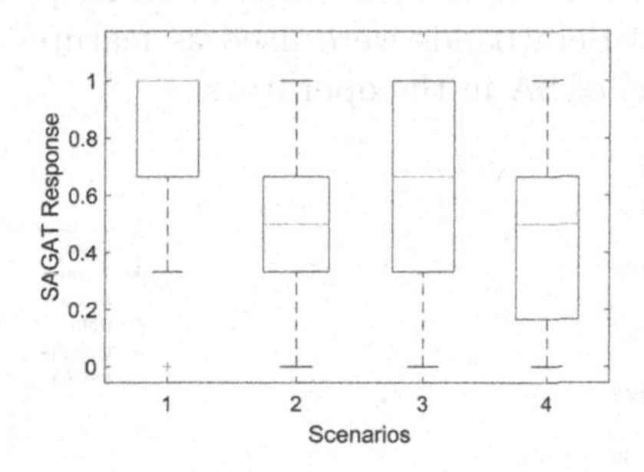

Fig. 3. Subjects response to SAGAT questionnaire.

A repeated measures ANOVA test was used to examine changes in mean scores under different conditions (scenarios). The research hypothesis is that the mean SAGAT scores are significantly different in at least two conditions (i.e., that the distribution of responses varies according to the scenario). Therefore, the null hypothesis is as follows, H_0: The mean SAGAT scores are the same at all conditions. The response rate of each question is measured and the observed means (the SAGAT scores) in each condition are obtained. The level of significance to reject H_0 is $\alpha = 0.05$. The results (Sphericity Assumed $p = 0.21$) showed that mean participants response to the SAGAT questionnaire differed significantly between conditions [$F(3, 27) = 4.541$, $p = .011$, *partial* $\eta^2 = 0.335$].

Tukey's multiple comparison was used as post-hoc test. The pairwise comparisons revealed significant differences between the communication conditions with good-quality(scenario 1), medium quality (scenario 2) ($p = 0.026$), and bad quality (scenario 4) ($p = 0.017$). Based on these results, we conclude that the difference in the mean scores of these three conditions (Scenario 1, Scenario 2, and Scenario 4) is significant and that the situation awareness assessment is significantly ($p < 0.05$) affected by the quality in communication presented in our teleoperated system.

3.2 Neural Response to Situation Awareness

We investigated the cognitive perception to SA using EEG. The majority of the neural response was observed in the frontal area, in particular in the Delta, Theta, and Alpha bands; however, in the beta and gamma bands the activation was also observed in channels located on the temporal lobe (such as, T7 and T9). The hypothesis is that channels with strong response will present better discrimination between different levels of SA.

First, we determined the frequency bands that showed the best response using all the available channels. Figure 4 presents the average response of each feature to different levels of SA. Overall, features from all frequency bands presented an increase between Scenario 1 and Scenario 4, and this increase was statistically significant ($p < 0.01$) only in Theta, Alpha, and Beta bands. These results suggest that the operators' perception to good-quality communication scenarios (Scenario 1) and bad-quality communication scenarios (Scenario 4) can be observed as an increase of PSD value in all frequency bands. Therefore, the Theta, Alpha, and Beta bands were used as features to train our classifier and to predict the level of SA in the operators.

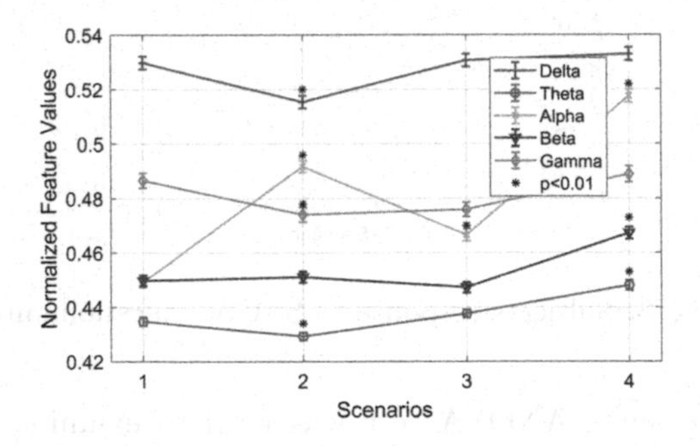

Fig. 4. Average response of all subjects for each frequency band to different levels of SA

Second, based on the observed frequency response (Fig. 4), the channels that best responded to each scenario were explored. The response distributions of Theta, Alpha, and Beta were explored in all channels, and the best data separation were observed in Channels F3, O2, and F4 (please refer to Fig. 2 for cortical locations). Figure 5 shows the distributions across the scenarios (1–4) using the average of the three bands. The overall trend among all the channels showed similar distributions, however, these channels (F3, O2, F4) presented better separation between good-quality communication scenarios (Scenario 1) and bad-quality communication scenarios (Scenario 4).

Third, two conditions were tested in the classification task using the LDA classifier to corroborate the observed distributions. These conditions are: using features of all the frequency bands, and using features of only Theta, Alpha, and Beta. Table 1 presents the classification results (accuracy ± std.) of these

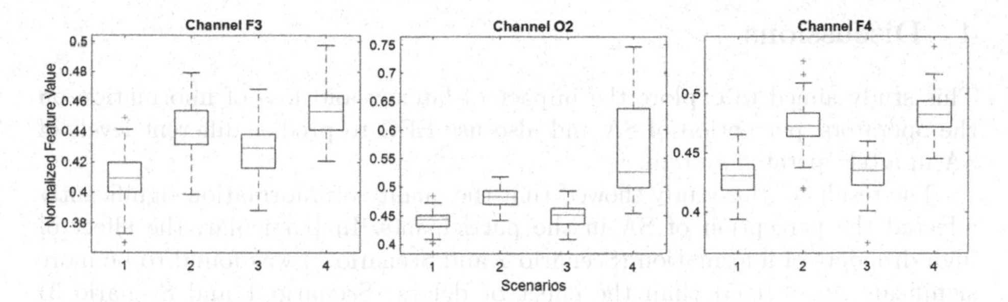

Fig. 5. Average response of all subjects to different levels of Situation Awareness using Theta, Alpha, and Beta bands.

two conditions. Overall, the LDA classifier using only the three PSD features (Theta, Alpha, Beta) showed slightly better results (76.22 ± 10.2) than the classifier using all PSD features (72.34 ± 8.2). These results are in line with the frequency response of each band presented in Fig. 4. For instance, the uniform (non-significant) response of Delta and Gamma bands between scenarios (except for Scenario 2) suggest that the discrimination between scenarios becomes more difficult when these two bands are included in the feature space, therefore the lower classification rate in most of the subjects.

Table 1. Performance (accuracy in %) results of the LDA classifier.

	Number of features	
	Theta, Alpha, Beta	All PSD features
Subject 1	92.03 ± 3.3	80.07 ± 4.7
Subject 2	75.22 ± 6.6	63.71 ± 6.6
Subject 3	76.99 ± 4.2	79.64 ± 5.9
Subject 4	61.94 ± 5.0	60.17 ± 6.0
Subject 5	87.61 ± 3.7	74.33 ± 5.6
Subject 6	81.41 ± 5.4	82.30 ± 4.4
Subject 7	75.39 ± 4.6	69.91 ± 5.2
Subject 8	59.13 ± 7.1	61.06 ± 7.3
Subject 9	80.53 ± 6.1	74.33 ± 6.8
Subject 10	71.94 ± 4.6	77.87 ± 5.9
Average	76.22 ± 10.2	72.34 ± 8.2

4 Discussions

This study aimed to explore the impact of latency and loss of information on the operators' perception of SA and also use EEG to predict different levels of SA in a teleoperated system.

The results of this study showed that the quality of information significantly affected the perception of SA in the participants. In particular, the effect of high dropouts of information (Scenario 2 and Scenario 4) was found to be more significant ($p < 0.05$) than the effect of delays (Scenario 1 and Scenario 3) in the subjective SAGAT scores. These results suggest that the loss of SA in scenarios with high dropouts reflects the operators' difficulty in understanding and identifying new information after the loss of an already-familiar scenario. These gaps of information also indicated that the operators might be missing important information to maintain a complete perception of the environment. This is in line with previous studies in the effect of automation in teleoperated system, where operators encounter new information after the use of automation and have reported poorer results in SAGAT scores [4].

In addition, EEG data provided objective confirmation of the perceived SA in all four scenarios. In particular, Theta, Alpha, and Beta bands showed a significant ($p < 0.01$) increase in the amplitude of their PSD values between the good-quality communication scenario (Scenario 1) and the low-quality communication scenario (Scenario 4). Increased activity in the Theta band has been related to increased mental load during focused attention [7]. Alpha band has been associated with the level of attention in visual tasks [9]. In addition, Theta, Alpha, and Beta bands have been linked to task engagement and attention [10]. In our study, the increase in PSD values in these bands might reflect the subjects' adaptation effect (or increased attention) to understand (or adapt to) new information after each loss of information during the simulation; and this increase is more evident in Scenarios 2 and 4, where the dropout of information is as long as 9 s.

The cortical location of the Channels F3, O2, and F4 might explain their distribution response, in particular during the loss of SA (e.g., Scenario 4). For instance, Catherwood et al. [2] identified F3 and F4 cortical areas (Brodmann area BA8) as one of the regions to be associated with decisions under uncertainty, conflict, or ambiguity; conditions that the operators might faced during the loss of SA in our simulation. On the other hand, Channel O2 corresponds to the occipital lobe, in particular this cortical region is responsible for processing visual stimuli [3,12]; in our experiment, during scenarios with bad-quality information, brain activity in the occipital region suggest that visual perception is actively occurring. This also demonstrates that the occipital lobe is constantly processing visual stimuli in particular after facing new or unexpected environments.

In summary, preliminary results presented in this study have provided evidence of the effect of delay and dropout of information in the perception of SA and the use of EEG as a possible objective indicator of SA. Future work will focus on using the identified neural response to develop an on-line tool for a real-time assessment of SA, which could improve the performance in teleoperations.

Other human factors (e.g., workload) should be investigated to understand their effect on the success of teleoperated systems and their interaction with situation awareness. In addition, the use of multiple sensors (e.g., heart rate, galvanic skin response) should be investigated [5].

References

1. Berka, C., Levendowski, D.J., Davis, G., Whitmoyer, M., Hale, K., Fuchs, K.: Objective measures of situational awareness using neurophysiology technology. Augment. Cogn.: Past, Present. Future 145–154 (2006)
2. Catherwood, D., et al.: Mapping brain activity during loss of situation awareness: an EEG investigation of a basis for top-down influence on perception. Hum. Factors **56**(8), 1428–1452 (2014)
3. Chao, L.L., Lenoci, M., Neylan, T.C.: Effects of post-traumatic stress disorder on occipital lobe function and structure. NeuroReport **23**(7), 412–419 (2012)
4. Chen, S.I., Visser, T.A., Huf, S., Loft, S.: Optimizing the balance between task automation and human manual control in simulated submarine track management. J. Exp. Psychol.: Appl. **23**(3), 240 (2017)
5. Debie, E., et al.: Multimodal fusion for objective assessment of cognitive workload: a review. IEEE Trans. Cybern. 1–14 (2019)
6. Endsley, M.R.: Toward a theory of situation awareness in dynamic systems. Hum. Factors **37**(1), 32–64 (1995)
7. French, H.T., Clarke, E., Pomeroy, D., Seymour, M., Clark, C.R.: Psycho-physiological measures of situation awareness. Decis. Mak. Complex Environ. **291** (2007)
8. Lingard, L., et al.: Communication failures in the operating room: an observational classification of recurrent types and effects. BMJ Qual. Saf. **13**(5), 330–334 (2004)
9. Mathewson, K.E., Lleras, A., Beck, D.M., Fabiani, M., Ro, T., Gratton, G.: Pulsed out of awareness: EEG alpha oscillations represent a pulsed-inhibition of ongoing cortical processing. Front. Psychol. **2**, 99 (2011)
10. Pope, A.T., Bogart, E.H., Bartolome, D.S.: Biocybernetic system evaluates indices of operator engagement in automated task. Biol. Psychol. **40**(1–2), 187–195 (1995)
11. Thornton, R.C.: The effects of automation and task difficulty on crew coordination, workload, and performance, 00019 (1992)
12. Wang, M., Hussein, A., Rojas, R.F., Shafi, K., Abbass, H.A.: EEG-based neural correlates of trust in human-autonomy interaction. In: 2018 IEEE Symposium Series on Computational Intelligence (SSCI), pp. 350–357. IEEE (2018)
13. Winkler, I., Brandl, S., Horn, F., Waldburger, E., Allefeld, C., Tangermann, M.: Robust artifactual independent component classification for BCI practitioners. J. Neural Eng. **11**(3), 035013 (2014)
14. Yeo, L.G., et al.: Mobile EEG-based situation awareness recognition for air traffic controllers. In: 2017 IEEE International Conference on Systems, Man, and Cybernetics (SMC), pp. 3030–3035. IEEE (2017)

Adaptive Estimation of Human-Robot Interaction Force for Lower Limb Rehabilitation

Xu Liang[1,2], Weiqun Wang[1(✉)], Zengguang Hou[1,2,3], Shixin Ren[1,2], Jiaxing Wang[1,2], Weiguo Shi[1,2], and Tingting Su[4]

[1] Institute of Automation, Chinese Academy of Sciences, Beijing 100190, China
{liangxu2013,weiqun.wang,zengguang.hou,renshixin2015,wangjiaxing2016, shiweiguo2017}@ia.ac.cn

[2] University of Chinese Academy of Sciences, Beijing 100149, China

[3] CAS Center for Excellence in Brain Science and Intelligence Technology, Beijing 100190, China

[4] North China University of Technology, Beijing 100144, China
sutingting@ncut.edu.cn

Abstract. Human-robot interaction force information is of great significance for realizing safe, compliant and efficient rehabilitation training. In order to accurately estimate the interaction force during human-robot interaction, an adaptive method for estimation of human-robot interaction force is proposed in this paper. Firstly, the dynamics of human-robot system are modeled, which allows to establish a state space equation. Then, the interaction force is described by a polynomial function of time, and is introduced into the state space equation as a system state. Meanwhile, the Kalman filter is adopted to estimate the extended state of system online. Moreover, in order to deal with the uncertainty of system noise covariance matrix, sage-husa adaptive Kalman filter is used to correct the covariance matrices of system noises online. Finally, experiments were carried out on a lower limb rehabilitation robot, and the results show that the proposed method can precisely estimate the interaction force and also has good real-time performance.

Keywords: Human-robot interaction · State estimation · Rehabilitation robot · Interaction force estimation

1 Introduction

Cerebral infarction, cerebral hemorrhage, brain trauma, acute myelitis and other neurological diseases can cause paralysis and limb weakness. Physical exercise is

W. Wang—This research is supported in part by the National Key R&D Program of China (Grant 2017YFB1302303), National Natural Science Foundation of China (Grants 61720106012, 91848110), and Strategic Priority Research Program of Chinese Academy of Science (Grant XDB32000000).

T. Gedeon et al. (Eds.): ICONIP 2019, CCIS 1142, pp. 540–547, 2019.
https://doi.org/10.1007/978-3-030-36808-1_59

extremely important for the recovery of paralyzed patients. Rehabilitation robot can be applied in various periods of stroke rehabilitation, since it can be used to promote the functional compensation and reorganization of central nervous system through specific training and improve their daily living activities [1].

Studies have shown that rehabilitation training with patients' active participation can effectively promote neuroplasticity and motor function recovery [2]. Precise recognition of motion intention is the premise and one of the key issues of active rehabilitation training [3], and meanwhile, human-robot interaction force is an intuitive manifestation of human motion intention. Therefore, whether the interaction force between human and robot can be estimated in real time is of great significance for active rehabilitation training [4].

Human motion intention can be recognized by two types of methods. One is physiological signals based method. Physiological electrical signals mainly include muscle electrical signals and brain electrical signals. In [5], human motion intention was detected by surface electromyography (sEMG) signals, and then the robot motion is controlled according to human limb impedance. The physiological signals directly reflect the human motion intention, but they are susceptible to the surroundings. The collected EEG signals are weakly and difficult to recognize, and they are also susceptible to external interference [6].

The alternative is motion signals based method. The motion signal sensor has the characteristics of convenient wear, strong versatility and good environmental adaptability. For example, in [7], the force/position sensor based method was used to establish the moment mapping model between human limbs and robot joints, by detecting the generated force and motion of human limbs, to determine human motion intention. Huang placed a force sensor on the robot end effector to estimate the wearer's motion intention in real time [8].

Kalman filter method has higher estimation accuracy, and can also achieve estimation of robot state at the same time. Reasonable assumptions of interaction force model can improve the estimation accuracy of interaction force [9]. At present, most of the research work on human-robot interaction force uses the constant value hypothesis [10]. Hu adopted the interaction force model with polynomial and sinusoidal variation expression [9], which improved the effect of dynamic hypothesis to some extent. However, in actual process of human-robot interaction, the model of interaction force is usually time-varying, so the assumption of fixed order cannot meet the practical demand.

Based on the dynamic model of human-robot interaction system, an extended state space equation can be established by introducing the interaction force model using polynomial function of time. The improved sage-husa adaptive Kalman filter (SHAKF) is used to correct statistical characteristics of system state noise in real time to optimize the estimation of interaction force. Finally, the effectiveness of the proposed method is verified by experiments.

2 Human-Robot System Dynamic Model

During the rehabilitation exercise with robot, the patient's lower limbs are usually attached to the mechanical legs, as shown in Fig. 1. The hip and knee joints of mechanical leg can be respectively corresponding to the joints of human leg by adjusting the length of each link. Meanwhile, the lower limb of the human body can be fixed on the mechanical leg by using velcro fastener. Since the ankle joint contributes less to the end motion range, the above human-robot system can be treated as a two-bar linkage mechanism.

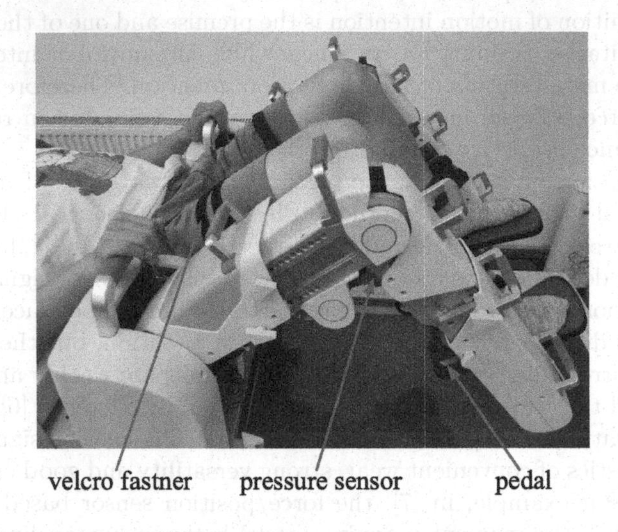

Fig. 1. Experiment platform for identification and validation of dynamics

The dynamic model of human-robot system can be obtained by Euler-Lagrange equation.

$$\frac{d}{dt}\left(\frac{\partial L}{\partial \dot{\theta}}\right) - \frac{\partial L}{\partial \theta} = \tau \tag{1}$$

where θ and $\dot{\theta}$ denote the joint angular and its velocity, τ denotes the joint moment, and L is Lagrangian equation of human-robot system.

$$L = \sum_{i=1}^{2} K_i - P_i \tag{2}$$

where K_i and P_i denote the human-robot system's kinetic and potential energy of linkage i respectively. In order to obtain the kinetic and potential energy of human-robot system, the two links are respectively considered to be composed of innumerable mass micro-elements. Firstly, the kinetic and potential energy are calculated for each micro-element, and then the integral operation is performed. As a result, the kinetic and potential energy of links 1 and 2 can be obtained as follows.

$$
\begin{cases}
K_1 = \int_{v_1} \frac{1}{2}\rho_v l_v^2 \dot{\theta}_1^2 dv \\
K_2 = \int_{v_2} \frac{1}{2}\rho_v l_1^2 \dot{\theta}_1^2 dv + \int_{v_2} \rho_v l_1 l_v \cos(\theta_2)\cos(\theta_v)\dot{\theta}_1(\dot{\theta}_1 + \dot{\theta}_2) dv \\
\quad + \int_{v_2} \frac{1}{2}\rho_v l_v^2 (\dot{\theta}_1 + \dot{\theta}_2)^2 dv - \int_{v_2} \rho_v l_1 l_v \sin(\theta_2)\sin(\theta_v)\dot{\theta}_1(\dot{\theta}_1 + \dot{\theta}_2) dv \\
P_1 = \int_{v_1} \rho_v g l_v \sin(\theta_1 + \theta_v) dv \\
P_2 = \int_{v_2} \rho_v g l_2 \sin(\theta_1) dv + \int_{v_2} \rho_v g l_v \sin(\theta_1 + \theta_2)\cos(\theta_v) dv \\
\quad + \int_{v_2} \rho_v g l_v \cos(\theta_1 + \theta_2)\sin(\theta_v) dv
\end{cases}
\tag{3}
$$

where $\dot{\theta}_i$ represents the angular velocity of joint i, dv represents the mass micro-element on the corresponding link, l_v represents the distance between dv and corresponding joint, θ_v represents the angle from the link's midline to connection between the joint and mass micro-element.

Combined with formulas one to three, the standard form of human-robot system dynamic equation can be derived.

$$D(\theta)\ddot{\theta} + C(\theta, \dot{\theta})\dot{\theta} + G(\theta) + \tau_f = \tau \tag{4}$$

where θ, $\dot{\theta}$ and $\ddot{\theta}$ denote the joint angular, its velocity and acceleration respectively. τ_f denotes the classical friction term, which consists of viscous friction and Coulomb friction. $D(\theta)$ is a symmetric positive definite matrix, $C(\theta, \dot{\theta})\dot{\theta}$ denotes the Coriolis and centripetal moment, $G(\theta)$ denotes the gravitational moment.

According to the linear characteristic of robot's dynamics, there is a parameter vector that makes them satisfy the following linear relationship.

$$D(\theta)\ddot{\theta} + C(\theta, \dot{\theta})\dot{\theta} + G(\theta) + \tau_f = Y\varphi \tag{5}$$

where Y is the regression matrix of joint variable, and φ is an unknown constant parameter vector.

3 Estimation of Human-Robot Interaction Force

3.1 Model of Human-Robot Interaction Force

In this paper, the joints of human lower limb are simplified into hip and knee joint, of which the interaction force models are similar to each other. Hence, the interaction force of hip joint is modeled below as an example.

Assuming that in the process of human-robot interaction, the change law of hip joint's interaction force, τ_a, is a polynomial function of time in a finite period. Then the dynamic expression of interaction force can be expressed as follows.

$$\begin{cases} \dot{\lambda} = \mathrm{L}\lambda \\ \tau_a = \mathrm{S}\lambda \end{cases} \tag{6}$$

where $\lambda = [\lambda_1 \ \cdots \ \lambda_{r+1}]$, r is the order of polynomial,

$$\mathrm{L} = \begin{bmatrix} \mathbf{0}_{r\times 1} & diag(\partial_1, \cdots, \partial_r) \\ 0 & \mathbf{0}_{1\times r} \end{bmatrix}$$
$$\mathrm{S} = \begin{bmatrix} 1 \ \mathbf{0}_{1\times r} \end{bmatrix} \tag{7}$$

where $diag(\cdot)$ is a diagonal matrix, $\partial_1, \cdots, \partial_r$ is partial coefficients of polynomial.

3.2 Extended State Space Equation

We can get the following extended state space equation by introducing λ into state vector.

$$\begin{aligned} \dot{x} &= Ax + Bu + V \\ z &= Hx + N \end{aligned} \tag{8}$$

Therefore, the human-robot interaction force can be achieved by Eq. 9 under the extended state space model.

$$\tau_a = \begin{bmatrix} \mathbf{0}_{2\times 4} \ \mathbf{I}_{2\times 2} \ \mathbf{0}_{2\times r} \end{bmatrix} x \tag{9}$$

So, the estimation of state vector can be gained by using Kalman filter.

3.3 Sage-Husa Adaptive Kalman Filter

Kalman filtering is an autoregressive optimal estimation algorithm [11], which principle consists of two parts: state prediction process and update process. In the prediction step, the current state is estimated by previous state value, while in the update step, Q and R are calculated, based on which the confidence of the estimated and the measured value are weighted. The optimal estimation of current moment is performed according to the predicted value of previous moment, the measured value and the error covariance of the current moment, then the state value of next time is predicted, thereby forming an iterative loop.

$$\begin{aligned}
\hat{x}_{k|k-1} &= A_k \hat{x}_{k-1|k-1} + B_k u_k \\
\hat{P}_{k|k-1} &= A_k \hat{P}_{k-1|k-1} A_k^T + Q_{k-1} \\
V_k &= Z_k - H_k \hat{x}_{k|k-1} \\
K_k &= \hat{P}_{k|k-1} H_k^T (H_k \hat{P}_{k|k-1} H_k^T)^{-1} \\
Q_k &= (1 - d_k) Q_{k-1} + d_k (K_k V_k V_k^T K_k^T + A_k \hat{P}_{k|k-1} A_k^T) \\
\hat{x}_{k|k} &= \hat{x}_{k|k-1} + K_k V_k \\
\hat{P}_{k|k} &= \hat{P}_{k|k-1} - K_k H_k \hat{P}_{k|k-1}
\end{aligned} \tag{10}$$

4 Experiments and Discussion

In order to verify the effectiveness of the estimation method of human-robot interaction force, experiments are carried out on the lower limb rehabilitation robot, as shown in Fig. 1.

4.1 Identification and Validation of System Parameter

The parameters of human-robot system dynamic model need to be identified at first, so that the extended state space Eq. 8 can be obtained.

The samples used for parameter identification are collected during the motion process of performing excitation trajectory driven by motors in the human-robot interaction system, while the subject is required to passively follow the mechanical leg to move without applying any active torque. Unknown dynamic parameters are recognized by the Eq. 5 using the least squares method.

The parameters identification results are shown in Table 1.

Table 1. Identification results of system parameter

Parameter	IU	Number
φ_1	$Kg * m^2$	23.1118
φ_2	Nm	8.2613
φ_3	$Kg * m^2$	−2.4986
φ_4	$Kg * m^2$	−0.9827
φ_5	$Kg * m^2$	−0.4954
φ_6	Nm	−7.1493
φ_7	Nm	17.9843
φ_8	Nm	−0.5973
φ_9	Nm	26.2949
φ_{10}	Nm	−1.5535

After the parameters are recognized, a reference trajectory different from optimal excitation trajectory is carried out to verify the accuracy of identification parameters. The root mean square errors of hip and knee joints are 0.4417 Nm and 0.6937 Nm respectively, which indicates that the identification method used in this paper can effectively recognize the parameters of human-robot system dynamic model.

4.2 Estimation of Human-Robot Interaction Force

In this section, the experiment for estimation of human-robot interaction force is performed on the lower limb rehabilitation robot to verify the effectiveness of the proposed method.

In the experiment, the rehabilitation robot performs treadmill trajectory in the vertical sagittal plane. Treadmill exercise is a common rehabilitation training mode, which can slow the muscle atrophy, promote the recovery of limb motor function and improve blood circulation. For patients with central nervous system injury, it also has the effect of reducing muscle tension and improving muscle strength. During the treadmill exercise, the subject applies an interaction force to the pedal through foot, which can be collected by force sensor mounted on the pedal. To illustrate the effectiveness and versatility of the proposed method, the force applied to pedal by the subject is required to be a reciprocating force that varies in magnitude and direction over time. This force can be transformed by Jacobian matrix J into torque of robot's joint space.

By comparing the measured value of human's active joint torque with the calculated torque obtained by the proposed method, the feasibility for method of estimating the human-robot interaction force can be verified. The experimental results are illustrated in Fig. 2. The root mean square errors of the hip and knee joint measurements and estimation torques are 0.1379 Nm and 0.2413Nm respectively, which indicates that the improved sage-husa adaptive Kalman filter method based on the force model using polynomial function of time can precisely estimate the human-robot interaction force, thus verifying the effectiveness of the proposed method.

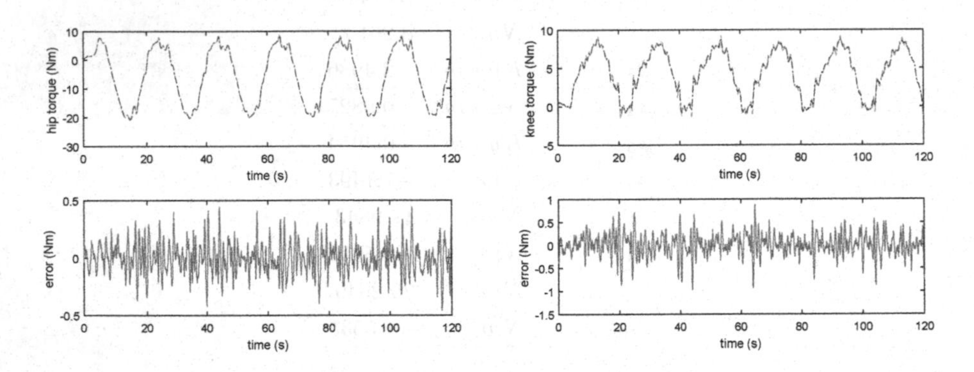

Fig. 2. Estimation of human-robot robot interaction force

5 Conclusion

In order to precisely estimate the interaction force during human-robot interaction, an adaptive method for estimation of human-robot interaction force is proposed in this paper. The interaction force is fitted by polynomial and then imported into the human-robot system dynamic model, in order to obtain the extended state space equation. To correct the time-varying covariance matrix of system noise, an improved sage-husa adaptive Kalman filter method is designed to estimate the state online. Experiments were carried out on the lower limb rehabilitation robot. The experimental results demonstrated that the proposed method can accurately estimate the human-robot interaction force and also has good real-time performance, which verifies the effectiveness of the proposed method. The experiments for paralyzed patients will be carried out in future research to test the clinical feasibility.

References

1. Pons, T.P., Garraghty, P.E., Ommaya, A.K., Kaas, J.H., Taub, E., Mishkin, M.: Massive cortical reorganization after sensory deafferentation in adult macaques. Science **252**, 1857–1860 (1991)
2. Warraich, Z., Kleim, J.A.: Neural plasticity: the biological substrate for neurorehabilitation. PM&R **2**(12), S208–S219 (2010)
3. Hou, Z.G., Zhao, X.G., Cheng, L., Wang, Q.N., Wang, W.Q.: Recent advances in rehabilitation robots and intelligent assistance systems. Acta Autom. Sin. **42**, 1765–1779 (2016)
4. Lee, K.H., Lee, H.K., Lee, J., Ji, S.H., Koo. J.C.: A simple method to estimate the impedance of the human hand for physical human-robot interaction. In: 2017 IEEE International Conference on Ubiquitous Robots and Ambient Intelligence, pp. 152–154 (2017)
5. Li, Z.J., Huang, Z.C., He, W., Su, C.Y.: Adaptive impedance control for an upper limb robotic exoskeleton using biological signals. IEEE Trans. Ind. Electron. **64**(2), 1664–1674 (2017)
6. Norman, S.L., Dennison, M., Wolbrecht, E., Cramer, S.C., Srinivasan, R., Reinkensmeyer, D.J.: Movement anticipation and EEG: implications for BCI-contingent robot therapy. IEEE Trans. Neural Syst. Rehabil. Eng. **24**(8), 911–919 (2016)
7. Chen, S.H., et al.: Assistive control system for upper limb rehabilitation robot. IEEE Trans. Neural Syst. Rehabil. Eng. **24**(11), 1199–1209 (2016)
8. Huang, J., Huo, W., Xu, W., Mohammed, S., Amirat, Y.: Control of upper-limb power-assist exoskeleton using a human-robot interface based on motion intention recognition. IEEE Trans. Autom. Sci. Eng. **12**(4), 1257–1270 (2015)
9. Hu, J., Xiong, R.: Contact force estimation for robot manipulator using semiparametric model and disturbance Kalman filter. IEEE Trans. Ind. Electron. **65**(4), 3365–3375 (2018)
10. Wahrburg, A., Bös, J., Listmann, K.D., Dai, F., Matthias, B., Ding, H.: Motor-current-based estimation of cartesian contact forces and torques for robotic manipulators and its application to force control. IEEE Trans. Autom. Sci. Eng. **15**(2), 879–886 (2017)
11. Auger, F., Hilairet, M., Guerrero, J.M., Monmasson, E., Orlowska-Kowalska, T., Katsura, S.: Industrial applications of the Kalman filter: a review. IEEE Trans. Ind. Electron. **60**(12), 5458–5471 (2013)

Effect of Incomplete Memorization in a Computational Model of Human Cognition

Toshihiko Matsuka[1](✉), Yoshiko Kawabbatas[2], and Kuangzhe Xu[1]

[1] Chiba University, Chiba, Japan
{matsuka,jo}@chiba-u.jp
[2] National Institute of Japanese Language and Linguistics, Tachikawa, Japan
ykawabat@gmail.com

Abstract. Dropout has been introduced as a simple yet effective method to prevent over-learning in deep learning. Although its mechanism, i.e., incapable of utilizing all memorized units, seems quite natural to human cognition, the effect of dropout on models of human cognition has not been addressed. In the present research, we apply dropout to a computational model of human category learning. We compared models with and without complete memorization abilities, and results showed that they differed acquired association weights, dimensional attention strengths, and how they handled exceptional exemplars.

Keywords: Cognitive modeling · Dropout · Categorization

1 Introduction

High-order human cognitive processes almost always involve highly abstracted categorical information [1,2]. For example, instead of describing every small detailed features of an animal, “dog” (e.g. a four-legged, brown haired animal with pointed teeth, etc), we use a word “dog” to categorically represent the object in communication. Categorization is a form of data compression that enables us to process, understand, and communicate complex thoughts and ideas by efficiently utilizing task- and context-relevant information while ignoring other types. Unlike many other methods of data compression (e.g. PCA), data compression with humans’ categorization processes are shown to produce highly interpretable results. We can infer several statistical properties about features of members of a particular category, say, dogs. These are reasons why cognitive scientists inarguably suggested that our high-order cognition is largely driven by categorically organized knowledge.

In cognitive science, research on human categorization has been strongly associated with computational modeling as a means to test theories of human categorization and category learning [3,4]. One important research question about

Supported by JSPS KAKENHI JP16H02835.

T. Gedeon et al. (Eds.): ICONIP 2019, CCIS 1142, pp. 548–556, 2019.
https://doi.org/10.1007/978-3-030-36808-1_60

human categorization has been how categories are represented in our mind. Among several theories and models, cognitive models built on the basis of the exemplar theory have shown promising results in computer simulations [3,4]. The exemplar theory of categorization assumes that humans utilize psychological similarities or conformities between the input stimulus and previously-seen, memorized exemplars as evidence to probabilistically assign the input stimulus to an appropriate category.

Although exemplar models of categorization could account for several behavioral data, they have been criticized for their assumptions about how they handle exemplars. In a typical implementation, the model is assumed to be capable of memorizing and utilizing all exemplars that an agent has encountered without any noise. Like categorization being suggested to play a central role in high-order human cognition, humans' memory is also inarguably suggested to be imperfect and erroneous, as almost every textbook on cognitive science describes limitations in the human memory system [2]. Given that the human memory system is limited, previous simulation studies on exemplar models with an unlimited memory capacity might not have been what cognitive scientists intended to model or describe. In the present research, we apply a limited memory system to a computational model of human category learning to examine its effects. In particular we apply dropout [5] (i.e., a simple yet effective method to prevent over-learning in deep learning) to an exemplar model of category learning and compared models with and without complete memorization capability to examine its effect on acquired association weights, dimensional attention weights, and how they handled exceptional exemplars.

1.1 A Cognitive Model of Category Learning - ALCOVE

The exemplar theory of categorization assumes that humans utilize previously-seen, memorized exemplars as reference points in categorization process. ALCOVE [3] is one of the most well-known models of category learning built on the basis of exemplar theory. One important cognitive process in ALCOVE and many other models is a selective attention operation that translates physical or logical distances between input stimuli and memorized exemplars into psychological similarities between them. In ALCOVE, the psychological similarity between a memorized exemplar R_j and input x is denoted as s_j and formulated as:

$$s_j = s(x, R_j) = \exp\left(-\beta \cdot \sum\nolimits_i a_i \left(R_{ji} - x_i\right)^2\right) \quad (1)$$

where $a_i \geq 0$ is a selective attention weight allocated to feature dimension i, and β is a constant that the experimenter can define in order to manipulate an overall similarity gradient. In ALCOVE, even a small physical difference on one dimension results in a significant psychological difference, when the corresponding dimension is attended. In contrast, a large physical difference on another dimension is perceived to be negligible when the dimension is weakly attended or unattended. The similarity is highest when every "attended" feature of an input matches that of a memorized exemplar.

ALCOVE uses the following function to model activations of category nodes:

$$O_k(x) = \sum_j w_{kj} s(x, R_j) \tag{2}$$

where w_{kj} is learnable association weights between exemplar j and category node k, representing how strongly or weakly category nodes and exemplars are coupled. The probability that x being classified as category A is calculated using the following choice rule:

$$P(A|x) = \frac{\exp(\phi \cdot O_A(x))}{\sum_k \exp(\phi \cdot O_k(x))} \tag{3}$$

where ϕ is a constant defining decisiveness of responses. The larger the ϕ, the more decisive a simulated human is. That is, when ϕ is large a small differences in category node activations would be psychologically perceived as large.

ALCOVE assumes that human is capable of and indeed memorizes and utilizes many if not all previously-seen exemplars. Thus, in this model, R_j represents all exemplars that one has encountered in the past, and those exemplar basis units have links to every category node. A categorization response will be made on the basis of the "collective" similarities between a input stimuli and exemplars from different categories. In the present paper, when dropout is applied, ALCOVE cannot utilize exemplars that are subject to dropout.

ALCOVE adjusts its selective attention and association weights in learning. Note that in ALCOVE's objective function is not minimization of crossentropy with softmax activation (Eq. 3) which is typically used in machine learning classification tasks. Rather, the objective of learning is a minimization of sum of squared differences between the target outputs (t_k) and activations of category nodes:

$$E = 1/2 \sum_k (t_k - O_k)^2 \tag{4}$$

A typical implementation of ALCOVE utilizes an online version of the gradient descent method for learning:

$$\Delta w_{kj} = -\lambda_W \frac{\partial E}{\partial w_{kj}}, \Delta a_i = -\lambda_a \frac{\partial E}{\partial a_i} \tag{5}$$

where λ_W and λ_a are learning rates for association weights and attention, respectively

2 Simulation Studies

We conducted three simulation studies to examine the effects of the incomplete memory system (i.e., dropout) of the exemplar model of category learning. In Simulations 1 and 2, we examined how association weights and attention strengths were affected by dropout. In Simulation 3, we compared how ALCOVE handled exceptional exemplars with and without dropout.

Overview of Simulation Setups. In all simulations, there were a total of 50 training epochs, and there were a total of 10000 replications, each with different combinations of hyperparameters. We randomly selected values of λ_W and λ_a from the uniform distribution ranging from 0.01 to 0.50. The other hyperparameters were set at constant values ($\beta = 3$, $\phi = 1$). At the beginning of each replication, we randomly initialized association weights, hyperparameters, and the order of training stimuli with the identical seed number for ALOVE with and without dropout. Thus, in each replication, ALCOVE with and without dropout had the identical setup. The probability of dropout was 0.5 for all simulations, and we randomly dropped out exemplars as there is no single study indicating which exemplars should be dropped out.

2.1 Study 1: Effects of Incomplete Memory on Association Weights

In Simulation 1, we examined the effects of incomplete memory on association weights. In so doing, we used very simple categories that consist of 20 exemplars with a single feature. The exemplars' feature values were equally spaced integer from 1 to 20. The exemplars with feature values less than or equal to 10 belong to Category A, one with feature values more than or equal to 11 belong to Category B.

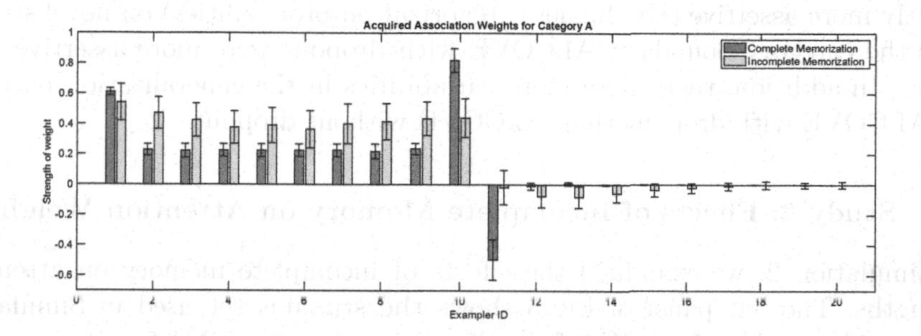

Fig. 1. Result of Simulation 1. Averaged acquired association weights to Category A for ALCOVE with (light gray) and without (dark gray) dropout. While ALCOVE with dropout formed evenly distributed association weights, ACOVE without dropout utilizes exemplars that are close to the category boundary more critically than prototypical exemplars

Results of Study 1. Figure 1 shows the averaged acquired association weights to Category A for ALCOVE with (light gray) and without (dark gray) dropout. There were noticeable differences. As compared with ALCOVE with dropout, the original ALCOVE acquired a very strong positive association between Category A and the boundary exemplar (i.e., one with feature value equals to 10). It also shows that original ALCOVE acquired a very strong negative association

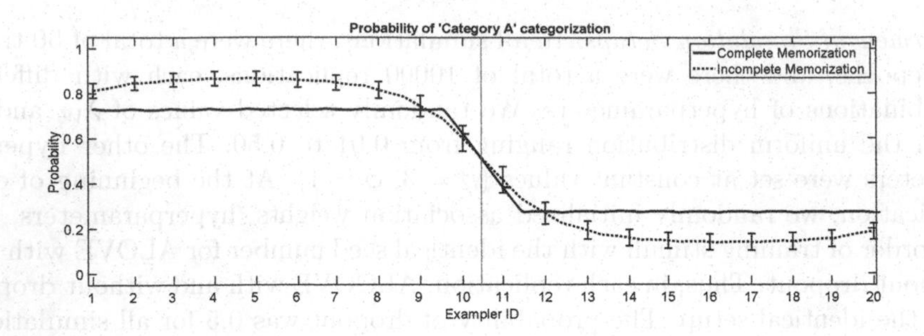

Fig. 2. The averaged generalization patterns for ALCOVE with (dotted line) and without dropout.

between Category A and another boundary exemplar (i.e., one that belong to Category B with feature value equals to 11). These two tendencies together indicate that the original ALCOVE utilizes exemplars that are close to the category boundary more critically than prototypical exemplars. In contrast, ALCOVE with dropout acquired more or less evenly distributed association weights, treating all exemplars equally.

Figure 2 shows the averaged generalization patterns for ALCOVE with (dotted line) and without (solid line) dropout. While the original ALCOVE could be slightly more assertive (i.e., higher categorization probabilities) on novel stimuli near the category boundary, ALCOVE with dropout were more assertive elsewhere. In addition, there were more variabilities in the generalization patterns for ALCOVE with dropout than ALCOVE without dropout.

2.2 Study 2: Effects of Incomplete Memory on Attention Weights

In Simulation 2, we examined the effects of incomplete memory on attention strengths. The left panel of Fig. 3 shows the stimulus set used in Simulation 2. Note that only information from dimension 1 was needed for correct categorization. Both types of ALCOVE were predicted to pay more attention to

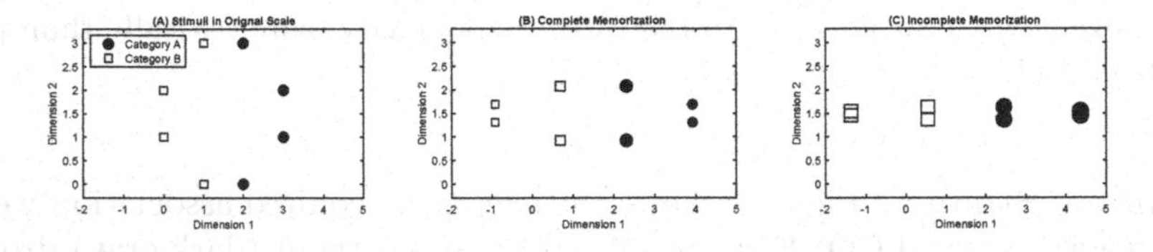

Fig. 3. Left panel: Stimulus set used in Simulation 2. Note that only information from dimension 1 was needed for the correct categorization. Middle panel: Internal representations (scaled by attention paid to dimensions 1 an 2) of categories for ALCOVE without dropout. Right panel: Scaled internal representations of categories for ALCOVE with dropout. The sizes of markers were proportional to positive association weights between exemplars and categories in Middle and Right panels.

this informative dimension, and we examined whether their attention learning processes differed.

Results of Study 2. Figures 3 and 4 show the results Simulation Study 2. The left panel of Fig. 4 shows the learning curves for ALCOVE with (dotted line) and without (solid line) dropout. The original ALCOVE resulted in high categorization accuracies than ALCOVE with dropout. This was simply because ALCOVE with dropout utilizes only a half of exemplars. If ALCOVE with dropout was capable of accessing all exemplar in predictions (but not in learning), its categorization accuracies were much higher than the original ALCOVE as shown in the left panel of Fig. 4 (line with black circles).

The right panel of Fig. 4 shows the attention allocation curves ALCOVE with (dotted line) and without (solid line) dropout. ALCOVE with dropout learned to allocate a greater amount of attention to the informative dimension than ALCOVE without dropout. The middle and right panels of Fig. 3 show the acquired internal representations of the stimulus set for ALCOVE without and with dropout, respectively. By internal representations, we mean that the stimulus set was scaled by learned attention strengths (cf. Eq. 1). The sizes of markers were proportional to positive association weights between exemplars and categories in the middle and right panels of Fig. 3. As in Simulation 1, while the association weights learned by ALCOVE with dropout were equally distributed, those of the original ALCOVE were not. The original ALCOVE put a stronger emphasis on exemplars near the category boundary. ALCOVE without dropout acquired more efficient representation properly ignoring the uninformative dimension than the original ALCOVE. The main reason why the original ALCOVE paid a weaker amount of attention to the informative dimension than ALCOVE with dropout was that it effectively formed strong associations between categories and exemplars near the category boundary (i.e., exemplars

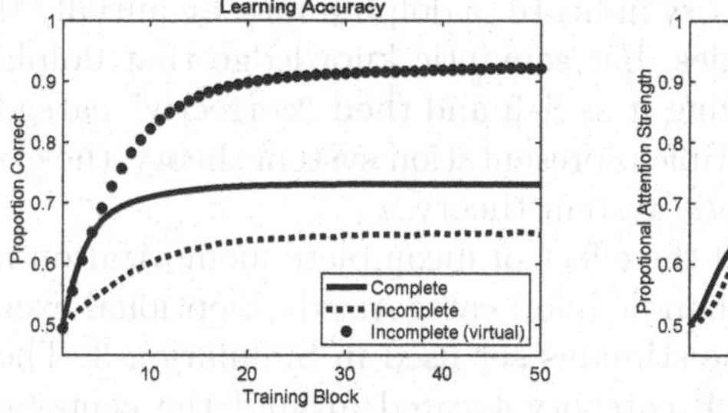

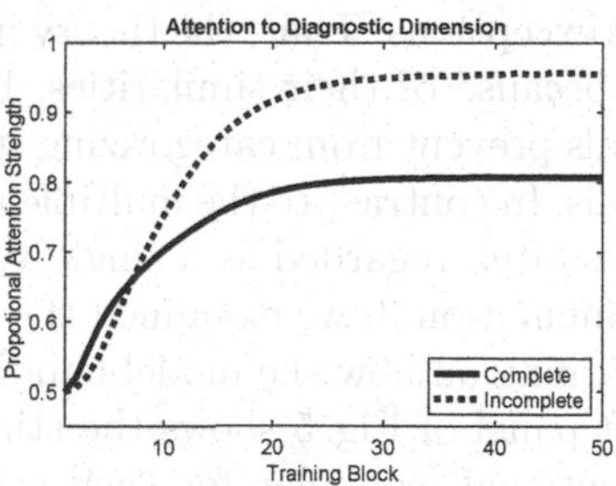

Fig. 4. Results Simulation 2. Left panel shows the learning curves for ALCOVE with (dotted line) and without (solid line) dropout. The line with black circles indicate the virtual learning curve (i.e., as if all exemplars were utilized in prediction) for ALCOVE with dropout. Right panel shows the attention allocation curves ALCOVE with (dotted line) and without (solid line) dropout.

that were more difficult to categorize) and thus there was no need for paying more attention to the informative dimension.

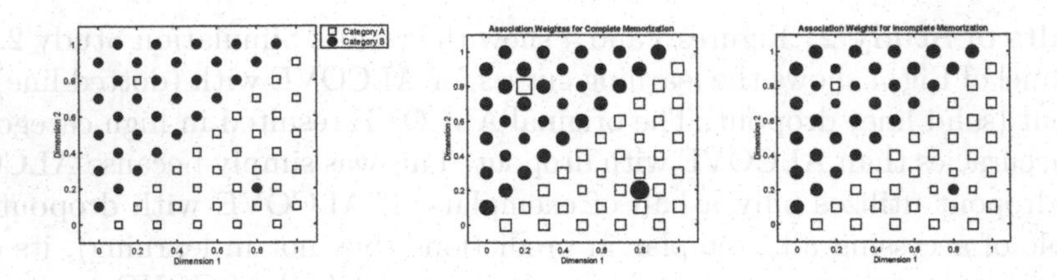

Fig. 5. Left panel: Stimulus set used in Simulation 3. There was an exceptional exemplar for each category located in the middle of each category. Middle panel: The averaged association weights learned by ALCOVE without dropout. The sizes of markers were proportional to positive association weights between exemplars and categories. Right panel: The averaged association weights learned by ALCOVE with dropout.

2.3 Study 3: Effects of Incomplete Memory on Exceptions

There have been some debates about how humans handle exceptions during categorization [6]. There are many cases where categories contain exceptions. Dolphins may be a good example of exceptions, as their physical appearances, behaviors, and the environment in which they live seem closer to those of fish than mammals. The exemplar theory suggests that exceptions are handled in the same manner that typical exemplars are handled. The other theory such as the multiple representation system theory [6] indicated there are at least two representation systems in human categorization processes, one for perceptual and the other for semantic knowledge. The multiple representation system theory suggests that typical exemplars trigger categorization based on perceptual representation, but semantic knowledge overwrites the categorization to accommodate exceptions. Thus, the theory indicates a dolphin may be initially thought as fish because of their similarities, but semantic knowledge that dolphins are mammals prevent from categorizing it as fish and then "correctly" categorize as mammals. In contrast to the multiple representation system theory, the exemplar theory is often regarded as a single system theory.

In Simulation 3, we examined the effect of incomplete memorization in category learning on how the model handled and categorized exceptional exemplars. The left panel of Fig. 5 shows the stimulus set used in Simulation 3. There was an exceptional exemplar for each category located around the center of each category.

Results of Study 3. Figures 5 and 6 show the results of Simulation Study 3. The middle and right panel of Fig. 5 shows the averaged association weights learned by ALCOVE without dropout and with dropout, respectively. The sizes of markers were proportional to positive association weights between exemplars and categories. As in Simulations 1 and 2, the original ALCOVE formed strong excitatory weights to positive exemplars (exemplars that belong to a category) and inhibitory weights to negative exemplars (exemplars that do not belong to a category) near the category boundaries. The strongest excitatory weights were formed for the exceptions and the strongest inhibitory weights were formed for exemplar near the exceptions. ALCOVE with dropout also formed the "correct" associations between the exceptions and categories and the exemplars near the exceptions and categories, but in a lesser magnitude.

The left and right panel of Fig. 6 shows the generalization patterns for ALCOVE without dropout and with dropout, respectively. While the original ALCOVE was able to exhibit "correct" categorizations for the exceptions, ALCOVE with dropout could not "correctly" categorize them, even though it had valid associations between the exceptions and categories. Thus, if humans memory system was indeed limited, capable of utilizing a limited number of previously encountered exemplar (i.e., if ALCOVE with dropout is a valid model), then the exemplar theory (or the single representation system theory) may not be able to account for human's ability to categorize exceptions. ALCOVE with dropout exhibit categorization similar to that of perceptual representation in the multiple representation system theory.

Do the results of Simulation 3 invalidate the exemplar theory or the single representation system theory? We argue it is not necessarily true. Given that, even with a limited memory capacity, ALCOVE with dropout formed the correct associations between the exceptions and categories as shown in Fig. 5. If ALCOVE somehow could access to this information, it can correctly categorize them. Let us assume that there are two exemplar referencing processes, say naive and targeted. What we have modeled with ALCOVE with dropout in the paper was naive exemplar referencing, randomly select exemplars in categorization. If there was such a process as targeted exemplar referencing, selecting a limited number of exemplars in categorization based on context, then ALCOVE with dropout could correctly categorize exceptions with only one representation system. Thus, like two representation systems theory, a single representation system with multiple exemplar referencing processes can account for both failure to categorize exceptions (which often happens in real life) and successful categorization of exceptions.

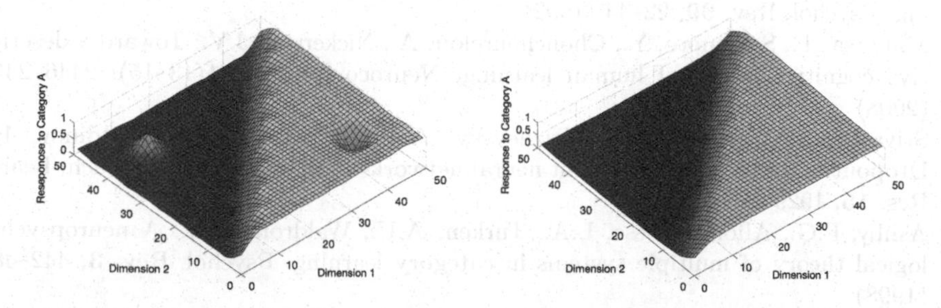

Fig. 6. Left panel: Averaged generalization pattern for ALCOVE without dropout. Right panel: Averaged generalization pattern for ALCOVE without dropout

3 Conclusion

In cognitive science, it is inarguably suggested that our high-order cognition is driven mainly by categorically organized knowledge. It is also inarguably suggested that the human memory system is limited. However, those two suggestions have not been simultaneously considered. The exemplar models of categorization have been successfully accounted for several behavioral data, but they have been criticized for their unrealistic assumption about human's memory system. In the present research, we apply a limited memory system to a computational model of human category learning to examine its effects. In particular, we apply dropout to an exemplar model of category learning and compared models with and without a complete memorization capability. The results of three simulation studies showed several differences between them. First, while a model with complete memorization formed strong excitatory associations between categories and positive exemplars near the category boundaries and strong inhibitory associations between categories and negative exemplars near the boundaries, a model with incomplete memorization tended to form equally distributed associations. Second, a model with incomplete memorization distributes its attention more efficiently than a model with complete memorization. Third, while a model with complete memorization could categorize exceptional exemplars "correctly," exceptions were generally ignored by a model with incomplete memorization. Although exceptions were ignored, a model with incomplete memorization acquired valid associations for exceptions.

In the present paper, we incorporated naive exemplar referencing (i.e., randomly select or omit exemplars) in categorization and category learning, because we do not know how exemplars were referenced. Both empirical and computational studied are much needed to clarify about this issue and the nature of human categorization and category learning.

References

1. Murphy, G.L.: The Big Book of Concepts. MIT Press, Cambridge (2002)
2. Medin, D.L., Ross, B.H., Markman, A.B.: Cognitive Psychology, 4th edn. Wiley, Hoboken (2005)
3. Kruschke, J.E.: ALCOVE: an exemplar-based connectionist model of category learning. Psychol. Rev. **99**, 22–44 (1992)
4. Matsuka, T., Sakamoto, Y., Chouchourelou, A., Nickerson, J.V.: Toward a descriptive cognitive model of human learning. Neurocomputing **71**(13–15), 2446–2455 (2008)
5. Srivastava, N., Hinton, G., Krizhevsky, A., Sutskever, I., Salakhutdinov, R.: Dropout: a simple way to prevent neural networks from overfitting. J. Mach. Learn. Res. **15**, 1929–1958 (2014)
6. Ashby, F.G., Alfonso-Reese, L.A., Turken, A.U., Waldron, E.M.: A neuropsychological theory of multiple systems in category learning. Psychol. Rev. **3**, 442–481 (1998)

Human Centred Computing and Medicine

Decoding Action Observation Using Complex Brain Networks from Simultaneously Recorded EEG-fNIRS Signals

Yi-chuan Jiang, Peng Wang, Hui Liu, and Sheng Ge(✉)

Key Laboratory of Child Development and Learning Science of Ministry of Education, Research Center for Learning Science, Southeast University, Nanjing 210096, Jiangsu, China
{scott_j,wang-peng,liu-hui,shengge}@seu.edu.cn

Abstract. In the current study, a novel brain-machine interaction was proposed, which incorporates action observation decoding into the traditional control circuit of a brain-machine interface. In this new brain-machine interaction, the machine can actively decode the user's action observation and stop immediately if it detects that the user does not understand the intention of the action correctly. We measured brain activation using electroencephalography (EEG)-functional near-infrared spectroscopy (fNIRS) bimodal measurement while 16 healthy participants observed three action tasks: drinking, moving a cup, and action with unclear intention. Complex brain networks were constructed for EEG and fNIRS data separately, and four network measures were chosen as features for classification. The obtained results revealed that the classification of three action observation tasks achieved accuracy of 72.3% using EEG-fNIRS confusion features, which was higher than that using fNIRS features (52.7%) or EEG features (68.6%) alone. Thus, the current findings suggested that our proposed method could provide a promising direction for brain-machine interface systems design.

Keywords: Brain-machine interface · Complex brain networks · Electroencephalography · Functional near-infrared spectroscopy · Action observation

1 Introduction

Brain-machine interfaces (BMIs), designed to provide direct functional interfaces between brains and artificial devices, have received increasing attention in the past several decades [1]. BMI systems are mainly composed of the following three parts: signal acquisition, feature extraction and device control. Moreover, many systems also incorporate feedback mechanisms for error correction. Users can learn about the execution of the command through auditory [2] or visual [3] feedback, and send instructions when misoperation is detected. For BMI systems with a relatively low information transfer rate (ITR), however, it may take several or more seconds to send a single command. During this process, the misoperation of the device is likely to cause

T. Gedeon et al. (Eds.): ICONIP 2019, CCIS 1142, pp. 559–569, 2019.
https://doi.org/10.1007/978-3-030-36808-1_61

accidents. To be effective, error correction must be automatic or require minimal user effort [4], and a more effective human-machine interaction mode could have useful applications.

One potentially promising solution is to construct a machine that is capable of decoding the user's action observation, and to stop immediately if it detects that the user did not correctly understand the intention of the action. For example, in a daily care situation, if a robot has the intention to feed a user, a user may not notice the action, or might misunderstand the intention of the action. At this time, the robot could detect a dangerous action by decoding the user's action observation activity, and cease the action immediately. Thus, it is important to enable BMIs to decode the user's brain activity during action observation. However, few studies have examined this issue.

Many neuroimaging methods have been used in BMI system design, including electroencephalography (EEG) and functional near-infrared spectroscopy (fNIRS). Because of its high temporal resolution and relatively low cost, EEG is currently the most popular neuroimaging method in BMI system design. However, EEG has several disadvantages, such as susceptibility to electromagnetic and motion interference, which makes it unsuitable for use in daily life. Unlike direct measurement of electrical potentials generated by cortical postsynaptic currents, fNIRS measures changes in the oxygenation correlates of neural activity and possesses several advantages, such as relatively low sensitivity to participant motion and ease of administration [5]. However, this method also has limitations, including low temporal resolution. Because EEG and fNIRS each have specific shortcomings, combining EEG and fNIRS measurement can provide an approach for overcoming the limitations of each method. A number of previous studies have used simultaneous recording of EEG and fNIRS to design bimodal BMIs [6, 7].

In the current study, we used combined EEG-fNIRS bimodal measurement to record neural activity during action observation. Simultaneously recorded EEG and fNIRS signals were then used to construct complex brain networks for actions with different potential intentions: drinking, moving a cup, and actions with unclear intentions. We sought to use complex brain networks to learn the spatio-temporal patterns for brain activities generated during action observation, and to extract useful features for intention classification. The main contributions of the current study can be summarized as follows:

1. A new method of human-machine interaction is proposed, which can greatly reduce the serious consequences of not correcting misoperation in time, and can improve the robust qualities of the BMI system.
2. Complex brain networks are used to decode the brain activities generated during action observation, and the nodal characteristics of different networks are discussed in detail.
3. The classification performance of EEG, fNIRS and EEG-fNIRS bimodal signals are compared. Based on this comparison, it can be seen that using EEG-fNIRS bimodal data was able to improve the performance of intention classification.

The remainder of this paper is organized as follows: Sect. 2 introduces the experimental setup, complex brain network construction, feature extraction, and feature classification. Section 3 presents the results of our analysis of action observation using a complex brain network. Finally, Sect. 4 discusses the conclusions that can be drawn from these findings.

2 Materials and Methods

2.1 Participants

Sixteen healthy participants (10 males and six females, mean age: 24.1 years) were recruited to participate in this study. All participants had normal or corrected-to-normal vision and provided informed consent. All participants provided written informed consent before enrolment in the study, which was approved by The Ethics Committee of the Affiliated Zhongda Hospital, Southeast University (2016ZDSYLL002.0 and 2016ZDSYLL002-Y01). Each of the participants received monetary compensation of 200 yuan after the experiment.

2.2 Experimental Procedure

Before the experiment, all participants were informed that pictures with three different kinds of actions (see Fig. 1) would be displayed randomly on the computer monitor. These actions and their potential intentions were as follows: grasping the handle of the cup with the intention of drinking (Sd), grasping the rim of the cup with the intention of moving it (Sm) and touching the rim of the cup with an unclear intention (Su). To avoid the influence of cup color, we used seven differently colored cups. Thus, there were 21 different pictures (three actions × seven colors) in the image set.

(a)

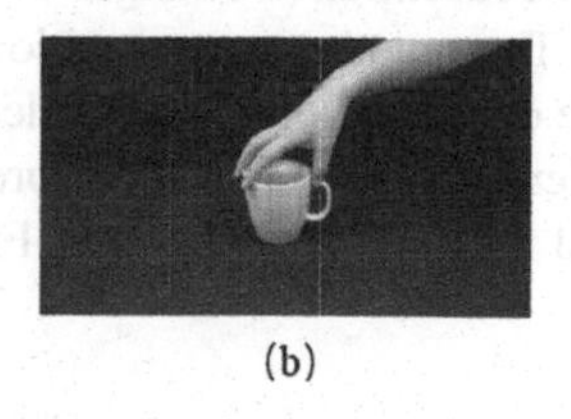

(b)

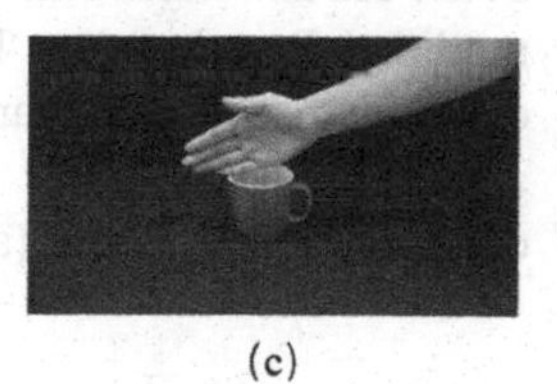

(c)

Fig. 1. Three kinds of hand-cup interaction stimuli corresponding to different potential intentions: (a) Sd, (b) Sm and (c) Su.

The experiment consisted of four sessions and each session contained 21 trials. Each trial started with a pre-rest period in which a fixation cross was presented for 6 s on the screen. Subsequently, the cross on the screen was replaced by a cup as a cue to indicate that the observation task was about to begin and lasted 0.5 s. During the observation period, the picture was displayed on the screen for 3.5 s and subjects were asked to interpret the potential intentions of the action in the picture. Finally, there was a post-rest period in which subjects could rest for 6 s. The experimental paradigm is shown in Fig. 2.

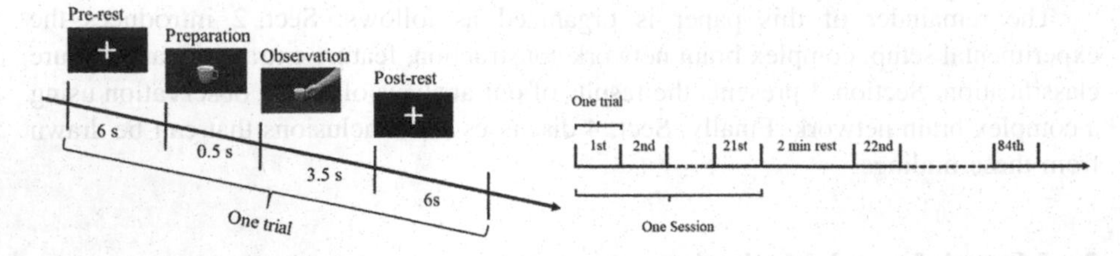

Fig. 2. Experimental procedure for action observation

Before the experiment, a training session was conducted in which participants were instructed to familiarize themselves with each action and its potential meaning, within 3.2 min. After the session, participants were debriefed to ensure they understood the experimental instructions and correctly understood the action intentions shown. During EEG-fNIRS measurement, participants received clear instructions to carefully observe the three different kinds of hand-object interaction stimuli, and to attempt to understand the intention behind the stimuli.

2.3 Data Acquisition

During the experiment, 64-channel EEG and 48-channel fNIRS bimodal signals were simultaneously recorded. The EEG data were measured using a Synamps2 EEG system (Neuroscan, USA) at a sampling rate of 1000 Hz. All electrodes were placed according to the international 10–20 system and the impedance of the electrodes was kept below 5 kΩ. Electrooculogram (EOG) signals were also be recorded for later removal of eye movement and blink artifacts. fNIRS data were recorded using a LABNIRS system (Shimadzu CO., LTD., Kyoto, Japan). The observation of three wavelengths (780, 805, and 830 nm) of continuous near infrared light was recorded at a sampling rate of 37.04 Hz and transformed into concentration changes of HbO, HbR and HbT using the modified Beer-Lambert Law. The optodes were positioned over the 64-channel EEG cap (Neuroscan, USA) and the optodes and the electrodes were placed at intervals with a distance between the emitters and detectors of approximately 3 cm. The channel configuration for EEG (a) and fNIRS (b) is shown in Fig. 3.

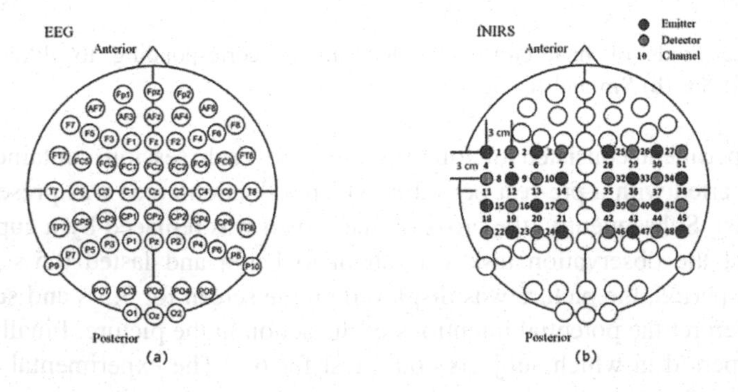

Fig. 3. Channel configuration for (a) EEG and (b) fNIRS.

2.4 Data Preprocessing

The EEG signals were band-pass filtered from 1 to 30 Hz using a finite impulse response (FIR) filter to remove some low frequency artifacts, such as eye blinks, eye movements, heart beat and breathing. In addition, an independent component analysis (ICA) method based on EEG and EOG data was performed to remove eye movement and blink artifacts. The fNIRS signals were also band pass filtered between 0.01 and 0.1 Hz. The baseline correction was performed for each trial by subtracting the mean value of the 1 s of data obtained before the trial.

2.5 Complex Brain Network Construction

Complex network modeling is based on graph theory, which uses a collection of nodes (vertices) and links (edges) between pairs of nodes to describe important properties of complex systems. As a large and complex system, the human brain shares a number of common features with networks from other biological and physical systems, and can thus be characterized by a complex network model [8]. Using complex network models to analyze the brain's structural and functional systems is referred to as a complex brain network method [9]. When using EEG-fNIRS bimodal measurement, the network nodes can be defined as EEG electrodes, or as fNIRS channels. Links usually represent various connections, for instance, anatomical connections, functional connections and effective connections. Because action observation is a dynamic and spatiotemporal process [10, 11], it is suitable to use functional connections to represent the magnitudes of temporal correlations in activity between pairs of electrodes or channels. In the current study, we constructed complex brain networks for EEG and fNIRS respectively.

The construction of complex brain networks consists of four major procedures: (a) defining the network nodes; (b) estimating the link between two nodes and generating a connectivity (adjacency) matrix; (c) binarizing the connectivity matrix; (d) choosing measures of the complex brain network. The details of building the complex brain networks are as follows:

(a) Defining the complex brain network nodes. In the current study, we used EEG-fNIRS bimodal measurement and constructed a network for each measurement method. Therefore, the nodes could be defined as the electrodes or channels, and the number of nodes in EEG network and fNIRS network was set to 62 (except for M1 and M2 electrodes) and 48, respectively.
(b) Estimating the link between two nodes and generate connectivity (adjacency) matrix. Here, we used Pearson's correlation coefficients to represent the temporal correlations between a pair of nodes, which can be computed as follows:

$$r_{ij} = \frac{\sum_{k=1}^{K}(x_i(k) - \overline{x_i})(x_j(k) - \overline{x_j})}{\sqrt{\sum_{k=1}^{K}(x_i(k) - \overline{x_i})^2 \sum_{k=1}^{K}(x_j(k) - \overline{x_j})^2}} \quad (1)$$

where x_i denotes the time series measured by the *ith* electrodes/channels, and $\overline{x_i}$ is its mean value. Rows and columns in the connectivity matrix denote the nodes, and each element in the matrix is the corresponding link between two nodes.

(c) Binarizing the connectivity matrix. First, the connectivity matrix was converted into Fisher's z maps using Fisher's r-to-z transformation to improve normality:

$$Z = \frac{1}{2}\ln(\frac{1+r}{1-r}) \quad (2)$$

We then applied a threshold to each element in the transformed connectivity matrix to form a binary connectivity matrix.

(d) Choosing measures of the complex brain network. As a graph, the topology of complex brain network can be quantitatively described using a variety of measures. In the current study, we chose the four most commonly used measures, which were (1) degree, (2) clustering coefficient, (3) local efficiency and (4) betweenness centrality:

1. Degree. Degree is the most fundamental measure of complex brain networks. The degree of node i is equal to the number of all links connected to that node:

$$k_i = \sum_{j \in N} a_{ij} \quad (3)$$

2. Clustering coefficient. The clustering coefficient of an individual node is equivalent to the fraction of triangles around it. In functional networks, a high clustering coefficient implies functional segregation. The clustering coefficient of a node i is defined as [12]:

$$C_i = \frac{2t_i}{k_i(k_i - 1)} \quad (4)$$

where k_i and t_i denote the degree and the number of triangles of node i respectively. The number of triangles can be calculated as follows:

$$t_i = \frac{1}{2}\sum_{j,h \in N} a_{ij}a_{ih}a_{jh} \quad (5)$$

3. Nodal local efficiency. The local efficiency of node i measures the extent of information transmission among the neighbors of the node, which can be calculated as follows [13]:

$$E_{loc,i} = \frac{\sum_{j,h \in N, j \neq i} a_{ij}a_{ih}[d_{jh}(N_i)]^{-1}}{k_i(k_i - 1)} \quad (6)$$

where $d_{jh}(N_i)$ is the length of the shortest path between node j and h, that contains only neighbors of i.

4. Betweenness centrality. The betweenness centrality of node i is defined as the number of all shortest paths in the network that pass through it [14]:

$$b_i = \frac{1}{(n-1)(n-2)} \sum_{h \neq j, h \neq i, j \neq i} \frac{\rho_{hj}^{(i)}}{\rho_{hj}} \tag{7}$$

where ρ_{hj} is the number of shortest paths between h and j, and $\rho_{hj}^{(i)}$ is the number of shortest paths between h and j that pass through i.

2.6 Feature Selection and Classification

As mentioned in Sect. 2.6, we chose four kind of measures to quantitatively describe each node in the network and calculated these nodal features for the EEG and fNIRS networks separately, and the process results in 440 features after bimodal feature fusion. However, redundant features not only waste computing resources but also impact the classifier performance. In the current study, the ReliefF algorithm proposed in a previous study [15] was used to choose a small subset of features, which is necessary and sufficient for describing the target. Compared with the Relief algorithm, ReliefF can deal with multiclass problems and is more robust. As a feature estimator, the ReliefF algorithm can estimate the quality of the feature according to how well their values distinguish between instances that are near to each other. ReliefF randomly selects an instance R_i. We searched its k ($k = 10$) nearest neighbors from the same class and different classes and named their nearest hits H_j and nearest misses M_j. The initial weight of feature A is set to zero, then updated iteratively, as follows [16]:

$$\begin{aligned} &\textit{for}\ i \in \{1..m\}\ \textit{do} \\ &\quad W(A) = W(A) - \sum_{j=1}^{k} \textit{diff}(A, R_i, H_j)/(m \times k) \\ &\quad\quad + \sum_{C \neq class(R_i)} \left[\frac{P(C)}{1 - P(class(R_i))} \sum_{j=1}^{k} \textit{diff}(A, R_i, M_j(C))\right]/(m \times k) \\ &\textit{end for} \end{aligned} \tag{8}$$

where m is the number of instances in each class, $\textit{diff}(A, I_1, I_2)$ denotes the difference between the values of the feature A for instances I_1 and I_2, and P denotes the prior probability of the class. All of the calculated feature weights will be sorted and only the features with weights greater than the threshold will be retained, leading to a relatively small subset of features. We used LIBSVM [17] to classify those features, and the classification accuracy was obtained using the averaged accuracies of 10 times 10-fold cross-validation.

3 Results and Discussion

In the current study, complex brain networks were used to decode the observation of three kinds of actions using simultaneously recording EEG-fNIRS data. Figure 4 shows the complex brain networks of three different actions constructed from EEG and fNIRS data for subject 9, separately. The nodes of the complex brain networks represent the EEG electrodes or fNIRS channels, while the edges of the networks represent functional connections. The resulting complex brain networks contain the spatio-temporal information generated in action observation and therefore, they can be used as spatio-temporal patterns for different action observation.

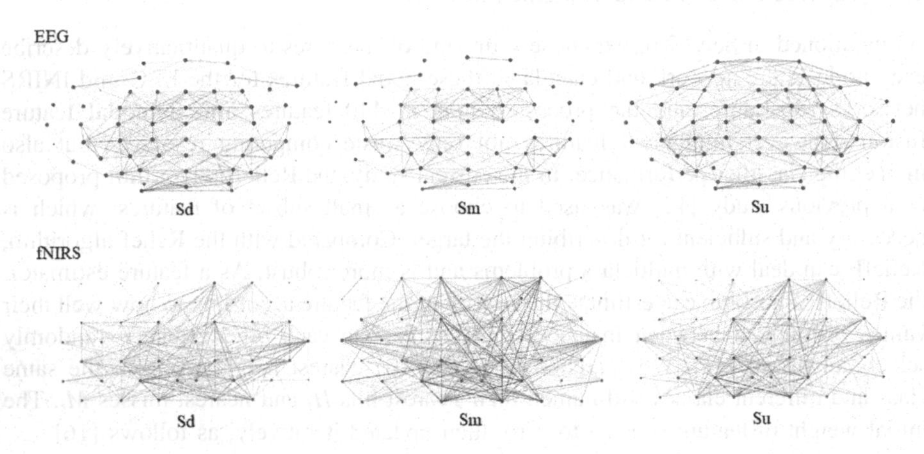

Fig. 4. Complex brain networks constructed from EEG and fNIRS data for S9.

Hubs are nodes with high degree or centrality, which plays an important role in the complex network [9]. In this study, we employed normalized betweenness centrality $b_{norm,i} = b_i/b_{ave}$ to identify hubs of different networks, where b_{ave} denotes the mean value of betweenness centrality for all nodes. The calculated $b_{norm,i}$ were averaged across all subjects and those nodes with high normalized betweenness centrality ($B_{norm,i} > 2$) were selected as functional hubs of a network. From Fig. 5, we can observe that there existed distinct differences of the functional hubs distribution for different complex brain networks.

Schippers and Keysers have proved that there was a flow of information within mirror neuron systems (MNS) during gesture observation, which goes from visual cortex → temporal cortex → parietal cortex → premotor cortex [18]. Functional hubs in the corresponding cortex are marked with dashed frames in Fig. 5. The distribution of EEG functional hubs within MNS for complex networks composed of different action observation are roughly the same, which demonstrates that the complex brain networks can reflect the feed-forward model of MNS during action observation. As shown in Fig. 1, both the action of drinking (Sd) and moving the cup (Sm) involve the grasping movement, while the action with unclear intention is just touching the rim of the cup. Previous study demonstrated that during grasping gesture observation,

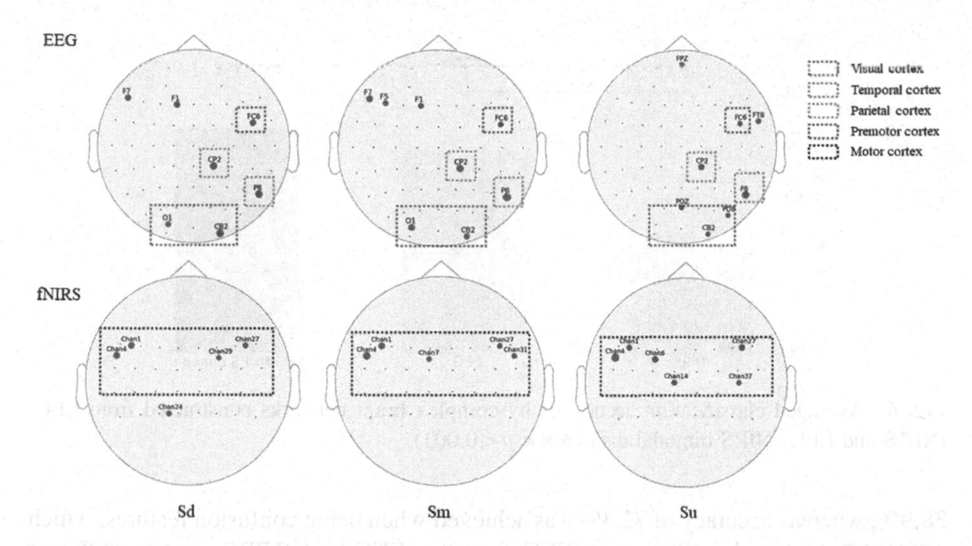

Fig. 5. Distribution of functional hubs for different complex brain networks. The corresponding cortex is marked with dashed frames.

significantly greater activation can be observed not only within above-mentioned MNS but also in left inferior frontal gyrus (IFG) [19]. From Fig. 5 we can observe that both Sd and Sm networks possessed the F7 hub, which is located in IFG, while it was absent in Su network. Moreover, Su network has another functional hub FPz, suggest the involvement of dorsolateral prefrontal cortex (DLPFC) in manipulating memory and high-level inference. The hubs calculated using fNIRS data are mostly distributed in motor cortex. The distribution of hubs coincides with the mechanism of action observation proves the feasibility of complex brain networks.

Four kinds of network measures were selected as features for further analysis. To build robust models, we used the ReliefF algorithm for feature selection before classification. In the current study, we not only used EEG or fNIRS features for classification, but also used EEG-fNIRS bimodal fusion features for classification. Figure 6 shows the average classification accuracies of fNIRS, EEG and fNIRS-EEG bimodal data, which were 52.7% (SD = 7.9%), 68.6% (SD = 6.8%) and 72.7% (SD = 4.4%), respectively. The one-way analysis of variance (ANOVA) results demonstrated that there was a statistically significant difference between these three methods ($F[2, 45] = 39.25$, $P < .001$). Post-hoc paired t-tests revealed that the accuracies calculated from EEG and EEG-fNIRS bimodal data were significantly higher than those calculated from fNIRS data ($P < .001$). Although the t-test failed to reveal a significant difference between classification accuracies calculated from EEG and EEG-fNIRS bimodal data ($P = 0.17$), the mean accuracy of EEG-fNIRS bimodal data was approximately 4% higher.

Compared with the classification accuracies calculated from EEG features, the accuracies of fNIRS features are relatively low. However, for most participants, the use of EEG-fNIRS confusion features can achieve higher classification accuracy. Particularly for participant No. 16, the classification accuracy of fNIRS features was only

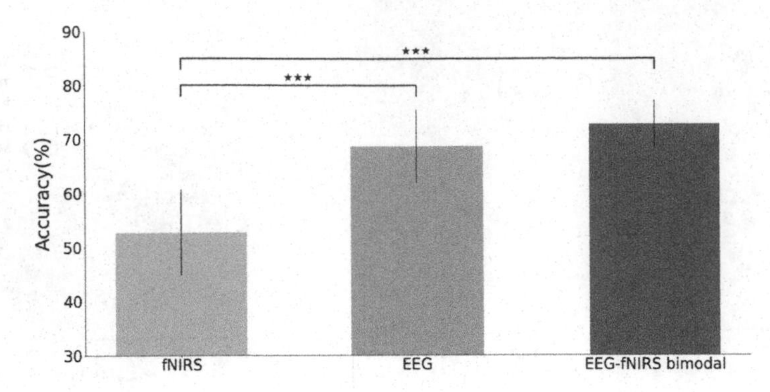

Fig. 6. Averaged classification accuracies by complex brain networks constructed from EEG, fNIRS and EEG-fNIRS bimodal data (★★★ $p < 0.001$).

38.9%, whereas accuracy of 72.3% was achieved when using confusion features, which was 8.6% higher than that using EEG features. EEG and fNIRS are two different functional brain activity measurements, with their own advantages and disadvantages. A previous study [7] reported that using simultaneous recording of EEG-fNIRS data could significantly improve the classification performance. In the current study, we constructed complex brain network models for EEG and fNIRS data separately and fused the calculated features at the feature level. The results demonstrated that using EEG-fNIRS confusion features was able to improve model robustness, and effectively reduce the misclassification.

4 Conclusion

In the current study, complex brain networks were used to decode simultaneously recorded EEG-fNIRS signals during action observation. The results revealed that a classification accuracy of 72.3% was achieved using EEG-fNIRS bimodal data, which was higher than that obtained using fNIRS data (52.7%) or EEG data (68.6%) alone. In addition, the results demonstrated the feasibility of incorporating action observation decoding into BMI system design. Detecting misoperation of a device using action observation decoding and responding to an action with unclear intention properly can improve the robustness of a BMI system. Although the current classification performance was not ideal, it could be improved using other feature extraction methods and novel machine learning algorithms. Future studies should attempt to use dynamic graph convolution neural networks (DGCNN) [20] for classification. In conclusion, our proposed method provides a promising direction for BMI systems design.

Acknowledgments. This work was supported in part by the National Basic Research Program of China under Grant 2015CB351704, the National Nature Science Foundation of China (61473221, 61773408), and in part by the Fundamental Research Funds for the Central Universities of China (CZY18047).

References

1. Lebedev, M.A., Nicolelis, M.A.L.: Brain–machine interfaces: past, present and future. Trends Neurosci. **29**, 536–546 (2006)
2. Klobassa, D.S., et al.: Toward a high-throughput auditory P300-based brain–computer interface. Clin. Neurophysiol. **120**, 1252–1261 (2009)
3. Chen, X., Wang, Y., Nakanishi, M., Gao, X., Jung, T.-P., Gao, S.: High-speed spelling with a noninvasive brain–computer interface. Proc. Natl. Acad. Sci. **112**, E6058 (2015)
4. Patil, P.G., Turner, D.A.: The development of brain-machine interface neuroprosthetic devices. Neurotherapeutics **5**, 137–146 (2008)
5. Cui, X., Bray, S., Bryant, D.M., Glover, G.H., Reiss, A.L.: A quantitative comparison of NIRS and fMRI across multiple cognitive tasks. NeuroImage **54**, 2808–2821 (2011)
6. Khan, M.J., Hong, M.J., Hong, K.-S.: Decoding of four movement directions using hybrid NIRS-EEG brain-computer interface. Front. Hum. Neurosci. **8**, 244 (2014)
7. Ge, S., et al.: A Brain-Computer Interface Based on a Few-Channel EEG-fNIRS Bimodal System. IEEE Access **5**, 208–218 (2017)
8. Rubinov, M., Sporns, O.: Complex network measures of brain connectivity: uses and interpretations. NeuroImage **52**, 1059–1069 (2010)
9. Bullmore, E., Sporns, O.: Complex brain networks: graph theoretical analysis of structural and functional systems. Nat. Rev. Neurosci. **10**, 186 (2009)
10. Ge, S., et al.: Temporal-spatial features of intention understanding based on EEG-fNIRS bimodal measurement. IEEE Access **5**, 14245–14258 (2017)
11. Gardner, T., Goulden, N., Cross, E.S.: Dynamic modulation of the action observation network by movement familiarity. J. Neurosci. **35**, 1561 (2015)
12. Watts, D.J., Strogatz, S.H.: Collective dynamics of 'small-world' networks. Nature **393**, 440–442 (1998)
13. Wang, J., et al.: Parcellation-dependent small-world brain functional networks: a resting-state fMRI study. Hum. Brain Mapp. **30**, 1511–1523 (2009)
14. Freeman, L.C.: Centrality in social networks conceptual clarification. Soc. Netw. **1**, 215–239 (1978)
15. Kononenko, I.: Estimating attributes: analysis and extensions of RELIEF. In: Bergadano, F., De Raedt, L. (eds.) ECML 1994. LNCS, vol. 784, pp. 171–182. Springer, Heidelberg (1994). https://doi.org/10.1007/3-540-57868-4_57
16. Robnik-Šikonja, M., Kononenko, I.: Theoretical and empirical analysis of ReliefF and RReliefF. Mach. Learn. **53**, 23–69 (2003)
17. Chang, C.-C., Lin, C.-J.: LIBSVM: A library for support vector machines. ACM Trans. Intell. Syst. Technol. **2**, 1–27 (2011)
18. Schippers, M.B., Keysers, C.: Mapping the flow of information within the putative mirror neuron system during gesture observation. NeuroImage **57**, 37–44 (2011)
19. Buxbaum, L.J., Kyle, K.M., Tang, K., Detre, J.A.: Neural substrates of knowledge of hand postures for object grasping and functional object use: evidence from fMRI. Brain Res. **1117**, 175–185 (2006)
20. Song, T., Zheng, W., Song, P., Cui, Z.: EEG emotion recognition using dynamical graph convolutional neural networks. IEEE Trans. Affect. Comput. 1 (2018)

Characterizing and Identifying Autism Disorder Using Regional Connectivity Patterns and Extreme Gradient Boosting Classifier

Thomas M. Epalle, Yuqing Song, Hu Lu, and Zhe Liu(✉)

School of Computer Science and Telecommunication Engineering, Jiangsu University, Zhenjiang 212013, Jiangsu, People's Republic of China
1000004088@ujs.edu.cn

Abstract. In this paper, we design and implement a procedure to capture and extract regional connectivity patterns from brain connectomics. Moreover, we assess the viability of such patterns as predictors for both childhood and adult autism. Finally, we investigate which regions and connections are significant for characterizing and predicting this psychiatric pathology. We use two publicly-available neuroimaging datasets and systematically train 90 extreme gradient boosting trees classifiers (XGBoost) for each set, each classifier receiving connectivity patterns extracted for one of the 90 regions of interest that form the automated anatomical labeling (AAL) atlas. Our most predictive regional connectivity pattern features achieved an accuracy of 78.95% (precision = 78.98%, recall = 78.75%) for the adult population and 75.01% accuracy for the pediatric dataset (precision = 75.00%, recall = 75.09 %) for the pediatric population. These classification accuracies are higher than those reported in prior studies that used the same datasets. Altogether, our results indicate that local connectivity around the lingual gyrus can predict both adult and childhood autism with relatively high accuracy.

Keywords: Autism · Brain connectivity · Classification · eXtreme Gradient Boosting · Functional magnetic resonance imaging

1 Introduction

Autism spectrum disorder (ASD) is a heterogeneous, persistent neurodevelopmental disorder with a range of symptom expression profiles, including deficits in communication and social interaction, along with repetitive patterns of behavior and interests. A growing body of literature suggests that persons with autism

This study was supported by the National Natural Science Foundation of China (Projects No. 61572239 and No. 61772242), and the Doctoral Fund of the Ministry of Education of China (Project No. 2017M611737).

T. Gedeon et al. (Eds.): ICONIP 2019, CCIS 1142, pp. 570–579, 2019.
https://doi.org/10.1007/978-3-030-36808-1_62

Table 1. Datasets

Dataset	ASD		CONTROL		Age($\mu \pm \sigma$)	Total
	M/F	Age	M/F	Age		
CAL	15/4	17.5–45.1	15/4	17–56.2	28.15 ± 0.41	$N = 38$
STA	16/4	7.5–12.9	16/4	7.8–12.4	9.9 ± 1.5	$N = 40$

M = male, F = female, μ = mean, σ = standard deviation.

exhibit altered functional brain connectivity, as well as altered anatomical connectivity, lending more credence to the dysconnectivity theory of this pathology [7]. A more in-depth investigation of the patterns of dysconnectivity might reveal clinically relevant diagnostic predictors or improve our understanding of subtypes of the spectrum, leading to new or improved detection approaches.

Recent research using pattern recognition and machine learning methods applied to whole-brain neuroimaging data has proved effective at diagnosing autism based on brain features computed from complex network methods. Graph theoretical approaches make it possible to extract network-based features for single-subject classification at different levels of granularity. With the development of machine learning technologies, alterations in network connectivity have been extensively leveraged for building predictive models of brain disorders. Currently, the majority of brain disorder classification studies made use of nodal pair-wise correlations as features that were fed to machine learning classifiers such as SVMs, discriminant analysis classifiers, and neural networks [7].

In this paper, resting-state brain networks are modeled as undirected, weighted graphs. Unlike previous works that focussed on global connectivity patterns, we extract and analyze regional connectivity patterns, and highlight how they differ between persons living with autism and healthy individuals. We examine the significance of autism-related variations in regional connectivity for all the brain regions described in the Automated Anatomical Labeling (AAL) atlas. Further, we address the classification problem by training gradient boosting trees classifiers with regional features. To the best of our knowledge, no study has systematically analyzed the whole range of regional connectivity for classification purposes. Since functional connectivity varies significantly between children and adults, we illustrate our approach on two resting-state functional magnetic resonance imaging (Rs-fMRI) datasets and show that local connectivity features can effectively diagnose both children and adults with autism.

2 Materials and Methods

2.1 Experimental Datasets and Preprocessing

Rs-fMRI scans in the current study were collected at the California Institute of Technology (CAL) and Stanford University (STA). Their corresponding preprocessed regional time-series were downloaded from the Autism Brain Imaging

Data Exchange (ABIDE) Preprocessed Connectomes Project [3]. Table 1 provides demographic information about the enrolled participants. All data contained in the ABIDE repository were previously anonymized, and private health information was protected according to the Health Insurance Portability and Accountability Act (HIPAA). Detailed information about imaging acquisition parameters, informed consent, and site-specific protocols are available on the consortium's website[1]. Selected data were already preprocessed according to the DPARSF pipeline [10] and warped into the Anatomical Automatic Labeling (AAL) atlas; the mean time-series for 90 regions of interest (ROI) were extracted for each subject.

2.2 Brain Network Construction

We modeled each brain imaging data as a network or undirected, weighted graph $G = (V, E)$, where $V = \{v_1, v_2, ..., v_n\}$ is the set of 90 regions of interest defined according to the Anatomical Automatic Labeling (AAL) atlas, and $E = \{e_{ij}\}_{i,j=1}^{n}$ with $e_{ij} = \{v_i, v_j, w_{i,j}\} \in V \times V \times \mathbb{R}$, a collection of connections among ROIs, with $w_{i,j}$ denoting Pearson's correlation coefficient that measures the strength of association between any possible pair of regional mean time-series v_i and v_j.

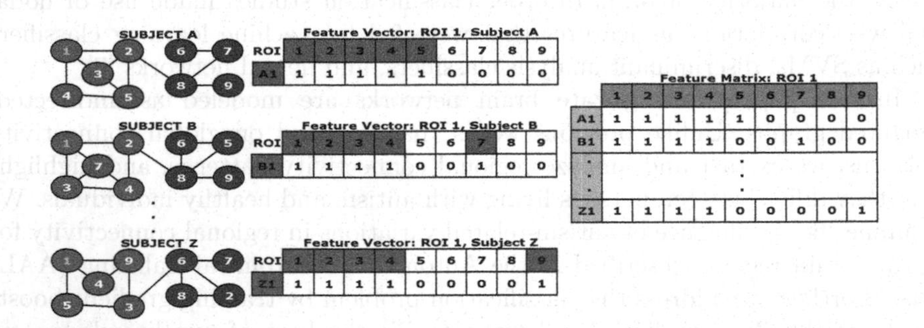

Fig. 1. An example schematic of our regional connectivity pattern extraction procedure. This figure illustrates how regional connectivity patterns were extracted based on Katz similarity to the reference ROI labeled "1". Left column: Toy graphs representing all subjects labeled "subject A", "subject B", ..., "subject Z", having the same set of ROIs but with different regional connectivity circuitry. Green and red colors are assigned to close and distant connections of the reference ROI, respectively. Middle column: binary feature vectors generated according to ROI's assignment to the long- or short-range classes of the reference ROI. Right column: feature matrix of the reference ROI, formed by aggregating the corresponding feature vectors across subjects. (Color figure online)

[1] http://fcon_1000.projects.nitrc.org/indi/abide/.

2.3 Regional Connectivity Features Estimation

Regional connectivity features were extracted for each region of interest (ROI) and diagnostic group as follows. Based on topological similarity with respect to a given reference ROI v_R, nodes in each network were grouped in two classes: the class of short-range connections, C_L^R, and the class of distant or long-range connections C_D^R as illustrated in Fig. 1. More formally, given any dyadic similarity metric $sim(.,.)$ and a threshold value $\tau \in]0,1[$, $C_L^R = \{v_l \mid sim(v_R, v_l) > \tau\}$, the set of ROIs that are structurally close to the reference node v_R, and $C_L^D = \{v_l \mid sim(v_R, v_l) \leq \tau\}$, the set of nodes that are distant to the reference node. Then, for each ROI and network, a 90 binary-valued feature vector was generated corresponding to whether for a given reference ROI, all other ROIs were assigned to the same connectivity group. ROIs were marked as either participating (1) and not participating (0) in the reference ROI's class of short-range connections. Finally, for each ROI, feature vectors were aggregated to include entries from all subjects in the same diagnostic group and form the regional connectivity matrix (feature matrix). This procedure was applied to each dataset and diagnostic group separately and yielded 90×2 feature matrices corresponding to each ROI and clinical group. The assessment of regional connectivity differences between healthy and pathological participants using statistical tests, as well as classification, were performed using feature matrices extracted for each ROI. The nodal similarity between each pair of nodes was computed using Katz's metric, which is based on network paths. While a variety of metrics could be employed for this purpose, the use of Katz index was motivated by a recent study which found Katz index highly correlates with the underlying neural activity of the brain at rest [4]. In essence, this metric plausibly captures how brain regions interact at rest, especially when they are not connected with direct links. Katz metric is defined as:

$$sim^{Katz}(v_i, v_j) = \sum_{l=0}^{\infty} \beta^l \left\| \sigma^l(v_i, v_j) \right\|, \tag{1}$$

where $\beta \ll 1$ is a computing parameter set to 0.001 in our experiments, $\|A\|$ the cardinal of set A, and $\sigma^l(v_i, v_j)$ the set of all paths of length l between v_i and v_j. The similarity threshold τ was set to 0.5.

2.4 Single Subject Classification

XGBoost (eXtreme Gradient Boosting) model was adopted for binary classification [2]. XGBoost is an implementation of the gradient boosting trees algorithm where gradient descent is used to minimize the loss. Briefly, a binary classifier tries to find a relationship between training inputs, $\boldsymbol{x}_i \in \mathbb{R}^m$, and their corresponding label, $\boldsymbol{y}_i = \{-1, +1\}$ (e.g. healthy controls and ASD), by estimating a classification function $f(\boldsymbol{x}_i) : \mathbb{R}^m \rightarrow \mathbb{R}$, where i is a training sample and m is the dimensionality of $\boldsymbol{x}$. For XGBoost model, assuming that the model is made of K trees, we have: $\hat{\boldsymbol{y}}_i = \sum_{k=1}^{K} f_k(x_i)$, where f_k belongs to the set of regression trees.

In order to additively train our model, we optimized the following L_2-regularized objective:

$$\xi(\boldsymbol{x}_i, \boldsymbol{y}_i) = \sum_{i=1}^{n} l(\boldsymbol{y}_i, \hat{\boldsymbol{y}}_i^{(k)}) + \sum_{k=1}^{K} \Omega(f_k), \quad (2)$$

where l denotes the binomial logistic loss function $\frac{1}{(1+e^{-t})}$ and $\Omega(f) = \gamma K + \frac{1}{2}\lambda\sum_{j=1}^{K}\theta_j^2$ is the regularization term which helps avoiding overfitting. In the regularization term, θ denotes the vector of scores on tree leaves, γ and λ are two regularization hyperparameters.

The Scikit-learn package was implemented to perform classification, and the hyperparameters were left to their default values [6]. We performed 90 different classification experiments, each classifier being fed with connectivity features extracted based on similarity with respect to a specific region. All the classifiers were trained using the same hyperparameters. To evaluate the performance of each classifier, we applied the leave-one-out cross-validation (LOOCV) strategy, and classification results are reported in terms of accuracy, precision, and recall. A statistical permutation test was used to assess whether the estimated accuracies outperformed chance.

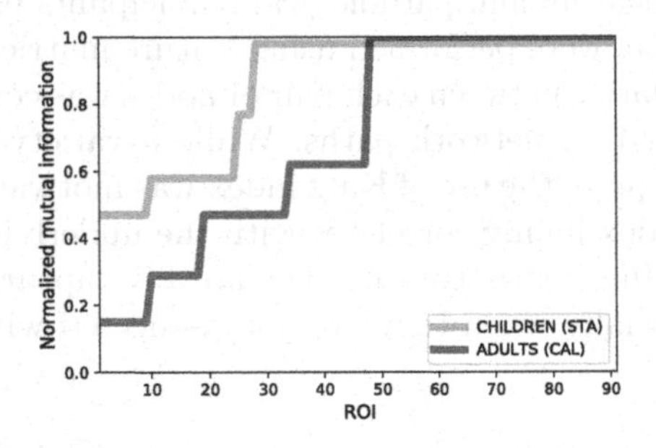

Fig. 2. Normalized mutual information (NMI) plots. NMI was computed between regional average connectivity sets derived from homologous reference ROIs in the two clinical groups (ASD and controls).

3 Results and Discussion

3.1 Regional Connectivity Differences

Group Level Analyses. To examine group-level regional connectivity differences between ASD subjects and healthy controls, group-level mean regional connectivity matrices were generated for each ROI. In addition, normalized mutual information (NMI) was computed between the diagnostic groups in each dataset. Figure 2 shows that regional connectivity patterns are well-preserved for many brain regions in both children and adult's autism ($NMI \approx 1$.) However, several ROIs exhibit less similar connectivity between the two diagnostic populations. Also, we note that for many regions, NMI was higher in the pediatric population, implying that regional connectivity is more atypical in the adult population.

Table 2. Significantly altered reference regions in autism

Label	Regions	Hemisphere	MNI coordinates	Pvalue
Adult dataset (CAL)				
7	Middle frontal gyrus	Left	[−33.43, 32.73, 35.46]	0.009**
11	Inferior frontal gyrus, opercular part	Left	[−48.43, 12.73, 19.02]	0.001**
20	Supplementary motor area	Right	[8.62, 0.17, 61.85]	0.001**
23	Superior frontal gyrus, medial	Left	[−4.8, 49.17, 30.89]	0.009**
27	Gyrus rectus	Left	[−5.08, 37.07, −18.14]	0.001**
30	Insula	Right	[39.02, 6.25, 2.08]	0.009**
39	Parahippocampal gyrus	Left	[−21.17, −15.95, −20.7]	0.001**
43	Calcarine fissure and surrounding cortex	Left	[−7.14, −78.67, 6.44]	0.001**
47	Lingual gyrus	Left	[−14.62, −67.56, −4.63]	0.001**
50	Superior occipital gyrus	Right	[24.29, −80.85, 30.59]	0.001**
52	Middle occipital gyrus	Right	[37.39, −79.7, 19.42]	0.001**
54	Inferior occipital gyrus	Right	[38.16, −81.99, −7.61]	0.001**
66	Angular gyrus	Left	[45.51, −59.98, 38.63]	0.001**
70	Paracentral lobule	Right	[7.48, −31.59, 68.09]	0.001**
85	Middle temporal gyrus	Left	[−55.52, −33.8, −2.2]	0.018*
Children dataset (STA)				
1	Precental gyrus	Left	[−38.65, −5.68, 50.94]	0.018*
4	Superior frontal gyrus, dorsolateral	Right	[21.9, 31.12, 43.82]	0.009**
11	Inferior frontal gyrus, opercular part	Left	[−48.43, 12.73, 19.02]	0.001**
20	Supplementary motor area	Right	[8.62, 0.17, 61.85]	0.018**
24	Superior frontal gyrus, medial	Right	[9.1, 50.84, 30.22]	0.001**
32	Anterior cingulate and paracingulate gyri	Right	[8.46, 37.01, 15.84]	0.009**
37	Hippocampus	Left	[−25.03, −20.74, −10.13]	0.036*
48	Lingual gyrus	Right	[16.29, −66.93, −3.87]	0.001**
51	Middle occipital gyrus	Left	[−32.39, −80.73, 16.11]	0.027*
72	Caudate nucleus	Right	[14.84, 12.07, 9.42]	0.001**
78	Thalamus	Right	[13, −17.55, 8.09]	0.001**

Subject Level Analyses. While normalized mutual information plots allow for measuring the degree of disagreement between regional connectivity patterns, they do not indicate if the difference is significant. To this end, we used an approach proposed by [1] to test for regional connectivity differences between the two clinical populations. This approach relies on the idea that if the clinical group irrefutably justifies the discrepancy in connectivity patterns of a specific ROI, then the mean NMI between all possible pairs of participants within a diagnostic group should be higher than the mean NMI of pairs of participants between randomized groups. The underlying distribution of group NMI being unknown, a null-distribution was generated through a permutation method (10,000 permutations). Thus, a set of 90 p-values was generated corresponding to whether each regional connectivity patterns were more similar for subjects in the same clinical group than in shuffled groups. The p-values were subsequently FDR-corrected for multiple comparisons ($p < 0.05$). As shown in Table 2, the adult population (CAL) displayed nineteen regions with significant alterations in connectivity, and the pediatric population (STA) twenty. However, only two identical brain regions, the superior frontal gyrus, and the supplementary motor area were found atypical in both demographic groups.

Table 3. Classification performance of significantly discriminating regional connectivity features

Label	Reference regions	Accuracy (%)	Precision (%)	Recall (%)	P-value
Adult's dataset (CAL)					
12	Inferior frontal gyrus, opercular part	72.05	71.59	72	0.010*
15	Inferior frontal gyrus, orbital part	71.05	71.01	71.08	0.020*
36	Posterior cingulate gyrus	76.32	76.48	76.07	0.010*
47	Lingual gyrus	78.95	78.98	78.75	0.020*
66	Angular gyrus	68.42	68.3	68.52	0.048*
70	Paracentral lobule	73.68	73.77	73.33	0.020*
Children's dataset (STA)					
11	Inferior frontal gyrus, opercular part	75.01	75.00	75.09	0.009**
20	Supplementary motor area	73.09	73.13	72.98	0.019*
28	Gyrus rectus	63.89	63.17	63.71	0.047*
47	Lingual gyrus	70.65	70.26	70.33	0.039*

3.2 Identifying ASD Patients

The results showing the discriminative reference regions are summarized in Table 3. The adult population displayed six significantly discriminative reference regions, while the pediatric population showed only four. As can be seen, using regional long- versus short-range connectivity patterns as features yielded a peak accuracy of 78.95% for the adult dataset (reference ROI: Lingual gyrus, 78.98% precision, 78.75% recall and p-value = 0.020, permutation test with 100 repetitions) and a peak accuracy of 75.01% for the pediatric dataset (reference ROI: inferior frontal gyrus, opercular part, 75.00% precision, 75.09% recall, and p-value = 0.009, permutation test with 100 repetitions). To the best of our knowledge, these are the highest classification accuracies reported for these two datasets using a LOOCV evaluation strategy. The peak accuracy reported for CAL in [5] using fine-grained correlations as features and a LOOCV strategy was as high as 50%. Also, the highest accuracy obtained by [8] for STA using 303 regional morphological features and 400 inter-regional functional features with a support vector classifier was about 69%. Although we did not seek to reproduce their results, the baselines used in these papers are the same as ours.

3.3 Deriving Significant Neural Patterns in ASD

Additional follow-up analyses were performed for the classifiers that yielded the highest classification accuracy in order to identify connections that were involved in the construction of decision trees. These connections are those that effectively contributed to the identification of autistic patients. We extended our permutation test to evaluate the predictive power of each connection in the connectivity sets. We re-ran the classification framework 100 times and computed the average over 100 runs of the total number of times a specific connection was involved in the decision process of boosted trees.

The most discriminative connections can be visualized in Fig. 3. In these Figures, significantly discriminative connectivity patterns between patients with ASD and healthy individuals involve only a small number of connections that

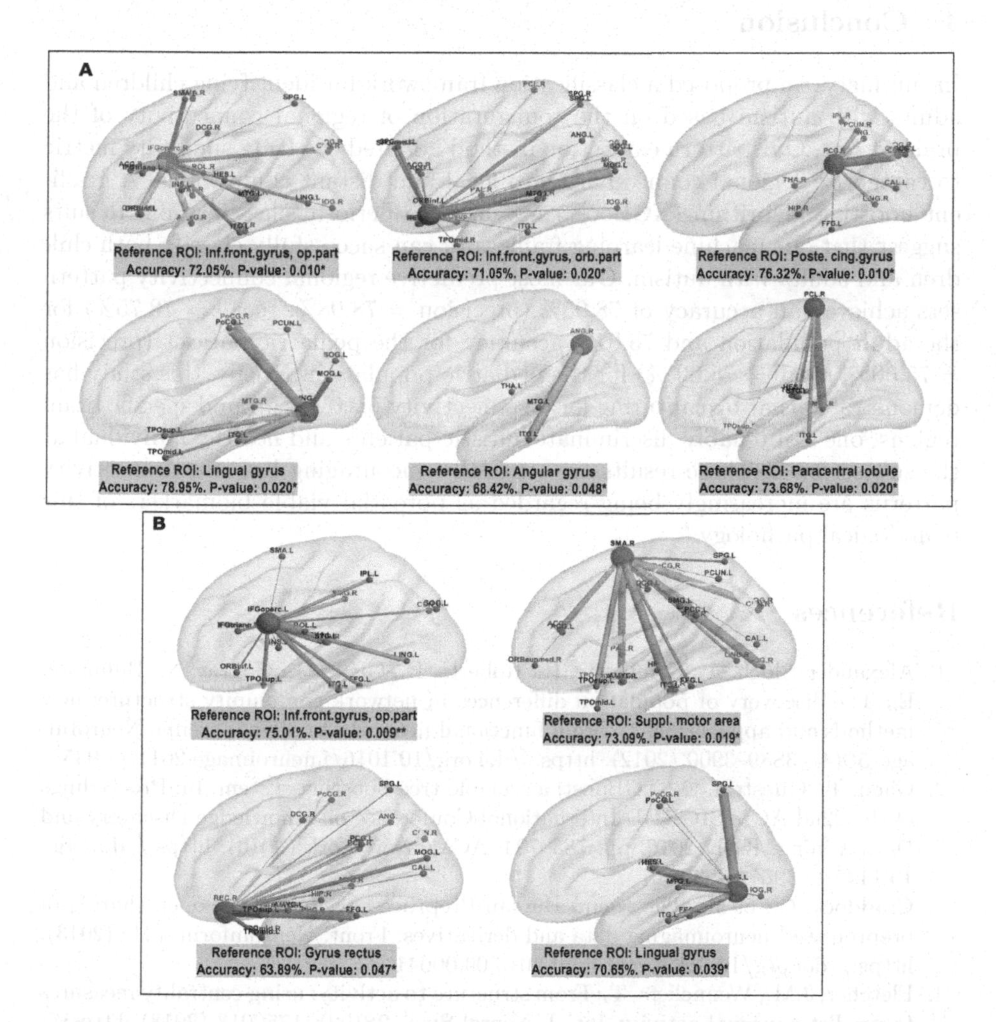

Fig. 3. Visualizing significantly discriminative reference ROIs and their connections for the adult cohort (A) and the pediatric cohort (B). The size of the reference nodes is increased only for distinction purpose. The width of each connection is proportional to the total number of times it contributed to the discrimination process. Visualizations were made possible using BrainNet Viewer [9].

yielded high classification accuracies. The most discriminative reference ROIs differ between children and adults, except for the lingual gyrus. Regional connections involving the lingual gyrus proved to be highly discriminative for both adults and children datasets, but not with an identical set of connections. Taken together, significantly altered regions and highly discriminative reference regions found in this study are mainly in line with what has been previously reported [5,8].

4 Conclusion

In summary, we proposed a classification framework for identifying children and adults with autism based on the configuration of regional connectivity of the brain at rest. Our pattern extraction method adopted the Katz similarity metric to categorize regional connectivity into local and distant connections. A gradient boosting algorithm (XGBoost) was used to perform classification. Results suggest that our machine learning framework can successfully classify both children and adults with autism. Our most predictive regional connectivity pattern sets achieved an accuracy of 78.95% (precision = 78.98%, recall = 78.75%) for the adult population and 75.01% accuracy for the pediatric dataset (precision = 75.00%, recall = 75.09 %) for the pediatric population. Briefly, this study has demonstrated that by capturing local connectivity patterns around specific brain regions, one can reliably discriminate autistic patients and healthy individual at the subject level. These results are especially encouraging because connectivity patterns are increasingly being regarded as potential viable biomarkers of this neurological pathology.

References

1. Alexander-Bloch, A., Lambiotte, R., Roberts, B., Giedd, J., Gogtay, N., Bullmore, E.: The discovery of population differences in network community structure: new methods and applications to brain functional networks in schizophrenia. NeuroImage **59**(4), 3889–3900 (2012). https://doi.org/10.1016/j.neuroimage.2011.11.035
2. Chen, T., Guestrin, C.: XGBoost: a scalable tree boosting system. In: Proceedings of the 22nd ACM SIGKDD International Conference on Knowledge Discovery and Data Mining, KDD 2016, pp. 785–794. ACM, New York (2016). https://doi.org/10.1145/2939672.2939785
3. Craddock, C., et al.: The Neuro Bureau Preprocessing Initiative: open sharing of preprocessed neuroimaging data and derivatives. Front. Neuroinform. (41) (2013). https://doi.org/10.3389/conf.fninf.2013.09.00041
4. Fletcher, J.M., Wennekers, T.: From structure to activity: using centrality measures to predict neuronal activity. Int. J. Neural Syst. **28**(02), 1750013 (2018). https://doi.org/10.1142/S0129065717500137, pMID: 28076982
5. Nielsen, J.A., et al.: Multisite functional connectivity MRI classification of autism: ABIDE results. Front. Hum. Neurosci. **7**, 599 (2013). https://doi.org/10.3389/fnhum.2013.00599
6. Pedregosa, F., et al.: Scikit-learn: machine learning in python. J. Mach. Learn. Res. **12**, 2825–2830 (2011)

7. Song, Y., Epalle, T.M., Lu, H.: Characterizing and predicting autism spectrum disorder by performing resting-state functional network community pattern analysis. Front. Hum. Neurosci. **13**, 203 (2019). https://doi.org/10.3389/fnhum.2019.00203
8. Wang, J., et al.: Multi-task diagnosis for autism spectrum disorders using multi-modality features: a multi-center study. Hum. Brain Mapp. **38**(6), 3081–3097 (2017). https://doi.org/10.1002/hbm.23575
9. Xia, M., Wang, J., He, Y.: BrainNet Viewer: a network visualization tool for human brain connectomics. PLoS ONE **8**, e68910 (2013)
10. Yan, C., Zang, Y.: DPARSF: a MATLAB toolbox for "pipeline" data analysis of resting-state fMRI. Front. Syst. Neurosci. **4**, 13 (2010). https://doi.org/10.3389/fnsys.2010.00013

TP-ADMM: An Efficient Two-Stage Framework for Training Binary Neural Networks

Yong Yuan[1,2], Chen Chen[1,2](✉), Xiyuan Hu[1,2], and Silong Peng[1,2,3]

[1] Institute of Automation, Chinese Academy of Sciences, Beijing, China
{yuanyong2015,chen.chen,xiyuan.hu,silong.peng}@ia.ac.cn
[2] University of Chinese Academy of Sciences, Beijing, China
[3] Beijing Visystem Co. Ltd., Beijing, China

Abstract. Deep Neural Networks (DNNs) are very powerful and successful but suffer from high computation and memory cost. As a useful attempt, binary neural networks represent weights and activations with binary values, which can significantly reduce resource consumption. However, the simultaneous binarization introduces the coupling effect, aggravating the difficulty of training. In this paper, we develop a novel framework named TP-ADMM that decouples the binarization process into two iteratively optimized stages. Firstly, we propose an improved target propagation method to optimize the network with binary activations in a more stable format. Secondly, we apply the alternating direction method (ADMM) with a varying penalty to get the weights binarized, making weights binarization a discretely constrained optimization problem. Experiments on three public datasets for image classification show that the proposed method outperforms the existing methods.

Keywords: Binary neural network · ADMM · Target propagation

1 Introduction

Recently, Deep Neural Networks (DNNs) have achieved state-of-the-art performance in various tasks such as speech recognition, computer vision, and natural language processing. However, the complexity of DNN increases dramatically, which hinders its deployment on embedded devices. To alleviate this problem, a number of approaches have been proposed [1,2], such as network quantization, weight pruning, low-rank decomposition, compact structure design, and knowledge distillation. Among these methods, binary neural networks have received great attention from both the research community and the industry. Binarizing weights and activations together, the computation will be replaced by XNOR and bitcount operations, and the network can get a 58× speedup in theory [3–5].

Unfortunately, binarizing weights and activations simultaneously incurs severe accuracy drop for large-scale classification tasks. The binarization of

T. Gedeon et al. (Eds.): ICONIP 2019, CCIS 1142, pp. 580–588, 2019.
https://doi.org/10.1007/978-3-030-36808-1_63

weights only leads to marginal accuracy loss. However, the binarization of activations degrades the performance significantly, which may be caused by the following two reasons: (1) The features are more sensitive to binarization comparing to weights and more bits are needed for the representation. Zhou et al. increased the bit-width of activations and can achieve obvious performance improvement than binary activations [6]. (2) The continuous approximations of the non-differentiable operators in backpropagation cause gradient mismatch, which leads to sub-optimal solutions [7]. Cai et al. tried to minimize the gradient mismatch of quantization networks with variants of ReLU activation [8]. In addition to the difficulty of binarizing activations, the simultaneous binarization of weights and activations introduces the coupling effect, which adds the optimization difficulty. Wang et al. decomposed the training of quantization networks into two steps, which can get a better result than simultaneous quantization [9]. Nevertheless, relevant efforts for binary neural networks are still scarce.

To alleviate the coupling effect of simultaneous binarization, we decompose the training of binary neural networks into two stages: (1) optimizing a network with binary activations; (2) binarizing the weights based on the network from the first stage. Inspired by target propagation that can be used to train the network with discrete outputs [10,11], we improve the target propagation method [11] to obtain a network with binary activations. With activations binarized, in order to binarize the weights further, we adopt a penalty varying alternating direction method of multipliers (ADMM). ADMM can convert the binarization of weights to a discretely constrained optimization problem [12,13], and the varying penalty parameter can reduce the dependence on the initial setting.

2 The Proposed Method

2.1 Target Propagation Based Activation Binarization

In the first stage, we utilize target propagation to obtain a network with binary activations. Consider an N-layer network and the i-th layer has n_i activation units. Let $Z_i(z_{i1}, ..., z_{in_i})$, $H_i(h_{i1}, ..., h_{in_i})$, and $T_i(t_{i1}, ..., t_{in_i})$ denote pre-activations, binary activations, and targets of the i-th layer, respectively. The pre-activation unit z_{ij} is binarized to -1 or 1 by a *sign* function. In target propagation, each activation unit h_{ij} is assigned with a binary target t_{ij}, and then the layer weights can be optimized based on the given targets as shown in Fig. 1. The key to target propagation is how to define the layer loss and how to set targets for each layer.

The binary targets based optimization can be considered as a binary classification problem, which can be optimized using *hinge loss* intuitively. However, *hinge loss* is sensitive to noisy data and outliers that cannot be used in this problem directly. As in [11], we adopt *soft hinge loss* as the layer loss, which can alleviate the effect of noisy data and outliers.

For targets setting, a natural method is to choose targets that can reduce layer loss. Only having two opposite values, the targets can be set by a heuristic method, which is based on the direction information of the derivatives [11].

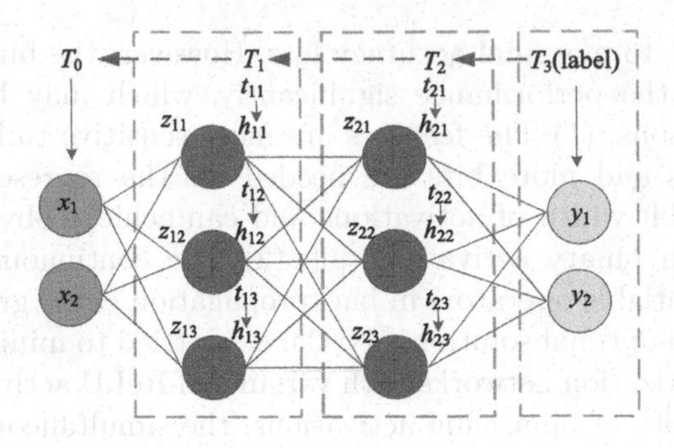

Fig. 1. Illustration of target propagation. The loss function of the last layer is *cross entropy loss*, and the targets of the last layer are set to the labels.

Nevertheless, this direction based method is sensitive to small derivatives. To reduce the permutation caused by small derivatives, we improve the targets setting method proposed in [11]. The targets can be set as:

$$t_{ij} = \begin{cases} sign(-\frac{\partial}{\partial h_{ij}} L_{i+1}(Z_{i+1}, T_{i+1})), & when \left|\frac{\partial}{\partial h_{ij}} L_{i+1}\right| \geq e_i, \\ sign(z_{ij}), & others, \end{cases} \tag{1}$$

where L_{i+1} is the loss of layer $i + 1$, and e_i denotes a threshold of the i-th layer. For the activation units whose derivatives are greater than the threshold, the targets are set to be consistent with the negative sign of the partial derivatives of the next layer's loss. This scheme is simple but effective, as the negative gradient usually indicates the optimization direction. For the activation units whose derivatives are smaller than the threshold, the targets are set to be consistent with the activations. The mismatch between the sign of small derivatives and binary activations leads to opposite targets, degrading the stability of training. A threshold is introduced to ignore the perturbation from small derivatives, which can improve the robustness of target propagation.

2.2 Penalty Varying ADMM Based Weight Binarization

Based on the activation-binarized network from the first stage, we employ ADMM to binarize the weights further. Let $f(W)$ denote the loss function of the activation-binarized network, where $W = \{W_1, W_2, ..., W_N\}$, and W_i denotes the full-precision weights of the i-th layer. Let m_i denote the number of output channels in the i-th layer. The weights are binarized channel-wisely, and the problem can be defined as:

$$\min_W f(W) \quad s.t. \quad W_{ij} \in C_{ij} = \{-\alpha_{ij}, \alpha_{ij}\}, \quad i = 1, ..., N, \quad j = 1, ..., m_i, \tag{2}$$

where W_{ij} represents the weights of i-th layer and j-th channel, and α_{ij} is a scaling factor. Defining $g(\cdot)$ as an indicator function of the set C, the augmented Lagrange can be formulated as:

$$L_\rho = f(W) + \sum_{i=1}^{N} g(Z_i) + \sum_{i=1}^{N} \frac{\rho}{2} \|W_i - Z_i + U_i\|^2, \tag{3}$$

where Z_i denotes the auxiliary variables, U_i denotes the dual variables, and ρ denotes the penalty parameter. The ADMM algorithm proceeds by repeating, for k = 0, 1,..., the following steps [12]:

$$W^{k+1} := \arg\min_W (f(W) + \sum_{i=1}^{N} \frac{\rho^k}{2} \|W_i - Z_i^k + U_i^k\|^2) \tag{4}$$

$$Z^{k+1} := \arg\min_Z (\sum_{i=1}^{N} \|W_i^{k+1} - Z_i^k + U_i^k\|^2) \tag{5}$$

$$U_i^{k+1} := U_i^k + W_i^{k+1} - Z_i^{k+1}, \tag{6}$$

which is proximal step, binarization projection step, and dual update, respectively.

To accomplish the binarization of weights, we need to optimize Eqs. (4) and (5). In Eq. (4), the first term is the loss function of the first stage, and the second term can be considered as a special regularizer. In fact, these two terms cannot be sufficiently optimized by stochastic gradient descent. Since the penalty parameter ρ is sensitive to initialization, the first term is difficult to optimize under a large penalty, and a small penalty will slow down the convergence. To overcome this challenge, an increasing penalty is introduced in ADMM, which can reduce the dependence on the initial setting. In the early iterations, the activation-binarized network converges rapidly, and the distance between full-precision weights and binary weights is very large. With the increase of the penalty parameter, the effect of the regularization term is enhanced, and the distance is optimized.

As in Eq. (5), the full-precision weights can be binarized using the Euclidean projection. The binarizaiton of weights in i-th layer and j-th channel can be formulated as:

$$B_{ij} = sign(W_{ij}^{k+1} + U_{ij}^k) \tag{7}$$

$$\alpha_{ij} = \frac{1}{c \times h \times w} \|W_{ij}^{k+1} + U_{ij}^k\|_1, \tag{8}$$

where $B_{ij} \in \{-1, 1\}^{c \times h \times w}$, and c, h, w denote input channels, kernel height, and kernel width, respectively.

Finally, we update the dual variables according to Eq. (6). As we adopt a varying penalty parameter, the dual variables U^{k+1} should be rescaled after updating ρ^{k+1}. This concludes one epoch of the ADMM algorithm. The regularization term in Eq. (4) varies with the training, and this ADMM based method can be considered as a special regularization method to achieve binarization.

The training of binary neural networks is decoupled into two stages as the above sections. By optimizing the activation-binarized network from the first stage under the ADMM framework iteratively, we can obtain an enhanced binary neural network. This two-stage optimization framework is named as TP-ADMM, and the detailed procedure is demonstrated in Algorithm 1.

Algorithm 1. TP-ADMM for training binary neural networks.

Input: An initialized network with binary activations.

1: **ADMM LOOP:**
2: **for** epoch = 1 to K **do**
3: **Step 1**: Proximal step
4: **TP LOOP:** Train the activation-binarized network
5: **for** iter = 1 to M **do**
6: **Forward propagation**:
7: Compute pre-activations Z and binary activations H.
8: **Backward propagation**:
9: Compute binary targets T_i and the layer loss.
10: Compute the gradients of W_i based on the layer loss.
11: Update the learning rate η.
12: **end for**
13: **Step 2**: Binarization projection: Update Z^{k+1} by Eq. (7) and Eq. (8).
14: **Step 3**: Dual update: Update U^{k+1} by Eq. (6).
15: Update the penalty parameter ρ^{k+1} and rescale U^{k+1}.
16: **end for**

3 Experiments

In this section, we compare our method with the following methods: (1) BNN [4]; (2) XNOR-Network (XNOR) [5]; (3) ADMM on three commonly used public datasets. Besides, we extend the bit-width of weights or activations and conduct some additional experiments.

3.1 Datasets and Experiments Setting

CIFAR10. It contains 32×32 color images from ten object classes, 50000 images for training, and 10000 images for testing. We adopt a 8-layer model as in [11] to validate the effectiveness of our approach: "(48C5) - MP2 - (2×64C3) - MP2 - (3×128C3) - (512C3) - 10", where C5 is a 5×5 convolution layer and MP2 is a 2×2 max-pooling layer.

CIFAR100. It contains 32×32 color images from 100 object classes, 50000 images for training, and 10000 images for testing. We use a VGG-like architecture as in [14]: "(2×128C3) - MP2 - (2×256C3) - MP2 - (2×512C3) - MP2 - (2×1024FC) - 100".

SVHN. It contains 32×32 color images from ten digit classes. We use 604388 images for training, and the remaining 26032 for testing. The model we use is the same with [4]: "(2×64C3) - MP2 - (2×128C3) - MP2 - (2×256C3) - MP2 (2×1024FC) - 100".

Experiments Setting. ADAM is used as the optimizer. The initial learning rate is set to 0.001, 0.002, and 0.002 on CIFAR10, CIFAR100, and SVHN, respectively. In target proportion, a threshold is proposed to improve the stability. The activation units are sorted based on the magnitude of the gradients, and the threshold is set by a given proportion. We gradually increase the proportion to evaluate the impact of the threshold. As shown in Fig. 2, when the proportion is set to 6%, the performance outperforms the original model by 0.4%, so the proportion is set to 6% in the following experiments. In ADMM, the step size of penalty parameter ρ is set to 2×10^{-6}.

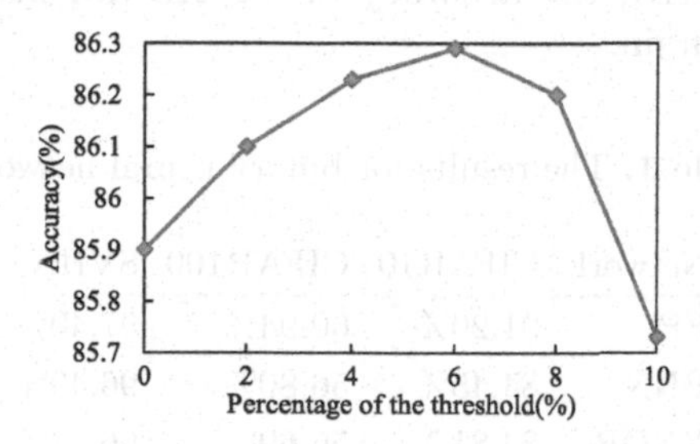

Fig. 2. The impact of the threshold. Experiments are implemented on CIFAR10 with only activations binarized.

3.2 Experimental Results

Experiments on Binary Neural Network. The training process of TP-ADMM is presented in Fig. 3. The $L2$ distance is used to measure the model difference between the two stages. As we can see, the accuracy of the binary neural network approaches the network from the first stage fast in the early iterations. With more epochs and heavier penalty, these two networks converge to the same accuracy finally. At the start of training, the regularization term in Eq. (4) has little effect on the activation-binarized network. The network converges rapidly, making the binary neural network difficult to follow. With the enhancement of the regularization term, the weights of the two stages approach to each other, and we can get an optimized binary neural network in the end.

We compare the full-precision network (FP), BNN, XNOR-Net, and ADMM with the proposed TP-ADMM. As shown in Table 1, training binary neural networks under the ADMM framework outperforms the existing binarization approaches. Combining target propagation with ADMM further, we can achieve the best performance on the three datasets. On CIFAR10 and CIFAR100, TP-ADMM outperforms BNN and XNOR by a significant margin with 1% accuracy

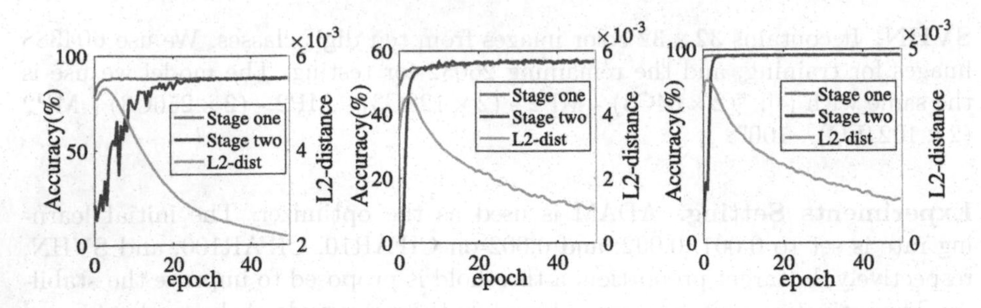

Fig. 3. The training curve on CIFAR10, CIFAR100, and SVHN (left to right). The red line represents the accuracy of the activation-binarized network, the blue line represents the accuracy of binary neural network, and the yellow line represents the average $L2$ distance between full-precision weights and binary weights. (Color figure online)

improvement. On SVHN, the accuracy of TP-ADMM also exceeds BNN and XNOR by a small margin.

Table 1. The results for binary neural networks.

Network	CIFAR10	CIFAR100	SVHN
FP	91.20%	60.94%	97.40%
BNN	84.45%	56.80%	96.49%
XNOR	84.81%	56.69%	96.52%
ADMM	85.71%	57.01%	96.52%
Ours	**85.85%**	**57.74%**	**96.58%**

Experiments on Ternary Weights. To improve the representation ability of binary weights, we represent the weights with $\{-\alpha, 0, \alpha\}$ as in Ternary Weight Network (TWN) [15] for further experiments. The additional zero value does not add computation consumption. The results are presented in Table 2. As we can see, TP-ADMM outperforms TWN on the three datasets. However, the performance only improves slightly comparing to binary weights.

Table 2. The results for ternary weights and binary activations.

Network	CIFAR10	CIFAR100	SVHN
FP	91.20%	60.94%	97.40%
TWN	85.80%	57.10%	96.55%
Ours	**86.07%**	**57.96%**	**96.60%**

Experiments on 2-Bit Activations. We extend the bit-width of activations to 2-bit for more experiments. Target propagation for multi-bit activations is processed as in [11]. As shown in Table 3, the performance improves apparently than binary neural networks. On CIFAR10, TP-ADMM outperforms $XNOR_2$ by 1.69%. On SVHN, the accuracy of TP-ADMM nearly approaches the full-precision network. Moreover, the proposed method even outperforms the full-precision network on CIFAR100, indicating the effectiveness of TP-ADMM. According to the experiments, activations are more sensitive to binarization and should be paid with more attention.

Table 3. The results for binary weights and 2-bit activations. $XNOR_2$ refers to replacing the activation of XNOR with 2-bit.

Network	CIFAR10	CIFAR100	SVHN
FP	91.20%	60.94%	97.40%
$XNOR_2$	87.75%	57.94%	97.14%
Ours	**89.44%**	**61.03 %**	**97.34%**

4 Conclusion

In this paper, we present a two-stage framework named TP-ADMM to optimize the training of binary neural networks. An improved target propagation method and a penalty varying ADMM are jointly employed to achieve the two-stage binarization, which can alleviate the coupling effect of simultaneous binarization, making the binarization process more stable. Experiments on image classification demonstrate the effectiveness of the proposed method. In addition, the proposed method can be easily applied to train quantization networks with more bits.

Acknowledgments. All correspondences should be forwarded to Chen Chen, the corresponding author, via chen.chen@ia.ac.cn. This work was supported by the National Science Foundation of China under Grant NSFC 61571438.

References

1. Cheng, Y., Wang, D., Zhou, P., et al.: Model compression and acceleration for deep neural networks: the principles, progress, and challenges. IEEE Signal Process. Mag. **35**(1), 126–136 (2018)
2. Cheng, J., Wang, P., Li, G., et al.: Recent advances in efficient computation of deep convolutional neural networks. Front. Inf. Technol. Electron. Eng. **19**(1), 64–77 (2018)
3. Courbariaux, M., Bengio, Y., David, J.P.: BinaryConnect: training deep neural networks with binary weights during propagations. In: International Conference on Neural Information Processing Systems, pp. 3123–3131 (2015)

4. Hubara, I., Soudry, D., Yaniv, R.E: Binarized neural networks. In: Advances in Neural Information Processing Systems (2016)
5. Rastegari, M., Ordonez, V., Redmon, J., Farhadi, A.: XNOR-Net: ImageNet classification using binary convolutional neural networks. In: Leibe, B., Matas, J., Sebe, N., Welling, M. (eds.) ECCV 2016. LNCS, vol. 9908, pp. 525–542. Springer, Cham (2016). https://doi.org/10.1007/978-3-319-46493-0_32
6. Zhou, S., Wu, Y., Ni, Z., et al.: DoReFa-Net: training low bitwidth convolutional neural networks with low bitwidth gradients. arXiv preprint. arXiv:1606.06160 (2016)
7. Bengio, Y., Léonard, N., et al.: Estimating or propagating gradients through stochastic neurons for conditional computation. arXiv preprint. arXiv:1308.3432 (2013)
8. Cai, Z., He, X., Sun, J., et al.: Deep learning with low precision by half-wave Gaussian quantization. In: IEEE Conference on Computer Vision and Pattern Recognition, pp. 5406–5414 (2017)
9. Wang, P., Hu, Q., Zhang, Y., et al.: Two-step quantization for low-bit neural networks. In: CVPR, pp. 4376–4384 (2018)
10. Lee, D.-H., Zhang, S., Fischer, A., Bengio, Y.: Difference target propagation. In: Appice, A., Rodrigues, P.P., Santos Costa, V., Soares, C., Gama, J., Jorge, A. (eds.) ECML PKDD 2015. LNCS (LNAI), vol. 9284, pp. 498–515. Springer, Cham (2015). https://doi.org/10.1007/978-3-319-23528-8_31
11. Friesen, A.L., Domingos, P.: Deep learning as a mixed convex-combinatorial optimization problem. In: International Conference on Learning Representations (2018)
12. Boyd, S., Parikh, N., Chu, E.: Distributed optimization and statistical learning via the alternating direction method of multipliers. Found. Trends Mach. Learn. **3**(1), 1–122 (2011)
13. Leng, C., Li, H., Zhu, S.: Extremely low bit neural network: squeeze the last bit out with ADMM. In: AAAI Conference on Artificial Intelligence (2018)
14. Hou, L., Yao, Q., Kwok, J.T.: Loss-aware weight quantization of deep networks. In: International Conference on Learning Representations (2018)
15. Li, F., Zhang, B., Liu, B.: Ternary weight networks. arXiv preprint. arXiv:1605.04711 (2016)

Fast and Accurate Lung Tumor Spotting and Segmentation for Boundary Delineation on CT Slices in a Coarse-to-Fine Framework

Shuchao Pang[1], Anan Du[2], Xiaoli He[3], Jorge Díez[4], and Mehmet A. Orgun[1(✉)]

[1] Department of Computing, Macquarie University, Sydney, Australia
pangshuchao1212@sina.com, mehmet.orgun@mq.edu.au

[2] School of Electrical and Data Engineering, University of Technology, Sydney, Australia
duanan2008@163.com

[3] Department of Internal Medicine, Qingdao Huikang Hospital, Qingdao, China
dochexiaoli@126.com

[4] Artificial Intelligence Center, University of Oviedo at Gijon, Gijon, Spain
jdiez@uniovi.es

Abstract. *Label noise* and *class imbalance* are two of the critical challenges when training image-based deep neural networks, especially in the biomedical image processing domain. Our work focuses on how to address the two challenges effectively and accurately in the task of lesion segmentation from biomedical/medical images. To address *the pixel-level label noise problem*, we propose an advanced transfer training and learning approach with a detailed DICOM pre-processing method. To address *the tumor/non-tumor class imbalance problem*, we exploit a self-adaptive fully convolutional neural network with an automated weight distribution mechanism to spot the Radiomics lung tumor regions accurately. Furthermore, an improved conditional random field method is employed to obtain sophisticated lung tumor contour delineation and segmentation. Finally, our approach has been evaluated using several well-known evaluation metrics on the Lung Tumor segmentation dataset used in the 2018 IEEE VIP-CUP Challenge. Experimental results show that our weakly supervised learning algorithm outperforms other deep models and state-of-the-art approaches.

Keywords: Boundary delineation · Lung tumor segmentation · Fully convolutional neural networks

1 Introduction

With the improvement of clinical diagnostic equipment in terms of their capability, quality and availability in hospitals, biomedical/medical image data analysis has attracted much attention. The Volume, Variety, and Velocity (3V) of these images make it impractical and infeasible for clinicians to analyze them manually without making subjective errors [1]. Among different imaging devices, computed tomography

T. Gedeon et al. (Eds.): ICONIP 2019, CCIS 1142, pp. 589–597, 2019.
https://doi.org/10.1007/978-3-030-36808-1_64

is the most popular imaging modality because of its high resolution, imaging sensitivity, and isotropic acquisition, e.g., locating the lung and its lesions. Moreover, with the successful application of imaging technology in clinical medicine, automated image segmentation has been playing an increasingly important role. The accuracy of lesion region segmentation can be improved further by the consideration of Radiomics feature extraction and its detailed qualification with the ultimate goal of developing predictive models for precise prognosis in clinical medicine [2]. However, due to the variability and diversity during medical imaging processing and the existence of noisy-labelled datasets as well as tumor/non-tumor class imbalance in images, it is very hard to train a discriminative model for a specific lesion spotting and segmentation task with accurate contour delineation. In particular, in real-world applications, many popular methods often fail to perform well or even completely fail on raw datasets. Because it is really difficult and time-consuming to obtain pure data sets and labels in many image processing applications, so we have to directly use the available raw dataset.

In this paper, we focus on how to overcome the two critical challenges of (i) noisy pixel-level labels and (ii) tumor/non-tumor class imbalance when training a robust pixel-level deep model for biomedical/medical lesion spotting and segmentation with sophisticated gross tumor contour delineation. Figure 1 highlights the diversity of, and challenges that arise from, the lung computed tomography (CT) dataset used in this work.

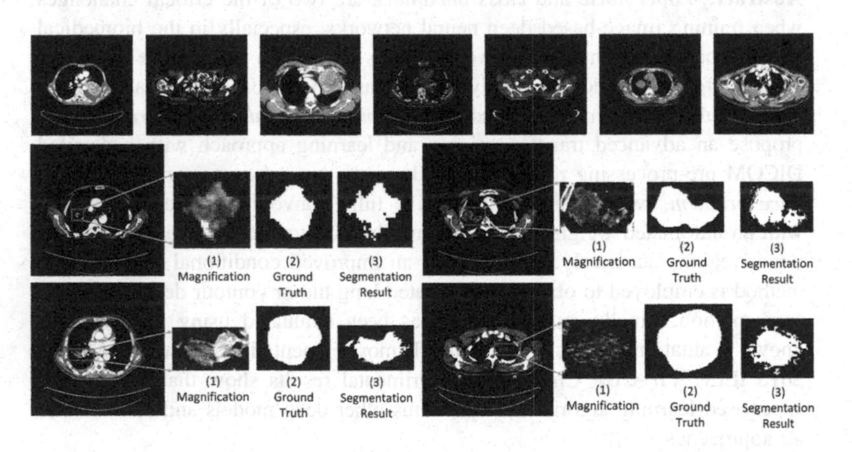

Fig. 1. Some Lung CT slices from the public NSCLC-Radiomics dataset. Note that the first row indicates the diversity of the dataset and the rest of the figure shows the obvious noisy pixel-level label problems and small tumor regions, where the lesion region in each lung CT slice is magnified in contrast with its rough manual ground truth and our segmentation result.

The main contributions of our work can be summarized as follows: We first propose a novel framework to simultaneously address noisy labels and class imbalance of raw datasets for accurate biomedical/medical image lesion segmentation and boundary delineation in real applications. We design an algorithm for reading from raw DICOM meta files as a preprocessing step for training deep neural networks. Then, a fully convolutional neural network is proposed to address pixel-level noisy labels using transfer learning. Furthermore, we develop a self-adapting weight allocation mechanism for addressing severe tumor/non-tumor imbalance problems to establish a differentiable cost function for spotting tumors. Finally, an improved conditional random field is used for accurate CT lung tumor segmentation and contour delineation.

Experimental results show that our weakly supervised learning framework outperforms some of the other deep models and state-of-the-art approaches in lesion spotting and boundary delineation from biomedical/medical images, based on several well-known evaluation criteria. Moreover, with fast processing time, our average dice coefficient result is higher than those of the winners of 2018 IEEE VIP-CUP Challenge by a large margin.

2 Related Work

The current biomedical/medical image segmentation methods can be grouped under two categories: the co-segmentation methods and the deep learning methods.

Co-segmentation Methods. Co-segmentation methods involve combining two segmentation methods or using two or more types of biomedical/medical image modalities. Many studies [3] have indicated that the co-segmentation method by combining different segmentation methods is treated as an energy minimization problem to delineate the gross tumor contours. Besides, due to the superior contrast of positron emission tomography (PET) images and high spatial resolution of CT images, more recent methods [4] in the field of clinic and lesion segmentation prefer to integrate PET and CT images.

Deep Learning Methods. In the past few years, deep learning has already swept through most research fields of computer vision and has achieved better results than traditional methods, and biomedical/medical image segmentation is no exception [5]. The existing methods [6, 7] strive to obtain more precise and comprehensive tumor features by taking the advantage of deep learning in its superior ability of hierarchical feature representation.

3 The Proposed Method

3.1 Problem Setup and Preprocessing

Suppose that there is a testing CT slice image (I) with a lung tumor from any 3D CT scans of a patient and the size of the image is 512×512 pixels. We denote each pixel in image I as $v_i, i = 1, 2, \ldots, N$, where N is the total number of pixels in I. And the set of possible labels for each pixel can be represented as $L = \{0, 1, \ldots, t\}$. Besides, for each pixel v_i, we define a variable $l_{v_i} \in L$ that indicates the assigned label. The probability of a pixel v_i belonging to label k in the given CT slice image I is formalized as $P(l_{v_i} = k|I)$ and it is calculated by our proposed fully convolutional neural networks (FCNNs). In this gross tumor segmentation task, we take the set $L = \{0, 1\}$ to denote the labels: 0 means non-tumor and 1 means tumor. In the training stage, our work is to train a deep neural network $\emptyset(X, \theta)$, where $\emptyset(\cdot)$ is the learned network on each training image X and θ indicates all network parameters, with the NSCLC-Radiomics dataset from 2018 IEEE VIP-CUP Challenge to compute the predicted probability maps $P(l_{v_i}|I)$ for any testing image I. In order to predict the result for each testing image, we use all the training data pairs $\left(X_q, Y_q\right), q = 1, 2, \ldots, Q$ for supervised training the neural network model $\emptyset(X, \theta)$ looking for the best network parameters θ^*. Note that $\left(X_q, Y_q\right)$ is the qth training image and its label, and Q is the total number of the images in the training dataset. In this way, $P(l_{v_i}|I)$ is equivalent to $\emptyset(I, \theta^*)$. To compute it, let $C\left(\hat{Y}, Y\right)$ be the loss function to minimize during the training phase, where $\hat{Y} = \emptyset(X, \theta)$ and Y is the label of each training image X. Now, the optimal parameter θ^* for gross tumor segmentation can be calculated with the following formula:

$$\theta^* = arg\min_{\theta} \sum\nolimits_{s=1}^{m} C\left(\hat{Y}_s, Y_s\right) = arg\,min_{\theta} \sum\nolimits_{s=1}^{m} C(\emptyset(X_s, \theta), Y_s), \tag{1}$$

where $m \ll Q$ stands for the number of images in a mini-batch. Then, the predicted pixel probability result $P(l_{v_i}|I, \theta^*)$ for each gross tumor slice can be further refined by using our improved dense conditional random fields as maximum a posteriori inference.

In addition, this subsection also shortly introduces the preprocessing steps from reading RTSTRUCT annotations through an extra DICOM metafile in clinics to transforming the original DICOM images in CT volumes as the inputs for deep neural networks (the workflow can be found in Fig. 2 ①). After this procedure, the whole NSCLC-Radiomics dataset [11] from 422 patients are further grouped into missing file cases with 104 patients, irrelevant labelling cases with 30 patients checked by our physician and roughly usable cases (in spite of some noisy labels in each slice) with 288 patients.

3.2 The Whole Architecture

Adaptive Fully Convolutional Neural Networks. As shown in the step ② of Fig. 2, a whole fully convolutional neural network model is illustrated with different stacked layers, which mainly comprises two key modules, one of which is the encoder part which aims to capture spatial and context information of tumors and non-tumors, while the other is the decoder model that is used to recover the details and localize the position of tumors and non-tumors.

To address the noisy pixel-level label problem from the provided NSCLC-Radiomics dataset, we adopt a transfer learning strategy among different but interrelated biomedical image segmentation datasets for alleviating noisy pixel-level label interruption. In particular, we resort to the Neuronal Structure Segmentation Dataset in Electron Microscopic Stacks at 2015 ISBI Challenge [9], which has an accurate neuronal structure segmentation gold standard in spite of only 30 training images.

When we train any classification network, the class imbalance could make the network recognize the classes with the vast majority of examples and ignore the rarely seen classes. To solve the severe tumor/non-tumor class imbalance problem from the

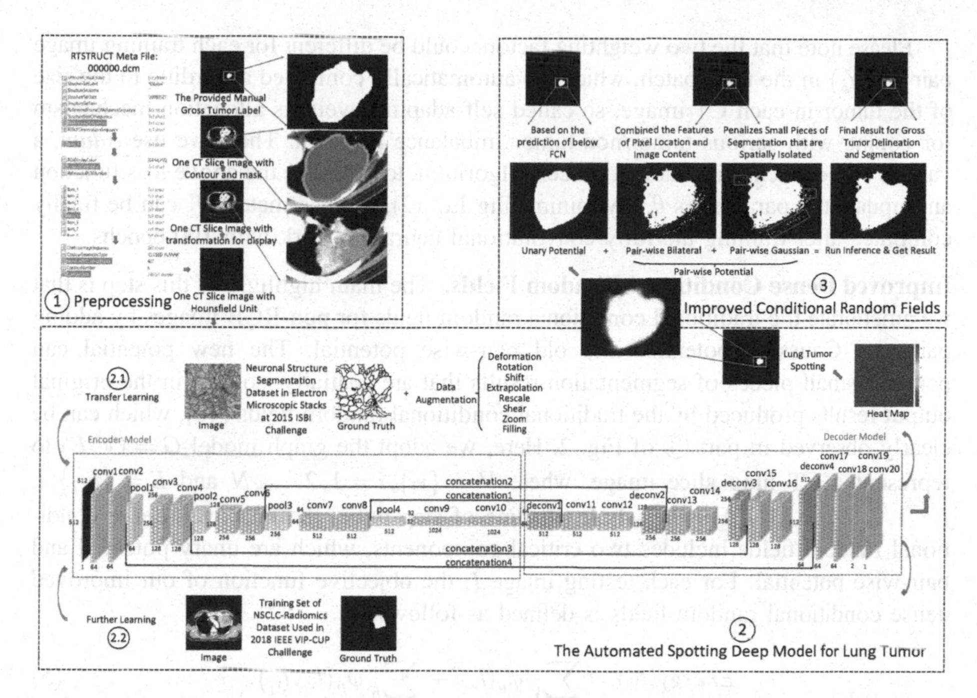

Fig. 2. Overview of the proposed gross tumor spotting and segmentation pipeline for Radiomics Lung CT images from different patients on NSCLC-Radiomics dataset.

dataset when training the deep model, we improve the binary loss function with a self-adapting weights allocation mechanism for these two categories in each mini-batch:

$$\begin{aligned} Loss_{mini-batch} &= \sum\nolimits_{s=1}^{m} C(\hat{Y}_s, Y_s) = \sum\nolimits_{s=1}^{m} C(\emptyset(X_s, \theta), Y_s) \\ &= -\frac{1}{m * N} \sum\nolimits_{S=1}^{m} \sum\nolimits_{i=1}^{N} \left[\omega_{tumour}^{class} Y_{s,v_i} log \hat{Y}_{s,v_i} + \omega_{non-tumour}^{class} (1 - Y_{s,v_i}) log (1 - \hat{Y}_{s,v_i})\right], \end{aligned} \tag{2}$$

where the weighting factors for tumor ω_{tumour}^{class} and non-tumor $\omega_{non-tumour}^{class}$ are defined in the following formulas respectively:

$$\omega_{tumour}^{class} = 1.0/\left(\sum\nolimits_{i=1}^{N} Y_{s,v_i} = 1\right) \Big/ \tfrac{1}{2}\left(1.0/\left(\sum\nolimits_{i=1}^{N} Y_{s,v_i} = 1\right) + 1.0/\left(\sum\nolimits_{i=1}^{N} Y_{s,v_i} = 0\right)\right), \tag{3}$$

$$\omega_{non-tumour}^{class} = 1.0/\left(\sum\nolimits_{i=1}^{N} Y_{s,v_i} = 0\right) \Big/ \tfrac{1}{2}\left(1.0/\left(\sum\nolimits_{i=1}^{N} Y_{s,v_i} = 1\right) + 1.0/\left(\sum\nolimits_{i=1}^{N} Y_{s,v_i} = 0\right)\right). \tag{4}$$

Please note that the two weighting factors could be different for each training image pair (X_s, Y_s) in the mini-batch, which are automatically computed according to the size of the tumor in each CT image, so called self-adapting weights allocation mechanism for coping with serious tumor/non-tumor imbalance problem. Then, we use *Adam*, a variant of the stochastic gradient descent algorithm, to optimize the above loss function and update the parameters θ. By minimizing Eq. (2), the parameters θ^* can be finally computed after training the fully convolutional neural network with 100 epochs.

Improved Dense Conditional Random Fields. The main highlight of this step is that we ameliorate the traditional conditional random fields for non-RGB images by adding pair-wise Gaussian potential into old pair-wise potential. The new potential can penalize small pieces of segmentation results that are spatially isolated in the original output results produced by the traditional conditional random fields [10], which can be clearly observed in part ③ of Fig. 2. Here, we adopt the graph model $G = (V, E)$ to represent a CT lung slice image, where $V = \{v_i\}, i = 1, 2, \ldots, N$ and $E = \{e_{i,j}\}, i,$ $j = 1, 2, \ldots, N, i < j$. For the whole pipeline of step three, the improved dense conditional random fields includes two critical components, which are unary potential and pair-wise potential. For each testing image I, the objective function of our improved dense conditional random fields is defined as follows:

$$Energy(l_V) = \sum\nolimits_V \varphi_u(l_{v_i}) + \sum\nolimits_E \varphi_p(l_{v_i}, l_{v_j}), \tag{5}$$

where the first term $\varphi_u(l_{v_i})$ denotes unary potential and it can be equal to the probabilistic output of our trained deep gross tumor spotting model with $P(l_{v_i}|I, \theta^*)$. And the second term $\varphi_p(l_{v_i}, l_{v_j})$ is the pair-wise potential, where we define the bilateral potential and Gaussian potential inside with the following Eq. (6). Here, $\mu(l_{v_i}, l_{v_j})$ is given with the Potts function that evaluates the label compatibility, and p_v and I_v denote pixel position and intensity content information respectively. Besides, δ_{bil} and δ_{gau} separately stand for the proportion of each kind of pair-wise potential with different effective range α, β, γ.

$$\varphi_p(l_{v_i}, l_{v_j}) = \mu(l_{v_i}, l_{v_j}) \left[\delta_{bil} \exp\left(-\frac{|p_{v_i} - p_{v_j}|^2}{2\alpha^2} - \frac{|I_{v_i} - I_{v_j}|^2}{2\beta^2} \right) + \delta_{gau} exp\left(-\frac{|p_{v_i} - p_{v_j}|^2}{2\gamma^2} \right) \right]. \tag{6}$$

Finally, to gain more precise gross tumor boundaries with $P(l_{v_i}|I)$, the best label result l_v^* can be computed by the following formula with the efficient approximation inference approach proposed in [10]:

$$l_v^* = argmin_{l_v \in L} Energy(l_v). \tag{7}$$

4 Experimental Evaluation

Datasets and Evaluation Metrics. As in [9], we also use different data augmentation techniques on the Neuronal Structure Segmentation Dataset for transfer learning, and NSCLC-Radiomics dataset [11] is randomly split into training, validation and test sets with the proportion of 7:1:2 for lung tumor spotting and segmentation. In the evaluation of our framework, several public and widely used image semantic segmentation evaluation metrics are utilized in our experiments, including Dice Coefficient, Hausdorff Distance, Jaccard Index, Precision, Sensitivity (Recall), Specificity and F1.

Experimental Results and Analysis. Our coarse-to-fine algorithm with three steps can achieve significant lung tumor spotting and segmentation results. In terms of qualitative analysis shown in Fig. 1, the second and third rows help check the performance on different sizes of gross tumor areas and we can also observe that the obtained boundaries of lesions are described in more detail and clearer than the given roughly manual ground truth. Furthermore, our algorithm not only accurately spots the position of tumors, but also carefully discriminates between tumors and non-tumors pixel by pixel.

In addition, the results in Fig. 1 also reveal the manual annotation errors in the provided dataset by a radiation oncologist [11]. Furthermore, in order to evaluate boundary location performance of our segmentation method with high quality, a clear tumor contour delineation is shown in Fig. 3 with several local enlargement patches.

We have also evaluated our model over the test set by removing the cases with obvious and unrealistic raw data errors by our physician. The results in Table 1 show that our method achieves a significant performance for lung tumor spotting and segmentation in CT slices, and it is especially competitive compared to other typical segmentation models and methods. Except the original U-Net without any prediction for tumors, we compare our method with SegNet and the latter can obtain a better recall. However, its precision is really worse than that of ours, which means SegNet classifies lots of non-tumor pixels as tumor regions. In particular, it might take more misclassified results under the situation of many noisy ground truths. Next, we improve the U-Net method with our techniques proposed in this paper and we can find it can roughly predict tumor regions better than before. Besides, we have also compared our improved conditional random fields with its naive model, and observed that our framework can obtain a 92.47% precision compared to traditional CRFs' 81.55%. Furthermore, by comparison, our approach can even achieve a much better average dice coefficient on the test set with 0.7767 than the reported results of 2018 VIP Cup (Winner: Team Markovian of 0.594 and Runner up: Team NTU_MiRA of 0.521) [12]. In our approach, the average processing time for a CT slice is 468 ms, including spotting time with 79 ms and segmentation time with 389 ms, which is rather fast for practical applications.

Table 1. Segmentation results on all the test data with different evaluation criteria. Note that the numbers in bold face indicate the best result under different criteria for the whole test set.

Methods\ Metrics	DICE	HD	JAC	Precision	Sensitivity	Specificity	F1
SegNet [13]	0.7518	50.935	0.6260	0.6666	**0.9256**	0.9958	0.7750
U-Net [8] -w.-Our-Tech.	0.6209	346.71	0.4662	0.5367	0.7869	0.9947	0.6382
Ours-w.-OldCRF [10]	0.5313	48.436	0.3953	0.8155	0.4425	0.9991	0.5737
Ours	**0.7767**	**15.492**	**0.6493**	**0.9247**	0.6951	**0.9995**	**0.7936**

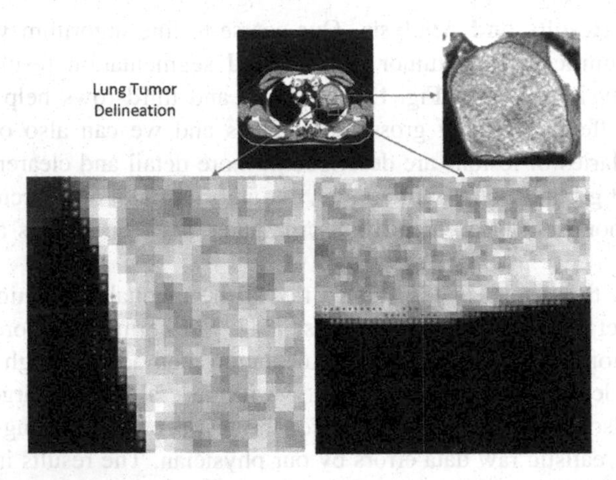

Fig. 3. Comparison of our proposed automatic delineation result (green) and the manual ground truth (red) in details. (Color figure online)

5 Conclusions

In this work, we have proposed a novel framework to leverage the integration benefits of the co-segmentation model and powerful discriminative capability of the deep learning method to localize the gross tumor boundaries in medical images precisely and automatically. Most importantly, we propose a unified framework to successfully overcome these two critical bottlenecks in training a deep model for pixel-level medical image segmentation tasks: (i) noisy pixel-level labels and (ii) tumor/non-tumor class imbalance problems. Finally, by designing a coarse-to-fine model via weakly supervised learning, the proposed Radiomics gross tumor segmentation approach can achieve more precise contour delineation and segmentation than those state of the art methods. Moreover, our proposed approach has fast response times for assisting a more accurate clinical diagnosis and a good radiation therapy treatment planning.

References

1. Pang, S., Yu, Z., Orgun, M.A.: A novel end-to-end classifier using domain transferred deep convolutional neural networks for biomedical images. Comput. Methods Programs Biomed. **140**, 283–293 (2017)
2. Litjens, G., et al.: A survey on deep learning in medical image analysis. Med. Image Anal. **42**, 60–88 (2017)
3. Ju, W., Xiang, D., Zhang, B., Wang, L., Kopriva, I., Chen, X.: Random walk and graph cut for co-segmentation of lung tumor on pet-ct images. IEEE Trans. Image Process. **24**(12), 5854–5867 (2015)
4. Song, Q., et al.: Optimal co-segmentation of tumor in pet-ct images with context information. IEEE Trans. Med. Imaging **32**(9), 1685–1697 (2013)
5. Pang, S., del Coz, J.J., Yu, Z., Luaces, O., Díez, J.: Deep learning to frame objects for visual target tracking. Eng. Appl. Artif. Intell. **65**, 406–420 (2017)
6. Dong, H., Yang, G., Liu, F., Mo, Y., Guo, Y.: Automatic brain tumor detection and segmentation using u-net based fully convolutional networks. In: Valdés Hernández, M., González-Castro, V. (eds.) MIUA 2017. CCIS, vol. 723, pp. 506–517. Springer, Cham (2017). https://doi.org/10.1007/978-3-319-60964-5_44
7. Christ, P.F., et al.: Automatic liver and lesion segmentation in CT using cascaded fully convolutional neural networks and 3D conditional random fields. In: Ourselin, S., Joskowicz, L., Sabuncu, Mert R., Unal, G., Wells, W. (eds.) MICCAI 2016. LNCS, vol. 9901, pp. 415–423. Springer, Cham (2016). https://doi.org/10.1007/978-3-319-46723-8_48
8. Ronneberger, O., Fischer, P., Brox, T.: U-Net: convolutional networks for biomedical image segmentation. In: Navab, N., Hornegger, J., Wells, William M., Frangi, Alejandro F. (eds.) MICCAI 2015. LNCS, vol. 9351, pp. 234–241. Springer, Cham (2015). https://doi.org/10.1007/978-3-319-24574-4_28
9. Arganda-Carreras, I., et al.: Crowdsourcing the creation of image segmentation algorithms for connectomics. Frontiers in Neuroanatomy **9**, 142 (2015)
10. Krähenbühl, P., Koltun, V.: Efficient inference in fully connected CRFs with gaussian edge potentials. In: Advances in Neural Information Processing Systems, pp. 109–117 (2011)
11. Aerts, H.J., Velazquez, E.R., Leijenaar, R.T., Parmar, C., Grossmann, P., et al.: Decoding tumour phenotype by noninvasive imaging using a quantitative radiomics approach. Nat. Commun. **5**, 4006 (2014)
12. Mohammadi, A., et al.: Lung cancer radiomics: highlights from the ieee video and image processing cup 2018 student competition [sp competitions]. IEEE Signal Process. Mag. **36**(1), 164–173 (2018)
13. Badrinarayanan, V., Kendall, A., Cipolla, R.: Segnet: a deep convolutional encoder-decoder architecture for image segmentation. IEEE Trans. Pattern Anal. Mach. Intell. **39**(12), 2481–2495 (2017)
14. Russakovsky, O., et al.: Imagenet large scale visual recognition challenge. Int. J. Comput. Vision **115**(3), 211–252 (2015)

Exploration of Different Attention Mechanisms on Medical Image Segmentation

Jie Tian, Kaijie Wu[✉], Kai Ma, Hao Cheng, and Chaocheng Gu

Shanghai Jiao Tong University, 800 Dongchuan Rd, Shanghai, People's Republic of China
{Tianjie_13, kaijiewu, makaay, jiaodachenghao, jacygu}@sjtu.edu.cn

Abstract. Nowadays, medical image segmentation plays an important role in computer-aided medical diagnosis. To realize effective segmentation, Attention Mechanism (AM) is widely adopted. It can be trained to automatically highlight salient features and integrated into convolution neural networks conveniently. However, many researchers choose the attention mechanism without sufficient theoretical interpretability. They ignore the differences and dominant characteristics between various datasets, which causes the failure to select the most appropriate one. In this paper, we explore the implementation and discrimination of four specific attention mechanisms. To evaluate their performances, we incorporate these mechanisms within the U-Net and make a comparison on three medical image datasets. The experimental results show that all these attention mechanisms can improve the value of Mean IoU. More significantly, we find the best AM for each type of dataset and analyze the reasons for different performances from underlying mathematical principles.

Keywords: Deep learning · Attention mechanism · Medical image segmentation

1 Introduction

Automated medical image segmentation aims to segment special parts, which is a key issue in determining if it can provide reliable basis for diagnosis. Medical image segmentation is difficult for images are too complex and lack simple linear features.

Recently, methods based on deep learning [5, 6] has made remarkable achievements. Fully Convolutional Networks (FCNs) [1] and the U-Net [2] are two typical architectures. However, they rely on multi-stage cascaded CNNs, which leads to redundancy of model parameters and repeated extraction of low-level features. To solve it, attention mechanisms are proposed. Generally, random selection of AM cannot receive the best results. Therefore, we do plenty of experiments to prove it and made a discussion on the differences between different attention mechanisms.

In this paper, we choose U-Net as the base model and compare the results when it is added with different AMs. The experimental data are several medical image sets, including segmentations of nuclear and lesions. Totally, there are four types of AMs been

T. Gedeon et al. (Eds.): ICONIP 2019, CCIS 1142, pp. 598–606, 2019.
https://doi.org/10.1007/978-3-030-36808-1_65

adopted: Position Attention Module (PAM), Channel Attention Module (CAM) [3], Region Attention Block (RAB) and Channel Attention Block (CAB) [4].

Generally speaking, our main contributions can be summarized as follows:

- We propose four AMs and five fusions of them with U-Net to do experiments on three different medical datasets.
- We achieve great improvement in Mean IoU after adding attention mechanisms to the original networks. This obviously proves the superiority of AMs.
- We discuss the results of the experiments and find that the same attention module has different promotion for different datasets. That is to say, there will be a most suitable module for each dataset according to its specific characteristics.

2 Related Work

2.1 Semantic Segmentation

Several networks [10, 11, 13–15] based on FCNs achieved improvement. As to medical images, U-Net shows great advantages. Wang et al. [20] proposed a wound image analysis system, which adopts the U-Net to segment the wound image and SVM classifier to classify. Milletari et al. [17] obtained V-Net by the deformation of U-Net, which uses the dice coefficient loss function instead of the cross-entropy loss function.

2.2 Attention Mechanism

AM learns a weight distribution of image features and apply it to the original features, which provides different effects of features. AM can be divided into soft attention [12] and hard attention [21]. The former one is to retain all components for weighting, and the latter is to select partial components by some strategy. AM can be weighted on the original image [21], the spatial scale, the channel scale [4] and combinations [7, 12, 22].

Therefore, AMs are applied in several tasks, including image captioning, segmentation [16] and object recognition [21]. Wang et al. [9] enhanced the receptive field of the underlying features and increased the depth of the network in disguise through the attention map. Chen et al. [8] constructed the attention model by two convolutional layers to automatically learn the weight of different scales and carry out the fusion.

3 Methods

3.1 Overview Framework

The main framework of our algorithm is shown in Fig. 1. The U-Net is the base model, for its superiority in medical image segmentation. We add different AMs to the U-Net: CAB, RAB, PAM and CAM. Besides, we also make five fusions. The positions of these AMs are a bit different. As is shown in Fig. 1, the yellow circle is where the PAM, CAM display, while the other four orange circles represent the insertion of CAB, RAB.

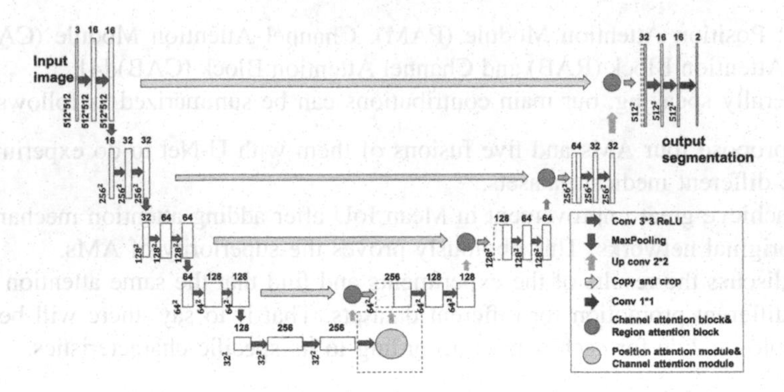

Fig. 1. A detailed framework of our method. The base network is the U-Net. The yellow circle is where PAM and CAM placed, while the four orange circles are for CAB and RAB. (Color figure online)

3.2 Attention Modules

As the name suggests, CAB and CAM are related to channels while RAB and PAM depend on positions. Besides, CAB and RAB are similar, for both of their weight distributions represent the influence on the final results. Meanwhile, PAM and CAM belong to self-attention and focus on the similarity. The more similar the semantic information of each position to the specified position, the greater the weight value is in PAM.

Channel Attention Block (CAB). Some channels are more significant, so CAB is proposed to make each channel have a corresponding weight. It is composed of spatial squeeze and channel excitation block, as is illustrated in Fig. 2(A). The first step is to squeeze the spatial information by a global average pooling layer, which transforms the size from $C*H*W$ to $C*1*1$. We set $\boldsymbol{A}=[\boldsymbol{a}_1,\boldsymbol{a}_2,\ldots,\boldsymbol{a}_c]$ $(\boldsymbol{a}_i \in R^{H\times W})$ as the original input maps, $\boldsymbol{b}=[b_1,b_2,\ldots,b_c]$ $(b_i \in R)$ as the results of squeezing:

$$b_k = \frac{1}{H\times W}\sum_{i=1}^{H}\sum_{j=1}^{W} a_k(i,j) \tag{1}$$

Then it is linked with two fully-connected (fc) layers. $\boldsymbol{W}_1$ reduces the dimension and $\boldsymbol{W}_2$ increases it. Thus, the feature map is transformed to $\boldsymbol{s}=\boldsymbol{W}_2(\delta(\boldsymbol{W}_1\boldsymbol{b}))$. Compared with only one fc layer, this has more nonlinearity. Then, a sigmoid layer σ is proposed:

$$\widetilde{\boldsymbol{R}}_1 = F_{se}(\boldsymbol{A}) = [\sigma(s_1)\boldsymbol{a}_1, \sigma(s_2)\boldsymbol{a}_2, \ldots, \sigma(s_c)\boldsymbol{a}_c] \tag{2}$$

In the formula (2) above, the value of $\sigma(s_i)$ represents the weight of the ith map. Finally, we multiply the original map $\boldsymbol{A}$ by $\widetilde{\boldsymbol{R}}_1$ to obtain the result map $\boldsymbol{R}_1$.

Region Attention Block (RAB). It focuses on spatial information. The size of the map is C and (i,j) means the location of the pixel. The channel is compressed by a convolution kernel $\boldsymbol{K}(\boldsymbol{K} \in R^{1\times 1\times C\times 1})$. We set the intermediate results as $\boldsymbol{b} = \boldsymbol{K} * \boldsymbol{A}(\boldsymbol{b} \in \boldsymbol{R}^{\boldsymbol{H}\times \boldsymbol{W}})$, $*$ means convolution operation. Then we employ a sigmoid activation σ:

$$\widetilde{\boldsymbol{R}}_2 = F_{ra}(\boldsymbol{A}) = [\sigma(b_{1,1})\boldsymbol{a}_{1,1}, \sigma(b_{1,2})\boldsymbol{a}_{1,2}, \ldots, \sigma(b_{H,W})\boldsymbol{a}_{\boldsymbol{H},\boldsymbol{W}}] \tag{3}$$

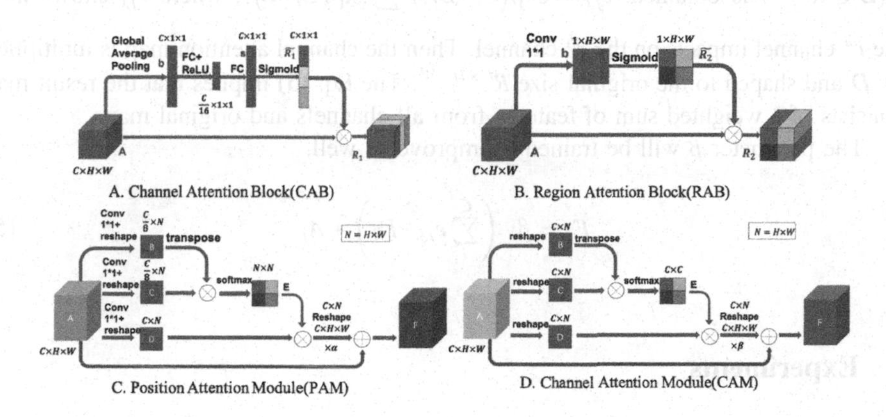

Fig. 2. Detailed network of four types of attention mechanisms.

Where $b_{i,j}$ means the weight of spatial position (i,j) for all channels. The larger the value of $b_{i,j}$ is, the greater the correlation between this position and the result.

Position Attention Module (PAM). PAM is adopted to obtain the similarity of pixels in different positions. For a specific location feature, it is combined of features of all pixels and their weights are determined by the degree of similarity. Similar features at different locations can promote each other's improvement.

As is shown in Fig. 2(C), we consider $\boldsymbol{A}$ as the input local feature. It experiences a convolutional layer and is reshaped into $\{\boldsymbol{B},\boldsymbol{C}\} \in R^{(C/8)\times N}, \mathbf{D} \in R^{C\times N}$ $(N = H \times W)$. Then we obtain the spatial attention map $\boldsymbol{E}(\boldsymbol{E} \in R^{N\times N})$ by a softmax layer: $e_{j,i} = \exp(B_i \cdot C_j) / \sum_{i}^{N} \exp(B_i \cdot C_j)$, where $e_{j,i}$ represents the weight of how the ith position works on the jth position. Besides, we obtain the map with original size by reshaping the multiplication between $\mathbf{D}$ and $\boldsymbol{E}$. It is multiplied by α and added to the input map:

$$F_j = \alpha \cdot (\sum_{i}^{N} e_{j,i} \cdot D_i) + A_j \tag{4}$$

Where $\boldsymbol{F}(\boldsymbol{F} \in R^{C \times H \times W})$ means the final map. After several epochs of training, the value of parameter α will raise to a suitable point so that the features of other positions can influence each position effectively.

Channel Attention Module (CAM). CAM is similar to the PAM, which aims to discover the relationship between the semantic responses of different channels. We also set $\boldsymbol{A}$ as the input feature and reshape it into $\{\boldsymbol{B}, \boldsymbol{C}, \boldsymbol{D}\} \in R^{C \times N}$. Then a matrix multiplication is performed between the transpose of $\boldsymbol{B}$ and $\boldsymbol{C}$. The channel attention map $\boldsymbol{E}(\boldsymbol{E} \in R^{C \times C})$ is obtained: $e_{j,i} = \exp(C_i \cdot B_j) / \sum_{i}^{C} \exp(C_i \cdot B_j)$, where $e_{j,i}$ shows how the ith channel impacts on the jth channel. Then the channel attention map is multiplied by $\boldsymbol{D}$ and shaped to the original size $R^{C \times H \times W}$. The Eq. (5) implies that the result map consists of a weighted sum of features from all channels and original maps.

The parameter β will be trained to improve as well.

$$F_j = \beta \cdot \left(\sum_{i}^{C} e_{j,i} \cdot D_i \right) + A_j \tag{5}$$

4 Experiments

4.1 Datasets

The first dataset [18] is from MoNuSeg 2018. It contains 30 images and about 22000 nuclear boundary annotations. The second dataset CVC-ClinicDB [19] is extracted from colonoscopy videos. It consists of 612 still images from 29 different sequences. The third dataset is made by ourselves. We cooperated with Shanghai International Peace Maternity and Child Health Hospital (IPMCH) and they provided a total of 168 cervical cancer sections in 2014–2016. This dataset is composed of 47 images.

4.2 Results and Analysis

The choice of combination is based on the similarity or consistency, which has been stated in the beginning of Sect. 3.2.

For the first dataset, it reaches the optimal result of 75.7% in Mean IoU when four AMs are all adopted, which brings 3.0% improvement. Employing CAB, RAB or their fusion improve the results slightly. But it can be raised by over two percent when added with PAM, CAM or their fusion. This gap is caused by the characteristics of the dataset. Each image is composed of large amounts of nuclei. They are distributed in various positions and have similar features. Consequently, the correlation among positions or channels is more important. Thus, PAM and CAM are superior to CAB and RAB.

Table 1. Performance of several attention mechanisms on three datasets.

Base model	CAB	RAB	PAM	CAM	Mean IoU		
					Dataset1	Dataset2	Dataset3
U-Net					0.727	0.692	0.712
U-Net	√				0.735	0.723	0.718
U-Net		√			0.731	0.706	0.721
U-Net	√	√			0.736	0.711	0.723
U-Net			√		0.750	0.721	0.723
U-Net				√	0.748	0.730	0.721
U-Net			√	√	0.751	0.726	0.723
U-Net	√			√	0.754	**0.739**	0.726
U-Net		√	√		0.743	0.729	0.724
U-Net	√	√	√	√	**0.757**	0.719	**0.728**

As to the second dataset, it is completely different from the first one. Mostly, there exists only one lesion area to be segmented. The correlation or similarity between different positions become less important than channels. Figure 3 proves the differences. The network with PAM generates more redundant information compared with CAM. Meanwhile, the result of CAB is closer to the ground truth than RAB. In a word, the model added by PAM or RAB performs worse than CAM or CAB. As is displayed in Table 1, CAB reaches 72.3% in Mean IoU, which is about 2 points better than RAB. Thence, this dataset achieves the best value 73.9% when added with both CAB and CAM.

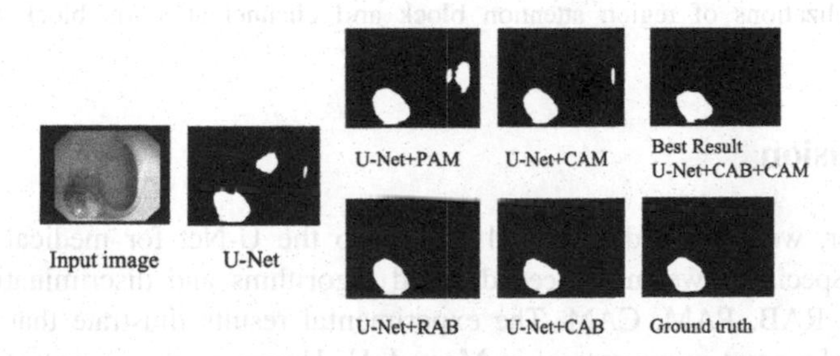

Fig. 3. Visualized results of PAM, CAM, RAB, CAB and the best result on an image chosen from the second dataset CVC-ClinicDB.

Finally, the third dataset also achieves the best performance with four AMs added. However, it only improves 1.6%. The difference between these attention mechanisms is tiny as well. We consider the reason is that the labeling of this dataset is lack of accuracy compared with others, for it is accomplished by our own comprehension.

4.3 Visualization of Attention Gates

We select one image in each dataset as examples to visualize the attention mechanisms. The results are all illustrated in Fig. 4. The second column is the attention map of RAB, which represents the colorful square in Fig. 2(B). When added with RAB, we choose the last one to be nearest to the results so the size is $1 \times \frac{H}{2} \times \frac{W}{2}$. It is then resized to the same size as the input image and added to the original one. We utilize a pseudo-color mapping on the map so that the larger the weight, the closer it is to red. Thus, the contours of the segmented regions are all close to red in Fig. 4. The third column is the maps of CAB and the size is $32 \times 1 \times 1$. Therefore, we transform the weight of each channel into a strip, in which the more it is close to blue, the less important the channel is. The last two columns demonstrate the results of RAB and the ground truth. It can be seen that the results are consistent with the reddish parts or contours in the RAB map.

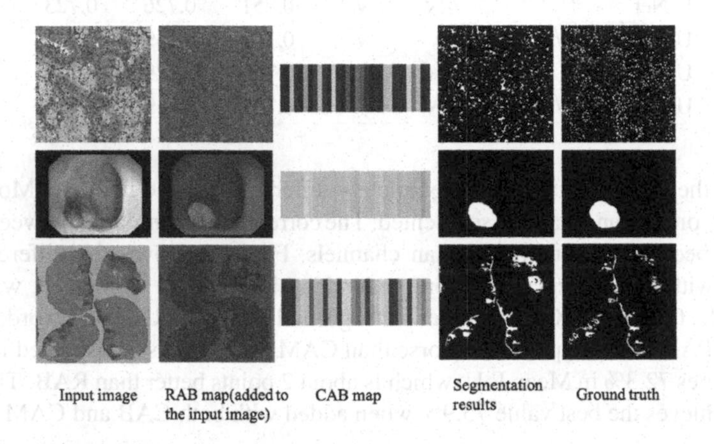

Fig. 4. Visualizations of region attention block and channel attention block (Color figure online).

5 Conclusion

In this paper, we have added several AMs into the U-Net for medical image segmentation. Specially, we introduced detailed algorithms and discriminations of four AMs: CAB, RAB, PAM, CAM. The experimental results illustrate that all of these AMs can make great improvement in Mean IoU. However, the choice of AM can be reasonable based on dominant or recessive features of specific datasets. For instance, when the image contains fewer lesion areas, the function of location mechanisms (PAM, RAB) perform worse than that of channel (CAM, CAB). Conversely, when there exist many lesions to be segmented, it is more meaningful to concentrate on the similarity between locations or channels (PAM, CAM).

Acknowledgment. This work is supported by National Key R&D Program of China (2017YFC0112705), National Key Scientific Instruments and Equipment Development Program of China (2013YQ03065101) and partially supported by National Natural Science Foundation (NNSF) of China under Grant 61503243.

References

1. Long, J., Shelhamer, E., Darrell, T.: Fully convolutional networks for semantic segmentation. In: Proceedings of the IEEE Conference on Computer Vision and Pattern Recognition, pp. 3431–3440 (2015)
2. Ronneberger, O., Fischer, P., Brox, T.: U-Net: convolutional networks for biomedical image segmentation. In: Navab, N., Hornegger, J., Wells, W.M., Frangi, A.F. (eds.) MICCAI 2015. LNCS, vol. 9351, pp. 234–241. Springer, Cham (2015). https://doi.org/10.1007/978-3-319-24574-4_28
3. Fu, J., et al.: Dual attention network for scene segmentation. In: Proceedings of the IEEE Conference on Computer Vision and Pattern Recognition, pp. 3146–3154 (2019)
4. Hu, J., Shen, L., Sun, G.: Squeeze-and-excitation networks. In: Proceedings of the IEEE Conference on Computer Vision and Pattern Recognition, pp. 7132–7141 (2018)
5. Wu, X.: An iterative convolutional neural network algorithm improves electron microscopy image segmentation. arXiv preprint arXiv:1506.05849 (2015)
6. Su, H., Liu, F., Xie, Y., Xing, F., Meyyappan, S., Yang, L.: Region segmentation in histopathological breast cancer images using deep convolutional neural network. In: 2015 IEEE 12th International Symposium on Biomedical Imaging (ISBI), pp. 55–58. IEEE, April 2015
7. Chen, L., et al.: SCA-CNN: spatial and channel-wise attention in convolutional networks for image captioning. In: Proceedings of the IEEE Conference on Computer Vision and Pattern Recognition, pp. 5659–5667 (2017)
8. Chen, L.C., Yang, Y., Wang, J., Xu, W., Yuille, A.L.: Attention to scale: scale-aware semantic image segmentation. In: Proceedings of the IEEE Conference on Computer Vision and Pattern Recognition, pp. 3640–3649 (2016)
9. Wang, F., et al.: Residual attention network for image classification. In: Proceedings of the IEEE Conference on Computer Vision and Pattern Recognition, pp. 3156–3164 (2017)
10. Chen, L.C., Papandreou, G., Kokkinos, I., Murphy, K., Yuille, A.L.: DeepLab: semantic image segmentation with deep convolutional nets, atrous convolution, and fully connected CRFs. IEEE Trans. Pattern Anal. Mach. Intell. **40**(4), 834–848 (2017)
11. Chen, L.C., Papandreou, G., Schroff, F., Adam, H.: Rethinking atrous convolution for semantic image segmentation. arXiv preprint arXiv:1706.05587 (2017)
12. Roy, A.G., Navab, N., Wachinger, C.: Concurrent spatial and channel 'squeeze & excitation' in fully convolutional networks. In: Frangi, A., Schnabel, J., Davatzikos, C., Alberola-López, C., Fichtinger, G. (eds.) MICCAI 2018. LNCS, pp. 421–429. Springer, Cham (2018). https://doi.org/10.1007/978-3-030-00928-1_48
13. Lin, G., Shen, C., Van Den Hengel, A., Reid, I.: Efficient piecewise training of deep structured models for semantic segmentation. In: Proceedings of the IEEE Conference on Computer Vision and Pattern Recognition, pp. 3194–3203 (2016)
14. Liu, S., Qi, X., Shi, J., Zhang, H., Jia, J.: Multi-scale patch aggregation (MPA) for simultaneous detection and segmentation. In: Proceedings of the IEEE Conference on Computer Vision and Pattern Recognition, pp. 3141–3149 (2016)
15. Pinheiro, P.H., Collobert, R.: Recurrent convolutional neural networks for scene labeling. In: No. CONF (2014)
16. Oktay, O., et al.: Attention U-Net: learning where to look for the pancreas. arXiv preprint arXiv:1804.03999 (2018)
17. Milletari, F., Navab, N., Ahmadi, S.A.: V-net: fully convolutional neural networks for volumetric medical image segmentation. In: 2016 Fourth International Conference on 3D Vision (3DV), pp. 565–571. IEEE, October 2016

18. Kumar, N., Verma, R., Sharma, S., Bhargava, S., Vahadane, A., Sethi, A.: A dataset and a technique for generalized nuclear segmentation for computational pathology. IEEE Trans. Med. Imaging **36**(7), 1550–1560 (2017)
19. Bernal, J., Sánchez, F.J., Fernández-Esparrach, G., Gil, D., Rodríguez, C., Vilariño, F.: WM-DOVA maps for accurate polyp highlighting in colonoscopy: validation vs. saliency maps from physicians. Comput. Med. Imaging Graph. **43**, 99–111 (2015)
20. Wang, C., et al.: A unified framework for automatic wound segmentation and analysis with deep convolutional neural networks. In: 2015 37th Annual International Conference of the IEEE Engineering in Medicine and Biology Society (EMBC), pp. 2415–2418. IEEE, August 2015
21. Mnih, V., Heess, N., Graves, A.: Recurrent models of visual attention. In: Advances in Neural Information Processing Systems, pp. 2204–2212 (2014)
22. Hu, Y., Li, J., Huang, Y., Gao, X.: Channel-wise and spatial feature modulation network for single image super-resolution. arXiv preprint arXiv:1809.11130 (2018)

Machine Learning Based Method for Huntington's Disease Gait Pattern Recognition

Xiuyu Huang[1], Matloob Khushi[1(✉)], Mark Latt[2], Clement Loy[3], and Simon K. Poon[1]

[1] The University of Sydney, Camperdown, NSW 2006, Australia
mkhushi@uni.sydney.edu.au
[2] Royal Prince Alfred Hospital, Camperdown, NSW 2050, Australia
[3] Westmead Hospital, Sydney, NSW 2145, Australia

Abstract. Huntington's disease (HD) is an inherited neurodegenerative disorder causing problems with mobility, cognition and mood. Gait abnormality is a potential diagnostic sign as it can occur even in the early stages of HD. We developed a machine learning method for detecting HD with gait dynamics as the model features. Concretely, standard deviation (SD) and interquartile range (IQR) were calculated for 6 gait time series sequences as 12 candidate features. An exhaustive feature and hyperparameter selector was then applied to optimize the features and hyperparameter subsets for 5 different machine learning models. Classification outcomes were determined by nested leave-one-out cross-validation (nested LOOCV) method. Support Vector Machines (SVM) achieved the highest accuracy (97.14%) without overfitting bias assumptions. Our result showed that the machine learning based method with gait dynamics features can be a complementary tool for HD diagnosis.

Keywords: Huntington's disease · Machine learning · Nested LOOCV

1 Introduction

Huntington's disease (HD) is an autosomal dominant neurodegenerative condition named after George Huntington, who described the disease as hereditary chorea in 1872 [1]. It is a neurodegenerative brain condition with a distinct clinical phenotype and diagnosis of HD is complicated [2, 3].

Gait analysis may be a means of identifying signs, estimating severity and monitoring progression of HD. People with HD gradually lose motor function as the disease progresses, leading to gait abnormalities even in the early stages of disease [4]. Walker [2] and Grimbergen et al. [5] found differences in gait between HD patients and a healthy control group. In recent related studies, many different automatic computer-aided methods have been used to differentiate between HD patients and healthy control group using gait variables, including time series stride, swing and stance interval [6–10]. Based on these parameters, Daliri [6] used Support Vector Machine (SVM) with different kernels for HD diagnosis. Similarly, Zeng and Wang [7] used a radial basis

T. Gedeon et al. (Eds.): ICONIP 2019, CCIS 1142, pp. 607–614, 2019.
https://doi.org/10.1007/978-3-030-36808-1_66

function neural network for HD diagnosis using time series stride-to-stride interval features. In addition, a meta-classifier for HD and other neurodegenerative gait pattern recognition was established by Sánchez-Delacruz et al. [8]. Aziz and Arif [9] transferred the stride time series into a specific symbol sequence and then applied threshold-dependent symbolic entropy in their analysis of gait differences between HD patients and healthy control groups. Klomsae et al. [10] converted left-foot stride interval into symbols sequence and then built a gait classification model with String Grammar Unsupervised Possibilistic Fuzzy C-Medians.

There are serval gaps in existing methods that need to be addressed. Firstly some machine-learning studies used low computational time feature selection strategies, which may have sacrificed model performance [6–8]. Secondly, the studies evaluated models in a biased way. Those studies used the k-fold cross-validation method to select either hyperparameters or features for models with the whole dataset [6–10], and used the best performance in the selection process to denote the model assessment. However, the models were developed and adjusted within the same dataset and should be evaluated with external data. Thirdly, the studies ignored features such as interquartile range (IQR) which can capture the difference in variance of healthy controls and HD subjects [11].

To fill in those gaps, we proposed a novel machine learning-based framework to differentiate HD subjects from healthy controls using stride-to-stride information. The framework integrates 3 steps including (1) unused features derivation, (2) exhaustive feature selection and hyperparameter tuning, and (3) nested cross-validation evaluation to maximize the reliability and validity of classification results. The remainder of this paper is organized as follows: Sect. 2, overview of common classification models and feature extraction process; Sect. 3, results; Sect. 4, discussion; and Sect. 5, conclusion.

2 Method

The focus of this study was to build a reliable machine learning model for differentiating Huntington's disease subjects from healthy controls. The steps of our analysis were summarized in Fig. 1 and presented in the following sections.

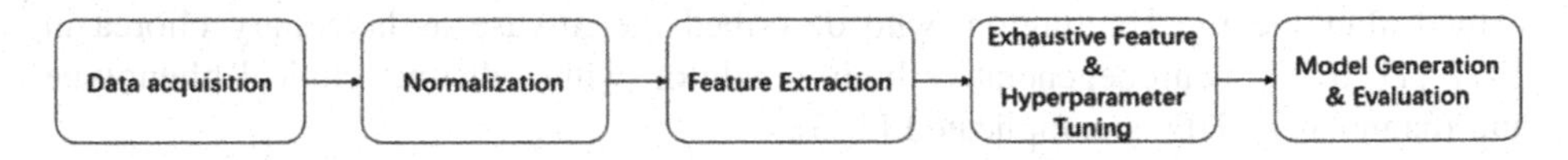

Fig. 1. An integrative framework to create a classification system for HD

2.1 Data Acquisition

Data sets for analysis and classification were taken from gait time series in a neurodegenerative database: http://www.physionet.org/physiobank/database/gaitndd. The database contains gait dynamics data of both healthy people and HD patients, and it was collected by a walking experiment [12]. The data set contained gait records of 16 healthy controls and 20 HD patients. We only included 16 healthy controls and 19 HD patients in the analysis. Data from one HD subject was excluded because all rows have same number, suggesting that this data was erroneous.

2.2 Normalization

Data normalization may improve pattern recognition and reduce computational time [13]. In the analysis, z-score normalization was implemented before feature selection and model evaluation steps [14].

2.3 Feature Extraction

We derived two types of features from the stride, swing and stance interval data to evaluate gait variability: standard deviation (SD) and interquartile range (IQR). These Features were summarized in Table 1.

Table 1. SD and IQR features derived from data

Sequence	Feature type	
	Standard deviation	Interquartile range
Left stride interval	LSTRSD	LSTRIQR
Right stride interval	RSTRSD	RSTRIQR
Left swing interval	LSWISD	LSWIIQR
Right swing interval	RSWISD	RSWIIQR
Left stance interval	LSTASD	LSTAIQR
Right stance interval	RSTASD	RSTAIQR

2.4 Feature Selection and Hyperparameter Tuning

An exhaustive selector for feature and hyperparameter was used to determine the best feature subsets from Table 1 as well as the best hyperparameters for the classifiers. The method was run over ten-fold cross-validation (ten-fold CV) for different commonly used machine learning classifiers including Support Vector Machine (SVM), Naïve Bayes (NB), Decision Tree (DT), Random Forest (RF) and Logistics Regression (LR).

2.5 Model Evaluation and Generation

Following feature selection, we determined the best performing classifier for our model, using the best performance in feature selection process of each algorithm in Sect. 2.4.

However, potential overfitting bias could occur from cross-validation in the same data set for both feature selection and model evaluation [15]. To control for the overfitting selection bias, we used the nested leave-one-out cross-validation (nested LOOCV) method (shown in Fig. 2) to divide the original data into training (n – 2 samples), validation (1 sample) and testing (1 sample) subsets n * n – 1 times (where n is the sample size of the original dataset) [7]. Feature selection was applied in the inner loop with training and validation sets. The best feature subset was then passed to the outer loop to assess model performance with the testing subset.

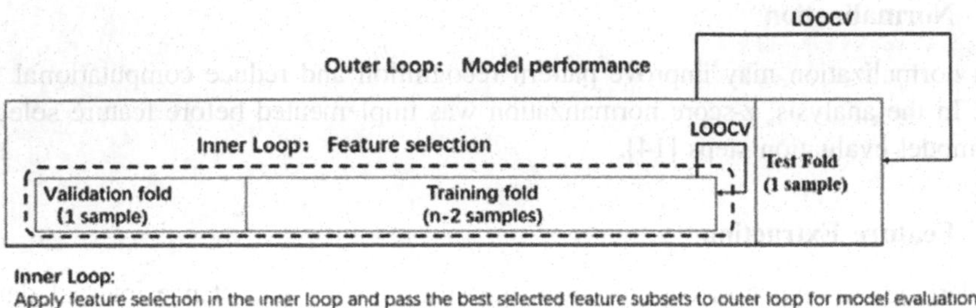
LOOCV
Outer Loop: Model performance
Inner Loop: Feature selection
LOOCV
Test Fold
(1 sample)
Validation fold
(1 sample)
Training fold
(n-2 samples)
Inner Loop:
Apply feature selection in the inner loop and pass the best selected feature subsets to outer loop for model evaluation
Outer loop:
Assess the model performance with the average testing accuracy
n is the sample size of original dataset

Fig. 2. Process of nested LOOCV

3 Result

3.1 The Classification Models

The best subset and performance of each classifier (Sect. 2.4) are presented in Table 2. The SVM method achieved 97.14% accuracy and 96.77% F1 score with a feature subset in size 3 obtained from the previous exhaustive feature selection method (RSTRIQR, LSWIIQR, RSWIIQR). The Decision Tree algorithm had a slightly lower average accuracy of 91.43%, and 91.43% F1 score with a feature subset in size 4 (RSTRSD, RSWISD, RSTASD, LSWIIQR). The Naïve Bayes algorithm also achieved a 91.43% average accuracy, and 91.43% F1 score with a feature subset in size 2 (RSTRSD, RSWISD). The Random Forest achieved the same accuracy and F1 score as DT and NB with a feature subset in size 3 (LSTRSD, RSWISD, LSWIIQR). The Logistics Regression, however, achieved the lowest accuracy and f1 score (both 85.71%) with a feature subset in size 2 (RSTRSD and RSTASD).

Table 2. Evaluation of the models by ten-fold CV with the highest performed feature subset

Classifier	Average accuracy	F1 score	Feature subset	Subset size
SVM with polynomial kernel (degree = 3)	97.14%	96.77%	RSTRIQR, LSWIIQR, RSWIIQR	3
Decision Tree (max_depth = 10)	91.43%	91.43%	RSTRSD, RSWISD, RSTASD, LSWIIQR	4
Naïve Bayes	91.43%	91.43%	RSTRSD, RSWISD	2
Random Forest (n_estimators = 110)	91.43%	91.43%	LSTRSD, RSWISD, LSWIIQR	3
Logistics Regression (penalty: l1)	85.71%	85.71%	RSTRSD, RSTASD	2

3.2 Nested Cross Validation Evaluation

The average accuracies of outer cross-validation loops were used to identify the unbiased performances of models (Table 3). The accuracy of SVM and Logistics Regression remained 97.14% and 85.71% respectively, but the accuracies of Decision Tree, Naïve Bayes and Random Forest classifier decreased to 77.14%, 85.72% and 82.85% respectively.

Table 3. Average accuracy in outer loop of nested LOOCV

Classifier	Average accuracy 10-fold CV	Average accuracy (LOOCV)
SVM with polynomial kernel (degree = 3)	97.14%	97.14%
Decision Tree (max_depth = 10)	91.43%	77.14%
Naïve Bayes	91.43%	85.72%
Random Forest (n_estimators = 110)	91.43%	82.85%
Logistics Regression (penalty: l1)	85.71%	85.71%

3.3 Comparison with Other Methods

We also compared our model performance with other existing methods (Table 4). Our proposed method achieved an accuracy of 97.14%, while accuracies of existing methods ranged from 83.33% to 100%.

Table 4. Comparison of our model and existing the methods

Model	Classification accuracy	Reference
Our model	**97.14%**	
Symbolic entropy	95%	[9]
Radial basis function (RBF) neural networks (All-training-all-testing)	100%	[7]
Radial basis function (RBF) neural networks (Leave-one-out)	83.33%	[7]
Radial basis function (RBF) support vector machines	90.23%	[6]
String grammar unsupervised possibilistic fuzzy C-Medians	97.22%	[10]
Meta-classifier	88.67%	[8]

4 Discussion

4.1 Summary

Our results revealed that all classifiers based on a features subset, obtained from feature selection procedures, achieved ranging from 85% to 97% accuracy. However, the CV error estimate for the classifiers with the optimal parameters was substantially biased (biased error can possibly achieve 20%), especially for a dataset where its overall size is small [15]. Therefore, a nested LOOCV was implemented to reduce possible overfitting biases caused by CV in feature selection and model assessment using the same data. The nested LOOCV procedure reduces bias and provides a true error estimate, as the testing data was not used in the feature selection process [16]. As expected, performances of different classifiers stayed the same or dropped. SVM, the best classifier identified by the ten-fold CV method, maintained its accuracy (97.14%) in nested LOOCV evaluation. Our strategy demonstrates that gait analysis may complement clinical, neuropsychological and genetic assessments in the differentiation between persons with HD and healthy individuals.

4.2 Comparison with Other Methods

Our method performed slightly worse than those of previous investigators [7, 10]. However, the evaluated accuracy of the model developed by Zeng and Wang [7] was determined using training data; and the performance of the model developed by Klomsae et al. [10] was evaluated by data used in hyperparameter selection. These models had a risk of overfitting bias. In contrast, our models were validated using a testing subset that had not been used in feature and model-selection processes, reducing the potential for overfitting bias.

4.3 Feature Selection and Selected Features

Feature selection as explained in [17, 18] is an important step in building a classifier for HD based on gait variables [19] and may avoid overfitting and redundant variables. Our study used wrapper strategy as a method of feature selection. As the number of features in our analysis was only 12, an exhaustive search could be implemented [20]. The exhaustive searching method naïvely tried all combinations of features to obtain the best model. As shown in the result part, all the sizes of the best feature subsets for different algorithms were no more than 4, which indicated that only a small part of gait dynamic variance features can denote the whole part of gait features for each classification algorithm. Additionally, the features used in the best classifier (SVM) are all IQR, indicating that IQR also can possibly be an indicator for detecting the gait dynamics difference between health people and HD patient.

4.4 Limitation and Further Development

One of the limitations in our finding was a small sample size, as the total number of HD and healthy subjects was only 35. A larger sample size is required for confirmation of our proposed method in HD classification.

In addition, we only used statistics stride-to-stride information as features in our study. However, many other techniques may also derive features useful for gait classification in HD. For example, Fourier Analysis [21] and Symbolic Aggregate Approximation (SAX) algorithm [22] can also be considered. They are used to transform time domain signal into other components. In the manner, we can have a different analysis scenario for the dataset currently used in the paper.

Lastly, the reason, why the accuracy gap between CV and nested LOOCV are different from each classifier, is still not clear. As shown in the result, the accuracy of SVM and LR stayed at the same level in both CV and nested LOOCV evaluation, while the accuracies of DT, NB, and RF classifier had different drops from CV to nested LOOCV. Further studies can be explored to investigate the cause of those results by analyzing the mechanism of each classifier.

5 Conclusion

In this paper, a novel classification stepwise framework was introduced. In particular, the statistic variables (STD and IQR) were extracted from gait time series data as candidate features. The exhaustive wrapper feature selection method was then implemented to find the best feature subset for 5 common machine learning algorithms. Finally, ten-fold CV and nested LOOCV were used to evaluate those 5 models. We found that the best model was SVM with 3 features which achieved 97.14% accuracy in the nested LOOCV result. From the indirect comparison, our results demonstrated that our proposed algorithm performs better than the existing algorithms on average. It can achieve very high accuracy without selection bias. External validity can be further tested by collecting new sample data.

References

1. Paulsen, J.S., et al.: Neuropsychiatric aspects of Huntington's disease. J. Neurol. Neurosurg. Psychiatry **71**(3), 310–314 (2001)
2. Walker, F.O.: Huntington's disease. Lancet **369**(9557), 218–228 (2007)
3. U.S. National Library of Medicine (2019). https://ghr.nlm.nih.gov/condition/huntington-disease#genes
4. Hausdorff, J.M., et al.: Altered fractal dynamics of gait: reduced stride-interval correlations with aging and Huntington's disease. J. Appl. Physiol. **82**(1), 262–269 (1997)
5. Grimbergen, Y.A., et al.: Falls and gait disturbances in Huntington's disease. Mov. Disord. Off. J. Mov. Disord. Soc. **23**(7), 970–976 (2008)
6. Daliri, M.R.: Automatic diagnosis of neuro-degenerative diseases using gait dynamics. Measurement **45**(7), 1729–1734 (2012)

7. Zeng, W., Wang, C.: Classification of neurodegenerative diseases using gait dynamics via deterministic learning. Inf. Sci. **317**, 246–258 (2015)
8. Sánchez-Delacruz, E., Acosta-Escalante, F., Wister, M.A., Hernández-Nolasco, J.A., Pancardo, P., Méndez-Castillo, J.J.: Gait recognition in the classification of neurodegenerative diseases. In: Hervás, R., Lee, S., Nugent, C., Bravo, J. (eds.) UCAmI 2014. LNCS, vol. 8867, pp. 128–135. Springer, Cham (2014). https://doi.org/10.1007/978-3-319-13102-3_23
9. Aziz, W., Arif, M.: Complexity analysis of stride interval time series by threshold dependent symbolic entropy. Eur. J. Appl. Physiol. **98**(1), 30–40 (2006)
10. Klomsae, A., Auephanwiriyakul, S., Theera-Umpon, N.: String grammar unsupervised possibilistic fuzzy C-Medians for gait pattern classification in patients with neurodegenerative diseases. Comput. Intell. Neurosci. **2018**, 10 (2018)
11. Upton, G., Cook, I.: Understanding Statistics. Oxford University Press, Oxford (1997)
12. Hausdorff, J.M., et al.: Gait variability and basal ganglia disorders: stride-to-stride variations of gait cycle timing in Parkinson's disease and Huntington's disease. Mov. Disord. **13**(3), 428–437 (1998)
13. Tahir, N.M., Manap, H.H.: Parkinson disease gait classification based on machine learning approach. J. Appl. Sci. **12**(2), 180–185 (2012)
14. Jain, A., Nandakumar, K., Ross, A.: Score normalization in multimodal biometric systems. Pattern Recogn. **38**(12), 2270–2285 (2005)
15. Varma, S., Simon, R.: Bias in error estimation when using cross-validation for model selection. BMC Bioinform. **7**(1), 91 (2006)
16. Palmerini, L., et al.: Feature selection for accelerometer-based posture analysis in Parkinson's disease. IEEE Trans. Inf. Technol. Biomed. **15**(3), 481–490 (2011)
17. Khushi, M., et al.: Automated classification and characterization of the mitotic spindle following knockdown of a mitosis-related protein. BMC Bioinform. **18**(16), 566 (2017)
18. Barlow, H., Mao, S., Khushi, M.: Predicting high-risk prostate cancer using machine learning methods. Data **4**(3), 129 (2019)
19. Snijders, A.H., et al.: Bioinformatics. Lancet Neurol. **6**(1), 63–74 (2007)
20. Guyon, I., Elisseeff, A.: An introduction to variable and feature selection. J. Mach. Learn. Res. **3**(Mar), 1157–1182 (2003)
21. Hörmander, L.: The Analysis of Linear Partial Differential Operators I: Distribution Theory and Fourier Analysis. Springer, Heidelberg (2003). https://doi.org/10.1007/978-3-642-61497-2
22. Lin, J., et al.: A symbolic representation of time series, with implications for streaming algorithms. In: Proceedings of the 8th ACM SIGMOD Workshop on Research Issues in Data Mining and Knowledge Discovery. ACM (2003)

Research on Deep Learning-Based Intelligent Diagnosis Algorithms for OCT Medical Images of Macular Edema

Ziwei Li(✉), Xuesong Zhao, Airu Yin(✉), Chuyang Guo, and Li Chen

College of Computer Science, KLMDASR, Nankai University, Tianjin, China
lzw_nku@mail.nankai.edu.cn, yinar@nankai.edu.cn

Abstract. Macular edema is the most important cause of visual impairment in the center of human eyes, which causes a lot of life problems for a large number of patients. Optical coherence tomography is a very important medical imaging material in the diagnosis and treatment of macular diseases. Firstly, on the basis of Faster R-CNN, this paper adjusts the processing strategy of the model by modifying the tag generation method to detect the lesions area of OCT images of fundus lesions. Then, using the U-Net basic model, the task of semantics segmentation of OCT images of fundus lesions is accomplished by fusing multi-attention modules in the decoding stage. Good results have been achieved in the OCT medical image dataset of the largest fundus lesions, which can help doctors quickly identify and locate the lesions areas in the image, and quantify the severity of specific fundus edema.

Keywords: Retinal macular edema · OCT images · Faster R-CNN · U-Net

1 Introduction

The general office of the State Council issued the opinions on promoting the development of "Internet + medical health". It pointed out that we should attach great importance to the development of "Internet + medical health". Fundus macular disease, as a disease with high blindness rate and difficult to reverse, has attracted wide attention in the medical field. In the aspect of target detection of medical images, Kim et al. [1] constructed an automatic noise reduction encoder on lung CT images, and characterized the potential non-linear correlation between morphological features. More than 3500 nodule images were studied by unsupervised method, and more than 90% of the high accuracy was obtained. In medical image segmentation, Venhuizen et al. [2] used U-Net model to segment age-related macular lesions. The systematic prediction error of macular thickness was $14.0 \pm 22.1\,\mu\text{m}$. By analyzing the structural changes of retinal vessels, we can directly or indirectly observe various diseases such as ophthalmic diseases, cardiovascular diseases, etc. Zhou [3] uses neural networks to extract vascular

T. Gedeon et al. (Eds.): ICONIP 2019, CCIS 1142, pp. 615–623, 2019.
https://doi.org/10.1007/978-3-030-36808-1_67

features, adds a set of filters to the micro-vessels, and uses dense conditional random fields to segment the blood vessels to assist medical personnel in observing the fundus blood vessels. Zhang Kang's research group [4] of Guangzhou Medical University Affiliated Medical Center used image recognition technology and applied it to more than 200,000 OCT images of retina acquired clinically. It applied the learned model to the diagnosis of children's pneumonia by using transfer learning technology, and achieved good results.

In the task of lesion area recognition and detection, based on the model Faster R-CNN of conventional target detection, the basic algorithm is reformed, and the steps of conventional target detection are reformed by different tag generation methods in this paper, so as to get a more suitable algorithm for this problem.

In the image segmentation task of fundus diseases, this paper designs and experiments several lesions segmentation models, using experimental data and visualization effect to verify the effectiveness of the segmentation task.

The main work of this paper includes: Sect. 2 introduces the OCT medical image dataset, Sect. 3 introduces the detection of ocular fundus OCT image lesions area, Sect. 4 introduces the segmentation of ocular fundus lesions image, Sect. 5 is a summary.

2 OCT Medical Image Dataset

In this paper, the largest OCT medical image dataset of fundus lesions in China is used. The dataset includes 100 OCT volumetric data, 128 OCT images of 512 * 1024 in size per volume data, of which 70 are trained and 15 are validated. The medical image samples in the dataset include three conditions: retinal edema (REA), pigmented epithelial detachment (PED) and subretinal fluid (SRF). In the lesions marking map, the black corresponds to the background; the white is the REA lesions area, the light grey is the SRF lesions area, and the dark grey is the PED lesions area. The samples in each category of the dataset are shown in Fig. 1.

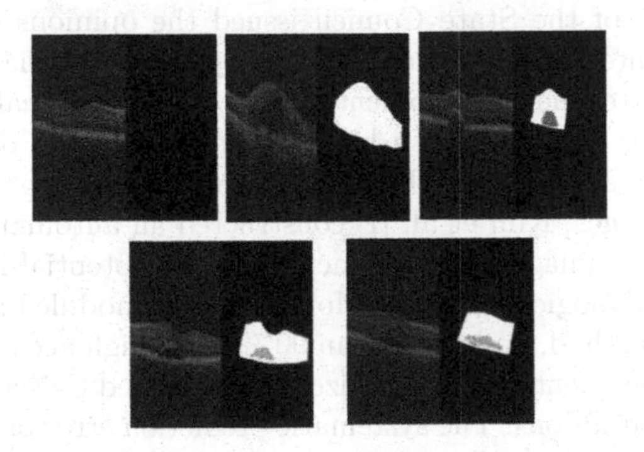

Fig. 1. The samples in each category of the dataset

3 Detection of Lesions Area

In order to help doctors quickly identify and locate the lesions area in the image, this paper adjusts the generation of the initial candidate boxes on the basis of Faster R-CNN, and adds the non-lesions samples to the training to detect the lesions area.

3.1 Generation and Mapping of Candidate Boxes

Faster R-CNN [5,6] firstly inputs images into CNN network for feature extraction, and the extracted features are shared in the whole network. Then RPN is used to generate candidate boxes of various sizes for each pixel. These candidate areas are put into two layers of network at the same time for border regression and classification. Finally, the two layers of branch network are aggregated to achieve the preliminary filtering of candidate boxes.

Table 1. Statistical analysis of the size and length-width ratio of the lesions area in dataset subsets.

	y/x = 0.5	y/x = 1	y/x = 2
(100, 255)	0	0	36
(255, 400)	0	24	35
(400, 625)	13	123	6
(625, 900)	35	172	1
(900, 1225)	33	150	13
(1225, 1600)	25	98	12
>1600	4	20	0

The lesions area of fundus in the dataset used in this paper has a certain range of changes, so RPN network needs to be reformed. Firstly, 800 samples were randomly selected from the training set, and the size and length-width ratio of the lesions area were analyzed, the results are shown in Table 1. Among them, the horizontal header represents the closer value of the length-width ratio of the lesions area in each sample, and the vertical header represents the area range of the lesions area calculated by every ten pixels in the sample. Among them, the horizontal header represents the closer value of the length-width ratio of the lesions area in each sample, and the vertical header represents the area range of the lesions area calculated by every ten pixels in the sample.

From the analysis of the subset of dataset in this paper, we can find that when the area is smaller (less than 400), the higher probability of the length-width ratio of samples is close to 2; when the area is larger (between 400 and 1225), the higher probability of the length-width ratio of samples is close to 1; In the case of larger area, the ratio of length-width is more likely to be 0.5 and

1. In order to match the size of the lesions area in the dataset, the window size generated by the initial candidate boxes of this model is set as Table 2.

According to the previous statistical analysis of the size range of the lesions area in the dataset, it can be found that the lesions area whose length-width ratio is close to 1 accounts for the largest proportion. Therefore, in the RoIPooling stage, the candidate boxes is pooled into 16 * 16.

3.2 Joint Training of Classification and Border Regression

Classification refers to putting a series of fixed-size feature maps obtained from the previous processing into a fully connected network to determine which category the candidate box belongs to and outputs probability vectors. Border Regression is a more refined adjustment of the target detection frame. As shown in Formula 1, the total loss of the model consists of three parts, namely, the loss of candidate box classification, the loss of fine-tuning regression and the loss of weight regularization.

$$L = L_{classification} + L_{regression} + L_{regularization} \tag{1}$$

Table 2. Window size generated by initial candidate box.

Initial candidate box window	Window size
Window1	(280, 140)
Window2	(180, 360)
Window3	(250, 250)
Window4	(360, 180)
Window5	(250, 500)
Window6	(350, 350)
Window7	(500, 250)
Window8	(140, 280)
Window9	(300, 600)

3.3 Experimental Settings

The Faster R-CNN model adopted in this paper is based on PyTorch platform. In the training stage, the 10 fold cross-validation is adopted. The optimizer chooses SGD, the learning rate lr is 4e−3, and the batch size is 2.

3.4 Metrics

For a certain category, the target box of an image output prediction is marked as y_predict, and the corresponding real target box is marked as y_true, which is measured by the overlapping ratio of y_predict and y_true, namely, the ratio of the intersection and union (IoU) of the predicted value area and the label area. For all samples in a certain category of data set, the accuracy of prediction is judged. Artificially setting a threshold of positioning accuracy, such as PASCAL VOC2007 is usually set to 0.5. When the IoU of the predicted target frame in this sample is larger than the threshold value, it is considered that the prediction is correct on this entity of the image, that is, True Positive (TP), otherwise is False Positive (FP), and the target of missing detection is recorded as False Negative (FN). The formulas for calculating the accuracy and recall rates are as follows:

After several recall and recision values are obtained, PR curves are drawn and the area enclosed by the curves is calculated. For each recall value, the maximum value of Precision is selected when the recall value is greater than or equal to the recall value, and these Precision values are averaged as the area under the PR curve.

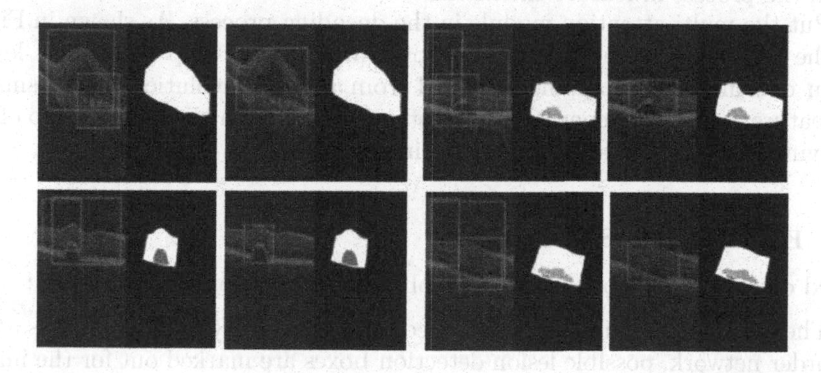

Fig. 2. Comparison of the predicted lesions areas before and after the improvement

3.5 Experimental Results

The accuracy of detection of ocular fundus lesions area is 79.79% by putting the dataset of OCT images into the model. For normal images without lesions, a “no_disease” label with an area of zero was designed, and the prediction accuracy of the model reached 93.64%. Figure 2 is a comparison of the predicted lesions areas before and after the improvement. After adding the training of non-lesions samples, the model greatly improves the ability to distinguish the normal and lesions parts of the fault zone and the classification and location of the detection frame are more accurate.

4 Image Segmentation of Fundus Edema Lesions in OCT

The effective information in OCT image samples of fundus lesions occupies a relatively small proportion in the complete image, Large area background images may increase noise. Therefore, In this paper, we proposed to use lesions area detection to reduce the input image, and add the self-attention mechanism of visual spatial scale and channel scale on the basis of u-net [7,8] to complete the task of OCT image semantic segmentation of fundus lesions.

4.1 Integration of Multi-attention Module and Full Convolution Network

The two types of self-attention are fused into multi-attention modules, and then merged with the model architecture of u-net in the following two ways:

I. Place the multi-attention module in the coding phase. As shown in Fig. 3. In the process of down sampling, each hidden layer contains two convolution operations. Multi-attention module is added between the two convolution operations, and the image is gradually screened according to its importance in the process of feature extraction.
II. Put the multi-attention module in the decoding process. As shown in Fig. 4, the model keeps the feature map of samples from deep to shallow levels. In decoding, the mapping obtained from the deconvolution of the smaller feature map of a deeper level is constantly fused with the feature map of the symmetric level to add the shallow information.

4.2 Experimental Settings

Based on the u-net depth model, the following experiments are designed:

I. The detection of fundus lesions based on the Faster R-CNN is used as a pre-order network, possible lesion detection boxes are marked out for the image, then the original image and its corresponding real label map are cut and put into the u-net network to generate corresponding predictions.
II. Add multi-attention module to u-net network and try to fuse the module in different ways in the coding stage and decoding stage respectively.

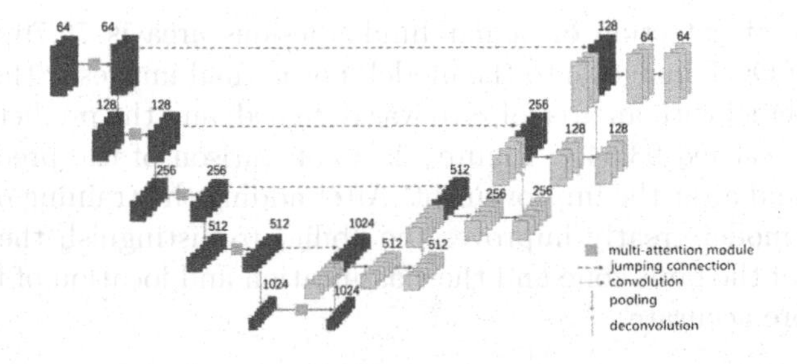

Fig. 3. Fusion of multi-attention module and U-Net in the coding phase

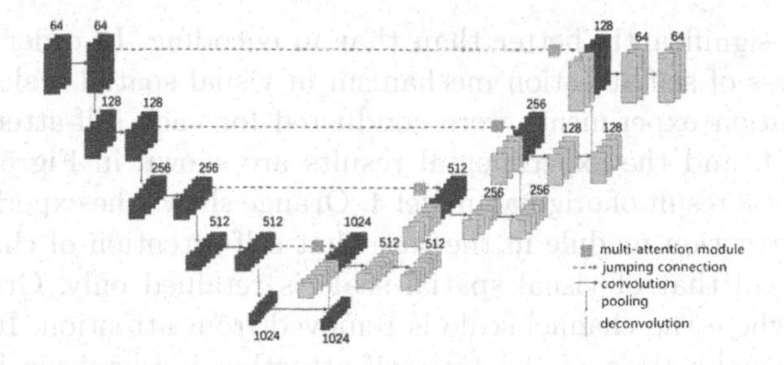

Fig. 4. Fusion of multi-attention module and U-Net in the decoding phase

4.3 Metrics

The commonly used evaluation index for this task is the overlap ratio, which is the ratio of predicted intersection to true area.

4.4 Experimental Results

The experimental results of this paper are shown in Table 3. Model 1 represents the original data and u-net network. Model 2 represents the regional box generated after detection of the lesion area to cut data and put it into u-net network for training results. Model 3 shows that the original data and u-net are integrated with multi-attention modules in the coding stage for training. Model 4 shows that the original data and u-net are combined with multi-attention module in decoding phase to train. In this paper, the segmentation accuracy of REA lesions, PED lesions and SRF lesions are calculated by these four models respectively, and the average accuracy is further obtained, based on which the following analysis is made:

Table 3. The experimental results of four models constructed on the data set of this paper.

	IoU_REA	IoU_PED	IoU_SRF	Avg_IoU
Model 1	92.12%	92.83%	75.60%	86.85%
Model 2	89.48%	87.85%	84.66%	84.72%
Model 3	89.79%	93.86%	71.94%	85.20%
Model 4	92.73%	92.83%	75.60%	86.85%

I. The segmentation effect of lesions with more data is significantly better: The effect on REA and PED is good, but the accuracy on SRF lesions is lower.
II. After using the detection results of the lesion area to cut data and greatly reduce the sample size, the segmentation accuracy of SRF is significantly improved, and the effect of integrating multi-attention module in decoding

process is significantly better than that in encoding. In order to prove the effectiveness of self-attention mechanism in visual spatial scale and channel scale, ablation experiments were conducted for each self-attention module in model 4, and the experimental results are shown in Fig. 5. Blue is the experimental result of original model 4. Orange shows the experimental effect of multi-attention module in the case that self-attention of channel scale is removed and that of visual spatial scale is retained only. Gray represents the case where the channel scale is removed from attention. It can be seen that the combination of the two self-attention mechanisms in this paper can improve the segmentation experiment. At the same time, the addition of visual spatial scale self-attention makes the segmentation effect of each lesion category more balanced.

5 Summary

In this paper, for OCT images of ocular fundus macular edema, firstly, On the basis of Faster R-CNN, the initial filtering strategy of model training is changed by modifying the tag generation method to detect the lesion area and assist doctors to quickly identify and locate the lesion area in the image. Then, in the U-Net decoding stage, multi-attention mechanism is fused to segment the fundus lesion image, which facilitates doctors to quantify the severity of specific fundus edema.

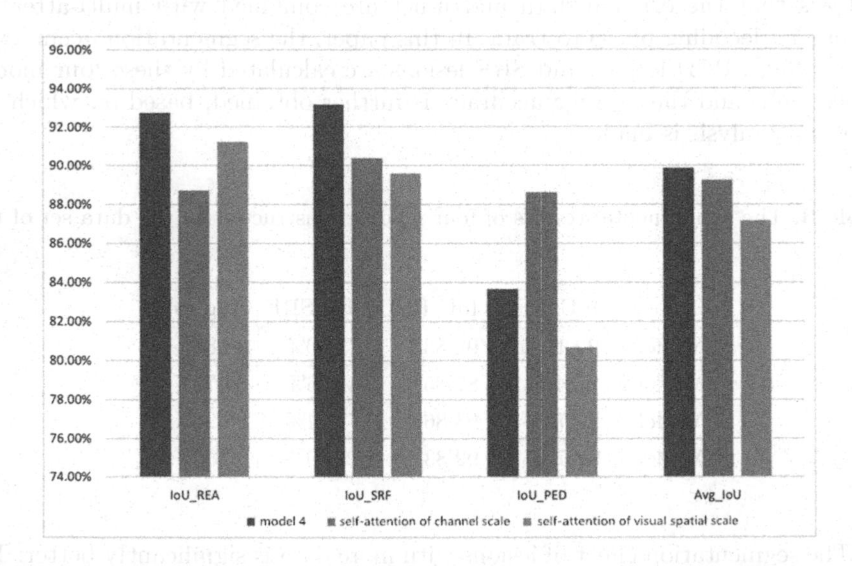

Fig. 5. Effectiveness comparison of self-attention module on model 4 (Color figure online)

Acknowledgement. This work is supported by the National Science Foundation of China (No. U1633103), the Open Project Foundation of Information Technology Research Base of Civil Aviation Administration of China (No. CAAC-ITRB-201502).

References

1. Kim, B.-C., Sung, Y.S., Suk, H.-I.: Deep feature learning for pulmonary nodule classification in a lung CT. In: 2016 4th International Winter Conference on Brain-Computer Interface (BCI). IEEE (2016)
2. Venhuizen, F.G., van Ginneken, B., Liefers, B., et al.: Robust total retina thickness segmentation in optical coherence tomography images using convolutional neural networks. Biomed. Opt. Express **8**(7), 3292–3316 (2017)
3. Zhou, L., Yu, Q., Xu, X., et al.: Improving dense conditional random field for retinal vessel segmentation by discriminative feature learning and thin-vessel enhancement. Comput. Methods Programs Biomed. **148**, 13–25 (2017)
4. Kermany, D.S., Goldbaum, M., Cai, W., et al.: Identifying medical diagnoses and treatable diseases by image-based deep learning. Cell **172**(5), 1122–1131 (2018)
5. Girshick, R.: Fast R-CNN. In: The IEEE International Conference on Computer Vision, pp. 1440–1448 (2015)
6. Ren, S., He, K., Girshick, R., et al.: Faster R-CNN: towards real-time object detection with region proposal networks. In: Advances in Neural Information Processing Systems, pp. 91–99 (2015)
7. Ronneberger, O., Fischer, P., Brox, T.: U-Net: convolutional networks for biomedical image segmentation. In: Navab, N., Hornegger, J., Wells, W.M., Frangi, A.F. (eds.) MICCAI 2015. LNCS, vol. 9351, pp. 234–241. Springer, Cham (2015). https://doi.org/10.1007/978-3-319-24574-4_28
8. Long, J., Shelhamer, E., Darrell, T.: Fully convolutional networks for semantic segmentation. In: Proceedings of the IEEE Conference on Computer Vision and Pattern Recognition, pp. 3431–3440 (2015)
9. Zhao, B., et al.: Diversified visual attention networks for fine-grained object classification. IEEE Trans. Multimed. **19**(6), 1245–1256 (2017)
10. Wang, F., et al.: Residual attention network for image classification (2017). arXiv preprint arXiv:1704.06904
11. Wang, F., Jiang, M., Qian, C., et al.: Residual attention network for image classification (2017). arXiv preprint arXiv:1704.06904

Using (Automated) Machine Learning and Drug Prescription Records to Predict Mortality and Polypharmacy in Older Type 2 Diabetes Mellitus Patients

Simon Kocbek[1(✉)], Primoz Kocbek[2], Tina Zupanic[3], Gregor Stiglic[2,4], and Bogdan Gabrys[1]

[1] Advanced Analytics Institute, FEIT, University of Technology Sydney, Sydney, Australia
skocbek@gmail.com, bogdan.gabrys@uts.edu.au
[2] Faculty of Health Sciences, University of Maribor, Maribor, Slovenia
{primoz.kocbek,gregor.stiglic}@um.si
[3] Health Data Center, National Institute of Public Health, Ljubljana, Slovenia
tina.zupanic@nijz.si
[4] Faculty of Electrical Engineering and Computer Science, University of Maribor, Maribor, Slovenia

Abstract. We analyse a large drug prescription dataset and test the hypothesis that drug prescription data can be used to predict further complications in older patients newly diagnosed with type 2 diabetes mellitus. More specifically, we focus on mortality and polypharmacy prediction. We also examine the balance between interpretability and predictive performance for both prediction tasks, and compare performance of interpretable models with models generated with automated methods. Our results show good predictive performance in the polypharmacy prediction task with AUC of 0.859 (95% CI: 0.857–0.861). On the other hand, we were only able to achieve the average predictive performance for mortality prediction task with AUC of 0.754 (0.747–0.761). It was also shown that adding additional drug related features increased the performance only in the polypharmacy prediction task, while additional information on prescribed drugs did not influence the performance in the mortality prediction. Despite the limited success in mortality prediction, this study demonstrates the added value of the systematic collection and use of Electronic Health Record (EHR) data in solving the problem of polypharmacy related complications in older age.

Keywords: Mortality prediction · Polypharmacy prediction · Diabetes · Drug prescription · Interpretability · Automated machine learning · Logistic regression

T. Gedeon et al. (Eds.): ICONIP 2019, CCIS 1142, pp. 624–632, 2019.
https://doi.org/10.1007/978-3-030-36808-1_68

1 Introduction

Drug prescription data contains valuable information about the patient's medical history and offers potentially rich source for predictive modelling. Therefore, the aim of this paper is to evaluate predictive power for the following two important use cases: mortality prediction and polypharmacy prediction, both for patients with chronic type 2 diabetes (T2D). It has been previously shown that T2D patients die earlier than patients without T2D and that T2D is a leading underlying or contributing cause of death in high-income countries [1, 2]. Therefore, we explore possibility to use drug prescription records for mortality prediction in older T2D patients. Mortality prediction is an active research field with several proposed methods [3].

Multimorbidity is becoming increasingly common, especially in older population [4]. Polypharmacy is defined as concurrent use of multiple medications by one individual and is becoming another major health concern and is tightly related to multimorbidity. World Health Organisation (WHO) estimates that more than half of all medicines are prescribed, or sold inappropriately, and that half of all patients fail to take them correctly. In the older population, the number of concurrent health conditions is directly related to a number of medications prescribed, eventually resulting in polypharmacy [5, 6]. Most of previous research has focused on potential negative consequences of polypharmacy, e.g., nonadherence, interactions, and adverse drug reactions [7]. In this study, we focused on medications taken in the last three months in older patients with newly diagnosed chronic T2D patients. Compared to our previous work on polypharmacy prediction [8], we use additional drug features, and experiment with automated ML methods as described later.

Although, systematic collection of healthcare data in hospital information systems creates opportunities for more powerful and accurate models, such predictive modelling often results in models that are difficult to interpret by domain experts and healthcare professionals, which hinders their decision making process [9]. Therefore, one of the main goals of the paper is also to examine the balance between interpretability and predictive performance in regularized Logistic Regression (LR) based predictive models. It has been argued that LR performs better or same as more complex Machine Learning (ML) models on clinical datasets [10].

To compare performance of our interpretable models with other, more complex models, we performed additional experiments using automated approach. Following the drive towards automation of predictive systems building, our recent work has resulted in various fundamental contributions to this area with some open source tools [11] which, in principle, allow automatic composition, optimisation and adaptation of multi- component predictive systems. The emergence of successful automated models, and the release of various, open-source state-of-the-art tools, has recently led to the establishment of a new field – automated machine learning (AutoML), which overlaps with terms such as One Button Data Mining [12]. The aim of the additional experiments was also to measure readiness of AutoML this in critical domain. AutoML should be considered as one of the key options and an opportunity for scaling up this kind of ML model development and deployment in the future.

2 Methods

2.1 Data, and Definition of Polypharmacy and Chronic Disease

In this study, drug prescription records collected by the National Institute of Public Health of Slovenia from 2008 to 2016 were used. Approximately 95 million records were obtained covering more than 750 thousand unique patients. The data contained 14 fields including patient information (e.g., patient id, patient gender, patient's geographical information), information about the prescribed medication, and the patient's doctor information. The data was collected centrally by the National Institute of Public Health and contains complete data for every patient. All records were properly anonymised by inclusion of randomly generated patient identifiers, lower fidelity of the date of prescription, and discretised dates of birth for all patients. The data was also linked to a death register which contained date of death for those patients who died.

Anatomical Therapeutic Chemical Classification System (ATC) codes which provided information about the prescribed drugs in 5 different detail levels. The first level indicates the main anatomical group, the second level indicates the main anatomical group, the third level indicates the therapeutic/pharmacological subgroup, the fourth level indicates chemical/therapeutic/pharmacological subgroup, and the final, fifth level indicates subgroup for chemical substance. In this study all five levels of ATC codes were provided and patients with at least one A10 L2 prescribed medication were selected (A10 is the code for drugs used to treat T2D patients).

The data was processed to filter patients based on age, presence of polypharmacy and chronic condition (as described in Subsect. 2.2). The final dataset contained 10,767 instances and 487 features consisting of class attribute (i.e., polypharmacy or mortality), age, gender, and L5 and L2 ATC codes. Table 1 summarizes number of positive/negative cases, average age (with 95% confidence interval), and number of females and males for both prediction tasks. The mortality prediction dataset contained only 584 positive cases, which represents only 5.4% of all instances. A patient was marked as positive when its death was recorded in the year following the prediction time point (PTP).

Table 1. Summary table for both prediction tasks.

Feature	Polypharmacy		Mortality	
	Pos (n = 3,993)	Neg (n = 6,774)	Pos (n = 584)	Neg (n = 10,183)
Age (years) [95% CI]	68.56 [68.30–68.84]	66.02 [65.85–66.20]	74.57 [73.82–75.30]	66.53 [66.39–66.68]
Female [n (%)]	1,945 (49%)	2,857 (42%)	227 (39%)	4,575 (45%)
Male [n (%)]	2,048 (51%)	3,917 (58%)	357 (61%)	5,608 (55%)

From the data, we removed entries with no ATC codes, since these were probably prescriptions for, e.g., medical appliances or were simply errors. Next, we selected only prescriptions for patients with at least one prescription for T2D. Finally, we selected only patients that are born before 1960. For AutoML experiments we also manually split the data into 10 training/testing stratified folds. These datasets were used for training and testing the AutoML models as described in Sect. 2.3.

Next, we identified patients with: (i) polypharmacy, and (ii) newly diagnosed chronic T2D condition. WHO defines polypharmacy as the *use of too many medicines per patient.* Five concurrent medications are mentioned most often [13], however, polypharmacy has also been defined as, e.g., the use of three or six [14] concurrent drugs. In addition, according to WHO, chronic diseases *are* not *passed from person to person* and *they are of long duration and generally slow progression.* The latter is not clearly defined, however, authors often characterize chronic disease as a condition that is expected to last or lasts at least 12 months or more [15].

We defined polypharmacy as a concurrent use of at least five medications, and a chronic disease as a condition for which a patient had been taking T2D medications for at least a year every 3 months. Concurrent use was defined as all medications that were prescribed in three consecutive months, for example, January, February, March.

Figure 1 illustrates four different possible chronic disease/polypharmacy scenarios. Each line presents a patient's medical history for one year. All data is partitioned into time periods of three consecutive months. Each three-month interval is used to: (i) define the number of concurrent use of medications, and (ii) check for T2D medications. For example, the first line on the left presents a non-chronic T2D patient, since the patient did not receive any T2D medication in July, August and September, while the second line on the left presents a chronic patient. Similarly, the two lines on the right present a patient: with detected (top line) and no polypharmacy (bottom line).

Both definitions (i.e., polypharmacy and chronic disease) were then used to filter data for training subset and to define the positive polypharmacy cases. Specifically, in the training data only non chronic patients with no polypharmacy before the PTP, who became chronic after the PTP, were kept. All patients that had polypharmacy detected in the year following the PTP were labelled as positive cases.

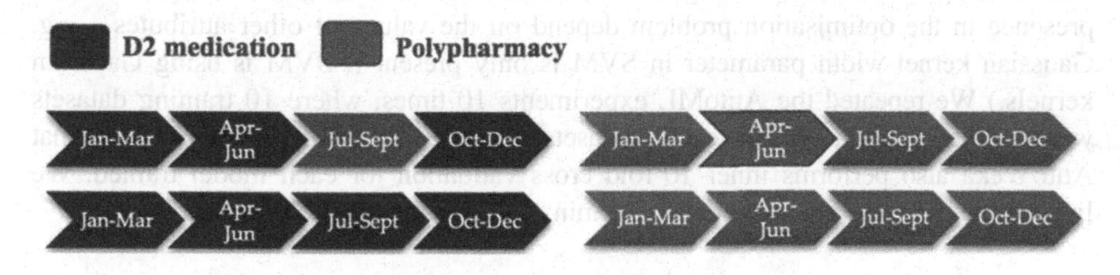

Fig. 1. Illustration of 4 different patient scenarios.

2.2 Logistic Regression Model (LASSO)

As one of our goals was to build interpretable models to increase usability, we restricted model building to regularized linear models, where it is possible to control the model complexity (dimensionality) to some extent. The latter also helps in avoiding overfitting, a problem in ML where models do not generalise well. The generalized linear model via penalized maximum likelihood LASSO regularization was used as defined by Friedman et al. [16]:

$$min_{\beta_0,\beta} \frac{1}{N} \sum_{i=1}^{N} w_i l(y_i, \beta_0 + \beta^T x_i) + \lambda \|x_i\|_1, \quad (1)$$

where $\boldsymbol{i}$ represents observations and its negative log-likelihood contribution is noted as $l(y, n)$ with w_i representing weights and tuning (shrinkage) parameter λ controlling the overall strength of the penalty.

We further controlled the complexity of the model with the Maximal number of dimensions (MND) parameter with values from 10 to 100 in steps of 10. Number of selected features is controlled by stopping the λ parameter tuning before the number of selected features exceeds MND. Each experiment was repeated 100 times.

To validate our models, we focused on their predictive performance measured by Area Under ROC Curve (AUC) and Area Under the Precision Recall Curve (AUPRC). The latter can often be more informative than AUC, especially for unbalanced datasets [17], which was the case for the mortality prediction in this work. In addition to AUC and AUPRC, we also report final numbers of selected features for both prediction tasks.

2.3 Auto ML Models

The AutoWeka tool [11] was used to prepare AutoML experiments. Due to the nature of our dataset (i.e., sparse, high dimensional and unbalanced dataset), we experimented with the following filters and models enabled in AutoWeka: attribute selection filters, balancing filters, Naïve Bayes (NB), Support Vector Machines (SVM), and Random Forest (RF). Bayesian optimisation, specifically Sequential Model Algorithm Configuration (SMAC) [18] was used to optimise models towards the highest AUC. SMAC method support continuous, categorical and conditional attributes (i.e. attributes whose presence in the optimisation problem depend on the values of other attributes – e.g. Gaussian kernel width parameter in SVM is only present if SVM is using Gaussian kernels.) We repeated the AutoML experiments 10 times, where 10 training datasets were used for training and 10 testing datasets were used for testing purposes. Note that AutoWeka also performs inner 10-fold cross validation for each model trained. We limited the training time a model to 10 min, while overall time limit was 2 h.

3 Results

Figure 2 shows AUC and AUPRC results for mortality and polypharmacy prediction for different number of selected features. The results are presented as violin plots, representing a combination of box plots and density plots, where different MND values are presented on the horizontal axis (NDR presents a model with default LASSO regularization parameters - i.e. no restriction of dimensionality). The main advantage of violin plots over simple box plots is that violin plots show the distribution shape of the data. Table 2 summarises final number of selected features for both prediction tasks for different MNDs. Please note, that the number of all selected variables in an experiment can be higher than the experiment's MND parameter, since we repeat each experiment 100 times and its selected variables do not necessarily always overlap.

Almost all of our AutoML experiments resulted in a Naïve Bayes model as a model with highest AUC performance. The only exception was one experiment which resulted in a RandomForest model on a polypharmacy dataset, however, the performance was lower to NB models in other folds. Interestingly, although AutoML optimisation methods searched through space using also filters (e.g., balancing filter), no filter has been selected in final models. In Table 3 we present AUC and AUPRC results for each training/test split, as well as their mean, maximum and minimum values.

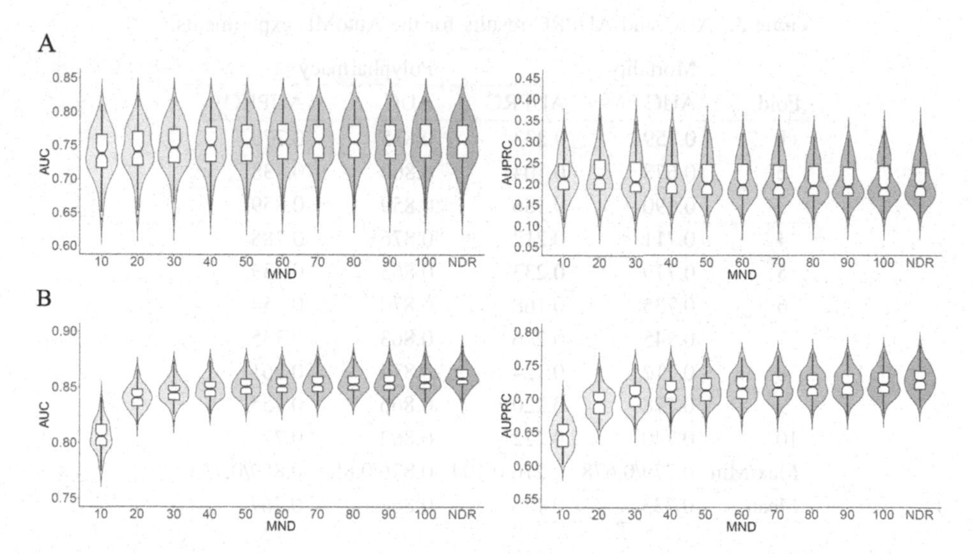

Fig. 2. AUC (left) and AUPRC (right) results for mortality (A) and polypharmacy (B) prediction with 100 iterations and different MND values.

Table 2. Number of selected features for both tasks (95% CI)

MND	Mortality	Polypharmacy
10	1.880 (1.694–2.066)	6.640 (6.320–6.960)
20	13.040 (12.480–13.600)	14.020 (13.782–14.258)
30	19.230 (18.606–19.854)	19.550 (19.111–19.989)
40	29.220 (28.630–29.810)	25.580 (25.027–26.133)
50	36.400 (35.771–37.029)	31.520 (30.882–32.158)
60	42.420 (41.717–43.123)	38.100 (37.387–38.813)
70	47.940 (47.023–48.857)	45.620 (44.851–46.389)
80	51.900 (50.608–53.192)	53.600 (52.317–54.883)

(continued)

Table 2. (*continued*)

MND	Mortality	Polypharmacy
90	56.910 (55.322–58.498)	59.770 (58.819–60.721)
100	56.980 (54.857–59.103)	67.300 (66.390–68.210)
NDR	58.540 (55.529–61.551)	183.770 (176.265–191.275)

Table 3. AUC and AUPRC results for the AutoML experiments.

	Mortality		Polypharmacy	
Fold	AUC	AUPRC	AUC	AUPRC
1	0.759	0.233	0.875	0.773
2	0.678	0.104	0.860	0.738
3	0.690	0.186	0.859	0.859
4	0.711	0.212	0.876	0.788
5	0.779	0.233	0.863	0.733
6	0.735	0.168	0.870	0.754
7	0.745	0.276	0.863	0.735
8	0.712	0.124	0.874	0.761
9	0.778	0.220	0.866	0.755
10	0.740	0.222	0.862	0.775
Max/Min	0.779/0.678	0.276/0.104	0.876/0.859	0.859/0.733
Mean	0.733	0.198	0.867	0.767

4 Discussion and Conclusion

We showed that it is possible to use prescription data to build models for polypharmacy and mortality prediction, and that it is feasible to find a balance between performance and interpretability. With LR and polypharmacy, we achieved the highest AUC and AUPRC values on the model with no dimension reduction. The values were 0.859 (95% CI: 0.857–0.861) and 0.729 (0.725–0.733) for AUC and AUPRC, respectively. Although the performance decreased with lower MND values, the difference was not significant, especially when MND increased to >=50. One can observe stabilisation of both performance metrics after the MND is increased from 10 to 20. Table 2 showed a large reduction in the complexity of the model with 31.520 (95% CI: 30.882–32.158) and 183.770 (176.265–191.275) selected features for MND = 50 and NDR respectively.

In the mortality prediction task, using LR, the maximum AUC was 0.754 (0.747–0.761), while AUPRC was 0.221 (0.211–0.231). The former was achieved with the NDR model, while the latter presents the result for the MND = 20 model. AUPRC is often more informative for skewed datasets and it is interesting to see that the highest AUPRC value was a result of a low dimensional model. Low AUPRC in general indicates that predicting mortality from drug prescription data represents a bigger challenge than predicting polypharmacy. Such conclusion can also be visually seen in not changing distribution plots with regard to MND, (complex models did not improve the model significantly). Figure 2 and Table 2 showed that adding drug related features increased the performance in the polypharmacy prediction task, while additional information on prescribed drugs did not influence the performance in the mortality prediction.

AutoML experiments showed comparable performance to those experiments with LR. The most successful model was NB. The reason for this might be high dimensionality of our data and the fact that NB often performs well on binary features such as ours. For mortality prediction, mean values for AUC and AUPRC were 0.733 and 0.198, respectively. These values are lower than those achieved with LR, which confirms our findings that mortality prediction using only prescription data is a challenging task. On the other hand, polypharmacy prediction resulted in mean AUC and AUPRC of 0.867 and 0.767, respectively. These results are higher than those achieved with the LR model, which indicates that there might be more room to improve results.

Even though our study showed a limited success in mortality prediction, we demonstrate the added value of the systematic collection and use of EHR data in solving the problem of polypharmacy related complications in older population. The fact that the data contained only months and years for prescription or death, influenced our definitions for concurrent use of drugs, polypharmacy and chronic disease. Since the day is removed from the data due to anonymisation purposes, it is impossible to address this issue. In the future, we plan to use additional data sources, investigate reasons for low performance of mortality prediction, and perform a deeper analysis of the most important features in our models. We also plan to experiment with AutoML in details.

References

1. Saely, C.H., Vonbank, A., Lins, C et al.: Type 2 diabetes, chronic kidney disease, and mortality in patients with established cardiovascular disease. Eur. Heart. J. 38 (2017)
2. Ogurtsova, K., da Rocha Fernandes, J.D., Huang, Y., et al.: IDF Diabetes atlas: global estimates for the prevalence of diabetes for 2015 and 2040. Diabetes Res. Clin. Pract. **128**, 40–50 (2017)
3. Gannon, W.D., Lederer, D.J., Biscotti, M., et al.: Outcomes and mortality prediction model of critically ill adults with acute respiratory failure and interstitial lung disease. Chest **153**(6), 1387–1395 (2018)
4. Calderón-Larrañaga, A., Santoni, G., Wang, H.X., et al.: Rapidly developing multimorbidity and disability in older adults: does social background matter? J. Intern. Med. **283**(5), 489–499 (2018)
5. Hajjar, E.R., Cafiero, A.C., Hanlon, J.T.: Polypharmacy in elderly patients. Am. J. Geriatr. Pharmacother. **5**, 345–351 (2007)
6. Good, C.B.: Polypharmacy in elderly patients with diabetes. Diabetes Spectr. **15**, 240–248 (2002)

7. Zelko, E., KlemencKetis, Z., TusekBunc, K.: Medication adherence in elderly with polypharmacy living at home: a systematic review of existing studies. Materia Socio-Medica **28**, 129 (2016)
8. Kocbek, S., Kocbek, P., Stozer, A., et al.: Building interpretable models for polypharmacy prediction in older chronic patients based on drug prescription records. PeerJ. **6**, e576 (2018)
9. Miotto, R., Wang, F., Wang, S., Jiang, X., Dudley, J.T.: Deep learning for healthcare: review, opportunities and challenges. Briefings Bioinform. **19**(6), 1236–1246 (2017)
10. Jie, M.A., Collins, G.S., Steyerberg, E.W., Verbakel, J.Y., van Calster, B.: A systematic review shows no performance benefit of machine learning over logistic regression for clinical prediction models. J. clin. epidemiol. (2019)
11. Martin Salvador, M., Budka, M., Gabrys, B.: Automatic composition and optimization of multicomponent predictive systems with an extended auto-WEKA. IEEE Trans. Autom. Sci. Eng. **16**(2), 946–959 (2019)
12. Stiglic, G., Brzan, P.P., Fijacko, N., et al.: Comprehensible predictive modeling using regularized logistic regression and comorbidity based features. PLoS ONE **10**(12), e0144439 (2015)
13. Geurts, M.M.E., Stewart, R.E., Brouwers, J.R.B.J., et al.: Implications of a clinical medication review and a pharmaceutical care plan of polypharmacy patients with a cardiovascular disorder. Int. J. Clin. Pharm. **38**, 808–815 (2016)
14. Jörgensen, T.M., Isacson, D.G.L., Thorslund, M.: Prescription drug use among ambulatory elderly in a swedish municipality. Ann. Pharmacother. **27**, 1120–1125 (1993)
15. Anderson, G.: Chronic Care: Making the Case for Ongoing Care. Robert Wood Johnson Found, pp. 1–43 (2010)
16. Friedman, J., Hastie, T., Tibshirani, R.: Regularization paths for generalized linear models via coordinate descent. J. Stat. Softw. **33**(1), 1 (2010)
17. Saito, T., Rehmsmeier, M.: The precision-recall plot is more informative than the ROC plot when evaluating binary classifiers on imbalanced datasets. PLoS ONE **10**(3), e0118432 (2015)
18. Hutter, F., Hoos, H.H., Leyton-Brown, K.: Sequential model-based optimization for general algorithm configuration. In: Coello, C.A.C. (ed.) LION 2011. LNCS, vol. 6683, pp. 507–523. Springer, Heidelberg (2011). https://doi.org/10.1007/978-3-642-25566-3_40

Deep Vision System for Clinical Gait Analysis in and Out of Hospital

Hosang Yu[1], Kyunghun Kang[2], Sungmoon Jeong[1,3](✉), and Jaechan Park[1,2]

[1] Center for Artificial Intelligence in Medicine,
Kyungpook National University Hospital, Daegu, South Korea
youhs4554@gmail.com, jparkmd@hotmail.com
[2] Department of Neurology, School of Medicine,
Kyungpook National University, Daegu, South Korea
kangkh@knu.ac.kr
[3] School of Medicine, Kyungpook National University, Daegu, South Korea
jeongsm00@gmail.com

Abstract. To follow-up Parkinson's disease (PD) progress, clinical gait analysis is performed with the precise measuring equipments (e.g. IMU, electric walkway, etc.). However, the existing gait analysis methods have a limitation such that patients must visit a certain space in hospital for the checkup. For clinical gait analysis in and out of hospital, we propose a baseline model of 'deepvision' system, which can estimate 15 clinical gait parameters measured from electric walkway named GAITRite. We constructed 3D convolution layers which have skip connections to grasp spatio-temporal characteristics of the walking behavior with an effective manner. Afterwards, we validated the method with scripted walking videos, and achieved the following results: error range of temporal and spatial parameters as 32–71 ms, 1.6–6.7 cm respectively, and error for cadence, velocity and functional ambulation profile as 7.0 steps/min, 4.1 cm/min, and 4.9 points respectively.

Keywords: Parkinson's disease · Contactless visual monitoring · Clinical gait analysis

1 Introduction

Parkinson's disease (PD) is a long-term degenerative disorder of the central nervous system that mainly affects the motor system. Early in the disease, the most obvious symptoms are shaking, rigidity, slowness of movement, and difficulty with walking. To follow-up the disease progress before and after a therapy intervention, accurate assessment procedure for gait parameters is performed. The procedure includes measuring the acceleration and angular velocities with wearable sensors [1,9], or force platforms [14], or motion capture systems [2,5]. In determining temporal gait parameters, detecting each time of initial foot contact (IC) and final foot contact (FC) timing is required, called gait events. [7,11,12,14] tried to detect IC/FC timings from a single/dual IMU positioned

T. Gedeon et al. (Eds.): ICONIP 2019, CCIS 1142, pp. 633–642, 2019.
https://doi.org/10.1007/978-3-030-36808-1_69

on the lower trunk, and have proposed for both normal and clinical gait analysis usages. [9] tried to estimate stride/step/stance time based on IC/FC timing obtained from IMUs, and validated free walking performance for each case attached to the shank and waist. As most of the existing systems require special equipment (e.g. IMU, GAITRite, etc.), patients must visit the hospital for checkup, and the medical process becomes inefficient and complicated.

As a preliminary study for developing 'contactless' deep vision clinical system, we insist that our main contribution is to validated baseline model of vision-based gait analyzer with normal scripted walking. In this paper, person detection is performed with YOLO v3 [6]. And we introduce an algorithm for localizing patient area in gait video, which is based on tracing patterns of each detected person. Features characterizing the motion of the patients are obtained from pre-trained C3DNet [10] which were trained on Sports-1M dataset [4]. Architecture of proposed regression model has residual blocks, containing skip connections in it, inflated 2D→3D, to interpret dynamic characteristics in both space and time effectively. To train the actual regression model, we used data including walking videos and corresponding parameters collected from medical gait checkup, and permission to use data was given from Institutional Review Board (IRB) of Kyungpook National University Hospital. We achieved the following results: error range of temporal and spatial parameters as 32–71 ms, 1.6-6.7 cm respectively, and error for cadence, velocity and functional ambulation profile as 7.0 steps/min, 4.1 cm/min, and 4.9 points respectively.

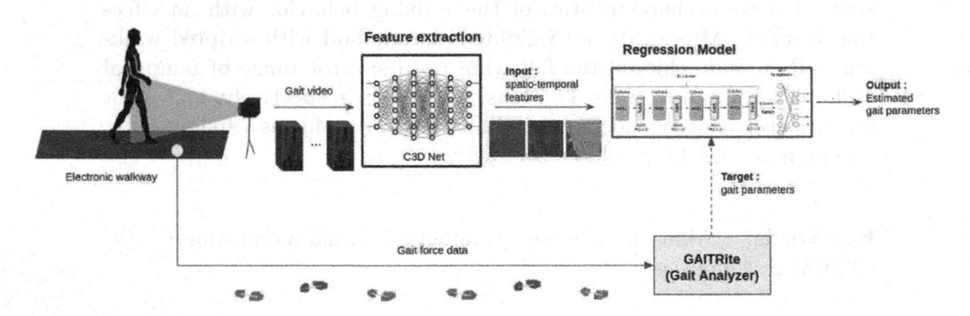

Fig. 1. Overview of data acquisition environment and pipelines for end-to-end deep learning system to estimate gait parameters.

2 Methods

Overview. Figure 1 shows the overview of our system. We conducted data collection for 640 patients, who were asked to walk straight for 5 m across a pressure electronic walkway. Gait force data is sampled from the walkway at 120 Hz and analyzed by GAITRite system, which is well-known gold standard gait analyzer, which provides the 50 gait parameters for clinicians to identify anomalies in gait patterns. Among the 50 parameters, 15 gait parameters, which are essential for medical diagnosis, were selected. Simultaneously, gait video is recorded using a single RGB camera while facing the patient, which has resolution of 640 × 480 and 24 FPS spec. In order to verify the performance of a normal scripted straight walking, we recorded video of 4 times gait trials at most per each person for 640 subjects in various disease groups including PD. Totally, 1563 video samples were used in this work. Descriptive statistics for entire parameters are shown in Table 1. All parameters in temporal/spatial group have left and right footage values each. Because each left/right pair has nearly similar statistical characteristics, averaged values are included in the table.

Brief descriptions of each gait parameter are as follows: 'Cycle Time' is elapsed time between the first contacts of two consecutive footfalls of the same foot. 'Stance Time' is the time elapsed between the First Contact and the Last Contact of two consecutive footfalls on the same foot. 'Double Supp. Time' is the time elapsed while both feet are on the floor. 'Swing Time' is the time elapsed between the last contact of the current footfall to the first Contact of the next footfall on the same foot. 'Stride Length' is distance on the line of progression between the heel points of two consecutive footprints of the same foot. 'HH Base Support' is the vertical distance from heel center of one footprint to the line of progression formed by two footprints of the opposite foot. 'Cadence' is the number of steps per minute. 'Velocity' is obtained after dividing the distance traveled by the ambulation time. 'Functional Amb. Profile' is a rating score which reflects functional aspects of gait and represents a quantification of patients' gait, and calculated by subtracting points from a maximum score of 100 for a self-selected velocity gait trial.

Table 1. Mean and SD values of gait parameters

Params group	Params name	$\mu \pm \sigma$
Temporal	Cycle Time (sec)	1.16 ± 0.25
	Stance Time (sec)	0.77 ± 0.23
	Double Supp. Time (sec)	0.38 ± 0.21
	Swing Time (sec)	0.39 ± 0.062
Spatial	Stride Length (cm)	81.08 ± 26.13
	HH Base Support (cm)	11.65 ± 3.61
Etc.	Cadence (steps/min)	106.02 ± 15.57
	Velocity (cm/min)	71.67 ± 26.61
	Functional Amb. Profile	77.24 ± 16.57

2.1 Preprocessing

Patient Localization. To analyze the gait patterns of patient, our vision system need to mainly focus on a patient under the checkup. Most of our collected gait video consists of single patient's scripted straight walking, but some cases one or more persons are appearing in the video frame because of difficulty in walking of patients, which makes it hard to focus on a patient. To remedy this, we conducted patient localization, which contains the process of finding where the persons are in the video frame and determination of who is a patient. For localization of spatial areas where patient is walking, deep learning based object detector YOLO v3 [6], a state-of-the-art real-time object detection system which can processes each frame at 30 FPS, is used. YOLO can find locations of human in the form of bounding box (bbox), but it is hard to determine which bbox contains patient among all the other bboxes, as YOLO does not have a tracking algorithm for specific person, general person tracking is possible instead. An example view from our camera is shown in Fig. 2. Most of our video data contains scripted walking, and we found that bbox of patient (label = 1) has much larger deviation in the y-axis with respect to x-axis, compared with the other (label = 0). Based on this characteristic, we applied a tracking algorithm as shown below:

Step 1: Compute spatio-temporal intersection over union (STIOU) for all bboxes at time $t-1$. STIOU between i-th bbox at time t and j-th bbox at time $t-1$, i.e. $STIOU(b_t^i, b_{t-1}^j)$, is defined as in Eq. 1. M, N denotes entire number of bbox at time $t, t-1$ respectively.

$$STIOU(b_t^i, b_{t-1}^j) = \frac{intersect(b_t^i, b_{t-1}^j)}{union(b_t^i, b_{t-1}^j)}, i \in 1...M, j \in 1...N \quad (1)$$

Step 2: Get maximum value of STIOU for all b_{t-1}^j, and it is defined as effective STIOU value,

$$STIOU^{eff}(b_t^i, b_{t-1}^{j...N}) = \max_{b_{t-1}^j, j \in 1...N} (STIOU(b_t^i, b_{t-1}^j)) \quad (2)$$

Step 3: If effective STIOU exceeds any threshold value, the current bbox b_t^i gets tracking label from the previous bbox which has maximum value of STIOU denoted as $b_{t-1}^{j_{max}}$. Otherwise, a new label is assigned to the current bbox b_t^i. The threshold value for branching of labeling assignment policy is determined empirically. In our case, $thresh = 0.1$ was best.

$$label_{b_t^i} = \begin{cases} label_{b_{t-1}^{j_{max}}}, & \text{if } STIOU^{eff}(b_t^i, b_{t-1}^{j...N}) > thresh. \\ label_{new}, & \text{otherwise.} \end{cases} \quad (3)$$

Step 4: Each label has array for analyzing its bbox traces. The bbox position of b_t^i is appended into array of the corresponding label. And step 1–4 is repeated for all bboxes for all video frames.

Step 5: Select a bbox that has maximum of δ_y/δ_x. And crop each video frame with the bbox coordinates. δ_x and δ_y denotes displacement for x- and y-axis positions for entire walking time respectively.

Fig. 2. Example of patient tracking results. At bottom left of each bbox, tracking label is displayed, and each bbox is distinguished by its color.

Input Data. To receive a series of features capturing spatio-temporal information from the video frames, a sequence of features is given as an input for the proposed model: $f_t = F(v_t : v_t + \delta)$ where δ is the time resolution of each feature f_t. In this paper, F extracts C3D features where $\delta = 16$ frames. The output of F is a tensor of size D × N × 4 × 4 where $D = 512$ dimensional features and $N = T/\delta$ discretizes the video frames and we set the maximum length of the sequence as 20, to consider $320(= 20 * 16)$ frames. In upper path of Fig. 1 shows pipeline of video data processing. Each cubic in the figure represents a video clip, and it is fed to pretrained C3DNet to capture well-organized spatio-temporal characteristics. To see the features intuitively, we visualized an example of conv5 layer activation maps in Fig. 1. We can see that the activated area are mainly located near the lower body, so it might be nice feature in interpreting the gait.

Target Data. Since all of target gait parameters have different scale and units, standardization is required. To make distributions of all gait parameters similar in statistical properties, quantile transformation which can force the data to follow normal distribution, is applied. Quantile transformation help to spread out the most frequent values, and reduces the impact of outliers.

2.2 Proposed Model

Residual Block. Residual networks (ResNet) was introduced by [3] to construct extremely deep nets. Authors of ResNet insist that identity skip connections can resolve the vanishing gradient problem by preserving gradient flow throughout the entire depth of deep networks. However, [13] showed that ResNet enables deep networks training by leveraging only the short paths. They measured how much gradient contributed the paths of different lengths in a ResNet. They proved it in their experiment by measuring changes in gradient magnitude at input with respect to various length of paths. Also, they showed only the short paths are needed during training: the longer paths is, the smaller contribution on gradients is. Finally, they concluded the paths through ResNet that contribute gradient during training are shorter than expected. In this paper, to capture spatio-temporal features from the sequence of 'conv5' layer output, proposed model has residual blocks inflatted 2D→3D. Also, in ResNet structure, the output of the previous layer contains multiple scales of receptive field. We can expect that, as the network gets deeper, an ensemble effect that will allow us to consider multiple scales. However, based on the fact that the longer the path is the less effect on the gradient at input layer, thus we implemented a shallow 3D ResNet as shown in Fig. 3.

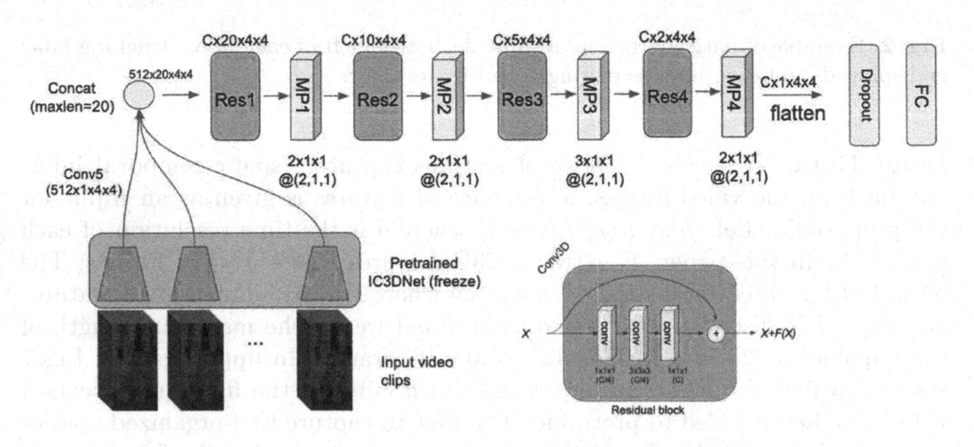

Fig. 3. Proposed model for gait parameter regression.

Regression Model. The proposed regression model has 3D ConvNet-MLP structure: In 3D ConvNet, we expect the networks to learn to represent spatio-temporal relations between each features from pre-trained C3DNet for nice regression results in MLP. Highly inspired by [13], we constructed shallow networks with only 16 residual blocks as shown in Fig. 3. Each green box contains 4 residual blocks in it. BN and ReLU layer has been omitted for simplicity: each CONV in the figure has placement of CONV-BN-ReLU. Exceptionally, last CONV in each residual block located in front of addition has placement of Conv-BN. After addition with X from skip connection, ReLU is applied. Between each group of residual block, we max-pooled outputs from previous layer along time axis. After 4 iterations of same operations, we finally flatten features to a vector of $C*1*4*4$ dimension. And we apply dropout and feed into MLP(i.e. fully connected layer), without applying any activation function, and each output node is trained to do regression for each gait parameters.

Training and Evaluation. We split the dataset 80% for train 20% for test. And we conduct 5-fold cross validation(CV), splitting train dataset 80% for train 20% for validation at every fold iteration. Mean squared error (MSE) is used as cost function. And during cross validation, we conduct random hyperparameter searching for 6 kinds of hyperparameters listed in Table 1. Every hyperprameters is sampled from normal distribution with reasonable ranges for each, and 20 times of sampling is performed for each CV iteration. We train the model with SGD optimizer and Cyclical learning rates scheduling [8] during 50 epochs. To prevent over-fitting, we applied early-stop technique with 5 epochs of patience. Based on R^2 score (or coefficient of determination) best model is selected. We implemented our model with PyTorch, and to accelerate training speed, we used Dask frameworks supporting parallelization in distributed computing. We used 4 NVIDIA TITAN V GPUs, to conduct parallelized hyperparameter searching. Finally, we achieved following best hyperparameters: `batch_size=4, base_lr=0.00639, lr_damping=0.762, drop_rate=0.22, output_residual=250, weightdecay_rate=0.0063.`

3 Results and Discussions

For evaluation, we measure the performances with widely used matrices in regression tasks: Mean Absolute Error (MAE) and R-squared regression score (R^2). Evaluation results are shown in Table 2. We conducted six times of training-testing sessions to evaluate performance. The evaluation was conducted for test dataset with 313 samples, and each mean and SD of error was estimated for entire experiments. R^2 provides a measure of how well target values can be predicted by the regression model, based on the proportion of total variation of data explained by the model.

Table 2. Evaluation results of gait parameters with best model.

Params group	Params name	MAE	R^2
Temporal	Cycle Time (sec)	0.071 ± 0.0011	0.47 ± 0.019
	Stance Time (sec)	0.048 ± 0.0015	0.61 ± 0.024
	Double Supp. Time (sec)	0.033 ± 0.0011	0.74 ± 0.018
	Swing Time (sec)	0.032 ± 0.00052	0.31 ± 0.014
Spatial	Stride Length (cm)	6.7 ± 0.22	0.84 ± 0.0049
	HH Base Support (cm)	1.6 ± 0.11	0.57 ± 0.038
Etc.	Cadence (steps/min)	7.0 ± 0.091	0.45 ± 0.023
	Velocity (cm/min)	4.1 ± 0.046	0.92 ± 0.0028
	Functional Amb. Profile	4.9 ± 0.11	0.73 ± 0.032

Since YOLO-based human detection method is applied to find a patient at the stage of preprocessing, if some of video frames are not detected by YOLO then it may embarrass our model in predicting the gait parameters. Especially in case of temporal parameters group, only small amount of video frames are used to extract their values, thus any omitted frames could have enormous influence on temporal parameter. We evaluated detection error rate of YOLO as $1.57 \pm 2.75\%$. In the worst detection error as 4.32%, 0.327 s error in temporal parameters is expected because an average gait travel time is 7.58 s. It is big enough to make temporal parameters worse than the other groups (see temporal parameters group's values in Table 1).

To show fitness of proposed model intuitively, we draw scatter plots of predicted values versus true values in Fig. 4. Each green dot denotes pair of [true, pred], and dotted red line denotes all points satisfying "*true == pred*". From the figure, we conclude that our regression model can follow the overall tendency of spatio-temporal gait parameters pretty well.

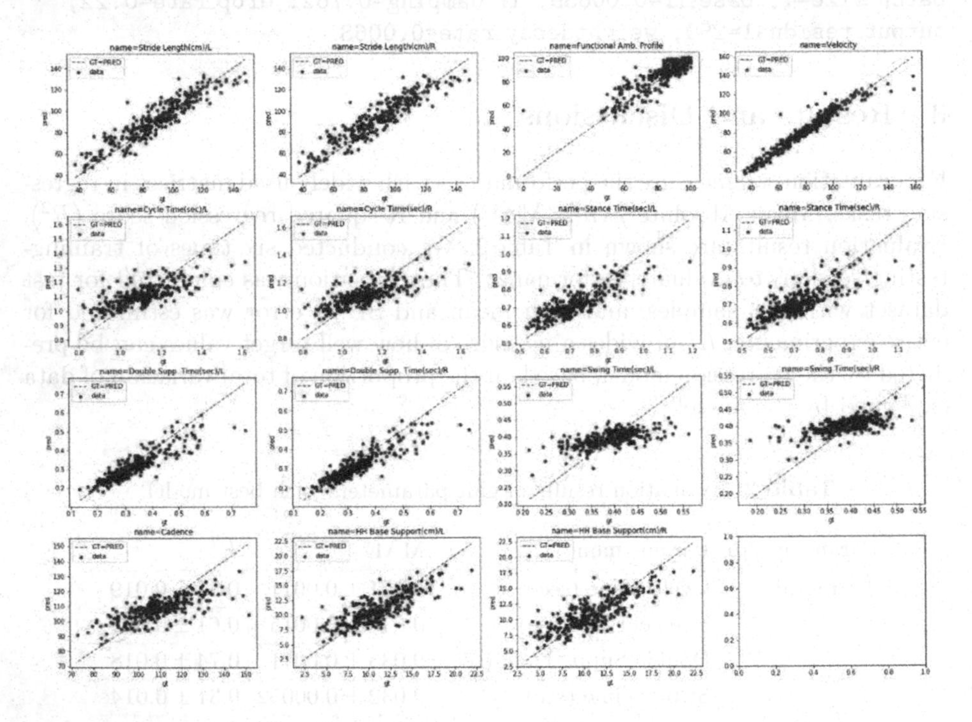

Fig. 4. Scatter plots of predicted values versus true values.

4 Conclusion and Future Works

The aim of this work was to develop a baseline of vision-based gait analyzer for 'contactless' deep vision clinical system. Our main contribution of this work is to build a visual framework, and verify the possibility of this approach by validating performances. Our model achieves an accuracy in the rage of 3%–7.1% and R^2 in the range of 0.31–0.92. In order to enhance the model accuracy, more accurate patient localization and key frame extraction methods will be considered to reduce the missing video frames including patients as a preprocessing stage.

Since our main objective is to assist clinical diagnosis of PD, the predictability of gait analyzer is of course important, but capability of distinguishing between normal group and disease group is also important in clinical point of view. In order words, it may have clinical meaning in diagnosis assistance, if our system can extract significant different parameters in gait analysis between the two groups. Therefore, the distinguishing ability should be also investigated between different groups of subjects based on clinical point of view in near future. We hope our system can help not only the diagnosis assistance for PD but also anomalies detection in everyday gait video.

Acknowledgement. This work was supported by Institute of Information & Communications Technology Planning & Evaluation(IITP) grant funded by the Korea government(MS-IT) (2016-0-00564, Development of Intelligent Interaction Technology Based on Context Awareness and Human Intention Understanding) and supported by the Korea government (MOTIE) (P0004794, Creating innovate ecosystem for Convergence medical industry of Daegu innovation city).

References

1. Benson, L.C., Clermont, C.A., Bošnjak, E., Ferber, R.: The use of wearable devices for walking and running gait analysis outside of the lab: a systematic review. Gait Posture **63**, 124–138 (2018)
2. Cloete, T., Scheffer, C.: Benchmarking of a full-body inertial motion capture system for clinical gait analysis. In: 2008 30th Annual International Conference of the IEEE Engineering in Medicine and Biology Society, pp. 4579–4582. IEEE (2008)
3. He, K., Zhang, X., Ren, S., Sun, J.: Deep residual learning for image recognition. In: Proceedings of the IEEE Conference on Computer Vision and Pattern Recognition, pp. 770–778 (2016)
4. Karpathy, A., Toderici, G., Shetty, S., Leung, T., Sukthankar, R., Fei-Fei, L.: Large-scale video classification with convolutional neural networks. In: Proceedings of the IEEE Conference on Computer Vision and Pattern Recognition, pp. 1725–1732 (2014)
5. Pfister, A., West, A.M., Bronner, S., Noah, J.A.: Comparative abilities of microsoft kinect and vicon 3D motion capture for gait analysis. J. Med. Eng. Technol. **38**(5), 274–280 (2014)
6. Redmon, J., Farhadi, A.: Yolov3: an incremental improvement. arXiv preprint. arXiv:1804.02767 (2018)
7. Salarian, A., et al.: Gait assessment in Parkinson's disease: toward an ambulatory system for long-term monitoring. IEEE Trans. Biomed. Eng. **51**(8), 1434–1443 (2004)
8. Smith, L.N.: Cyclical learning rates for training neural networks. In: 2017 IEEE Winter Conference on Applications of Computer Vision (WACV), pp. 464–472. IEEE (2017)
9. Storm, F.A., Buckley, C.J., Mazzà, C.: Gait event detection in laboratory and real life settings: accuracy of ankle and waist sensor based methods. Gait Posture **50**, 42–46 (2016)
10. Tran, D., Bourdev, L.D., Fergus, R., Torresani, L., Paluri, M.: C3D: generic features for video analysis. CoRR **2**(7), 8 (2014). arXiv:1412.0767
11. Trojaniello, D., et al.: Estimation of step-by-step spatio-temporal parameters of normal and impaired gait using shank-mounted magneto-inertial sensors: application to elderly, hemiparetic, parkinsonian and choreic gait. J. Neuroeng. Rehabil. **11**(1), 152 (2014)

12. Trojaniello, D., Ravaschio, A., Hausdorff, J.M., Cereatti, A.: Comparative assessment of different methods for the estimation of gait temporal parameters using a single inertial sensor: application to elderly, post-stroke, Parkinson's disease and Huntington's disease subjects. Gait Posture **42**(3), 310–316 (2015)
13. Veit, A., Wilber, M.J., Belongie, S.: Residual networks behave like ensembles of relatively shallow networks. In: Advances in Neural Information Processing Systems, pp. 550–558 (2016)
14. Zijlstra, W., Hof, A.L.: Assessment of spatio-temporal gait parameters from trunk accelerations during human walking. Gait Posture **18**(2), 1–10 (2003)

Human Centred Computing for Emotion

Recognizing Facial Expressions of Occluded Faces Using Convolutional Neural Networks

Mariana-Iuliana Georgescu[1,2] and Radu Tudor Ionescu[1(✉)]

[1] University of Bucharest, 14 Academiei Street, Bucharest, Romania
raducu.ionescu@gmail.com
[2] Novustech Services, 12B Aleea Ilioara, Bucharest, Romania

Abstract. In this paper, we present an approach based on convolutional neural networks (CNNs) for facial expression recognition in a difficult setting with severe occlusions. More specifically, our task is to recognize the facial expression of a person wearing a virtual reality (VR) headset which essentially occludes the upper part of the face. In order to accurately train neural networks for this setting, in which faces are severely occluded, we modify the training examples by intentionally occluding the upper half of the face. This forces the neural networks to focus on the lower part of the face and to obtain better accuracy rates than models trained on the entire faces. Our empirical results on two benchmark data sets, FER+ and AffectNet, show that our CNN models' predictions on lower-half faces are up to 13% higher than the baseline CNN models trained on entire faces, proving their suitability for the VR setting. Furthermore, our models' predictions on lower-half faces are no more than 10% under the baseline models' predictions on full faces, proving that there are enough clues in the lower part of the face to accurately predict facial expressions.

Keywords: Facial expression recognition · Convolutional neural networks · Severe face occlusion · Virtual reality headset

1 Introduction

Facial expression recognition from images is an actively studied problem in computer vision, having a broad range of applications including human behavior understanding, detection of mental disorders, human-computer interaction, among others. Our particular application is to recognize the facial expressions of a person wearing a virtual reality (VR) headset and use the recognition result in order to provide feedback to the VR system, which can automatically change the VR experience according to the user's emotions.

Research supported by Novustech Services through Project 115788 (Innovative Platform based on Virtual and Augmented Reality for Phobia Treatment) funded under the POC-46-2-2 by the European Union through FEDR.

T. Gedeon et al. (Eds.): ICONIP 2019, CCIS 1142, pp. 645–653, 2019.
https://doi.org/10.1007/978-3-030-36808-1_70

Fig. 1. Images (of people wearing VR headsets) with corresponding Grad-CAM [28] explanation masks and labels from a VGG-face model trained on lower-half images.

In the past few years, most works [2,6,8,11,13,16,18–21,23,25,31,34–36] have focused on building and training deep neural networks in order to achieve state-of-the-art results. Engineered models based on handcrafted features [1,14,29,30] have drawn very little attention, since such models usually yield less accurate results compared to deep learning models. As most recent works, we adopt deep convolutional neural networks (CNNs) due to their capability of attaining state-of-the-art results in facial expression recognition. However, we have to train the neural networks for a very difficult setting, in which the person to be analyzed wears a VR headset. The currently available VR headsets essentially occlude the upper part of the face, posing significant problems for a standard (not adapted) facial expression recognition system. Our goal is to adapt the facial expression recognition system for this specific setting. To achieve this goal, we train two CNN models, VGG-f [3] and VGG-face [26], on modified training images in which the upper half of the face is completely occluded. This forces the neural networks to find discriminative clues in the lower half of the face, as shown in Fig. 1.

We perform experiments showing that our models (trained with occluded faces) obtain better accuracy rates than models trained on the entire faces, when the test set contains occluded faces. The experiments are conducted on two benchmark data sets, FER+ [2] and AffectNet [24]. Our empirical results show that our CNN models' trained on lower-half faces are up to 13% higher than the baseline CNN models trained on entire faces, when the test set includes images of lower-half faces. Furthermore, our models trained and tested on lower-half faces are about 10% (or even less) under the baseline models, when the baseline models are tested on full faces, thus proving that there are enough clues in the lower part of the face to accurately predict facial expressions. Overall, our empirical results indicate that learning and inferring facial expressions solely on the lower half of the face is a viable option for recognizing facial expression of persons wearing VR headsets. We thus conclude that our final goal, that of providing feedback to the VR system in order to automatically change the VR experience based on the user's emotions, is achievable.

We organize the rest of this paper as follows. We discuss related work in Sect. 2. We present the convolutional neural networks in Sect. 3. We describe the empirical results in Sect. 4. Finally, we draw our conclusions in Sect. 5.

2 Related Art

The early works on facial expression recognition are mostly based on handcrafted features [32]. After the success of AlexNet [17] in the ImageNet Large Scale Visual Recognition Challenge (ILSVRC) [27], deep learning has been widely adopted in the computer vision community. Perhaps some of the first works to propose deep learning approaches for facial expression recognition were presented at the 2013 Facial Expression Recognition (FER) Challenge [9]. Interestingly, the top scoring system in the 2013 FER Challenge is a deep convolutional neural network [31], while the best handcrafted model ranked only in the fourth place [14]. With only a few exceptions [1,29,30], most of the recent works on facial expression recognition are based on deep learning [2,6–8,11,13,16,18–21,23,25,34–36]. Some of these recent works [13,16,18,34,35] proposed to train an ensemble of convolutional neural networks for improved performance, while others [4,15] combined deep features with handcrafted features such as SIFT [22] or Histograms of Oriented Gradients (HOG) [5]. Works that combine deep and handcrafted features usually employ a single CNN model and various handcrafted features, e.g. Connie et al. [4] employed SIFT and dense SIFT, while Kaya et al. [15] employed SIFT, HOG and Local Gabor Binary Patterns (LGBP). While most works studied facial expression recognition from static images as we do here, some works approached facial expression recognition in video [11,15].

Different from these mainstream works [2,4,6,8,11,13–16,18,19,21,23,25,29, 30,34–36], we focus on recognizing facial expressions of occluded faces. More closely related to our work, Li et al. [20] proposed an end-to-end trainable Patch-Gated CNN that can automatically perceive occluded region of the face, making the recognition based on the visible regions. To find the visible regions of the face, their model decomposes an intermediate feature map into several patches according to the positions of related facial landmarks. Each patch is then reweighted by its importance, which is determined from the patch itself. Different from Li et al. [20], we consider a more difficult setting in which half of the face is occluded.

To our knowledge, the only work that studies facial expression recognition for people wearing VR headsets is that of Hickson et al. [12]. In their work, Hickson et al. [12] presented an algorithm that automatically infers expressions from images of the user's eyes captured from an infrared gaze-tracking camera mounted inside the VR headset, while the user is engaged in a VR experience. While their approach is applicable to VR headsets that have an infrared camera mounted inside, our approach, which uses an external camera, is cheaper and applicable to all VR headsets, thus being more generic.

3 Method

In this work, we choose two pre-trained CNN models, namely VGG-f [3] and VGG-face [26]. We proceed by fine-tuning the networks in two stages. In the first stage (see details in Sect. 3.1), we fine-tune the CNN models on images with full faces. In the second stage (see details in Sect. 3.2), we further fine-tune the models on images in which the upper half of the face is occluded. All models are trained using data augmentation, which is based on including horizontally flipped images. To prevent overfitting, we employ Dense-Spare-Dense (DSD) training [10] to train our CNN models. The training starts with a dense phase, in which the network is trained as usual. When switching to the sparse phase, the weights that have lower absolute values are replaced by zeros after every epoch. A sparsity threshold is used to determine the percentage of weights that are replaced by zeros. The DSD learning process, typically ends with a dense phase. It is important to note that DSD can be applied several times in order to achieve the desired performance. We next describe in detail each of the three training stages.

3.1 Training on Non-occluded Faces

VGG-face. VGG-face [26] is a deep neural network that is pre-trained on a closely related task, namely face recognition. The architecture is composed of 16 layers. We keep its 13 convolutional (conv) layers, replacing the fully-connected (fc) layers with a single max-pooling layer for faster inference. We also replace the softmax layer of 1000 units with a softmax layer of 8 units, since FER+ [2] and AffectNet [24] contain 8 classes of emotion. We randomly initialize the weights in the softmax layer, using a Gaussian distribution with zero mean and 0.1 standard deviation. We add 6 dropout layers after each conv layer, starting from the fourth convolutional block. The first dropout layer has a dropout rate of 0.3. For each subsequent dropout layer, the dropout rate increases by 0.05. Thus, the last dropout layer has a dropout rate of 0.55. We set the learning rate to 10^{-4} and we decrease it by a factor of 10 when the validation error stagnates for more than 10 epochs. In order to train VGG-face, we use stochastic gradient descent using mini-batches of 64 images and set the momentum rate to 0.9. We fine-tune VGG-face using DSD training [10] for a total of 50 epochs. For DSD training, we set the sparsity rate to 0.2 for the second conv layer, increasing the rate up to 0.7 with each additional conv layer. We refrain from pruning the weights of the first conv layer. Since VGG-face is pre-trained on a closely related task (face recognition), it converges in only 50 epochs.

VGG-f. We also fine-tune the VGG-f [3] network with 8 layers, which is pre-trained on the ILSVRC benchmark [27]. As for VGG-face, we keep the conv layers of VGG-f, replacing the fc layers with a single max-pooling layer. We also replace the softmax layer of 1000 units with a softmax layer of 8 units. We add 3 dropout layers after each conv layer, starting from the third convolutional layer. In each dropout layer, we set the dropout rate to 0.2. We set the learning rate to 10^{-3} and we decrease it by a factor of 10 when the validation error stagnates for more than 10 epochs. At the end of the training process, the learning rate drops to 10^{-5}. In order to train VGG-f, we use stochastic gradient descent using mini-batches of 512 images and set the momentum rate to 0.9. As for VGG-face, we use the DSD training method to fine-tune the VGG-f model. However, we refrain from pruning the weights of the first conv layer during the sparse phases. We set the sparsity rate to 0.2 for the second conv layer, increasing the rate up to 0.5 with each additional conv layer. We train this network for a total of 800 epochs. Since VGG-f is pre-trained on a distantly related task (object recognition), it converges in a higher number of epochs than VGG-face.

3.2 Training on Occluded Faces

VGG-face. We fine-tune VGG-face on images containing only occluded faces, using, in large part, the same parameter choices as in the first training stage. The architecture is the same as in the first training stage. We hereby present only the different parameter choices. We applied DSD training for 40 epochs, starting with a learning rate of 10^{-3}, decreasing it by a factor of 10 each time the validation error stagnated for more than 10 epochs. At the end of the training process, the learning rate drops to 10^{-4}.

VGG-f. In a similar fashion, we fine-tune VGG-f on images containing only occluded faces, preserving the architecture and most of the parameter choices. We next describe the differences from the first training stage. We applied DSD training for 80 epochs using a learning rate of 10^{-3}. This time we did not have to decrease the learning rate during training, as the validation error drops after every epoch.

4 Experiments

4.1 Data Sets

FER+. The FER+ data set [2] is a curated version of FER 2013 [9] in which some of the original images are relabeled, while other images, e.g. not containing faces, are completely removed. Barsoum et al. [2] add *contempt* as the eighth class of emotion along with the other 7 classes: *anger*, *disgust*, *fear*, *happiness*, *neutral*, *sadness*, *surprise*. The FER+ data set contains 25045 training images, 3191 validation images and another 3137 test images. All images are of 48×48 pixels in size.

AffectNet. The AffectNet [24] data set contains 287651 training images and 4000 validation images, which are manually annotated. Since the test set is not publicly available, researchers [24,36] evaluate their approaches on the validation set containing 500 images for each of the 8 emotion classes.

4.2 Implementation Details

The input images in both data sets are scaled to 224×224 pixels. We use the MatConvNet [33] library to train the CNN models. Each CNN architecture is trained on the joint AffectNet and FER+ training sets. On AffectNet, we adopt the down-sampling setting proposed in [24], which solves, to some extent, the imbalanced nature of the facial expression recognition task. As Mollahosseini et al. [24], we select at most 15000 samples from each class for training. This leaves us with a training set of 88021 images.

4.3 Results

In Table 1, we present the empirical results conducted on AffectNet [24] and FER+ [2] using either VGG-f or VGG-face. The models are trained and tested on various combinations of full or occluded images. We include a bag-of-visual-words baseline [14] and two state-of-the-art CNN models (trained on full faces), VGG-13 [2] and AlexNet [24], for reference. First, we note that our models, VGG-f and VGG-face, attain results that are on par with the state-of-the-art CNNs, when full faces are used for training and testing. However, the CNN models trained on full faces give poor results when lower-half faces are used for testing. These results indicate that the CNN models (trained on full faces, as usual) are not particularly designed to handle severe facial occlusions, justifying our idea of fine-tuning the models on images containing such severe occlusions. Indeed, we observe significant accuracy improvements when VGG-f and VGG-face are fine-tuned on occluded images. For instance, the fine-tuning of VGG-face on lower-half images (occluded upper half) brings an improvement of 11.53% (from 37.70% to 49.23%) on AffectNet and an improvement of 13.39% (from 68.89% to 82.28%) on FER+. Interesting, our final models trained and tested on lower-half faces attain results that are not very far from the CNN models trained and tested on full faces. For example, our VGG-face model is 8.77% under the state-of-the-art AlexNet [24] on AffectNet, and 2.71% under the state-of-the-art VGG-13 [2] on FER+. Despite being tested on lower-half images, both VGG-f and VGG-face surpass the bag-of-visual-words model [14], which is tested on full images. We thus conclude that our CNN models can provide reliable results, despite being tested on faces that are severely occluded.

Table 1. Accuracy rates of various VGG-f and VGG-face models on AffectNet [24] and FER+ [2], using full faces or lower-half faces (occluded upper half) for training and testing. A bag-of-visual-words model [14] and two state-of-the-art CNN models (trained on full faces), VGG-13 [2] and AlexNet [24], are also included for reference.

Model	Train set	Test set	AffectNet	FER+
Bag-of-visual-words [14]	full faces	full faces	48.30%	80.65%
VGG-13 [2]	full faces	full faces	-	84.99%
AlexNet [24]	full faces	full faces	58.00%	-
VGG-f	full faces	full faces	57.37%	85.05%
VGG-face	full faces	full faces	59.03%	84.79%
VGG-f	full faces	lower-half faces	41.58%	70.00%
VGG-face	full faces	lower-half faces	37.70%	68.89%
VGG-f	lower-half faces	lower-half faces	47.58%	78.23%
VGG-face	lower-half faces	lower-half faces	49.23%	82.28%

5 Conclusion

In this paper, we proposed a learning approach based on fine-tuning CNN models in order to recognize facial expressions of severely occluded faces. The empirical results indicate that our learning framework can bring significant performance gains, leading to models that provide reliable results in practice, even surpassing a bag-of-visual-words baseline [14] tested on images with fully visible faces.

References

1. Al Chanti, D., Caplier, A.: Improving bag-of-visual-words towards effective facial expressive image classification. In: Proceedings of VISIGRAPP, pp. 145–152 (2018)
2. Barsoum, E., Zhang, C., Ferrer, C.C., Zhang, Z.: Training deep networks for facial expression recognition with crowd-sourced label distribution. In: Proceedings of ICMI, pp. 279–283 (2016)
3. Chatfield, K., Simonyan, K., Vedaldi, A., Zisserman, A.: Return of the devil in the details: delving deep into convolutional nets. In: Proceedings of BMVC, pp. 1–12 (2014)
4. Connie, T., Al-Shabi, M., Cheah, W.P., Goh, M.: Facial expression recognition using a hybrid CNN-SIFT aggregator. In: Proceedings of MIWAI, vol. 10607, pp. 139–149 (2017)
5. Dalal, N., Triggs, B.: Histograms of oriented gradients for human detection. In: Proceedings of CVPR, vol. 1, pp. 886–893 (2005)
6. Ding, H., Zhou, S.K., Chellappa, R.: FaceNet2ExpNet: regularizing a deep face recognition net for expression recognition. In: Proceedings of FG, pp. 118–126 (2017)
7. Georgescu, M.I., Ionescu, R.T., Popescu, M.: Local learning with deep and handcrafted features for facial expression recognition. IEEE Access **7**, 64827–64836 (2019)
8. Giannopoulos, P., Perikos, I., Hatzilygeroudis, I.: Deep learning approaches for facial emotion recognition: a case study on FER-2013. In: Hatzilygeroudis, I., Palade, V. (eds.) Advances in Hybridization of Intelligent Methods. SIST, vol. 85, pp. 1–16. Springer, Cham (2018). https://doi.org/10.1007/978-3-319-66790-4_1
9. Goodfellow, I.J., et al.: Challenges in representation learning: a report on three machine learning contests. In: Lee, M., Hirose, A., Hou, Z.-G., Kil, R.M. (eds.) ICONIP 2013. LNCS, vol. 8228, pp. 117–124. Springer, Heidelberg (2013). https://doi.org/10.1007/978-3-642-42051-1_16
10. Han, S., et al.: DSD: dense-sparse-dense training for deep neural networks. In: Proceedings of ICLR (2017)
11. Hasani, B., Mahoor, M.H.: Facial expression recognition using enhanced deep 3D convolutional neural networks. In: Proceedings of CVPRW, pp. 2278–2288 (2017)
12. Hickson, S., Dufour, N., Sud, A., Kwatra, V., Essa, I.: Eyemotion: classifying facial expressions in VR using eye-tracking cameras. In: Proceedings of WACV, pp. 1626–1635 (2019)
13. Hua, W., Dai, F., Huang, L., Xiong, J., Gui, G.: HERO: human emotions recognition for realizing intelligent internet of things. IEEE Access **7**, 24321–24332 (2019)

14. Ionescu, R.T., Popescu, M., Grozea, C.: Local learning to improve bag of visual words model for facial expression recognition. In: Proceedings of ICML Workshop on Challenges in Representation Learning (2013)
15. Kaya, H., Gürpınar, F., Salah, A.A.: Video-based emotion recognition in the wild using deep transfer learning and score fusion. Image Vis. Comput. **65**, 66–75 (2017)
16. Kim, B.K., Roh, J., Dong, S.Y., Lee, S.Y.: Hierarchical committee of deep convolutional neural networks for robust facial expression recognition. J. Multimodal User Interfaces **10**(2), 173–189 (2016)
17. Krizhevsky, A., Sutskever, I., Hinton, G.E.: ImageNet classification with deep convolutional neural networks. In: Proceedings of NIPS, pp. 1106–1114 (2012)
18. Li, D., Wen, G.: MRMR-based ensemble pruning for facial expression recognition. Multimedia Tools Appl. **77**(12), 1–22 (2017)
19. Li, S., Deng, W., Du, J.: Reliable crowdsourcing and deep locality-preserving learning for expression recognition in the wild. In: Proceedings of CVPR, pp. 2584–2593 (2017)
20. Li, Y., Zeng, J., Shan, S., Chen, X.: Patch-gated CNN for occlusion-aware facial expression recognition. In: Proceedings of ICPR, pp. 2209–2214 (2018)
21. Liu, X., Kumar, B., You, J., Jia, P.: Adaptive deep metric learning for identity-aware facial expression recognition. In: Proceedings of CVPRW, pp. 522–531 (2017)
22. Lowe, D.G.: Distinctive image features from scale-invariant keypoints. Int. J. Comput. Vis. **60**(2), 91–110 (2004)
23. Meng, Z., Liu, P., Cai, J., Han, S., Tong, Y.: Identity-aware convolutional neural network for facial expression recognition. In: Proceedings of FG, pp. 558–565 (2017)
24. Mollahosseini, A., Hasani, B., Mahoor, M.H.: AffectNet: a database for facial expression, valence, and arousal computing in the wild. IEEE Trans. Affect. Comput. **10**(1), 18–31 (2019)
25. Mollahosseini, A., Hassani, B., Salvador, M.J., Abdollahi, H., Chan, D., Mahoor, M.H.: Facial expression recognition from world wild web. In: Proceedings of CVPRW, pp. 1509–1516 (2016)
26. Parkhi, O.M., Vedaldi, A., Zisserman, A., et al.: Deep face recognition. In: Proceedings of BMVC, pp. 6–17 (2015)
27. Russakovsky, O., et al.: ImageNet large scale visual recognition challenge. Int. J. Comput. Vis. **115**(3), 211–252 (2015)
28. Selvaraju, R.R., Cogswell, M., Das, A., Vedantam, R., Parikh, D., Batra, D.: Grad-CAM: visual explanations from deep networks via gradient-based localization. In: Proceedings of ICCV, pp. 618–626 (2017)
29. Shah, J.H., Sharif, M., Yasmin, M., Fernandes, S.L.: Facial expressions classification and false label reduction using LDA and threefold SVM. Pattern Recogn. Lett. (2017)
30. Shao, J., Gori, I., Wan, S., Aggarwal, J.: 3D dynamic facial expression recognition using low-resolution videos. Pattern Recogn. Lett. **65**, 157–162 (2015)
31. Tang, Y.: Deep learning using linear support vector machines. In: Proceedings of ICML Workshop on Challenges in Representation Learning (2013)
32. Tian, Y., Kanade, T., Cohn, J.F.: Facial expression recognition. In: Handbook of Face Recognition, pp. 487–519. Springer, Cham (2011). https://doi.org/10.1007/978-0-85729-932-1_19
33. Vedaldi, A., Lenc, K.: MatConvNet - convolutional neural networks for MATLAB. In: Proceeding of ACMMM (2015)

34. Wen, G., Hou, Z., Li, H., Li, D., Jiang, L., Xun, E.: Ensemble of deep neural networks with probability-based fusion for facial expression recognition. Cogn. Comput. **9**(5), 597–610 (2017)
35. Yu, Z., Zhang, C.: Image based static facial expression recognition with multiple deep network learning. In: Proceedings of ICMI, pp. 435–442. ACM (2015)
36. Zeng, J., Shan, S., Chen, X.: Facial expression recognition with inconsistently annotated datasets. In: Proceedings of ECCV, pp. 222–237 (2018)

Analysis of Key Face Parts to Detect Emotional Expression Using a Neural Network Model

Katsuki Izumi(✉), Osamu Araki, and Tomokazu Urakawa

Department of Applied Physics, Tokyo University of Science,
6-3-1 Niijuku, Katsushika-ku, Tokyo 125-8585, Japan
1515007@alumni.tus.ac.jp

Abstract. Psychological studies on recognition of facial expression reported that local parts of a face image such as eyes or mouth are important to detect the facial expression. On the other hand, artificial neural network technology has progressed greatly in facial recognition. However, the neural mechanism behind the recognition remains to be unknown. The purpose of this study is to extract important features from the neural network model after learning emotions from facial expression and to clarify which face parts are key to detect emotional expression. First, we trained a 2-layered neural network model with backpropagation for recognition of 7 kinds of emotional faces. Then, we found more weighted input pixels for the recognition by tracing and accumulating the synaptic weights linearly from the output layer toward the hidden layer. By this method, we extracted the 6 face-image filters for each emotional expression. Using the face-image filters divided into 10 parts, we designed a new analytical method to evaluate which facial part or parts are important for the discrimination of emotional expression. From this analysis, it can be concluded that key parts are very different for each emotion. For example, nose and mouth are effective for happy smile while strained cheek for sad face.

Keywords: Neural network model · Facial expression · Emotions · Backpropagation

1 Introduction

How do we read other's emotion from the face? A behavioral experiment to answer emotions using masked facial images, which divided into 3 regions (each eye, nose and cheek, and mouth) reported that the mouth and eye regions showed highest correct rate and shortest reaction time after the onset of a face image in reporting its effect [1]. Another eye-tracking experiment using facial images to detect their emotion showed longer duration that staying subject's eye gaze at eyes and mouth regions [2]. So far the facial expression from the eyes and mouth region has been thought to or interpreted to affect strongly the emotional discrimination. Whereas, the influence from the other regions such as nose, cheek,

T. Gedeon et al. (Eds.): ICONIP 2019, CCIS 1142, pp. 654–661, 2019.
https://doi.org/10.1007/978-3-030-36808-1_71

and jaw has not been paid attention. It remains unclear whether detection of facial expressions would be based on a part of face image or information on integrated parts. On the other hand, artificial neural network technology has made great process in facial recognition [3–5]. Whereas these studies aimed at higher recognition accuracy, they have been less interested in the reason why the learned neural model answered correctly. The neural network which have learned emotional expressions from human facial images may give us significant hints about the information processing in the brain. In fact, it is a crucial problem to clarify the reason for the model's output and to extract the learned knowledge in neural network models, and some techniques to visualize important features for discrimination have been proposed [6,7]. In this study, we adopted a simpler method among them because we focus on evaluation of what the features are rather than the method. Thus, the purpose of this study is to clarify the important parts of a face for discrimination of each emotion by extracting and analyzing the stimulus patterns required for the discrimination. First, we trained a 2-layered neural network model by backpropagation to distinguish 6 kinds of basic emotions [8] (Happiness, Sadness, Surprise, Anger, Disgust, and Fear) and Neutral from facial images. Then, we found more weighted input pixels for the recognition through tracing and accumulating the synaptic weights linearly from the output layer toward the hidden layer. Consequently, we extracted the 6 face-image filters for each emotional expression.

2 Methods

2.1 A Neural Network Model

The neural network model we used has 2 layers, a hidden layer and an output layer [9] (Fig. 1). The synaptic weights between layers are fully connected. The number of inputs is $28 \times 28 = 784$ which corresponds to the pixels of input images, and number of neurons in the hidden layer is 50. The output layer has 7 neurons corresponding to 6 emotions plus the neutral. The output x_j of the j-th neuron in hidden layer is calculated as follows:

$$x_j = f(\sum_{i=1}^{784} w_{ji} I_i - \theta_j), \tag{1}$$

where I_i is an i-th input, w_{ji} is a connected weight from i-th input value to j-th neuron in the hidden layer, θ_j is a threshold, and f is the ReLU function. The input value is a pixel value of a facial image normalized to the range 0–1. Each initial value of synaptic weight follows a Gaussian distribution $N(0, 0.01)$. And, the output y_k of the k-th neuron in the output layer is represented as follows:

$$y_k = \sum_{j=1}^{50} W_{kj} x_j - \theta_k, \tag{2}$$

where W_{kj} is a connection weight from j-th neuron in hidden layer to k-th neuron in the output layer, and θ_k is a threshold. After these calculations, this network chooses the index k that maximizes to $g(y_k)$ as an answer. g is the softmax function as follows:

$$g(y_k) = \frac{\exp y_k}{\sum_{i=0}^{6} \exp y_i}. \tag{3}$$

Learning of connection weights is performed by the backpropagation algorithm to minimize the energy function E as follows through the gradient decent method.

$$E = -\sum_{k=0}^{6} t_k \log g(y_k), \tag{4}$$

where t_k is the one-hot label, in which $t_k = 1$ if k is the answer and $t_k = 0$ otherwise. The learning rate is set to be 0.01. For 1 epoch training period, a mini-batch learning using 50 images randomly selected from training data was executed. Totally 30,000 epochs were performed. During the learning, we evaluated the average accuracy of emotional discrimination using test images every 1,000 epochs.

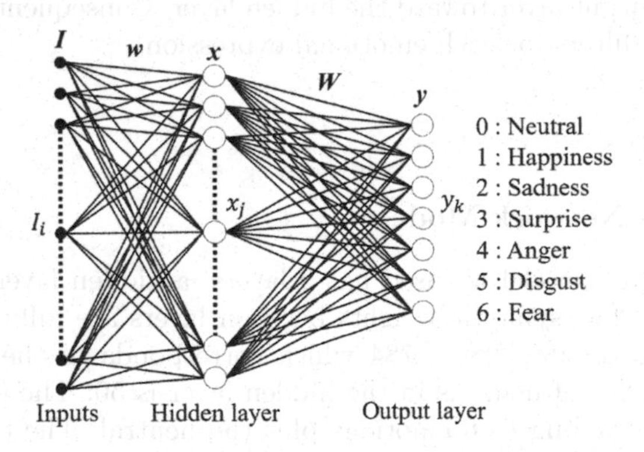

Fig. 1. Structure of the neural network model.

2.2 Analytical Methods

First, we show the method to calculate the input patterns which strongly affect the discrimination of facial emotion from the connection weights after learning. Assuming that $\boldsymbol{y}$, a 7-dimentional vertical vector, represents the result whose answer is set to 1 and the others are 0, we assign proportionally the synaptic weights toward the inputs as follows:

$$\boldsymbol{x} = \boldsymbol{W}^T \boldsymbol{y}, \tag{5}$$

$$\boldsymbol{I} = \boldsymbol{w}^T \boldsymbol{x}, \tag{6}$$

where $\boldsymbol{W}^T$ is the 50×7 transposed weight matrix (W_{kj}), $\boldsymbol{w}^T$ is the 784×50 transposed weight matrix (w_{ji}). $\boldsymbol{x}$ is a 50-dimentional vertical vector, and $\boldsymbol{I}$ is a 784-dimentional vertical vector. $\boldsymbol{I}$ is calculated for each emotion. $\boldsymbol{I}$ is considered to contain elements that play an important role in determining the emotion from input images. To extract important elements from $\boldsymbol{I}$, we created filters consisting of parts where the element value in $\boldsymbol{I}$ is larger than the threshold value ξ. ξ is set for each emotion as follows: Happiness: 5, Sadness: 3, Surprise: 1, Anger: 5, Disgust: 1, and Fear: 1. Using the face images which are masked by the filter for each emotion face, we checked whether the model discriminated the inputs correctly. In addition, in order to examine the influence of each facial region on judging facial expressions, the masking images are spatially divided into 10 regions (both eyebrows, eyes, both cheeks, nose, mouth, and jaws on both sides), and they are input to the learned model. Specifically, we selected 1 part or combination of 2 parts and more parts of the divided masked images, and other areas are filled in black.

2.3 Face Images

We used Picture of Facial Affect (POFA) [10] image database and ATR Facial Expression Image Database (DB99) [11]. POFA is a black-and-white image database with a total of 110 images: full faces of 6 males and 8 females, with emotional expressions. DB99 is a color image database with a total of 629 images: full faces of 6 males and 4 females, with emotional expressions. Among DB99 images, 39 images other than 7 expressions were included, and 590 images were used. The total 700 images are evaluated by emotional impression into 6 basic emotions plus neutral ones through a psychological procedure. We labeled each image as the emotion with the highest evaluation, and we used them as teaching signals for the learning of the model. Face images were normalized and compressed to 28×28 pixels size, with adjusted brightness. Finally, 700 image datasets are randomly divided into 70 for testing (each 10 for 7 kinds of emotions) and 630 for training. The number of each expression in the database are as follows: (0) Neutral 78, (1) Happiness 187, (2) Sadness 102, (3) Surprise 45, (4) Anger 192, (5) Disgust 47, and (6) Fear 49.

3 Results

3.1 Learning of Emotion in Faces

The accuracy rate for the training images increased to nearly 100% after 20,000 epochs, while the rate for the test images was saturated at 70–80% (Fig. 2). Since the accuracy rate depends on the randomly selected images and initial values, we executed 10 trials assuming that 30,000 epochs training for 1 trial. The results are shown in Table 1.

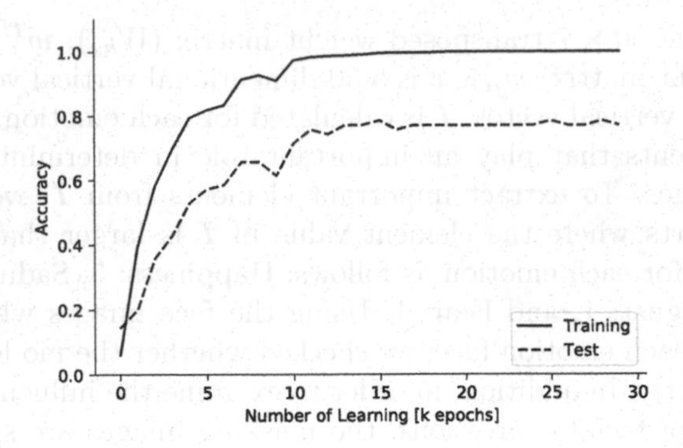

Fig. 2. Learning process of facial emotions. The solid line indicates accuracy for the training set, and the dashed line indicates accuracy for 70 test images.

Table 1. Average correct rate for discrimination of emotional face.

Emotion	Neutral	Happiness	Sadness	Surprise	Anger	Disgust	Fear	Training	Test
Correct rate	0.84	0.96	0.77	0.78	0.83	0.44	0.62	1.00	0.75

3.2 Important Features for Detection of Facial Emotion

Figure 3 shows the value of the calculated of $\boldsymbol{I}$ for each facial emotion (Sect. 2.2) with heatmaps as follows: 1. Happiness, 2. Sadness, 3. Surprise, 4. Anger, 5. Disgust, 6. Fear. The region with large value strongly influences the discrimination of emotion. Because the regions with larger values were distributed in the images, the discriminations did not depend on a specific face part like a mouth or eyes. When the face images masked by the filters were input, correct answers were output. It confirmed that the filter plays an important role for the discrimination. In Fig. 3, the correspondence between these filters and face images is entirely vague, though they seem to reflect facial parts such as eyes and mouth. In Sect. 3.3, we examine the effects that face parts have on the emotional discrimination.

3.3 Key Facial Parts for Emotional Detection

The filters were divided into 10 parts, numbered from 0 to 9 (Fig. 4). Each region includes facial parts as follows: 0, 1: both eyebrows, 2, 3: both eyes, 4, 6: both cheeks, 5: nose, 8: mouth, 7, 9: both sides of jaw. First, either 1 filter or a combination of partial filters was selected from the 10 filters, and face images which were masked by the filters were input to the learned model. Then, we calculated the rate of correct outputs including the target partial filter for each emotion (Table 2). In the case of one part, face images masked with the target filter were input to the learned model. In the case of a combination of 2 parts, face images masked with a target filter plus another one were input and evaluated. Similar procedures were applied to the combination more than 2. Since the

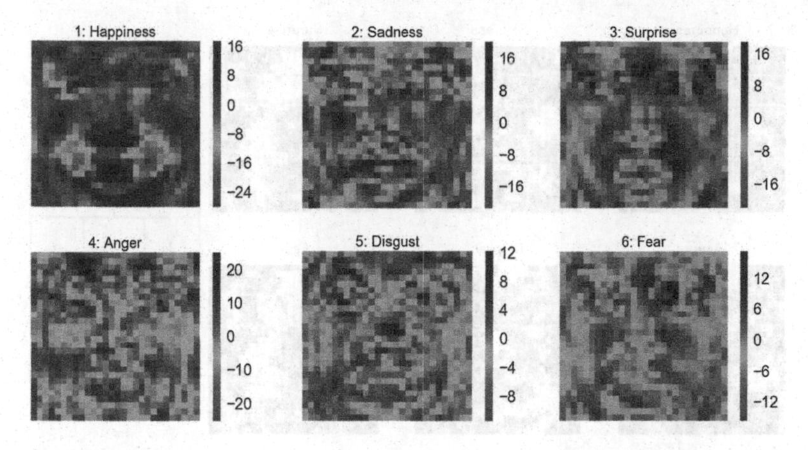

Fig. 3. Important filters for emotional discrimination in the neural network model.

correct rate tends to increase as the combination increase, the results with higher rates were omitted from Table 2. For example for Happiness, the correct rate was relatively high if the input combination includes #5, for example (1, 5), (2, 5), and (3, 5). Thus, #5 part is considered to be important for the detection of happy emotion. Happiness expression was correctly answered 100% in response to #5, #8, and #9 partial parts regardless of single or combination. It means that facial parts around a nose and a mouth are very effective to the discrimination of a happy face. This may be partially because the bright part of nasal bridge becomes wider when we smile. Also mouth region reflects the contour of the lower lip and teeth in the smiling face. For Sadness expression, #4 filter was relatively strong. This may reflect a spacial shape of a cheek in the sad expression. For Surprise expression, #2 and #5 were found to be effective facial parts. It is considered to reflect widely opened eyes which are characteristic to the surprise expression. Additionally, the wide outline of nose seems to be reflected. For Anger expression, the effective filters are special compared with others because any partial filter led the model to the correct answer. This implies that individual part includes cues for anger emotion. For example, the right eyebrow (#0) seemed to reflect the wrinkles in the middle forehead and its direction. For Disgust expression, the nose (#5) and the right side jaw (#7) showed higher rates. The upper part of the nose is thought to reflect wrinkles on the nose showing disgust. This result matches the hypothesis that disgust reminds a human the bad primitive memory of a smell such as rotten food [12]. For Fear expression, both sides of jaw (#7 and #9) have a large influence. It may reflect the contour of the opened mouth.

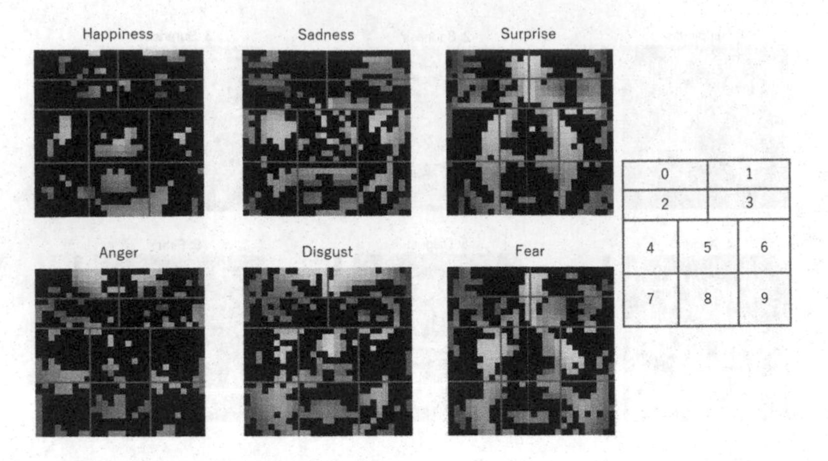

Fig. 4. Key filters $\boldsymbol{F}$ divided into 10 parts.

Table 2. Correct discrimination rates with some filter(s) including the target filter for each emotion using one or some combination area as input.

Happiness

Number of Combination	Target Filter 0	1	2	3	4	5	6	7	8	9
1 Part	0.00	0.85	0.00	0.00	0.00	1.00	0.00	0.00	1.00	1.00
2 Parts	0.98	1.00	0.78	0.78	0.99	1.00	0.76	0.98	1.00	1.00
3 Parts	1.00	1.00	0.99	0.99	1.00	1.00	0.99	1.00	1.00	1.00

Sadness

Number of Combination	Target Filter 0	1	2	3	4	5	6	7	8	9
1 Part	0.00	0.85	0.00	0.05	1.00	0.00	0.00	0.05	0.50	0.00
2 Parts	0.56	0.96	0.68	1.00	1.00	0.71	0.48	0.99	0.94	0.64
3 Parts	0.90	0.99	0.97	1.00	1.00	0.94	0.88	1.00	0.99	0.94

Surprise

Number of Combination	Target Filter 0	1	2	3	4	5	6	7	8	9
1 Part	0.00	0.00	0.00	0.00	0.00	0.00	0.00	0.00	0.00	0.00
2 Parts	0.00	0.08	0.13	0.01	0.00	0.13	0.00	0.00	0.00	0.01
3 Parts	0.26	0.39	0.46	0.26	0.10	0.44	0.05	0.17	0.08	0.29
4 Parts	0.57	0.66	0.65	0.61	0.35	0.65	0.28	0.47	0.33	0.51
5 Parts	0.76	0.80	0.75	0.84	0.61	0.80	0.54	0.72	0.59	0.66
6 Parts	0.88	0.89	0.84	0.95	0.80	0.88	0.75	0.86	0.79	0.80
7 Parts	0.96	0.96	0.94	0.99	0.93	0.96	0.92	0.95	0.92	0.92

Anger

Number of Combination	Target Filter 0	1	2	3	4	5	6	7	8	9
1 Part	1.00	1.00	1.00	1.00	1.00	1.00	0.95	1.00	1.00	1.00
2 Parts	1.00	1.00	1.00	1.00	1.00	1.00	1.00	1.00	1.00	1.00

Disgust

Number of Combination	Target Filter 0	1	2	3	4	5	6	7	8	9
1 Part	0.00	0.00	0.00	0.00	0.00	0.50	0.00	0.90	0.00	0.00
2 Parts	0.01	0.28	0.24	0.23	0.27	0.73	0.23	0.86	0.10	0.41
3 Parts	0.26	0.49	0.46	0.62	0.56	0.84	0.55	0.88	0.36	0.76
4 Parts	0.48	0.63	0.69	0.82	0.72	0.90	0.75	0.90	0.60	0.86
5 Parts	0.65	0.73	0.76	0.86	0.80	0.93	0.82	0.92	0.70	0.89
6 Parts	0.82	0.84	0.87	0.92	0.88	0.96	0.89	0.96	0.83	0.94

Fear

Number of Combination	Target Filter 0	1	2	3	4	5	6	7	8	9
1 Part	0.00	0.00	0.00	0.00	0.00	0.00	0.00	0.00	0.00	0.00
2 Parts	0.00	0.00	0.00	0.26	0.14	0.11	0.11	0.49	0.16	0.35
3 Parts	0.43	0.29	0.27	0.56	0.71	0.44	0.51	0.84	0.70	0.77
4 Parts	0.70	0.60	0.55	0.74	0.83	0.64	0.70	0.96	0.84	0.90
5 Parts	0.84	0.78	0.75	0.86	0.91	0.79	0.82	0.98	0.92	0.94
6 Parts	0.94	0.92	0.92	0.94	0.97	0.92	0.93	0.99	0.97	0.97

4 Conclusion

First, we extracted discrimination filters for 6 kinds of emotional expressions in human faces from the 2-layered neural network model after learning face images through backpropagation. Then, we examined which facial part of each filter is more effective for the discrimination using the modified face images masked by combinations of the partial filters. This is our proposed analytical method.

Consequently, important face expressions for the emotional discrimination are quite different as follows: nose and mouth for smile in happiness, cheek for tension in sadness, eyes and nose for surprise, all for anger, nose and jaw for disgust expression, and contour of mouth for fear. These suggest that the model discriminated the emotion using some characteristics specific for each expression.

It has been shown that we can find essential features of face parts for discrimination of each emotional expression using a neural network model with the proposed method. However, these are superficial features that can be mapped on a face. A more difficult problem, how to extract the more abstract emotional information from the higher hidden layers, remains to be solved in the future. In addition, whether our brain utilizes these features should be examined by psychological experiments.

References

1. Blais, C., Roy, C., Fiset, D., Arguin, M., Gosselin, F.: The eyes are not the window to basic emotions. Neuropsychologia **50**(12), 2830–2838 (2012)
2. Wells, L.J., Gillesp, S.M., Rotshtein, P.: Identification of emotional facial expressions: effects of expression, intensity, and sex on eye gaze. PLoS One **11**(12), e0168307 (2016). https://doi.org/10.1371/jounal.pone.0168307
3. Lawrence, S., Giles, C.L., Tsoi, A.C., Back, A.D.: Face recognition: a convolutional neural-network approach. IEEE Trans. Neural Netw. **8**(1), 98–113 (1997)
4. Lopes, A.T., Aguiar, E., Souza, A.F., Santos, O.T.: Facial expression recognition with convolutional neural networks: coping with few data and the training sample order. Pattern Recogn. **61**, 610–628 (2017)
5. Ding, C., Tao, D.: Robust face recognition via multimodal deep face representation. IEEE Trans. Multimedia **17**(11), 2049–2058 (2015)
6. Zeiler, M.D., Fergus, R.: Visualizing and understanding convolutional networks. In: Fleet, D., Pajdla, T., Schiele, B., Tuytelaars, T. (eds.) ECCV 2014. LNCS, vol. 8689, pp. 818–833. Springer, Cham (2014). https://doi.org/10.1007/978-3-319-10590-1_53
7. Simonyan, K., Vedaldi, A., Zisserman, A.: Deep inside convolutional networks: visualising image classification models and saliency maps. In: CVPR (2014)
8. Ekman, P., Friesen, V.W.: Universals and cultural differences in the judgements of facial expressions of emotion. J. Pers. Soc. Psychol. **54**(4), 712–717 (1987)
9. Hertz, J., Krogh, A., Palmer, R.G.: Introduction to the Theory of Neural Computation. Wylde, A.M., Santa Fe Institute (1991)
10. Picture of Facial Affect (POFA). https://www.paulekman.com/product/pictures-of-facial-affect-pofa/. Accessed 29 June 2019
11. ATR Facial Expression Image Database (DB99). http://www.atr-p.com/products/face-db.html. Accessed 29 June 2019
12. Rozin, P., Lowery, L., Ebert, R.: Varieties of disgust faces and the structure of disgust. J. Pers. Soc. Psychol. **66**(5), 870–881 (1994)

Multi-task Gated Contextual Cross-Modal Attention Framework for Sentiment and Emotion Analysis

Suyash Sangwan(✉), Dushyant Singh Chauhan, Md. Shad Akhtar, Asif Ekbal, and Pushpak Bhattacharyya

Indian Institute of Technology, Patna, Bihta, India
{suyash.mtmc17,1821CS17,shad.pcs15,asif,pb}@iitp.ac.in

Abstract. Multi-modal sentiment and emotion analysis have been an emerging and prominent field nowadays at the intersection of natural language processing, deep learning, machine learning, computer vision, and speech processing. Sentiment and emotion prediction model finds the attitude of a speaker or writer towards any discussion, debate, event, document or topic. It can be expressed in different ways like the words spoken, energy and tone while delivering words, accompanying facial expressions, gestures, etc. Moreover related and similar tasks generally depend on each other and are predicted better if solved through a joint framework. In this paper, we present a multi-task gated contextual cross-modal attention framework which considers all the three modalities *(viz. text, acoustic and visual)* and multiple utterances for sentiment and emotion prediction together. We evaluate our proposed approach on CMU-MOSEI dataset for sentiment and emotion prediction. Evaluation results depict that our proposed approach extracts co-relation among the three modalities and attains an improvement over the previous state-of-the-art models.

1 Introduction

Microblogging websites and social media platforms like Twitter, YouTube, etc. have evolved and shown a stupendous growth to become a source of varied kinds of information *(like images, audios and videos)*. People post real-time messages on these platforms about their opinions on different topics, products, discussions, etc. They use these social media platforms as an open and comfortable environment to express and discuss current issues or to raise their voice to complain about the products they use in day-to-day life and various organizations utilize the users' inputs as feedbacks.

Thus, multi-modal analysis has been an emerging field of study. The main motivation of using multi-modalities in sentiment and emotion prediction lies in the fact that videos are quite a rich source of information as we can have all the three modalities *(viz. visual, acoustic and text)* from a video. The key

First two authors have equal contributions.

T. Gedeon et al. (Eds.): ICONIP 2019, CCIS 1142, pp. 662–669, 2019.
https://doi.org/10.1007/978-3-030-36808-1_72

challenge in the multi-utterance multi-modality framework is to utilize and fuse the relevant information for the prediction.

Sentiment [1–3] and emotion [4,5,7] are closely related and depend on each other in a way that we can classify '*sad*', '*fear*', '*anger*' and '*disgust*' emotions to have '*negative*' sentiment, whereas '*surprise*' and '*happy*' to reflect '*positive*' sentiment. Therefore, motivated by these advantages of multitasking, we present an effective approach that jointly predicts the expressed emotions and sentiment in a video. Multi-task learning (MTL) paradigm provides advantages over the single-task learning (STL) paradigm as they can leverage the relatedness of each task in a joint framework and helps to achieve the generalization across the multiple tasks.

Further, we employ a gated architecture (i.e. Gated Multi-modal Unit (GMU)) to refine an input representation *w.r.t.* all the other participating inputs i.e. it evaluates the importance of an individual modality based on its role in final prediction. We apply this GMU module at both the raw inputs and generalized attentive representations. The GMU at raw inputs helps the model to filter out any noise in the data for effective learning i.e. based on the role of a modality in final prediction, it is either suppressed or passed. On the other hand, GMU at the generalized attentive representations aims to extract the importance of various attentive modalities in accordance with a specific task (i.e. sentiment and emotion in our case).

Our model is different in the sense that it predicts both sentiment and emotions through a single model and applies gated attention over multi-modalities present and contextual utterances present in only one step.

The main contributions of this paper are: **(a)** *we propose a multi-task framework to leverage the inter-dependence of two related tasks (i.e. sentiment and emotion)*; **(b)** *we introduce a contextual cross-modal attention mechanism to assign weights to the contributing contextual utterances and/or to different modalities simultaneously*; **(c)** *we apply GMU module to refine the input representations as well as attention outputs for the specific tasks (i.e. the refinement for sentiment is performed independent from emotion and hence different attention outputs are selected for sentiment and emotion predictions)* (Fig. 1).

2 Proposed Methodology

In this section, we describe our proposed methodology where we aim to leverage the multi-modal and contextual information for solving multiple tasks (i.e. sentiment and emotion) together. Utterances *(i.e. an uninterrupted chain of spoken languages)* of a video are time-dependent *(i.e. they must be serially connected in time)*. Emotion and sentiment of an utterance are generally dependent on the other neighboring utterances *(i.e. its contextual utterances)* and more importantly all the modalities of contextual utterances may not contribute equally for final prediction. Therefore to model these relationships, we propose an RNN (Recurrent Neural Network) based multi-modal multi-utterance gated attention framework. The proposed model takes multi-modal *(text, visual and acoustic)*

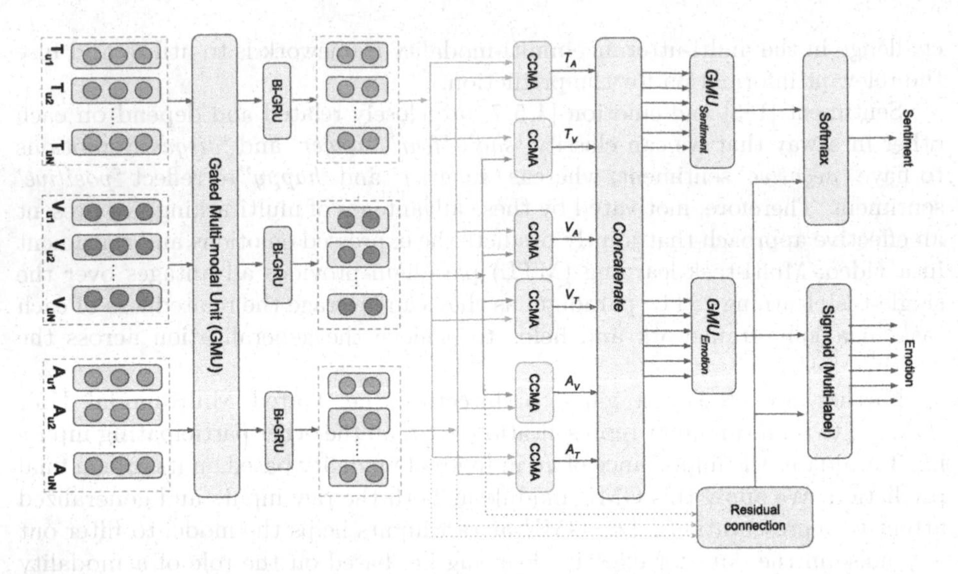

Fig. 1. Overall Architecture of the proposed Multi-task Gated Contextual Cross-Modal Attention framework

information for a sequence of utterances of a video and process them through three separate bi-directional Gated Recurrent Units (GRUs) for capturing the contextual information. Subsequently, we extract the relationships among the contextual modalities through an attention framework. The objective is to learn the joint-association among the utterances and their modality information and to emphasize the contributing features by putting more attention to the contextual utterances. The contextual cross-modal attention framework learns the importance of one modality (e.g. '*acoustic*') *w.r.t.* the other modality (e.g. '*text*' or '*visual*') for all the utterances in a video. We term this '*text-aware acoustic*' (A_T) or '*visual-aware acoustic*' (A_V) attentive representations and so on. Similarly, we compute attention for all the combinations of text, acoustic & visual modalities (i.e. A_T, T_A, A_V, V_A, V_T & T_V). Finally, we concatenate all these contextual cross-modal attentive representations along with the residual connections of *text, acoustic* & *visual* representations for the final prediction. We append residual connection representations to boost the gradient flow to the lower layers.

Our multi-task framework shares the concatenated representation up to the attention layer, therefore, they help each other in better predictions. The shared representation will receive gradients of errors from both the branches (sentiment & emotion) and accordingly adjust the weights of the model. Thus, the shared representations will not be biased to any particular task, and it will assist the model to achieve generalization for multiple tasks.

Since the concatenated representations (A_T, T_A, A_V, V_A, V_T & T_V) aim to achieve the generalization, not all these attentive representations are equally important to both sentiment and emotion. In other words, some of these representations might be more significant than others for sentiment classification, whereas the same might be less important for emotion prediction. For example, if we have to classify a video having 6 utterances then our model decides that to classify 'u1' utterance, textual features of 'u4' and 'u6', acoustic features of 'u2' and visual features of 'u4' are important for sentiment classification, whereas textual features of 'u2', 'u4' and 'u6', acoustic features of 'u6' and visual features of 'u2' and 'u4' are important for emotion detection. Hence, only selected modalities are sent for final prediction. Therefore, for final prediction, we introduce a gated multi-modal unit (GMU) to assign weights to these representations according to their importance for the respective tasks.

Contextual Cross-Modal Attention Framework (CCMA): In our proposed attention framework, we calculate the cross-modality attention scores for each utterance in a video. As outputs of Bi-GRU already contain contextual information of utterances for each modality separately, we, at first, compute a matching matrix $M_x \in \mathbb{R}^{u \times u}$ to capture the cross-modality information. So for the *text-aware acoustic* (A_T), we calculate $M_{A_T} = A.T^{'}$. In the next step, we compute probability distribution score ($P_x \in \mathbb{R}^{u \times u}$) over each utterance of the matrix M_x using a Softmax function. Probability distribution scores calculated here are the weights (or attention scores) for the contextual utterances. Then, we apply soft attention for computing cross-modality aware representation i.e. S_x. Finally, a multiplicative gating mechanism [8] (G_x) is introduced to attend the important components of cross-modalities and utterances. Similarly, we compute cross-modal attentions for all the combinations (i.e. A_T, T_A, A_V, V_A, V_T & T_V).

GMU (Gated Multi-modal Unit): The module, called Gated Multi-modal Unit (GMU) [8] is shown below. It works like the flow control in recurrent architectures like Long Short Term Memory (LSTM) and GRU. Traditionally, the GMU module is used to find a representation of input modality based on the combination of all other modalities. Each input representation (X_i), corresponding to a modality 'i', is passed through an activation function (i.e. *tanh*) to encode the input representation. Further, the encoded representation is filtered through a gating mechanism computed on all the available input sources. The gating mechanism is controlled through a *sigmoid* (σ). This essentially computes the significance of an input source with respect to other available sources of information. For N sources of information, equations for computing GMU for i^{th} source (X_i) is as follows:

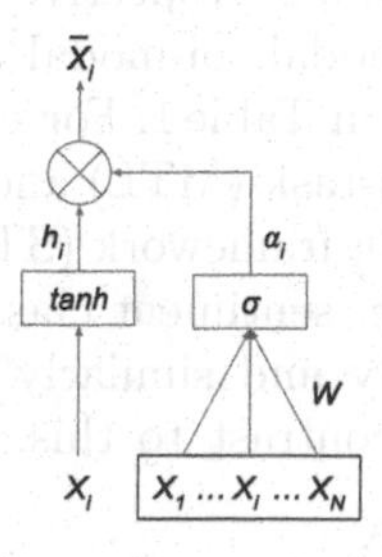

$$h_i = tanh(W_i.X_i)$$
$$X = concat(X_1, ., X_i, ., X_N)$$
$$\alpha_i = \sigma(W.X)$$
$$\overline{X_i} = h_i * \alpha_i$$

3 Experimental Results and Analysis

Dataset: We use the benchmark dataset of sentiment and emotion analysis, namely CMU Multi-modal Opinion Sentiment and Emotion Intensity (CMU-MOSEI) [9] to evaluate our proposed approach. This dataset consists of 3,229 videos spanning over 23,000 utterances. The training, validation, and test set consist of 16216, 1835 and 4625 utterances, respectively. Each utterance has a degree of emotion *(i.e. emotion intensity)* for all the six emotion classes *i.e. anger, disgust, fear, happy, sad and surprise* which depicts the intensity of each emotion. All non-zero intensity values of an utterance are considered as its emotion labels, representing multi-labels of an utterance. If an utterance has no emotion label, then it is considered as of '*No class*'. In contrast, the sentiment values for each utterance are disjoint i.e. $value < 0$ represents *negative* sentiment and $value \geq 0$ represents *positive* sentiment.

We use CMU-Multi-Modal data SDK[1] for downloading and for extracting the features. In MOSEI dataset, word-level features were given where *textual* features were extracted using *GloVe Embeddings*, *acoustic* features were extracted using *CovaRep* and *visual* features were extracted using *Facets.* We compute word-level average of these features to obtain the utterance level features. For each word, the dimension of the feature vector is 300 for *text*, 74 for *acoustic* and 35 for *visual.*

Experiments: We use the Python-based Keras library for its implementation. For evaluation, we compute F1-Score and accuracy values to measure the performance of sentiment classification. For emotion prediction, we use F1-Score and weighted-accuracy [10]. We choose weighted accuracy as a metric because samples are unbalanced across various emotions.

We use grid search to find the optimal hyper-parameters for our experiment. For consistency, we use same hyper-parameters for training all the models ($model_1$, $model_2$ and $model_3$). We use three Bi-GRUs with 300 neurons, one for each modality (*i.e. text, visual and acoustic*). We set dropout to 0.3 and epochs to 50. We use *ReLu* as an activation and *Adam* as an optimizer. We use *softmax* for sentiment classification and *sigmoid* for emotion prediction and *binary cross-entropy* as the loss function. As the dataset is suited for multi-label classification (i.e. more than one emotions possible), we choose a threshold value and consider all the emotions as present whose respective values are above that threshold value. We cross-validate and optimize both the evaluation metrics i.e. weighted accuracy and F1- score and set the threshold as 0.3 and 0.2, respectively.

We experiment our proposed model with all uni-modal, bi-modal and tri-modal input combinations, and their results are shown in Table 1. For comparison, we experiment and report the results of both multi-task (MTL) and single-task (STL) learning framework in Table 1. In Single task framework (STL), two models are built for sentiment and emotion analysis i.e. sentiment class is predicted by learning the model on sentiment labels only and similarly disjoint framework is learned for emotion prediction. But in contrast to this, we also

[1] https://github.com/A2Zadeh/CMU-MultimodalDataSDK.

Table 1. Single-task learning (STL) and Multi-task learning (MTL) frameworks for the proposed approach. W-Acc: Weighted-Accuracy.

Tasks		F1-Scores							Acc (Sent) & W-Acc (Emo)						
		T	A	V	TV	TA	AV	TAV	T	A	V	TV	TA	AV	TAV
Sent	STL	76.26	70.18	63.53	77.16	77.24	74.58	77.82	79.28	70.94	74.77	79.10	79.57	73.83	79.66
	MTL	76.38	71.09	73.96	77.25	77.44	74.99	78.30	79.86	77.73	75.52	79.89	79.97	77.96	80.15
Emo	STL	77.76	75.78	76.47	78.17	78.10	73.49	78.79	60.98	57.36	57.94	62.43	61.54	53.68	62.80
	MTL	77.88	76.26	76.48	78.38	78.21	77.05	79.06	61.19	57.75	58.19	62.53	61.98	59.87	63.16

Table 2. Comparison of our proposed multi-task framework with state of the are systems. *Values are taken from [9]

	Emotion														*Sentiment*	
	Anger		*Disgust*		*Fear*		*Happy*		*Sad*		*Surprise*		*Average*†			
System	*F1*	*W-Acc*	*F1*	*W-Acc*	*F1*	*W-Acc*	*F1*	*W-Acc*	*F1*	*W-Acc*	*F1*	*W-Acc*	*F1*	*W-Acc*	*F1*	*Acc*
Blanchard et al. [1]	-	-	-	-	-	-	-	-	-	-	-	-	-	-	*63.2*	*60.0*
Nojavanasghari et al. [4]*	71.4	-	-	67.0	-	-	-	-	-	-	-	-	-	-	-	-
TFN [5]*	-	60.5	-	-	-	-	66.6	**66.5**	-	58.9	-	52.2	-	-	-	-
Random Forest [6]*	72.0	-	73.2	-	**89.9**	-	-	-	61.8	-	85.4	-	-	-	-	-
Zadeh et al., [7]*	-	-	-	-	-	-	71.0	-	-	-	-	-	-	-	-	-
EF-LSTM [9]*	-	-	-	-	-	56.7	-	57.8	-	59.2	-	-	-	-	-	-
Rajagopalan et al. [11]*	-	56.0	-	-	-	-	-	-	-	-	-	-	-	-	76.4	76.4
Zadeh et al. [13]*	-	-	71.4	65.2	**89.9**	-	-	-	60.8	-	85.4	53.3	-	-	76.0	76.0
Zadeh et al. [9]	72.8	62.6	76.6	69.1	**89.9**	62.0	66.3	66.3	66.9	60.4	85.5	53.7	76.3	62.3	77.0	76.9
Proposed (STL)	73.98	67.15	**82.05**	73.25	87.95	60.48	70.08	57.16	**72.87**	61.84	**86.06**	**57.7**	78.79	62.80	77.82	79.66
Proposed (MTL)	**74.92**	**67.25**	81.99	**73.45**	87.96	**60.50**	**71.33**	60.94	70.97	**63.57**	86.01	57.02	**79.06**	**63.16**	**78.30**	**80.15**
T-test *w.r.t.* SOTA	-	-	-	-	-	-	-	-	-	-	-	-	***0.024***	***0.042***	***0.001***	***0.004***
T-test *w.r.t.* STL	-	-	-	-	-	-	-	-	-	-	-	-	***0.017***	***0.031***	***0.001***	***0.027***

perform multi-task (MTL) framework where a joint-model is learned for both sentiment and emotion i.e. both sentiment and emotion help each other in learning in a better way and hence giving better predictions than STL.

Moreover, better results are obtained while we consider tri-modal input features i.e. all three modalities *(i.e. textual, visual and acoustics)*.

Comparative Analysis: We compare the results obtained from our proposed model against the various other existing models [1,4,7,9,11–13] which also use the same dataset. For each case, we report the results of the top three existing systems (as reported in [9]) and the comparative analysis is shown in Table 2. Our proposed multi-modal multi-task contextual framework reports the best F1-score of 79.06% and weighted accuracy of 63.16% for emotion classification as compared to F1-score of 76.3% and weighted accuracy of 62.3% of the state-of-the-art system. Similarly, for sentiment classification, our proposed model obtains an F1-score of 78.30% and accuracy of 80.15%, whereas the state-of-the-art system obtains the F1-score of 77% and accuracy of 76.9%. Hence, these results show significant performance improvement over the state-of-the-art model. We also perform statistical significance test (*paired T-test*) on the obtained results and observe that performance improvement in the proposed model over the state-of-the-art is significant with 95% confidence (i.e. p-value < 0.05).

Table 3. Comparison with MTL and STL frameworks. Few error cases where MTL framework performs better than STL

	Utterances	Sentiment			Emotion		
		Actual	STL	MTL	Actual	STL	MTL
1.	This information had been brought to me so i filed a case	Pos	**Neg**	Pos	Hp, Sr	**An, Dg**, Hp, **Sd**	Hp, **Sd**, Sr
2.	When the judge asked so are you suggesting dr shiva	Pos	**Neg**	Pos	No Class	**An, Dg, Hp, Sd**	**Sd**
3.	Stars because i have the previous two	Pos	**Neg**	Pos	An, Hp	**Dg, Sd**	Hp
4.	Remembered seeing the previews for it speaker	Neg	**Pos**	Neg	An, Dg, Sd	An, Dg, **Hp**, Sd, **Sr**	An, Sd, Dg

STL vs. MTL Error Analysis: We compare STL and MTL frameworks as shown in Table 3. We have shown a few cases to show where and how multi-task helps for better prediction over the single-task framework. For example, the first utterance has gold sentiment label as *positive* which was misclassified by STL as *negative*. Similarly, in emotion predictions, gold labels are *happy and surprise*, whereas STL predicts *anger, disgust, happy and sad*, but MTL predicts *happy, sad and surprise*. Precision and recall for STL are (1/4) and (1/2), respectively, whereas, we observe improved precision (2/3) & recall (2/2) for MTL. In the second utterance, MTL predicts the correct label for sentiment class i.e. *positive* and in gold emotion label, no class is present for that given utterance. So MTL predicts the presence of only one class i.e. *sad* and STL misclassifies its sentiment and emotion labels both. Similarly, for utterance 3, precision and recall for STL are (0/2) and (0/2), respectively whereas precision and recall for MTL are (1) and (1/2), respectively. In 4th utterance, correct emotion labels i.e. *anger, disgust and sad* for MTL framework help in predicting the correct sentiment label i.e. *negative*. Hence these examples show inter-dependence of two related tasks i.e. sentiment and emotion and also show how MTL framework predicts better than the STL framework.

4 Conclusion

In this paper, we have proposed an RNN based multi-task framework that aims to reveal and utilize the inter-dependence of two related tasks i.e. sentiment and emotion. Our proposed approach learns a joint-representation for both the tasks but selects a different combination of modalities for sentiment and emotion using the GMU model at attention. We evaluate our proposed approach on the recently released benchmark dataset on CMU-MOSEI i.e. the largest available dataset for multi-modal sentiment and emotion analysis having multi-label data. Experimental results suggest that sentiment and emotion help each other for better predictions when learned in a joint framework. In the future, we would like to explore the other dimensions of our multi-task framework.

Acknowledgement. Asif Ekbal acknowledges the Young Faculty Research Fellowship (YFRF), supported by Visvesvaraya Ph.D. scheme of MeiTY, Government of India. The research reported here is also partially supported by "Skymap Global India Private Limited".

References

1. Blanchard, N., Moreira, D., Bharati, A., Scheirer, W.: Getting the subtext without the text: scalable multimodal sentiment classification from visual and acoustic modalities. In: Proceedings of Grand Challenge and Workshop on Human Multimodal Language, Melbourne, Australia, pp. 1–10 (2018)
2. Zhang, Y., et al.: A quantum-inspired multimodal sentiment analysis framework. Theor. Comput. Sci. **752**, 21–40 (2018)
3. Poria, S., Cambria, E., Hazarika, D., Mazumder, N., Zadeh, A., Morency, L.: Multi-level multiple attentions for contextual multimodal sentiment analysis. In: 2017 IEEE International Conference on Data Mining (ICDM), New Orleans, LA, pp. 1033–1038, November 2017
4. Nojavanasghari, B., Gopinath, D., Koushik, J., Baltrušaitis, T., Morency, L.-P.: Deep multimodal fusion for persuasiveness prediction. In: Proceedings of the 18th ACM International Conference on Multimodal Interaction, Tokyo, Japan (2016)
5. Deng, D., Zhou, Y., Pi, J., Shi, B.E.: Multimodal utterance level affect analysis using visual, audio and text features. arXiv preprint arXiv:1805.00625 (2018)
6. Breiman, L.: Random forests. Mach. Learn. **45**(1), 5–32 (2001)
7. Zadeh, A., Liang, P.P., Poria, S., Vij, P., Cambria, E., Morency, L.P.: Multi-attention recurrent network for human communication comprehension. In: Thirty-Second AAAI Conference on Artificial Intelligence (AAAI-2018), New Orleans, USA, pp. 5642–5649 (2018)
8. Arevalo, J., Solorio, T., Montes-y-Gómez, M., González, F.A.: Gated multimodal units for information fusion. arXiv preprint arXiv:1702.01992 (2017)
9. Zadeh, A., Liang, P.P., Poria, S., Cambria, E., Morency, L.-P.: Multimodal language analysis in the Wild: CMU-MOSEI dataset and interpretable dynamic fusion graph. In: Proceedings of the 56th Annual Meeting of the ACL, Melbourne, Australia, pp. 2236–2246 (2018)
10. Tong, E., Zadeh, A., Jones, C., Morency, L.-P.: Combating human trafficking with multimodal deep models. In: Proceedings of the 55th Annual Meeting of the ACL, Vancouver, Canada, pp. 1547–1556. ACL (2017)
11. Rajagopalan, S.S., Morency, L.-P., Baltrušaitis, T., Goecke, R.: Extending long short-term memory for multi-view structured learning. In: Leibe, B., Matas, J., Sebe, N., Welling, M. (eds.) ECCV 2016. LNCS, vol. 9911, pp. 338–353. Springer, Cham (2016). https://doi.org/10.1007/978-3-319-46478-7_21
12. Zadeh, A., Chen, M., Poria, S., Cambria, E., Morency, L.-P.: Tensor fusion network for multimodal sentiment analysis. In: Proceedings of the 2017 Conference on Empirical Methods in Natural Language Processing, Copenhagen, Denmark, pp. 1103–1114, September 2017
13. Zadeh, A., Liang, P.P., Mazumder, N., Poria, S., Cambria, E., Morency, L.-P.: Memory fusion network for multi-view sequential learning. In: Proceedings of the 32nd AAAI, New Orleans, Louisiana, USA, 2–7 February 2018

A Cross-Culture Study on Multimodal Emotion Recognition Using Deep Learning

Lu Gan[1], Wei Liu[1], Yun Luo[1], Xun Wu[1], and Bao-Liang Lu[1,2,3](✉)

[1] Center for Brain-Like Computing and Machine Intelligence, Department of Computer Science and Engineering, Shanghai Jiao Tong University, Shanghai, China
{ganlu_paristech,liuwei-albert,angeleader,stephanie_wx,bllu}@sjtu.edu.cn

[2] Key Laboratory of Shanghai Education Commission for Intelligent Interaction and Cognition Engineering, Shanghai Jiao Tong University, Shanghai, China

[3] Brain Science and Technology Research Center, Shanghai Jiao Tong University, Shanghai, China

Abstract. In this paper, we aim to investigate the similarities and differences of multimodal signals between Chinese and French on three emotions recognition task using deep learning. We use videos including positive, neutral and negative emotions as stimuli material. Both Chinese and French subjects wear electrode caps and eye tracking glass while doing experiments to collect electroencephalography (EEG) and eye movement data. To deal with the problem of lacking data for training deep neural networks, conditional Wasserstein generative adversarial network is adopted to generate EEG and eye movement data. The EEG and eye movement features are fused by using Deep Canonical Correlation Analysis to analyze the relationship between EEG and eye movement data. Our experimental results show that French has higher classification accuracy on beta frequency band while Chinese performs better on gamma frequency band. In addition, EEG signals and eye movement data of French participants have complementary characteristics in discriminating positive and negative emotions.

Keywords: Emotion recognition · EEG · Eye movement · Deep learning · Cross-culture · Chinese · French

1 Introduction

Facial expressions, speech and non-verbal vocalizations are often used as input to recognize different emotions. Recent research found that facial expressions of emotion are not culturally universal [1]. People from different cultures can reach an agreement on the most intense emotion in judging facial expressions. However, culture differences are found when people judge the absolute level of emotional intensity [2]. Differences of non-verbal emotion cognition between western culture and remote tribe were also studied [3]. Cross-cultural similarities and differences

T. Gedeon et al. (Eds.): ICONIP 2019, CCIS 1142, pp. 670–680, 2019.
https://doi.org/10.1007/978-3-030-36808-1_73

appear in music mood perception as well. Research and the experimental results showed that listeners from different cultural backgrounds behaved differently in their selection of mood clusters and agreement ratio in each mood cluster. The similar result was found in Shuar hunter-horticulturalists from Amazonian Ecuador and American native English speakers [4]. However, it is widely agreed that cross-cultural agreement levels are lower than intra-cultural one [5,6].

With the quick development of brain-computer interface (BCI), many researches start to use neural signals to study the relationship between emotion and brain activities. EEG signals are proved to be effective in the field of emotion recognition. Recent researches indicated that there exists a stable neural pattern of EEG signals for positive, neutral and negative emotions [7]. Researchers also used EEG to investigate the differences of neural patterns between Chinese and Germans [8]. Combining EEG modality with other modalities provided an efficient way to recognize human emotions [9].

Eye movements have been widely used in studying attention, perceptions and emotion. Eye tracking data allow researchers to find users' areas of interest, attention track and subconscious behaviors. Therefore, more and more studies start to focus on the relationship between emotion and the movements of eyes. It was proved that higher trait emotional intelligence was associated with more attention to positive emotional stimuli [10]. The increase of gaze to eye region in children with autism spectrum disorders led to higher emotion recognition accuracy [11]. Furthermore, the characteristics of eye movements and EEG are complementary to emotion recognition [12]. Using modality fusion methods can significantly enhance the accuracy on emotion recognition task [13].

In this paper, we focus on investigating the similarities and differences of EEG and eye movement signals between Chinese and French on emotion recognition task using deep learning. The task is to classify positive, neutral and negative emotions. We evaluate the performance of emotion classification with different features and different frequency bands. Functional brain connectivity patterns are adopted to visualize the similarities and differences between Chinese and French. Since the complementary characteristics of EEG and eye movements in Chinese subjects have already been proven [12], we focus on the results for French participants. Multi-modality fusion algorithm is also used to reveal the relationship between EEG signals and eye movement data.

2 Methods

2.1 Functional Brain Connectivity Patterns

Functional brain connectivity patterns are used to visualize the neural patterns of Chinese and French participants instead of focusing on single-channel analysis [14]. Each EEG channel represents one node and the connections between pairs of channels are the links. To construct the functional brain network, we use spectral coherence to calculate the connectivity indices between two EEG channels under different frequency bands. Thus, one connectivity matrix can represent one sample's brain network. Then we use critical subnetwork selection to choose the emotion-related subnetworks.

Critical subnetwork selection can be divided into several steps. Firstly, we calculate the average matrices for each emotion. The brain connectivity matrix of subjects under the same culture background are used to calculate the mean connectivity matrix. Secondly, we sort each mean connectivity matrix based on the absolute value of the connection weights. Since some weak connections between electrodes are not relevant to emotion and they may obscure the profile for the network topology, we discard the connections based on a proportional threshold. The connectivity matrices of positive, neutral and negative emotions are processed respectively. The intersection of connections under three emotions is considered to be less relevant to the specific emotion. Hence, these connections are removed from brain connectivity matrix in the visualization. The choice of threshold is based on the performance of classification. The topological feature strength is extracted from three critical subnetworks of each subject with different thresholds and then fed into a classifier. The threshold who can obtain the highest accuracy is considered to have remained the most emotion-related connections.

2.2 Augmentation of EEG and Eye Movement Data

To overcome the problem of lacking training data for deep neural network, we use Conditional Wasserstein Generative Adversarial Network (CWGAN) to generate both EEG and eye movement data [15]. CWGAN consists of two components. The generator G produces realistic-like data X_g by giving real data distribution X_r and generated data distribution X_g. The objective of generator is to confuse discriminator D which tries to distinguish whether a sample comes from X_r or X_g. The target is to solve the minimax problem during the adversarial training procedure. The formula is defined as:

$$\begin{aligned}\min_{\theta_G}\max_{\theta_D} L(X_r, X_g) = \mathbb{E}_{x_r\sim X_r}[log(D(x_r))] \\ + \mathbb{E}_{x_g\sim X_g}[log(1-D(x_g))]\end{aligned} \quad (1)$$

where θ_g and θ_d represent the parameters of the generator and discriminator, respectively.

In CWGAN, the Earth-Mover distance (EMD, also known as Wasserstein-1 distance) is used to replace Jensen-Shannon divergence to calculate the distance between probability distribution of real data and generated data. Compared with Jensen-Shannon divergence, EMD is continuous and differentiable almost everywhere, which ensures the convergence of GAN and avoids the problem of mode collapse. To make training procedure more stable and convergence faster, a gradient penalty is added instead of using weight clipping [16].

In order to generate samples for multiple classes, label information is used. An auxiliary label Y_r is fed into both generator and discriminator. In the generator, X_z is concatenated with Y_r. In discriminator, X_r and X_g are concatenated with Y_r to construct a hidden representation. The final objective function is defined as:

$$\min_{\theta_G} \max_{\theta_D} L(X_r, X_g, Y_r) = \mathbb{E}_{x_r \sim X_r, y_r \sim Y_r}[D(x_r|y_r)] - \mathbb{E}_{x_g \sim X_g, y_r \sim Y_r}[D(x_g|y_r)] - \lambda \mathbb{E}_{\hat{x} \sim \hat{X}, y_r \sim Y_r}[||\nabla_{\hat{x}|y_r} D(\hat{x}|y_r)||_2 - 1)^2] \quad (2)$$

where λ is a hyperparameter controlling the trade-off between the original objective and gradient penalty, and $\hat{x}$ is defined as:

$$\hat{x} = \alpha x_r + (1-\alpha)x_g, \alpha \sim U[0,1], x_r \sim X_r, x_g \sim X_g \quad (3)$$

The loss of discriminator is the maximum term, and the loss of generator is the minimum term. They are optimized simultaneously. The discriminator loss is updated for critic times in each adversarial training iteration.

2.3 Multi-modality Fusion Approach

To analyze the characteristics of eye movements and EEG data, Deep Canonical Correlation Analysis (DCCA) is used [13]. For each modality, a neural network is constructed to realize nonlinear feature transformation which aims to represent original modality features in another feature space supposed to be related with emotion. The layer sizes for both modalities are the same. Then Canonical Correlation Analysis (CCA) is used to calculate the correlation between transformed features of two modalities. The back-propagation algorithm is adopted to update parameters of network in order to get higher correlation in CCA layer. The extracted features are fused by using the formula defined as follows:

$$F_{fusion} = \alpha M_1 + \beta M_2 \quad (4)$$

where M_1 and M_2 represent the extracted features for each modality, respectively, and α and β are the parameters to control the weight of each modality. Since we consider that EEG and eye movement features have an equivalent importance here, we choose $\alpha = \beta = 0.5$.

3 Experiment Setup

3.1 The SEED Dataset

The SEED[1] dataset is a public dataset for emotion recognition. Fifteen Chinese healthy subjects participated in the experiments to watch 15 Chinese film clips. Each subject was invited to participate in 3 sessions of experiments. The stimuli material contains positive, neutral and negative emotions. During the experiment, subjects were demanded to watch film clips attentively. 62-channel EEG signals based on international 10–20 system and eye movement signals were recorded at the same time.

[1] http://bcmi.sjtu.edu.cn/~seed/index.html.

3.2 Experiment for French Participants

To compare the results of Chinese with those of French, we have to keep consistency in experiment design and data collection. Thus, we choose film clips as stimuli material as well. Since French participants may not understand the expressions of emotion in Chinese films, film clips used in the experiments for French subjects are chosen from a large database of emotion-eliciting films developed by Schaefer *et al.* [17]. All the film excerpts were nominated by 50 experts and evaluated by 364 Belgian French-speaking undergraduates. We add film clips with highest Positive And Negative Affect Schedule (PANAS) into our stimuli material. Due to the lack of neutral excerpts, extra neutral excerpts are chosen from calm landscape films, which are consistent with SEED dataset. Finally, 21 film excerpts are chosen.

Six healthy subjects aged from 22 to 41 participated in the experiments. All of the subjects come from France and their native language is French. Since all the subjects are exchange students and professors on the campus, the number of subjects are limited. Each participant was required to perform the experiments for two sessions. During experiment, participants were asked to immerse in the film clips. 62-channel EEG signals based on international 10–20 system and eye movement signals were recorded simultaneously.

3.3 Feature Extraction and Classification

To keep balance between the number of Chinese subjects and French subjects, we randomly choose 6 subjects from the SEED dataset. In order to keep consistency with the number of sessions each French subject participated, two sessions of a Chinese subject are chosen. We apply the same data preprocessing and feature extraction methods on Chinese and French subjects.

The EEG data are downsampled to 200 Hz and transformed by a Short-Term Fourier Transform (STFT) with an 1-s Hamming window. By using a band-pass filter from 1 to 50 Hz, it allows us to filter out a large part of artifacts. Power Spectral Density (PSD), Differential Entropy (DE), Rational Asymmetry (RASM), Differential Asymmetry (DASM), Asymmetry (ASM) and Differential Causality (DCAU) features are extracted from five frequency bands: δ: 1–3 Hz, θ: 4–7 Hz, α: 8–13 Hz, β: 14–30 Hz, and γ: 31–50 Hz. The data recorded from the same film excerpt are labeled as the same label. The features extracted from EEG usually contain noises which cannot be thoroughly filtered. Therefore, we use linear dynamic system (LDS) approach to filter out the unrelated features for emotion recognition.

As the eye movement data contain different parameters, every eye movement parameter is processed separately. We adopt the same extracted features of eye movement in the work of Lu *et al.* [12] since these features were proven to be effective in emotion recognition. We also apply LDS to filter out unrelated features for eye movement data. The total number of dimension of eye movement features is 33. The details of eye movement features are presented in Table 1.

We use an SVM with linear kernel as a classifier. All the results are obtained by a 5-fold cross validation. The parameter c is searched from 2^{-10} to 2^{9}.

Table 1. Details of extracted features from Eye Movement

Eye movement parameters	Extracted features
Pupil diameter (X and Y)	Mean, standard deviation and DE in four bands: 0–0.2 Hz, 0.2–0.4 Hz, 0.4–0.6 Hz, 0.6–1 Hz
Dispersion (X and Y)	Mean, standard deviation
Fixation duration (ms)	Mean, standard deviation
Blink duration (ms)	Mean, standard deviation
Saccade	Mean, standard deviation of saccade duration (ms) and saccade amplitude (°)
Event statistics	Blink frequency, fixation frequency, fixation duration maximum, fixation dispersion total, fixation dispersion maximum, saccade frequency, saccade duration average, saccade amplitude average, saccade latency average

4 Experiment Results

4.1 Comparison on Features

In this part, we compare the performance of emotion classification on different features. Figure 1(a) shows the classification accuracy for Chinese and French subjects.

We can see that the mean accuracy of Chinese reaches 72.93%, which is much higher than the mean accuracy of French (47.39%). The gap of accuracy between Chinese and French shows that the emotions of Chinese have been stimulated effectively while the emotions of French are relatively difficult to stimulate. The unfamiliar environment may make French subjects feel difficult to relax and immerse in the films. The standard deviation (SD) of Chinese (5.98) is close to the SD of French (6.08), indicating that the individual differences exist on both datasets. We also use two-way analysis of variance to study the statistical significance of nation and features. The p-values for the nation (0.0000), the features (0.0000), and the interaction between nation and features (0.3695) indicates that the nation and features affect the accuracy, but there is no evidence of an interaction effect of the two.

Among different features, DE feature achieves the highest classification accuracy on both datasets, 79.37% with Chinese subjects and 49.65% with French subjects. DE feature gets the lowest SD on Chinese dataset which means DE feature is a relatively stable feature for emotion recognition for Chinese subjects.

4.2 Comparison on Frequency Bands

We also compare the classification accuracy on five non-overlapping frequency bands. The results are shown in Fig. 1(b). The mean accuracy of Chinese subjects achieves 72.92% (SD = 7.23) and that of French is 47.38% (SD = 7.52).

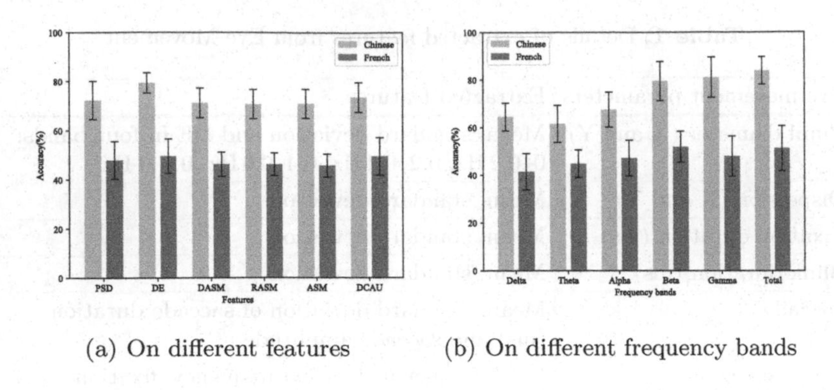

(a) On different features (b) On different frequency bands

Fig. 1. Classification accuracies on different features and bands

By using two-way analysis of variance, the p-values for the nation (0.0000), the frequency bands (0.0000), and the interaction between nation and frequency bands (0.2887) indicates that the nation and the frequency bands affect the accuracy, but there's no interaction between nation and the frequency bands. For Chinese, the performance on higher frequency bands, such as beta and gamma, is better than that of lower frequency bands. The finding is consistent with the existing work [18]. Total frequency band, which means to concatenate all frequency bands together, gets the highest accuracy (83.77%) with regards to Chinese subjects. For French, we find that on beta frequency band the best result (51.89%) is obtained. Unlike Chinese subjects, gamma frequency band has a relatively poor for French subjects performance (47.86%) compared with that of Chinese subjects (80.98%).

4.3 Functional Brain Connectivity Patterns

Figure 2 shows the functional brain connectivity patterns of Chinese and French with three emotions in five frequency bands. There are more connections of Chinese than those of French. It is because that French has a larger number of intersections shared by three emotions, which have been removed from visualization. Here, we choose 0.2 as threshold, which means 20% of total connections have been discarded. We get the highest mean accuracy for Chinese (71.24%) and French (44.25%) when threshold equals to 0.2.

For both Chinese and French, we can observe higher coherence connectivity of frontal lobes in positive emotion on alpha, beta and gamma frequency bands. The connectivity patterns on neutral and negative emotions are relatively similar on beta and gamma frequency bands. For Chinese, we find higher coherence connectivity on temporal and occipital lobes. For French, the higher coherence is found especially on left hemisphere. Watching positive film excerpts, the temporal and occipital sites of Chinese subjects show higher coherence while French subjects show higher coherence at frontal and temporal sites. Watching neutral film excerpts, higher coherence connectivities are located at frontal sites for

Chinese but at occipital sites for French. For both Chinese and French, higher coherence is found on lower frequency bands. However, unlike Chinese subjects, who have a relatively symmetry distribution of connectivities, French are relatively asymmetry and higher coherence connectivities appear in left hemisphere.

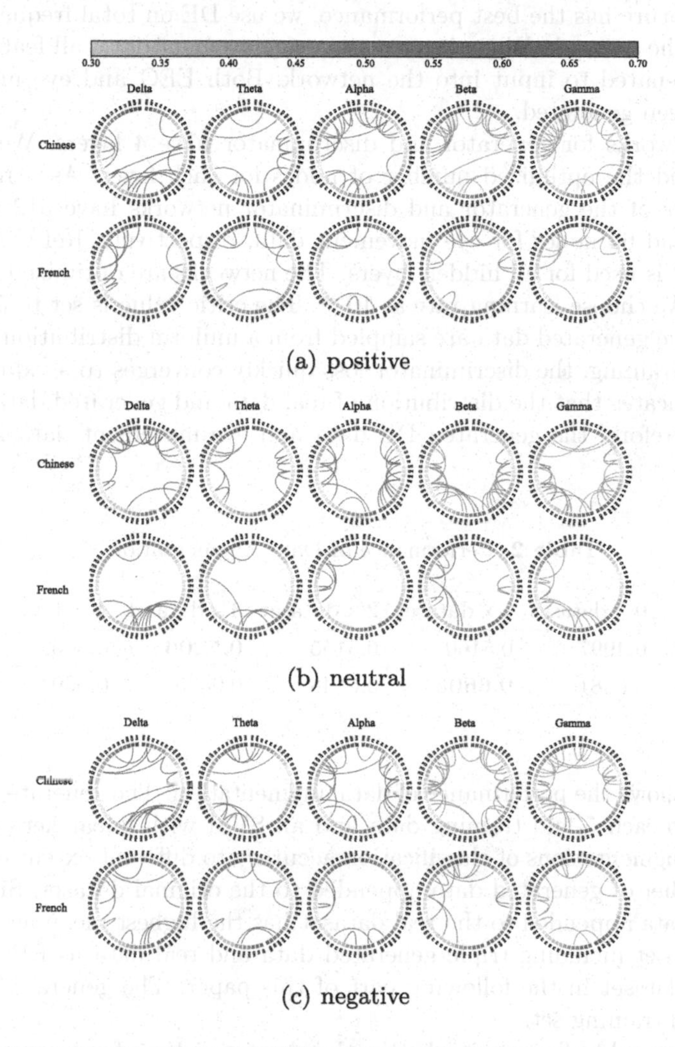

(a) positive

(b) neutral

(c) negative

Fig. 2. The functional brain connectivity patterns for three emotions in five frequency bands with coherence as the connectivity index. The text on each node means the name of electrode. The nodes from top to bottom represent EEG channels from the frontal, temporal, parietal to the occipital lobes. Here, the maps from first row of each emotion represent the results of Chinese and those from second row represent the results of French.

4.4 Multi-modality Fusion

Considering the lower sample rate of eye tracking glass, an STFT with a 4-second non-overlapping window is used to compute both EEG and eye movement features. Because of lack of data, we use CWGAN as data augmentation method. Since DE feature has the best performance, we use DE on total frequency band as input to the network. When it comes to eye movement data, all features have been concatenated to input into the network. Both EEG and eye movements data have been generated.

Both networks for generator and discriminator have 4 layers. We use grid search to find the optimized number of nodes for each layer. As a result, the hidden layers of the generator and discriminator networks have 512 nodes for EEG data and 64 nodes for eye movement data, respectively. ReLU (Rectified Linear Unit) is used for all hidden layers. The networks are optimized by Adam optimizer. We choose learning rate as 10^{-3}. The critic value is set to 5 and λ is set to 10. The generated data are sampled from a uniform distribution $U[-1, 1]$. During the training, the discriminator loss quickly converges to a value close to 0, which indicates that the distribution of real data and generated data are very similar. Therefore, the generated DE data and eye movement data have high quality.

Table 2. Performance of Data Augmentation

	$0 \times$ dataset	$1 \times$ dataset	$2 \times$ dataset	$3 \times$ dataset	$4 \times$ dataset
EEG	0.4997	0.5160	0.5155	**0.5206**	0.5202
Eye	0.6381	**0.6603**	0.6448	0.6595	0.6504

Table 2 shows the performance of data augmentation. The generated data are appended to each 5-fold training data and an SVM with linear kernel is used. There are augmentations of classification accuracy to different extent depending on the number of generated data appended to the original dataset. Since triple generated data appended to the real dataset has the highest mean accuracy, we use the dataset including triple generated data and real data as EEG and eye movement dataset in the following part of this paper. The generated data are only used in training set.

DCCA is used to figure out whether the characteristics of eye movements are complementary with EEG. Each modality is constructed by three full connected layers. We use random search between 50 and 200 to find the optimal number of layer nodes. The learning rate is set to 10^{-3}. Batch size is set to 100 and regulation parameter is set to 10^{-7}. We choose the output dimension of features for each modality as 20.

The mean accuracy by using EEG data only is 55.35% and the mean accuracy by using eye movements only is 60.98%. When we combine two modalities and project them into another feature space with lower dimension, we get an augmentation of classification accuracy to 64.22%. Figure 3 shows the confusion matrices of classification results. From Fig. 3, we have found that eye movement and EEG modalities have complementary characteristics. By using EEG features solely, it's very likely to confuse negative emotion with other two emotions while using eye movements alone shows a better performance. When it comes to discriminate positive emotions, using EEG features solely shows a better performance. After combining two modalities, we find that the negative emotion can be recognized with higher accuracy (64.71%).

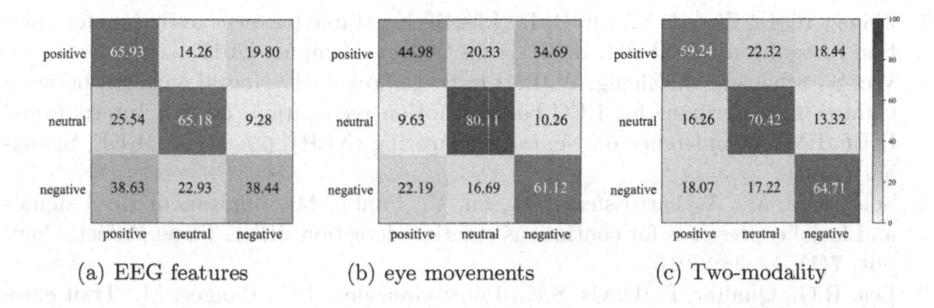

(a) EEG features (b) eye movements (c) Two-modality

Fig. 3. The confusion matrices of classification results by using different features.

5 Conclusions and Future Work

In this paper, we have compared the neural patterns between Chinese and French on a task of recognizing three emotions (positive, neutral and negative). We have found that French has higher mean accuracy on beta frequency band while Chinese tends to perform better on gamma frequency band. The functional brain connectivity patterns indicate the coexistence of similarities and differences of neural patterns between Chinese and French subjects. The results of classification by using DCCA reveal that EEG and eye movement data of French subjects are complementary in discriminating positive and negative emotions.

As future work, we will recruit more number of subjects to participate in the experiments and use different multi-modality fusion methods to investigate the relationship between EEG signals and eye movement data.

Acknowledgements. This work was supported in part by the grants from the National Key Research and Development Program of China (Grant No. 2017YFB100 2501), the National Natural Science Foundation of China (Grant No. 61673266), and the Fundamental Research Funds for the Central Universities.

References

1. Jack, R.E., Garrod, O.G.B., Yu, H., Caldara, R., Schyns, P.G.: Facial expressions of emotion are not culturally universal. Proc. Natl. Acad. Sci. **109**(19), 7241–7244 (2012)
2. Ekman, P., et al.: Universals and cultural differences in the judgments of facial expressions of emotion. J. Pers. Soc. Psychol. **53**(4), 712–717 (1987)

3. Sauter, D.A., Eisner, F., Ekman, P., Scott, S.K.: Cross-cultural recognition of basic emotions through nonverbal emotional vocalizations. Proc. Natl. Acad. Sci. **107**(6), 2408–2412 (2010)
4. Bryant, G., Barrett, H.C.: Vocal emotion recognition across disparate cultures. J. Cogn. Culture **8**(1–2), 135–148 (2008)
5. Elfenbein, H.A., Ambady, N.: On the universality and cultural specificity of emotion recognition: a meta-analysis. Psychol. Bull. **128**(2), 203 (2002)
6. Hutchison, A.N., Gerstein, L.H.: The impact of gender and intercultural experiences on emotion recognition. Revista De Cercetare Si Interventie Sociala **54**, 125 (2016)
7. Zheng, W.-L., Zhu, J.-Y., Lu, B.-L.: Identifying stable patterns over time for emotion recognition from EEG. IEEE Trans. Affect. Comput. (2017)
8. Wu, S., Schaefer, M., Zheng, W.-L., Lu, B.-L., Yokoi, H.: Neural patterns between Chinese and Germans for EEG-based emotion recognition. In: 8th International IEEE/EMBS Conference on Neural Engineering (NER), pp. 94–97. IEEE, Shanghai (2017)
9. Soleymani, M., Asghari-Esfeden, S., Fu, Y., Pantic, M.: Analysis of EEG signals and facial expressions for continuous emotion detection. IEEE Trans. Affect. Comput. **7**(1), 17–28 (2016)
10. Lea, R.G., Qualter, P., Davis, S.K., Pérez-González, J.C., Bangee, M.: Trait emotional intelligence and attentional bias for positive emotion: an eye tracking study. Pers. Individ. Differ. **128**, 88–93 (2018)
11. Bal, E., Harden, E., Lamb, D., Van Hecke, A.V., Denver, J.W., Porges, S.W.: Emotion recognition in children with autism spectrum disorders: relations to eye gaze and autonomic state. J. Autism Dev. Disord. **40**(3), 358–370 (2010)
12. Lu, Y., Zheng, W.L., Li, B., Lu, B.L.: Combining eye movements and EEG to enhance emotion recognition. In: IJCAI 2015, pp. 1170–1176 (2015)
13. Qiu, J.-L., Liu, W., Lu, B.-L.: Multi-view emotion recognition using deep canonical correlation analysis. In: Cheng, L., Leung, A.C.S., Ozawa, S. (eds.) ICONIP 2018. LNCS, vol. 11305, pp. 221–231. Springer, Cham (2018). https://doi.org/10.1007/978-3-030-04221-9_20
14. Wu, X., Zheng, W.-L., Lu, B.-L.: Identifying functional brain connectivity patterns for EEG-based emotion recognition. In: 9th International IEEE/EMBS Conference on Neural Engineering. IEEE, San Francisco (2019)
15. Luo, Y., Lu, B.-L.: EEG data augmentation for emotion recognition using a conditional wasserstein GAN. In: 40th Annual International Conference of the IEEE Engineering in Medicine and Biology Society (EMBC), pp. 2535–2538. IEEE, Honolulu (2018)
16. Gulrajani, I., Ahmed, F., Arjovsky, M., Dumoulin, V., Courville, A.C.: Improved training of Wasserstein GANs. In: Advances in Neural Information Processing Systems, NIPS, Long Beach, pp. 5767–5777 (2017)
17. Schaefer, A., Nils, F., Sanchez, X., Philippot, P.: Assessing the effectiveness of a large database of emotion-eliciting films: a new tool for emotion researchers. Cogn. Emot. **24**(7), 1153–1172 (2010)
18. Zheng, W.-L., Lu, B.-L.: Investigating critical frequency bands and channels for EEG-based emotion recognition with deep neural networks. IEEE Trans. Auton. Ment. Dev. **7**(3), 162–175 (2015)

Time-Frequency Deep Representation Learning for Speech Emotion Recognition Integrating Self-attention

Jiaxing Liu[1], Zhilei Liu[1(✉)], Longbiao Wang[1(✉)], Lili Guo[1], and Jianwu Dang[1,2]

[1] Tianjin Key Laboratory of Cognitive Computing and Application, College of Intelligence and Computing, Tianjin University, Tianjin, China
{jiaxingliu,zhileiliu,longbiao_wang,liliguo}@tju.edu.cn
[2] Japan Advanced Institute of Science and Technology, Nomi, Ishikawa, Japan
jdang@jaist.ac.jp

Abstract. Learning efficient deep representations from spectrogram for speech emotion recognition still represents a significant challenge. Most existing spectrogram feature extraction methods empowered by deep learning have demonstrated great success, but the respective changing information of time and frequency exhibited by the spectrogram is ignored. In this paper, a speech emotion recognition method integrating self-attention is proposed by considering the interactive and respective changing information of time and frequency. To learn the deep representations from spectrogram, a time-frequency convolutional neural network (TFCNN) is proposed at first. After that, a Multi-head Self-attention layer inspired by Transformer proposed by Google is introduced to fuse deep representations more efficiently. Finally, extreme learning machine (ELM) and bidirectional long short term memory (BLSTM) models are adopted as emotion classifiers. Experiments conducted on IEMOCAP dataset demonstrate the effectiveness of our proposed methods showing better visual illustrations and classification results.

Keywords: Speech emotion recognition · Time-frequency · Self-attention

1 Introduction

As the fundamental research of emotion artificial intelligence, speech emotion recognition (SER) has become an active research area [1]. The SER systems consist of two stages, one is feature extraction and the other is classification. Finding effective emotional features representation in feature extraction stage is the key to the success of SER systems [2–4].

The SER methods can be categorized as traditional methods and deep learning methods. In traditional methods, segment features, such as Mel-Frequency Cepstral Coefficients (MFCC) [5], Linear Prediction Cepstral Coefficients (LPCC), prosodic features and the statistics of segment features, perform

T. Gedeon et al. (Eds.): ICONIP 2019, CCIS 1142, pp. 681–689, 2019.
https://doi.org/10.1007/978-3-030-36808-1_74

well in Automatic Speech Recognition (ASR) tasks, but may not be suitable for SER tasks to get satisfactory performance.

In recent years, deep learning methods have gained outstanding performances in vision and speech recognition. Deep neural networks (DNNs) can extract emotion related features from a large amount of data, which make deep learning based SERs achieve competitive results. The high-level features learned by DNN can overcome the pre-defined limitations of hand-crafted features used in traditional methods. Han et al. [6] proposed a model based on DNN to obtain the emotion state probability distribution, and the extreme learning machine (ELM) was used as the classifier; Satt et al. [7] proposed the famous model which uses CNN to learn emotional features directly from spectrogram, and followed by a BLSTM to learn the contextual information; Guo et al. [8] improved Satt's model, employing ELM instead of the complicated structure BLSTM, and got an outstanding result. The models mentioned above have been regarded as the state-of-the-art (SOA) models in the field of speech emotion recognition. Among them CNN shows more powerful performance in representation learning and the usage is also the same as image processing. The traditional CNN only considers the interactive information in the receptive field. However, the respective changing information of time and frequency is also highly related to speech emotions.

Attention mechanism was first used in machine translation to solve the bottleneck of information loss [9]. This attention model and its variants were quickly introduced to various research fields such as computer vision, neutral language processing (NLP) [10] and automatic speech recognition (ASR). Attention mechanism was also introduced in SER, Li et al. [11] used the second-order attention as a pooling layer instead of max-pooling and average-pooling layer; Gorrostieta et al. [12] added attention mechanism to emphasize salient regions of the audio clip. The success achieved in those research fields demonstrated the effectiveness of attention mechanism. Among those attention models used in various research fields, the most successful one was Transformer [13] proposed by Google.

To study both interactive and respective changing information between time and frequency in spectrogram, we propose a time-frequency deep representation learning method which could be called Time-Frequency CNN (TFCNN) for SER with self-attention as shown in Fig. 1. Firstly, three groups of filters with different shapes are designed to extract three kinds of representations, including time, frequency interactive and respective changing information from spectrogram directly. Secondly, the extracted three information representations are fed to properly designed CNNs to learn deep representations, respectively. Then, the three deep representations are further concatenated into a fusion representation. Thirdly, inspired by Transformer, Multi-head Self-attention is introduced to explore relations between three deep representations to learn a more efficiently fusion representation. Finally, the new fusion representation is sent to a classifier to get the classification result.

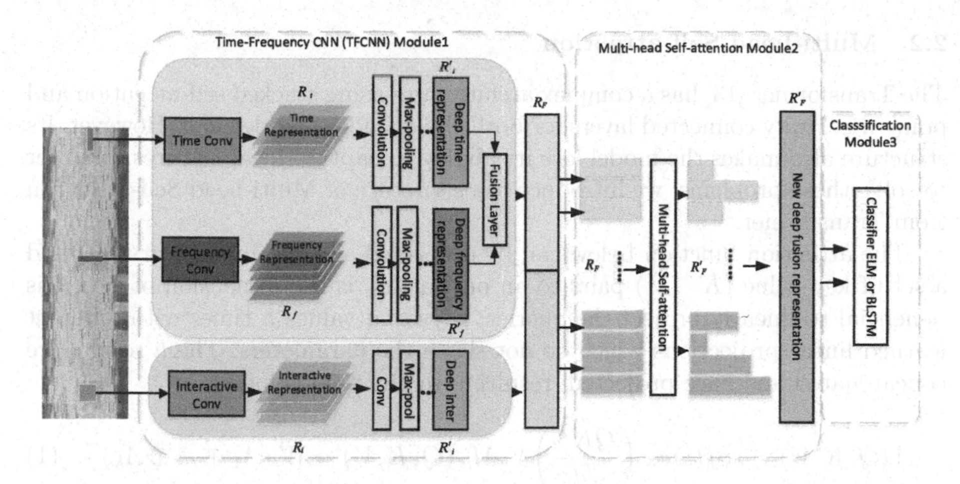

Fig. 1. Time-Frequency CNN (TFCNN) integrating self-attention.

2 Time-Frequency CNN (TFCNN) Integrating Self-attention

2.1 Time-Frequency CNN (TFCNN)

To study the interactive and respective changing information of time and frequency from spectrogram, a TFCNN representation learning module $Module1$ is proposed as shown in Fig. 1. At first, three groups of specially designed filters are introduced, the first one is along the time dimension of spectrogram to capture time changing information R_t and frequency dimension is set to 1; the second one is along the frequency dimension of spectrogram to capture frequency information R_f and time dimension is set to 1. These two groups are designed for minimizing the interaction of time and frequency to get only one kind of information. The third group is the traditional rectangle filter to learn the interactive representation R_i. Secondly, the time changing representation R_t, frequency changing representation R_f and the interactive representation R_i are fed to three CNNs to learn deep time changing representation R'_t, deep frequency changing representation R'_f and deep interactive representation R'_i. Thirdly, we concatenate R'_t, R'_f and R'_i to get the fusion representation R_F. Considering the difference of three kinds of representations, three CNNs are designed to fit their own characteristics. The respective changing representations of time and frequency are concatenated at deep level rather than after first layer [11] which treats time and frequency representation equally. Researching on deep interactive and respective changing representations carefully, there still exists two problems. One is the lack of adaptive adjustment of linear connection between the three deep representations. The other is ignoring the inner relation in fusion representation. To address this two problems, we introduce the solution in the following part.

2.2 Multi-head Self-attention

The Transformer [13] has a complex architecture using stacked self-attention and point-wise, fully connected layers for both the encoder and decoder. However, its structure also makes the model lose its ability to capture local features. In order to solve these problems, we introduce the structure of Multi-head Self-attention from Transformer.

The attention function below can be described as mapping a query (Q) and a set of key-value ($K-V$) pairs to an output. d_k is the dimension of key. it is beneficial to linearly project the queries, keys and values h times with different learned linear projections which do not share the parameters. The h results are concatenated and once projected, resulting in the final values.

$$A\left(Q,K,V\right)=softmax\left(\frac{QK^T}{\sqrt{d_k}}\right)V, M_h\left(Q,K,V\right)=W\left(A_1\oplus...\oplus A_h\right) \quad (1)$$

Where the A_i means ith attention, $\oplus$ means concatenation and W means parameter matrices.

When $Q==K==V==X$, the mechanism is called as self-attention. X represents the input of the fusion representation R_F. The fusion representation R_F will be computed to a new representation R'_F by Multi-head Self attention which is $Module2$ proposed as shown in Fig. 1.

$$R'_F=M_h\left(R_F\right) \quad (2)$$

This Multi-head Self-attention is introduced to enhance adaptation and learn the inner relations in fusion representation. The following experiments also prove the effectiveness of the new fusion representation R'_F.

2.3 Classification

In this paper, we choose three state of the art (SOA) models [6–8] as the benchmarks. In order to compare the classification results with them fairly and verify the effectiveness of the new representation R'_F for SER, we use two classifiers, namely ELM and BLSTM which are as the one used in the SOA models.

3 Experiments

3.1 Experimental Setup

Dataset: Interactive Emotional Dyadic Motion Capture database (IEMOCAP) [14] is a well-known database which contains about 12 h audiovisual data performed by 10 skilled actors. We only use the audio data. The 5531 utterances are as Atypical Affect Challenge selected from IEMOCAP. It consists of four emotion categories: Neutrality, Anger, Sadness and Happiness. According to the recording scenarios, the data can be divided into improvised speech section and scripted speech section. The scripted section may lead to a bad influence on the results [15]. However, considering the richness of scenarios and robustness of the model, all improvised and scripted sections in 5531 utterances are used.

Preprocessing: Each speech signal of the 5,531 utterances in dataset is sampled at 16 HKz. The length of each segment is thus an open problem. Fortunately, some researchers have found that more than 250 ms speech segment can contain efficient and effective emotional information [16]. The utterance is converted into frames using a 25-ms window with an overlapping of 15-ms. The size of each segment is set to 25 frames. The time of each segment is 265 ms. The input spectrogram has the following $time \times frequency : 32 \times 129$.

Experimental Setup: In this paper, we focus our attention on performance of TFCNN and Multi-head Self-attention to validate the effectiveness of our proposed method. Therefore, we select three SOA models, DNN+ELM [6], CNN+ELM [8] and CNN+BLSTM [7] as the benchmarks. We also design two groups of ablation studies with different classifiers.

Benchmarks: We choose three SOA models as the benchmark algorithms. These experiments are designed to evaluate the effectiveness of proposed TFCNN followed by ELM and BLSTM as classifiers.

(1) *DNN+ELM* [6]: This experiment is design to show the performance of DNN. Input of DNN is a vector of low-level descriptors (384 dimensions LLDs) which are extracted by openSMILE tools.
(2) *CNN+ELM* [8]: This experiment is set to evaluate the performance of CNN. The input is spectrogram mentioned above. The structure of CNN contains three convolutional layers and two max-pooling layers.
(3) *CNN+BLSTM* [7]: This structure of CNN is same as in experiment (2) and the classifier is BLSTM. The BLSTM has two hidden layers each with 64 units.

Ablation studies: We design two groups ablation studies to evaluate the effectiveness of TFCNN and Multi-head Self-attention with two different classifiers.

(4) *TFCNN+BLSTM*: This experiment is design to evaluate our proposed TFCNN representation learning method comparing with experiment (3) with the same classifier BLSTM. The proposed TFCNN consists of two parts as shown in Fig. 1. The first part time representation: the first convolutional layer has 32 filters with size of 8×1 to catch time changing representation followed a max-pooling with size of 2×1; The first part frequency representation: the first convolutional layer has 32 filters with size of 1×8 to catch frequency changing representation followed a max-pooling with size of 1×2. The second part is interactive representation: the structure of CNN is same as in experiment (3).
(5) *TFCNN_att+BLSTM*: This experiment is design to evaluate the effectiveness of Multi-head Self-attention comparing with experiment (4). The structure of TFCNN is same as in experiment (4) and with a 16-head Self-attention.
(6) *TFCNN+ELM*: This experiment is design to evaluate the effectiveness of proposed TFCNN comparing with experiment (2). The structure of TFCNN

is same as in experiment (4) and the classifier is ELM which is same as in experiment (2).

(7) *TFCNN_att+ELM*: This experiment is design to evaluate the effectiveness of Multi-head Self-attention comparing with experiment (6). The structure of TFCNN is same as in experiment (6) with a 16-head Self-attention.

3.2 Experiments Results

We design two groups of comparative experiments to validate the effectiveness of our proposed method. The first group uses visualization analysis and the second group uses emotional classification results.

Visualization Analysis: Four deep representations are considered, respectively the output of DNN in experiment (1), output of CNN in experiment (2), output of TFCNN in experiment (4) and output of TFCNN_att in experiment (5). The t-Distributed Stochastic Neighbor Embedding (t-SNE) [17] is introduced to visualize the four deep representations in Fig. 3. Compared with Fig. 2(a), the points which represent Anger and Sadness in Fig. 2(b) can be distinguished more easily. The points which represent Neutrality and Happiness have similar performance. Compared with Fig. 2(b), all the four emotion points have better distinguishment which means the deep representations extracted by TFCNN have better performance. Compared with Fig. 2(c), the points which represent Neutrality, Anger and Sadness in Fig. 2(d) can be distinguished more easily, but the Happiness points spread across the other three emotion points. The performance of visualization analysis shows that the attention leads to some side effects. We will discuss this phenomenon in the following part.

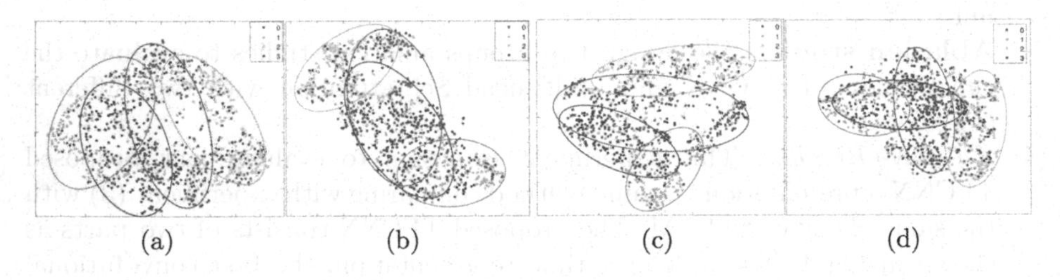

Fig. 2. The t-SNE visualization of four deep representations. (a): DNN, (b): CNN, (c): TFCNN, (d): TFCNN_att. 0: Neutrality, 1: Anger, 2: Sadness, 3: Happiness

Classification Results: In order to quantitatively evaluate the performance of the proposed model, the classification results are provided in Table 1. Compared with the DNN+ELM, the CNN+ELM shows a better accuracy; Compared with the SOA model CNN+BLSTM, the improvements of the proposed TFCNN+BLSTM and TFCNN_att+BLSTM are 3.16% and 5.33%; Compared with the SOA CNN+ELM, the improvements of the proposed TFCNN+ELM

Table 1. The classification accuracy

	Model	Accuracy(%)
Benchmarks	DNN+ELM [6]	62.66
	CNN+ELM [8]	66.09
	CNN+BLSTM [7]	60.49
Ablation studies	TFCNN+BLSTM	63.65
	TFCNN_att+BLSTM	**65.82**
	TFCNN+ELM	67.27
	TFCNN_att+ELM	**68.99**

and TFCNN_att+ELM are 1.18% and 2.90%. The comparisons in Table 1 demonstrate the effectiveness of the proposed TFCNN integrating Self-attention.

Observing comparisons above carefully, we find that not only the feature extraction networks influence the classification accuracy, but also the classifiers and the Multi-head Self-attention have certain impacts on the classification results. To further explore this question, four confusion matrices which represent TFCNN+BLSTM, TFCNN_att+BLSTM, TFCNN+ELM, TFCNN_att+ELM corresponding to Table 1 are shown in Fig. 3. In comparisons between Fig. 3(a) with (c), Fig. 3(b) with (d), ELM is more sensitive to Neutrality and Anger, and BLSTM is more sensitive to Happiness. But accuracy of Happiness in both classifiers gets worse when attention mechanism is added through comparing Fig. 3(a) with (b), Fig. 3(c) with (d). The results verify the phenomenon we find in visualization analysis. However, error rates of Neutrality, Anger and Sadness get lower, especially these three emotions classified into Happiness. The measurements present in Fig. 3 and Table 1 demonstrate the attention mechanism's ability to improve overall accuracy, although it is not sensitive to specific one emotion.

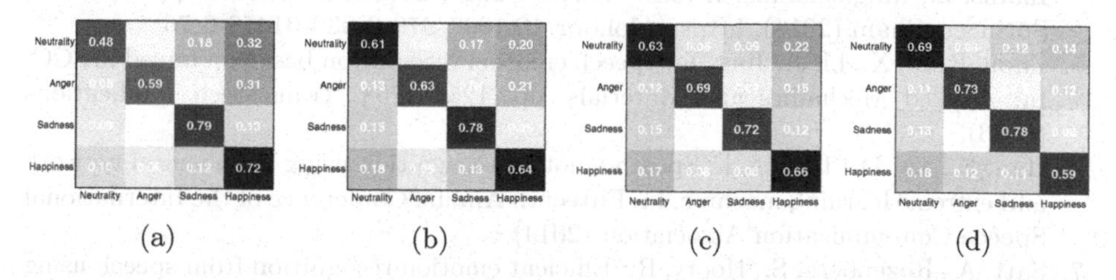

Fig. 3. The confusion matrices of four classification results. (a): TFCNN+BLSTM, (b): TFCNN_att+BLSTM, (c): TFCNN+ELM, (d): TFCNN_att+ELM.

4 Conclusions

In this paper, we studied the interactive and respective changing information of time and frequency from spectrogram and proposed a time-frequency deep representation learning method integrating self-attention for SER. The effectiveness of the proposed method has been verified under both representation visualization and classification results on IEMOCAP. Compared with SOA models, the classification accuracies achieve 65.82% and 68.99% with absolute increments about 5.3% and 3.0%. The proposed model shows high sensitivity to all four emotions, highlighting great potential for the fusion of traditional methods and deep learning methods. In the future, we plan to investigate the performance of the proposed model on different databases. Furthermore, multi-task deep representation learning method will also be studied to improve the performance of SER task.

Acknowledgments. This work was supported in part by the National Natural Science Foundation of China under Grant 61771333 and the Tianjin Municipal Science and Technology Project under Grant 18ZXZNGX00330.

References

1. Ramakrishnan, S., El Emary, I.M.: Speech emotion recognition approaches in human computer interaction, pp. 937–940. Kluwer Academic Publishers (2013)
2. Guo, L., Wang, L., Dang, J., Zhang, L., Guan, H., Li, X.: Speech emotion recognition by combining amplitude and phase information using convolutional neural network. In: Proceedings of the Interspeech, pp. 1611–1615 (2018)
3. Guo, L., Wang, L., Dang, J., Liu, Z., Guan, H.: Exploration of complementary features for speech emotion recognition based on kernel extreme learning machine. IEEE Access **7**, 75798–75809 (2019)
4. Zhang, L., Wang, L., Dang, J., Guo, L., Yu, Q.: Gender-aware CNN-BLSTM for speech emotion recognition. In: Kůrková, V., Manolopoulos, Y., Hammer, B., Iliadis, L., Maglogiannis, I. (eds.) ICANN 2018. LNCS, vol. 11139, pp. 782–790. Springer, Cham (2018). https://doi.org/10.1007/978-3-030-01418-6_76
5. Zhou, P., Li, X., Li, J., Jing, X.: Speech emotion recognition based on mixed MFCC. In: Applied Mechanics and Materials, pp. 1252–1258. Trans Tech Publications (2013)
6. Han, K., Yu, D., Tashev, I.: Speech emotion recognition using deep neural network and extreme learning machine. In: Fifteenth Annual Conference of the International Speech Communication Association (2014)
7. Satt, A., Rozenberg, S., Hoory, R.: Efficient emotion recognition from speech using deep learning on spectrograms. In: Proceedings of the Interspeech, pp. 1089–1093 (2017)
8. Guo, L., Wang, L., Dang, J., et al.: A feature fusion method based on extreme learning machine for speech emotion recognition. In: IEEE International Conference on Acoustics, Speech and Signal Processing (ICASSP), pp. 2666–2670. IEEE (2018)
9. Bahdanau, D., Cho, K., Bengio, Y.: Neural machine translation by jointly learning to align and translate. Comput. Sci. (2014)

10. Young, T., Hazarika, D., Poria, S., Cambria, E.: Recent trends in deep learning based natural language processin. IEEE Comput. Intell. Mag. **13**, 55–75 (2018)
11. Li, P., Song, Y., McLoughlin, I., Guo, W., Dai, L.: An attention pooling based representation learning method for speech emotion recognition. In: Proceedings of the Interspeech 2018, pp. 3087–3091 (2018)
12. Gorrostieta, C., et al.: Attention-based sequence classification for affect detection. In: Proceedings of the Interspeech 2018, pp. 506–510 (2018)
13. Vaswani, A., et al.: Attention is all you need. In: NIPS, pp. 5998–6008 (2017)
14. Busso, C., et al.: IEMOCAP: interactive emotional dyadic motion capture database. Lang. Resour. Eval. **42**, 335 (2008)
15. Ma, X., Wu, Z., Jia, J., Xu, M., Meng, H.: Emotion recognition from variable-length speech segments using deep learning on spectrograms. In: Proceedings of the Interspeech 2018, pp. 3683–3687 (2018)
16. Kim, Y., Provost, E.M.: Emotion classification via utterance-level dynamics: a pattern-based approach to characterizing affective expressions. In: IEEE ICASSP, pp. 3677–3681 (2013)
17. Maaten, L.V.D.: Accelerating t-SNE using tree-based algorithms. J. Mach. Learn. Res. **15**, 3221–3245 (2014)

Sparse Graphic Attention LSTM for EEG Emotion Recognition

Suyuan Liu, Wenming Zheng(✉), Tengfei Song, and Yuan Zong

Southeast University, Nanjing, China
{syl,wenming_zheng,songtf,xhzongyuan}@seu.edu.cn

Abstract. In this paper, a novel multichannel EEG emotion recognition method based on sparse graphic attention long short-term memory (SGA-LSTM) is proposed. The basic idea of SGA-LSTM is to adopt graph structure modeling EEG signals to enhance the discriminative ability of EEG channels carrying more emotion information while alleviate the importance of the EEG channels carrying less emotion information. To this end, we employ two graphic branches. One branch generates global features reflecting the intrinsic relationship between EEG channels and the other generates an attention vector guiding the global features to focus on specific EEG channels. Researches on brain emotion show that different brain regions may be related to different brain functions and the contribution of each EEG channel to one specific brain function are possibly sparse such that ℓ_1-norm penalty is applied. Extensive experiments are conducted on our dry electrodes EEG database and MPED database. The experimental results show that the proposed method is superior to the state-of-the-art methods.

Keywords: EEG emotion recognition · SGA-LSTM · Graph convolution · Attention mechanism · Sparse constraint

1 Introduction

EEG emotion recognition has drawn an increasing attention recently due to its potential applications to human-machine interaction. Generally, EEG emotion recognition contains two parts: feature extraction and recognition methods.

Basically, EEG feature can be divided into three categories, i.e., time domain (Hjorth, HOC), frequency domain(PSD) and time-frequency domain(STFT, HHS) [6]. Before the EEG feature extraction, EEG signals are usually decomposed into several frequency bands, e.g., δ (1–3 Hz), θ (4–7 Hz), α (8–13 Hz), β (14–30 Hz) and γ (30–50 Hz) [20]. To cope with EEG emotion recognition, deep learning methods, especially long short-term memory (LSTM) [5], had shown to be powerful and had been widely adopted in recently years [7,15]. In [7], Li et al. proposed the method of combining both CNN and LSTM for EEG emotion recognition. In [15], Tang et al. proposed a bimodal-LSTM model for EEG emotion recognition. Recently, Song et al. [14] provide a novel DGCNN method to

T. Gedeon et al. (Eds.): ICONIP 2019, CCIS 1142, pp. 690–697, 2019.
https://doi.org/10.1007/978-3-030-36808-1_75

explore the relationship between EEG channels and brain functions and achieve a good performance on EEG emotion recognition.

Neuroscience research has proved that human emotion is closely related to some brain subregions [9]. For EEG signals, not all channels are helpful to recognize emotion states. Although there have been many algorithms designed for channel selection, the relationships between EEG channels are rarely considered due to the imperceptible neuromechanism, which are significant for EEG emotion recognition. Based on the above considerations, motivated by [14] and [7,15], we propose to combine graphic model and LSTM [5] to deal with EEG emotion recognition. Additionally, inspired by [17], we provide a graph-based attention structure to produce an attention vector to select EEG channels for extracting more discriminative features. Moreover, we suppose the contribution of different EEG channels may be possibly sparse according to [2]. So ℓ_1-norm penalty of the attention vector is proposed to adopt on loss function, so as to obtain sparse weight parameters for measuring the contributions of different EEG channels. Take into account the above considerations, we propose a novel sparse graphic attention long short-term memory method (SGA-LSTM).

To evaluate the proposed method, extensive experiments are conducted on our DEED database and MPED database [13]. The experimental results demonstrate that SGA-LSTM is superior to state-of-the-art methods.

2 Proposed Approach

SGA-LSTM consists of a graph attention structure and LSTM, as illustrated in Fig. 1. The graph attention structure containing trunk branch and attention branch, is applied for extracting discriminative features. LSTM is adopted for modeling the spatial information in EEG channels to futher improve the performance.

2.1 Graph Attention Structure

Graph attention structure consists of two branches, i.e. trunk branch and attention branch, which are both based on graph convolution layers. The trunk branch is employed to extract global features. The attention branch is adopted to select useful channels.

Graph Convolution Layer: Let $G = \{\mathcal{V}, \mathbf{A}\}$ denote a directed and weighted graph, where $\mathcal{V}$ is a vertex set of N EEG channels, $\mathbf{A} \in R^{N \times N}$ is the adjacency matrix of G with entries $a_{ij} \geq 0$ denoting the degree of relation from channel v_i to v_j. For graph filtering, let $g(\mathbf{A})$ be a filtering function, signal $\tilde{\mathbf{x}}$ filtered from $\mathbf{x}$ by $g(\mathbf{A})$ can be expressed as: $\tilde{\mathbf{x}} = g(\mathbf{A})\mathbf{x}$, where $x \in R^{N \times B}$ is the EEG feature, and B is the number of feature bands. To capture the multi-hop information, we present K-order polynomial filter, which has following form: [3],

$$\tilde{\mathbf{x}} = g(\mathbf{A})\mathbf{x} = \sum_{k=0}^{K-1} \theta_k \varphi_k(\mathbf{A})\mathbf{x} \tag{1}$$

where θ_k is polynomial coefficients, $\varphi_k(\mathbf{A}) = \mathbf{A}^k$ is the K-hop filtering, and $\tilde{\mathbf{x}}$ has the size of $N \times (B*K)$. Generally, $\mathbf{A}$ can be normalized by $\mathbf{A}^{norm} = \mathbf{D}^{-\frac{1}{2}}\mathbf{A}\mathbf{D}^{-\frac{1}{2}}$, where $\mathbf{D} \in R^{N \times N}$ is a diagonal matrix with entries $d_{ii} = \sum_j a_{ij}$.

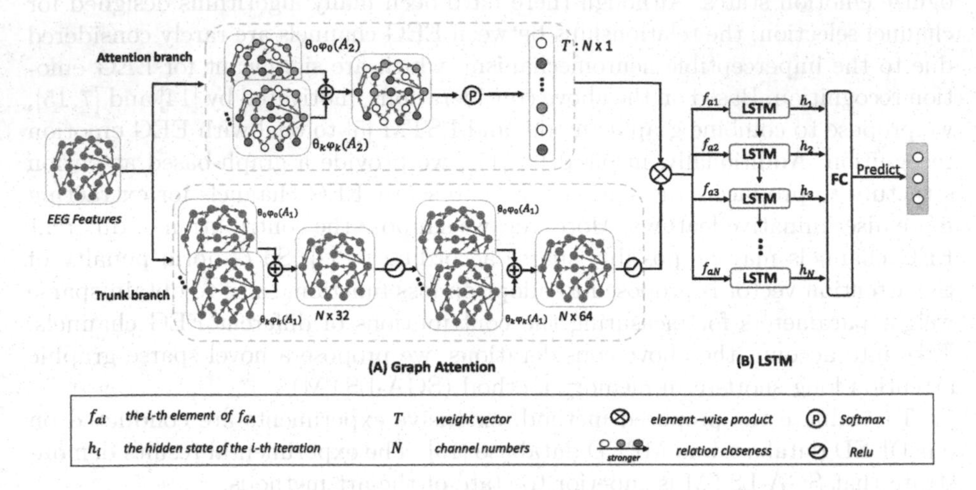

Fig. 1. The framework of SGA-LSTM method.

After graph filtering, a matrix $W \in R^{(B*K) \times O}$ is used for dimension transformation, where O is the expected output size.

Graph Attention: The trunk branch adopts two graph convolution layers with output size of $N \times 32$ and $N \times 64$ respectively. Let $f_{GCN}(\mathbf{X}, \mathbf{A_1})$ denote the output of the trunk branch, it can be calculated as following:

$$f_{GCN}(\mathbf{X}, \mathbf{A_1}) = \Phi(\Phi(\mathbf{X}, \mathbf{A_1}), \mathbf{A_1}) \tag{2}$$

where $\mathbf{A_1}$ denotes the adjacency matrix used in the trunk branch, $\mathbf{\Phi}(\cdot)$ denotes the graph convolution procedure formulated by Eq. (1). Relu [11] is adopted after each graph convolution layer, to increase the nonlinearity and make the output of graph convolution layers non-negative.

The attention branch adopts one graph convolution layer to generate an attention vector $\mathbf{T} \in R^{N \times 1}$, which is formulated as following:

$$\mathbf{T} = \Phi(\mathbf{X}, \mathbf{A_2}) = [\tau_1, \cdots, \tau_n]^T \tag{3}$$

where $\mathbf{A_2}$ denotes the adjacency matrix used in the attention branch, and τ_i indicates the contribution of the $i\text{-}th$ channel in the emotion recognition. Moreover, softmax is adopted on $\mathbf{T}$ to generate a normalized attention vector $\tilde{\mathbf{T}}$.

We obtain the output of graph attention structure by weighting the graph convolution results of each EEG channel with the corresponding weight parameters in attention vector. To this end, we expand $\tilde{\mathbf{T}}$ to a diagonal matrix

$\text{diag}(\tilde{\mathbf{T}}) \in R^{N\times N}$. Let f_{GA} denote the output of the graph attention, the weighted procedure can be formulated as following:

$$f_{GA} = \text{diag}(\tilde{\mathbf{T}}) \cdot f_{GCN}(\mathbf{X}, \mathbf{A_1}) \tag{4}$$

2.2 Spatial LSTM

The use of LSTM in the SGA-LSTM framework aims to capture the additional emotional features produced by the spatial topographic distribution of the EEG channels. Hence, we take the output of graph attention, i.e., $f_{GA} = [f_{a1}^T, \dots, f_{an}^T]$ as the input of LSTM.

Let i_t, g_t, c_t and o_t denote input gate, forget gate, cell activation and output gate, respectively, of LSTM. Then, they can be iteratively calculated via the following formulations:

$$\begin{cases} i_t = \sigma(W_{xi} f_{at} + W_{hi} h_{t-1} + W_{ci} c_{t-1} + b_i), \\ g_t = \sigma(W_{xg} f_{at} + W_{hg} h_{t-1} + W_{cg} c_{t-1} + b_g), \\ c_t = g_t c_{t-1} + i_t \tanh(W_{xc} f_{at} + W_{hc} h_{t-1} + b_c), \\ o_t = \sigma(W_{xo} f_{at} + W_{ho} h_{t-1} + W_{co} c_t + b_o), \\ h_t = o_t \tanh(c_t), \end{cases} \tag{5}$$

where σ denotes the logistic sigmoid function, h_t is the hidden vector, W_{xi}, W_{hi}, W_{ci}, W_{xg}, W_{hg}, W_{cg}, W_{xc}, W_{hc}, W_{xo}, W_{ho} and W_{co} are model parameters. The hidden layer $H = [h_1, \dots, h_n]$ is served as the input of the fully connected layer.

2.3 Optimization of SGA-LSTM

The loss function of SGA-LSTM is formulated as the following one:

$$Loss = \Psi(I, I^p) + \lambda \|\Theta\|_2, \tag{6}$$

where $\Psi(I, I^p)$ denotes cross entropy of predicted label I^p with ground truth label I, Θ denotes all trainable parameters, and λ is a trade-off parameter.

To obtain sparse weight parameters for measuring the contributions of different EEG channels, we impose a ℓ_1-norm penalty [12] of $\mathbf{T}$ onto the loss function of (6), resulting in the following regularized loss function:

$$Loss = \Psi(I, I^p) + \lambda \|\Theta\|_2 + \mu \|\mathbf{T}\|_1, \tag{7}$$

where μ is another trade-off parameter.

To learn the optimal parameters Θ, back propagation (BP) is adopted to update the network parameters. To this end, the partial derivatives of the loss function with respect to $\mathbf{A_1}$ and $\mathbf{T}$ are calculated respectively, i.e. $\partial Loss/\partial \mathbf{A_1}$ and $\partial Loss/\partial \mathbf{T}$. After calculating the partial derivatives of the loss function with respect to $\mathbf{A_1}$ and $\mathbf{T}$, we can update them by using the following rules:

$$\mathbf{A_1} = (1-\eta)\mathbf{A_1} + \eta \frac{\partial Loss}{\partial \mathbf{A_1}}, \; \mathbf{T} = (1-\eta)\mathbf{T} + \eta \frac{\partial Loss}{\partial \mathbf{T}},$$

where η denotes the learning rate.

3 Experiments and Results

In this part, we will give a brief introduction on DEED and MPED database, the implement details and the experimental results.

3.1 Database

DEED: We build an emotional EEG database at the Southeast University by using dry electrodes (DEED) to evaluate our proposed method. DEED contains 15 subjects with 7 males and 8 females aging from 22 to 29 years old. We collect 15 trails of EEG data for each subject, which contains three emotional states averagely, i.e, positive, neutral and negative emotion.

MPED: MPED [13] contains 23 subjects with 10 males and 13 females aging from 18 to 24 years old. For each subject, 28 trails data containing seven emotional states averagely, i.e., joy, funny, anger, disgust, fear, sadness and neutrality, are collected. In our experiments, we only use the EEG data to evaluate our method, obeying the protocols proposed in [13] strictly.

3.2 Implement Details

The raw EEG feature is decomposed into five frequency bands using the method proposed in [20]. EEG features are extracted according to each frequency band. Then five frequency bands of each kind of features are concatenated for model evaluation. For DEED, Hjorth, HOC, PSD, STFT and HHS features are chosen for both subject-dependent and subject-independent experiments. While for MPED, only subject-dependent experiments using STFT feature are conducted, for being consistent with former studies. In our experiments, the number of EEG channels N is set to 18 for DEED and 62 for MPED. For the model part, the order of graph convolution K is set to 3, the hidden layers of LSTM is set to 64. For the trade-off parameter of the attention vector in loss function, i.e., μ, is searched in range of [1e−4, 5].

3.3 Experiments

Subject-Dependent Experiments on DEED: We adopt five-folder cross-validation experimental strategy in this experiment. Experiments on state-of-the-art methods like SVM [1], CCA [16], GSCCA [8] and DGCNN [14] are conducted for comparison purpose. In addition, experiments using LSTM and G-LSTM (a simplified SGA-LSTM by removing the attention branch) are also conducted for ablation study.

From Table 1, we can see the following points. Among these EEG features, time domain features (i.e. Hjorth and HOC) perform better while frequency domain feature (i.e. PSD) achieves the least accuracies, which indicates frequency feature is inferior to other features. Besides, the LSTM-based methods achieve much higher accuracies than the others in most cases. This is very

Table 1. The results on DEED for subject-dependent experiments. G-LSTM means removing the attention branch from the SGA-LSTM method.

Method	Hjorth ACC/STD(%)	HOC ACC/STD(%)	PSD ACC/STD(%)	STFT ACC/STD(%)	HHS ACC/STD(%)
SVM	79.14/19.18	80.36/20.37	48.73/11.01	66.42/18.22	67.42/16.80
CCA	82.16/19.04	59.90/22.36	46.33/09.50	73.27/18.81	72.58/15.92
GSCCA	55.57/16.50	44.72/18.52	38.38/08.33	52.36/14.57	52.05/12.41
DGCNN	77.17/16.31	43.71/13.77	54.96/10.81	81.79/15.32	84.25/15.54
LSTM	86.90/10.40	86.50/12.68	54.27/09.69	84.44/12.04	83.58/11.19
G-LSTM	89.21/09.59	89.28/11.29	55.78/09.22	85.47/12.02	86.95/11.66
SGA-LSTM	**90.19/08.91**	**90.38/10.70**	**56.35/09.98**	**86.91/11.22**	**88.01/10.24**

likely because LSTM is advantageous to capture the additional discriminative information from the spatial relationships among the various EEG channels. Moreover, among the LSTM-based methods, G-LSTM performs better than LSTM, indicating the superiority of catching the intrinsic connections between EEG channels by graphic structure. SGA-LSTM achieves higher accuracies than G-LSTM, which demonstrates the use of attention branch is useful to improve the performance, indicating the rationality of our method. Additionally, the standard deviation is high in these experiments, which owing to the great difference between individuals.

Subject-Independent Experiments on DEED: We adopt leave-one-subject-out (LOSO) strategy in this experiment. Experiments using SA [4], TKL [10], DGCNN [14], LSTM, G-LSTM and SGA-LSTM have been conducted.

Table 2. The results on DEED for subject-independent experiments.

Method	Hjorth ACC/STD(%)	HOC ACC/STD(%)	PSD ACC/STD(%)	STFT ACC/STD(%)	HHS ACC/STD(%)
SA	59.70/07.22	58.56/06.73	47.08/03.27	63.80/11.31	67.38/10.08
TKL	57.70/13.89	54.37/08.97	45.66/03.06	61.42/10.31	63.13/10.22
DGCNN	55.61/10.79	49.60/14.56	45.88/05.92	60.83/11.07	54.58/10.92
LSTM	61.58/14.19	61.14/11.9	47.23/05.21	69.53/08.00	66.35/09.26
G-LSTM	62.80/10.26	62.17/12.10	47.90/05.39	70.46/08.30	68.10/08.54
SGA-LSTM	**63.93/11.57**	**63.40/11.28**	**48.00/05.37**	**72.14/07.05**	**70.65/10.95**

From Table 2, we can observe the following points. Among these EEG features, time-frequency domain features (i.e. STFT and HHS) perform better, which is different from that in subject-dependent experiments. This may owing to that emotion changes quite different for different subjects, thus time domain features can't capture the generality well for subject-dependent experiments. Moreover, the LSTM-based methods perform better than the other methods. And SGA-LSTM method still achieves the highest accuracies among these methods.

Subject-Dependent Experiments on MPED: In this experiment, we obey the protocol proposed in [13]. Since data of each categories are imbalanced for protocol two, f1 score is calculated for evaluation. Experiments on SVM [1], DBN [19], STRNN [18], DGCNN [14], A-LSTM [13], LSTM, G-LSTM and SGA-LSTM are conducted with these three protocols.

Table 3. The results on MPED for subject-dependent experiments.

Method	Protocol one ACC/STD(%)	Protocol two ACC/F1(%)	Protocol three ACC/STD(%)
SVM	59.86/16.29	57.06/24.43	31.14/08.06
DBN	65.83/13.20	65.95/59.19	29.26/09.19
STRNN	65.38/13.20	66.84/60.57	35.64/09.57
DGCNN	71.13/15.77	68.02/61.11	36.92/12.78
A-LSTM	72.93/13.19	71.57/67.74	38.74/07.75
LSTM	72.09/14.94	71.92/65.12	38.55/08.43
G-LSTM	73.79/12.71	72.20/66.64	39.33/11.41
SGA-LSTM	**74.74/12.46**	**73.00/67.48**	**40.69/11.12**

Table 3 shows the results, we can observe that, among the three protocols, the LSTM-based methods achieve higher accuracies than the other methods. Besides, G-LSTM performs better than LSTM, and SGA-LSTM achieves the highest accuracies, which is consistent with the performance in experiments on DEED.

4 Conclusion and Discussion

In this study, a novel method SGA-LSTM was proposed for EEG emotion recognition. Extensive experiments had been conducted on both DEED and MPED database with five kinds EEG features for evaluation. The experimental results had demonstrated that SGA-LSTM method achieved the highest recognition accuracies among these methods, which are very likely due to the fact of using attention mechanism in building the learning network. The ℓ_1-norm in the loss function helps to generate a sparse attention vector to enhance the discriminative ability of EEG channels carrying more emotion information.

Acknowledgment. This work was supported in part by the National Key Research and Development Program of China under Grant 2018YFB1305200, in part by the National Natural Science Foundation of China under Grant 61921004, Grant 81971282, Grant 61572009, Grant 61902064, and Grant 61906094, and in part by the Fundamental Research Funds for the Central Universities under Grants 2242018K3DN01, Grant 2242019K40047, and Grant 30919011232.

References

1. Cortes, C., Vapnik, V.: Support vector machine. Mach. Learn. **20**(3), 273–297 (1995)
2. Dalgleish, T.: The emotional brain. Nat. Rev. Neurosci. **5**(7), 583 (2004)
3. Defferrard, M., Bresson, X., Vandergheynst, P.: Convolutional neural networks on graphs with fast localized spectral filtering. In: Advances in Neural Information Processing Systems, pp. 3844–3852 (2016)
4. Fernando, B., Habrard, A., Sebban, M., Tuytelaars, T.: Unsupervised visual domain adaptation using subspace alignment. In: Proceedings of the IEEE International Conference on Computer Vision, pp. 2960–2967 (2013)
5. Hochreiter, S., Schmidhuber, J.: Long short-term memory. Neural Comput. **9**(8), 1735–1780 (1997)
6. Jenke, R., Peer, A., Buss, M.: Feature extraction and selection for emotion recognition from EEG. IEEE Trans. Affect. Comput. **5**(3), 327–339 (2014)
7. Li, Y., Huang, J., Zhou, H., Zhong, N.: Human emotion recognition with electroencephalographic multidimensional features by hybrid deep neural networks. Appl. Sci. **7**(10), 1060 (2017)
8. Lin, D., Zhang, J., Li, J., Calhoun, V.D., Deng, H.W., Wang, Y.P.: Group sparse canonical correlation analysis for genomic data integration. BMC Bioinform. **14**(1), 245 (2013)
9. Lindquist, K.A., Barrett, L.F.: A functional architecture of the human brain: emerging insights from the science of emotion. Trends Cogn. Sci. **16**(11), 533–540 (2012)
10. Long, M., Wang, J., Sun, J., Philip, S.Y.: Domain invariant transfer kernel learning. IEEE Trans. Knowl. Data Eng. **27**(6), 1519–1532 (2015)
11. Nair, V., Hinton, G.E.: Rectified linear units improve restricted Boltzmann machines. In: Proceedings of the 27th International Conference on Machine Learning (ICML 2010), pp. 807–814 (2010)
12. Nasrabadi, N.M.: Pattern recognition and machine learning. J. Electron. Imaging **16**(4), 049901 (2007)
13. Song, T., Zheng, W., Lu, C., Zong, Y., Zhang, X., Cui, Z.: MPED: a multi-modal physiological emotion database for discrete emotion recognition. IEEE Access **7**, 12177–12191 (2019)
14. Song, T., Zheng, W., Song, P., Cui, Z.: EEG emotion recognition using dynamical graph convolutional neural networks. IEEE Trans. Affect. Comput. (2018)
15. Tang, H., Liu, W., Zheng, W.L., Lu, B.L.: Multimodal emotion recognition using deep neural networks. In: Liu, D., Xie, S., Li, Y., Zhao, D., El-Alfy, E.S. (eds.) ICONIP 2017. LNCS, vol. 10637, pp. 811–819. Springer, Cham (2017). https://doi.org/10.1007/978-3-319-70093-9_86
16. Thompson, B.: Canonical correlation analysis. Encyclopedia of Statistics in Behavioral Science (2005)
17. Woo, S., Park, J., Lee, J.Y., So Kweon, I.: CBAM: convolutional block attention module. In: The European Conference on Computer Vision (ECCV), September 2018
18. Zhang, T., Zheng, W., Cui, Z., Zong, Y., Li, Y.: Spatial-temporal recurrent neural network for emotion recognition. IEEE Trans. Cybern. (99), 1–9 (2018)
19. Zheng, W.L., Lu, B.L.: Investigating critical frequency bands and channels for EEG-based emotion recognition with deep neural networks. IEEE Trans. Auton. Ment. Dev. **7**(3), 162–175 (2015)
20. Zheng, W.: Multichannel EEG-based emotion recognition via group sparse canonical correlation analysis. IEEE Trans. Cogn. Dev. Syst. **9**, 281–290 (2016)

Dynamic Facial Stress Recognition in Temporal Convolutional Network

Sidong Feng(✉)

Research School of Computer Science, Australian National University, Canberra, Australia
u6063820@anu.edu.au

Abstract. Stress is a major problem that infiltrates our society in countless ways. We cannot eliminate stress, but can recognize stress and manage it. Automatically recognizing stress through facial expressions has been extensively studied in the past decades. Recent research indicates that certain architectures can reach state-of-the-art accuracy in stress recognition. However, they recognise facial stress in view of static expressions, while only a few papers identify the fundamental limitations of static facial expression. This paper adapts ANUStressDB database in dynamic and develops a Temporal Convolutional Network to recognize continuous facial stress problem. We further apply Bimodal Distribution Removal to improve our result. The experimental results show that our system achieves 67.56% classification accuracy.

Keywords: Stress recognition · Temporal convolutional networks · Bimodal Distribution Removal

1 Introduction

Stress is defined as a state of mental or emotional strain. It is important to recognize stress so that it can be effectively managed. Automatically recognizing human's emotions through facial expressions (a.k.a. facial expression recognition or FER) has emerged as a key problem of human-computer interaction and psycho-physiology analysis [2]. We used ANUStressDB [10] to identify facial stress. In this problem, stress is identified based upon the signals acquired in real time from contact-less sensors such as RGB and thermal modalities. Given a time sequence signals, the goal is to simultaneously segment every emotion in time and classify each constituent segment as stress or not.

Deep learning practitioners commonly regard recurrent architecture as the default starting point for sequence modelling tasks [8]. In past decades Long Short-Term Memory (LSTM) [9] and Gated Recurrent Unit (GRU) [5] occupy time sequence problems. However, they take too long to process, because they read and interpret the time sequence one frame at a time, the neural network

T. Gedeon et al. (Eds.): ICONIP 2019, CCIS 1142, pp. 698–706, 2019.
https://doi.org/10.1007/978-3-030-36808-1_76

must wait to process the next frame until the current frame processing is completed. This means that RNNs cannot take advantage of massive parallel processing (MPP) [17] in the same way the CNNs can. Temporal Convolutional Net (TCN) [3] solve this problem.

By mid-2017, Bai et al. published a new architecture called TCN, which distills the best practices in convolutional (e.g. Causal Convolutions, Dilated Convolutions) network design into a simple architecture. It outperforms canonical recurrent networks such as LSTMs across a diverse range of sequence modeling tasks. Our task is thus to evaluate the performance on real time facial stress recognition problem using TCN.

Stress varies among individuals. Some people are naturally more sensitive and reactive to stress. Different kinds of stress have different symptoms and physiological signs [18]. It is a subjective topic, data could be fuzzy and vague, leading outliers in database. A number of methods for cleaning up noisy data has been proposed, such as Least Median Squares (LMS) by Rousseeuw [16] and Least Trimmed Squares (LTS) by Alfons [1]. These methods perform well on synthetic noisy data, but not well on real world data. Since our data is collected in real world, a more reliable method is required. Bimodal Distribution Removal (BDR) by Slade et al. [20] is a well-known outliers removal method, proved to perform well on both added artificial outliers and real noisy data [20].

Our contributions can be summarized below:

- We propose a convolutional based technique to automatically recognize temporal dynamic facial stress problem.
- We analyze the feasibility of BDR technique on the basis of ANUStressDB, which contains added artificial outlier and real world noisy data.
- We identify the fundamental limitations of static processing characteristics of stress recognition problem in previous work and propose to exploit continuity of stress to address these limitations.

2 The Proposed System

The proposed system is shown in Fig. 1. It involves four procedures. (1) prepare ANUStressDB data; (2) apply techniques like data augmentation, dimensionality reduction and data scaling to help manage the data; (3) train the model; (4) apply BDR on pretrained model for further improvement. A halting condition is provided to decide termination.

2.1 Data Preparation and Preprocessing

In this paper, we use ANUStressDB as benchmark dataset to evaluate our model. The dataset involves 24 participants. Instructors played a film with a collection of negative and positive clips as stress stimulator. The clips are separated by displaying few seconds blank screen in between the clips to neutralize the participants' emotion before playing the next clip. Two cameras are working at 30 frames per second to capture thermal and RGB modalities. Then, facial features are extracted by using Linear Spectral Clustering (LSC) [11] and Local Binary Patterns (LBP) [14], respectively. As a result, we extract 36 features for each frame. In the ground truth data, we assign the patterns in the time series as stressed or not when the label of the clip is stressed or not.

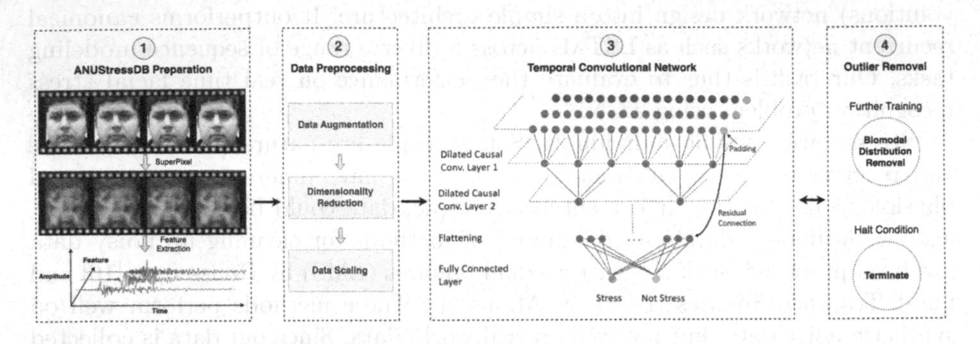

Fig. 1. Illustration of the proposed system.

Data Augmentation. We have 24 participants and 12 clips, results in 288 time series, whereas too small to train a deep learning neural network. Hence, data augmentation is applied. We split time series into fixed length sub time series. As a trade off, sub time series may lose some information on the origin time series. To balance the quality and quantity, length is set to 10. We round length to encounter aliquant time series. Although we carefully select the time series length, we still suffer from noise. For example, a 17 s time series would be split into 0–10, 7–17 sub time series. If there is no emotional disclosure in the first 10 s, 0–10 will be an artificial noisy data. Whereas, BDR in Sect. 2.3 can solve this implicit problem.

Dimensionality Reduction. As discussed in Sect. 2.1, a 10 s time series involves 300×36 features. Training model on high-dimensional data greatly increases the number of weights, making the training infeasible [21]. We reduce the data in two approaches. First, in time series dimension, we observe that the difference between each frame is small. Therefore, we take the average value of 30 frames (1 s) as one time sequence. Second, we reduce feature dimension by feature selection. We remove irrelevant features. By observation some features are slightly different. Therefore, we remove one of these features so we're left with only features with distinct values. Thus, each time series reduces to 10×16.

Data Scaling. Standardization [6] (i.e., rescaling with 0 mean and unit variance) that changes the values of numeric columns in dataset to a common scale ([−1,1]) is applied to improve neural network stability and training efficiency.

2.2 Temporal Convolutional Networks

This architecture is informed by convolutional architecture for sequential data (e.g., WaveNet [13]), but is deliberately kept simple. It combines the best practices of modern convolutional architectures, such as Dilated Causal Convolutions, Residual Connections. There are two major characteristics of TCN. (1) the convolutions in the architecture are causal, meaning that there is no information leakage from future to past; (2) the architecture can take a sequence of any length and map it to an output sequence of the same length, just as with an RNN.

2.3 Bimodal Distribution Removal

The outliers in training set will have larger errors relative to the rest of the training set. First, we calculate the errors of each training pattern by using cross entropy. Then, calculate the mean of errors ($\overline{\Delta}_{ts}$). We define the error greater than $\overline{\Delta}_{ts}$ as high error peak and calculate the mean and standard deviation of errors of high error peak ($\overline{\Delta}_{ss}$) (σ_{ss}). Since $\overline{\Delta}_{ss}$ will be heavily influenced by outliers, it will be relatively high. It is possible to decide which patterns to permanently remove from the set. If the error follows the pattern:

$$error \geq \overline{\Delta}_{ss} + \alpha\sigma_{ss} \quad (1)$$

α is to control how many outliers need to be removed. Since our dataset is not large enough, we decide to set the removal factor a to 1, so the least outliers are removed. To avoid removing all the data, a halting condition is set by variance v_{ts} and the size of the remaining set. Low variance means the network is well trained and small size of training set means the network could easily overfit.

Dilated Causal Convolutions. As mentioned, the TCN is based on two principles. One is the convolutions in the architecture are causal. To accomplish this point, the TCN uses causal convolutions. Causal simply means a filter at time step t can only see inputs that are no later than t. However, a major problem of this design is when the history is long. This is because, a causal convolution needs to look back at history with size linear to the time. For example, to predict output at time 1000, network needs to look back 1000 previous inputs. It requires an extremely deep network, which is inefficient and infeasible.

In the previous work WaveNet by Oord et al. [13], they employ dilated convolutions to allow the receptive field to increase exponentially [22]. Receptive field is the implicit area captured on the initial input by each input to the next layer. In TCN, it makes use of dilated convolution which is just a convolution applied to input with defined gaps. The kernel size k is to filter and the dilation factor d is to control the gaps. In common, dilation factor grows exponentially (i.e., $d = 2^i$ at depth i). This ensures that filter can hit each input within the effective history, while also allowing for an extremely large effective history using deep networks [3]. Takes the advantages of both techniques, the integration of causal and dilated is able to conquer long history problem.

Table 1. Performance on different models

	Training (%)	Validation (%)	Testing (%)
Epoch 300	73.95	59.56	60.98
Epoch 500	81.21	55.73	57.92
Early Stop	69.06	64.30	67.56
Early Stop + BDR	84.61	54.71	53.45
Sharma (GA-SVM) [19]	-	-	86
Irani (SVM) [10]	-	-	89
Prasetio (CNN) [15]	-	-	95.9

Table 2. Optimal hyperparameter settings of TCN and BDR

TCN		BDR	
Input features	16	Further train epochs	50
Sequence length	10	α	1
Kernel size k	7	Variance υ_{ts}	0.01
Hidden neurons	[10,10]	Min train size	1000
Learning rate	1e−4		

Fully Connection. To achieve another principle of TCN, we use a fully convolutional network (FCN) established by Long et al. [12]. Fully connection layers is added after the output of the TCN to address binary classification problem. Determining a certain number of hidden layers and neurons is crucial and difficult in the research community. For our problem, we perform several trials on different numbers of hidden layers and neurons, finding that the best performance appears when there are no hidden layers after the TCN layers. Thus, we apply one fully connected layer at the end. An illustration is provided in Fig. 1.

3 Experiments

In this section, we begin by discussing our hyperparameter settings. Then, we evaluate and analyze on the result of the system in detail. Finally, we discuss the comparison between our model and previous works. A synopsis of the result is shown in Table 1.

Hyperparameter Settings. Table 2 lists the hyperparameters we used when applying the TCN. The most crucial factor for the TCN is k. They determine whether the receptive field is large enough to capture the sufficient context to predict. As previous work suggested, larger kernel size k helps network to converge faster. By several trials, $k = 7$ performs best.

As discussed in Sect. 2.3, thresholds on variance v_{ts} and size of training dataset are defined in Table 2. Early Stop [4] is applied to prevent overtraining. The stopping point depends on either validation accuracy or validation loss. We decide to use validation accuracy as the driving metric since it is the most vital factor in our problem. Since accuracy oscillates, we set patience value to 30 to determine whether it reaches the end or just floating. All threshold values are carefully selected through manually check.

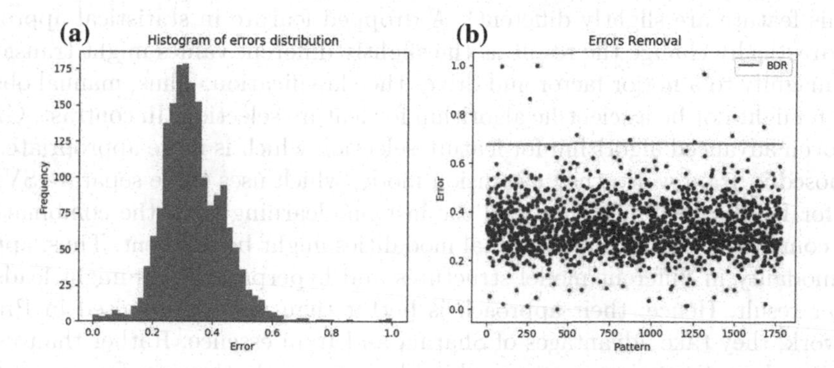

Fig. 2. (a) is the histogram of normalized error distribution at early stop checkpoint. (b) is the diagram of normalized error distribution of each patterns, green scatter point represents as each error. The line represents as the BDR line where the pattern above the line considered as outlier and will be removed. (Color figure online)

Model Analysis and Discussion. As we can see, the testing accuracy at early stop is 6.58% and 9.64% higher than training model at 300 epochs and 500 epochs. Therefore, early stop is an effective technique to use. We assume the model using this technique as the pretrained network.

As mentioned in Sect. 2.3, BDR can clean up noisy data. Hence it may help improve our network. To test the usefulness of BDR, we implement BDR on pretrained network. As we observe in Table 1, BDR boosts 15.55% on training accuracy. Contrastly, validation accuracy and testing accuracy decreases 9.59%, 14.11% respectively. It is possibly overfit. There are two reasons for this problem. In Fig. 2(a), we can observe that the errors distribution after pretraining is not bimodal distribution. Thus, the algorithm to calculate high error peak is not accurate and precise any more. Another reason might be outliers can be legitimate data, representing an accurate observation of a rare case. Removal decreases generalization ability in neural network. Thus, BDR is not an effective approach for our model.

Comparison with Previous Works. We further compare our best result 67.56% appears at Early Stop, with the previous works by Sharma et al. [19], Irani et al. [10] and Prasetio et al. [15]. From Table 1 we can see that previous works outperform our model. Next, we discuss possible reasons for the difference.

First, instead of using deep learning, Sharma and Irani use SVM as classifier. SVMs are originally designed for binary classification. On the basis of our problem, SVM has dominant position. Also, our dataset contains a small amount of training data. SVMs have advantages to predict in less training data. In such a case, SVM might be better than our model.

Apart from the benefits of using SVM, the model architecture is different. As proposed in Sharma work, they use GA [7] for feature selection. On the contrary, we use feature selection in statistical way (removing features if the values of this feature are slightly different). A dropped feature in statistical approach can drastically change the result as the slightly different values might transform and magnify to a major factor and drives the classification. Thus, manual observation might not be a scientific algorithm for feature selection. In contrast, GA is a proven advanced algorithm for feature selection which is more appropriate. As proposed in Irani work, they use fusion model which uses three separate SVMs, one for RGB, one for thermal and the last one learning from the combination. The complexity of RGB and thermal modalities might be different. Thus, applying modality in different model structures and hyperparameters might leads to better result. Hence, their approach is better than us. As proposed in Prasetio work, they take advantages of Sharma and Irani essence. Rather than using GA to reduce dimension as proposed in Sharma's work, they use feature extraction. Rather than fusing RGB and thermal modalities, they fuse Eye, Nose and Mouth, which is the intuition of Irani's work. In conclusion, the previous model architecture is more effective.

However, there is a fundamental limitation on training input in the previous works. Instead of time series data, they use frame data. Each frame considers as a pattern and labels as stress or not. Then randomly select some patterns into training set. However, as observation, many patterns in sequence have minimal differences, especially in one film. It is highly possible that the patterns in the testing set are mostly the same in the training set. For example, assume two patterns A and B are in sequence. The difference between them is slight. After shuffling, A is divided into training set and B into testing set. This causes a crucial problem. As long as the network can classify A, it can classify B. In other word, network can simply memorize patterns, not learn, and still performs well.

4 Conclusion

In this paper, we proposed a Temporal Convolutional Network (TCN), whose core is a dilated and causal convolution method for facial expression recognition. Rather than using the canonical recurrent neural networks such as LSTMs and GRUs, we have presented convolutional neural network which can also be used in a way of solving sequence modeling tasks. The type of input data has a tremendous impact on the results. Our experiments on ANUStessDB confirm this claim, showing that the results by using static input outperform that by dynamic input. We intend to extend our work by applying Bimodal Distribution

Removal (BDR) method to remove noise in artificial and real-world data. Contrastly, BDR worsens our neural network. The improvement on outlier removal suggests our proposed system has the potential to improve the performance of other methods, which will be investigated in future work.

References

1. Alfons, A., Croux, C., Gelper, S.: Sparse least trimmed squares regression for analyzing high-dimensional large data sets. Ann. Appl. Stat. **7**(1), 226–248 (2013)
2. Andreassi, J.L.: Psychophysiology : Human Behavior and Physiological Response. Oxford University Press, New York (1980)
3. Bai, S., Kolter, J.Z., Koltun, V.: An empirical evaluation of generic convolutional and recurrent networks for sequence modeling. arXiv:1803.01271 (2018)
4. Caruana, R., Lawrence, S., Giles, C.L.: Overfitting in neural nets: backpropagation, conjugate gradient, and early stopping. In: Advances in Neural Information Processing Systems, pp. 402–408 (2001)
5. Cho, K., et al.: Learning phrase representations using RNN encoder-decoder for statistical machine translation. arXiv preprint arXiv:1406.1078 (2014)
6. Cowan, R.: High technology and the economics of standardization (1992)
7. Gen, M., Lin, L.: Genetic Algorithms. Wiley Encyclopedia of Computer Science and Engineering, pp. 1–15 (2007)
8. Goodfellow, I., Bengio, Y., Courville, A.: Deep Learning. MIT Press, Cambridge (2016)
9. Hochreiter, S., Schmidhuber, J.: Long short-term memory. Neural Comput. **9**(8), 1735–1780 (1997)
10. Irani, R., Nasrollahi, K., Dhall, A., Moeslund, T.B., Gedeon, T.: Thermal super-pixels for bimodal stress recognition. In: IPTA, pp. 1–6. IEEE (2016)
11. Li, Z., Chen, J.: Superpixel segmentation using linear spectral clustering. In: Proceedings of the IEEE Conference on Computer Vision and Pattern Recognition, pp. 1356–1363 (2015)
12. Long, J., Shelhamer, E., Darrell, T.: Fully convolutional networks for semantic segmentation. In: Proceedings of the IEEE Conference on Computer Vision and Pattern Recognition, pp. 3431–3440 (2015)
13. Oord, A.V.D., et al.: WaveNet: a generative model for raw audio. arXiv:1609.03499 (2016)
14. Pietikäinen, M., Hadid, A., Zhao, G., Ahonen, T.: Local binary patterns for still images. In: Computer Vision Using Local Binary Patterns, pp. 13–47. Springer (2011). https://doi.org/10.1007/978-0-85729-748-8_2
15. Prasetio, B.H., Tamura, H., Tanno, K.: The facial stress recognition based on multi-histogram features and convolutional neural network. In: 2018 IEEE International Conference on Systems, Man, and Cybernetics (SMC), pp. 881–887. IEEE (2018)
16. Rousseeuw, P.J.: Least median of squares regression. J. Am. Stat. Assoc. **79**(388), 871–880 (1984)
17. Sankaradas, M., et al.: A massively parallel coprocessor for convolutional neural networks. In: 2009 20th IEEE International Conference on Application-Specific Systems, Architectures and Processors, pp. 53–60. IEEE (2009)
18. Schneiderman, N., Ironson, G., Siegel, S.D.: Stress and health: psychological, behavioral, and biological determinants. Annu. Rev. Clin. Psychol. **1**(1), 607–628 (2005)

19. Sharma, N., Dhall, A., Gedeon, T., Goecke, R.: Thermal spatio-temporal data for stress recognition. EURASIP J. Image Video Process. **2014**(1), 28 (2014)
20. Slade, P., Gedeon, T.D.: Bimodal distribution removal. In: Mira, J., Cabestany, J., Prieto, A. (eds.) IWANN 1993. LNCS, vol. 686, pp. 249–254. Springer, Heidelberg (1993). https://doi.org/10.1007/3-540-56798-4_155
21. Wójcik, P.I., Kurdziel, M.: Training neural networks on high-dimensional data using random projection. Pattern Anal. Appl. **22**(3), 1221–1231 (2019)
22. Yu, F., Koltun, V.: Multi-scale context aggregation by dilated convolutions. arXiv preprint arXiv:1511.07122 (2015)

A Comparison of CasPer Against Other ML Techniques for Stress Recognition

Jack Michael Harding Sekoranja(✉)

Australian National University, Canberra, Australia
jack.sekoranja@gmail.com

Abstract. When developing multi-layer neural networks (MLNNs), determining an appropriate size can be computationally intensive. Cascade Correlation algorithms such as CasPer attempt to address this, however, associated research often uses artificially constructed data. Additionally, few papers compare the effectiveness with standard MLNNs. This paper takes the ANUstressDB database and applies a genetic algorithm autoencoder to reduce the number of features. The efficiency and accuracy of CasPer on this dataset is then compared to CasCor, MLNN, KNN, and SVM. Results indicate the training time for CasPer was much lower than the MLNNs at a small cost to prediction accuracy. CasPer also had similar training efficiency to simple algorithms such as SVM, yet had a higher predictive ability. This indicates CasPer would be a good choice for difficult problems that require small training times. Furthermore, the cascading feature of the network makes it better at fitting to unknown problems, while remaining almost as accurate as standard MLNNs.

Keywords: CasCor · CasPer · Neural networks · KNN · SVM · Real world data set · Classification · Autoencoder · Genetic algorithm · Evolutionary algorithm

1 Introduction

A common feature of neural networks is the layering of interconnected nodes. However, if a network is too small or large, it will have poor predictive capabilities [1]. Furthermore, typical neural network training algorithms such as back-propagation are slow due to the *moving target problem*. This states that the constant changing of all weights in the network make it difficult for individual nodes to learn [2]. One solution, termed Cascade Correlation (CasCor) [2], is to dynamically increase the number of hidden units and layers until an arbitrary accuracy is reached. This algorithm was further improved upon to develop the CasPer algorithm, which uses the Progressive RPROP algorithm to train the entire network [3, 4].

While CasPer and CasCor are successful in complex artificial problems [2, 3], few studies have evaluated their performance on non-artificial datasets. Furthermore, there is little research that compares the performance of Casper and CasCor with multi-layer neural networks (MLNN). Therefore, this paper aims to determine the efficiency and effectiveness of CasPer against CasCor and MLNNs in the training and testing phases. This paper adds two additional benchmarks of simple yet successful algorithms – K-Nearest Neighbours (KNNs) and Support Vector Machines (SVMs).

T. Gedeon et al. (Eds.): ICONIP 2019, CCIS 1142, pp. 707–714, 2019.
https://doi.org/10.1007/978-3-030-36808-1_77

The dataset that was studied in this paper was the ANUstressDB database [6]. This database consists of time series data of 34 scalp electrode recordings from 24 participants. Subjects were shown 12 video clips that were separated by five seconds of blank screen to neutralise the participants' emotion. Each video was either attributed as being stressful or non-stressful, with the purpose of this paper being to predict these labels based on the electrode recordings. This dataset was used as it isn't artificially created like other CasPer and CasCor datasets, making it a realistic measure of effectiveness. Furthermore, there is a lot of variation in this kind of data, as it has been argued that individuals will react differently to the stimulus [6]. Therefore, the applied models will be required to sort through the nuance in the data, meaning the problem should be relatively difficult.

One potential caveat in using this data is that the different parts of the brain are responsible for different functions. This means that any machine-learning algorithm could potentially get a high accuracy by just reading off a single electrode. To remove this independence and therefore increase the complexity of the problem, the data in this paper was compressed using an autoencoder. However, since autoencoders utilise backpropagation in their training, they are prone to getting stuck in local minimums [7]. Therefore, this paper uses a genetic algorithm to select weight combinations that result in a more linearly separable data.

2 Method

2.1 Pre-processing Dataset

Each Model was tested on a pre-processed version of the aforementioned dataset. Participant responses were collected at the middle frame of each video, and were compressed using a genetic algorithm autoencoder. The autoencoder itself was set up in three layers with tanh activation functions, where the first and last layers both had 34 features and the middle (output) layer contained six. The weight and bias pairs in the network were sampled from a population of 50 candidates, and were trained for 100 epochs using mean squared error loss. Fitness of these candidates was determined by how accurately logistic regression could predict the labels. This results in the favouring of encoders that produce more linearly separable data. The top five candidates were placed in the next iteration, and were used as the mean of five multivariate normal distribution to sample the remaining 45 candidates. This population was then re-evaluated 30 times before the best candidate was chosen.

2.2 Implemented Models

Cascade Correlation (CasCor). CasCor is a Cascade Network algorithm that starts by connecting all inputs to a single output layer. Each iteration, a single hidden unit is trained and then inserted into the network, with all of its input weights being fixed. These new nodes are connected to all inputs and all previous hidden nodes, with the output node adjusting to accommodate the added node as part of its inputs. The CasCor architecture after adding two hidden nodes can be seen in Fig. 1 [2].

The implementation of CasCor in this paper trains a pool of neurons as suggested in the original paper [2], only adding the best one to the network. The Adam optimiser was used to train each candidate unit along with the output. Tanh activation functions were used for hidden nodes, while a sigmoid function was used for the output.

Cascade Correlation with Progressive RPROP (CasPer). CasPer is a Cascade Network Algorithm that is heavily based off CasCor. However, rather than freezing weights, they are separated into three regions with different learning rates. These rates are set such that new nodes will quickly minimise the network error, while changes in older nodes or the new node's output occur much more slowly. As a result, inputs to the new node are given a learning rate L1, the output is given rate L2, and all remaining weights are given rate L3, such that L1 ≫ L2 > L3. The CasPer architecture and location of the weights can be seen in Fig. 2 [3].

The implementation of CasPer in this paper followed the implementation suggested in the original paper [3]. A loss threshold parameter was added as an early exit feature to improve overall convergence properties [3]. The model initialised each weight to be within the range −0.7 to 0.7 and used the recommended RPROP values of $\eta^+ = 1.2$, $\eta^- = 0.5$, $\Delta_{max} = 50$, $\Delta_{min} = 1 \times 10^{-6}$, and $\Delta_0 = 0.2$. Tanh activation functions were used for hidden nodes, while a sigmoid function was used to normalise the output between 0 and 1, with values over 0.5 being taken as the class 'Stressed'.

Multi-Layer Neural Network (MLNN). The implementation MLNNs used in this paper again used the Adam optimiser as well as tanh activation functions for each hidden layer and a sigmoid function for the output.

K-Nearest Neighbour (KNN). KNN classifies targets by selecting the mode of the K data points that are closest (according to Euclidean distance) to the target [11]. This paper used the version of KNN provided in the scikit-learn python library.

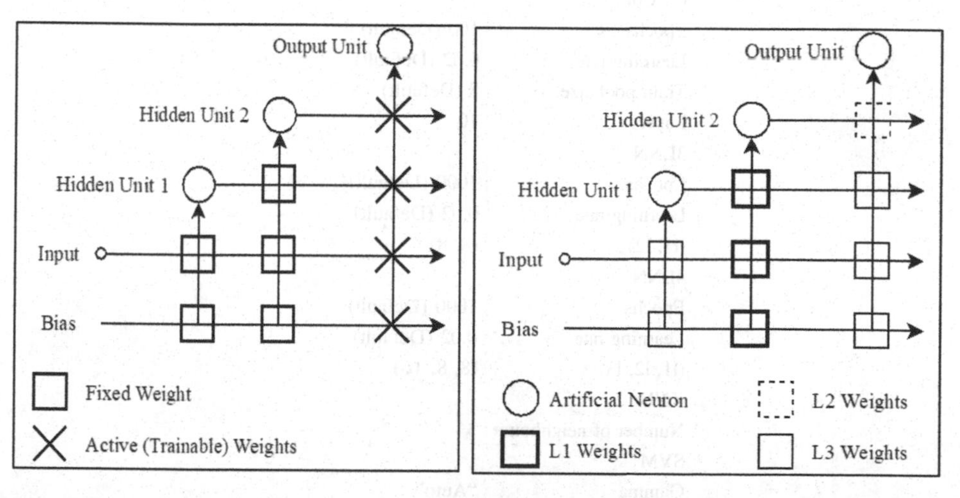

Fig. 1. The CasCor Architecture after two hidden units are added. Vertical lines indicate the sum of inputs.

Fig. 2. The CasPer Architecture after a single hidden node is added. Vertical lines indicate the sum of inputs

Support Vector Machine (SVM). SVMs work by mapping the input space on to a higher dimensional space that is easier to classify. The algorithm will then choose a decision boundary that maximises the distance of the points on either side. Classes are assigned based on the side of the boundary that the target appears [8]. This paper used the version of SVM from in the scikit-learn python library.

2.3 Parameter Selection

To ensure the models correctly fit the dataset, a mixture of Grid Search and Cross Validation was used. Different values were chosen for each hyperparameter, and every combination of these was tested for each model using 5-fold Cross Validation. This allows both thorough cross validation and efficient calculation of the average accuracy. Each cross validation was performed 5 times to prevent models from being more accurate as a result of the initial random weights. The average of this data was saved, and the hyperparameters that resulted in the highest average score over the five trials were used for generating the results. All network models used a Mean Squared Error loss function. The results from Grid Search can be seen in Table 1.

Table 1. Results of Grid Search for each model. "*(Default)*" indicates that the value of the hyperparameter was not run through grid search, and the default model value was used

Model hyperparameter	Value
CasPer	
Max epochs	1000 (Default)
P	5 (Default)
Loss threshold	0.01 (Default)
Layers	8
(L1, L2, L3)	(0.2, 0.005, 0.001)
CasCor	
Epochs	100 (Default)
Learning rate	0.02 (Default)
Train pool size	8 (Default)
Layers	10
3LNN	
Epochs	1000 (Default)
Learning rate	0.02 (Default)
(l1, l2)	(8, 8)
4LNN	
Epochs	1000 (Default)
Learning rate	0.02 (Default)
(l1, l2, l3)	(8, 8, 16)
KNN	
Number of neighbours	3
SVM	
Gamma	"Auto"
C	1
Degree	3
Kernel	Sigmoid
Shrinking	False

2.4 Result Gathering

Results were gathered by randomly shuffling all participants' responses and using 4/5 of the data for training, with the remainder used for testing. The 4:1 ratio ensures an ample selection of both train and test data. The same data split was used for each model to ensure any trends in the data (class imbalance, etc.) are learned by all models. This was to prevent one model from getting worse data than the rest. The accuracy on the train and test set was then recorded alongside the time taken, and the data was re-shuffled. This was repeated 500 times to reduce the likelihood of outliers occurring, with mean and standard deviation for all results being recorded.

3 Results and Discussion

The results of the testing accuracy indicate a lack of ability for any model to reliably predict the data. This was even the case for SVM, which was proven to be successful in a large number of cases with the exception of imbalanced classes [9]. Since we can verify the classes were not imbalanced, the optimal decision function of the data may be too complex or random for basic machine learning algorithms. We can also see that the accuracy from 3LNN and 4LNN on the training set was much higher than on the test set, which typically indicates overfitting. However, model parameters were selected using Grid Search, meaning other models with fewer nodes were tested. This indicates that it is unlikely a reduction in the number of hidden nodes would have generated better results. Therefore, the dataset itself may simply have too much noise, making it difficult to learn an appropriate decision function (Figs. 3 and 4).

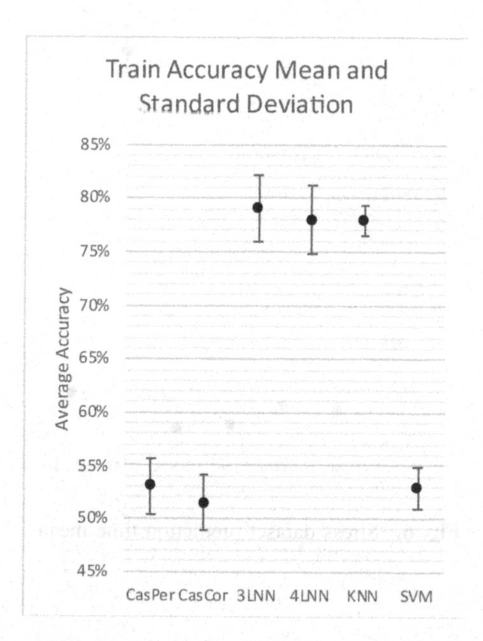

Fig. 3. Stress dataset training accuracy mean

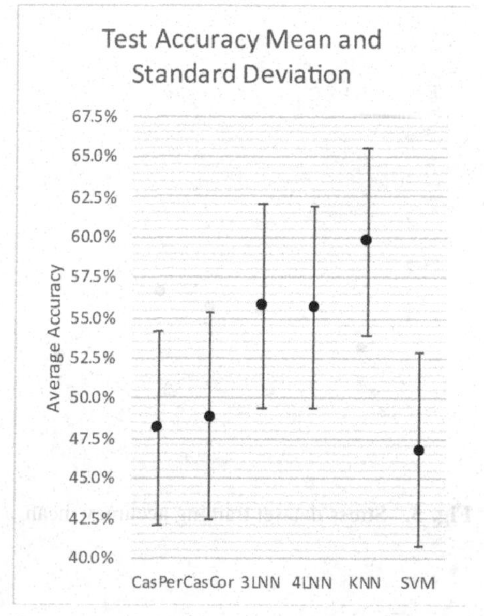

Fig. 4. Stress dataset prediction accuracy mean

This idea is further supported given the best-performing model was KNN, which relies on spatial similarities rather than data separability. One possible reason for this issue is the use of logistic regression in the autoencoder's genetic algorithm. The autoencoder likely chose a less accurate representation of the original dataset in order to maximise the linear separability for logistic regression. This idea is further supported by the poor performance of SVM, which again is successful in a large number of cases.

Regarding timing, the results show that CasCor takes the longest time to train, which is due to the training of a pool of 8 candidate nodes. Therefore, we can approximate how long it would take if it was training a single node by finding the average time taken per node (Approximate since there is a slight overhead that is unaccounted for due to caching). This was represented in the graph as the label *CasCor/Node*. After doing this, we can see that 3LNN and 4LNN taken the longest time to train, followed closely by the averaged CasCor. CasPer, KNN, and SVM are significantly faster than the other models, with all being at least six times faster than the next fastest model. When testing, KNN takes the most time, as it has to calculate the Euclidean distance between all the training nodes. SVM is the second slowest, predicting nearly 2.5 times slower than the next fastest model. This is then followed by CasCor and CasPer, with 3LNN and 4LNN being the most efficient at predicting. This is likely due to the fact that CasPer and CasCor had eight and ten nodes respectively, as the MLNN models had fewer iterations to compute (Figs. 5 and 6).

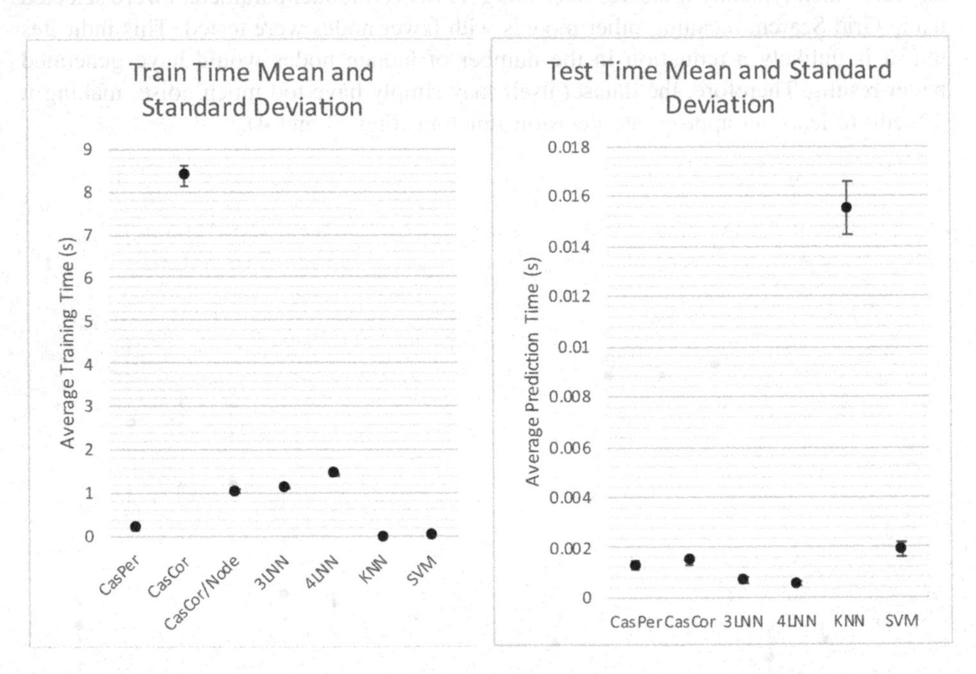

Fig. 5. Stress dataset training accuracy mean **Fig. 6.** Stress dataset prediction time mean

4 Conclusion and Future Work

This paper compares the use of CasPer with several other machine learning models with similar capabilities. To properly determine their effectiveness, these models were tested against a complex real-world dataset for stress prediction. Despite the lack of trainable data, this paper showed CasPer performs slightly worse than MLNNs, while being significantly less likely to overfit. CasPer was also significantly faster to train than MLNNs, while only being slightly slower when predicting. Given CasPer grows to match the size of the network, there is less finetuning required to set it up. This makes it useful when the problem space is unknown, as it combines the variability of models such as KNN with the accuracy of neural networks. Additionally, CasPer is able to exceed the accuracy of techniques such as SVM, while removing the large overhead associated with prediction on these models. CasPer is also similar in overall performance to CasCor, but has significantly reducing training time by avoiding the need to train multiple nodes at once. KNN was found to produce the best accuracy on the used dataset, however, this was likely a result of the noise introduced into the dataset. None of the MLNNs could learn the decision function despite the fact that they are universal approximators [5].

The results from this paper could be improved by modifying the pre-processing procedures used on the data as well as the raw data itself. This would allow the accuracy of each model to be fully taken into account. One approach would be to use the original ANUstressDB dataset that includes the participants' facial recordings. By using a convolutional neural network on this data an accuracy of 89% can be achieved [6]. Alternatively, the current pre-processing techniques could be improved by using all the time series data in the dataset, rather than just the middle frame. Furthermore, in the autoencoder genetic algorithm, the use of linear separability for the fitness function may have raised issues as previously discussed. Therefore, the results could be improved by either using a non-linear separation model such as SVC, or by removing the genetic component entirely and relying solely on the autoencoder to determine an acceptable feature mapping.

There have also been several suggestions for improvements to CasPer. These include training on a pool of neurons as CasCor does [4], or introducing new nodes in the same layer until a certain threshold is reached [10]. Therefore, the work in this paper could be extended by testing the efficiency and effectiveness of these new techniques with the approach used in this paper.

References

1. Geman, S., Bienenstock, E., Doursat, R.: Neural networks and the bias/variance dilemma. Neural Comput. **4**(1), 1–58 (1992)
2. Fahlman, S.E., Lebiere, C.: The cascade-correlation learning architecture. In: Advances in Neural Information Processing Systems, pp. 524–532 (1990)
3. Treadgold, N.K., Gedeon, T.D.: A cascade network algorithm employing progressive RPROP. In: Mira, J., Moreno-Díaz, R., Cabestany, J. (eds.) IWANN 1997. LNCS, vol. 1240, pp. 733–742. Springer, Heidelberg (1997). https://doi.org/10.1007/BFb0032532

4. Treadgold, N.K., Gedeon, T.D.: Extending casper: a regression survey. In: Proceedings of the International Conference on Neural Information Processing, pp. 310–313, November 1997
5. Hornik, K., Stinchcombe, M., White, H.: Multilayer feedforward networks are universal approximators. Neural Netw. **2**(5), 359–366 (1989)
6. Irani, R., Nasrollahi, K., Dhall, A., Moeslund, T.B., Gedeon, T.: Thermal super-pixels for bimodal stress recognition. In: 2016 Sixth International Conference on Image Processing Theory, Tools and Applications (IPTA), pp. 1–6. IEEE, December 2016
7. Ding, S., Su, C., Yu, J.: An optimizing BP neural network algorithm based on genetic algorithm. Artif. Intell. Rev. **36**(2), 153–162 (2011)
8. Chang, C.C., Lin, C.J.: LIBSVM: a library for support vector machines. ACM Trans. Intell. Syst. Technol. (TIST) **2**(3), 27 (2011)
9. Akbani, R., Kwek, S., Japkowicz, N.: Applying support vector machines to imbalanced datasets. In: Boulicaut, J.F., Esposito, F., Giannotti, F., Pedreschi, D. (eds.) ECML 2004. LNCS (LNAI), vol. 3201, pp. 39–50. Springer, Heidelberg (2004). https://doi.org/10.1007/978-3-540-30115-8_7
10. Shen, T., Zhu, D.: Layered_CasPer: layered cascade artificial neural networks. In: The 2012 International Joint Conference on Neural Networks (IJCNN), pp. 1–7. IEEE, June 2012
11. Wettschereck, D., Aha, D.W., Mohri, T.: A review and empirical evaluation of feature weighting methods for a class of lazy learning algorithms. Artif. Intell. Rev. **11**(1–5), 273–314 (1997)

Hybrid Models

On Explainable Flexible Fuzzy Recommender and Its Performance Evaluation Using the Akaike Information Criterion

Tomasz Rutkowski[1,2(✉)], Krystian Łapa[3], Maciej Jaworski[3], Radosław Nielek[2], and Danuta Rutkowska[4]

[1] Senfino, 1412 Broadway 21st floor, New York City, NY 10018, USA
XAI@senfino.com, tomasz.rutkowski@senfino.com
[2] Polish-Japanese Academy of Information Technology, Koszykowa 86, 02-008 Warsaw, Poland
[3] Institute of Computational Intelligence, Czestochowa University of Technology, Al. Armii Krajowej 36, 42-200 Czestochowa, Poland
[4] Information Technology Institute, University of Social Sciences, Lodz, Poland

Abstract. In the paper, fuzzy recommender systems are proposed based on the novel method for nominal attribute coding. Several flexibility parameters - subjects to learning - are incorporated to their construction, allowing systems to better represent patterns encoded in data. The learning process does not affect the initial interpretable form of fuzzy recommenders rules. Using the Akaike Information Criterion allows evaluating the trade-off between a number of rules and interpretability which is crucial to provide proper explanations for users.

Keywords: Recommender system · Explainable AI · Akaike information criterion

1 Introduction

In the past decade, recommender systems (also called recommendation systems or recommenders) have been successfully applied in many areas of our daily life, including books or movies recommendations, tourism services, and financial investments; see e.g. [1,8,12,13,15].

In this paper, we propose a novel explainable fuzzy recommender. The explainability is assured by generating a moderate number of interpretable fuzzy IF-THEN rules. A new method, well justified by mathematical statistics, for transforming nominal values of data into a numerical form is presented.

The paper is organized as follows. In Sect. 2, a new method for nominal attributes coding is proposed. Section 3 presents four explainable fuzzy recommender systems. In Sect. 4, we show exemplary simulation results illustrating the performance of the proposed recommender, by use of the MovieLens 10M benchmark [4]. Section 5 outlines conclusions and directions of future research.

T. Gedeon et al. (Eds.): ICONIP 2019, CCIS 1142, pp. 717–724, 2019.
https://doi.org/10.1007/978-3-030-36808-1_78

2 Nominal Attributes Coding for Recommender Systems

Let us consider a database, $S = \{\mathbf{o}_1, \ldots, \mathbf{o}_M\}$, of M objects, $\mathbf{o}_j$, $j = 1, \ldots, M$, characterized by n attributes $A_{1,j}, \ldots, A_{n,j}$, and d_j that is the decision attribute. Hence, every object is expressed as follows: $\mathbf{o}_j = (A_{1,j}, \ldots, A_{n,j}, d_j)$.

Values of $A_{i,j}$, for $i = 1, \ldots, n$ and $j = 1, \ldots, M$, can be numerical or nominal, from set $V_i = \{v_{i,1}, \ldots, v_{i,K_i}\}$, for i-th attribute. It should be noted that the attribute values of particular objects, $\mathbf{o}_j$, can be a subset of V_i.

Let us assume that i-th attribute of object $\mathbf{o}_j$, for $i = 1, \ldots, n$ and $j = 1, \ldots, M$, has nominal values. In the first step of the proposed method, we apply K_i-dimensional one-hot vector $X_{i,j} = [x_{i,j,1}, \ldots, x_{i,j,K_i}]^T$ where $x_{i,j,h} = 1$ if $v_{i,h}$ is a value of attribute $A_{i,j}$, and 0 otherwise, for $h = 1, \ldots, K_i$.

Let us consider the movie data, where i-th attribute of object $\mathbf{o}_j$ is *genre*, and $V_i = \{comedy, drama, fiction, action\}$. Hence, for example, $X_{i,j} = [0,1,0,0]^T$ if $A_{i,j} = \{drama\}$ but $X_{i,j} = [1,0,0,1]^T$ if $A_{i,j} = \{comedy, action\}$.

With regard to the movie data, values of d_j, for $j = 1, \ldots, M$, can be, for example, natural numbers from set $\{1, \ldots, 5\}$, expressing ratings of the movies. Alternatively, the rating values can be taken from $\{-1, 1\}$, representing negative and positive rates, respectively.

In the next step, transforming the nominal attributes into numerical ones, we propose to apply the Pearson's correlation coefficients between appropriate one-hot vector $X_{i,j}$, corresponding to i-th attribute of randomly chosen object $\mathbf{o}_j$, and the ratings (decision attribute), d_j, for $j = 1, \ldots, M$ and $i = 1, \ldots, n$.

Thus, with the assumption of the random variables, the Pearson's correlation coefficients, for $X_{i,j}$ and d_j, are determined, for $j = 1, \ldots, M$, $i = 1, \ldots, n$, and $h = 1, \ldots, K_i$, as follows:

$$\rho_{i,j,h} = \frac{Cov\left(x_{i,j,h}, d_j\right)}{\sqrt{Var\left(x_{i,j,h}\right) Var\left(d_j\right)}}. \tag{1}$$

Then, the correlation coefficients $\rho_{i,j,h}$, for $i = 1, \ldots, n$, $j = 1, \ldots, M$, and $h = 1, \ldots, K_i$, are estimated, based on dataset S, by use of the unbiased estimators of the covariance and variances:

$$\widehat{Cov}\left(x_{i,j,h}, d\right) = \frac{1}{M-1} \sum_{j=1}^{M} \left(x_{i,j,h} - \overline{x}_{i,h}\right)\left(d_j - \overline{d}\right), \tag{2}$$

$$\widehat{Var}\left(x_{i,j,h}\right) = \frac{1}{M-1} \sum_{j=1}^{M} \left(x_{i,j,h} - \overline{x}_{i,h}\right)^2; \tag{3}$$

$$\widehat{Var}(d) = \frac{1}{M-1} \sum_{j=1}^{M} \left(d_j - \overline{d}\right)^2; \tag{4}$$

where

$$\overline{x}_{i,h} = \frac{1}{M} \sum_{j=1}^{M} x_{i,j,h}; \qquad \overline{d} = \frac{1}{M} \sum_{j=1}^{M} d_j. \tag{5}$$

The estimators of correlation coefficients, $\widehat{\rho}_{i,j,h}$, obtained by replacing the covariance and variances in (1) by their estimates (2), (3) and (4), respectively, can be used in order to transform particular values of i-th attribute of object $\mathbf{o}_j$ to corresponding numerical values $a_{i,j}$, for $i = 1, \ldots, n$, and $j = 1, \ldots, M$.

Thus, the proposed procedure is composed of two steps. At first, the values of $A_{i,j}$ are expressed as the one-hot vector, $X_{i,j} = \left[x_{i,j,1}, \ldots, x_{i,j,N_j}\right]^T$. Then, applying the estimators of correlation coefficient, $\widehat{\rho}_{i,j,h}$, numerical values, $a_{i,j}$, are obtained from vector $X_{i,j}$, in the following way:

$$a_{i,j} = \frac{\sum_{h=1}^{K_i} x_{i,j,h}\ \overline{x}_{i,h}\ \widehat{\rho}_{i,j,h}}{\sum_{h=1}^{K_i} x_{i,j,h}\ \overline{x}_{i,h}}. \tag{6}$$

for $i = 1, \ldots, n$, $j = 1, \ldots, M$, and $h = 1, \ldots, K_i$, where $\overline{x}_{i,h}$ is given by (5).

This means that instead of $\mathbf{o}_j = (A_{1,j}, \ldots, A_{n,j}, d_j)$, the object is described as $\mathbf{o}_j = (a_{1,j}, \ldots, a_{n,j}, d_j)$, for $i = 1, \ldots, n$, and $j = 1, \ldots, M$.

The numerical values, $a_{i,j}$, determined according to formula (6), are applied in the recommender systems presented in Sect. 3.

3 Description of the Proposed Recommender Systems

In this paper, we propose four recommenders, marked as WM, WM+W, WM+D and WM+W+D, based on the Wang-Mendel method for fuzzy rule generation [14]. Each recommender works as a fuzzy system with n inputs and one output. Let $x_1, x_2, \ldots, x_n$ and y be linguistic variables corresponding to input and output variables, respectively, of the fuzzy system. The input vector $\mathbf{x} = [x_1, x_2, \ldots, x_n]^T$ in the space $\mathbf{X} = X_1 \times X_2 \times \cdots \times X_n$, as well as $y \in Y$, can take crisp values, denoted as $\overline{\mathbf{x}} = [\overline{x}_1, \overline{x}_2, \ldots, \overline{x}_n]^T$ and $\overline{y}$, respectively. In this case, each universe of discourse can be the space of real numbers. The nominal attributes coding is described in Sect. 2. The crisp values $\overline{\mathbf{x}}$ can be obtained from nominal values by use of formula (6); for $j = 1, \ldots, M$, we consider data pairs $(\overline{\mathbf{x}}_j, y_j)$ where $\overline{\mathbf{x}}_j = [a_{1,j}, a_{2,j}, \ldots, a_{n,j}]^T$ and $y_j = d_j$.

Applying the Wang-Mendel method, we get N fuzzy IF-THEN rules, R^j, of the following form:

$$\textbf{IF } x_1 \text{ is } A_1^j \text{ AND } x_2 \text{ is } A_2^j \text{ AND} \ldots \text{ AND } x_n \text{ is } A_n^j \textbf{ THEN } y \text{ is } B^j \tag{7}$$

where $x_1, x_2, \ldots, x_n, y$ are linguistic variables, A_i^j, B^j, for $i = 1, \ldots, n$, and $j = 1, \ldots, N$, are fuzzy sets – fuzzy (linguistic) values – defined in the universe of discourse (space $\mathbf{X}$) and Y, by membership functions, e.g. of Gaussian shape.

The maximal number of the rules depends on the number of the fuzzy regions determined by the fuzzy sets A_i^j, B^j, and is equal or less than the number of the fuzzy regions in space $\mathbf{X} \times Y$. Moreover, for M data pairs (objects), the Wang-Mendel algorithm produces the rule base R^j, of the form (7), for $j = 1, \ldots, N$, where $N \leq M$.

In the process of generating the rules, the antecedent matching degree, also called the degree of rule activation (or the rule firing level), expressed as:

$$\tau_j = T\left\{\mu_{A_1^j}(\overline{x}_1), \mu_{A_2^j}(\overline{x}_2), \ldots, \mu_{A_n^j}(\overline{x}_n)\right\}, \tag{8}$$

for $j = 1, 2, \ldots, N$, is used, with $\mu_{B^j}(\overline{y}^j)$ being included in this t-norm as additional argument of T; for details see [14], as well as [10] and [11].

The Mamdani type of a fuzzy system with inference based on the N fuzzy IF-THEN rules, generated by the Wang-Mendel method, can be described by the following mathematical models:

$$\overline{y} = \frac{\sum_{j=1}^{N} \overline{y}^j \tau_j}{\sum_{j=1}^{N} \tau_j} \quad \text{or} \quad \overline{y} = \frac{\sum_{j=1}^{N} w_j \overline{y}^j \tau_j}{\sum_{j=1}^{N} w_j \tau_j}, \tag{9}$$

denoted as WM and WM+W, respectively, where $\overline{y}^j$, for $j = 1, 2, \ldots, N$, is a point in which membership function $\mu_{B^j}(y)$ takes the maximal value, and τ_j is given by (8). The latter (WM+W), studied in this paper, differs from the former (WM) that is a classical approach, by introducing to antecedents of rules (7) their importance weights (see e.g. [5]).

The WM and WM+W systems refer to the case where the algebraic t-norm, $T(x, y) = xy$, most often used, is applied; see e.g. [10].

Apart from the WM and WM+W, we propose another method for tuning the fuzzy system, based on the parameterized triangular norms. Thus, we use the parametric Dombi t-norm, which in the simplest case is defined as follows:

$$T(x, y) = \begin{cases} 0 & \text{if } \; x = 0 \; \text{ or } \; y = 0, \\ \left(1 + \left(\left(\frac{1-x}{x}\right)^q + \left(\frac{1-y}{y}\right)^q\right)^{1/q}\right)^{-1} & \text{otherwise,} \end{cases} \tag{10}$$

where q is the Dombi t-norm parameter, and $q > 0$; see e.g. [11].

In the systems based on the Dombi t-norm, called WM+D and WMD+W+D, respectively, in this paper, it is assumed that each rule has its own q_j parameter, for $j = 1, 2, \ldots, N$.

Of course, a proper selection of such parameters can improve the performance of the fuzzy system. We employ evolutionary strategies (ES), as an optimization method, (see e.g. [11]), in order to optimize the system parameters.

We also implement a simple mechanism for further reduction of the fuzzy rules, by removal the least beneficial ones that increase the system error - Root Mean Square Error (RMSE); for every variant of the systems. It should be noted that in the literature there are several other methods for reduction, designing and visualization of systems given by (9), see e.g. [7,9].

Adding parameters to the fuzzy system increases their degree of freedom but at the same time their complexity. Therefore, it is important to check how much the increase in the number of parameters improves system performance. In our case, it is worth checking whether the additional rule reduction allows significantly improving system performance. To evaluate the solutions, from this point of view, we apply the Akaike information criterion (AIC), expressed as follows [3]: $AIC = M \ln Q + 2p$, where M is a number of items in a dataset (in our case, M objects in the database, S; see Sect. 2), Q denotes the system error (the RMSE), and p is a number of parameters - that in the systems under consideration includes weights and the Dombi t-norm parameters for every rule; see (9) and (10), respectively.

4 Illustrations of the Systems Performance

The MovieLens 10M dataset [4] has been used in order to illustrate the performance of the proposed systems. Six attributes of the movies have been considered: *genre*, *year*, *keywords*, as well as *country*, *actors*, *directors*. As a matter of fact, we compare the performance of the systems with three inputs (corresponding to the first tree attributes), and the systems with six inputs (all attributes). In addition, the user rate of the movies, included in this database, have been applied as the decision attribute that refers to the output of the systems. From this dataset, 200 users that rated more than 30 movies have been selected. Values of the *genre*, *country*, *actors*, *directors* have been coded according to Eq. (6).

For optimization of the weights and Dombi parameters, the evolutionary strategy $(\mu + \lambda)$ has been applied, with the following parameters: (a) population size: 100, (b) number of iterations: 200, (c) crossover probability: 0.9, (d) mutation probability: 0.3, (e) mutation range: 0.2.

For the system evaluation, k-fold cross-validation ($k = 5$) has been employed, with 80% of the data samples used for learning and 20% for testing.

Simulation results concerning the RMSE (system error), for all users, are presented in Table 1, for each system. As mentioned above, two versions of the recommenders are distinguished: with 3 and 6 inputs (3 and 6 attributes of the movies, respectively). Values of the RMSE are determined for learning and testing data. Comparison of the performance of these systems is illustrated in Fig. 1. We observe how the RMSE (denoted as rmse) depends on the percentage of rules reduced.

Table 1. Average RMSE for all users; It is obvious that the average RMSE error of the recommendation systems with 6 inputs (corresponding to 6 attributes of the movies) has lower values than the RMSE of the systems with 3 inputs (only 3 attributes considered): there more attributes characterize the movies there better recommendations (better performance of the recommender).

System	Three inputs		Six inputs	
	Learning	Testing	Learning	Testing
WM	0.431	0.601	0.224	0.385
WM+W	0.329	0.562	0.167	0.361
WM+D	0.312	0.563	0.158	0.364
WM+W+D	0.312	0.562	0.152	0.359

Figure 2 portrays isocriterial lines that represent constant values of the Akaike criterion, with different values of the system error, Q, and the number of

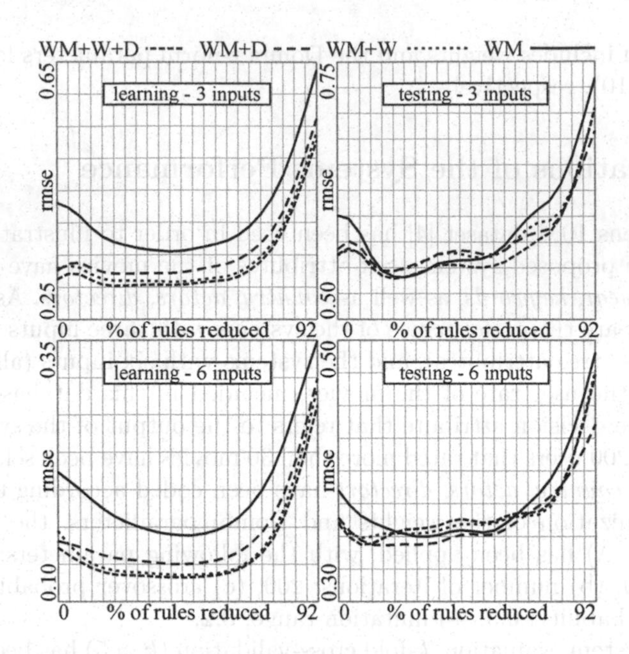

Fig. 1. Simulation results illustrating the impact of the reduction of fuzzy rules.

parameters, p, for the systems under consideration. We see that the optimal number of parameters should be low, for all considered systems (8–16 for 3-inputs and 12–24 for 6-inputs).

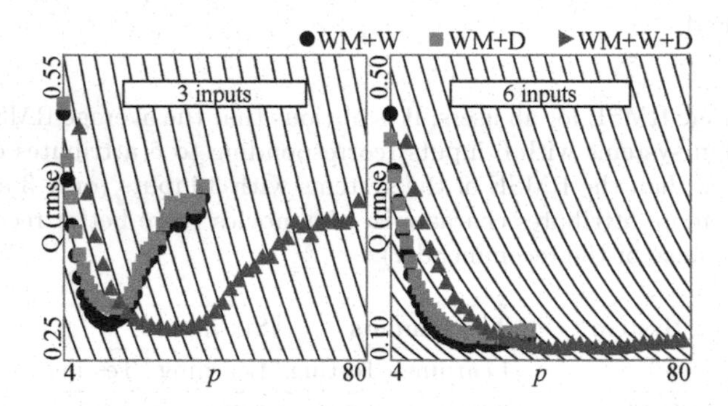

Fig. 2. Isocriterial lines representing the Akaike criterion for the recommender systems.

Examples of fuzzy rules obtained for the fuzzy systems applied as the recommenders are shown in Table 2. In addition, values of the importance weights and the t-norm parameters, for particular rules, are presented. In this case, the recommender system with 6 inputs (6 attributes) is considered.

Table 2. An example of fuzzy rules in the recommender system, e.g. the first rule should be formulated as follows: IF x_1 is Medium AND x_2 is Very High AND x_3 is Very Low AND x_4 is Very Low AND x_5 is Medium AND x_6 is Medium THEN y is Medium.

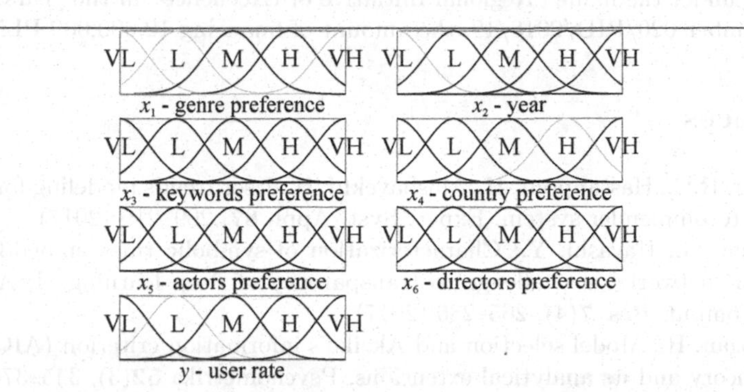

VL - Very Low, L - Low, M - Medium, H - High, VH - Very High

IF						THEN	w_j	p_j
x_1	x_2	x_3	x_4	x_5	x_6	y	=	=
M	VH	VL	VL	M	M	M	0.196	0.895
H	VH	H	L	M	M	M	0.625	0.575
M	VH	M	L	M	L	H	0.078	0.557
H	H	L	L	L	VL	VL	0.339	0.855
M	H	M	L	VL	L	VL	0.393	0.576
M	M	M	L	M	M	M	0.267	1.149
H	L	VH	M	VH	M	VH	0.515	0.750
M	M	L	L	VL	M	VL	0.979	0.689
M	VH	H	L	L	M	M	0.738	0.581
M	L	H	VH	H	H	H	0.633	1.806
H	M	M	M	L	M	M	0.140	1.565
H	VH	VH	L	VH	VH	VH	0.903	0.617

5 Conclusions

Explainability of the recommenders is realized by fuzzy IF-THEN rules with fuzzy sets that are semantically interpretable by the linguistic labels, e.g. Low, Medium, with regard to particular attributes of recommended objects (Table 2).

The recommendation system, proposed in this paper, is a flexible fuzzy recommender. The flexibility is realized by incorporating various parameters into its construction and optimizing by use of the Akaike criterion. The AIC allows finding the best trade-off between a number of rules and interpretability.

Figure 1 shows that applying both the rule importance weights and the Dombi t-norm parameters allows obtaining better results on testing datasets (for 3-input systems and 50% reduction of fuzzy rules), and an additional reduction of fuzzy rules improves the performance of the recommendation systems.

In future research, we plan to adopt several other rule-based methods (see e.g. [2,6]) for designing explainable recommender systems.

Acknowledgments. – This research was supported by the Polish National Science Center grants 2015/19/B/ST6/03179.

– The project financed under the program of the Minister of Science and Higher Education under the name "Regional Initiative of Excellence" in the years 2019-2022, project number 020/RID/2018/19, the amount of financing 12,000,000 PLN.

References

1. Bagher, R.C., Hassanpour, H., Mashayekhi, H.: User trends modeling for a content-based recommender system. Expert Syst. Appl. **87**, 209–219 (2017)
2. Bologna, G., Hayashi, Y.: Characterization of symbolic rules embedded in deep DIMLP networks: a challenge to transparency of deep learning. J. Artif. Intell. Soft Comput. Res. **7**(4), 265–286 (2017)
3. Bozdogan, H.: Model selection and Akaike's information criterion (AIC): the general theory and its analytical extensions. Psychometrika **52**(3), 345–370 (1987)
4. Harper, F.M., Konstan, J.A.: The movielens datasets: history and context. ACM Trans. Interact. Intell. Syst. **5**(4), 19 (2016)
5. Ishibuchi, H., Yamamoto, T.: Rule weight specification in fuzzy rule-based classification systems. IEEE Trans. Fuzzy Syst. **13**(4), 428–435 (2005)
6. Liu, H., Gegov, A., Cocea, M.: Rule based networks: an efficient and interpretable representation of computational models. J. Artif. Intell. Soft Comput. Res. **7**(2), 111–123 (2017)
7. Prasad, M., Liu, Y.-T., Li, D.-L., Lin, C.-T., Shah, R.R., Kaiwartya, O.P.: A new mechanism for data visualization with TSK-type preprocessed collaborative fuzzy rule based system. J. Artif. Intell. Soft Comput. Res. **7**(1), 33–46 (2017)
8. Ricci, F., Rokach, L., Shapira, B. (Eds.): Recommender Systems Handbook. Springer (2015)
9. Riid, A., Preden, J.-S.: Design of fuzzy rule-based classifiers through granulation and consolidation. J. Artif. Intell. Soft Comput. Res. **7**(2), 137–147 (2017)
10. Rutkowska, D.: Neuro-Fuzzy Architectures and Hybrid Learning. Springer, New York (2002). https://doi.org/10.1007/978-3-7908-1802-4. Physica
11. Rutkowski, L.: Computational Intelligence: Methods and Techniques. Springer, Heidelberg (2008). https://doi.org/10.1007/978-3-540-76288-1
12. Rutkowski, T., Romanowski, J., Woldan, P., Staszewski, P., Nielek, R.: Towards interpretability of the movie recommender based on a neuro-fuzzy approach. In: Rutkowski, L., Scherer, R., Korytkowski, M., Pedrycz, W., Tadeusiewicz, R., Zurada, J.M. (eds.) ICAISC 2018. LNCS (LNAI), vol. 10842, pp. 752–762. Springer, Cham (2018). https://doi.org/10.1007/978-3-319-91262-2_66
13. Rutkowski, T., Romanowski, J., Woldan, P., Staszewski, P., Nielek, R., Rutkowski, L.: A content-based recommendation system using neuro-fuzzy approach. In: 2018 IEEE International Conference on Fuzzy Systems (FUZZ) (2018)
14. Wang, L.X., Mendel, J.M.: Generating fuzzy rules by learning from examples. IEEE Trans. Syst. Man Cybern. **22**(6), 1414–1427 (1992)
15. Year, R., Martnez, L.: Fuzzy tools in recommender systems: a survey. Int. J. Comput. Intell. Syst. **10**(1), 776–803 (2017)

A Hybrid Evolutionary Algorithm with Taboo and Competition Strategies for Minimum Vertex Cover Problem

Gang Yang, Daopeng Wang, and Jieping Xu(✉)

Key Lab of Data Engineering and Knowledge Engineering,
Renmin University of China, Beijing, China
xjieping@ruc.edu.cn

Abstract. The Minimum Vertex Cover (MVC) problem is a prominent NP-hard combinatorial optimization problem, which is of great significance in both theory and application. Evolutionary algorithms and local search algorithms have proved to be two important methods to solve this problem. However, the combination of these two methods does not perform well. In order to acquire an effective hybrid evolutionary algorithm, two new control strategies are proposed, which are taboo of solution-distance and intensive competition of individuals. A hybrid evolutionary algorithm for the MVC problem, referred to HETC, is proposed in this paper using these two strategies. The effectiveness of the proposed scheme is validated by conducting deep simulations. The results obtained by the proposed scheme are compared with results obtained by EWSL, the state-of-the-art algorithm, and NuMVC.

Keywords: Evolutionary algorithms · Estimation of Distribution Algorithms · Local search · Guiding strategy · Minimum Vertex Cover (MVC)

1 Introduction

A vertex cover of an undirected graph $G = (V, E)$ is a subset $S \subseteq V$ such that each edge in E is incident to at least one vertex in S. The Minimum Vertex Cover (MVC) problem is to find the minimum sized vertex cover. The MVC problem is a classical and typical example of combinatorial optimization problems and of great importance with many real-world applications, such as network security, industrial machine assignment, and so on. The MVC problem is also a well-known NP-complete problem of combinatorial optimization, which is included in the famous Karp's 21 NP-complete problems [11], i.e., there is no polynomial-time algorithm for approximating the MVC within any factor

J. Xu—This work was supported by the Beijing Natural Science Foundation (No. 4192029), and the National Natural Science Foundation of China (61773385, 61672523). We gratefully acknowledge the support of NVIDIA Corporation with the donation of the Titan Xp used for this research.

T. Gedeon et al. (Eds.): ICONIP 2019, CCIS 1142, pp. 725–732, 2019.
https://doi.org/10.1007/978-3-030-36808-1_79

smaller than 1.3606 [8] unless $P = NP$, but generally, we believe that $P \neq NP$. Effective algorithms including EA/G [18], ULSA [17], and MIMIC [4] have been proposed to solve this kind of problems. Moreover, evolutionary algorithms and local search are considered as two important methods for these questions. Local search is a meta-heuristic strategy for solving computationally hard optimization problems. A local search algorithm starts from a candidate solution and then iteratively moves to a neighbor solution. It means to search within some local area instead of the whole area, and it aims to decrease the time to search in unpromising area. There are a lot of efficient algorithms using this thinking, such as NuMVC [6]. Furthermore, evolutionary algorithms (EAs) simulating the evolution of the nature work with a group of solutions and combine them to generate new solutions (offsprings), and the traditional genetic algorithms (GAs) are the most representative evolutionary algorithms. Many researchers devote themselves to GAs, and the strengths and weaknesses of simple GAs have been studied theoretically [9]. There are also some improved methods to guide the offspring by using heuristic strategies, such as Estimation of Distribution Algorithms (EDAs) [12].

From the aspect of algorithm optimization, some researchers intend to take both the advantages of EDAs and local search, but the results are not promising [18]. To get an effective algorithm, we propose a hybrid algorithm combining the two methods to solve the MVC problem. In this paper, we mix EDAs, an evolutionary algorithm, and NuMVC, a local search algorithm to form the new algorithm, noted as hybrid evolutionary algorithm with novel taboo and competition (HETC). Our algorithm applies two novel control strategies, taboo of solution-distance and intensive competition among individuals. In basic local search, its main shortcoming is that it is easy to fall into local optimal solution, but EDAs working with a group of individuals could ensure the diversity and release this question. However, there is a great possibility that different individuals search around the same area. That would be a waste of time if this situation happens, so we take the strategy of taboo of solution-distance to avoid this. As for the second strategy, by dropping the bad ones during the evolution and giving the left ones more time to do local search, the strength of the better individuals bursts out. In other words, individuals would compete with each other on the aspect of computing time in every epoch. In this paper, we carry out some experimental analyses about the two control strategies in HETC. The results indicate that HETC has shown superior performance on the MVC problem, compared with other excellent algorithms, such as EWSL [5], and NuMVC. We believe the new strategies would be useful for others who research this question later.

2 Two New Control Strategies

2.1 Taboo of Solution-Distance

Some local search algorithms such as NuMVC apply the tabu search for the vertex to avoid unnecessary exchange, but in our algorithm, given the genetic algorithms, point-tabu-search is not enough. When combining EDAs and the

NuMVC, a group of individuals, *gen[population]*, are working, generating and going on local search. In our strategy, a queue, $TabuSolPer$, with the maximum size of $TabuSolListSize$, consisting of the labels of the individuals that the solutions found recently belong to is maintained, and when we get a better solution, $better_{label}$ the label of the individual getting it, would be added to $TabuSolPer$. If $better_{label}$ is in $TabuSolPer$, it will be updated and avoided to be dropped next time dropping is needed. When the queue is so long that its size is over the given maximum size, the FIFO strategy will be used and the oldest one added will be dropped.

If a *candidate solution* is around the best solutions the individuals in the $TabuSolPer$ have found, it is considered as an illegal one, and we use the function *AroundTabu* to check whether a *candidate solution* is illegal. In this function, a parameter θ is the reference standard. If $dis(X, Y)$ is less than θ, we think that they are near to each other. Instead of setting a constant to θ, it is calculated by the following formula.

$$size = TabuSolPer.size \tag{1}$$

$$\theta = \rho \frac{\sum_{i=1}^{size-1} dis(gen_{TabuSolPer_{last}}.best, gen_{TabuSolPer_i}.best)}{size - 1} \tag{2}$$

$\rho \in (0, 1)$ is a parameter to control the formula. $\frac{\theta}{\rho}$ reflects the average distance between the old taboo solutions and the best solutions found recently, and it could better represent the normal distance of two candidate solutions of this graph. The value get from this formula would be more reasonable to judge whether two vertex covers is in the same area.

2.2 Intensive Competition of Individuals

This strategy is proposed to solve the following dilemma: on the one hand, if the time for local search is set too much, the frequency of mutation will be very limited, which slows down the evolution process; on the other hand, if local search doesn't get enough time, a single individual is hard to get a better solution due to its limited searching time, which also hinders the generation to get a better solution.

In HETC, better individuals refer to those with smaller size of solutions. HETC calculates the importance of each individual and decreases the population after several generations. The competition among individuals is mainly for surviving and time. Only the winners could be left and given the chance to take part in the later competition, and then be given more time. The worst individual would be dropped termly. When the size of the group, *population*, is less than a given threshold γ, the time for local search for each individual will begin increasing. In fact, when *population* reaching the threshold γ, each generation will be given the same total time for all individuals. As the *population* decreases, the time each one gets to do local search is more, but only those who win in the competition could get into this stage, and the final one left will be given as much time as possible to allow it finish it local search.

3 HETC

In this section, we present the HETC algorithm, which utilizes the strategies of taboo of solution-distance and intensive competition of individuals.

Firstly, to adapt to the strategy of taboo of solution-distance, we need to modify the two basic methods: EDAs and NuMVC. The changes is shown in the following two functions, Function *GuideAlgorithms*(*gen*, *p*), and Function *LS*(*gen*). In Function *GuideAlgorithms*, the group is sorted according to its own best size and then let the better half of the individuals be the template for the rest individuals. Besides, the process of generating a new *candidate solution* according to the template and the possibility information, will continue going unless the new *candidate solution* is legal. In Function *LS*(*gen*), when we get a new *vertex cover* that is not worse than the current best found solution, we would add its label to the *TabuSolPer*. We try to update the queue by this to avoid the old solutions influence the search all the time, and when we get a new *candidate solution*, we would check if it breaks into the forbidden zone to stop the invalid search timely.

Class *Individual* is the structure used to represent the individuals of population. It consists of the basic information of a *candidate solution* and the attribute *id* is the identification of it. Its attribute *best* is the best solution it has found most recently and attribute *cur* is the *candidate solution* it gets most recently, and attribute *delv* is the selected vertex with the maximal value. *cBest* refers to the best solution found of the whole.

```
Function GuideAlgorithms(Individual gen[ ], Double p[ ])
  sort gen_i according to gen_i.best.size .
  remove the individuals whose serial number is
  greater that population/2 from the TabuSolPer.
  for i = 1 to population/2 do
    repeat
      for j = 1 to |vnum| do
        if random() < β then
          if random() < p_j then
            set the jth vertex in gen_{i+population/2} as 1
          else
            set the jth vertex in gen_{i+population/2} as 0
          end if
        else
          set the jth vertex
          in gen_{i+population/2} the same as gen_i.
        end if
      end for
    until AroundTabu(i + population/2) is not true
  end for
```

```
Function LS(Individual gen)
while true do
  if the time reaches the limit-time then
    return
  end if
  if the candidate solution is a vertex cover then
    update gen.best and cBest if needed
    if gen.best.size ≤ cBest.size then
      Add gen_id to TabuSolPer
    end if
    remove gen.delv from gen.cur
    update the value and the weight of each
    edge if the threshold is reached,
    continue
  end if
  if AroundTabu(gen_id) then
    return
  end if
  update gen.delv and delete it from gen.cur.
  find the uncovered edge, and
  choose one of its endpoints,
  add this vertex to gen.cur
end while
```

To realize the strategy of intensive competition of individuals, *population*, the size of the living individuals, is changed in the algorithm HETC, which is also the kernel of HETC. In this function, the bad individuals is dropped gradually. The attribute *flag* of the class *Individual* indicates whether it has the right to calculate the influential factor, and *BestIndividual* refers to the label of the individual that owns the best solution.

Algorithm 1. HETC

```
initiate all the individuals gen[]
step = 0
repeat
  step = step + 1
  initiate all sump_i = 0
  for i = 1 to population do
    if i != BestIndividual then
      set gen_i.flag = 0 to allow gen_i calculate its influential factor
    end if
  end for
  for i = 1 to |vnum| do
    LS(gen_i)
  end for
  if population!=1 then
    sort gen[] according to the current best solution in increasing order
    if step mod α == 0 then
      population = population − 1
      delete the last individual from array gen
      CutoffTime = calc(population)
    end if
  end if
until population == 1
```

The variable $sump_i$, indicating the weight of the vertex i, is used to guide the evolution. We get the possibility vector *p*[], which is needed in the function *Guide*, through the variable vector *sump*[]. To increase the stability of the algorithm, we adopt the widely used Population-Based Incremental Learning algorithm (PBIL) [1], as shown in the following formula, where *totp* is the sum of the influential factor calculated already.

$$p_i = (1 - \lambda)p_i + \lambda\frac{sump_i}{totp}$$

Different from other EDAs, when one individual finds its own better solution, HETC calculates its influential factor according to the quality of its solution and adds the value to $sump_i$. However, when the time goes, HETC, like other algorithms, will slow down its pace to update the best solution. If the frequency of updating the best individual is not controlled, the best individual will count for guidance all the time. Therefore, to avoid some individual updating $sump_i$ too much, the frequency to update for those individuals whose solution size is the same as global best solution is set as once a generation. In particular, the best individual calculates its influential factor only once until other individuals find better solution. We use the attribute $flag$ to control the calculation of influential factor. In HETC, function $calc$ calculates the fitness of a new solution and accumulating the fitness to $sump_i$. It executes each time an individual gen_i updates its own best solution.

4 Simulations

Our algorithm is implemented in C++, compiled by the g++ compiler. All simulations executed on an Intel 1.6 GHz × 4 machine with 3.6 GB RAM under LINUX. In order to make comparison impartially, hardware differences are taken into consideration. The instances chosen in our simulations come from the famous BHOSLIB benchmark.

To show the effectiveness of intensive competition, we take the algorithm without it into consideration. That is the naive combination of EDAs and NuMVC, without considering the two control strategies proposed previously. Firstly, we find that if time is spent on insufficient local search, after the local search operator, the improvement of local search for each would be very small. So, the cutoff time for each individual to do local search is set as 0.01 s. We can obtain one generation within one second. As the whole program is primarily controlled by EDAs, we call this method as EDAs-oriented control (EDAOC), and the performance is unpleasing. For example, in instance frb56-25-1, EDAOC can only find 1346 or worse in 3 h while other methods usually find the best solution 1344 within 1 h. Then, in another test, the cutoff time is set as 1 s. Since the whole program is nearly controlled by local search, we call this method as local-search-oriented control (LSOC). Besides, other cutoff time, such as 0.05 s, and 0.2 s has also been tried. However, the results are all unpleasing. Table 1 shows the performance of naive combination of EDAs and local search with different cutoff time. The computational results shown in all tables include the following information: avg - the average size of the vertex cover found in ten runs; time - the run-time (in seconds) to find best solution. k - the minimum known vertex cover size.

As shown in Table 1, naive combination of EDAs and local search with cutoff time of 1 s does the best among the tested cutoff time. On the one hand, more cutoff time improves the performance of the simple combination of EDAs and local search. Besides, Lin-Kernighan method shows that the structure of any two local optimal solutions contains nearly 85% similar part on average [13]. So, it will be hard to find a solution with fresh new structure if time for local search is limited too much. On the other hand, if we continue to enlarge the cutoff time, the algorithm would be almost the same as running local search respectively, and it cann't outperform the original local search.

Table 1. Comparison among different cutoff time of naive combination of EDAs and local search

Graph		0.01 s		0.2 s		1 s	
Instance	k^*	avg	CPU(s)	avg	CPU(s)	avg	CPU(s)
frb50-23-1	1100	1101.7	n/a	1100	429.6	1100	996.0
frb53-24-1	1219	1221	n/a	1220	n/a	1219.8	5312.6
frb56-25-1	1344	1346.2	n/a	1345.9	n/a	1344.5	2599.3
frb59-26-1	1475	1476.8	n/a	1476.7	n/a	1475.5	6970.7
frb100-40	3900	3906.2	n/a	3906.5	n/a	3906.3	n/a

Table 2. Comparison among naive combination of EDAs & local search, EWLS, NuMVC and BIOC

Graph		Combination of EDAs & local search (1 s)		EWLS		NuMVC		HETC	
Instance	k^*	avg	CPU(s)	avg	CPU(s)	avg	CPU(s)	avg	CPU(s)
frb50-23-1	1100	1100	996.0	1100	790.7	1100	429.6	1100	**211.6**
frb53-24-1	1219	1219.8	5312.6	1219.4	3339.0	1219.6	2437.2	1219	3015.1
frb56-25-1	1344	1344.5	2599.3	1344.8	9175.4	1344	1657.2	1344	**1179.1**
frb59-26-1	1475	1475.5	6970.7	1475.6	5507.0	1475.6	4117.9	**1475.2**	**2371.2**
frb100-40	3900	3906.3	n/a	3903.8	1610.3	3903.6	5258.1	**3903.4**	2425.9

We test another excellent MVC solver called EWLS as comparison [5]. Table 2 shows the results. EWSL is the abbreviation for edge weight local search. EWLS established a new record for a challenging instance frb100-40 in 2010. We also add one of the original method, NuMVC, and the naive combination of the two methods as comparison. From Table 2, we can see that HETC outperforms naive combination of EDAs and NuMVC in all instances. The naive combination of EDAs and NuMVC usually finds two or more nodes larger than the best solution. HETC also shows improvement, both in consuming time and size of solution, comparing to NuMVC. In frb59-26-1, the average time of NuMVC to find the best solution is 4118 s, and EWLS uses 9175 s on average, while HETC consumes merely 2371 s on average. HETC can find the best solution 1219 for frb53-24-1 in all runs, while the success rate for EWLS is 60%, and for NuMVC it is 40%.

5 Conclusions

In this paper, we analyzed the reasons why naive combination of EDAs and local search is unpromising, and then proposed our new algorithm called HETC to make EDAs cooperate with local search more effectively. HETC achieves two new control strategies, taboo of solution-distance and intensive competition of individuals. Taboo of solution-distance effectively avoids different individuals searching around the same area, and promising the diversity and improves the effectiveness. Intensive competition of individuals solves the dilemma meet in deciding the time for local search. HETC is used to solve MVC problems with large size, and it shows a great increase in performance comparing with the original local search method used in the paper. HETC also shows excellent performance in graph with large size and better stability in different graphs. In the future, we intend to implement HETC to other local search solvers and put it into practice.

References

1. Baluja, S.: Population-based incremental learning: a method for integrating genetic search based function optimization and competitive learning. Technical report CMU-CS-94-163, Computer Science Department, Pittsburgh, PA (1994)
2. Bauer, M., Golinelli, O.: Core percolation in random graphs: a critical phenomena analysis. Eur. Phys. J. B **24**, 339 (2001)
3. Bollob, B.: Random Graphs, 2nd edn. Cambridge University Press, Cambridge (2001)
4. Bonet, J.S.D., Isbell, Jr, C.L., Viola, P.: 1996. MIMIC: finding optima by estimating probability densities. In: Advances in Neural Information Processing Systems, vol. 424. The MIT Press (1997)
5. Cai, S., Su, K., Luo, C., Sattar, A.: NuMVC: an efficient local search algorithm for minimum vertex cover. J. Artif. Int. Res. **46**(1), 687–716 (2013)
6. Cai, S., Su, K., Chen, Q.: EWLS: a new local search for minimum vertex cover. In: AAAI Conference on Artificial Intelligence (2010)
7. Chickering, D., Heckerman, D., Meek, C.: A Bayesian approach to learning Bayesian networks with local structure. Technical report MSR-TR-97-07, Microsoft Research, Redmond, WA (1997)
8. Dinur, I., Safra, S.: On the hardness of approximating minimum vertex cover. Ann. Math. **162**(1), 439–485 (2005)
9. He, J., Yao, X.: A study of drift analysis for estimating computation time of evolutionary algorithms. Nat. Comput. **3**(1), 21–35 (2004)
10. Karloff, H.: Linear Programming. Birkhauser, Boston (1991)
11. Karp, R.: Reducibility among combinatorial problems. In: Miller, R., Thatcher, J., Bohlinger, J. (eds.) The IBM Research Symposia Series. Complexity of Computer Computations, pp. 85–103. Springer, US (1972). https://doi.org/10.1007/978-1-4684-2001-2_9
12. Larraanaga, P., Lozano, J.A.: Estimation of Distribution Algorithms: A New Tool for Evolutionary Computation. Kluwer Academic Publishers, Norwell (2001)
13. Lin, S., Kernighan, B.W.: An effective heuristic algorithm for the traveling-salesman problem. Oper. Res. **21**(2), 498–516 (1973)
14. Mckee, T., Mcmorris, F.: Topics in intersection graph theory. SIAM, Philadelphia (1999)
15. Nemhauser, G.L., Trotter, L.E.: Vertex packings: structural properties and algorithms. Math. Progam. **8**, 232–248 (1975)
16. Papadimitriou, C., Yannakakis, M.: Optimization, approximation, and complexity classes. J. Comput. Syst. Sci. **43**, 425–440 (1991)
17. Rosin, C.D.: Unweighted stochastic local search can be effective for random CSP benchmarks. CoRR abs/1411.7480 (2014)
18. Zhang, Q., Sun, J., Tsang, E.: An evolutionary algorithm with guided mutation for the maximum clique problem. Trans. Evol. Comp **9**(2), 192–200 (2005)

Ant Colony System for Carpool Service Problem with High Seating Capacity

Zhi-Min Huang[1], Wei-Neng Chen[1(✉)], Wen Shi[1], and Xiao-Min Hu[2]

[1] School of Computer Science and Engineering, South China University of Technology, Guangzhou 510006, China
Cwnraul634@aliyun.com
[2] School of Computers, Guangdong University of Technology, Guangzhou 510006, China

Abstract. Carpool Service Problem (CSP), which aims at providing the ridesharing plans in order to alleviate traffic congestion, has attracted high attention in the past years. A considerable amount of efforts have been devoted to solving CSP with low vehicle capacity and small travel demand, e.g., ridesharing by private cars. However, there are few studies involving CSP with high vehicle capacity, e.g., ridesharing by buses or microbuses. In view of the special high capacity characteristic, this paper proposes an ant colony system (ACS) with novel heuristic information and pheromone calculation strategies (HC-ACS). First, we redesign the heuristic information of each edge, both the length of the specific edge and the estimation of the total travel are considered. Second, a summation rule is applied to the usage of pheromone to maintain the searching diversity. Our experiments on datasets with different spatial distribution show that the proposed HC-ACS is promising in the environment with relatively high seating capacity.

Keywords: Ant colony algorithm · Carpool service problem · Ridesharing

1 Introduction

In the past few years, the growth of population and economy increased the number of vehicles in cities, leading to serious traffic problems [1]. Since ridesharing can help not only lower the citizens' travel expenses but also alleviate the problem of congestion, considerable amounts of research efforts has been made to tackle the carpool service problem (CSP), which is an NP-hard problem [2]. Two important themes have emerged from the studies discussed so far: one is the optimization of the carpool system, the other is the optimization of the algorithm for matching between passengers and drivers. Huang *et al.* set up an intelligent carpool system to provide carpooling service for passengers via a smart portable device [2]. Similarly, QADIR *et al.* proposed a highest aggregated score vehicular recommendation framework, which is based on five parameters including average time delay, vehicle's capacity, fare reduction, driving distance and profit increment [3]. In spite of the optimization of the carpool system, more researchers concentrate on making full use of resources in CSP. Based on the assumption that some passengers get off and new passengers get on during the journey

T. Gedeon et al. (Eds.): ICONIP 2019, CCIS 1142, pp. 733–740, 2019.
https://doi.org/10.1007/978-3-030-36808-1_80

of the bus, Duan *et al.* removed the static constraints and assigned passengers more than the vehicle's capacity. Moreover, a multi-round matching based greedy algorithm and geometry partition strategy is adopted to improve the performance of the algorithm [4]. However, it is worthwhile mentioned that the current studies concentrate more on CSP in an environment with low seating capacity. For example, the sum of the seat demand of each passenger assigned to a driver or the seating capacity of each driver provided to passengers is mainly in the interval from one to four. The study of the CSP with high seating capacity should be further investigated.

Inspired by the foraging behavior of ant colony, ACS, a variant of ant colony optimization (ACO), is a meta-heuristic searching algorithm which has been widely used in solving combinational optimization problems (COPs) [5–7]. Compared with other evolutionary algorithms such as genetic algorithm (GA) [2] and particle swarm optimization (PSO) [3, 8], the application of ACS in solving CSP remains to be explored.

In this paper, we redesigned the heuristic information and pheromone calculation rule of the original ACS, making it suitable for CSP in the context of high seating capacity. We compared the proposed HC-ACS with four algorithms including binary particle swarm optimization (BPSO) [9], GA [10], hill climbing algorithm [11], and simulated annealing algorithm [12]. Experimental results show that HC-ACS has better performance on solving CSP with high seating capacity.

The remainder of this paper is organized as follows. Section 2 formulates the mathematical model for the carpool service problem (CSP). Section 3 describes the proposed heuristic information and pheromone calculation rule. Section 4 carries out a series of experiments to verify the effectiveness of the proposed strategies. Conclusions are presented in Sect. 5.

2 Mathematical Formulation

As a COP, CSP considers both the driver-passenger assignments and their internal path planning [13]. Although the existing studies concentrate on CSP with low capacity, it is a fact that the internal path planning is of little importance, because many algorithms can solve it with low time complexity such as bellman optimality [14]. Given a specific context that the maximum capacity available for passengers is relatively larger, the internal path planning becomes much more important because of dimension explosion. In this paper, we aim at redesigning the details of original ACS, making it more suitable to solve the internal path planning problem for large capacity.

2.1 CSP Model

In this paper, we only consider the situation of CSP without time window. There are two important definitions including the task definition identified by passengers and the driver definition identified by drivers [15]. A tuple $<w, u, s, d, p>$ is used to define the requirement of a passenger, namely a task t. It should be mentioned that a driver will be assigned to a task once the task is created. The terms w, u, s, d and p refer to driver, passenger, starting location, destination location, and path, respectively.

Another tuple $<s, T, P>$ is used to define the order information of a driver. Concretely, s is the location of driver, T is the task sets completed by the driver, and P is the path set of the driver. For the CSP with high seating capacity discussed in this paper, we extend the capacity of each vehicle to 30, which corresponds to buses or microbuses.

2.2 Objective Function

In our study, the objective function can be stated mathematically as follow:

$$\begin{aligned} \min f = & m \cdot c_d \\ & + \sum_{i=1}^{m} \left(\sum_{j=1}^{p-1} L_{s_j^i s_{j+1}^i} \right) \cdot \omega_d \\ & + \sum_{i=1}^{m} \left(\sum_{j=1}^{p-1} L_{s_j^i s_{j+1}^i} \right) \cdot \omega_f + \sum_{i=1}^{m} c_v^i \end{aligned} \quad (1)$$

where m and p represent the number of drivers and the maximum number of locations, respectively. L_{xy} represents the length of routes between two neighboring locations x and y. ω_d and ω_f represent the fixed cost per kilometer of driver and fuel. c_d and c_v represent the fixed cost per driver and per vehicle. The objective of HC-ACS is to minimize the total cost of CSP when given the information of drivers and passengers.

2.3 Constraints

It should be mentioned that there are still lots of constraints for the CSP in the real world, including the maximum passenger capacity constraint, the maximum mileage of vehicle constraint and the unique driver-passenger matching constraint. The details of the above three constraints are list as follows. In the maximum passenger capacity constraint, the total passenger number in each vehicle should not exceed the maximum passenger capacity, which is set to be 30 as mentioned above. In the maximum mileage of vehicle constraint, the total mileage of each vehicle should not exceed the maximum mileage, which is set to be 50 km in this paper. In the unique driver-passenger matching constraint, each passenger should only be assigned to one specific driver, but each driver can complete more than one tasks.

3 The Proposed HC-ACS

3.1 Heuristic Information with Prediction Strategy

With the increase of maximum capacity of each vehicle, the internal path planning becomes much more complicated. When the vehicle makes a decision about the next visiting location, the result brought by the decision needs a prediction. In the original ACS, the heuristic information of an edge is the reciprocal of its length. However, in the CSP with a high seating capacity, we should focus on the total length of the route. The heuristic information of an edge is calculated by:

$$\eta_{ij} = \frac{A}{L_{ij} + \sum_{h \in N_i} L_{jh} + \sum_{k \in M_i} L_{ki}} \tag{2}$$

where η_{ij} denotes the heuristic information between location i and location j. N_i and M_i denote the set of unvisited locations and the set of visited locations respectively. A is a constant, which is used adjust the magnitude of heuristic information. We can observe that the probability of a remote location being chosen will be relatively small. Based on this heuristic information, the ant will prone to construct a route in areas with high travel demand.

3.2 Pheromone with Summation Rule

In the vehicle routing problem especially with a large-scale network, the pheromone on each edge is easy to evaporate because of the sparse distribution of ants. Due to the evaporation phenomenon, the ant colony will lose searching diversity quickly, making it prone to repeat the former routes and liable to trap into local optima. To address these issues, we applied the summation rule [16] to decrease the negative effect. The summation rule of pheromone can extend the spatio-temporal influence of pheromone, ensuring the high concentrations of pheromone continues to be a great impetus to the latter path selection. The possibility of visiting location j when the vehicle is at location i, namely p_{ij}, is shown in (4), where γ is a coefficient which determines the influence intensity of the pheromone at the former location. Based on the special pheromone calculating rule, the state transition rule is defined as follows.

$$j = \begin{Bmatrix} \arg\max_{j \in J_k(i)} \left\{ [\tau(i,j)], [\eta(i,j)]^{\beta} \right\}, & q \le q_0 \\ S, & \text{otherwise} \end{Bmatrix} \tag{3}$$

$$p_k(i,j) = \begin{Bmatrix} \dfrac{\left[\sum_{k=1}^{i} \gamma^{i-k} \tau_{kj}\right]^{\alpha} [\eta_{ij}]^{\beta}}{\sum_{h \in N_i} \left(\left[\sum_{k=1}^{i} \gamma^{i-k} \tau_{kh}\right]^{\alpha} [\eta_{ih}]^{\beta} \right)}, & j \in J_k(i) \\ 0, & \text{otherwise} \end{Bmatrix} \tag{4}$$

Before making node selection, a random number q is produced to compare with parameter q_0. The ant will select the next point to visit, namely exploitation, only if $q \le q_0$. As in (3). Otherwise, the ant will make roulette selection based on heuristic information and pheromone information, the detail of which is shown in (4).

3.3 Pheromone Management

Pheromone Initialization. The initial pheromone should be delicately designed because it can directly affect the searching capacity in the initial state. Plenty of experiments have been used to investigate the appropriate setting of initial pheromone [17]. The initial pheromone has been proven to be related to the characteristic of the network such as length in route scheduling problems. As a result, the initial pheromone in solving CSP is defined as follow,

$$\tau_0 = \sum_{o \in O} \sum_{d \in D} 1/(n_{od} C_{od}^{g}) \tag{5}$$

where C_{od}^{g} is the length of the path between the starting bus location o and the ending bus location d constructed by the greedy algorithm. This path should contain all the locations of passengers at least on time. n_{od} is the number of nodes along the path between the starting point o and the ending point d. It should be highlighted that the maximum number of locations that one vehicle can visit is fixed to be 20.

Local Pheromone Update. The local pheromone updating rule in ACS is used to promote colony exploration. The local pheromone updating rule for solving CSP in this paper is defined as follow,

$$\tau(i,j) = (1 - \xi) \cdot \tau(i,j) + \xi \cdot \tau_0 \tag{6}$$

where $\tau(i,j)$ denotes the amount of pheromone along the edge (i,j), τ_0 denotes the initial pheromone. The parameter ξ denotes the local pheromone volatilization rate, where $0<\xi<1$. The local pheromone updating rule will decrease the amount of pheromone of the historical edges, which encourages the latter ants to explore the unvisited edges.

Global Pheromone Update. The global pheromone updating rule in ACS is used to reallocate the pheromone at the global level. For solving CSP with high seating capacity in this paper, the global pheromone updating rule in HC-ACS is defined as follows,

$$\begin{aligned} \tau(i,j) = (1 - \rho) \cdot \tau(i,j) \\ + \Delta_b(i,j) \end{aligned} \tag{7}$$

$$\Delta_b(i,j) = \begin{cases} (T_b)^{-1}, & (i,j) \text{ on the route } R_b \\ 0, & \text{otherwise} \end{cases} \tag{8}$$

where ρ is the evaporation rate of pheromone on the edge. The global pheromone updating rule only happens on the best path. T_b and Δ_b are the consumption of ant colony with the best solution so far and the amount of pheromone released by that ant colony, respectively. When the global updating rule works, the shorter path will be assigned with more pheromone, which will attract more ants in the next generation.

4 Experiments

In this section, contrast experiments between the proposed HC-ACS and four compared algorithms are carried out to verify the effectiveness of the heuristic information with prediction strategy and pheromone with summation rule in HC-ACS. The proposed HC-ACS and the compared algorithms are implemented in C ++, run on a PC with a Pentium Dual CPU i7 and 4.00 GB RAM. All the experiments are repeated 30 times for each test case for the statistical credibility of data.

4.1 Test Benchmarks

In general, there are three moving configurations in metropolises, including inward radiating movement (CI) configuration, lateral drifting movement (CL) configuration and outward radiating movement (CO) configuration. Two different instances are generated for each kind of dataset respectively in this paper.

4.2 Compared Algorithms

Table 1. Comprehensive result of HC-ACS and BPSO, Hill Climbing Algorithm, Genetic Algorithm, Simulated Annealing Algorithm.

Dataset		HC-ACS	BPSO	GA	HC	SA
CI-1	Mean	**22178.8**	28850.2	25912.8	26444.6	26692.6
	Std.	219.7	524.1	139.0	304.7	277.5
CI-2	Mean	**24654.4**	25628.4	26253.3	27057.6	27007.7
	Std.	160.4	432.9	153.3	336.9	422.4
CL-1	Mean	**22035.6**	22115.4	25883.0	26907.6	26861.5
	Std.	**328.5**	885.7	101.12	400.0	563.3
CL-2	Mean	**22951.6**	24661.4	26180.0	27034.3	27408.0
	Std.	**196.1**	508.7	40.2	345.4	547.4
CO-1	Mean	**24661.4**	26084.8	25744.5	26720.6	26501.7
	Std.	**148.4**	1492.9	32.12	489.3	400.8
CO-2	Mean	25789.0	31687.2	**25774.5**	26782.6	26514.6
	Std.	107.4	768.8	**70.8**	308.1	399.6

We compare the HC-ACS with four compared algorithm including binary particle swarm optimization (BPSO), GA, hill climbing algorithm and simulated annealing algorithm.

The compared algorithms can be divided into two parts: heuristic method including hill climbing algorithm, simulated annealing algorithm, and meta-heuristic method including BPSO, GA. Firstly brought to extend particle swarm optimization (PSO) to solve discrete problems, binary particle swarm optimization (BPSO) has been widely adopted to solve the combinational problems (COP) [9]. Meanwhile, GA, an algorithm based on chromosome crossover and chromosome mutation, is also an effective approach for solving COP [18]. Hill Climbing algorithm is a kind of local searching method. In order to improve its global searching capacity, we adopt a regular restart strategy in this paper. Simulated annealing is a simple but efficient global optimization algorithm, which makes uphill moves with a certain probability.

4.3 Results

The experimental results including the mean value and the standard deviation are shown in Table 1. The mean value illustrates the performance of a specific algorithm, and standard deviation represents the stability of a specific algorithm. It can be seen that HC-ACS performs better than BPSO on all instances, and HC-ACS performs better than GA on 5 out of 6 instances. Based on these results, we concluded that HC-ACS performs better than BPSO and GA in the context of high seating capacity. What is more, HC-ACS performs better than hill climbing algorithm and simulated annealing algorithm on all instances. Similarly, we can conclude that HC-ACS performs better than hill climbing algorithm and simulated annealing algorithm. Moreover, the results obtained from Table 1 reveals that HC-ACO can qualified for solving the CSP under different moving configurations including CI, CL and CO.

5 Conclusion

This paper proposes an HC-ACS for solving CSP in the environment with high seating capacity. While the seating capacity per vehicle increases, the scheduling for internal routing planning will become much more important. For this reason, the delicate redesign of heuristic information calculation rule and pheromone usage rule guarantee effective guidance and the global searching ability when ACS is applied to solve CSP with high seating capacity. In this paper, we apply HC-ACS to solve CSP without time window constraints. In the future, it is interesting to apply distributed and parallel ACO algorithms [19–22] to solve large-scale CSPs.

Acknowledgement. This work was supported in part by the National Natural Science Foundation of China under Grant 61622206, Grant 61976093 and Grant 61873097, in part by the Science and Technology Plan Project of Guangdong Province under Grant 2018B050502006, and in part by Guangdong Natural Science Foundation Research Team 2018B030312003.

References

1. Barrero, R., Van Mierlo, J., Tackoen, X.: Energy savings in public transport. IEEE Veh. Technol. Mag. **3**(3), 26–36 (2008)
2. Huang, S., Jiau, M., Lin, C.: A genetic-algorithm-based approach to solve carpool service problems in cloud computing. IEEE Trans. Intell. Transp. Syst. **16**(1), 352–364 (2015)
3. Qadir, H., Khalid, O., Khan, M.U.S.: An optimal ride sharing recommendation framework for carpooling services. IEEE Access **6**, 62296–62313 (2018)
4. Duan, Y., Mosharraf, T., Wu, J., Zheng, H.: Optimizing carpool scheduling algorithm through partition merging. IEEE Int. Conf. Commun **2018**, 1–6 (2018)

5. Yang, Q., et al.: Adaptive multimodal continuous ant colony optimization. IEEE Trans. Evol. Comput. **21**(2), 191–205 (2017)
6. Chen, W.-N., et al.: Ant colony optimization for the control of pollutant spreading on social networks. IEEE Trans. Cybern. in press (2019)
7. Yu, X., et al.: ACO-A*: ant colony optimization plus A* for 3D traveling in environments with dense obstacles. IEEE Trans. Evol. Compt. **23**(4), 617–631 (2019)
8. Yang, Q., Member, S., Chen, W., Member, S., Da Deng, J.: A level-based learning swarm optimizer for large-scale optimization. IEEE Trans. Evol. Comput. **22**(4), 578–594 (2018)
9. Mirjalili, S., Lewis, A.: S-shaped versus V-shaped transfer functions for binary particle swarm optimization. Swarm. Evol. Comput. **9**, 1–14 (2013)
10. Deb, K., Pratap, A., Agarwal, S., Meyarivan, T.: A fast and elitist multiobjective genetic algorithm: NSGA-II. IEEE Trans. Evol. Comput. **6**(2), 182–197 (2002)
11. Davis, L.: Bit-climbing, representational bias, test suite design. In: Transportation Research (1991)
12. van Laarhoven, P.J., Aarts, E.H..: Simulated Annealing: Theory Applications, pp. 7–15 (1987)
13. Huang, S.C., Jiau, M.K., Liu, Y.P.: An ant path-oriented carpooling allocation approach to optimize the carpool service problem with time windows. IEEE Syst. J. **13**, 1–12 (2018)
14. Chou, S.K., Jiau, M.K., Huang, S.C.: Stochastic set-based particle swarm optimization based on local exploration for solving the carpool service problem. IEEE Trans. Cybern. **46**(8), 1771–1783 (2016)
15. Wang, B.I.N., Zhu, R.U.I., Zhang, S., Zhao, Z., Yang, X., Wang, G.: PPVF: a novel framework for supporting path planning over carpooling. IEEE Access **7**, 10627–10643 (2019)
16. Daniel, M., Middendorf, M., Schmeck, H.: Ant colony optimization for resource-constrained project scheduling. IEEE Trans. Evol. Comput. **6**(4), 333–346 (2002)
17. Mazzeo, S., Loiseau, I.: An ant colony algorithm for the capacitated vehicle routing. Electron. Notes Discret. Math **18**, 181–186 (2004)
18. Fujita, K., Akagi, S., Hirokawa, N.: Hybrid approach for optimal nesting using a genetic algorithm and a local minimization algorithm. In: Proceedings of the 19th Annual ASME Design Automation Conference (1993)
19. Chen, W.-N., et al.: A cooperative co-evolutionary approach to large-scale multisource water distribution network optimization. IEEE Trans. Evol. Comput. in press (2019)
20. Jia, Y.-H., et al.: Distributed cooperative co-evolution with adaptive computing resource allocation for large scale optimization. IEEE Trans. Evol. Comput. **23**(2), 188–202 (2019)
21. Yang, Q., et al.: A distributed swarm optimizer with adaptive communication for large scale optimization, IEEE Trans. Cybern. in press (2019)
22. Song, A., Chen, W.-N., Gong, Y.-J., Luo, X., Zhang, J.: A divide-and-conquer evolutionary algorithm for large-scale virtual network embedding. IEEE Trans. Evol. Comput. in press (2019)

Enhancing Artificial Bee Colony Algorithm with Directional Information

Qiyu Cai, Xinyu Zhou(✉), Anquan Jie, Maosheng Zhong, and Mingwen Wang

School of Computer and Information Engineering, Jiangxi Normal University, Nanchang, China
xyzhou@jxnu.edu.cn

Abstract. Artificial bee colony (ABC) algorithm is a swarm intelligence based optimization technique, which has attracted wide attention from different research fields. In the basic ABC, however, the same solution search equation is used in both of the employed bee phase and onlooker bee phase, which performs well in exploration but poorly in exploitation. To address this concerning defect, in this paper, we propose an improved ABC variant by designing a mechanism of utilizing directional information. In this mechanism, we first construct a pool of differential vectors in the employed bee phase, and then utilize a differential vector randomly selected from the pool as directional information to guide search in the onlooker bee phase. Furthermore, we propose two novel solution search equations based on the current best solution and some good solutions with the aim of balancing exploration and exploitation. Experiments are conducted on a set of 22 well-known benchmark functions, and the results demonstrate that our proposed approach shows promising performance.

Keywords: Artificial bee colony · Directional information · Solution search equation · Exploration and exploitation

1 Introduction

To solve complicated optimization problems, many evolutionary optimization techniques have been developed in recent decades, such as the genetic algorithm (GA) [1], particle swarm optimization (PSO) [2], differential evolution algorithm (DE) [3] and artificial bee colony algorithm (ABC) [4]. Among these techniques, ABC is a relatively new one which simulates the intelligent foraging behavior of the honey bee colony. In comparison with other optimization techniques, ABC has some advantages, such as fewer control parameters and simpler structure. Although ABC has shown good performance, it still has a deficiency concerning its solution search equation which is good at exploration but poor at exploitation.

To overcome this deficiency, numerous improved ABC variants have been proposed. For example, motivated by PSO, Zhu et al. [5] proposed a gbest-guided ABC (GABC) algorithm in which the global best solution is integrated into the solution search equation to improve exploitation. Gao et al. [6] presented a novel solution search equation like the crossover operation of GA in their proposed CABC algorithm, which has no bias to any

T. Gedeon et al. (Eds.): ICONIP 2019, CCIS 1142, pp. 741–749, 2019.
https://doi.org/10.1007/978-3-030-36808-1_81

search direction. Zhou et al. [7] designed a Gaussian bare-bones ABC (GBABC) which utilizes the global best solution as well, and the reported experimental results showed that GBABC can offer higher solution quality. Recently, Cui et al. [8] introduced a depth-first search framework and elite-guided solution search equation (DFSABC_elite) by using some good solutions to balance exploration and exploitation.

From the above representative ABC variants, it is not difficult to observe that the global best solution or some good solutions are usually used in the modified solution search equation. While these solutions can effectively improve exploitation, if they are improperly used, it is possible to cause a problem that the algorithm becomes too greedy. To solve this issue, we propose a mechanism of utilizing directional information in which a pool of differential vectors is constructed. In the employed bee phase, a differential vector can be obtained when a food source is improved, which aims to preserve the beneficial information among different good solutions. Then in the onlooker bee phase, one differential vector is randomly selected from the pool to be used to guide search. The mechanism of utilizing directional information is helpful to properly use some good solutions without sacrificing diversity. It is necessary to point out that this proposed mechanism is inspired by the concept of directional mutation operator for DE (DMDE) [9]. Furthermore, we designed two new solution search equations based on the current best solution and some good solutions with the aim of balancing exploration and exploitation. In the experiments, 22 well-known benchmark functions and five relatively new ABC variants are used. The comparative results indicate that our approach offers promising performance.

2 Basic ABC Algorithm

The ABC algorithm is inspired by the intelligent foraging behavior of bee colony. Its search process is divided into three phases: employed bee phase, onlooker bee phase and scout bee phase. Generally, ABC begins with an initial population of *SN* food sources which are randomly generated according to the Eq. (1).

$$x_{i,j} = x_j^L + rand_j \cdot (x_j^U - x_j^L) \tag{1}$$

where $X_i = (x_{i,1}, x_{i,2}, \cdots, x_{i,D})$ represents the *i*th food source, $i \in \{1, 2, \cdots, SN\}$, $j \in \{1, 2, \cdots, D\}$, and D denotes the problem dimension size. Note that a food source corresponds to a candidate solution of the problem. x_j^L and x_j^U are the lower and upper bounds for the *j*th dimension, respectively. The three phases are described as follows.

(1) Employed bee phase

In this phase, each employed bee generates a new food source V_i by the Eq. (2). If the fitness value of V_i is better than its parent X_i, then X_i is replaced by V_i.

$$v_{i,j} = x_{i,j} + \Phi_{i,j} \cdot (x_{i,j} - x_{k,j}) \tag{2}$$

where X_k is a randomly selected food source and has to be different from X_i. $j \in \{1, 2, \cdots, D\}$ is a randomly selected dimension, and $\Phi_{i,j}$ is a uniformly distributed random number within the range $[-1, 1]$.

(2) Onlooker bee phase

After all of the employed bees finish their work, they will share information about nectar with the onlooker bees. Then the onlooker bees will continue to search for new food sources with the same solution search equation listed in the Eq. (2). However, being different from the search patter of the employed bees, the onlooker bees select the existing food source based on the probability which is calculated by the Eq. (3), and fit_i denotes the fitness value of the ith food source.

$$p_i = \frac{fit_i}{\sum_{j=1}^{SN} fit_j} \quad (3)$$

(3) Scout bee phase

If a food source cannot be improved for at least consecutively *limit* times, it will be considered to be exhausted and requires to be reset. In this case, the Eq. (1) is used to reset the food source.

3 The Proposed Method

3.1 The Mechanism of Utilizing Directional Information

In the basic ABC, its solution search equation performs well in exploration but poorly in exploitation. Although many modified solution search equations utilizing good solutions have shown better performance, it may cause the problem of making the algorithm become too greedy. To overcome this issue, we propose a mechanism of utilizing directional information, which attempts to properly utilize good solutions without sacrificing diversity. This mechanism is inspired by the concept of directional mutation operator for DE (DMDE) [9]. In DMDE, if a child individual is better than its parent individual, then the difference vector between these two individuals are considered as directional information which will be utilized in the mutation operator for guiding search. Motivated by the DMDE, we propose a mechanism of utilizing directional information from DE to ABC, in which two steps are included.

First, in the employed bee phase, we will check the quality of new food sources, if V_i is better than its parent food source X_i, then we will continue check whether V_i is better than the current best food source X_{best}. If V_i is indeed better than X_{best}, we consider that the difference vector $V_i - X_i$ represents a promising search direction, and it is worth preserving this difference vector. As a result, we construct a pool to contain all of these difference vectors. Let δ denotes the pool, and it can be formally represented as $\delta = \{\lambda_1, \lambda_2, \ldots \lambda_N\}$, where λ_i is the ith difference vector and N is the pool size which meets the condition: $0 \leq N \leq SN$. Note that being different from other evolutionary algorithms, such as the DE algorithm, only one dimension of a parent food source is updated to generate a new food source in the basic ABC, and it implies that in fact the difference vector is a scalar in this case. In addition, the pool is reset at every generation and it may be empty when all of the new food sources are worse than X_{best}.

Second, in the onlooker bee phase, the directional information is utilized to guide search. If the pool of difference vector is not empty, we will randomly select one difference vector from the pool and then incorporate it into the solution search equation. To maximize the performance of the directional information, the following new solution search equation is defined.

$$v_{e,j} = x_{best,j} + \Phi_{e,j} \cdot \lambda_i \tag{4}$$

where V_e is the new food source for the corresponding parent food source X_e, X_e is an elite food source. Note that the top 10%•*SN* food sources are regarded as elite food sources. X_{best} is the current best food source, $\Phi_{e,j}$ is a uniformly distributed random number within [−1, 1], and λ_i is a randomly selected difference vector. If the pool is empty, however, another new solution search equation listed in the following Eq. (8) will be used for the onlooker bees. In addition, the roulette selection mechanism is removed for the onlooker bees.

3.2 Two New Solution Search Equations

As pointed out by many other researchers, the original solution search equation does well in exploration but badly in exploitation, and this may result in slow convergence speed for ABC. To solve this issue, being inspired by the ABC_elite algorithm [8], we further design a new solution search equation for the employed bees and onlooker bees, respectively. In the ABC_elite, two modified solution search equations are designed as follows.

$$v_{i,j} = x_{e,j} + \Phi_{i,j} \cdot \left(x_{e,j} - x_{k,j}\right) \tag{5}$$

$$v_{e,j} = \frac{1}{2} \cdot \left(x_{e,j} + x_{best,j}\right) + \Phi_{e,j} \cdot \left(x_{best,j} - x_{k,j}\right) \tag{6}$$

where X_e is an elite food source selected from the top 10%•*SN* food sources, X_k is a randomly selected food source and has to be different from X_i. The Eq. (5) is used for the employed bees, while the Eq. (6) is for the onlooker bees. Although these two equations have shown good performance, it can be observed that the base vectors in these two equations only include the current best food sources or some elite food source, and this may bias the search direction only towards them and easily trigger the premature problem.

To avoid this issue of ABC_elite, we design two new solution search equations listed in the following Eqs. (7) and (8).

$$v_{i,j} = \frac{1}{2} \cdot (x_{best,j} + x_{k,j}) + \Phi_{i,j} \cdot (x_{e,j} - x_{k,j}) \tag{7}$$

$$v_{e,j} = \frac{1}{2} \cdot (x_{best,j} + x_{k,j}) + \Phi_{i,j} \cdot (x_{e,j} - x_{k,j}) \tag{8}$$

The Eq. (7) is designed for the employed bees, while the Eq. (8) is used for the onlooker bees. In comparison with the Eq. (5), the base vector in the Eq. (7) includes a randomly selected food source, which is helpful to enhance diversity. Compared with the Eq. (6), although the base vector in the Eq. (8) changes X_e to X_{best}, the difference vector replaces X_{best} with X_e, and it is beneficial to add more perturbation.

3.3 The Pseudocode of Our Approach

To better clarify our approach, abbreviated as DIABC, the pseudocode is described in the Algorithm 1. In there, *FEs* denotes the used number of fitness function evaluations, and *MaxFEs* is the maximum number of fitness function evaluations.

Algorithm 1. The pseudocode of DIABC

```
1:  Randomly generate SN food sources as an initial population according to the Eq. (1)
2:  while FEs ≤ MaxFEs do
3:      Initialize a pool δ to preserve the difference vectors
4:      for i=1 to SN do            //The employed bee phase
5:          Generate a new food source V_i according to the Eq. (7)
6:          Evaluate V_i and set FEs = FEs + 1
7:          if f(V_i) < f(X_i) do
8:              if f(V_i) < f(X_best) do
9:                  Insert the difference vector λ_i = V_i − X_i into the pool δ
10:             end if
11:             Replace X_i with V_i
12:         else
13:             count_i = count_i + 1
14:         end if
15:     end for
16:     for i=1 to SN do            //The onlooker bee phase
17:         if the pool δ is empty do
18:             Generate a new food source V_e according to the Eq. (8)
19:         else
20:             Generate a new food source V_e according to the Eq. (4)
21:         Evaluate V_e and set FEs = FEs + 1
22:         if f(V_e) < f(X_e) do
23:             Replace X_e with V_e
24:         else
25:             count_e = count_e + 1
26:         end if
27:     end for
28:     if max(count_i) ≥ limit do  //The scout bee phase
29:         Reset X_i according to the Eq. (1)
30:     end if
31: end while
```

4 Experiments and Discussion

To verify the performance of our approach, 22 well-known benchmark functions are used, and they are widely used in other literatures as well. Among these functions, the first 11 ones are unimodal types and the remaining ones are multimodal types. The global optimum of all of these functions are zero. Due to the limited paper space, the definitions about these functions can refer to [7] and [12]. In the experiments, five well-established ABC variants are compared with our approach, they are: CABC [6], GBABC [7], ILABC [11], MGPABC [10], DFSABC_elite [8].

Table 1. The comparison results of DIABC with other five ABC variants for D = 30

Function	CABC	GBABC	ILABC	MPGABC	DFSABC_elite	DIABC
F_1	3.00E-50 +	1.33E-33 +	5.60E-56 +	8.07E-53 +	2.67E-83 +	4.23E-90
F_2	1.41E-26 +	3.11E-21 +	1.23E-29 +	6.83E-29 +	5.03E-43 +	1.26E-46
F_3	1.15E+04 +	1.77E+03 +	8.15E+03 +	5.05E+03 =	4.14E+03 -	5.53E+03
F_4	2.05E+00 +	2.04E-01 +	6.27E-01 +	1.15E+00 +	8.78E-02 -	5.74E-02
F_5	1.63E-01 =	7.36E+00 +	2.90E-02 -	3.17E+00 =	4.85E-01 =	4.66E-01
F_6	0.00E+00 =	0.00E+00 =	0.00E+00 =	0.00E+00 =	0.00E+00 =	0.00E+00
F_7	1.63E-02 +	1.54E-02 +	1.53E-02 +	2.46E-02 +	1.26E-02 +	9.78E-03
F_8	7.33E-42 +	5.40E-26 +	1.42E-48 +	2.62E-50 +	7.87E-80 +	3.90E-87
F_9	6.02E-52 +	1.70E-34 +	5.67E-57 +	1.55E-53 +	9.61E-85 +	4.27E-93
F_{10}	5.20E-31 +	1.45E-71 -	1.16E-43 +	6.74E-58 -	2.44E-54 +	2.80E-55
F_{11}	7.18E-66 =	7.18E-66 =	7.18E-66 =	7.18E-66 =	7.18E-66 =	7.18E-66
F_{12}	1.53E-12 +	1.16E-12 +	1.02E-12 +	1.16E-12 +	7.28E-13 +	2.18E-13
F_{13}	0.00E+00 =	7.11E-16 =	0.00E+00 =	0.00E+00 =	0.00E+00 =	0.00E+00
F_{14}	2.94E-14 +	1.98E-14 =	2.82E-14 +	3.50E-14 +	2.52E-14 +	2.06E-14
F_{15}	6.73E-11 =	0.00E+00 =	1.84E-13 =	4.12E-04 =	0.00E+00 =	0.00E+00
F_{16}	1.57E-32 =	1.83E-32 =	1.57E-32 =	1.57E-32 =	1.57E-32 =	1.57E-32
F_{17}	1.35E-32 =	5.53E-32 -	1.35E-32 =	1.35E-32 =	1.35E-32 =	1.35E-32
F_{18}	0.00E+00 =	0.00E+00 =	0.00E+00 =	0.00E+00 =	0.00E+00 =	0.00E+00
F_{19}	1.49E-27 -	7.53E-09 +	8.72E-30 -	5.23E-08 +	7.55E-17 =	8.33E-16
F_{20}	1.35E-31 =	2.83E-31 =	1.35E-31 =	1.35E-31 =	1.35E-31 =	1.35E-31
F_{21}	0.00E+00 =	0.00E+00 =	0.00E+00 =	0.00E+00 =	0.00E+00 =	0.00E+00
F_{22}	0.00E+00 -	6.14E-03 =	0.00E+00 -	4.50E-03 =	1.16E-03 -	2.77E-03
+ / = / -	10/10/2	10/10/2	10/9/3	9/12/1	8/11/3	–

To make a fair comparison, the specific parameters of these five ABC variants are kept the same with their original literatures. For the proposed DIABC, the elite food sources are set to the top 10%·*SN* food sources, and *limit* is set to *SN*·*D*. For the other common parameters, they are set as follows: $SN = 50$, $D = 30$ or 50, and $MaxFEs = 5000 \cdot D$. Furthermore, the paired Wilcoxon signed-rank test with $\alpha = 0.05$ is used to compare the statistically significant difference of two algorithms. The signs "+", "=", and "−" indicate DIABC is significantly better than, equals to, and worse than the competitor, respectively. All of the algorithms are run 30 times on each function independently, and the average best fitness value of each function is recorded in the Tables 1 and 2.

Table 2. The comparison results of DIABC with other five ABC variants for D = 50

Function	CABC	GBABC	ILABC	MPGABC	DFSABC_elite	DIABC
F_1	1.48E-49 +	1.84E-26 +	1.78E-54 +	7.77E-51 +	7.66E-83 +	6.77E-89
F_2	4.48E-26 +	1.09E-18 +	1.19E-28 +	7.09E-28 +	7.12E-43 +	1.50E-46
F_3	3.10E+04 +	1.77E+04 =	2.60E+04 +	1.78E+04 =	1.35E+04 -	1.78E+04
F_4	8.80E+00 +	3.45E+00 +	4.57E+00 +	5.91E+00 +	7.12E-01 +	5.53E-01
F_5	1.56E-01 =	7.16E+01 +	2.64E-02 -	7.97E-02 =	9.47E-01 =	3.49E+00
F_6	0.00E+00 =	0.00E+00 =	0.00E+00 =	0.00E+00 =	0.00E+00 =	0.00E+00
F_7	3.82E-02 +	3.86E-02 +	3.25E-02 +	4.87E-02 +	2.53E-02 +	1.73E-02
F_8	4.11E-41 +	3.27E-36 +	4.72E-47 +	1.26E-47 +	2.20E-79 +	2.11E-86
F_9	7.83E-51 +	1.18E-26 +	9.90E-55 +	3.50E-52 +	2.18E-83 +	1.23E-89
F_{10}	7.12E-31 +	1.09E-73 -	2.99E-43 +	1.21E-57 -	2.23E-54 +	1.78E-54
F_{11}	2.67E-109 =	2.67E-109 =	2.67E-109 =	2.67E-109 =	2.67E-109 =	2.67E-109
F_{12}	2.24E-11 +	4.74E+00 +	1.97E-03 +	1.83E-11 =	4.74E+00 +	1.83E-11
F_{13}	0.00E+00 =	1.22E-01 +	0.00E+00 =	0.00E+00 =	0.00E+00 =	0.00E+00
F_{14}	5.44E-14 +	9.20E-13 +	5.22E-14 +	6.74E-14 +	4.81E-14 +	3.92E-14
F_{15}	1.86E-13 =	0.00E+00 =	0.00E+00 =	0.00E+00 =	0.00E+00 =	0.00E+00
F_{16}	9.42E-33 =	1.27E-28 +	9.42E-33 =	9.42E-33 =	9.42E-33 =	9.42E-33
F_{17}	1.35E-32 =	4.24E-29 +	1.35E-32 =	1.35E-32 =	1.35E-32 =	1.35E-32
F_{18}	0.00E+00 =	4.00E-02 =	0.00E+00 =	0.00E+00 =	0.00E+00 =	0.00E+00
F_{19}	3.85E-27 -	1.97E-07 +	3.33E-17 -	4.22E-07 +	4.22E-17 -	1.67E-15
F_{20}	1.35E-31 =	2.32E-25 +	1.35E-31 =	1.35E-31 =	1.35E-31 =	1.35E-31
F_{21}	0.00E+00 =	0.00E+00 =	0.00E+00 =	0.00E+00 =	0.00E+00 =	0.00E+00
F_{22}	1.71E-15 -	1.43E-02 +	2.84E-15 =	1.81E-02 +	0.00E+00 -	1.83E-15
+ / = / -	10/10/2	15/6/1	10/10/2	9/12/1	9/10/3	–

For D = 30, in the Table 1, it is clear that DIABC achieves the best overall performance among the involved six algorithms. To be specific, DIABC performs better than CABC, GBABC, ILABC, MPGABC and DFSABC_elite on 10, 10, 10, 9 and 8 test functions, respectively. Compared with MPGABC, DIABC defeats it on 9 functions and only lose one on the SumPower function (F_{10}). For D = 50, in the Table 2, we can see that similar conclusions can be drawn as in the case of D = 30. As seen, although the complexity increases with the dimension size, the performance of DIABC is not always affected, and it is also superior to the other five ABC variants. In addition, we conduct the Friedman test on the final results for the involved six algorithms. The Table 3 shows the average rankings for D = 30 and 50, and the best results are marked in **boldface**. It can be seen that both the best average rankings are obtained by DIABC for two dimension sizes.

Table 3. Average rankings of the involved six algorithms

Algorithms	Average rankings	
	$D = 30$	$D = 50$
CABC	4.11	3.93
GBABC	4.34	4.82
ILABC	3.21	3.32
MPGABC	4.02	3.66
DFSABC_elite	2.73	2.77
DIABC	**2.59**	**2.52**

5 Conclusions

In the basic ABC, the solution search equation is good at exploration but poor at exploitation, and this may result in slow convergence for ABC. To solve this concerning issue, many modified solution search equations have been proposed in recent years. Among these modified solution search equations, most of them focus on utilizing the global best food source or some good food sources to enhance exploitation. Although the performance of ABC is indeed improved by this way, it may also cause a problem that ABC would easily be too greedy. Thus how to design a mechanism of utilizing good food sources to enhance exploitation is a challenging topic.

In this paper, we propose a mechanism of utilizing directional information to meet the challenge. In the proposed mechanism, we construct a pool to preserve difference vectors between a promising new food source and its parent food source, then the difference vectors are utilized in a new designed solution search equation. Furthermore, in order to maximize the performance of the proposed mechanism, we design another two new solution search equations for the employed bees and onlooker bees. The experiments are conducted on 22 benchmark functions, and five well established ABC variant are compared with our approach. The comparison results indicate that our approach can offer better performance.

Acknowledgments. This work is supported by the National Natural Science Foundation of China (Nos. 61603163, 61966019, 61877031 and 61876074), the Science and Technology Foundation of Jiangxi Province (No. 20192BAB207030).

References

1. Tang, K., Man, K., Kwong, S., He, Q.: Genetic algorithms and their applications. IEEE Signal Process. Mag. **13**(6), 22–37 (1996)
2. Wang, F., Zhang, H., Li, K., Lin, Z., Yang, J., Shen, X.: A hybrid particle swarm optimization algorithm using adaptive learning strategy. Inf. Sci. **436**, 162–177 (2018)
3. Zhou, X., Wu, Z., Wang, H., Rahnamayan, S.: Enhancing differential evolution with role assignment scheme. Soft Comput. **18**(11), 2209–2225 (2014)

4. Karaboga, D.: An idea based on honey bee swarm for numerical optimization, Kayseri: Engineering Faculty Computer Engineering Department, Ereiyes University (2005)
5. Zhu, G.P., Kwong, S.: Gbest-guided artificial bee colony algorithm for numerical function optimization. Appl. Math. Comput. **217**(7), 3166–3173 (2010)
6. Gao, W.: A novel artificial bee colony algorithm based on modified search equation and orthogonal learning. IEEE Trans. Cybern. **43**(3), 1011–1024 (2013)
7. Zhou, X., Wu, Z., Wang, H., Rahnamayan, S.: Gaussian bare-bones artificial bee colony algorithm. Soft Comput. **20**(3), 907–924 (2016)
8. Cui, L., Li, G., Lin, Q., Du, Z., Gao, W., Chen, J., Lu, N.: A novel artificial bee colony algorithm with depth-first search framework and elite-guided search equation. Inf. Sci. **367–368**, 1012–1044 (2016)
9. Zhang, X., Yuen, S.Y.: A directional mutation operator for differential evolution algorithms. Appl. Soft Comput. **30**, 529–548 (2015)
10. Cui, L., Zhang, K., Li, G., Fu, X., Wen, Z.: Modified Gbest-guided artificial bee colony algorithm with new probability model. Soft Comput. **22**(7), 2217–2243 (2018)
11. Gao, W.F., Huang, L.L., Liu, S.Y., Dai, C.: Artificial bee colony algorithm based on information learning. IEEE Trans. Cybern. **45**(12), 2827–2839 (2017)
12. Zhou, X., Wang, H., Wang, M., Wan, J.: Enhancing the modified artificial bee colony algorithm with neighborhood search. Soft Comput. **21**(10), 2733–2743 (2017)

Using Evolutionary Algorithms for Hyperparameter Tuning and Network Reduction Techniques to Classify Core Porosity Classes Based on Petrographical Descriptions

Tommy Liu(✉) and Jo Plested

Research School of Computer Science, The Australian National University, Canberra, Australia
{tommy.liu,jo.plested}@anu.edu.au

Abstract. Classifying the porosity of sedimentary information is an important field of study with applications to tasks such as oil reservoir characterisation. Classifying porosity into groups based on Petrographical characteristics has been attempted in the past using: expert systems, supervised clustering techniques and neural networks. In this paper, we expand upon the usage of neural networks for this classification task by applying Evolutionary Algorithms to determine optimal parameters. Despite recent advances in techniques to select hyperparameters it is still difficult to determine the optimal parameters for a given dataset. We further apply network reduction techniques to further improve classification accuracy. We produce results similar to the work done by Gedeon et al. [1] on this dataset.

Keywords: Artificial neural networks · Petrographical features · Network reduction · Porosity · Evolutionary Algorithms

1 Introduction

In this paper we develop a classification task to classify the core porosity of minerals into 4 categories (very poor, poor, fair, good) based upon a number of petrographical descriptions. We expand upon Gedeon et als work [1] which compares different methods of classifying the porosity classes using expert systems, supervised clustering and neural networks. Their paper compared these algorithms and came to the conclusion that neural networks are the best solution for this problem because of the relatively good results and ease of reproducing the results, we will refer to their work as the original paper. The aim of this paper is to expand upon the usage of neural networks to solve this porosity classification problem. We apply a genetic algorithm to determine the optimal

T. Gedeon et al. (Eds.): ICONIP 2019, CCIS 1142, pp. 750–757, 2019.
https://doi.org/10.1007/978-3-030-36808-1_82

Table 1. Description of the Petrographical data set

	Number of attributes	Description
% porosity	Range 2–22%	The percentage porosity of the sediment or rock
Grain size	12	The dimensions such as diameter, volume and density of the particles in the given mineral layer
Sorting	8	Similarity measures between characteristics of particles
Matrix	16	A descriptor of the type of material which encloses or fills the smaller gaps between larger particles of the layer
Roundness	8	The degree that the particles found in the given layer are rounded with smooth edges and corners
Bioturbation	4	How much the layer has been disturbed by living organisms
Laminae	10	The smallest unit found by inspection in the given layer

hyper-parameters for the neural network like in [2]. We apply the network reduction technique of relevance [3] to our model produced to improve classification performance and generalisation of the model.

We use the same data-set as found in [1]. This data set is obtained from an oil well located in the North-West Shelf, offshore Australia. This data set consists of 226 samples, each sample is described by six porosity related petrographical descriptions along with the percentage porosity. More detailed descriptions of the data set can be found in Table 1. The descriptions of the features are categorical linguistic pieces of information obtained by the people working in the field. This makes the task of classification difficult since we must pre-process the data and convert it to a quantitative representation which can be used to train our neural network with.

We divide the porosity percentage into four classes: very poor [0, 5%), poor [5, 10%), fair [10, 15%) and good [15, 100%]. Our neural network takes an encoded version of the petrographical descriptions and predicts this class. A simple one hidden layer feed-forward neural network has been demonstrated to perform well on this data set [1] and achieves a classification accuracy of about 60%. Instead of training a simple fully connected neural network like in [1], we run a genetic algorithm to determine what choice of hyper-parameters we will use for our network. We then apply network reduction techniques to the final model determined by our algorithm. Network reduction techniques can increase the performance of our network and reduce the size of the network which may result in better generalisation [3] of our model.

2 Methodology

Missing Data. Some of the data points in our data set have one or more missing features. We keep the points with more than three feature values present, this leaves us with a data set of 159 points. The original paper [1] uses 140 samples so we believe that techniques to increase our available data such as imputation are not required. We judge our sample size to be sufficient in order to reproduce and improve upon the results presented in the original paper. There also are some mismatches in the features between our data set and the one discussed in the original paper.

Data Preprocessing. We represent each feature value as a number based on their ordering. For most of the features (Grain size, Sorting, Matrix, Bioturbation, Laminae) we represent them as ordered numerical values from from 0 up to the number of different classes (i.e. 0–15 for Matrix). For the roundness attribute we implement a circular encoding like [1] which encodes the feature into two features based on their circular distances from each other in terms of sine and cosine. All feature values are then normalised into the range [0, 1]. Previous studies have found that normalisation of inputs prior to training is crucial to obtaining good results and increases the speed of convergence [8].

We split our data set into training (109) and testing (50) sets. We do this for better robustness of training results since a sample size of 79 (50/50 split) may be too small to train a network that generalises well. To obtain our training and testing sets, we shuffle our original set of 159 randomly then select the first 109 values as our training set.

Evaluation Metrics. For each of the four porosity classes we calculate precision, recall and F1-score and take the averages across classes. We will mostly look at the average F1-Score to determine differences in model performance [5]. We will use the term accuracy to broadly refer to any of these evaluation metrics. We will use precision when comparing with the model present in [1] as that is the metric that they use for percentage correct values.

3 Evolutionary Algorithm for Hyper-parameter Tuning

Representation. We develop a representation of a neural network, this representation contains six hyper parameters which we tune (layers and no. neurons in layers, dropout percentage between layers, learning rate, epochs to train for, whether to learn bias or not, activation function). An example of a representation of an ANN is seen:

"{'lr': 0.08, 'neurons': array([25, 30]), 'dropout': array([0., 0.25]), 'bias': True, 'epochs': 2500, 'inputsize': 7, 'activation': 4, 'outputsize': 4, 'lasthidden': 30}"

Initialisation. We develop a function that generates random representations of an ANN. When we initialise each ANN, the hyper-parameter values are restricted within reasonable ranges based on empirical observations, we provide an upper bound to ranges so that our training time is not too large.

Fitness Function. To assess the fitness of a model, we use the average validation accuracy across three rounds of training and validation. Our validation sets consist of 29 random values selected from our 109 values and as a result the same value could be present in each round of validation. We change this validation set at every generation of our algorithm.

Operators. We define our cross over operator as a function that takes two sets of neural network hyperparameters and returns a new set of network hyperparameters. For each hyper-parameter in the parent networks we have an equal (50%) chance to have the same hyper-parameter in the child network. We define a mutation operator which takes a single neural network and modifies either one or two of the hyper-parameters randomly. Each mutation has an equal chance to be selected, the mutation operators are the following: Increase or decrease the learning rate by a number in the range [0, 0.02]; randomise the value of a dropout layer; modify the number of neurons within a given layer, increases or decreases by [0, (no. of neurons in layer/3)]; adds or removes a layer in the network; change the number of epochs the network trains for by [250, 500]; randomises the activation function of the network. The values present for much of these operations are kept relatively small in order for each mutation to affect a given network too much.

Selection. We use a simple form of selection where the worst five performers according to the fitness function are eliminated. We also apply elitism so that top two individuals goes through to the next generation unchanged [7]. At every generation we apply a chance of mutation to each of the models (except the top ranking elite models). This chance to mutate starts off high and falls linearly as the generation increases, down to 2% at the last generation. At each generation we keep track of the best model in that generation in a 'hall of fame'.

Evolutionary Algorithm Testing and Results. We apply our selection criteria to a population of size 8 over the period of 50 generations. We evaluate and plot our models according to our fitness criteria and compare them against the test set, Fig. 1. We observe the upward trend in mean fitness scores as the generation increases. We note there is no significant increase in the F-scores on the test set as the generation increases. This suggests that the fitness score does not have a strong correlation with the optimisation problem we are solving. There are several possible causes for this such as our training/testing set random split simply being unfavourable resulting in week generalisation from our training set. The performance could also increase if we have more iterations. The final model that we select performs the best in testing at generation 14.

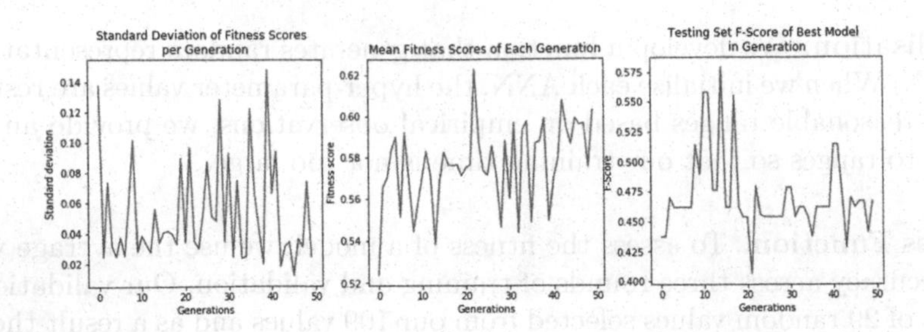

Fig. 1. Diagram of standard deviation plotted against generation, mean fitness against generation and testing set accuracy against generation

Performance of Selected Model. The model we obtain from our evolutionary algorithm consists of two layers, the first with 7 neurons and the second with 31 neurons, it uses the Leaky ReLu activation function with near 0 dropout between any layers and is trained for 2666 epochs with a learning rate of 0.1197. We record the results of this model in Table 2.

Table 2. Test set and training set results for best model (14)

Training data set (109)							Testing data set (50)						
		VP	PR	FR	GD	%Correct			VP	PR	FR	GD	%Correct
Actual class	VP	17	7	1	7	68.00%	Actual class	VP	6	4	1	0	54.54%
	PR	1	33	2	1	89.19%		PR	0	6	3	1	60.00%
	FR	0	3	15	4	68.18%		FR	2	4	6	3	40.00%
	GD	1	0	1	23	92.00%		GD	0	1	2	11	78.57%
F-1 score	79.93%						Average F-1 score	57.87%					

4 Relevance Network Measure

The papers [3] and [10] mention Relevance as a metric to determine neurons to remove. Removing neurons has two effects: improving generalisation; and speeding up learning and prediction [10]. It is difficult to determine the optimal number of neurons in any given hidden layer. Too few neurons will result in too little fitting power to capture patterns in the dataset and too many will result in overfitting occurring in the dataset. Given that our network in the previous section achieves a train/test accuracy of around 80%/58% we believe that our model may be overfitting, resulting in poor generalisation. Even with our evolutionary algorithm, we cannot determine the optimal number of neurons to have in our network because of the computational complexity and time required for our evolutionary algorithm to converge. We only run our evolutionary algorithm for 50 epochs which means our model can still be improved. Therefore we apply the network reduction technique of relevance in order to prune them in such

Table 3. Final performance results of our model

Training data set (109)							Testing data set (50)						
		VP	PR	FR	GD	%Correct			VP	PR	FR	GD	%Correct
Actual class	VP	16	7	2	0	64.00%	Actual class	VP	5	4	2	0	45.45%
	PR	1	33	2	1	89.19%		PR	0	6	3	1	60.00%
	FR	0	3	15	4	68.18%		FR	0	4	8	3	53.33%
	GD	0	0	1	24	96.00%		GD	0	1	2	11	78.57%
F-1 score	79.75%						Average F-1 score	59.92%					

a way to reduce overfitting [11]. The paper [10] first defines a straightforward measure of relevance p_i as:

$$p_i = E_{withneuron} - E_{withoutneuron}$$

where E is the evaluation of the model. The paper voices concerns that it is far too computationally expensive to compute p_i for all neurons in the network and goes on to develop more complex schemes and measures of Relevance. However with modern hardware this process is relatively quick and hence this is what we will implement. To determine if a neuron is relevant or not, we compare the F1-Score of our model on the testing data set with and without the single neuron. If the removal of the neuron results in an increase in the F1-Score then we consider it a candidate neuron for removal.

5 Results and Discussion

The results of our best neural network can be seen in Table 3. We observe that previously in Table 2 the model produced by our evolutionary algorithm produces a test/train F-1 score of 79.93%/57.87%. After applying our relevance procedure this changes to 79.75%/59.92%, Table 3. We compare our results to the original paper where they use precision as a metric [1], they demonstrate a train/test precision of 60.00%/62.80% compared to our 79.34%/59.33%. Our training set precision is far higher but our testing set precision is lower.

There may be several factors which cause this difference in performance. First, our raw data-set differs from the original paper. Second, we do not know how some of the variables have been prepossessed and what procedure was used to remove data. Thirdly, we used a different split of train/test (109:50) compared to the original paper (70/70). Fourth, our network is more complex than the one employed by the original paper [1], this may lead to overfitting on the training set and resulting in poor generalisation to the testing set.

Given that our training set score is far higher than our testing set score, we believe that it is the case that the model is overfitting. Yet it is curious that our testing set score cannot be further improved using relevance alone, this suggests that the weights themselves are overfitting on the training data. Given that we employed a form of validation in the training of the model itself, it may be the case that we randomly have a poor training/test split or that our data-set is too small. We propose that increasing the number of folds in our validation and increasing the number of generations would see overall better results as we do not achieve convergence in only 50 generations.

Our evolutionary algorithm did not produce an optimal model at the last generation. The models at the earlier generations tended to perform better. This suggests to us we can further tune our evolutionary algorithm. A major issue may be that there is not a strong correlation between the validated training f-score and the testing f-score. What is interesting is that for the optimal model we selected (14), the epoch and learning rates were optimal for the testing set f-score. We observe for this given model with a set number of neurons, activation function and learning rate, that around 2666 epochs results in the optimal value for the testing score. This actually provides support for our evolutionary algorithm because this model cannot improve further than this score.

6 Conclusion and Future Work

In this paper we have reproduced the problem found in [1] of classifying the core porosity of minerals using petrographical descriptions. There are however some slight differences as our dataset is not completely the same as the one found in [1].

We use an evolutionary algorithm to determine what may be good hyper-parameters to select. We run this algorithm for 50 epochs and test the model which performs the best. Our model performs better on the training set and slightly worse on the training set than in the original paper [1]. We then apply the network reduction technique of removing the least relevant neuron(s). Our model improves yet is still slightly worse than the original paper when using the testing set (59.92% vs 62.8%).

In the future we plan on integrating more network pruning techniques such as taking into account: complementary neurons, similar neurons, badness and sensitivity measures to further reduce our network [3]. This will hopefully result in better generalisation to the testing set. Our evolutionary algorithm can be further improved. The biggest improvement would be to run it for many more generations so that we can better observe the effects of accuracy trends over time. We can also devise better fitness functions such as: adding more rounds of validation on randomised subsets; or using a function based on a combination of factors. We plan on sourcing more computational power to explore different number of generations, possible mutations and rates along with bigger networks in general. We also plan on exploring various other encoding schemes for our petrographical features. It would also be interesting to compare and contrast the results from different petrographical data sets other than the one in the original paper. A broader range of data sources from groups who collect it in the field should provide deeper insights into the predictive capability of our model.

References

1. Gedeon, T.D., et al.: Use of linguistic petrographical descriptions to characterise core porosity: contrasting approaches. J. Petrol. Sci. Eng. **31**, 193–199 (2001)
2. Bergstra, J., Bardenet, R., Bengio, Y., Kegl, B.: Algorithms for hyper-parameter optimization. In: Neural Information Processing Systems (2011)
3. Gedeon, T.D.: Network reduction techniques. In: Proceedings International Conference on Neural Networks Methodologies and Applications, vol. 1, pp. 119–126 (1991)
4. Zhang, G.P.: Neural networks for classification: a survey. IEEE Trans. Syst. Man Cybern. Part C (Appl. Rev.) **30**(4), 451–462 (2000)
5. Buckland, M., Gey, F.: The relationship between recall and precision. J. Am. Soc. Inf. Sci. Banner **45**(1), 12–19 (1994)
6. Kambauer, G., Unterthiner, T., Mayr, A.: Self-normalising neural networks. Neural Inf. Process. Syst. (2017)
7. Vasconcelos, J.A., Ramirez, J.A., Takahashi, R.H.C., Saldanha, R.R.: Improvements in genetic algorithms. IEEE Trans. Magn. **37**(5), 3414–3417 (2001)
8. Sola, J., Sevilla, J.: Importance of input data normalization for the application of neural networks to complex industrial problems. IEEE Trans. Nucl. Sci. **44**(3), 1464–1468 (1997)
9. Holt, M.J.J., Semnani, S.: Convergence of back-propagation in neural networks using a log-likelihood cost function. Electron. Lett. **26**(23), 1964–1965 (1990)
10. Mozer, M.C., Smolensky, P.: Using Relevance to reduce network size automatically. Conn. Sci. **1**(1), 3–16 (2007)
11. Sietsma, J., Dow, R.J.F.: Creating artificial neural networks that generalize. Neural Netw. **4**(1), 67–79 (1991)

Using an Evolutionary Algorithm to Optimize the Hyper-parameters of a Cascading Neural Network

Angus Vos(✉) and Jo Plested

Research School of Computer Science, Australian National University, Canberra, Australia
u5581956@anu.edu.au

Abstract. This paper describes the implementation of an evolutionary algorithm to optimize the ability of a Casper neural network to classify the porosity of a reservoir into groups using the linguistic petrographical characteristics of the data. The Casper neural network technique is implemented on the petrographical data gained from a paper from Gedeon et al. [3]. We used an evolutionary algorithm to optimize the hyper-parameters of the network, specifically the learning rates of the different regions in the network and the period of time between the additions of new neurons. Several methods of producing offspring were tested, and each was able to improve the accuracy of the Casper neural network. Future work is suggested to improve the optimization of the evolutionary algorithm, such as testing for more generations, and utilizing other selection methods. Ultimately the results achieved suggest that there is potential for optimizing the hyper-parameter of a cascading neural network using evolutionary algorithms.

1 Introduction

This report describes the implementation of an evolutionary algorithm to optimize a Casper neural network's ability to solve a classification problem. The dataset used was the same dataset utilized by Gedeon et al. [3] ('Dataset Paper'). The Dataset Paper tested whether a neural network could be used to characterize the porosity of petrographical samples using linguistic petrographical descriptions. Gedeon et al. noted that there was an ad-hoc approach by experts to the problem and that a standardised algorithm which can effectively derive the porosity from linguistic petrographical descriptions could be very useful [3].

While the same classification problem is approached by the networks described in this paper, the aim of is not to solve this classification problem, but to identify whether an evolutionary algorithm can be used to optimize the hyperparameters of a Casper neural network [7]. The evolutionary algorithm used is based on the genetic algorithms first popularized by Holland, who suggested them as a method of searching for optimal solutions [4]. Whitley et al. demonstrated that evolutionary algorithms have the potential to optimize more standard neural networks [8], and this paper is attempting to extend on that. A flaw which has been identified in Casper neural networks is that the

T. Gedeon et al. (Eds.): ICONIP 2019, CCIS 1142, pp. 758–765, 2019.
https://doi.org/10.1007/978-3-030-36808-1_83

hyperparameters used to describe their behavior are often estimated initially. This means the optimal values are unknown without extensive testing of the networks [7]. We hypothesized that an evolutionary algorithm may shorten this search for the optimal parameters, allowing for more effective Casper neural networks to be built. Our investigations tested whether evolutionary algorithms can do this, with three different reproductive methods tested.

2 Method

The methodology comprised four stages; encoding of the data, implementing a Casper neural network on the dataset, building a genetic algorithm to generate learning rates and time periods for the Casper neural network, and finally testing different methods for producing offspring. The problem being modelled was the classification of the porosity of linguistic petrographical data. There are four possible categories of porosity that each sample can be in. The category of porosity into which an entry falls is the output of the neural network. The inputs are comprised of the linguistic petrographical data from the Dataset Paper [3].

2.1 Encoding the Data

We followed a similar method as used in the Dataset paper [3] to encode the data. Empty columns and columns containing strings were removed from the dataset. Then, the porosity data was normalized between 0 and 1. These values where placed into four categories, depending on the normalized value using the same categorization implemented by Gedeon et al. [3]. The columns with values for bioturbation and sorting were encoded into new columns so that each of these groups were represented by a single value. The other groups of values were not encoded to allow the neural network to discover any patterns in these inputs rather than risk losing valuable data. If an entry had no value in the Grain Size or Matrix columns, the row was dropped from the dataset.

When the initial neural network algorithm was run, the data was split into seven parts to enable k-fold validation to be done. Essentially, six parts of the dataset were used for training at any one time, and the seventh part was used for testing the dataset. This was done to prevent overfitting of the data and to improve the robustness of the test results.

2.2 Creating the Casper Neural Network

A Casper neural network was implemented, which is a cascade correlation network utilizing progressive resilient backpropagation (RPROP) for it's optimizer function. This technique comes from Treadgold and Gedeon [7]. It builds on the original Cascade Correlation algorithm from Fahlman and Lebiere [2]. This is a network which trains an initial layer of hidden neurons until the change in error reaches a certain point. At this point another hidden neuron is added which takes in as inputs all the inputs, and all previous outputs. A normal cascade correlation network freezes the weights of previous neurons after adding a new neuron, but Casper does not freeze these weights. Instead it uses a progressive RPROP algorithm to decay the learning rate used on the weights. Furthermore, different regions of the network have different initial learning rates, a learning rate which is reset whenever another neuron is added. These regions are classified as L1, L2 and L3. L1 covers the output of newly added hidden neuron. L2 refers to the outputs of previously added hidden neurons to the new hidden neuron, while L3 refers to every other weight. In the paper from which the technique is drawn from, the learning rate at L1 is significantly greater than at L2, which in turn is higher than the learning rate at L3 [7]. This scale was used for the control Casper network, but the learning rates generated by the genetic algorithm used were not limited by this constraint. to do this to explore whether the algorithm would find that the most effective learning rates were the same as those proposed in the technique paper. The learning rates used for the control Casper Neural network were (L1 = 0.05, L2 = 0.01, L3 = 0.005).

The learning rates are re-initialised whenever a new neuron is added. Essentially this allowed the network to modify itself when new neurons are added to account for them without losing all the information it has previously learned. A new neuron is added after the error falls by less than 1% when compared to a previous value attained a given number of epochs previously. The genetic algorithm attempted to optimize this given number of epochs, or the 'time period'. The formula for calculating this time period was, where P was the parameter being optimized:

$$Time\ Period = 15 + P * No.\ of\ Neurons\ currently\ in\ the\ Network$$

In the control Casper network, P was initialized as 15. The control was run 10 times, and the average accuracy from these tests was obtained.

2.3 Testing and Building the Evolutionary Algorithm

The evolutionary algorithm we used aims to optimize both the learning rates used for the different regions of the Casper network and the time period controlling the rate at which hidden neurons are added to the network. This algorithm starts by randomly generating a population of twenty 'chromosomes', each being a list of each value. Each chromosome is used to create and train a Casper neural network 10 times, and the test accuracy is measured at the end of each of these trials. This test accuracy is averaged to get a final score for that chromosome. The population is then sorted by that score, and a reproduction method is applied. I tested three reproduction methods; a simple asexual reproduction, sexual reproduction crossover and a combination of asexual and sexual reproduction. All the reproduction methods used the same mutation method, which applied a 10% chance for each allele to be mutated, which either increased or decreased the value by 20%. These values were chosen as they were large enough to potentially help escape local minima, while not being so large that any mutation would destroy progress made by the algorithm.

The asexual reproduction method consisted of getting the top half of the population, replicating it and passing that on as the next generation after mutation was applied.

Sexual reproduction was done by choosing the parent candidates from the population randomly, but with a strong weighting towards the chromosomes which obtained a higher score. This meant that it was far more likely that a more successful parent would be chosen. The two chromosomes were then crossed over and the mutation function was applied to produce an offspring.

The combined reproduction method preserved the top two chromosomes from each generation, and then used sexual reproduction as described above to fill out the rest of the population, and then applied the mutation algorithm as above to the population.

Each of these methods was tested by getting the average accuracy attained by the Casper networks described by chromosomes in that generation and the best accuracy attained by a chromosome of each generation and examining how this changed over time. These results were then compared with the control's accuracy and the accuracy achieved on the original dataset.

3 Results

3.1 Final Results

Table 1. Results of each reproduction method, the control and the results in the original dataset paper. Average Accuracy is the average accuracy of all the chromosomes in the final generation. Top Accuracy is the average accuracy of the highest scoring chromosome in the final generation.

Method used	Average accuracy	Top accuracy
Asexual reproduction	62.88%	68.4%
Sexual reproduction	62.68%	67.2%
Combined reproduction	63.28%	69.2%
Control	65.2%	n/a
Original dataset	61.4%	n/a

The Top Accuracy column demonstrates the potential of the use of evolutionary algorithms to optimize a Casper network. The best results achieved by each reproduction method are higher than the Control, indicating that the algorithm was able to find more optimal hyper-parameters for the network. The accuracy achieved by the control was higher than the final average accuracy of each of the reproductive methods, suggesting that the population in each generation had not converged yet. The high Top Accuracy, particularly with the results for Combined Reproduction.

In each of the reproduction methods the average accuracy remaining relatively consistent between generations. This suggests that there is potential for an improved selection algorithm to be used or that each chromosome was not sufficiently tested. It may also be because the search space is very large, and requires a lot of generations for the populations to converge.

An unexpected aspect of the results shown in Table 1 is the slight difference between the randomly chosen learning rates and the ones drawn from Gedeon and Treadgold's paper [7]. The average accuracy of the final generation and the earlier generations are lower than the Control, but are higher than were expected, suggesting that there is a limited range for optimizing the network over this dataset.

It should also be noted that running the evolutionary algorithm took up significant computational resources. While training the control Casper network took less than a second of time, running the evolutionary algorithm took several hours. Therefore, while the results suggest that the potential for optimization of the Casper network is there, using an evolutionary algorithm to attain this optimization may not be the most efficient method of doing so.

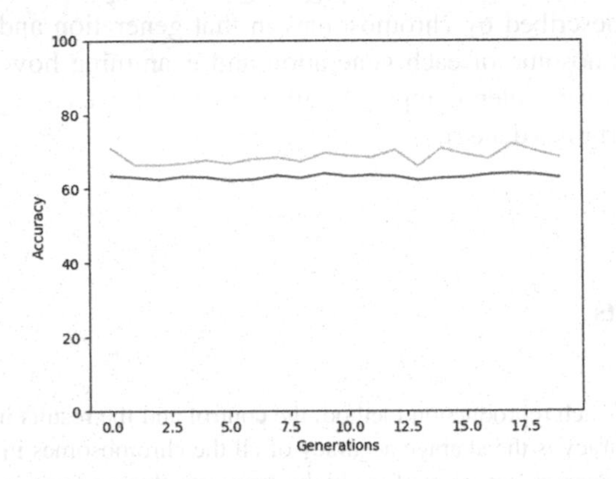

Fig. 1. Accuracy over the generations of asexual reproduction **Orange:** top score in generation, **Blue:** average score in generation (Color figure online)

3.2 Asexual Reproduction

Figure 1 contains the accuracy obtained over 20 generations of running the evolutionary algorithm using asexual reproduction. The results show that there was little variation in the accuracies achieved between the generations, but that there were chromosomes able to achieve results substantially higher than the average. This demonstrates that there is potential for the evolutionary algorithm to optimize the Casper networks hyper-parameters. However, the inability of the higher scores to proliferate through the generations suggests that the measure used to classify the higher scoring chromosomes could be improved. This could also be due to a need for running more than 10 tests on each chromosome to gauge a more accurate measure of its performance, or simply that more generations need to be run to allow the two measures to converge.

The reproductive method may have hindered the optimization of the neural network, as asexual reproduction is likely to get caught in local minima. It has less ability to create novel chromosomes but is reliant on mutation to generate new chromosomes. As such, the asexual reproduction used for the above test is unlikely to have been able to generate the optimal solution if it got caught in a local minima, as it may have.

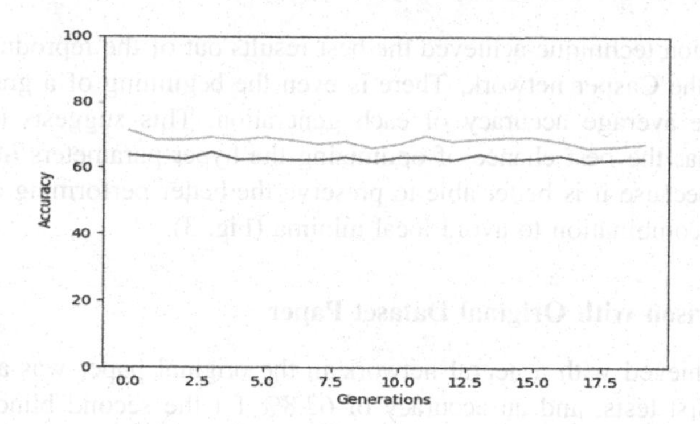

Fig. 2. Accuracy over generations of a sexual reproduction evolutionary algorithm **Orange:** top score in generation, **Blue:** average score in generation (Color figure online)

3.3 Sexual Reproduction

Figure 2 demonstrates that the use of an evolutionary algorithm using sexual reproduction was able to achieve marginal improvement and little variance over time. This may be for similar reasons as asexual reproduction. However, the fact that there was initially a high top accuracy again suggests that there is scope for the hyper-parameters of the Casper network to be optimized.

Sexual reproduction would potentially be better at escaping local minima than asexual reproduction. The same method which brings this benefit may also have lead to a greater chance for it to lose any gains it makes each generation. This can be seen by the noticeably flatter line in Fig. 2 when compared to Fig. 1.

3.4 Combined Reproduction

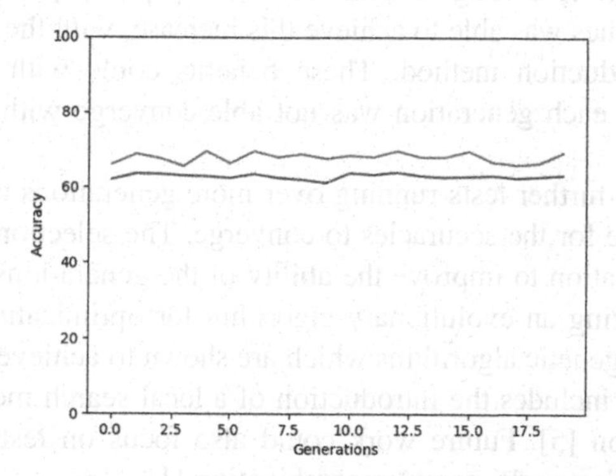

Fig. 3. Accuracy achieved over 20 generations of an evolutionary algorithm using a combined reproductive technique **Orange:** top score in generation, **Blue:** average score in generation (Color figure online)

This reproduction technique achieved the best results out of the reproductive methods at optimizing the Casper network. There is even the beginning of a gradual upwards increase in the average accuracy of each generation. This suggests that combined reproduction has the best chance of optimizing the hyper-parameters of the network. This may be because it is better able to preserve the better performing chromosomes, while using recombination to avoid local minima (Fig. 3).

3.5 Comparison with Original Dataset Paper

The results achieved with a neural network in the original paper was an accuracy of 60% for the first tests, and an accuracy of 62.8% for the second blind test [3]. The average accuracy achieved by the Casper neural networks after the evolutionary algorithms ranged from 62.68–63.28%, and the top accuracy achieved ranged from 68.4–69.2%. The control achieved an average accuracy of 65.2%. These results are noticeably higher than the results from the original dataset paper [3]. This suggests both that a Casper neural network is better than a standard neural network at solving this classification problem, and that there is potential for the hyper-parameters of the network to be optimized further to improve the results. Potentially if data is gathered about the number of neurons that the Casper networks deem optimal, then the standardized algorithm sought by Gedeon et al. [3] may be improved, although this is beyond the scope of the current report.

4 Conclusion and Future Work

The use of evolutionary algorithms in this paper demonstrates their potential for optimizing the hyper-parameters of a Casper neural network. It was able to generate hyper-parameters which proved to create a more accurate Casper network than the control developed from Treadgold and Gedeon's paper [7]. Each of the differing reproduction techniques was able to achieve this increase, with the most effective being the combined reproduction method. These benefits come with the caveat that the average accuracy of each generation was not able converge with the top accuracy in twenty generations.

For future work, further tests running over more generations could be done to test whether it is possible for the accuracies to converge. The selection algorithm may also need further optimization to improve the ability of the generations to converge. While this method of utilizing an evolutionary algorithm for optimization isn't efficient this can be mitigated by genetic algorithms which are shown to achieve the optimum results at a faster rate. This includes the introduction of a local search method to improve the speed of optimization [5]. Future work could also focus on testing other cross-over methods, such as using multi-parent recombination [1].

Ultimately, this paper demonstrates that an evolutionary algorithm can optimize the hyper-parameters of a Casper network. Further work is needed to test this and to see the extent of the optimization possible, but this paper demonstrates that the potential for this optimization is present.

References

1. Bremermann, H., Rogson, M., Salaff, S.: Global properties of evolution processes. In: Natural Automata and Useful Simulations, pp. 3–41. Spartan Books, Washington DC (1966)
2. Fahlman, S.E., Lebiere, C.: The cascade-correlation learning architecture. In: Touretzky, D.S (ed.) Advances in Neural Information Processing II, pp. 524–532. Morgan Kauffman, San Mateo (1990)
3. Gedeon, T.D., Tamhane, D., Lin, T., Wong, P.M.: Use of linguistic petrographical descriptions to characterise core porosity: contrasting approaches. J. Petrol. Sci. Eng. **31**(2–4), 193–199 (2001)
4. Holland, J.H., Goldberg, D.E.: Genetic algorithms and machine learning. Mach. Learn. **3**(2–3), 95–99 (1988)
5. Lara, A., Sanchez, G., Coello, C.A.C., Schutze, O.: HCS: a new local search strategy for memetic multiobjective evolutionary algorithms. IEEE Trans. Evolut. Comput. **14**(1), 112–132 (2009)
6. Sousa M.O.L., Suarez del Rio, L.M., Calleja, L., Argandona, V.G.R., Rey, A.R.: Influence of microfractures and porosity on the physico-mechanical properties and weathering of ornamental granites. Eng. Geol. **77**(1–2), 153–168 (2005)
7. Treadgold, N.K., Gedeon, T.D.: A cascade network algorithm employing Progressive RPROP. In: Mira, J., Moreno-Díaz, R., Cabestany, J. (eds.) IWANN 1997. LNCS, vol. 1240, pp. 733–742. Springer, Heidelberg (1997). https://doi.org/10.1007/BFb0032532
8. Whitley, D., Starkweather, T., Bogart, C.: Genetic algorithms and neural networks: optimizing connections and connectivity. Parallel Comput. **14**, 347–361 (1990)

Artificial Intelligence and Cybersecurity

Instruction Cognitive One-Shot Malware Outbreak Detection

Sean Park[2], Iqbal Gondal[1(✉)], Joarder Kamruzzaman[1], and Jon Oliver[2]

[1] ICSL-Federation University, University Dr, Mount Helen, VIC 3350, Australia
{iqbal.gondal,joarder.kamruzzaman}@federation.edu.au
[2] Trend Micro, 15/1 Pacific Highway, North Sydney, NSW 2060, Australia
{spark,jon_oliver}@trendmicro.com

Abstract. New malware outbreaks cannot provide thousands of training samples which are required to counter malware campaigns. In some cases, there could be just one sample. So, the defense system at the firing line must be able to quickly detect many automatically generated variants using a single malware instance observed from the initial outbreak by statically inspecting the binary executables. As previous research works show, statistical features such as term frequency-inverse document frequency and n-gram are significantly vulnerable to attacks by mutation through reinforcement learning. Recent studies focus on raw binary executable as a base feature which contains instructions describing the core logic of the sample. However, many approaches using image-matching neural networks are insufficient due to the malware mutation technique that generates a large number of samples with high entropy data. Deriving instruction cognitive representation that disambiguates legitimate instructions from the context is necessary for accurate detection over raw binary executables. In this paper, we present a novel method of detecting semantically similar malware variants within a campaign using a single raw binary malware executable. We utilize Discrete Fourier Transform of instruction cognitive representation extracted from self-attention transformer network. The experiments were conducted with in-the-wild malware samples from ransomware and banking Trojan campaigns. The proposed method outperforms several state of the art binary classification models.

Keywords: Deep learning · Self-attention transformer · One to many malware · Outbreak detection · Instruction recognition · Raw binary executable

1 Introduction

The majority of approaches use a large number of samples for training, which tends to overfit without generalising individual sample's characteristics. Many approaches also assume the features extracted by the model possess the meaningful signals of the raw executable files without verifying it. Given that it is crucial to provide detection on initial outbreak for raw binary executables with packed data, this paper will focus on one-shot learning and instruction-cognitive feature extraction. To the best of our knowledge, no work has been conducted providing these attributes in malware detection.

T. Gedeon et al. (Eds.): ICONIP 2019, CCIS 1142, pp. 769–778, 2019.
https://doi.org/10.1007/978-3-030-36808-1_84

Saxe et al. [1] and Vinayakumar et al. [2] implemented a deep feed forward network using statistical features derived from the executable file metadata. Anderson et al. [3] showed that it is easy to defeat these statistical features largely from the executable header metadata such as import table entries, sections, entropy, and other relevant metadata. Byte n-gram has been considered an attractive approach when dealing with highly structured data such as raw executable files. However, Zak et al. [4] showed that byte n-gram learns little information from code sections contrary to common hypotheses in machine learning. Then Raff et al. [5] discovered a potential that neural network models can learn useful representation from uninterpreted sequence of executable bytes that helps classification. This paper makes following contributions:

1. A method that learns a representation directly correlated with the legitimate instructions embedded within the raw binary executable file.
2. A method to use the model to detect malware variants that possess instruction-wise similarity by performing one-shot training.

2 Related Works

The idea of one-to-many detection over raw binary executables poses a number of challenges such as producing sufficient accuracy with one-shot training, correctly identifying valid instructions from raw sequence of bytes, detecting diverse variations, and staying resilient against adversarial attacks.

2.1 One-Shot Training

Park et al. [6] demonstrated that malware can be detected through one-shot learning using adversarial autoencoder [7] trained over prepared instructions, which finds Nash equilibrium in a non-cooperative minmax game.

2.2 Identifying Legitimate Sequence of Instructions

Kan et al. [8] and HaddadPajouh et al. [9] implemented a binary classifier using convolutional neural network (CNN) and Recurrent Neural Network (RNN), respectively. However the instruction feature used was prepared instead of automatically being derived by the model. Treating the executable as an image, Le et al. [10] created a Multiclass classifier using a combination of CNN and Long Short Term Memory (LSTM). Raff et al. [11] also attempted to detect malware with CNN using the features extracted from raw binary executables. Pascanu et al. [12] created binary classifier using LSTM with Max-Pooling, which is impractical for real world training on a large sequence such as binary executables.

However, a problem arises when the trained model is not designed to distinguish legitimate instructions from data. A significant number of malware samples deliberately insert arbitrary amount of high entropy data in between the code fragments (see Fig. 1). This randomized data scattered within the code section works as a significant amount of noise that contributes to the incorrect decision. The models without instruction cognitive capability will make a coin-flip decision for the malware samples with this tactic.

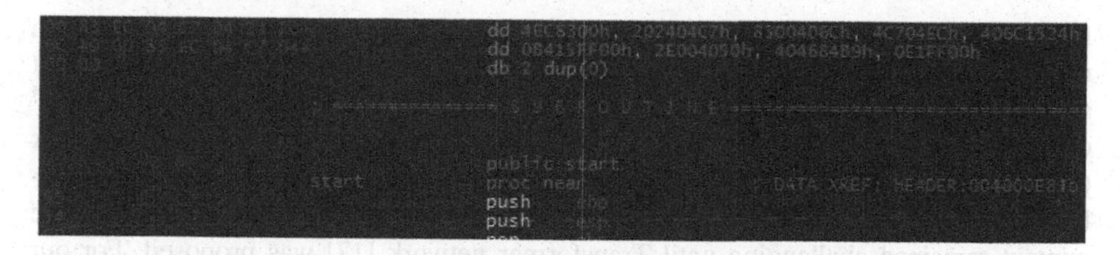

Fig. 1. Disassembly of a malware sample with packed data embedded in the code section.

Although the majority of the previous researches report respectable detection accuracy, none of them has demonstrated the capability to understand instructions, thereby vulnerable to the overfitting and the adversarial attacks as described in Sect. 2.4. As pointed out by Zak et al. [4], disambiguating instructions by their binary opcode is critical for model generalization. One of the major contributions of this paper is to develop a novel method to create instruction cognitive representations from the raw binary executables.

2.3 Detecting Diverse Variations

Metamorphism has been one of the major tactics to defeat detection [13]. The crux of various metamorphic techniques is in its global spatial translation with local context intact. In order to deal with these variations, the model must be able to coherently detect variations. Park et al. [6] have achieved this through global average pooling of CNN over prepared instructions. However, having raw byte sequence as the input, arbitrary amount of noise created by high entropy data in the input can cripple the detection accuracy as shown in Sect. 2.2. Detecting diverse variations on raw binary sequence poses yet another challenge.

2.4 Resiliency Against Adversarial Attacks

Kolosnjaji et al. [14] created adversarial malware binaries by injecting padded bytes and training with gradient descent. They showed the models utilizing raw executable bytes are vulnerable to this simple adversarial attack, including the Raff's model [11]. Grosse et al. [15] also introduced a method to induce misclassification of the detection model by perturbation of the malware binary.

3 Proposed Method

3.1 Transformer Network

Figuring out valid instructions from raw sequence of bytes in executables requires a model that understands the semantic relationship between the elements of the input sequence. A plethora of deep learning techniques have been produced in early 21st century. Notably CNN and RNN along with Generative Adversarial Network (GAN) [16] have been the base platform for language modelling and machine translation. Despite its success, correctly identifying the instructions purely based on the context remained challenging until Transformer network [17] was proposed. For our purpose, Transformer network is trained by providing raw sequence of bytes from an executable as an input and by setting the desired instructions as a target (see Fig. 2).

No	Input (raw bytes)	Instruction disassembly	Output (opcodes)
07	**83** ec 3c	sub esp, 0x3c	**83** 64 64
10	**8b** 35 03 6c 40 00	mov esi, dword ptr [0x406c03]	**8b** 64 64 64 64 64
16	**56**	push esi	**56**
17	ff **15** c4 50 40 00	call dword ptr [0x4050c4]	**ff 15** 64 64 64 64
23	2e ba c2 37 40 00	mov edx, 0x4037c2	**2e** 64 **c2** 64 64 64
29	ff e2	jmp edx	**64 64**
31	00 00	add byte ptr [eax], al	64 64
33	00 8b 3d ff 6b 40	add byte ptr [ebx + 0x406bff3d], cl	64 64 64 64 64 64

Fig. 2. The first column is line number. The second column shows input raw bytes to transformer model. The third column is the disassembly for the input. The last column shows the output of the transformer. All numbers are in hexadecimal while the line numbers in the first column is in decimal. Legitimate instructions are shown until line 29, and the following bytes are data bytes. (Color figure online)

The goal of the model is to produce correct opcodes at its output while padding the rest of the bytes with 64, which indicates INVALID. As highlighted in blue, the model correctly identifies opcodes until it starts outputting 64 (INVALID) from line 31. The model correctly disregards invalid instructions as highlighted in the disassembly in red from line 31, by filling output bytes with 64 (INVALID). Although the model's output (last column) is not correct at line 29 when transitioning from the end of the code block to the beginning of data block, the model mostly produces accurate outputs. In short, self-attention enables Transformer to find out the relationship between different positions of the input sequence. With LSTM model [18], the experiment shows that output instruction sequence is far from accurate in the presence of packed data, which suggests self-attention plays a key role in predicting the opcodes.

3.2 Model Architecture

Transformer network [17] is used as a base model to produce instruction-cognitive signals for the raw input sequence. The model is trained using off-the-dataset samples from both malicious and benign samples. Let $\boldsymbol{x} = (x_1, \ldots, x_n) \in \mathbb{R}^n$ be an input sequence of symbols, $\boldsymbol{z} = (z_1, \ldots, z_n) \in \mathbb{R}^{d \times n}$ be latent representation of dimension $\boldsymbol{d}$ retrieved from the encoder output, and $\boldsymbol{y} = (y_1, \ldots, y_n) \in \mathbb{R}^n$ be the desired output sequence with the opcode placed at the beginning of each valid instruction and INVALID symbol in the rest of the positions. $\boldsymbol{z}$ is learned by minimizing the softmax cross-entropy loss of the decoder output, $\hat{\boldsymbol{y}}$, against the label $\boldsymbol{y}$. The model architecture is shown in Fig. 3. The trained $\boldsymbol{z}$ has instruction cognitive signals that can directly transform the raw byte sequences into a sequence of legitimate instructions.

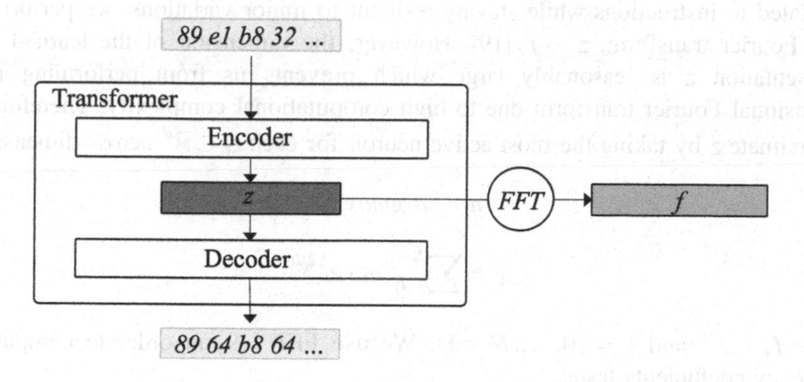

Fig. 3. Model architecture using transformer. Frequency spectrum of approximated encoded latent representation is used as the feature for malware detection.

Key hyperparameters for Transformer network and training process are:

number_of_layers: 2	epochs: 100
number_of_heads: 8	batch_size: 32
d_model: 128	optimizer: 'adam' [21]
d_k: 8 d_v: 8	learning_rate: 0.0001

Figure 4 illustrates an example of $\boldsymbol{z}$ for a malware sample with the majority of the executable occupied by high entropy packed data. This demonstrates that the model is resilient to adversarial perturbations modifying the binary executable by inserting arbitrary bytes without caring about the legitimacy of instructions.

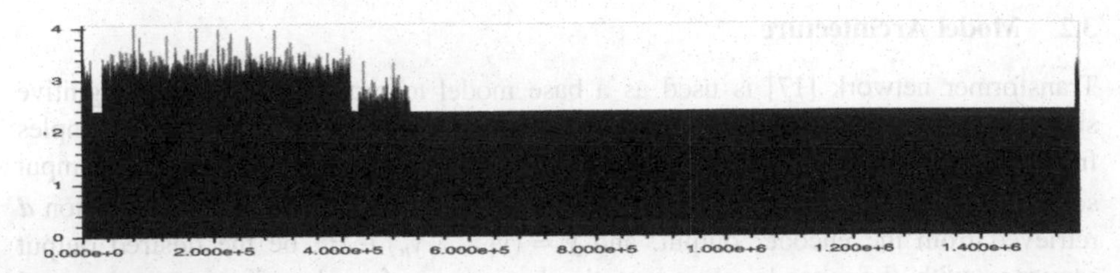

Fig. 4. Maximum activation for backdoor TORFSEE.SMF. There are several intervals where the activation strength is flat where no valid instruction was found.

As described in Sect. 2.3, it is critical to detect diverse malware variants deploying metamorphism. Given that frequency spectrum exhibits a coherent view of the features correlated to instructions while staying resilient to minor variations, we perform discrete Fourier transform, $z \rightarrow f$ [19]. However, the dimension of the learned latent representation z is reasonably large which prevents us from performing multi-dimensional Fourier transform due to high computational complexity. Therefore, we approximate z by taking the most active neuron for each $z_t \in \mathbb{R}^d$ across dimension $\boldsymbol{d}$.

$$\boldsymbol{a}_t = argmax\, z_t \tag{1}$$

$$f_k = \sum_{t=0}^{N-1} a_t \cdot e^{-\frac{2\pi i}{N}kt} \tag{2}$$

where $\boldsymbol{f}_k \in \mathbb{R}^1$ and $k = (0, \ldots, N-1)$. We use FFT [19] in order to compute the frequency coefficients faster.

We discovered that the samples sharing similar instruction-wise characteristics exhibit similar spectrum distributions. Therefore, we use Pearson Correlation Coefficient [20] against spectral density as a distance metric between samples, which is defined in Eq. (3).

$$\rho_{a,b} = \frac{E[(a-\mu_a)(b-\mu_b)]}{\sigma_a \sigma_b} \tag{3}$$

where $\boldsymbol{\sigma}_a$ is the standard deviation of a, $\boldsymbol{\sigma}_b$ is the standard deviation of b, $\boldsymbol{\mu}_a$ is the mean of a, $\boldsymbol{\mu}_b$ is the mean of b, and E is the expectation.

A sample is detected as malicious if the correlation, defined by the Eq. (3), to a known malware instance in a malware campaign is within the threshold. The threshold is empirically decided to cope with variations in the training dataset.

3.3 Analysis

Figure 5 shows the FFT of two separate malware campaigns captured in the wild. Each graph contains two variants exhibiting their frequency spectra overlapped to each other. Variants from the same campaign show similar spectral characteristics while the difference in spectra from different campaigns is distinct enough to distinguish them.

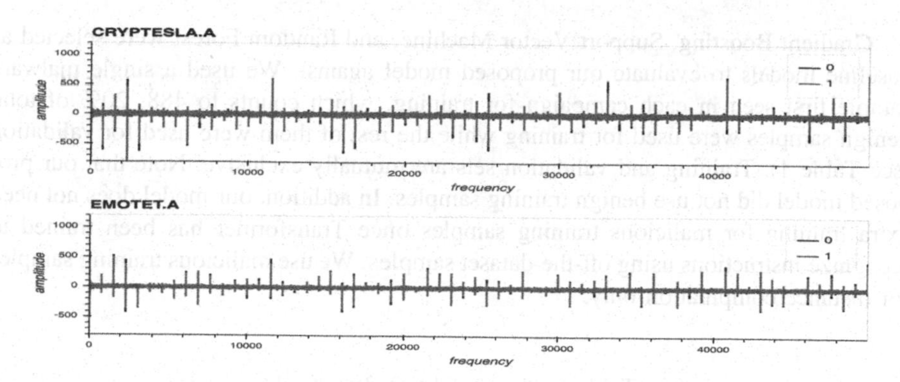

Fig. 5. The top graph shows FFT of CRYPTESLA variants whereas the bottom graph displays that of EMOTET variants.

4 Evaluation

4.1 Dataset

As stated in the introduction, one-shot training is used to evaluate the model's performance on one-to-many detection capability. There is no publicly available dataset for this problem setting. Repurposing public datasets for our problem setting is not optimal because some datasets come without binary samples, and others contain imbalanced samples with no campaign information, which makes it difficult to derive an accurate evaluation of the model's capability to detect malware variants originated from the same campaign. Besides most datasets are old and are not annotated with first-seen timestamp. For these reasons, we use a proprietary dataset provisioned by a commercial vendor that contains major ransomware and banking Trojans campaigns of 2017 and 2018. Each individual malware outbreak has been recorded along with its time and the binary sample. The largest campaigns are shown in Fig. 6.

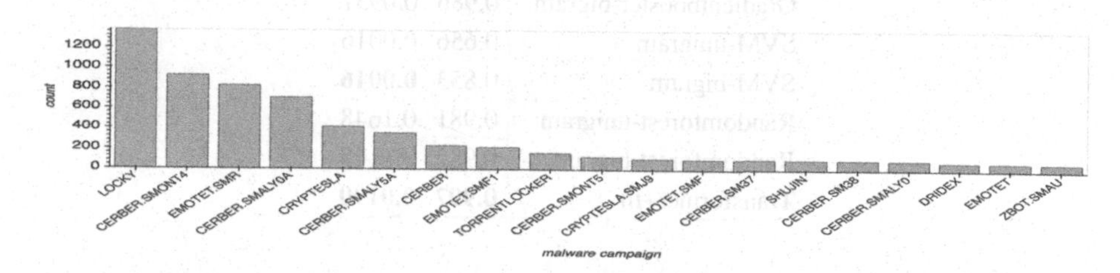

Fig. 6. A snapshot of the malware campaign distribution of the dataset used in the evaluation. The X-axis is the name of the malware campaign and the Y-axis is the number of samples within each malware campaign. Shown from the largest campaign (left) to the smaller ones (right).

Gradient Boosting, Support Vector Machine, and Random Forest were selected as baseline models to evaluate our proposed model against. We used a single malware sample first seen in each campaign for training, which counts to 488. 20% of total benign samples were used for training while the rest of them were used for validation (see Table 1). Training and validation sets are mutually exclusive. Note that our proposed model did not use benign training samples. In addition, our model does not need extra training for malicious training samples once Transformer has been trained to recognize instructions using off-the-dataset samples. We use malicious training samples for distance computation only.

Table 1. Train/validation dataset split

	Malicious	Benign	
		Baselines	Our model
Train	488	1365	0
Validation	3085	5461	5461

4.2 Model Performance

As shown in Table 2, our proposed model (*transformer+fft*) outperforms all models in TP (True Positive) despite the fact that no benign sample was used for training. Our model marginally comes in the second place for FP (False Positive) following SVM. However, SVM records a poor TP, which is sub-optimal to be used as a production model.

Table 2. Model performance comparison

Model	TP	FP
Gradientbooster-unigram	0.967	0.1518
Gradientbooster-bigram	0.986	0.0957
SVM-unigram	0.656	0.0016
SVM-bigram	0.853	**0.0016**
Randomforest-unigram	0.981	0.1648
Randomforest-bigram	0.982	0.1168
Transformer+fft	**0.997**	**0.0190**

ROC of the decision threshold for the Eq. (3) is shown in Fig. 7.

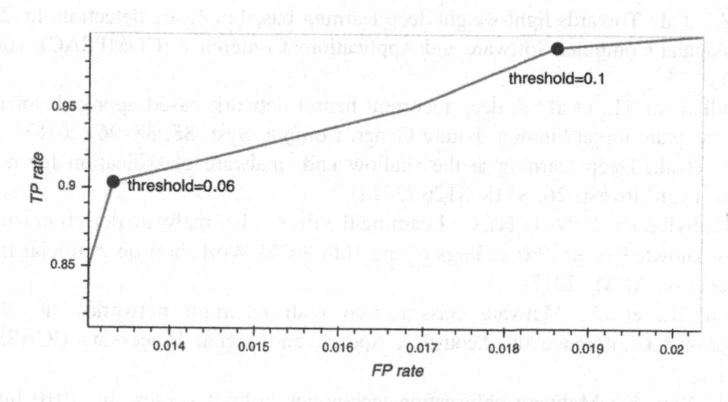

Fig. 7. ROC of decision threshold of the model as defined by Eq. (3).

5 Conclusion

In this work, we presented a novel method that extracts instruction cognitive representation from uninterpreted raw binary executables, which can be used for one-to-many malware detection via one-shot training against frequency spectrum of the Transformer's encoded latent representation. The method works regardless of the presence of diverse malware variations while remaining resilient to adversarial attacks that mostly use random perturbation against raw binaries.

One significant advantage of the method is that no computationally expensive training is required each time a new malware sample is added once Transformer is fully trained to produce the representation sufficient to recognize instructions within the binary sequence.

References

1. Saxe, J., Berlin, K.: Deep neural network based malware detection using two dimensional binary program features. In: 2015 10th International Conference on Malicious and Unwanted Software (MALWARE). IEEE (2015)
2. Vinayakumar, R., Soman, K.P.: DeepMalNet: evaluating shallow and deep networks for static PE malware detection. ICT Express **4**(4), 255–258 (2018)
3. Anderson, H.S., Kharkar, A., Filar, B., Evans, D., Roth, P.: Learning to evade static PE machine learning malware models via reinforcement learning (2018). arXiv:1801.08917
4. Zak, R., Raff, E., Nicholas, C.: What can N-grams learn for malware detection? In: 2017 12th International Conference on Malicious and Unwanted Software (MALWARE). IEEE (2017)
5. Raff, E., et al.: Malware detection by eating a whole exe. In: Workshops at the Thirty-Second AAAI Conference on Artificial Intelligence (2018)
6. Park, S., Gondal, I., Kamruzzaman, J., Oliver, J.: Generative malware outbreak detection. In: IEEE International Conference on Industry Technology ICIT, Melbourne (2019)

7. Makhzani, A., Shlens, J., Jaitly, N., Goodfellow, I., Frey, B.: Adversarial autoencoders (2015). arXiv:1511.05644
8. Kan, Z., et al.: Towards light-weight deep learning based malware detection. In: 2018 IEEE 42nd Annual Computer Software and Applications Conference (COMPSAC), vol. 1. IEEE (2018)
9. HaddadPajouh, H., et al.: A deep recurrent neural network based approach for internet of things malware threat hunting. Future Gener. Comput. Syst. **85**, 88–96 (2018)
10. Le, Q., et al.: Deep learning at the shallow end: malware classification for non-domain experts. Digit. Invest. **26**, S118–S126 (2018)
11. Raff, E., Sylvester, J., Nicholas, C.: Learning the PE header, malware detection with minimal domain knowledge. In: Proceedings of the 10th ACM Workshop on Artificial Intelligence and Security. ACM (2017)
12. Pascanu, R., et al.: Malware classification with recurrent networks. In: 2015 IEEE International Conference on Acoustics, Speech and Signal Processing (ICASSP). IEEE (2015)
13. You, I., Yim, K.: Malware obfuscation techniques: a brief survey. In: 2010 International Conference on Broadband, Wireless Computing, Communication and Applications (BWCCA), 4 November 2010, pp. 297–300. IEEE (2010)
14. Kolosnjaji, B., et al.: Adversarial malware binaries: evading deep learning for malware detection in executables. In: 2018 26th European Signal Processing Conference (EUSIPCO). IEEE (2018)
15. Grosse, K., et al.: Adversarial perturbations against deep neural networks for malware classification (2016). arXiv preprint arXiv:1606.04435
16. Goodfellow, I., et al.: Generative adversarial nets. In: Advances in Neural Information Processing Systems (2014)
17. Vaswani, A., et al.: Attention is all you need. In: Advances in Neural Information Processing Systems (2017)
18. Shin, E.C.R., Song, D., Moazzezi, R.: Recognizing functions in binaries with neural networks. In: 24th USENIX Security Symposium (USENIX Security 2015) (2015)
19. Cooley, J.W., Tukey, J.W.: An algorithm for the machine calculation of complex Fourier series. Math. Comput. **19**(90), 297–301 (1965)
20. https://en.wikipedia.org/wiki/Pearson_correlation_coefficient. Accessed 21 Jun 2019
21. Kingma, D.P., Ba, J.: Adam: a method for stochastic optimization (2014). arXiv preprint arXiv:1412.6980

Author Index

Printed in the United States
By Bookmasters